Crystal Optics

Crystal Optics

Properties and Applications

Ashim Kumar Bain

WILEY-VCH

Author

Prof. Ashim Kumar Bain
University of Birmingham
Electronic, Electrical and Systems Engineering
B15 2TT Edgbaston
United Kingdom

Library of Congress Card No.:
applied for

British Library Cataloguing-in-Publication Data
A catalogue record for this book is available from the British Library.

Bibliographic information published by the Deutsche Nationalbibliothek
The Deutsche Nationalbibliothek lists this publication in the Deutsche Nationalbibliografie; detailed bibliographic data are available on the Internet at <http://dnb.d-nb.de>.

Print ISBN: 978-3-527-41385-0
ePDF ISBN: 978-3-527-82303-1
ePub ISBN: 978-3-527-82302-4
oBook ISBN: 978-3-527-82301-7

Typesetting SPi Global, Chennai, India
Printing and Binding

Printed on acid-free paper

10 9 8 7 6 5 4 3 2 1

Contents

Preface

Since the last world war there has been a growing interest in all aspects of solid-state studies; inevitably, this has stimulated the teaching of crystallography and the physical properties of crystalline solids as a part of courses covering a wide range of academic disciplines and technological interests. Crystal optics is one of the most widely used practical techniques employed in the study of the physical properties of crystalline materials. The effect of electric and magnetic fields, mechanical stress, and ultrasound waves on the optical properties of crystals are studied in electro-optics, magneto-optics, photoelasticity, acousto-optics, and photorefractivity, which are based on the fundamental laws of crystal optics. The present book aims to provide the basic physical properties and applications of photoelastic, acousto-optic, magneto-optic, electro-optic, and photorefractive materials.

The first chapter deals with the basic concepts of crystal optics, such as index ellipsoid or optical indicatrix, crystal symmetry, wave surface, birefringence, polarization of light, changing the polarization of light, effects of reflection and transmission of polarization, and light polarizing devices. The second chapter provides an understanding of the physical phenomenon of the photoelastic effects in isotropic and crystalline materials. It describes in detail research information on modern photoelastic materials and reviews the up-to-date photoelastic device applications. The third chapter develops the underlying theory of acousto-optics from first principles, formulating results suitable for subsequent calculations and design. Special attention is given to designing procedures for the entire range of acousto-optic devices, and various applications of these devices are also described. The fourth chapter describes the basic principles of magneto-optic effects and mode of interaction with magnetic materials. It also describes in detail research information on modern magneto-optic materials and reviews the up-to-date magneto-optic device applications up to terahertz (THz) regime. The fifth chapter provides an understanding of the physical phenomenon of the linear and quadratic electro-optic effects in isotropic and crystalline materials. It describes in detail modern electro-optic materials and it also reviews the up-to-date electro-optic device applications in both bulk and plasmonic waveguide technologies. The sixth chapter is a collection of many of the most important recent developments in photorefractive effects and materials. Special attention has been paid to describe the up-to-date review of recent scientific findings and advances in photorefractive materials and devices.

I sincerely hope that this book will be of real value to the students and researchers moving into the wide field of crystal optics.

Finally, I would like to thank the Wiley-VCH publishing team for their outstanding support.

Birmingham, UK
January 2019

Ashim Kumar Bain

Overview

Crystal optics is the branch of optics that describes the behavior of electromagnetic waves in anisotropic media, that is, media (such as crystals) in which light behaves differently depending on which direction the light is propagating in. The phenomena characteristics of crystals that are studied in crystal optics include double refraction (birefringence), polarization of light, rotation of the plane of polarization, etc. The effect of electric, magnetic field, mechanical stress, and ultrasound waves on the optical properties of crystals are studied in electro-optics, magneto-optics, photoelasticity, acousto-optics, and photorefractivity, which are based on the fundamental laws of crystal optics.

This book deals with the basic physical properties and applications of photoelastic, acousto-optic, magneto-optic, electro-optic, and photorefractive materials. It also provides up-to-date information on design and applications of various optoelectronic devices based on these materials, such as photoelastic devices (modulator, Q-switches, accelerometer, sensor), acousto-optic devices (modulators, beam deflector, frequency shifter, Q-switch, tunable filter), magneto-optic devices (modulator, circulator, isolator, sensor, and magneto-optical recording), electro-optic devices (modulator, dynamic wave retarder, scanner, directional, coupler, deflector, tunable filter, Q-switch, attenuator, polarization controller, sensor), and photorefractive devices (waveguides, Q-switch, tunable filter, holographic interferometers, and holographic 3D stereograms). This book will be very useful for the scientific community including students, teachers, and researchers working in these fields. It will also find readership with non-experts of the subject.

1 Crystal Optics

1.1 Introduction

Crystal optics is the branch of optics that describes the behavior of electromagnetic waves in anisotropic media, that is, media (such as crystals) in which light behaves differently depending on the direction in which the light is propagating. The characteristic phenomena of crystals that are studied in crystal optics include double refraction (birefringence), polarization of light, rotation of the plane of polarization, etc.

The phenomenon of double refraction was first observed in crystals of Iceland spar by the Danish scientist E. Bartholin in 1669. This date is considered the beginning of crystal optics. Problems of the absorption and emission of light by crystals are studied in crystal spectroscopy. The effect of electric and magnetic fields, mechanical stress, and ultrasound waves on the optical properties of crystals are studied in electro-optics, magneto-optics, photoelasticity, acousto-optics, and photorefractivity, which are based on the fundamental laws of crystal optics.

Since the lattice constant (of the order of 10 Å) is much smaller than the wavelength of visible light (4000–7000 Å), a crystal may be regarded as a homogeneous but anisotropic medium. The optical anisotropy of crystals is caused by the anisotropy of the force field of particle interaction. The nature of the field is related to crystal symmetry. All crystals, except crystals of the cubic system, are optically anisotropic.

1.2 Index Ellipsoid or Optical Indicatrix

In isotropic materials, the electric field displacement vector **D** is parallel to the electric field vector **E**, related by $\mathbf{D} = \varepsilon_0 \varepsilon_r \mathbf{E} = \varepsilon_0 \mathbf{E} + \mathbf{P}$, where ε_0 is the permittivity of free space, ε_r is the unitless relative dielectric constant, and **P** is the material polarization vector.

The optical anisotropy of transparent crystals is due to the anisotropy of the dielectric constant. In an anisotropic dielectric medium (a crystal, for example), the vectors **D** and **E** are no longer parallel; each component of the electric

Crystal Optics: Properties and Applications, First Edition. Ashim Kumar Bain.

flux density **D** is a linear combination of the three components of the electric field E.

$$\mathbf{D}_i = \sum_j \varepsilon_{ij} E_j \tag{1.1}$$

where $i, j = 1, 2, 3$ indicate the x, y, and z components, respectively. The dielectric properties of the medium are therefore characterized by a 3×3 array of nine coefficients $\{\varepsilon_{ij}\}$ forming a tensor of second rank known as the electric permittivity tensor and denoted by the symbol ε. Equation (1.1) is usually written in the symbolic form $\mathbf{D} = \varepsilon\mathbf{E}$. The electric permittivity tensor is symmetrical, $\varepsilon_{ij} = \varepsilon_{ji}$, and is therefore characterized by only six independent numbers. For crystals of certain symmetries, some of these six coefficients vanish and some are related, so that even fewer coefficients are necessary.

Elements of the permittivity tensor depend on the choice of the coordinate system relative to the crystal structure. A coordinate system can always be found for which the off-diagonal elements of ε_{ij} vanish, so that

$$D_1 = \varepsilon_1 E_1, \quad D_2 = \varepsilon_2 E_2, \quad D_3 = \varepsilon_3 E_3 \tag{1.2}$$

where $\varepsilon_1 = \varepsilon_{11}$, $\varepsilon_2 = \varepsilon_{22}$, and $\varepsilon_3 = \varepsilon_{33}$. These are the directions for which E and D are parallel. For example, if E points in the x-direction, D must also point in the x-direction. This coordinate system defines the principal axes and principal planes of the crystal. The permittivities ε_1, ε_2, and ε_3 correspond to refractive indices.

$$n_1 = \left(\frac{\varepsilon_1}{\varepsilon_0}\right)^{1/2}, \quad n_2 = \left(\frac{\varepsilon_2}{\varepsilon_0}\right)^{1/2}, \quad n_3 = \left(\frac{\varepsilon_3}{\varepsilon_0}\right)^{1/2} \tag{1.3}$$

are known as the principal refractive indices and ε_0 is the permittivity of free space.

In crystals with certain symmetries two of the refractive indices are equal ($n_1 = n_2 \neq n_3$) and the crystals are called uniaxial crystals. The indices are usually denoted $n_1 = n_2 = n_o$ and $n_3 = n_e$. The uniaxial crystal exhibits two refractive indices, an "ordinary" index (n_o) for light polarized in the x- or y-direction, and an "extraordinary" index (n_e) for polarization in the z-direction. The crystal is said to be positive uniaxial if $n_e > n_o$ and negative uniaxial if $n_e < n_o$. The z-axis of a uniaxial crystal is called the optic axis. In other crystals (those with cubic unit cells, for example) the three indices are equal and the medium is optically isotropic. Media for which the three principal indices are different (i.e. $n_1 \neq n_2 \neq n_3$) are called biaxial. Light polarized at some angle to the axes will experience a different phase velocity for different polarization components and cannot be described by a single index of refraction. This is often depicted as an index ellipsoid.

The optical properties of crystals are described by the index ellipsoid or optical indicatrix. It is generated by the equation

$$\frac{x_1^2}{n_1^2} + \frac{x_2^2}{n_2^2} + \frac{x_3^2}{n_3^2} = 1 \tag{1.4}$$

where x_1, x_2, and x_3 are the principal axes of the dielectric constant tensor and n_1, n_2, and n_3 are the principal dielectric constants, respectively. Figure 1.1 shows the

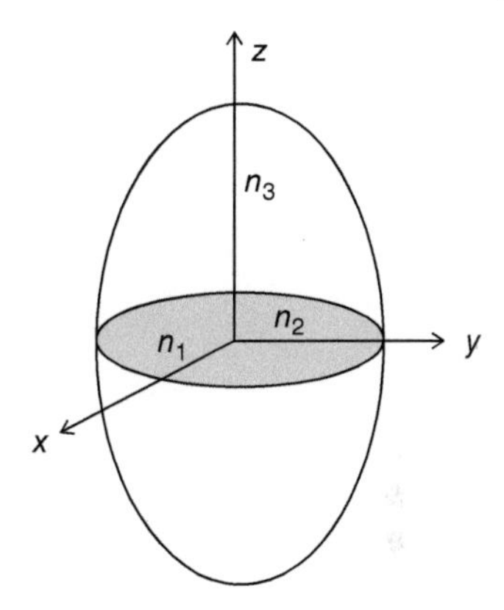

Figure 1.1 The index ellipsoid. The coordinates (x, y, z) are the principal axes and (n_1, n_2, n_3) are the principal refractive indices of the crystal.

optical indicatrix of a biaxial crystal. It is a general ellipsoid with $n_1 \neq n_2 \neq n_3$ representative of the optical properties of triclinic, monoclinic, and orthorhombic crystals.

1.3 Effect of Crystal Symmetry

In the case of cubic crystals, which are optically isotropic, ε is independent of direction and the optical indicatrix becomes a sphere with radius n. In crystals of intermediate systems (trigonal, tetragonal, and hexagonal), the indicatrix is necessarily an ellipsoid of revolution about the principal symmetry axis (Figure 1.2). The central section is perpendicular to the principal axis, and only this central section is a circle of radius n_0. Hence, only for a wave normal along the principal axis is there no double refraction. The principal axis is called the optic axis and the crystals are said to be uniaxial. A uniaxial crystal is called optically positive (+) when $n_e > n_0$ and negative (−) when $n_e < n_0$.

For crystals of the lower systems (orthorhombic, monoclinic, and triclinic), the indicatrix is a triaxial ellipsoid. There are two circular sections (Figure 1.3) and hence two privileged wave normal directions for which there is no double refraction. These two directions are called the primary optic axes or simply the optic axes, and the crystals are said to be biaxial.

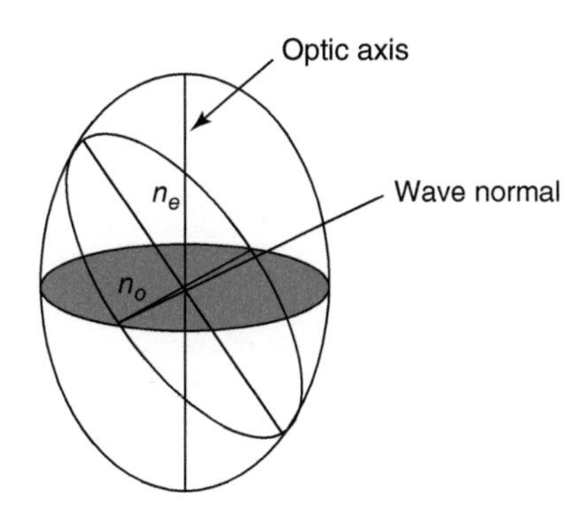

Figure 1.2 The indicatrix for a (positive) uniaxial crystal.

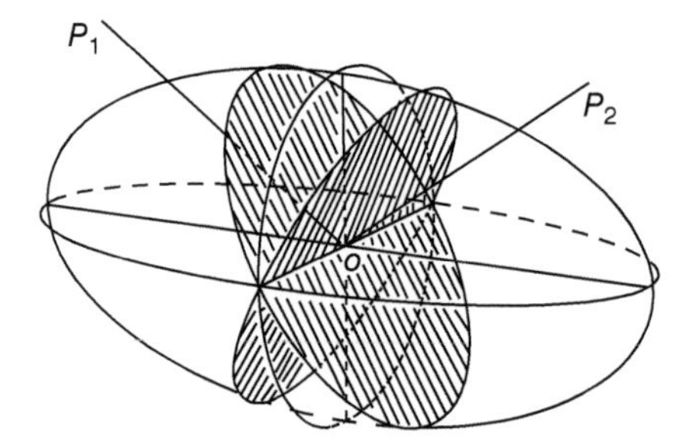

Figure 1.3 The two circular sections of the indicatrix and the two primary optic axes OP_1, OP_2 for a biaxial crystal.

1.4 Wave Surface

If a point source of light is situated within a crystal the wave front emitted at any instant forms a continuously expanding surface. The geometric locus of points at a distance v from a point O is called the ray surface or wave surface. Actually, the wave surface is a wave front (or pair of wave fronts) completely surrounding a point source of monochromatic light. This is also a double-sheeted surface. In most crystalline substances, however, two wave surfaces are formed; one is called the ordinary wave surface and the other is called the extraordinary wave surface. In both positive and negative crystals, the ordinary wave surface is a sphere and the extraordinary wave surface is an ellipsoid of revolution.

1.4.1 Uniaxial Crystal

In uniaxial crystals, one surface is a sphere and the other is an ellipsoid of revolution touching one another along the optical axis – OZ as shown in Figure 1.4a,b. In positive (+) crystals ($n_e > n_0$) the ellipsoid is inscribed within the sphere (Figure 1.4a), whereas in negative (−) crystals ($n_e < n_o$) the sphere is inscribed within the ellipsoid (Figure 1.4b).

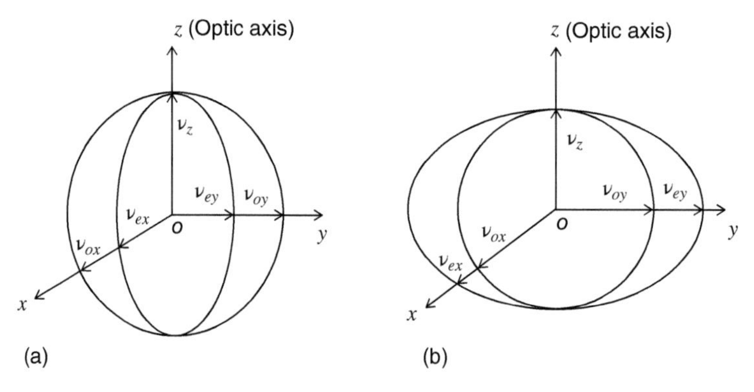

Figure 1.4 Ray surfaces of uniaxial crystals: (a) positive, (b) negative, (OZ) optical axis of the crystal, (v_0) and (v_e) phase velocities of ordinary and extraordinary waves propagating in the crystals.

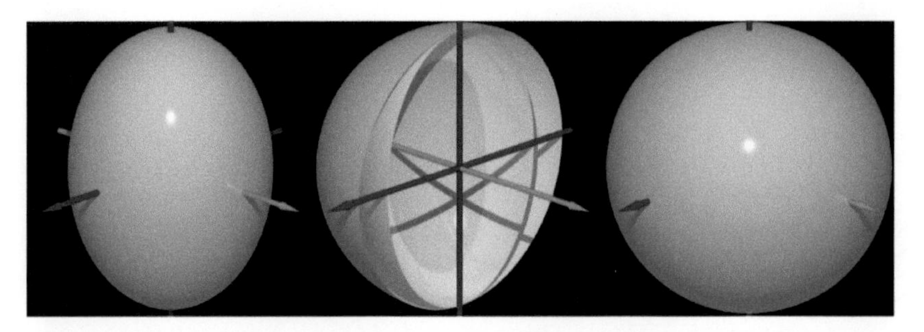

Figure 1.5 Positive uniaxial medium wave surface. Inner fold (left), vertical section (center), and outer fold (right). Source: Latorre et al. 2012 [1]. Reproduced with permission of Springer Nature.

Figure 1.6 Negative uniaxial medium wave surface. Inner fold (left), vertical section (center), and outer fold (right). Source: Latorre et al. 2012 [1]. Reproduced with permission of Springer Nature.

The dependence of the ray velocity of a plane wave propagating in a crystal on the direction of propagation and the nature of polarization of the wave leads to the splitting of light rays in crystals. In a uniaxial crystal, one of the refracted rays obeys the usual laws of refraction and is therefore called the ordinary ray, whereas the other ray does not (it does not lie in the plane of incidence) and is called the extraordinary ray. The three-dimensional view of positive and negative uniaxial medium wave surfaces can be seen in Figures 1.5 and 1.6.

1.4.2 Biaxial Crystal

It is very difficult to imagine what shapes the biaxial wave surface will have. For wave normals that lie in any of the three principal planes of the indicatrix the situation is very similar to that described for the uniaxial crystal. Three cross-sectional views of the wave surfaces for a biaxial crystal are given in Figure 1.7.

In biaxial media, the wave surface equations are fourth-order polynomials with even powers only; that is, the surface is symmetrical with respect to the origin. Both surface folds intersect only at four symmetrical points, as can be seen in Figure 1.8. Note that this intersection does not yield a curve but only the four points.

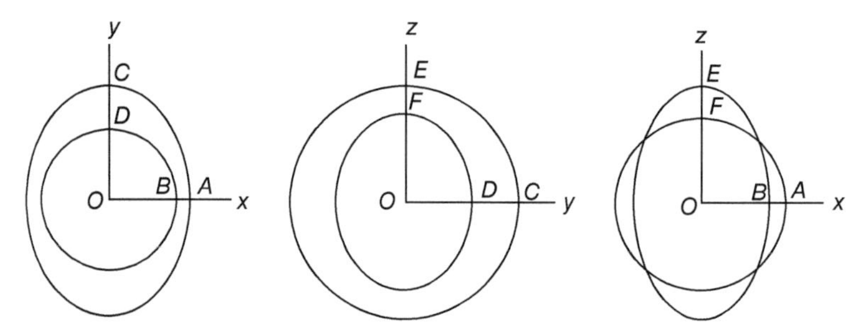

Figure 1.7 Principal sections of the wave surface for a biaxial crystal.

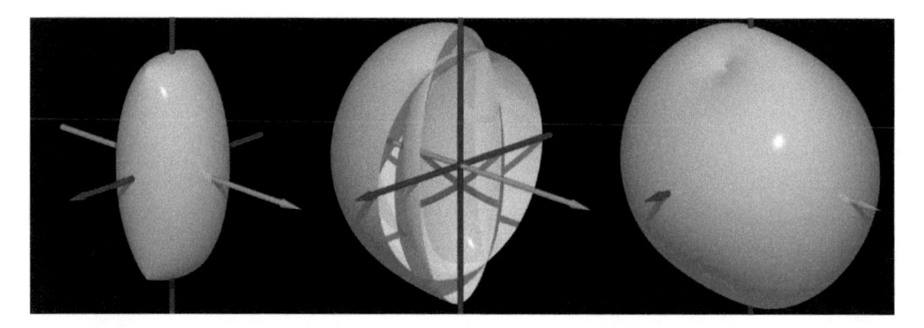

Figure 1.8 Wave surface for a biaxial medium. Inner fold (left), vertical section (center), and outer fold (right). Source: Latorre et al. 2012 [1]. Reproduced with permission of Springer Nature.

1.5 Birefringence

When a beam of nonpolarized light passes into a calcite or quartz crystal, the light is decomposed into two beams that refract at different angles. This phenomenon is called birefringence or double refraction. The ray for which Snell's law holds is called the ordinary or O-ray, and the other is called the extraordinary or E-ray (Figure 1.9).

Birefringent materials are optically anisotropic (their properties depend on the direction a light beam takes across them) because their molecules do not respond to the incident light evenly in all directions. This arises from their molecular (bond strengths) and crystal (arrangement) structures. However, in all such materials there is at least one optic axis (some materials have two) along which propagating light can travel with no consequences to either (any) component of its electric vector. This axis serves as a kind of reference. Light traveling in any other direction through the crystal experiences two different refractive indices and is split into components that travel at different speeds and have perpendicular polarizations.

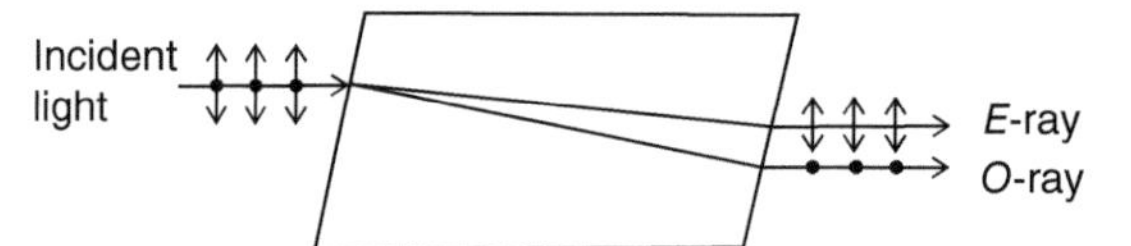

Figure 1.9 Side view of the double refraction of light by a calcite crystal.

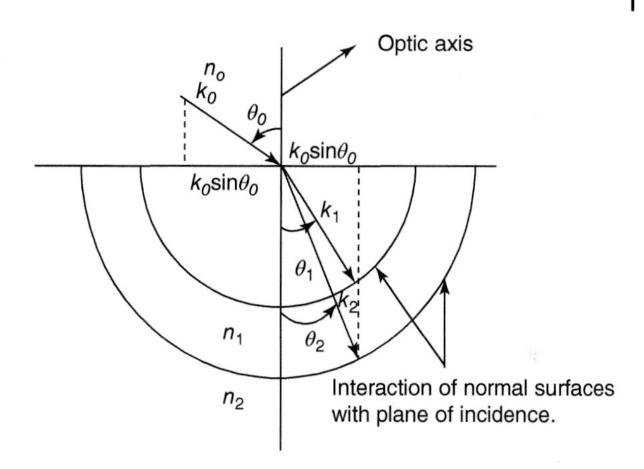

Figure 1.10 Double refraction at the boundary of an anisotropic medium.

This effect of double refraction or birefringence is further demonstrated in Figure 1.10 [2]. In Figure 1.10, subscript 0 indicates the incident wave, while 1 and 2 indicate the refracted waves. The refractive index for ordinary wave is denoted by n_o and is independent of the direction of propagation. The refractive index for extraordinary wave is denoted by $n_e(\theta)$ and depends on the direction of propagation (θ) relative to the optic axis.

The behavior of refractive index is usually described in terms of the refractive index surface, i.e. the index ellipsoid. In the case of the ordinary ray it is a sphere, while for the extraordinary ray it is an ellipsoid. That is, in terms of ellipsoid, this effect becomes a three-dimensional body with cylindrical symmetry. The two indices of refraction are then identical ($n_x = n_y$), so that the plane intersecting perpendicular to the optical axis forms a circle. If z-axis is considered as the axis of cylindrical symmetry (the optical axis of a uniaxial crystal), then for uniaxial crystal, the principal indices of refraction are

$$n_0^2 = \frac{\varepsilon_x}{\varepsilon_0} = \frac{\varepsilon_y}{\varepsilon_0} \quad \text{and} \quad n_e^2 = \frac{\varepsilon_z}{\varepsilon_0} \tag{1.5}$$

where ε_o is the dielectric constant in free space (~8.85×10^{-12} F/m); ε_x, ε_y, and ε_z are the dielectric constants along x, y, and z-axes. For uniaxial crystals $\varepsilon_x = \varepsilon_y$. It is also a known fact that refractive index and critical angle of materials are related by

$$\sin c_o = \frac{1}{n_0} \text{ and } \sin c_e = \frac{1}{n_e} \tag{1.6}$$

where c_o and c_e are the critical angles at the ordinary and extraordinary axes respectively. The critical angle of a material determines whether an internal ray will be reflected back into the material. As shown in Eq. (1.6), it is a function of the refractive index, and hence, the higher the refractive index the lower the critical angle.

In Figure 1.11a, the incident light rays giving rise to the ordinary and extraordinary rays enter the crystal in a direction that is oblique with respect to the optical axis and are responsible for the observed birefringent character. When an incident ray enters the crystal perpendicular to the optical axis, it is separated into ordinary and extraordinary rays, but instead of taking different pathways, the

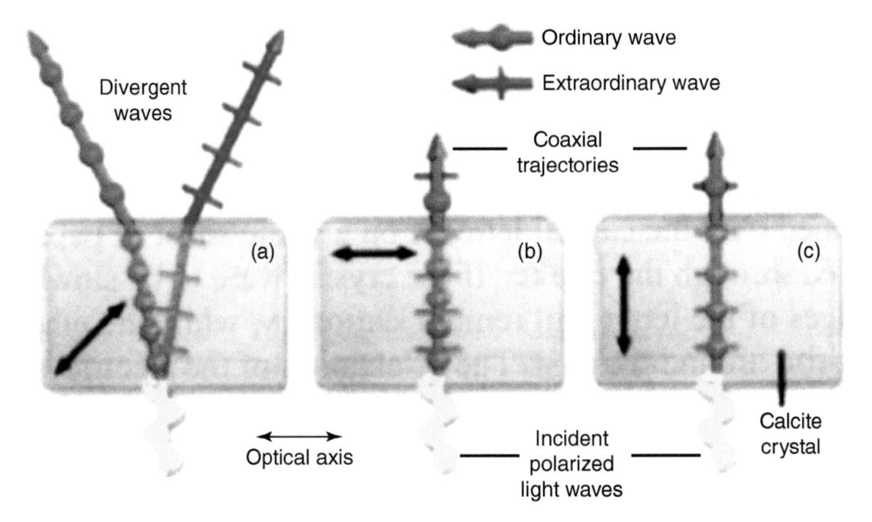

Figure 1.11 Separation of light waves by a birefringent crystal [3]. Source: Courtesy of Nikon.

trajectories of these rays are coincident. Even though the ordinary and extraordinary rays emerge from the crystal at the same location, they exhibit different optical path lengths and are subsequently shifted in phase relative to one another (Figure 1.11b). In the case where incident light rays impact the crystal in a direction that is parallel to the optical axis (Figure 1.11c), they behave as ordinary light rays and are not separated into individual components by an anisotropic birefringent crystal. Calcite and other anisotropic crystals act as if they were isotropic materials (such as glass) under these circumstances. The optical path lengths of the light rays emerging from the crystal are identical, and there is no relative phase shift.

1.6 Polarization of Light

Polarization generally just means "orientation." It comes from the Greek word *polos*, for the axis of a spinning globe. Wave polarization occurs for *vector* fields. For light (electromagnetic waves) the vectors are the electric and magnetic fields, and the light's polarization direction is by convention along the direction of the electric field. Generally, we should expect fields to have three vector components, e.g. (x, y, z), but light waves only have two non-vanishing components: the two that are perpendicular to the direction of the wave.

Electromagnetic waves are the solutions of Maxwell's equations in a vacuum:

$$\begin{aligned}
&\nabla \times \mathbf{E} = 0 \\
&\nabla \times \mathbf{B} = 0 \\
&\nabla \times \mathbf{E} = -\frac{\partial \mathbf{B}}{\partial t} \\
&\nabla \times \mathbf{B} = \varepsilon_0 \mu_0 \frac{\partial \mathbf{E}}{\partial t}
\end{aligned} \tag{1.7}$$

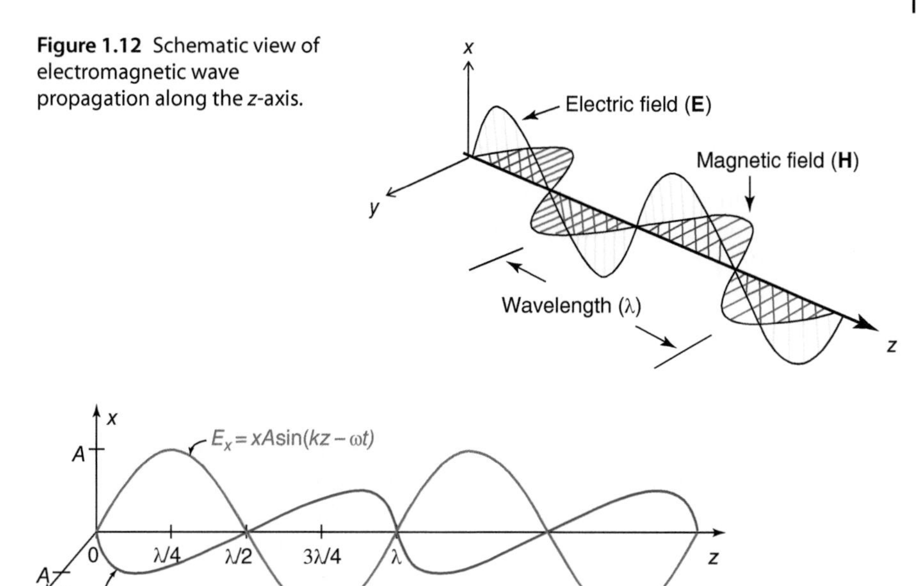

Figure 1.12 Schematic view of electromagnetic wave propagation along the *z*-axis.

Figure 1.13 Propagation of E_x and E_y components along the *z*-axis [4]. Source: Courtesy of Semrock.

In order to satisfy all four equations, the waves must have the **E** and **B** fields transverse to the propagation direction. Thus, if the wave is traveling along the positive z-axis, the electric field can be parallel to the x-axis and **B**-field parallel to the y-axis (Figure 1.12).

We shall call the two distinct waves E_x and E_y, where we denote these by vectors to remind that they point in (or oscillate along) a certain direction (the x- and y-directions, respectively) as shown in Figure 1.13. The amplitude of the light wave describes how the wave propagates in position and time. Mathematically, we can write it as a "sine wave" where the angle of the sine function is a linear combination of both position and time terms:

$$E(x,t) = A\sin\left(2\pi\frac{z}{\lambda} \pm 2\pi\nu t\right) \tag{1.8}$$

where A is called the "amplitude factor," the variable λ is the wavelength, and the variable ν is the frequency. If a snapshot of the wave could be taken at a fixed time, λ would be the distance from one wave peak to the next. If one sits at a fixed point in space and counts the wave peaks as they pass by, ν gives the frequency of these counts, or $1/\nu$ gives the time between peaks. The sign between the position and time terms determines the direction the wave travels: when the two terms have the opposite sign (i.e. the "−" sign is chosen), the wave travels in the positive z-direction.

For convenience, we often use two new variables called the wave number $k = 2\pi/\lambda$ and the angular frequency $\omega = 2\pi\nu$, which absorb the factor of 2π, so that the wave amplitude can now be written more compactly as

$$E(x,t) = A\sin(kz \pm \omega t) \tag{1.9}$$

Using this description of a single transverse orientation of a light wave, we can now consider multiple orientations to describe different states of polarization.

1.6.1 Linear Polarization – Equal Amplitudes

Here is an example of two waves E_x and E_y viewed in a "fixed time" picture (say at $t = 0$). The amplitude E or the potential for a charged particle to feel a force is vibrating along both the x- and y-directions. An actual charged particle would feel both of these fields simultaneously, or it would feel

$$E = E_x + E_y = (x + y)A \sin(kz - \omega t) \tag{1.10}$$

If we look down the propagation axis in the positive z-direction, the vector **E** at various locations (and at $t = 0$) appears as in Figure 1.14.

That is, **E** appears to oscillate along a line oriented at 45° with respect to the x-axis. Hence this situation is called linear polarization.

Equivalently, we could view the wave at a particular location ("fixed position") and watch its amplitude evolve with time. Suppose we sit at the position $z = 0$. Then we see that

$$E = E_x + E_y = -(x + y)A \sin(2\pi vt) \tag{1.11}$$

which appears as in Figure 1.15.

1.6.2 Linear Polarization – Unequal Amplitudes

If the two components E_x and E_y have unequal amplitude factors, we can see that the light wave is still linearly polarized (see Figure 1.16).

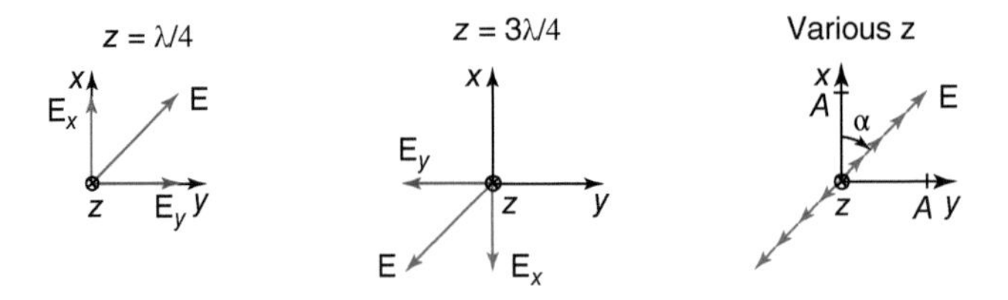

Figure 1.14 Orientation of **E** vector at various locations along the z-axis [4]. Source: Courtesy of Semrock.

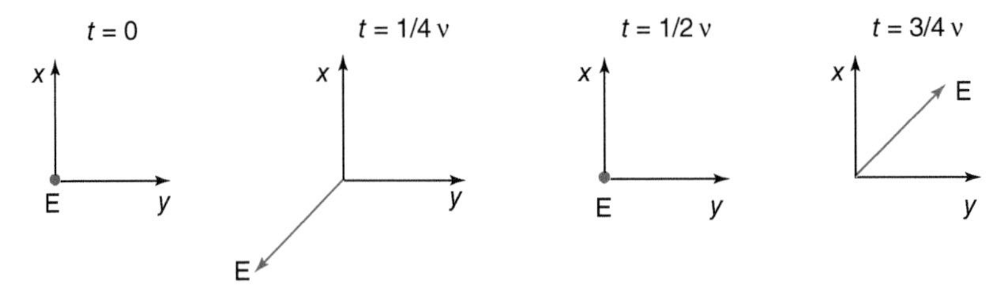

Figure 1.15 The oscillation of **E** vector back and forth along the same 45° line as time evolves [4]. Source: Courtesy of Semrock.

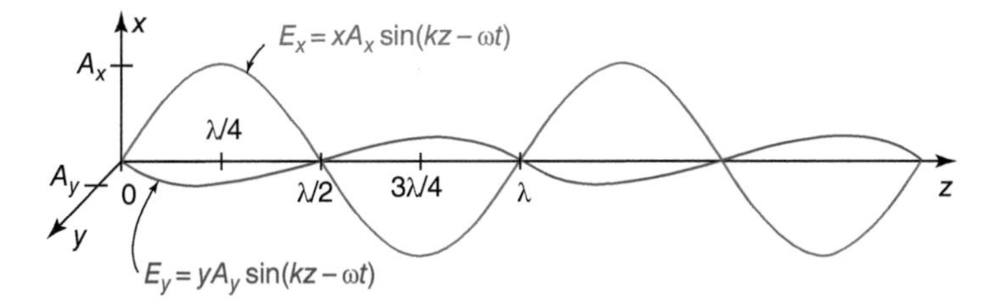

Figure 1.16 Propagation of the unequal E_x and E_y components along the z-axis [4]. Source: Courtesy of Semrock.

Figure 1.17 Orientation of **E** vector at various locations along the z-axis (when $A_x \neq A_y$) [4]. Source: Courtesy of Semrock.

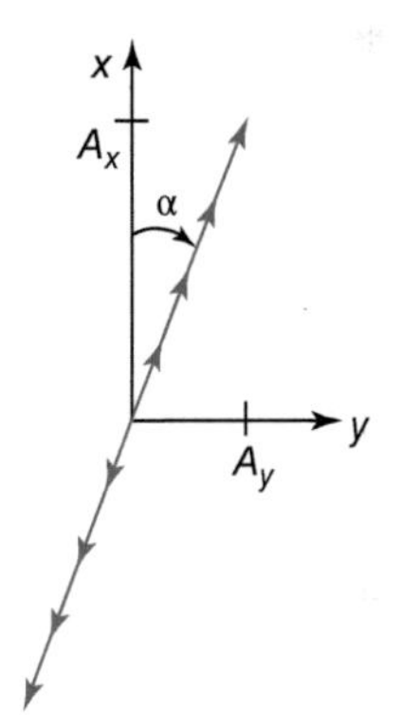

If $A_x \neq A_y$, the total wave E is linearly polarized, but it is no longer oriented at 45° with respect to the x-axis. In fact, we can see that it is oriented at an angle α where

$$\tan \alpha = \frac{A_y}{A_x} \tag{1.12}$$

In other words, if we look down the propagation axis in the positive x-direction, the vector **E** at various locations (and at $t = 0$) appears as in Figure 1.17.

1.6.3 Circular Polarization

Suppose the two components have equal amplitudes again, but now consider the case where these two components are not in phase, such that the angles of the "sine functions" are different. In particular, suppose there is a constant phase difference of $\pi/2$ between them, which corresponds to a distance of $\lambda/4$ in the "fixed time" picture.

The x-component is

$$E_x = xA \sin\left(kz - \omega t - \frac{\pi}{2}\right) \tag{1.13}$$

while the y-component is as before

$$E_y = yA \sin(kz - \omega t) \tag{1.14}$$

This case appears as in Figure 1.18.

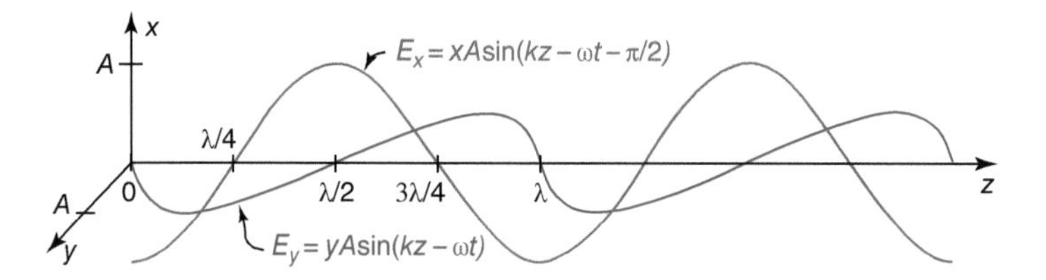

Figure 1.18 Propagation of E_x and E_y components along the z-axis when they are at a constant phase difference of $\pi/2$ [4]. Source: Courtesy of Semrock.

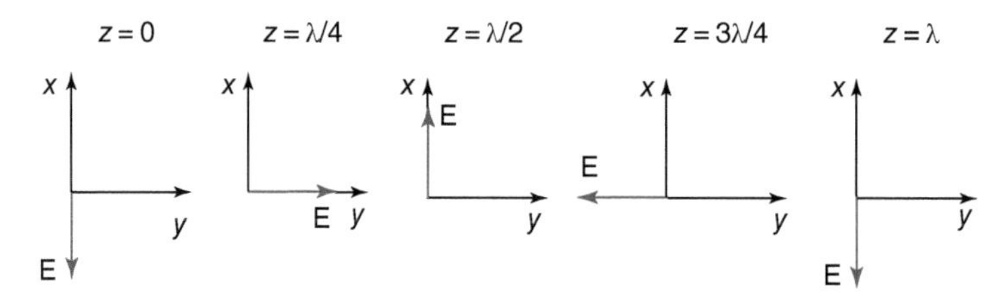

Figure 1.19 Orientation of **E** vector in the x–y plane at a fixed time [4]. Source: Courtesy of Semrock.

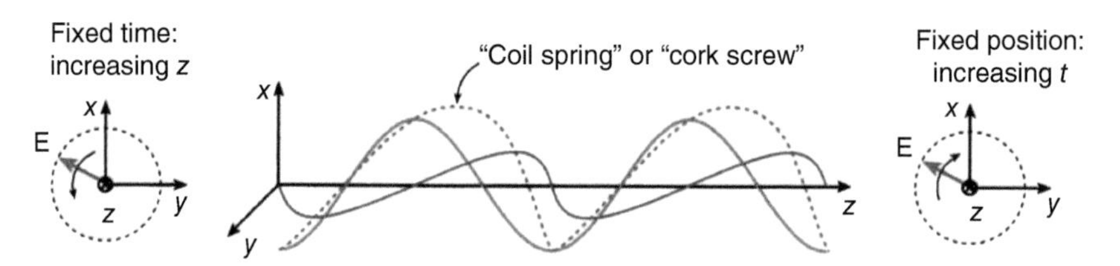

Figure 1.20 Circular orientation of **E** vector along the z-axis [4]. Source: Courtesy of Semrock.

To understand what is going on physically, again look down the z-axis (at time $t = 0$). We can see that the tip of E traces out a circle as we follow the wave along the z-axis at a fixed time (Figure 1.19).

Similarly, if we sit at a fixed position, the tip of E appears to trace out a circle as time evolves. Hence this type of polarization is called circular polarization (Figure 1.20).

1.6.4 Elliptical Polarization

All the states of polarization described above are actually special cases of the most general state of polarization, called elliptical polarization, in which the tip of the electric field vector **E** traces out an ellipse in the x–y plane. The two components might have unequal amplitudes $A_x \neq A_y$, and also might contain a different relative phase, often denoted δ. That is, we may write generally the x-component as follows:

$$E_x = xA_x \sin(kz - \omega t + \delta) \tag{1.15}$$

while the y-component is as before:

$$E_y = yA_y \sin(kz - \omega t) \tag{1.16}$$

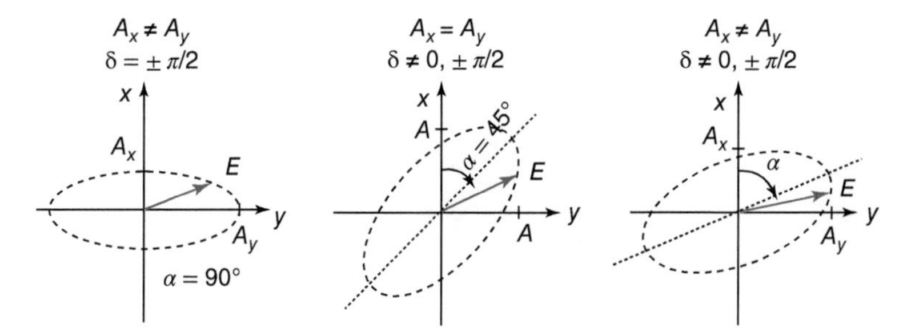

Figure 1.21 Elliptical polarization of **E** vector in the x–y plane [4]. Source: Courtesy of Semrock.

where, as before, $E = E_x + E_y$. The three special cases described in Sections 1.1, 1.2, and 1.3 thus correspond to $A_x = A_y$ and $\delta = 0$ (linear polarization; equal amplitudes); $A_x \neq A_y$ and $\delta = 0$ (linear polarization; unequal amplitudes); and $A_x = A_y$ with $\delta = -\pi/2$ (circular polarization), respectively. Some other examples of more general states of elliptical polarization are shown below (Figure 1.21).

1.7 Changing the Polarization of Light

Unpolarized light can be polarized using a polarizer or polarizing beam splitter (PBS) and the state of already polarized light can be altered using a polarizer and/or optical components that are phase retarders. In this section, we explore some examples of these types of components.

1.7.1 Polarizer and Polarizing Beam Splitters

A polarizer transmits only a single orientation of linear polarization and blocks the rest of the light. For example, a polarizer oriented along the x-direction passes E_x and blocks E_y (see Figure 1.22).

Some polarizer eliminates the non-passed polarization component (E_y in the above example) by absorbing it, while others reflect this component. Absorbing polarizers are convenient when it is desirable to completely eliminate one polarization component from the system. A disadvantage of absorbing polarizers is that they are not very durable and may be damaged by high-intensity light (as found in many laser applications).

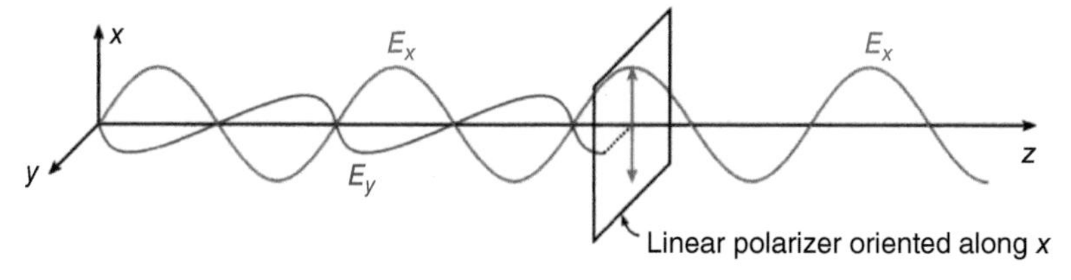

Figure 1.22 Linear polarizer transmits E_x component of the light oriented along the x-axis [4]. Source: Courtesy of Semrock.

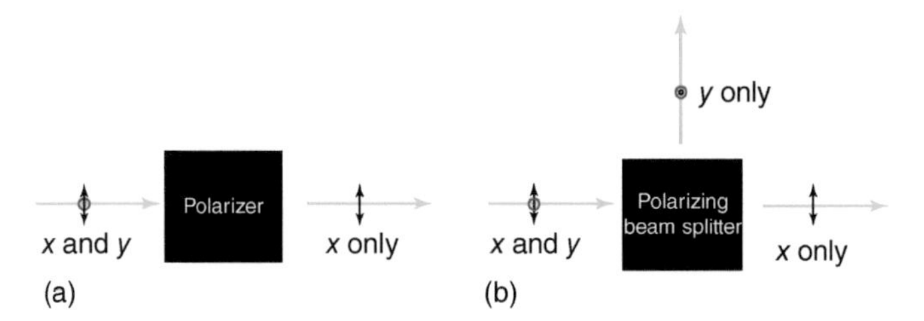

Figure 1.23 Polarization of light (y-component) by the absorbing polarizer (a), and by the reflective beam splitter (b) [4]. Source: Courtesy of Semrock.

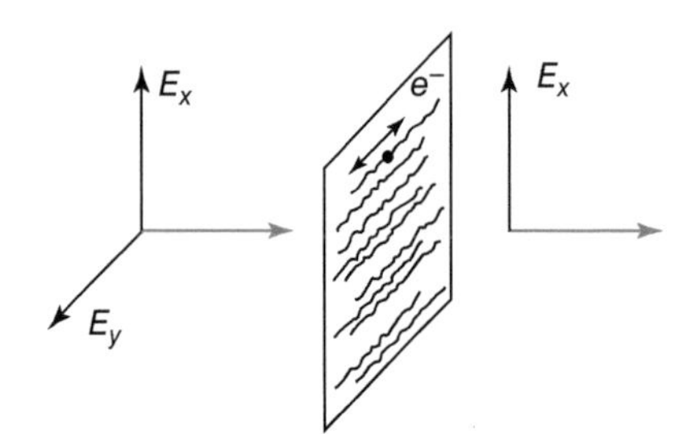

Figure 1.24 Transmission of E_x component of light through the PVA sheet Polaroid [4]. Source: Courtesy of Semrock.

When a reflective polarizer is operated in such a way that the blocked (i.e. reflected) polarization component is deflected into a convenient direction, such as 90° relative to the transmitted polarization component, then the polarizer acts like a PBS, as shown in Figure 1.23. Most PBSs are very efficient polarizers for the transmitted light (i.e. the ratio of desired to undesired polarization is very high); however, the reflected light generally contains some of both polarization components.

There are different ways of making a polarizer. However, as an example, consider one of the most popular absorbing polarizers: the well-known Polaroid "H-Sheet." This polarizer, invented by E. H. Land in 1938, is a plastic, poly-vinyl alcohol (PVA) sheet that has been heated and then stretched in one direction, forming long, nearly parallel hydrocarbon molecule chains. After dipping the sheet into an iodine-rich ink, long iodine chains form along the hydrocarbon molecules. Electrons freely move along the iodine chains, but do not easily move perpendicular to the chains. This ability for electrons to move freely in one direction but not the perpendicular direction is the key principle upon which most absorbing polarizers are based (see Figure 1.24).

When the electric field of a light wave encounters the sheet, the component parallel to the chains causes electrons to oscillate along the direction of that component (E_y in the above example), thus absorbing energy and inhibiting the component from passing through the sheet. Because electrons cannot respond to the other component (E_x), it is readily transmitted.

1.7.2 Birefringent Wave Plate

Some materials have a different index of refraction for light polarized along different directions. This phenomenon is called birefringence. For example, suppose that light polarized along the x-direction sees an index of n_x while light

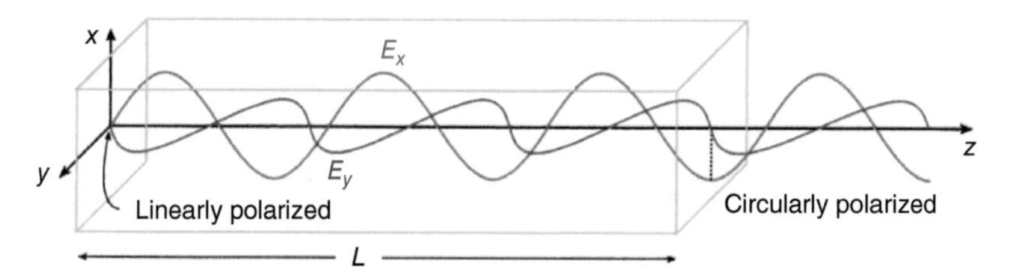

Figure 1.25 Transformation of linearly polarized light into circularly polarized light by the birefringent material [4]. Source: Courtesy of Semrock.

polarized along the y-direction sees an index n_y. Now, suppose linearly polarized light passes through a piece of such a birefringent material of length L, where the linear polarization axis is oriented at 45° angle with respect to the x- and y-axes. The fixed time picture thus looks as in Figure 1.25.

We can see that in general the light emerges in a different state of elliptic polarization. In fact, for the example illustrated above, the particular choice of L for a given difference between n_x and n_y causes the linearly polarized light at the input end to be converted to circularly polarized light at the other end of the birefringent material.

Consider the phases accumulated by the two component waves as they travel through the birefringent material. The waves can be described by

$$E_x = xA\sin\left(\frac{2\pi}{\lambda}n_x z - \omega t\right) \quad \text{and} \quad E_y = yA\sin\left(\frac{2\pi}{\lambda}n_y z - \omega t\right) \tag{1.17}$$

After traveling a length L, the waves have accumulated the respective phases of

$$\theta_x = \frac{2\pi}{\lambda}n_x L \quad \text{and} \quad \theta_y = \frac{2\pi}{\lambda}n_y L \tag{1.18}$$

If the difference between the two phase values is $\pi/2$, then the wave emerging from the material (say into air) will be circularly polarized. This occurs when

$$\theta_y - \theta_x = \frac{2\pi}{\lambda}(n_y - n_x)L = \frac{2\pi}{\lambda}\Delta nL = \frac{\pi}{2} \tag{1.19}$$

or when

$$\Delta nL = \frac{\lambda}{4} \tag{1.20}$$

Because of this relationship, a material with birefringence Δn of the appropriate thickness L to convert linear polarization to circular polarization is called a quarter-wave plate.

Some materials, especially crystals, are naturally anisotropic at microscopic (sub-wavelength) size scales. For example, calcite ($CaCO_3$) is shown in Figure 1.26. The structure, and hence the response to polarized light, along the c-direction is markedly different than that along the a- and b-directions, thus leading to a different index of refraction for light polarized along this direction (see Figure 1.26).

Other materials are nominally isotropic, but when they are bent or deformed in some way, they become anisotropic and therefore exhibit birefringence. This effect is widely used to study the mechanical properties of materials with optics.

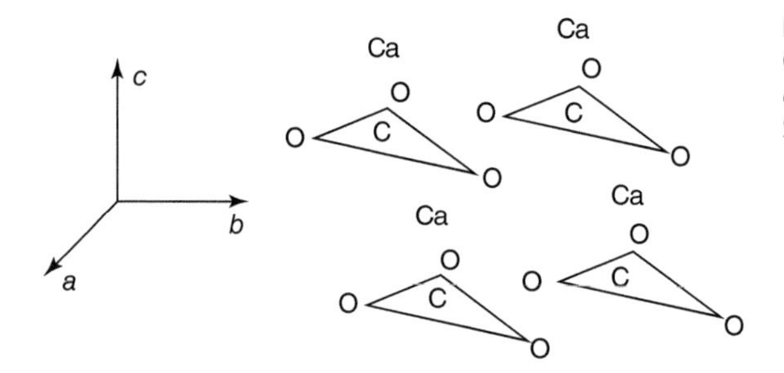

Figure 1.26 Structure of $CaCO_3$ along the *a*-, *b*-, and *c*-axis [4]. Source: Courtesy of Semrock.

1.8 Effects of Reflection and Transmission on Polarization

The polarization of light reflected and transmitted at an interface between two media or at a thin-film multilayer coating can be altered dramatically. These two cases are considered below.

1.8.1 Interface Between Two Media

When light is incident on an interface between two different media with different indexes of refraction, some of the light is reflected and some is transmitted. When the angle of incidence is not normal, different polarizations are reflected (and transmitted) by different amounts. This dependence was first properly described by Fresnel and hence it is often called "Fresnel reflection." It is simplest to describe the polarization of the incident, the reflected, and transmitted (refracted) light in terms of a vector component perpendicular to the plane of incidence, called the "*s*" component, and a component parallel to the plane of incidence, called the "*p*" component. The "plane of incidence" is the plane that contains the incident ray and the transmitted and reflected rays (i.e. all of these rays lie on one plane). In the example in Figure 1.27, the plane of incidence is the plane containing the *x*- and *z*-axes. That is, $\boldsymbol{E}_s \| y$, while $\boldsymbol{E}_p$ lies in the *x*–*z* plane.

The angle of the reflected ray, θ_r, is always equal to the angle of the incident ray, θ_i; this result is called the "law of reflection." The angle of the transmitted (or refracted) ray, θ_t, is related to the angle of incidence by the well-known "Snell's law" relationship: $n_i \sin\theta_i = n_t \sin\theta_t$.

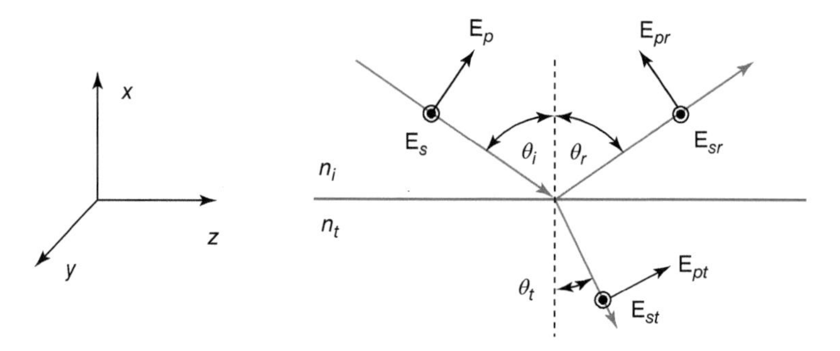

Figure 1.27 Polarization of the incident, reflected, and transmitted (refracted) light in terms of vector components ("*s*" and "*p*" components) [4]. Source: Courtesy of Semrock.

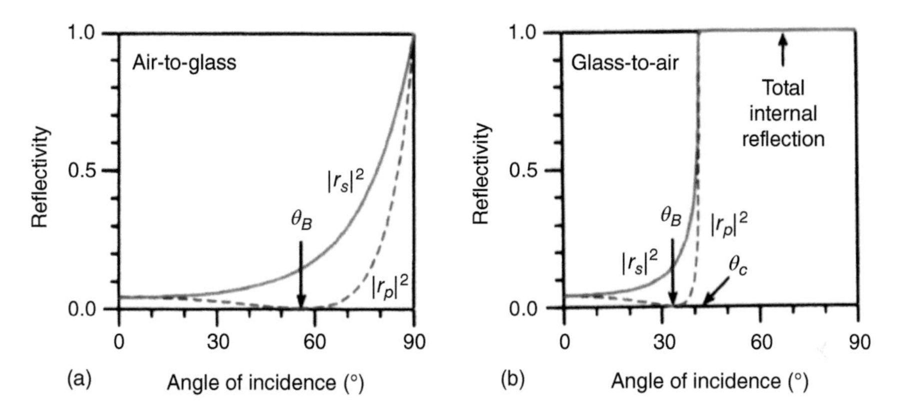

Figure 1.28 The intensity of reflection coefficient for a light wave for air-to-glass (a) and glass-to-air (b) [4]. Source: Courtesy of Semrock.

It turns out that s-polarized light is always more highly reflected than p-polarized light. In fact, at a special angle called "Brewster's angle," denoted θ_B, the p-polarized component sees no reflection, or is completely transmitted. Brewster's angle is given by $\theta_B = \arctan(n_t/n_i)$. The power or intensity reflection coefficients for a light wave (i.e. the squares of the amplitude reflection coefficients) for air-to-glass and glass-to-air polarization appear as in Figure 1.28a and b, respectively.

The Fresnel reflection coefficients for non-normal incidence are given by the equations:

$$r_s = \frac{n_i \cos\theta_i - n_t \cos\theta_t}{n_i \cos\theta_i + n_t \cos\theta_t} \quad \text{and} \quad r_p = \frac{n_t \cos\theta_i - n_i \cos\theta_t}{n_t \cos\theta_i + n_i \cos\theta_t} \tag{1.21}$$

Notice from the graph above on the right that for the case of reflection from a higher index region to a lower index region (in this case glass-to-air, or $n_i = 1.5$ and $n_t = 1.0$), the reflectivity becomes 100% for all angles greater than the "critical angle" $\theta_c = \arcsin(n_t/n_i)$ and for both polarizations. This phenomenon is known as "total internal reflection" (TIR).

For angles of incidence below the critical angle only the amplitudes of the different polarization components are affected by reflection or transmission at an interface. Except for discrete changes of π (or 180°), the phase of the light is unchanged. Thus, the state of polarization can change in only limited ways. For example, linearly polarized light remains linearly polarized, although its orientation (angle α) may rotate. However, for angles greater than θ_c, different polarizations experience different phase changes, and thus TIR can affect the state of polarization of a light wave in the same way birefringence does. Thus, linearly polarized light may become elliptical or vice versa, in addition to changes in the orientation.

1.8.2 Multilayer Thin-Film Filters

Multilayer thin-film coatings have a large number of interfaces, since they are generally comprised of alternating layers of high- and low-index layer materials. The fraction of incident light intensity I_{in} that is reflected (I_R) and transmitted

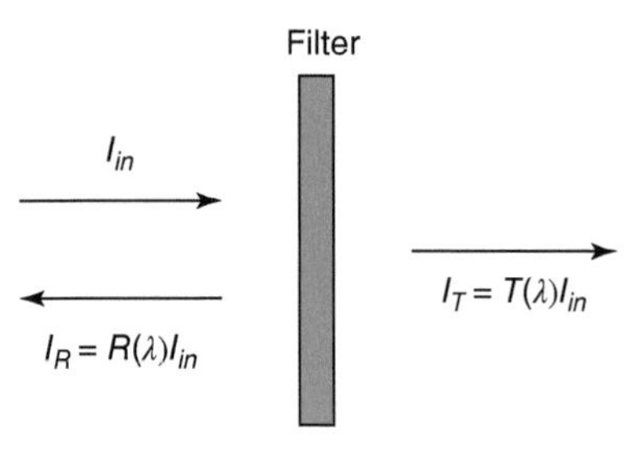

Figure 1.29 The fraction of intensity reflection (I_R) and transmission (I_T) for normal incidence of light on a multilayer thin film filter [4]. Source: Courtesy of Semrock.

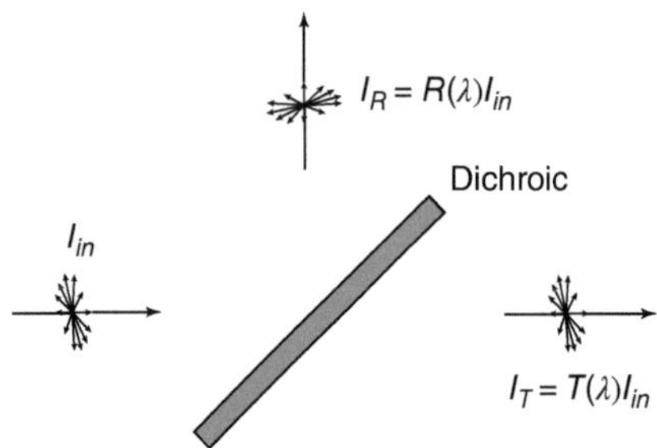

Figure 1.30 The fraction of intensity reflection (I_R) and transmission (I_T) for the non-normal, incoherent, and unpolarized incidence of light [4]. Source: Courtesy of Semrock.

(I_T) through a thin-film coating can be calculated from the indexes of refraction and the precise thicknesses of each layer. These intensity reflection and transmission functions $R(\lambda)$ and $T(\lambda)$, respectively, generally depend strongly on the wavelength of the light, because the total amount of light reflected from and transmitted through the coating comes from the interference of many individual waves that arise from the partial reflection and transmission at each interface. That is why optical filters based on thin-film coatings are called "interference filters" (see Figure 1.29).

When an optical filter is used at a non-normal angle of incidence, as is common with the so-called "plate beam splitters," the filter can impact the polarization of the light. If the incident light is incoherent and unpolarized and the optical system is "blind" to polarization, the standard intensity reflection and transmission functions $R(\lambda)$ and $T(\lambda)$ may be determined for the new angle of incidence, and they are sufficient to characterize the two emerging beams (see Figure 1.30).

However, if the optical system is in any way sensitive to polarization, even when the incident light is unpolarized, it is important to recognize that the beam splitter can transmit and reflect different amounts of the "s" and "p" polarization states, as shown in Figure 1.31.

The amount of light output in each polarization state can be determined by simply breaking up the incident light into its two polarization components (s and p) and then calculating how much of each intensity is transmitted and reflected. For systems based on incoherent light, this level of detail is usually sufficient to keep track of the impacts of components such as optical filters on polarization.

For some optical systems – particularly those based on coherent light and that utilize or are sensitive to interference effects – the complete state of polarization should be tracked at every point through the system. In that case, it is important to understand that optical filters based on multilayer thin-film coatings not only reflect and transmit different amounts of intensity for the "s" and "p" polarization

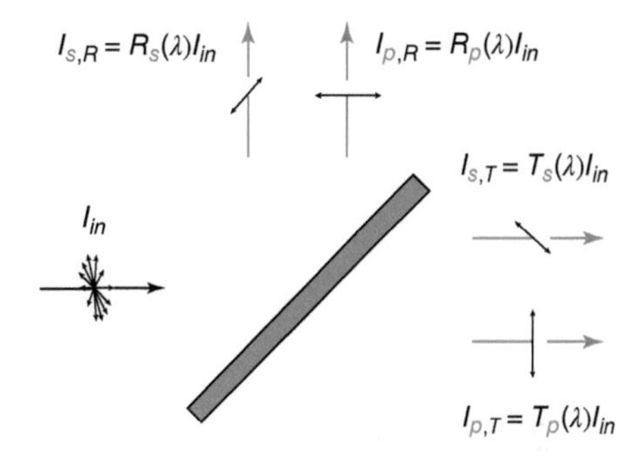

Figure 1.31 The fraction of intensity reflection (I_R) and transmission (I_T) in terms of "s" and "p" polarization states [4]. Source: Courtesy of Semrock.

states but also impart different phases to the two different states. Both the amplitude and phase contributions can depend strongly on the wavelength of light. Thus, in general, an optical filter can act like a combination of a partial polarizer and a birefringent wave plate, for both reflected and transmitted light.

To determine the effect of an optical filter on the light in such a system, the incident light should first be broken up into the two fundamental components associated with the plane of incidence of the filter (*s* and *p* components). Then, the amplitude and phase responses of the filter for the "*s*" and "*p*" components should be applied separately to each of the incident light components to determine the amplitudes and phases of the reflected and transmitted light components. Finally, the reflected "*s*" and "*p*" components can be recombined to determine the total reflected light and its state of polarization, and likewise for the transmitted light. These steps are illustrated in Figure 1.32.

Because the polarization response of a tilted multilayer thin-film coating can be very strong, optical filters can make excellent polarizers. For example, a basic edge filter at a high angle of incidence exhibits "edge splitting" – the edge wavelength for light at normal incidence shifts to a different wavelength for p-polarized light than it does for s-polarized light. As a result, there is a range of wavelengths

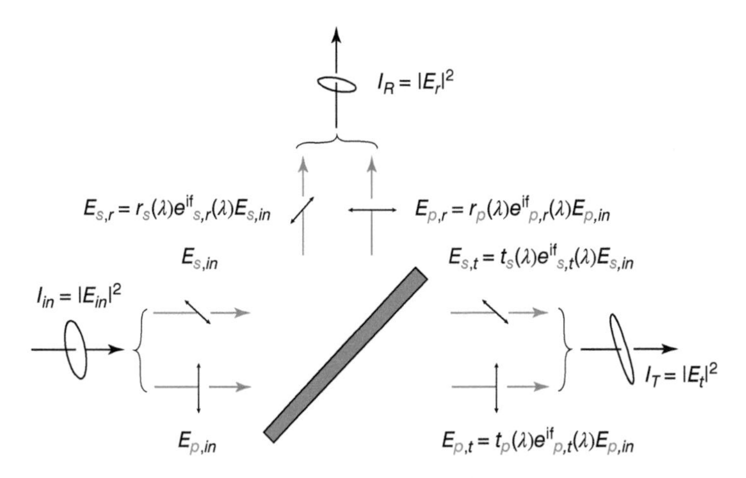

Figure 1.32 Distribution of I_R and I_T in terms of E_s and E_p of the incident light [4]. Source: Courtesy of Semrock.

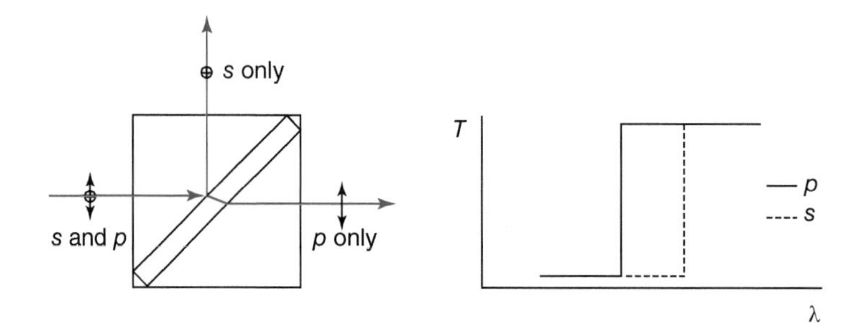

Figure 1.33 Thin-film plate polarizer [4]. Source: Courtesy of Semrock.

for which p-polarized light is highly transmitted while s-polarized light is highly reflected, as shown below (see Figure 1.33).

It is also possible to take advantage of an appreciable difference in reflected or transmitted phase for p- and s-polarized light over a region of the spectrum where the reflected and transmitted intensities are essentially equal, thus forming a wave plate.

1.9 Light Polarizing Devices

This section is a brief description of a number of devices that are used to modify the state of polarization of light. Their respective features are given below.

1.9.1 Polarizing Plate

A polarizing plate is a piece of film by itself or a film being held between two plates of glass. Adding salient iodine to preferentially oriented macromolecules will allow this film to have dichroism. Dichroism is a phenomenon in which discrepancies in absorption occur due to the vibration in the direction of incident light polarization. Since the polarizing plate absorbs the light oscillating in the arranged direction of the macromolecule, the transmitted light rays become linearly polarized. Despite its drawbacks of (i) limited usable wavelength band (visible to near infrared light) and (ii) susceptibility to heat, the polarizing plate is inexpensive and easy to enlarge.

1.9.2 Polarizing Prism

When natural light enters a crystal having double refraction the light splits into two separate linearly polarized beams. By intercepting one of these, linearly polarized light can be obtained; this kind of polarizing device is called a polarizing prism, and among those we find Glan–Thompson prism (Figure 1.34a) and Nicol prism (Figure 1.34b). A polarizing prism has higher transmittance than a polarizing plate and provides high polarization characteristics that cover a wide wavelength band. However, its angle of incidence is limited and it is expensive. In addition, when used in a polarizing microscope, this prism takes up more space than a polarizing plate and may cause image deterioration when placed in an

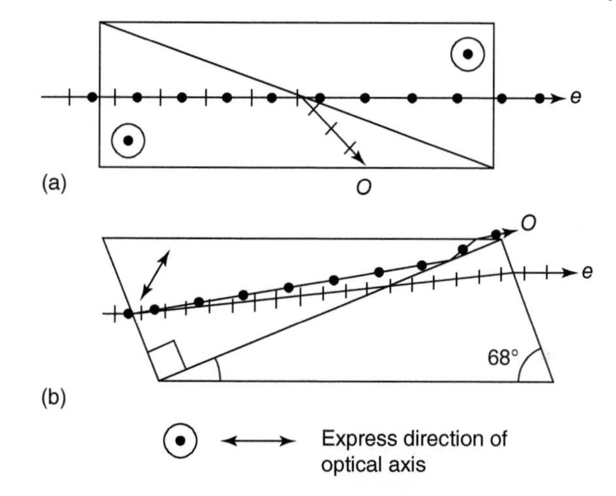

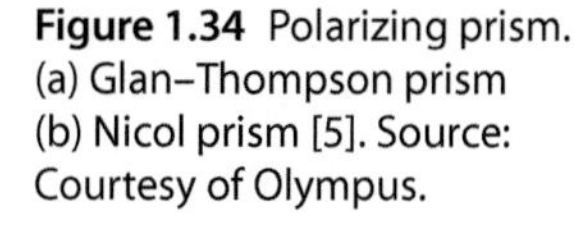
Figure 1.34 Polarizing prism. (a) Glan–Thompson prism (b) Nicol prism [5]. Source: Courtesy of Olympus.

image forming optical system. For these reasons, a polarizing plate is generally used except when brightness or high polarization is required.

1.9.3 Phase Plate

A phase plate is used in the conversion of linearly polarized light and circularly polarized light, and in the conversion of the vibration direction of linearly polarized light. A phase plate is an anisotropic crystal that generates a certain fixed amount of retardation, and based on that amount, several types of phase plates (tint plate, quarter-wave plate, and half-wave plate) are made. When using a quarter-wave plate, a diagonally positioned optical axis direction can convert incident linearly polarized light into circularly polarized light and vice versa (Figure 1.35).

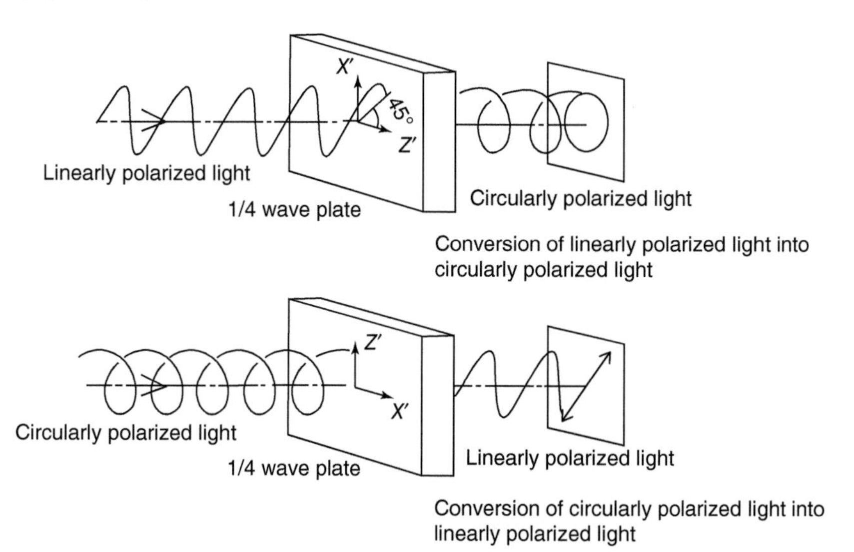

Figure 1.35 Quarter-wave plate conversion of linearly polarized light into circularly polarized light [5]. Source: Courtesy of Olympus.

A half-wave plate is mainly used for changing the vibration direction of linearly polarized light, and for reversing the rotating direction of circularly polarized and elliptically polarized light. Quarter-wave plates, half-wave plates, and tint plates are usually thin pieces of mica or crystal sandwiched in between the glass.

1.9.4 Optical Isolator

The optical isolator is a combination of a linearly PBS and a quartz quarter-wave plate as shown in Figure 1.36. Incident light is linearly polarized by the polarizer and converted to circular polarization by the quarter-wave plate. If any portion of the emerging beam is reflected back into the isolator, the quarter-wave plate produces a beam that is linearly polarized perpendicular to the input beam. This beam is blocked by the linear polarizer and not returned to the input side of the system. Two types of optical isolator are offered: a monochromatic optical isolator and a broadband optical isolator.

1.9.5 Optical Attenuators

An optical attenuator is built by combining two linear polarizers and a half-wave plate. The input and output polarizers are crossed so that no light passes through them; however, inserting the half-wave plate allows light to pass through the device. The amount of light is determined by the angle between the optical axis of the incoming polarizer and the half-wave plate. Placing the half-wave plate's optical axis at 45° to the incoming polarizer achieves maximum transmission; aligning the optical axis of the half-wave plate with either of the input or output polarizer optical axes gives the minimum transmission. How close the minimum is to zero transmission depends on the quality of the polarizer and the half-wave plate used in the device.

Replacing the half-wave plate with a liquid crystal (LC) variable retarder creates a variable attenuator. This configuration is shown in Figure 1.37. When we align the fast axis of the variable retarder at 45° to the input polarizer and modulate the retardance between half wave and full wave, transmission varies between the maximum and the minimum, creating an optical shutter chopper.

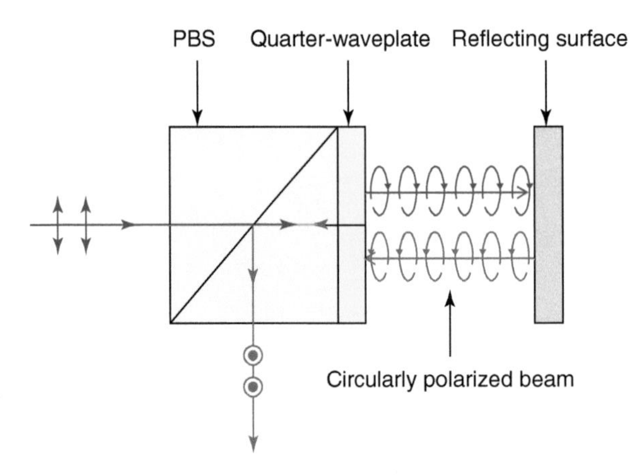

Figure 1.36 Demonstration of optical isolation [6]. Source: Courtesy of Union Optic.

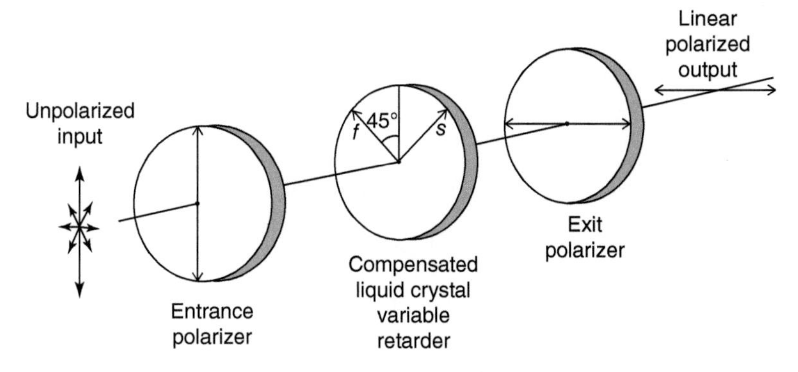

Figure 1.37 The variable attenuator configuration [7]. Source: Courtesy of Meadowlark Optics, Inc.

1.9.6 Polarization Rotator

A polarization rotator is an optical device that rotates the polarization axis of a linearly polarized light beam by an angle of choice. A simple polarization rotator consists of a half-wave plate in linear polarized light. Rotating the half-wave plate causes the polarization to rotate to twice the angle of the half-wave plate's fast axis with the polarization plane, as shown in Figure 1.38.

The polarization rotation induced by optically active crystals, such as wave plates, is reciprocal. If the polarization direction is rotated from right to left (say) on forward passage (as viewed by a fixed observer) it will be rotated from left to right on backward passage (as viewed by the same observer), so that back-reflection of light through an optically active crystal will result in light with zero final rotation, the two rotations having canceled out. This is best shown in the Figure 1.39.

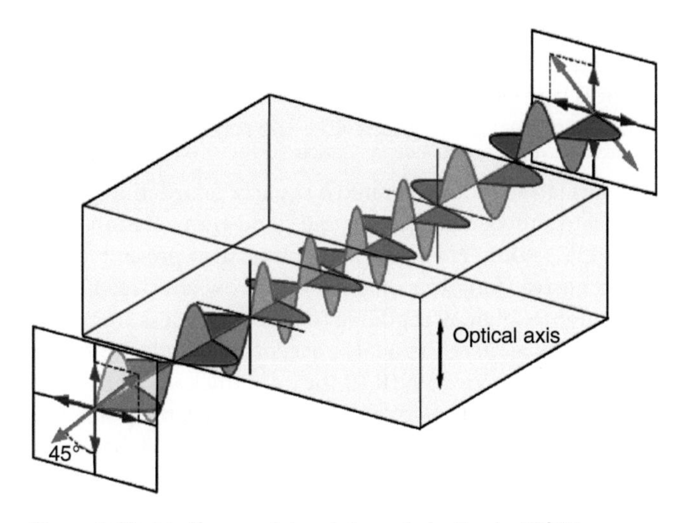

Figure 1.38 A half-wave plate rotates polarization by 90° [8].

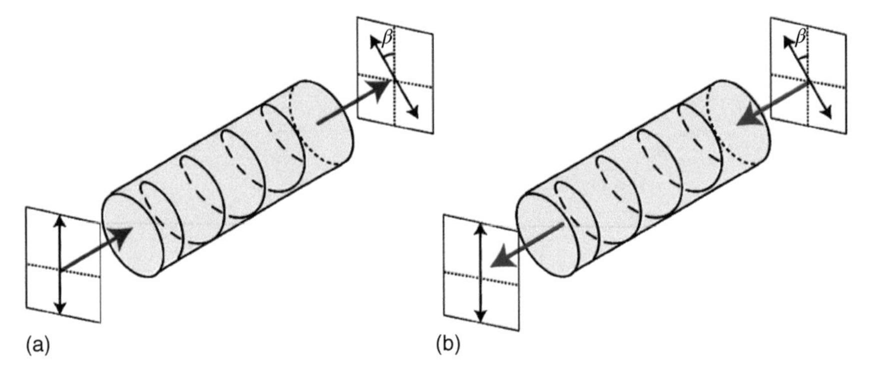

Figure 1.39 (a) Left to right polarization and (b) right to left polarization [9]. Source: Courtesy of Fosco.

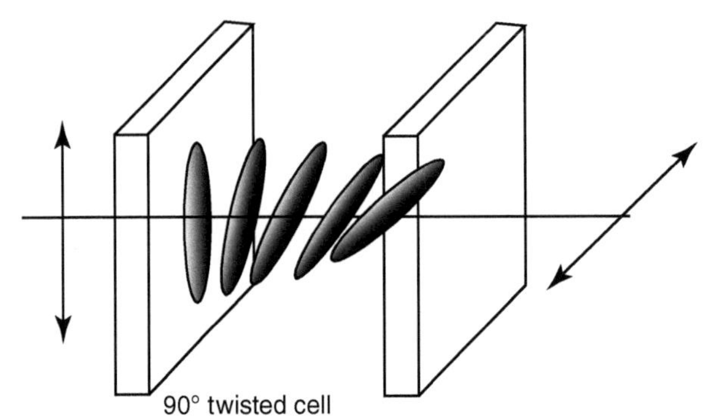

Figure 1.40 Liquid crystal twisted nematic polarization rotator cell [10]. Source: Courtesy of ARCoptix.

This is because the polarization rotation that occurs here is a result of a longitudinal spirality (represented as a helix) in the crystal structure. Hence, rotation following the handedness of the spiral in the forward direction will be opposed by the spiral in the backward direction.

Liquid crystal twisted nematic polarization rotator (TN cell) is very useful when one wants to rotate the orientation of linear polarization by a fixed amount, typically 45° or 90°. When light is traversing an LC twisted nematic cell its polarization follows the rotation of the molecules (Figure 1.40). The screen of any laptop computer is based on the same effect.

In optical systems, the polarization is often rotated by quartz retardation plates (1/2 or 1/4 plates). Quartz plate shows high quality and good transmission performances especially in the UV region. However, such plates also present some disadvantages: They are expensive, function only for a narrow spectral bandwidth, and have a small incidence angle acceptance (field of view less than 2°). The liquid crystal nematic cells have therefore a large acceptance angle, function over a very large spectral range from Vis to NIR (if they are thick enough), and are less expensive. Optionally, by applying a voltage on the TN cell, the polarization rotation can be "switched off." Also, when placing a 90° twisted cell between crossed polarizers, it can be used as a shutter.

References

1 Latorre, P., Seron, F.J., and Gutierrez, D. (2012). Birefringency: calculation of refracted ray paths in biaxial crystals. *Visual Comput.* 28 (4): 341–356.
2 Ubachs, W. (2001). *Nonlinear Optics*, Lecture Notes, Department of Physics and Astronomy. Amsterdam: University of Amsterdam.
3 Murphy, D.B., Spring, K.R., Fellers, T.J., and Davidson, M.W. *Principles of Birefringence: Introduction to Optical Birefringence*. Nikon Instruments, Inc. https://www.microscopyu.com/techniques/polarized-light/principles-of-birefringence (accessed 22 January 2019).
4 Erdogan T. A new class of polarization optics designed specifically for lasers. Semrock technical note Series: Understanding polarization, https://www.semrock.com/Data/Sites/1/semrockpdfs/whitepaper_understandingpolarization.pdf (accessed 22 January 2019).
5 Basics of polarizing microscopy – Production manual, Olympus. http://research.physics.berkeley.edu/yildiz/Teaching/PHYS250/Lecture_PDFs/polarization%20microscopy.pdf.
6 Product Specification-Optical Isolator, Union Optic. http://www.u-optic.com/gid_52_prdid_68_pid_300_l_1.
7 Meadowlark Optics, Inc. (2005). Basic polarizing Techniques and Devices. http://www.meadowlark.com/store/applicationNotes/Basic%20Polarization%20Techniques%20and%20Devices.pdf (accessed 22 January 2019).
8 Polarization Rotator, Wikipedia. https://en.wikipedia.org/wiki/Polarization_rotator (accessed 22 January 2019).
9 Fosco. Fiber optics for sale Co. https://www.fiberoptics4sale.com/blogs/wave-optics/99205446-faraday-effect (accessed 22 January 2019).
10 ARCoptix Production manual – polarization rotator. http://www.arcoptix.com/polarization_rotator.htm (accessed 22 January 2019).

2
Photoelasticity

2.1 Introduction

Photoelasticity is the property of some transparent solid materials, such as glass, plastic, or crystals, to become doubly refracting (i.e. a ray of light will split into two rays at entry) under the influence of external stress. When photoelastic materials are subjected to pressure, internal strains develop that can be observed in polarized light – i.e. light vibrating normally in two planes, which has had one plane of vibration removed by passing through a substance called a polarizer. Two polarizers that are crossed ordinarily do not transmit light, but if a stressed material is placed between them and if the principal axis of the stress is not parallel to this plane of polarization, some light will be transmitted in the form of colored fringes.

This phenomenon was first discovered by Sir David Brewster in the year 1815 [1], who presented a paper before the Royal Society of London where he reported the effect. In 1816, he reported the same effect in amorphous solids such as glass and in cubic crystals such as fluorspar and diamond. In 1818, he also studied the photoelastic behavior in some uniaxial and biaxial crystals. In his earlier experiments, he placed a piece of glass in between two crossed polarizers. He found that when the glass was stretched transversely to the direction of propagation of light, the field of view brightened up, thereby showing that an artificial birefringence was induced in the glass strip by the mechanical stress. Furthermore, he found in the case of solids that are initially birefringent that the initial birefringence is altered by the stress. Thereafter, photoelastic techniques were developed to study crystals and other transparent solids. It has become an important experimental method for the measurement of internal stress. The work of E. Coker and L. Filon at the University of London enabled photoelasticity to be developed rapidly into a viable technique for qualitative stress analysis. It found widespread use in many industrial applications, as in two dimensions it exceeded all other techniques in reliability, scope, and practicability. No other method had the same visual appeal or covered so much of the stress pattern.

2.2 Principle of Photoelasticity

Photoelasticity is an experimental method to determine stress distribution in a material. The method is based on the birefringence property exhibited by

Crystal Optics: Properties and Applications, First Edition. Ashim Kumar Bain.

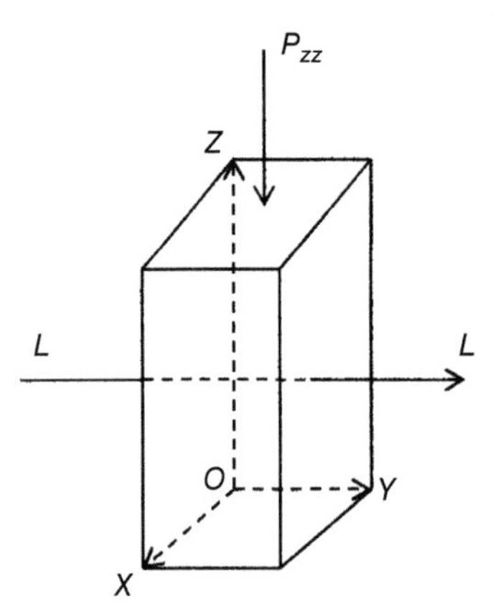

Figure 2.1 A solid under a linear stress of stress-optical measurements (P_{zz} is the applied stress and LL is the direction of light propagation and observation).

transparent materials. In particular, the phenomenon of stress-induced birefringence is utilized where the material becomes birefringent under the influence of external loading. The property of birefringence (or double refraction) is observed in many optical crystals. Upon the application of stresses, photoelastic materials exhibit the property of birefringence and the magnitude of the refractive indices at each point in the material is directly related to the state of stresses at that point. When a ray of light passes through a photoelastic material, its electromagnetic wave components get resolved along the two principal stress directions and each of these components experiences different refractive indices due to the birefringence. The difference in the refractive indices leads to a relative phase retardation between the two components.

If a rectangular parallelepiped with edges parallel to $x[100]$, $y[010]$, and $z[001]$ axes is stressed along z-axis and observation is made along y-axis, as shown in Figure 2.1; then the path retardation δ_{zy} introduced per unit length due to the stress-induced birefringence is given by

$$\delta_{zy} = (\Delta n_z - \Delta n_x) = C_{zy} P_{zz} \tag{2.1}$$

where Δn_z and Δn_x are the changes in the corresponding refractive indices, $(\Delta n_z - \Delta n_x)$ is the corresponding stress-induced birefringence, P_{zz} is the stress along z-axis, and C_{zy} is a constant called the Brewster constant or the relative stress-optical coefficient. In general, the Brewster constant is related to the stress-optical and strain-optical tensors of fourth rank [2] and is a measure of the stress-induced (piezo-optic) birefringence. It is conveniently expressed in the unit of $10^{-13}\,\text{cm}^2/\text{dyne}$ per centimeter thickness along the direction of observation, called a Brewster [2].

2.3 History of Photoelasticity

The photoelastic phenomenon was first observed by Sir David Brewster [1]. His first studies were concerned with the stress-optical effect in jellies, but in 1818, he had investigated the effect in amorphous solids, and isotropic, uniaxial, and biaxial crystals [3, 4]. Brewster was the first to discuss stress birefringence in diamonds. The work of Brewster was entirely qualitative but it inspired Fresnel [5, 6] and several scientists, to attempt a theoretical explanation. Neumann in 1840s founded a mathematical explanation for the observed behavior of noncrystalline substances, based on the strain dependence of the birefringence [7]. In 1853, Maxwell related the birefringence to stress and developed the stress-optical laws [8]. Although developed independently, both theories arrived at analogous results. Coker and Filon applied this technique to structural engineering in 1902 [9]. This technique was, however, systematically developed into a viable experimental stress analysis tool only in the second quarter of this century. The works of Frocht [10], Jessop and Harris [11] are particularly noteworthy in this regard. The potential of 3D stress analysis was also developed by Oppel [12] and Hetenyi [13] and to this day remains one of the very few experimental methods for 3D stress analysis. With this background, photoelasticity found widespread application in many industrial applications. In particular, determination of stress concentration in front of notches and holes was and still is one of the premier forte of photoelasticity in design of machine elements [14].

In the past two decades, photoelasticity has been overrun by the finite element method (FEM). The FEM has become the dominant technique, overshadowing many traditional techniques for stress analysis. Despite FEM advances, photoelasticity, one of the oldest methods for experimental stress analysis, has been revived through recent developments and new applications. When using FEM, it is crucial to assess the accuracy of the numerical model, and ultimately this can only be achieved by experimental verification. For example, a threaded joint experiences nonuniform contact, which is difficult to incorporate accurately into a computer model. Idealized models therefore tend to underestimate the actual maximum stress concentration at the root of the thread. Therefore, photoelasticity remains a major tool in modern stress analysis.

A new school of thought is the use of hybrid methods where advantages of both experimental and numerical methods are exploited. Nevertheless, recent needs such as continuous online monitoring of structures, residual stresses in glass (plastics) and microelectronics material, rapid prototype products, and dynamic visualization of stress waves have brought photoelasticity into the limelight once again.

With the advent of digital polariscope using light-emitting diodes, continuous monitoring of structures under load became possible. This led to the development of dynamic photoelasticity. Dynamic photoelasticity has contributed greatly to the study of complex phenomena such as fracture of materials. Many excellent scientists and research institutions have been involved in research, development, and application of the phenomenon of photoelasticity. A chronological listing of many of the more notable specific events in the history of photoelasticity is given in Table 2.1.

Table 2.1 Notable events in the history of photoelasticity.

1815	David Brewster (discovery of photoelasticity) [1]
1853	James Maxwell (one-dimensional photoelastic theory) [8]
1900	Coker polariscope (designed the standard polariscope) [15, 16]
1935	H. Muller (theory of the photoelastic effect of cubic crystals) [17]
1960	J. Badoz (first photoelastic modulator) [18]
1962	G. U. Oppel (photoelastic device for indicating principal strain directions) [19]
1967	A. Roberts (photoelastic stress indicating devices) [20]
1969	Kemp-modulator (conventional photoelastic modulators) [21]
1979	Muller and Saackel (photoelasticity + image processing) [15]
1980s	Digital image processing on photoelastic fringes [15]
1983	J. Bodez and J. C. Canit (new design for a photoelastic modulator) [22]
1984	J. C. Canit and C. Pichon (low-frequency photoelastic modulator) [23]
1994	A. K. Asundi and M. R. Sajan (low-cost digital polariscope for dynamic photoelasticity) [24]
2000	New polariscopes (cheaper cameras, better image quality, algorithms for automation) [15]
2007	F. Bammer (single crystal photoelastic modulator) [25]
2007	F. Bammer and R. Petkovsek (Q-switching of a fiber laser with a single crystal photoelastic modulator) [26]
2008	G. Ivtsenkov (fiber-optic strain sensor and associated data acquisition system) [27]
2009	D. Tang et al. (three-component hybrid-integrated optical accelerometer based on $LiNbO_3$ photoelastic waveguide) [28]
2015	N. E. Khelifa and Himbert (photoelastic force sensor) [29]

2.4 Phenomenological Theory of Photoelasticity

Fresnel developed the well-known elastic solid theory, in order to describe the phenomenon of double refraction in crystalline solids. According to this, to any given crystal there corresponds, for a given wavelength of incident light of radiation and at a given temperature, a certain ellipsoid of elasticity with its axes uniquely oriented with respect to the crystal lattice. This led him to the discovery of laws of governing the propagation of light in crystalline media.

The optical properties of crystals can be expressed in terms of the refractive index ellipsoid, whose semi axes are proportional to the three principal refractive indices as shown in Figure 2.2. The equation of refractive index ellipsoid for any crystal can be represented as

$$\frac{x^2}{n_1^2} + \frac{y^2}{n_2^2} + \frac{z^2}{n_3^2} = 1 \tag{2.2}$$

where, n_1, n_2, and n_3 are the three principal refractive indices for the wavelength λ. This equation is not valid for all wavelengths for crystals exhibiting monoclinic and triclinic symmetry. For these two cases, however,

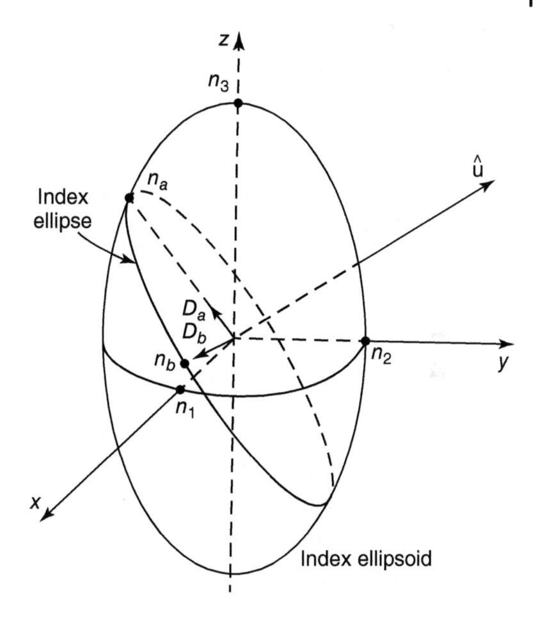

Figure 2.2 Optical indicatrix (index ellipsoid) of a positive crystal.

a transformation is necessary, leading to an equation for the index ellipsoid in terms of six parameters. For the sake of completeness, we shall use the equation for the index ellipsoid, in the general form

$$\frac{x^2}{n_{11}^2}+\frac{y^2}{n_{22}^2}+\frac{z^2}{n_{33}^2}+\frac{2yz}{n_{23}^2}+\frac{2zx}{n_{31}^2}+\frac{2xy}{n_{12}^2}=1 \tag{2.3}$$

The equation of this refractive index ellipsoid or optical index ellipsoid can be written in the form

$$\sum_{i,j} B_{ij}x_ix_j = 1 \quad (i,j = 1,2,3) \tag{2.4}$$

where the three optical parameters B_{ij} ($B_{ij} = \frac{1}{n_{ij}^2}$) are equal to the reciprocals of the squares of the respective refractive indices.

The choice of the x, y, z Cartesian axes, coinciding with the crystallographic axes of all crystal systems, except triclinic and monoclinic, is now followed to represent the phenomenological theory of photoelasticity as developed by Pockels. Pockels theory is based on the following two assumptions:

(1) In a homogeneous deformed solid, the effect of deformation is only to alter the optical parameters of the optical index ellipsoid.
(2) When the strain is within elastic limits, the change in the optical parameter (ΔB_{ij}) of the solid due to deformation can be expressed as the function of all nine stress components P_{kl} or all nine strain components ε_{kl} through the homogeneous linear relations.

$$\Delta B_{ij} = -q_{ijkl}P_{kl} \tag{2.5}$$

or

$$\Delta B_{ij} = p_{ijkl}\varepsilon_{kl} \tag{2.6}$$

where i, j, k, and l take the values 1, 2, 3. The coefficients q_{ijkl} and p_{ijkl} are fourth polar rank tensors, called the stress-optical (piezo-optic) and strain-optical (elasto-optic) coefficients, respectively. Both the optical parameter coefficients and the strain tensors are symmetric, so that $B_{ij} = B_{ji}$, $\varepsilon_{kl} = \varepsilon_{lk}$, and $p_{ijkl} = p_{jikl} = p_{ijlk} = p_{jilk}$. So, Eqs. (2.5) and (2.6) can be expressed in the matrix or two-suffix notation as given below:

$$\Delta B_i = -q_{ij}P_j \quad (i,j = 1,\ldots,6) \tag{2.7}$$

$$\Delta B_i = p_{ij}\varepsilon_j \quad (i,j = 1,\ldots,6) \tag{2.8}$$

Equation (2.8), when expanded, yields the following:

$$\begin{aligned}
\Delta B_1 &= p_{11}\varepsilon_1 + p_{12}\varepsilon_2 + p_{13}\varepsilon_3 + p_{14}\varepsilon_4 + p_{15}\varepsilon_5 + p_{16}\varepsilon_6\\
\Delta B_2 &= p_{21}\varepsilon_1 + p_{22}\varepsilon_2 + p_{23}\varepsilon_3 + p_{24}\varepsilon_4 + p_{25}\varepsilon_5 + p_{26}\varepsilon_6\\
\Delta B_3 &= p_{31}\varepsilon_1 + p_{32}\varepsilon_2 + p_{33}\varepsilon_3 + p_{34}\varepsilon_4 + p_{35}\varepsilon_5 + p_{36}\varepsilon_6\\
\Delta B_4 &= p_{41}\varepsilon_1 + p_{42}\varepsilon_2 + p_{43}\varepsilon_3 + p_{44}\varepsilon_4 + p_{45}\varepsilon_5 + p_{46}\varepsilon_6\\
\Delta B_5 &= p_{51}\varepsilon_1 + p_{52}\varepsilon_2 + p_{53}\varepsilon_3 + p_{54}\varepsilon_4 + p_{55}\varepsilon_5 + p_{56}\varepsilon_6\\
\Delta B_6 &= p_{61}\varepsilon_1 + p_{62}\varepsilon_2 + p_{63}\varepsilon_3 + p_{64}\varepsilon_4 + p_{65}\varepsilon_5 + p_{66}\varepsilon_6
\end{aligned} \tag{2.9}$$

where $\Delta B_1 = \Delta B_{11}$, $\Delta B_4 = \Delta B_{23}$, etc. and $\varepsilon_1 = \varepsilon_{11}$, $\varepsilon_4 = \varepsilon_{23}$, etc. Similarly, Eq. (2.7) can be expanded in the form

$$\begin{aligned}
\Delta B_1 &= -(q_{11}P_1 + q_{12}P_2 + q_{13}P_3 + q_{14}P_4 + q_{15}P_5 + q_{16}P_6)\\
\Delta B_2 &= -(q_{21}P_1 + q_{22}P_2 + q_{23}P_3 + q_{24}P_4 + q_{25}P_5 + q_{26}P_6)\\
\Delta B_3 &= -(q_{31}P_1 + q_{32}P_2 + q_{33}P_3 + q_{34}P_4 + q_{35}P_5 + q_{36}P_6)\\
\Delta B_4 &= -(q_{41}P_1 + q_{42}P_2 + q_{43}P_3 + q_{44}P_4 + q_{45}P_5 + q_{46}P_6)\\
\Delta B_5 &= -(q_{51}P_1 + q_{52}P_2 + q_{53}P_3 + q_{54}P_4 + q_{55}P_5 + q_{56}P_6)\\
\Delta B_6 &= -(q_{61}P_1 + q_{62}P_2 + q_{63}P_3 + q_{64}P_4 + q_{65}P_5 + q_{66}P_6)
\end{aligned} \tag{2.10}$$

where $P_1 = P_{11}$, $P_4 = P_{23}$, etc. The negative sign in Eqs. (2.5), (2.7), and (2.10) arises from Pockels convention that positive stress produces negative strain. Here, the stress is taken as positive when compressional and negative when extensional. The strain, however, is considered positive for an extension and negative for a compression.

The stress and strain components are related by the following equations (in single suffix notation):

$$P_k = -c_{kj}\varepsilon_j, \quad \varepsilon_k = -s_{kj}P_j \quad (j,k = 1-6) \tag{2.11}$$

Similarly, the piezo-optic and elasto-optic components are interrelated through the elastic constants (in two-suffix notation):

$$p_{ij} = \sum_{k=1}^{6} q_{ik}c_{kj}, \quad q_{ij} = \sum_{k=1}^{6} p_{ik}s_{kj}, \quad (i,j = 1-6) \tag{2.12}$$

where c_{kj} and s_{kj} are elastic stiffness constants and elastic compliance constants, respectively. The set of 36 constants composed of the p_{ij}s and q_{ij}s completely define the behavior of a crystal when subjected to known strains or stresses. Only in the crystals of lowest symmetry do all the 36 constants have values different from zero. The number decreases with increasing symmetry, becoming 20 for monoclinic, 12 for orthorhombic, and so on.

The first three equations in formulae (2.9) and (2.10) express changes in the principal velocities that the crystal experiences on deformation and these can be measured by the usual methods by making observations with the incident light having its electric vectors parallel to the three axes of the undeformed crystal. The last three equations represent a rotation of the principal axes of the ellipsoid on deformation. This rotation is determined by measuring the change in refractive index under stress with the incident electric vector perpendicular to any one of the axes and bisecting the angle between the other two. From the change in length of the corresponding radius vector of the ellipsoid, the amount of rotation can be calculated.

Thus, by measuring the changes in the principal refractive indices and the rotation of the principal axes for various directions of pressure and observation, one gets a system of linear equations in q_{ij} in terms of measurable quantities, from which q_{ij} and hence p_{ij} can be evaluated. From knowledge of q_{ij} and p_{ij}, the three principal refractive indices and the orientation of the principal axes of the deformed crystal can be evaluated. For details information, the reader is referred to the study on the book: *Photoelastic and Electro-optic Properties of Crystals by T.S. Narasimhamurty* [2].

2.5 Atomic Theory of Photoelasticity

In 1935, H. Muller developed the atomic theory of photoelasticity [17] of which earlier attempts were made by several scientists [30–32]. When a solid is subjected to a mechanical stress, its symmetry is altered and the calculation of photoelastic constants is based on the evaluation of the changes in the Coulomb field, the Lorentz–Lorentz field, and the intrinsic polarizability of the ions. As these computations are extremely complicated for crystals of lower symmetry, the theory has been worked out only for isotropic solids and cubic crystals.

The contribution of the Coulomb field to the refractive index of an undeformed cubic crystal is zero by virtue of its symmetry and would be significant only when the crystal is deformed anisotropically. When a cubic crystal is unidirectionally stressed along the cube normal, the deformation reduces the crystal symmetry to a tetragonal one, causing anisotropy in the Coulomb and the Lorentz fields. In photoelastic experiments, the lattice distances parallel to the stress are decreased and the distances normal to the stress are increased. Muller pointed out that a deformation of the lattice changes the energy levels and transition probabilities of the optical electrons and hence gives rise to anisotropy of the atomic refractions. By taking into account all these effects, Muller was able to give a fairly satisfactory explanation of the observed photoelasticity of some cubic crystals, with D (diagonal) lattices.

The change in polarizability of ions with strain follows readily from the classical electromagnetic theory. The interaction of light with crystalline matter under a mechanical stress can be represented by the influence of the electromagnetic field of the incident light wave on a charged oscillator, which in this case is an electron. The incident electric vector of light radiation causes the electron to get displaced from its equilibrium position. Since we are dealing with an ionic crystal, the oscillator is influenced not only by the incident electric vector, but also by the Coulomb fields, due to other ions and the Lorentz–Lorentz field. The displacement of the electron (the charged particle) results in the production of a dipole moment of value

$$\mu_i = \alpha_i E_{eff} \tag{2.13}$$

where α_i is the polarizability of the ith ion and E_{eff} is the effective electric field (Coulomb field and Lorentz–Lorentz fields). If N_i is the number of oscillators of type i in a cubic centimeter and μ_i is the dipole moment per oscillator, then the total dipole moment per unit volume is given by

$$P = \sum_i N_i \mu_i = \sum_i N_i \alpha_i E_{eff} \tag{2.14}$$

Now, combining the relations $D = kE$ and $D = E + 4\pi P$ with the relation $k = n^2$, we can have

$$(n^2 - 1)E = 4\pi P \tag{2.15}$$

where D is the dielectric displacement caused by an external electric field E, acting on a medium of dielectric constant k, P is the dipole moment induced per unit volume, and n is the refractive index of the medium.

Combining Eqs. (2.14) and (2.15), we can have

$$(n^2 - 1)E = 4\pi \sum_i N_i \alpha_i E_{eff} \tag{2.16}$$

We are concerned with the change in the refractive index caused by the mechanical strain developed in the medium. Such a relation can be obtained by differentiating Eq. (2.16) with respect to the strain ε.

$$2n\frac{dn}{d\varepsilon}E = 4\pi \sum_i \left[\frac{dN_i}{d\varepsilon}\alpha_i E_{eff} + N_i \frac{d\alpha_i}{d\varepsilon}E_{eff} + N_i \alpha_i \frac{dE_{eff}}{d\varepsilon}\right] \tag{2.17}$$

The first term inside the brackets describes the effect due to the change in the number of oscillators of type i per cubic centimeter due to the strain. The third term describes the effect due to the change in the total effective electric field due to strain. The second term $d\alpha_i/d\varepsilon$ describes the effect due to the change of polarizability of the ions of type i per cubic centimeter due to the strain.

The strain-optical constants p_{ij} in the phenomenological theory of Pockels can be related to the terms in Eq. (2.17) by the following equations:

$$\begin{aligned} p_x &= p_0 + p_x^L + p_x^C + p_x^A \\ p_z &= p_0 + p_z^L + p_z^C + p_z^A \\ p_{x'} &= p_{x'}^L + p_{x'}^C + p_{x'}^A \end{aligned} \tag{2.18}$$

where

$$p_x = \frac{n^4}{(n^2-1)^2}p_{12} = \frac{-2n}{(n^2-1)^2}\frac{\partial n_1}{\partial \varepsilon_3}$$
$$p_z = \frac{n^4}{(n^2-1)^2}p_{11} = \frac{-2n}{(n^2-1)^2}\frac{\partial n_3}{\partial \varepsilon_3}$$
$$p_{x'} = \frac{2n^4}{(n^2-1)^2}p_{44} = \frac{-2n}{(n^2-1)^2}\frac{\partial n_4}{\partial \varepsilon_3} \tag{2.19}$$

and p_0 is the contribution due to the change of oscillator density

$$p_0 = \frac{n^2+2}{3(n^2-1)} \tag{2.20}$$

The term p_x^L is due to the anisotropy of the Lorentz force, and p_x^C gives the effect produced by the Coulomb field of the ions. Hence p_x^L and p_x^C together arise out of $dE_{eff}/d\varepsilon$. Finally, p_x^A is the contribution from the change in polarizability $d\alpha_i/d\varepsilon$.

Here n is the refractive index for the unstrained state and n_1, n_3, and n_4 are for the strained state of the crystal. We see from Eq. (2.18) that the three constant terms p_x, p_z, and $p_{x'}$ introduced by Muller are all related to the strain-induced changes in the oscillator density, the polarizability, and the effective field.

The birefringence produced by a strain ε in the direction [001] is given by

$$p = p_z - p_x = \frac{n^4(p_{11}-p_{12})}{(n^2-1)^2} \tag{2.21}$$

Similarly, the birefringence introduced by a shear is given by

$$p' = p_{x'} = \frac{2n^4}{(n^2-1)^2}p_{44} = \frac{-2n}{(n^2-1)^2}\frac{\partial n_4}{\partial \varepsilon_3} \tag{2.22}$$

Equations (2.21) and (2.22) represent the final results. They give p and p', which are proportional to the elasto-optic constants. For detailed information, the readers are referred to follow the books [2, 33].

2.6 Photoelastic Devices

Photoelasticity was first observed during the 1800s when certain glasses and ceramics were seen to exhibit fringe patterns when the materials were under stress and viewed under polarized light. The principle of photoelasticity was not quantified until Brewster's law was applied to the fringe patterns to describe the relative change in the index of refraction as being proportional to the difference in strain. The photoelastic phenomenon soon became a useful engineering tool that could provide a quantitative analysis of residual strains or material deformation of manufactured components.

Many transparent ferroelectric materials (e.g. $LiNbO_3$ and $LiTaO_3$) have widely been employed to be used as hosts of photoelastic devices due to their interesting and much studied photoelastic properties. Some of the photoelastic devices made of ferroelectric materials are described here.

2.6.1 Photoelastic Modulator

A photoelastic modulator (PEM) is an optical device used to modulate the polarization of a light source. The photoelastic effect is used to change the birefringence of the optical element in the PEM. The PEM was first invented by J. Badoz in the 1960s and originally called a "birefringence modulator." It was initially developed for physical measurements including optical rotary dispersion and Faraday rotation, polarimetry of astronomical objects, strain-induced birefringence, and ellipsometry. Later developers of the PEM include J.C. Kemp, S.N. Jasperson, and S.E. Schnatterly.

The typical PEM consists of a piece of glass (e.g. SiO_2, CaF_2, ZnSe) on which a piezoelectric transducer made of SiO_2 is glued [21, 34]. Both pieces are tuned to the same longitudinal eigenfrequency such that when the transducer is electrically excited on this frequency, the whole assembly starts to oscillate with high amplitude in a longitudinal mode. Now due to the photoelastic effect an artificial birefringence is induced in the initial isotropic glass such that it acts as a wave plate with temporally varying retardation. Hence, the polarization of light will be modulated. For example, when the induced oscillation amplitude corresponds to quarter-wave retardation and the incident polarization is inclined 45° to the direction of oscillation, the output polarization will oscillate between left- and right-circularized light. If the oscillation amplitude corresponds to a half-wave retardation and the input is again 45°-linear polarized, the output polarization will oscillate between −45° and 45° linear polarized light with intermediate states 45°-linearly, left- and right-circularly and elliptically polarized.

A single crystal photoelastic modulator (SCPEM) is made of a ferroelectric optical transparent crystal $LiTaO_3$ that is electrically excited on one of its resonance frequencies (Figure 2.3) [26, 35]. A necessary feature is that the polarization of light of any wavelength passing this crystal must not be changed when the crystal is at rest. This, for example, cannot be fulfilled with the quartz crystal of a conventional PEM, due to its optical activity, which would need to be compensated by a second reversely oriented quartz crystal as proposed in an early patent on PEMs [36].

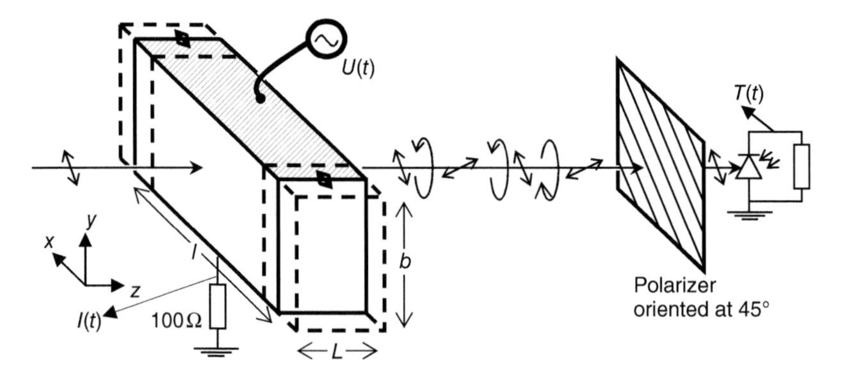

Figure 2.3 Single crystal photoelastic modulator (SCPEM) made of a 3 m crystal. Source: Bammer et al. 2011 [35]. Reproduced with permission of Intech.

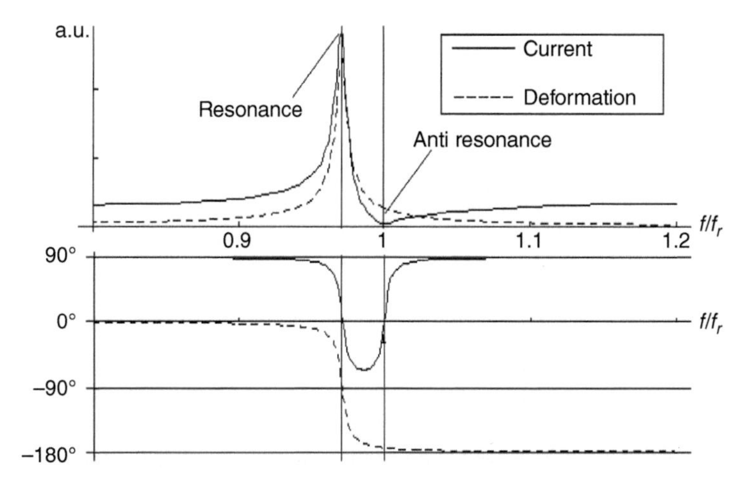

Figure 2.4 Course of amplitude (top) and phase (bottom) of current and deformation at one typical resonance of a piezoelectric element against normalized frequency. Source: Bammer et al. 2011 [35]. Reproduced with permission of Intech.

Figure 2.3 shows one favorable configuration of a PEM based on a crystal with symmetry 3m. The light travels along the optical axis (z-axis), the exciting electrical field points in the y-direction, and in most cases the longitudinal x-eigenmode is used. Two further important eigenmodes are the longitudinal y-eigenmode and the shear yz-eigenmode. Of course, there exist infinitely many higher eigenmodes and frequencies. In most cases, a polarizer (analyzer) oriented at 45° is placed behind the modulator. Figure 2.3 further shows a photodiode to get a transmission signal T and a resistor (here with 100 Ω) to generate a measure for the piezoelectric current I generated by the crystal.

Figure 2.4 shows typical resonance curves for current and deformation, which are common to all piezoelectric elements. It can be seen that outside the range of resonance the crystal acts electrically as a capacitor, i.e. the phase shift of the resulting current against the driving voltage is 90°. At resonance, current and deformation take a significant maximum while at the slightly higher frequency of antiresonance the current amplitude is extremely small (zero in case of a lossless device). The deformation amplitude in resonance is usually 2 or 3 orders of magnitude higher than the static deformation. The important advantage now is that the amplitude of the driving voltage is in the order of 10 V, i.e. 2 orders of magnitude lower than for electro-optic Q-switches.

However, this utilization of the resonance behavior of the SCPEM leads to a limited frequency range. A measure of the possible frequency shift is the 3 dB-bandwidth of the crystal resonance, which is usually below 0.1% of the resonance frequency. Depending on the available driving amplitude the driving frequency can be shifted within this range. Another option is to use other resonance frequencies. In their experiment, Bammer and Petkovsek [26] used the first resonance frequency at 199 kHz (x-longitudinal oscillation). The next useful frequencies can be found at 348 kHz (y-thickness oscillation) and 377 kHz (yz-shear oscillation).

2.6.2 Photoelastic Q-Switch

Pulsed laser operation is of utmost importance for many applications in manufacturing and medicine. One important method to produce short and strong laser pulses is Q-switching. This method uses inside the laser cavity an optical switch, which blocks the optical light path while energy is pumped to the gain medium. When this switch suddenly opens, all stored energy is rapidly converted to a strong light pulse. Owing to the very fast development of lasers based on an active fiber core there is a profound interest in various techniques for Q-switching of fiber lasers.

Bammer et al. introduced a new kind of Q-switching device based on the SCPEM [26, 35]. This modulator relies like acousto-optics modulators on photoelasticity, but modulates polarization instead of beam direction. It is made of $LiTaO_3$, which is electrically excited to oscillate in a longitudinal eigenmode. An important advantage of using eigenmode oscillations is that there is no need for a high voltage driver as in the case of standard electro-optic modulators.

The effect of any kind of PEM on the polarization of light with wavelength λ can be described by the so-called retardation defined as

$$\delta = \frac{L}{\lambda}(n_h - n_v) \tag{2.23}$$

where L is the optical interaction length and n_h, n_v are the temporally varying refractive indices for horizontally and vertically polarized light. These can be calculated via the time-dependent strains $\varepsilon_1,\ldots,\varepsilon_6$ and the photoelastic coefficients $p_{11},\ldots,p_{66}$ [36, 37]. For harmonic excitation at a resonance frequency f_r the harmonic retardation course is defined as

$$\delta = A\sin(\omega_r t) \tag{2.24}$$

where $\omega_r = 2\pi f_r$ and A is the amplitude of the retardation course. For the polarization course (sketched in Figure 2.3) $A = \lambda/2$ holds. If a polarizer oriented at 45° is placed behind the SCPEM and 45° linear polarized light is sent on the configuration, the transmission can be calculated by [37]

$$T_r(\delta) = \cos^2\left(\frac{\pi}{2}\delta\right) \tag{2.25}$$

A typical transmission curve with rather sharp transmission peaks achievable with a harmonic retardation course is shown in the third graph of Figure 2.6. Similar considerations can be found in Ref. [38], where a similar type of polarization modulator was used for time multiplexing of laser pulses emitted by high power laser diodes (LDs).

The main idea of the experimental setup is to test an SCPEM incorporated into the laser cavity as a Q-switch. The SCPEM is made of $LiTaO_3$ with dimensions $13.15\times 7.15\times 5.5\,\mathrm{mm}^3$ in x-, y-, and z-directions. The electrodes are on the zx-surfaces and the light travels along the z-axis through the xy-surfaces. The first resonance frequency is found at $f_r = 199\,\mathrm{kHz}$ and corresponds to a longitudinal oscillation in x-direction.

A schematic diagram of the experimental setup of Q-switching fiber laser is shown in Figure 2.5. It is based on a Nd-doped D-shaped double clad fiber

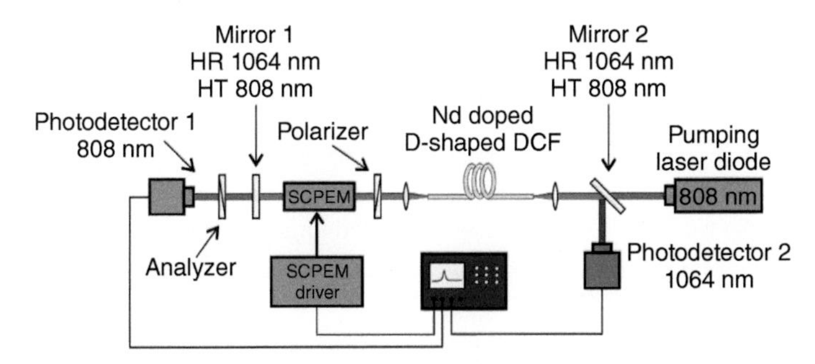

Figure 2.5 Setup of a fiber laser with SCPEM Q-switch. Two photodetectors are used simultaneously in order to detect both the transmittance of the SCPEM and the output pulses of the laser. Source: Bammer et al. 2011 [35]. Reproduced with permission of Intech.

(DCF). The concentration of Nd_2O_3 is 1300 mol/ppm. The length of the fiber is 5 m and the diameter of the active core and inner cladding is 13 and 400 μm, respectively. The corresponding numerical aperture is 0.13 and 0.37.

The measured laser output pulses and the transmission course of the modulator are as follows: Pumping light with 808 nm from a 5 W laser diode is guided through a coupler to the fiber end that also acts as an output mirror (4% of reflectance) of the laser cavity. The output laser light is reflected by the wavelength-sensitive mirror 2 and measured by photodiode 2. On the other end of the fiber the emerging laser light is collimated and then led through a polarizer and an SCPEM and is finally reflected by a highly reflective interference mirror 1 to travel again through the SCPEM and the polarizer back to the fiber. Pumping light that is not absorbed into the fiber and goes through the SCPEM is transmitted by mirror 1 and goes through an analyzer and falls onto photodiode 1.

Figure 2.6 shows the results of the measurements. The first graph shows the voltage on the electrodes (note the low amplitude: ~3 V), and the second one shows the resulting current generated by the crystal. It is measured with a 100 Ω resistor between the lower crystal electrode and ground (100 Ω are much lower as the absolute value of the crystal impedance in resonance and hence has negligible influence on the resonance behavior). The voltage amplitude is adjusted in order to optimize the output laser pulses. The current signal is in phase with the voltage signal, indicating a perfect adjustment to the resonance frequency. The third graph contains the transmittance of the pumping light through the polarizer/SCPEM/analyzer configuration measured by photodiode 1 (blue line). The red line is the estimated transmittance for the laser light when going through the optical elements in the sequence: polarizer/SCPEM (reflection) SCPEM/polarizer. This double pass configuration doubles the retardance induced by the SCPEM.

Now for the transmission of the pumping light Eq. (2.26) holds with a pumping light retardation δ_p. With estimated retardation amplitude of 0.70 a fit of Eq. (2.26) to the measured transmission curve (blue line in third graph of Figure 2.6) can be obtained. Hence the retardation is

$$\delta_p(t) = 0.70 \sin(\omega_r t) \tag{2.26}$$

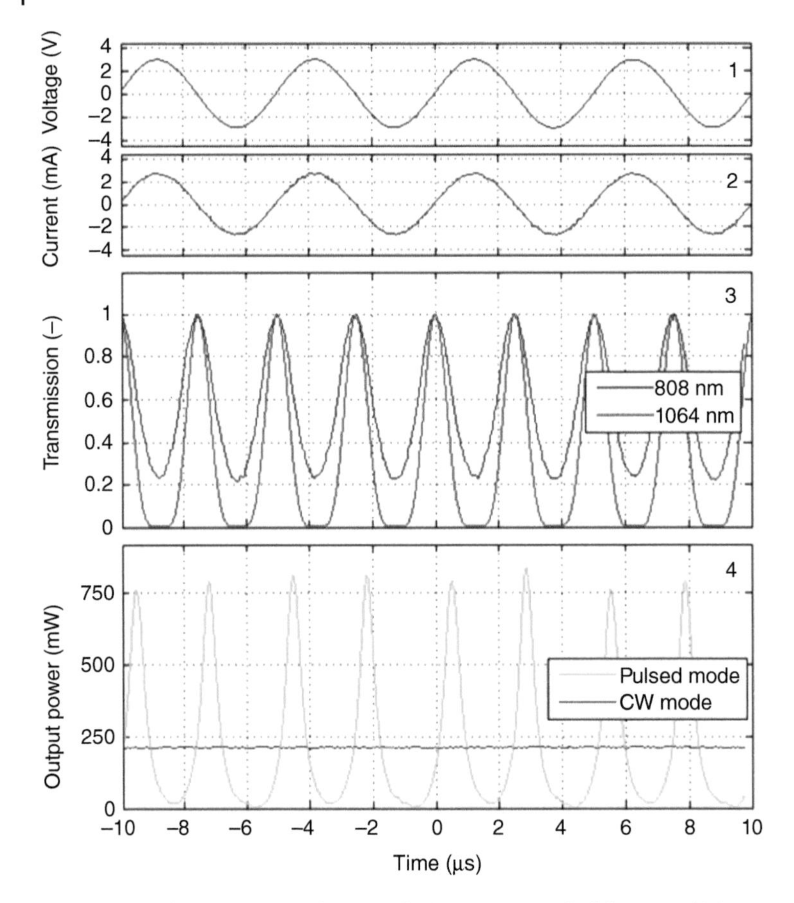

Figure 2.6 Voltage course on the crystal (1), current signal of the crystal (2), measured transmission for the pumping light (black), and estimated transmission for the laser light (gray) (3), optical output of the laser (4). Source: Bammer et al. 2011 [35]. Reproduced with permission of Intech.

Now Eq. (2.23) shows that δ scales with $1/\lambda$. Further the laser light retardation δ_l is doubled by the double pass configuration. Hence with the wavelengths $\lambda_1 = 1064$ nm for the laser light and $\lambda_p = 808$ nm for the pumping light, the estimated (neglecting the dispersion of refractive index and photoelastic coefficients) laser light retardation is

$$\delta_1(t) = 2\frac{\lambda_p}{\lambda_1}\delta_p(t) \tag{2.27}$$

The resulting estimated transmission curve for the laser light (red curve in third graph of Figure 2.6) shows between the transmission peaks for ~800 ns a complete block of the light path.

The resulting pulsed output of the laser (fourth graph) offers laser pulses with peak powers three to four times above the average power, which is obtained when no voltage is applied on the modulator. It is worth mentioning that the average output of the laser during pulsed mode is very close to that in *cw*-mode, namely

~200 mW. Hence the modulation scheme introduces no significant decrease in efficiency.

It can be seen in the fourth graph of Figure 2.6 that the amplitude of the laser pulses varies. It is believed that this is related to a phenomenon explained in Ref. [39], where certain repetition rates of externally imposed switching leads to a variation of the pulse energy.

The investigation presented allows a simple modulation of the output power of a laser, without significant loss of efficiency when compared to *cw*-operation. Possible applications may be found in fields where the simple laser structure is an important advantage and where the fixed pulse frequency does not represent a deficiency such as laser marking and engraving. This could be micro-welding of metals, where pulsed operations sometimes help to smooth the welding seam surface; rapid prototyping of plastics, where some polymers need high optical peak powers to harden; and laser range finding, where only small, simple, cheap, and robust solutions are possible.

2.6.3 Photoelastic Accelerometer

There have been several proposals for optical accelerometers based on photoelastic effects using fiber optics by Su et al. [40], the advantages of which are wider frequency band and dynamic range, excellent ability to withstand overload, and simple compact conformation. However, it can just detect one-component signal. In order to enhance the accuracy of three-component seismic acceleration detection, a novel three-component micro-optoelectronic mechanical system (MOEMS) accelerometer based on photoelastic of $LiNbO_3$ crystal has been designed by Tang et al. for the first time [28, 41].

Figure 2.7 shows a schematic diagram of the harmonic vibrator of the combined three-component photoelastic waveguide accelerometer. It consists of three separated dual Mach–Zehnder interferometers on X-cut and Y-propagation

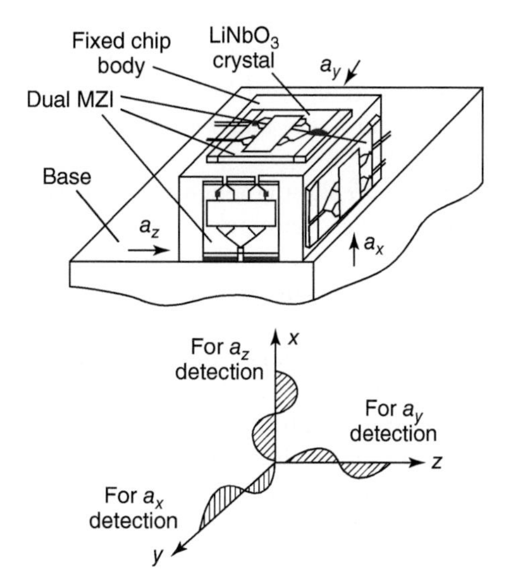

Figure 2.7 Harmonic vibrator of the combined three-component photoelastic waveguide accelerometer. Source: Tang et al. 2009 [28]. Reproduced with permission of Chinese Optics Letters.

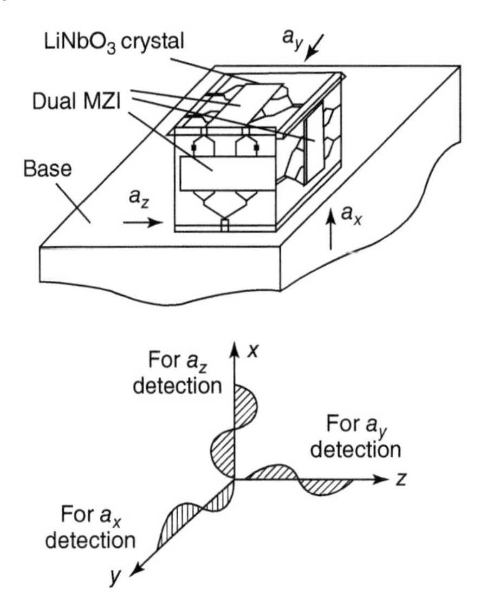

Figure 2.8 Harmonic vibrator of the integrated three-component photoelastic waveguide accelerometer. Source: Tang et al. 2009 [28]. Reproduced with permission of Chinese Optics Letters.

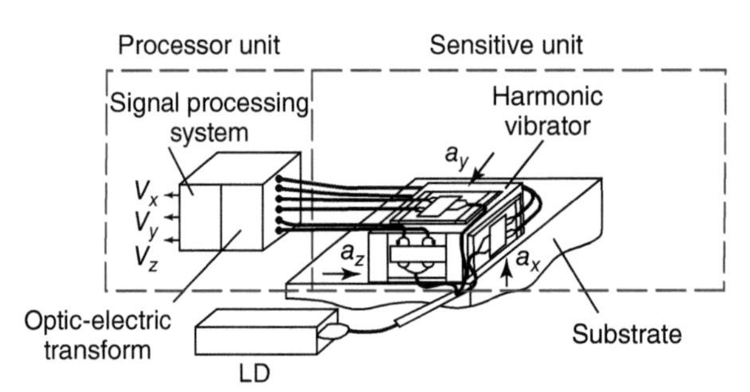

Figure 2.9 A schematic drawing of the three-component photoelastic waveguide accelerometer. Source: Tang et al. 2009 [28]. Reproduced with permission of Chinese Optics Letters.

$LiNbO_3$, respectively. With the separated dual Mach–Zehnder interferometers (MZIs) fixed on a body, a harmonic vibrator of the combined three-component photoelastic waveguide accelerometer is designed. For the integrated structure, three separated dual MZIs are monolithically integrated on three vertical surfaces of an X-cut and Y-propagation $LiNbO_3$ crystal respectively, as shown in Figure 2.8.

Figure 2.9 shows the schematic of the designed three-component photoelastic waveguide accelerometer with combined structure, which is mainly composed of LD laser, three-dimensional simple harmonic oscillator mounted on the basement, and outside processor. The sensitive unit is the three-component simple harmonic oscillator, which is made of three Mach–Zehnder interferometers arranged in the x-, y-, z-directions. The lights from LD-laser, which is divided into three beams by beam splitter, are input into three Mach–Zehnder interferometers, respectively. By the action of acceleration in an arbitrary direction, the three Mach–Zehnder interferometers are capable of sensing the three components in x-, y-, and z-axis (a_x, a_y, a_z) of acceleration a due to photoelastic effect, and transforming the three components of acceleration into optical phase-changes, which are then transmitted into the outside processor to execute the differential

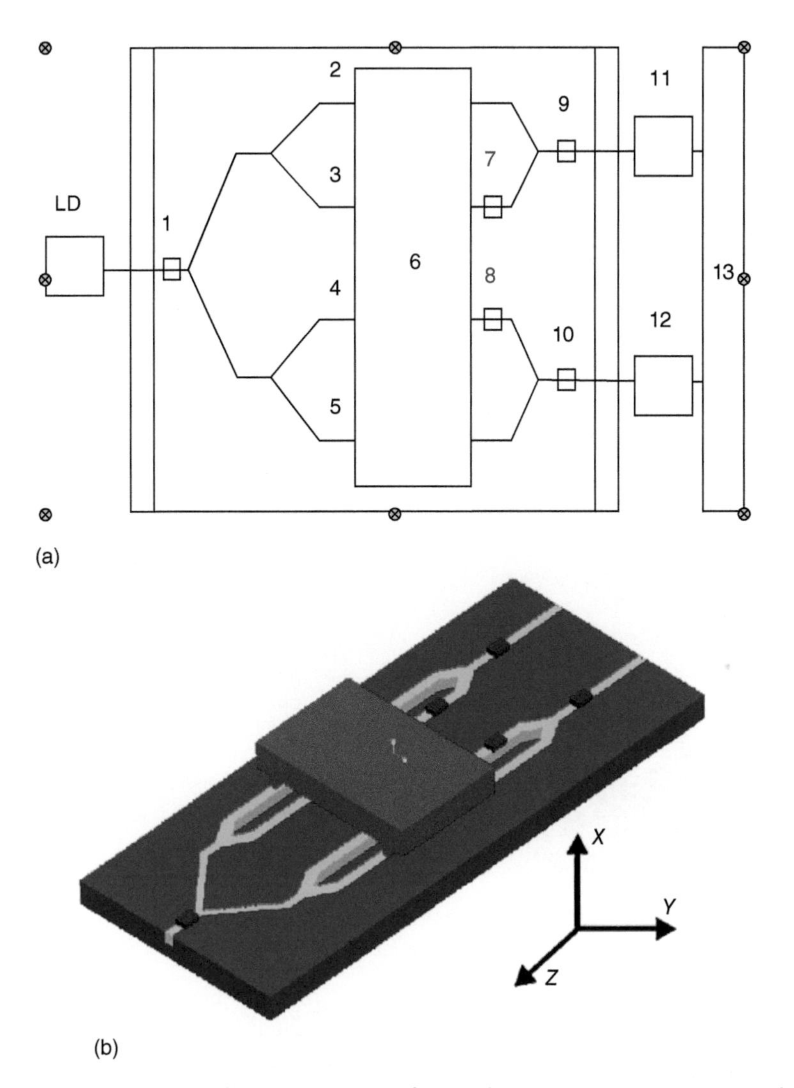

Figure 2.10 (a) The detection principle of photoelastic sensitive system in y-axis direction. 1, 9, 10 – polarimeters; 2, 5 – measurement waveguides; 3, 4 – reference waveguides; 6 – mass; 7, 8 – phase modulators; 11, 12 – photodetectors; 13 – signal processing unit. (b) Schematic of integrated optical chip of Mach–Zehnder microinterference accelerometer. Source: Tang et al. 2011 [41]. Reproduced with permission of Elsevier.

processing, and the relation between voltage and acceleration could be obtained afterwards; thus the acceleration testing is completed.

As shown in Figure 2.10a, light from the LD-laser molds into the horizontal polarized one via polarizer (1), and enters the waveguide in four parts after two times of equal strength division; when the mass (6) undergoes an acceleration in any direction, inertia force F is generated and then transmitted into waveguides (2) (5), which are attached to the mass. Owing to photoelastic effect, the index of refraction of waveguides changes correspondingly, which will lead to a change of optical phase. Because the light in waveguides (3) (4) are unaffected by the influence of acceleration from outside, they can be the reference arms that are modulated by the modulators (7) (8) to have a 90° phase shift. The lights, which are affected by outside acceleration and modulated by phase modulator, are coupled by branch waveguide into analyzers (9) (10), which are employed to realize and maintain the polarization state of the output light waves. The photoelectric cells (11) (12) translate the lights coming out from the analyzers into electric signals, which will be processed by the processing unit (13), and hence the acceleration that needs to be measured can be calculated.

The prototype accelerometer to be characterized is fixed onto the top of a mechanical vibrator, which provides a simulating acceleration signal, as shown in Figure 2.11. A spectrum analyzer is used to record the frequency spectrum. The sensitivity of the accelerometer as a function of vibration frequency is demonstrated in Figure 2.12. The acceleration is in the vertical direction of the light axis (a_z). The cross-axis sensitivity is lower than the on-axis sensitivity,

Figure 2.11 Photograph of the accelerometer on the mechanical vibrator. Source: Tang et al. 2009 [28]. Reproduced with permission of Chinese Optics Letters.

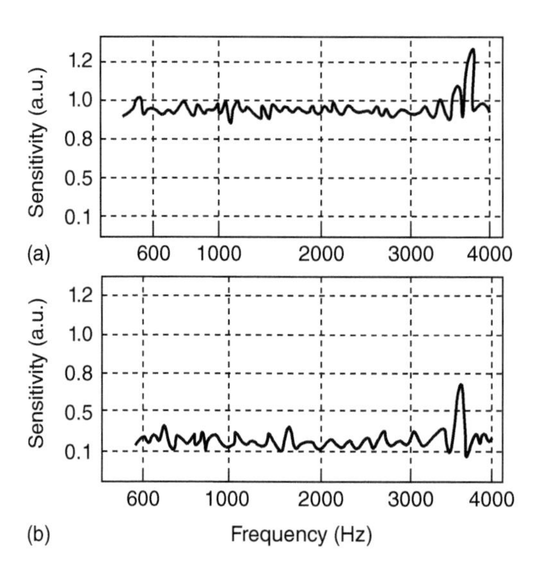

Figure 2.12 (a) On-axis sensitivity and (b) cross-axis sensitivity versus vibration frequency for acceleration in the vertical direction of light axis. Source: Tang et al. 2009 [28]. Reproduced with permission of Chinese Optics Letters.

and the corresponding resonance frequency is about 3.5 kHz and the transverse sensitivity ratio (TSR) = 0.1 (frequency range: 150–3000 Hz). It can be seen that the accelerometer has a good frequency responding characteristic when the frequency is below 3 kHz, which accords with the operation frequency in high-accuracy seismic exploration. This accelerometer can be employed to monitor or accurately measure vibrations in various areas such as seismic measurement in geophysical survey.

2.6.4 Photoelastic Force Sensor

A photoelastic force sensor is a solid-state laser in which the effect of the force is an induced birefringence. This kind of sensor was introduced first by Holzapfel et al. [42, 43]. In fact, a birefringent crystal is an optical support that has the property to resolve the field associated with a light wave into two orthogonal components, relative to the direction of propagation, and to transmit them with different speeds. A transparent and optically isotropic material under normal conditions, such as a crystal, can become birefringent when subjected to a constraint. The phenomenon, known as the photoelastic effect, appears and disappears almost instantaneously with the application of the force onto the crystal.

The birefringence is induced in an Nd:YAG laser rod under mechanical stress generated by external force applied on the crystal, as illustrated in Figure 2.13. In order to link the force intensity to the induced birefringence (based on photoelastic effect), Khelifa and Himbert [29] assumed that the stress distribution over the length of the laser rod is uniform. The stress components σ_y and σ_x along the principal directions of the rod, induced by the applied force F, are represented by [10]:

$$\sigma_x = \frac{\alpha}{\pi l d} \times F \quad \text{and} \quad \sigma_y = -\frac{\beta}{\pi l d} \times F \tag{2.28}$$

where l and d are the length and diameter of the cylindrical crystal respectively, while the parameters α and β depend on the nature of the contact between the

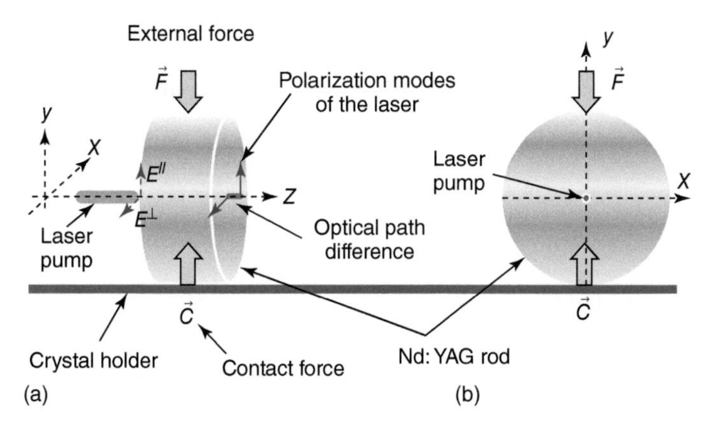

Figure 2.13 Orientation of the photoelastic rod relative to the direction of the applied force. (a) Front view; (b) side view. Source: Khelifa and Himbert 2015 [29]. Reproduced with permission of Sensors & Transducers.

laser crystal and its support and on the orientation of the pumping beam at 808 nm relative to the principal axis of the Nd:YAG crystal.

The relative stress, along the orthogonal directions x and y, induced in the center of the rod is then given by relation:

$$\sigma_x - \sigma_y = \frac{(\alpha + \beta)}{\pi l d} \times F \tag{2.29}$$

The induced frequency shift between the frequencies of the orthogonal polarizations $E^{\parallel}$ and $E^{\perp}$ of the oscillating laser mode is expressed as a function of external force by

$$\Delta \nu = \nu_q^{\perp} - \nu_q^{\parallel} = \frac{(\alpha + \beta) C_{PE}^{\lambda_q}}{\pi n} \times \frac{\nu_q}{l d} \times F \tag{2.30}$$

where $C_{PE}^{(\lambda_q)}$ is the photoelastic constant of the Nd:YAG crystal. For laser light of wavelength $\lambda_q = 1064\,\text{nm}$ (frequency $\nu_q \cong 281.76\,\text{THz}$), the theoretical value of the relative stress-optic coefficient is given by [44]

$$C_{PE}^{(\lambda_q)}[111] = 1.25 \times 10^{-12}\,\text{m}^2 \times \text{N}^{-1}$$

The sensitivity of the force sensor is then approximated by the relation:

$$S(l, d) \cong \frac{(\alpha + \beta) C_{PE}^{\lambda_q}}{\pi n} \times \frac{\nu_q}{l d} \tag{2.31}$$

In an ideal configuration, one edge of the cylindrical rod is in contact with a flat surface and the rod is illuminated by the laser beam along its axis of revolution. In this case α and β are related via [45]

$$(\alpha + \beta) \cong \frac{8}{\pi} \tag{2.32}$$

For a value of $(\alpha + \beta)$ given by Eq. (2.32), the sensitivity is reduced to

$$S(l, d) \cong 4928 \times \frac{1}{l d} \tag{2.33}$$

In Eq. (2.33), the sensitivity is in $\text{MHz}\,\text{N}^{-1}$ if l and d are in millimeters.

For technical reasons, Khelifa and Himbert [29] used a cylindrical crystal of length 2 mm and diameter 3 mm with parallel end faces (Figure 2.14). The

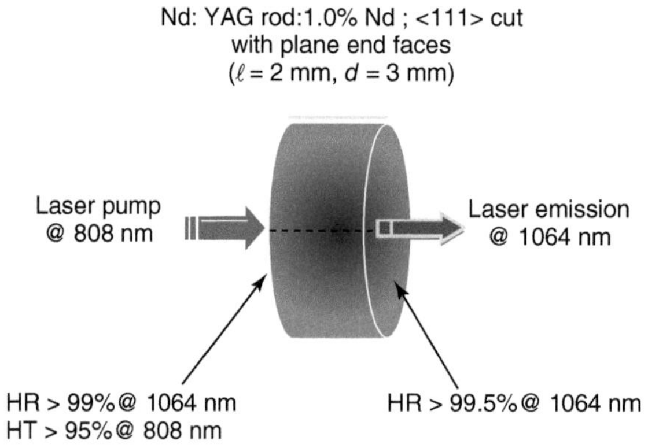

Figure 2.14 Monolithic configuration of the Nd:YAG laser used as sensing element of the force sensor. Source: Khelifa and Himbert 2015 [29]. Reproduced with permission of Sensors & Transducers.

pumping face is coated to have HR@1064 nm and HT@808 nm and the second one is coated with HT@808 nm and HR(99.5%)@1064 nm. The laser is bonded to its holder formed by a rectangular channel of width 3.5 mm and depth 3 mm, machined in an aluminum part. The temperature of the rod and the holder assembly is stabilized to better than ±0.02 °C by using a proportional–integral–derivative (PID) controller.

The response and the sensitivity of this photoelastic sensor were analyzed by measuring the deviation of the beat note frequency when applying, on the top of the crystal laser, deadweight linked to a mass standard, m_e. The mass standard values of 0.1 g (≈1 mN) to 20 g (≈200 mN) were used to study the response of the photoelastic sensor. As shown in Figure 2.15a, the response is almost linear over the range studied and is checked on a wider range. In fact, same results

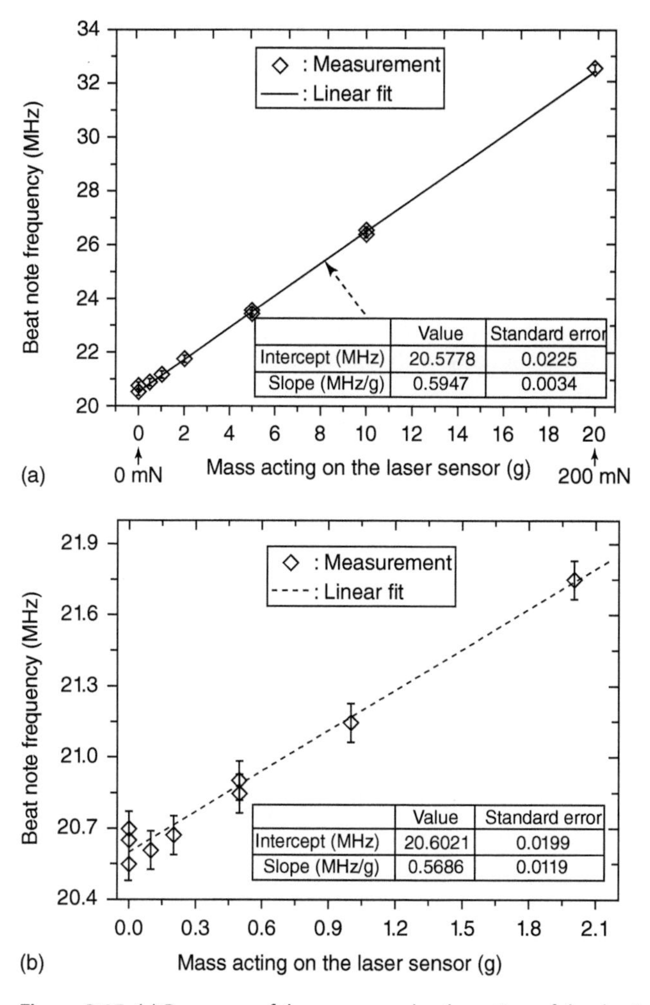

Figure 2.15 (a) Response of the sensor under the action of deadweight. (b) Zoomed in view of (a) at low loads. Source: Khelifa and Himbert 2015 [29]. Reproduced with permission of Sensors & Transducers.

were reported before for another size of photoelastic Nd:YAG crystal of the force sensor [43, 46, 47]. Using the least squares method, Khelifa and Himbert [29] deduced, from results reported in Figure 2.15a, the mean value of the sensitivity and its uncertainty to

$$S_{exp} \cong 0.5947\,(0.0034)\ \mathrm{MHz} \times \mathrm{g}^{-1}$$

In terms of force this value is equivalent to

$$S_{exp} \cong 60.63\,(0.35)\ \mathrm{MHz} \times \mathrm{N}^{-1}$$

To examine the response of the sensor to the low forces, Khelifa and Himbert [29] have considered only the beat frequency measurements observed when the crystal of the force sensor is loaded by mass standards smaller than 2 g (about 20 mN). From Figure 2.15b, which represents a part of Figure 2.15a, they deduced the sensitivity of the sensor.

$$S \cong 0.5686\,(0.0119)\ \mathrm{MHz} \times \mathrm{g}^{-1}$$

Or, in terms of sensitivity to a force

$$S = 57.97\,(1.21)\ \mathrm{MHz} \times \mathrm{N}^{-1}$$

Comparing these two results, Khelifa and Himbert [29] concluded that the sensor sensitivity to weak forces does not change significantly (lower by about 3% compared with the value deduced from a wider range of loading). However, they obtained an uncertainty on the derived sensitivity three times larger, because of the level of uncertainty on the frequency of the observed beat note between the two orthogonal polarizations.

The uncertainty of reproducibility is evaluated from the measurement of several repeated cycles of applying and removing the mass standard of 5 g. During successive cycles of loading and withdrawal of the force transducer, the mean frequency of the beat signal changes from 20.55 to 23.51 MHz. These values are obtained by averaging the observed frequencies measured alternately when the mass standard is inserted and then removed. For a given situation of the sensitive element of the transducer, the frequency of the beat note is measured with a repeatability better than 50 kHz. However, the reproducibility of the frequencies observed during a series of about 15 successive loadings and removals is of the order of 100 kHz. This large scatter is mainly generated by the reproducibility of internal stress distribution, induced in the center of the laser sensor, after each cycle of measurement when the sensing element is loaded by mass standard. A priori, one could reduce this limitation by improving the system of loading and unloading the sensitive element of the transducer.

Experimental sensitivity of the transducer, determined here by using only one mass standard, is given by

$$S^{*}_{exp} = \frac{\Delta\nu^{(load)} - \Delta\nu^{(free)}}{m_e} \cong 0.592\ \mathrm{MHz} \times \mathrm{g}^{-1}$$

Or, $S^{*}_{exp} \cong 60.35\ \mathrm{MHz} \times \mathrm{N}^{-1}$; this value is very close to $S_{exp} = 60.63\ \mathrm{MHz} \times \mathrm{N}^{-1}$, deduced from linear fit of results reported in Figure 2.15a.

The associated uncertainty of repeatability is evaluated as

$$u_{repeat}(S_{exp}) \cong 1.30\ \text{MHz} \times \text{N}^{-1} \tag{2.34}$$

If one considers the most pessimistic situation related to the reproducibility of frequency measurement in a series of a mass standard insertions and removals, one obtains

$$u_{reprod}(S_{exp}) \cong 2.61\ \text{MHz} \times \text{N}^{-1} \tag{2.35}$$

This last uncertainty is the main limitation in terms of measurement for this kind of force sensor.

The new values of sensitivity and its uncertainty are reported in Figure 2.16 and compared to the experimental measurements available for different force transducers reported earlier by other authors [45, 48]. In this figure, only results obtained with a photoelastic force sensor using monolithic Nd:YAG laser are considered for comparison. It can be seen from this figure that the sensitivities predicted by Eq. (2.33) corresponding to the ideal situation are clearly superior to experimental values when the dimensions ($l \times d$) of the laser sensor are small. As one can see from Figure 2.16, this is the case of the two available experimental values of sensitivity, corresponding to ($l \times d$) = 3 and 6 mm^2.

In Figure 2.16, the solid line is given by $S_1 = 4928/(ld)$ and corresponds to a simple theoretical model, while the dotted one, $S_2 = 310/(ld) + 3ld/4$, is an empirical relation that provides sensitivity values close enough to experimental results. Thus, the new results show that when the sizes of the sensor are small, the sensitivity of the photoelastic force sensor depends, in a relatively complex way, on the nature of the contact and on the localization of the applied force associated with the calibrated weight. Therefore, for the small size of the sensitive element, it is difficult to obtain high sensitivity with good reproducibility required for the measurement of small forces at the micro-newton level with acceptable accuracy.

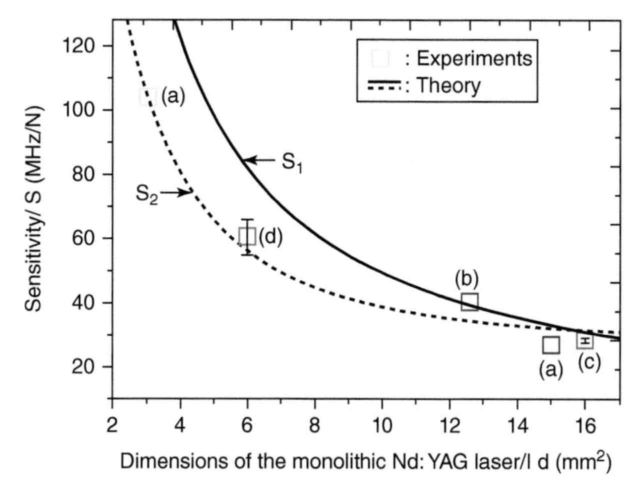

Figure 2.16 Sensitivity of photoelastic force sensor versus dimensions of the monolithic sensing element Nd:YAG laser. (a) Ref. [45]; (b) Ref. [48]; (c) Ref. [47]; (d) Ref. [29].

However, this type of sensor should prove useful for force measurement over a range of 5 orders of magnitude, namely, from less than 100 μN to more than 10 N. This wide range is very useful when trying to make comparisons with results obtained with other devices used in small force measurements to the International System of Units.

2.7 Photoelastic Materials and Applications

Photoelasticity has been used for a variety of stress analyses and even for routine use in design, particularly before the advent of numerical methods, such as, for instance, finite elements or boundary elements. Digitization of polariscopy enables fast image acquisition and data processing, which allows its industrial applications to control the quality of manufacturing process for materials such as glass and polymer. Dentistry utilizes photoelasticity to analyze strain in denture materials.

Photoelasticity can successfully be used to investigate the highly localized stress state within masonry or in proximity of a rigid line inclusion (stiffener) embedded in an elastic medium. Dynamic photoelasticity integrated with high-speed photography is utilized to investigate fracture behavior in materials. Another important application of the photoelasticity experiments is to study the stress field around bi-material notches. Bi-material notches exist in many engineering application such as welded or adhesively bonded structures.

R.W. Dixon studied the photoelastic tensor components and figures of merit of several materials for possible application in acoustic light modulators and scanners [49]. The most important results were that GaP is superior to fused quartz by nearly 2 orders of magnitude for modulation and scanning of visible light of wavelength longer than 0.58 μ and that GaAs is similarly superior at wavelengths longer than about 1 μ.

The use of the photoelastic effect to change the refractive index of semiconductor materials for lateral light confinement in double heterostructures without both chemical etching and regrowth is a relatively simple technique for fabricating planar optical waveguide and planar optoelectronic devices and was demonstrated in the mid-1970s [50–52]. Recently, a stable distribution of stress in AlGaAs/GaAs double heterostructures has been obtained by either metal–semiconductor interfacial reactions or a radio frequency sputtered refractory metal film with *dc* bias on the substrate to form stable stress distribution [53]. By using this technique, planar photoelastic AlGaAs/GaAs quantum well lasers [54], PEMs [55] and photoelastic couplers [56] are created. W.S. Gao et al. [57] demonstrated the form of photoelastic waveguide structure in InGaAsP/InP double heterostructure induced by a $W_{0.95}Ni_{0.05}$ metal thin-film stripe window. The strength of photoelastic waveguide caused by a 20-μm $W_{0.95}Ni_{0.05}$ compressive strain thin-film stripe window is 2.9×10^{-2} to 0.7×10^{-2} in the depth range from 1 to 2 μm of the InGaAsP/InP double heterostructure. The strength of narrow photoelastic waveguide induced by a 1.8 μm wide stripe window is 1.1×10^{-3} at 2.5 μm depth. The theoretical results demonstrate that the photoelastic waveguide structure may confine well the lateral light of

Figure 2.17 Photoelastic modulator. Hinds Instruments Inc. [34].

the InGaAsP/InP double heterostructure when the width of the $W_{0.95}Ni_{0.05}$ compressive strain thin-film stripe window is 5 times greater or 0.93 times smaller than the depth of the photoelastic waveguide structure.

Hinds Instruments Inc. [34] has designed a different model of PEMs using fused silica/ZnSe as the isotropic optical material and they operate at a wide range of frequency from vacuum UV to terahertz (Figure 2.17). They have high power handling capability – up to 5 GW/cm^2 for some applications.

2.7.1 $LiNbO_3$ and $LiTaO_3$ Crystals

Many ferroelectric materials, such as $LiNbO_3$ and $LiTaO_3$, have widely been employed as hosts for photoelastic waveguides and other optical components due to their interesting and much studied photoelastic properties. In 1985, A.R. Nelson designed a photoelastic optical waveguide utilizing an optically transparent crystal such as $LiTaO_3$ or $LiNbO_3$, which forms a channel for guiding light wave energy by applying a bias voltage to electrodes on the material, thereby effecting a change in the refractive index of the material [58]. In 2007, Bammer et al. [26, 35] designed a single crystal $LiTaO_3$ PEM for active Q-switching of a fiber laser. The modulator, which oscillates in a longitudinal eigenmode, was realized with $LiTaO_3$ crystals. This induces, due to the photoelastic effect, a modulated artificial birefringence, which modulates the polarization of passing light. When used together with a polarizer inside a laser cavity, the laser photon lifetime is strongly modulated and the laser may start to emit laser pulses.

On the basis of the work on Michelson fiber-optic accelerometer [59, 60], D. Tang et al. [28] designed a novel three-component hybrid-integrated optical accelerometer based on $LiNbO_3$ photoelastic waveguide. The resonant frequency of the accelerometer is 3.5 kHz. It can be employed to monitor or accurately measure vibrations in various areas such as seismic measurement in geophysical survey.

Precise determination of piezo-optic coefficients (POCs) is important in many aspects of materials science and optical engineering, especially when one searches for new efficient acousto-optic materials. With a complete set of POCs and elastic compliances the photoelastic constants p_{ij} may be evaluated [36, 61, 62]. Thus, the diffraction efficiency defined by the figure of merit $M_2 = n^6p^2/\rho V^3$ [63–65] can be determined for each particular geometry of acousto-optic interaction. Here V is the sound velocity, n is the refractive index, p is the effective photoelastic constant represented as a combination of p_{ij}

depending on sample geometry, and ρ is the crystal density. Optimized geometries of acousto-optic interaction, which provide the best diffraction efficiency of acousto-optic cells, can be found by analyzing the spatial anisotropy of $M_2(\theta,\varphi)$.

B.G. Mytsyk et al. [66] determined a complete set of POCs of pure and MgO-doped $LiNbO_3$ crystals as shown in Table 2.2.

Pure and MgO-doped $LiNbO_3$ crystals reveal nearly the same magnitudes of POCs. However, $LiNbO_3$:MgO exhibits about four times higher resistance with respect to powerful light radiation, making it more suitable for application in acousto-optic devices that deal with super powerful laser radiation.

B.G. Mytsyk et al. [68] also determined the POCs and the elasto-optic coefficients (EOCs) of calcium tungstate ($CaWO_4$) crystals as shown in Tables 2.3 and 2.4.

Table 2.3 shows that the value POCs of $CaWO_4$ crystals are more than two to seven times higher than the corresponding values of POCs of $LiNbO_3$ crystals, and the sum of absolute values of π_{im} is three times higher.

Table 2.4 shows the EOCs of $CaWO_4$ and $LiNbO_4$ crystals. It is seen that the largest P_{in} of $CaWO_4$ excels the P_{in} of $LiNbO_3$ considerably and the value of p_{11} is atypically large. The sum of absolute values of P_{in} for $CaWO_4$ crystals is also larger (in two times) than $LiNbO_3$ crystals.

Table 2.2 POCs of $LiNbO_3$:MgO and $LiNbO_3$ crystals in Brewster.

π_{im}	π_{11}	π_{12}	π_{13}	π_{31}	π_{33}	π_{14}	π_{41}	π_{44}
$LiNbO_3$:MgO	−0.43	0.15	0.78	0.50	0.32	−0.80	−0.90	2.0
$LiNbO_3$	−0.38	0.09	0.80	0.50	0.20	−0.81	−0.88	2.25
$LiNbO_3$ [67]	−0.47	0.11	2.0	0.47	1.6	0.7	−1.9	0.21

Table 2.3 The main POCs of $CaWO_4$ and $LiNbO_3$ crystals (in Brewster).

π_{im}	π_{11}	π_{12}	π_{13}	π_{31}	π_{33}
$CaWO_4$	1.76	−0.60	1.52	0.88	1.01
	±0.22	±0.03	±0.20	±0.23	±0.14
$LiNbO_3$ [66]	−0.38	0.09	0.80	0.50	0.20
	±0.03	±0.03	±0.10	±0.08	±0.05

Table 2.4 The main EOCs of $CaWO_4$ and $LiNbO_3$ crystals (in Brewster).

P_{in}	P_{11}	P_{12}	P_{13}	P_{31}	P_{33}
$CaWO_4$	0.41	−0.05	0.24	0.22	0.20
	±0.004	±0.02	±0.03	±0.04	±0.02
$LiNbO_3$ [69]	−0.02	0.06	0.17	0.14	0.12
	±0.02	±0.02	±0.03	±0.02	±0.02

The acousto-optic quality estimation of the $CaWO_4$ are given on the basis of the M_2 coefficient for the highest value of the p_{in} (0.41) and typical velocity of acoustic waves fort these crystals in the case of isotropic light diffraction ($V = 4.5 \times 10^3$ m/s):

$$M_2 = \frac{n^6 p^2}{\rho V^3} = 154 \times 10^{-15}\ \mathrm{c^3/kg}$$

where $\rho = 6.15\,\mathrm{g/cm^3}$ denotes the crystal density.

The presented value M_2 is greater than the greatest coefficient of acousto-optic quality M_2 for the isotropic light diffraction in $LiNbO_3$ crystals by 20 times. Thus, the $CaWO_4$ crystals are a prospective acousto-optic material in comparison with $LiNbO_3$, which is widely used in acousto-optic devices.

2.7.2 $Li_2Ge_7O_{15}$ Crystals

The study of piezo-optic dispersion of ferroelectric lithium heptagermanate $Li_2Ge_7O_{15}$ crystals (un-irradiated and X-irradiated) in the visible region of the spectrum of light at room temperature (RT = 298 K) shows an optical

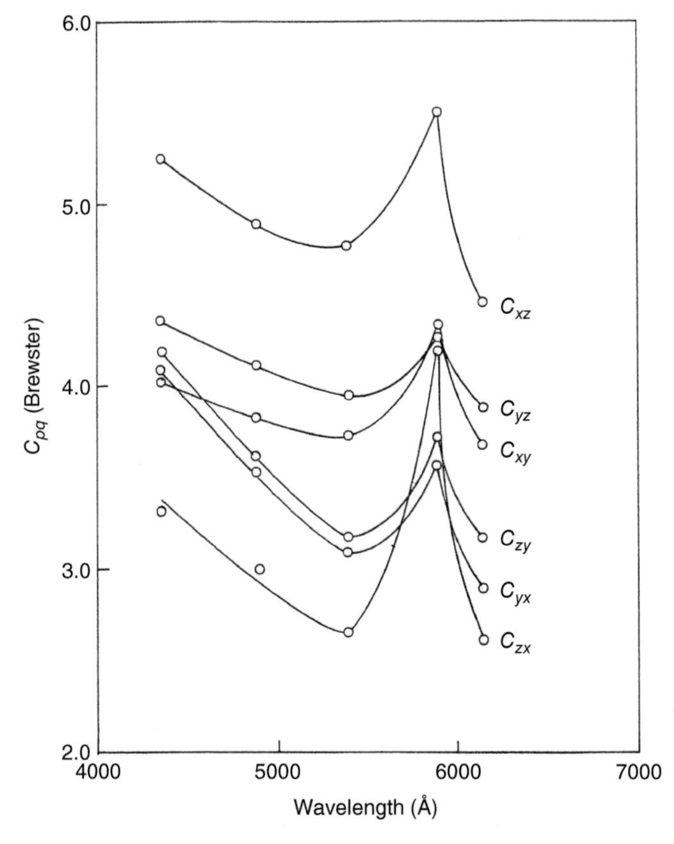

Figure 2.18 Stress-optical dispersion of $Li_2Ge_70_{15}$ crystals (un-irradiated) with wavelength at room temperature (298 K). Source: Bain et al. 2011 [70]. Reproduced with permission of Intech.

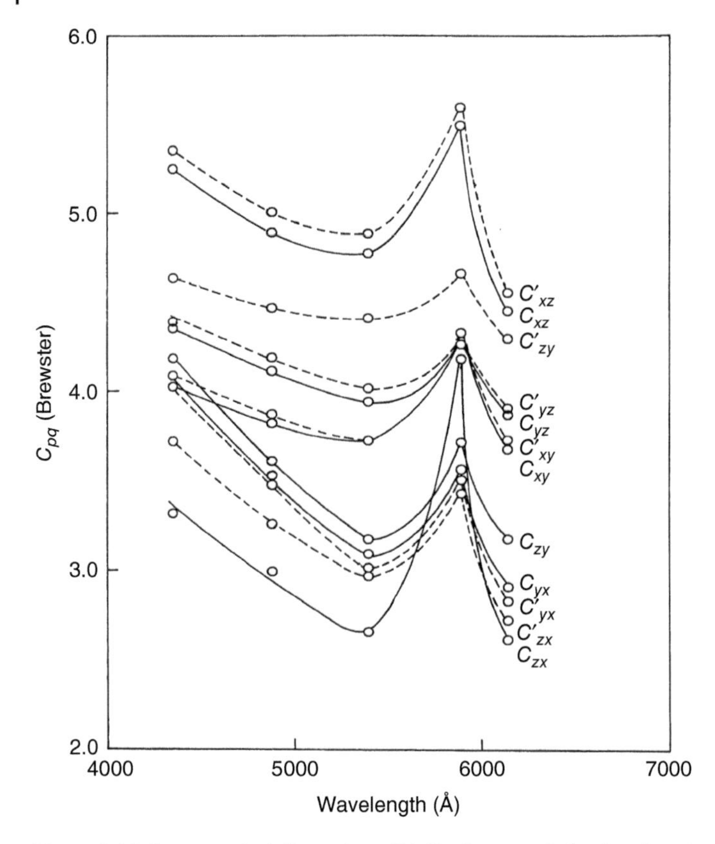

Figure 2.19 Stress-optical dispersion of $Li_2Ge_70_{15}$ crystals (un-irradiated and X-irradiated) with wavelength at room temperature (298 K). Source: Bain et al. 2011 [70]. Reproduced with permission of Intech.

zone/window between 5400 and 6200 Å with an enhanced piezo-optical behavior [70]. The values of POC obtained at different wavelengths are plotted in Figures 2.18 and 2.19. Here C_{pq} is the POC with the stress direction being p and observation direction being q.

The observed "optical zone or optical window" between 5400 and 6200 Å can have technical importance. This window region can act as an optical switch for acousto-optical devices. From the studies undertaken it may be concluded that $Li_2Ge_7O_{15}$ (LGO) is an attractive acousto-optic material, which deserves further probe.

The stress-optical coefficients C_{pq} of the crystals $Li_2Ge_7O_{15}$ at the paraelectric phase (RT = 298 K) and at T_c = 279 K are presented in Table 2.5 [70]. It is important to compare the values of C_{pq} for $Li_2Ge_7O_{15}$ with other ferroelectric crystals given in Table 2.6 particularly with Rochelle salt (RS), which belongs to the orthorhombic class like LGO [70].

The values of C_{pq} are significantly higher for LGO as compared to these ferroelectric systems. So, the large photoelastic coefficients and other properties such as good mechanical strength, a transition temperature close to room

Table 2.5 Stress-optical coefficients C_{pq} (in Brewsters) of $Li_2Ge_7O_{15}$ at RT = 298 K and at T_c = 279 K.

Obs.	C_{pq}	Paraelectric (PE) phase (RT = 298 K)	At T_c = 279 K
1	C_{xy}	4.38	3.85
2	C_{xz}	5.55	5.85
3	C_{yx}	3.60	4.46
4	C_{yz}	4.26	5.50
5	C_{zy}	3.71	4.83
6	C_{zx}	4.19	5.45

Source: Bain et al. 2011 [70]. Reproduced with permission of Intech.

Table 2.6 Piezo-optic coefficients C_{pq} (in Brewsters) for some ferroelectric crystals in their paraelectric (PE) phases.

Obs.	C_{pq}	Rochelle salt (RS) [71]	KDP [72]	ADP [73]
1	C_{xz}	3.74	0.28	1.25
2	C_{yz}	4.29	0.28	1.25
3	C_{yx}	3.56	1.04	4.30
4	C_{zx}	0.85	1.54	3.50
5	C_{zy}	2.61	1.54	3.50
6	C_{xy}	3.04	1.04	4.30

Source: Bain et al. 2011 [70]. Reproduced with permission of Intech.

temperature, and stability in ambient environment favor LGO as a potential candidate for photoelastic applications.

References

1 Brewster, D. (1815). *Philos. Trans. A* 105: 60–64.
2 Narasimhamurty, T.S. (1981). *Photoelasticity and Electro-optic Properties of Crystals*, vol. 514. New York: Plenum press.
3 Brewster, D. (1816). *Phil. Tras.* 156–178.
4 Brewster, D. (1818). *Trans. Roy. Soc. Edinburgh* 8: 281–369.
5 Fresnel, A. (1820). *Ann. Chim. Phys.* l5: 379.
6 Fresnel, A. (1822). *Ann. Chim. Phys.* 20: 376.
7 Neumann, F.E. (1841). Über Gesetze der Doppelbrechung des Lichtes in comprimierten oder ungleichformig Erwamten unkrystallischen Körpern. *Abh. Kon. Akad. Wiss. Berlin*, part II 1–254; see also Ann. Der Phys. Chem. (Germany) II-54, 449–476 (1841), in Germany.

8 Maxwell, J. (1853). *Trans. Roy. Soc. Edinburgh* 20, part 1: 87.
9 Coker, E.G. and Filon, L.N.G. (1931). *A Treatise on Photoelasticity*. New York: Cambridge University Press.
10 Frocht, M.M. (1941). *Photoelasticity*, vol. 1. New York: Wiley.
11 Jessop, H.T. and Harris, F.C. (1950). *Photoelasticity: Principles and Methods*. New York: Dover Publications, Inc.
12 Oppel, G. (1936). Polarisationoptische Untersuchung raumlicher Spannungs und Dehungszustande. *Forsch. Geb. Ingenieurw.* 7: 240–248.
13 Hetenyi, M. (1938). The fundamentals of three dimensional photoelasticity. *J. Appl. Mech.* 5 (4): 149–155.
14 Peterson, R.E. (1953). *Stress Concentration Design Factors*. New York: John Wiley & Sons.
15 Woolard, D. Obstacles Encountered in Calculating the Stress Tensor Using Thermoelasticity and Photoelasticity, CS-AAPT Fall 2000 Meeting.
16 Coker, E.G. and Thompson, S.P. (1912). *The Design and Construction of Large Polariscopes*. London: Optical Convention.
17 Muller, H. (1935). Theory of the photoelastic effect of the cubic crystals. *Phys. Rev.* 47: 947–957.
18 Photoelastic modulator-wikipedia. https://en.wikipedia.org/wiki/Photoelastic_modulator (accessed 22 January 2019)
19 Oppel George, U. Publication Number: US3067606 A, 11 December 1962.
20 Roberts Albert, Hawkes Ivor and Williams Frederik Trevor (1967). Photoelastic stress indicating devices. US Patent 3313,205A, filed 30 April 1963 and issued 11 April 1967.
21 kemp, J.C. (1969). Piezo-optical birefringence modulators. *J. Opt. Soc. Am.* 59: 950.
22 Canit, J.C. and Badoz, J. (1983). New design for a photoelastic modulator. *Appl. Opt.* 22 (4): 592–594.
23 Canit, J.C. and Pichon, C. (1984). Low frequency photoelastic modulator. *App. Opt.* 23 (13): 2198.
24 Asundi, A.k. and Sajan, M.R. (1994). Low-cost polariscope for dynamic photoelasticity. *Opt. Eng.* 33 (9): 3052–3055.
25 Bammer, F., Holzinger, B., and Schumi, T. (2007). A single crystal photo-elastic modulator. *Proc. SPIE* 6469: 1–8.
26 Bammer, F. and Petkovsek, R. (2007). Q-switching of a fiber laser with a single crystal photo-elastic modulator. *Opt. Express* 15 (10).
27 Ivtsenkov, G. (2008) Fiber optic strain sensor and associated data acquisition system. US Patent 7,359,586B2, filed 18 May 2006 and issued 15 April 2008.
28 Tang, D., Liang, Z., Zhang, X. et al. (2009). Three-component hybrid-integrated optical accelerometer based on $LiNbO_3$ photoelastic waveguide. *Chin. Opt. Lett.* 7 (1): 32–35.
29 Naceur-Eddine, K.H.E.L.I.F.A. and Marc, H.I.M.B.E.R.T. (2015). Sensitivity of miniaturized photo-elastic transducer for small force sensing. *Sens. Trans.* 184 (1): 19–25.
30 Banerjee, K. (1927). Theory of photoelasticity. *Indian J. Phys.* 2: 195–242.
31 Herzfeld, K.F. (1928). On the theory of forced double refraction. *J. Opt. Soc. Am.* 17: 26–36.

32 Herzfeld, K.F. and Lee, R. (1933). Theory of forced double refraction. *Phys. Rev.* 44: 625–631.
33 Krishnan, R.S. (ed.) (1958). *Progress in Crystal Physics*. Academic Press.
34 Hinds Instruments Inc. Photoelastic modulator-production manual. https://www.hindsinstruments.com/products/photoelastic-modulators/ (accessed 22 January 2019).
35 Bammer, F., Schumi, T., Carballido Souto, J.R. et al. (2011). Q-Switching with single crystal photo-elastic modulators. In: *Laser Systems for Applications* (ed. K. Jakubczak). InTech. ISBN: 978-953-307-429-0.
36 Nye, J.F. (1985). *Physical Properties of Crystals*. Oxford University Press.
37 Wong, K. (2002). *Properties of Lithium Niobate*, Chapter 8. London: INSPEC.
38 Bammer, F., Holzinger, B., and Schumi, T. (2006). Time multiplexing of high power laser diodes with single crystal photo-elastic modulators. *Opt. Express* 14: 3324–3332.
39 Dörring, J., Killi, A., Morgner, U. et al. (2004). Period doubling and deterministic chaos in continuously pumped regenerative amplifiers. *Opt. Express* 12: 1759–1768.
40 Su, W., Gilbert, J.A., Morrissey, M.D., and Song, Y. (1997). General-purpose photoelastic fiber optic accelerometer. *Opt. Eng.* 36 (1): 22–28.
41 Tang, D., Zhao, D., Wang, Y. et al. (2011). A MOEMS accelerometer based on photoelastic effect. *Optik* 122: 635–638.
42 Holzapfel, W. and Settgast, W. (1989). Force to frequency con-version by intracavity photoelastic modulation. *Appl. Opt.* 28 (21): 4585–4594.
43 Holzapfel, W. and Finnemann, M. (1993). High-resolution force sensing by a diode-pumped Nd:YAG laser. *Opt. Lett.* 18 (23): 2062–2064.
44 Holzapfel, W., Hou, L., and Neuschaefer-Rube, S. (2000). Error effects in microlaser sensors. In: *Proceedings of the XVI IMEKO World Congress*, vol. 3, 85–90. Austria.
45 Holzapfel, W., Neuschaefer-Rube, S., and Kobusch, M. (2000). High-resolution, very broadband force measurement by solid-state laser transducers. *Measurement* 28 (4): 277–291.
46 Khelifa, N. (2014). Small-force measurement by photo-elastic transducer. *Opt. Photonics J.* 4 (1): 14–20.
47 Khelifa, N.-E. and Himbert, M. (2014). Sensitivity of photo-elastic Nd-YAG laser for small force sensing. In: *Proceedings of the fifth International Conference on Sensor Device Technologies and Applications (SENSORDEVICES)*, Lisbon, Portugal (16–20 November 2014). IARIA. ISBN: 978-1-61208-375-9.
48 Ding, J., Zhang, L., Zhang, Z., and Zhang, S. (2010). Frequency splitting phenomenon of dual transverse modes in Nd:YAG laser. *Opt. Laser Technol.* 42: 341–346.
49 Dixon, R.W. (1967). Photoelastic properties of selected materials and their relevance for applications to acoustic light modulators and scanners. *J. Appl. Phys.* 38 (13): 5149–5153.
50 Yamamoto, Y., kamiya, T., and Yanai, H. (1975). *Appl. Opt.* 14: 322.
51 Campbell, J.C., Blum, F.A., Skaw, D.W., and Lawley, K.L. (1975). *Appl. Phys. Lett.* 27: 202.

52 Kirkby, P.A., Selway, P.R., and Westbrook, L.D. (1979). *J. Appl. Phys.* 50: 4567.
53 Jiang, X.S., Liu, Q.Z., Yu, L.S. et al. (1994). *Mater. Chem. Phys.* 38: 195.
54 Liu, Q.Z., Chen, W.X., Li, N.Y. et al. (1998). *J. Appl. Phys.* 83: 7442.
55 Pappert, S.A., Xia, W., Jiang, X.S. et al. (1994). *J. Appl. Phys.* 75: 4352.
56 Yu, L.S., Guan, Z.F., Xia, W. et al. (1993). *Appl. Phys. Lett.* 62: 2944.
57 Gao, W.S., Xing, Q.J., Yuan, Z.J., and Hu, T. (2003). *Chin. Phys. Lett.* 20 (8): 1296.
58 Nelson, A.R. (1985). Photoelastic effect optical waveguides. US Patent 4561718A, filed 22 August 1983 and issued 31 December 1985.
59 Chen, C., Zhang, D., Ding, G., and Cui, Y. (1999). *Appl. Opt.* 38: 628.
60 Chen, C., Wu, B., Ding, G., and Cui, Y. (2002). *Proc. SPIE* 4919: 1.
61 Sonin, A.S. and Vasilevskaya, A.S. (1971). *Electro-Optic Crystals*, 327. Atomizdat [in Russian].
62 Kloos, G. (1997). On photoelastic and quadratic electrostrictive effect. *J. Phys. D* 30: 1536–1539.
63 Uchida, N. and Niizeki, N. (1973). Acousto-optic deflection materials and techniques. *Proc. IEEE* 61: 1073–1094.
64 Silvestrova, I.M., Vinogradov, A.V., Shygorin, A.V. et al. (1990). Elastic, piezoelectric, acousto-optic and non-linear optical properties of caesium ortosulphobenzoat crystals. *Kristallografiya* 35: 906–911. [in Russian].
65 Nelson, D.F. (1979). *Electric, Optic and Acoustic Interactions in Dielectrics*, 539. Wiley-Interscience.
66 Mytsyk, B.G., Andrushchak, A.S., Demyanyshyn, N.M. et al. (2009). *Appl. Opt.* 48 (10): 1904–1911.
67 Mytsyk, B.H., Pryriz, Y.V., and Andrushchak, A.S. (1991). The lithium niobate piezo-optic features. *Cryst. Res. Technol.* 26: 931–940.
68 Mytsyk, B., Kost, Y., Andrushchak, A., and Solskii, I. (2010). Calcium tungstate as a perspective acousto-optic material, photoelastic properties. In: *2010 International Conference on Modern Problems of Radio Engineering, Telecommunications and Computer Science (TCSET)*, Lviv-Slavske, Ukraine (23–27 February 2010). IEEE.
69 Andrushchak, A.S., Mytsyk, B.G., Laba, H.P. et al. (2009). Complete sets of elastic constants and photoelastic coefficients of pure and MgO-doped lithium niobate crystals at room temperature. *J. Appl. Phys.* 106: 073510–073516.
70 Bain, A.K., Chand, P., and Veerabhadra Rao, K. (2011). Piezo-optic and dielectric behavior of the ferroelectric lithium heptagermanate crystals. In: *Ferroelectrics – Physical Effects* (ed. M. Lallart). InTech. ISBN: 978-953-307-453-5.
71 Narasimhamurty, T.S. (1969). *Phys. Rev.* 186: 945.
72 Veerabhadra Rao, K. and Narasimhamurty, T.S. (1975). *J. Mater. Sci.* 10: 1019.
73 Narasimhamurty, T.S., Veerabhadra Rao, K., and Petterson, H.E. (1973). *J. Mater. Sci.* 8: 577.

3
Acousto-Optics

3.1 Introduction

Acousto-optics is a branch of physics that studies the interactions between sound waves and light waves, especially the diffraction of laser light by ultrasound (or sound in general) through an ultrasonic grating. In general, acousto-optic effects are based on the change in the refractive index of a medium caused by the mechanical strain produced by an acoustic wave. Since the mechanical strain varies periodically in the acoustic wave, the refractive index of the medium also varies periodically leading to a refractive index grating. When a light beam is incident on such a refractive index grating, diffraction takes place and this produces either multiple-order diffraction or only single-order diffraction. The former is referred to as Raman–Nath diffraction and is usually observed at low acoustic frequencies. The latter is analogous to Bragg diffraction of X-rays in crystals and is referred to as Bragg diffraction. It is usually observed at high acoustic frequencies.

The acousto-optic effect is extensively used in a number of applications such as in acousto-optic modulators (AOMs), deflectors, frequency shifters for heterodyning, spectrum analyzers, *Q*-switching, delay lines, image processing, general and adaptive signal processing, tomography transformations, optical switches, neural networks, optical computing, and mode locking in lasers. Along with the current applications, acousto-optics presents interesting possible applications. It can be used in nondestructive testing, structural health monitoring, and biomedical applications, where optically generated and optical measurements of ultrasound give a noncontact method of imaging.

3.2 Short History of Acousto-optics

The acousto-optic effect has had a relatively short history, beginning with the French physicist Léon Brillouin predicting the diffraction of light by an acoustic wave, being propagated in a medium of interaction, in 1922 [1]. In 1932, Debye and Sears [2] and Lucas and Biquard [3] observed diffraction of light by an ultrasonic wave launched in an interaction medium. These were the first experimental verifications of the light–sound interaction phenomenon predicted

Crystal Optics: Properties and Applications, First Edition. Ashim Kumar Bain.

by Brillouin [1]. Since then a great number of theoretical and experimental studies have been performed to understand the phenomenon. The most useful theoretical treatment made in the early stage of work appeared in a series of papers written by the Indian physicists Raman and Nath [4]. The simple phase-grating theory developed by them explained well the multiple-order diffraction phenomenon, contrary to Brillouin's expectation [1], observed at a low-acoustic-frequency region. The quantitative agreement of diffracted light intensities with the Raman–Nath theory was demonstrated by Sanders [5] and Nomoto [6] for the case of normal incidence of light with respect to the acoustic wave vector and by Nomoto [7] for the case of oblique incidence. However, it should be noted that the Raman–Nath theory can be applied only to the case where the sound intensity is not so strong and the interaction length is not so long that at most 10 orders of diffraction are observed.

When the frequency of the acoustic wave is raised, the diffraction to higher orders is eliminated and, at certain angles of incidence of the light, energy exchange between zeroth- and first-order light beams becomes predominant. This type of diffraction, exactly the one predicted by Brillouin [1], was observed by Rytow [8], Bhagavantam and Rao [9]. Starting with the generalized theory of Raman and Nath [4], Phariseau [10] extended the coupled-wave theory, in which only two light beams were taken into account. It may be considered that the theory is practically applicable to the light–sound interaction in high-frequency Bragg devices [11], in which diffraction to a higher order is negligibly weak.

The preceding two theories of Raman–Nath and Phariseau give the diffraction features occurring at the two extreme and idealized cases, and many attempts have been made to treat the problem in a more general form [12–15]. However, we do not review the rather complicated results obtainable from the generalized theories because the practical acousto-optic devices do not require them in most cases. A treatment made by Klein and Cook [16] is of general interest from the practical point of view. They calculated the diffracted light intensity at a moderate-frequency region, where the Raman–Nath equation has no analytic solutions. The aspect of diffraction appearing in this "transition" region is characterized as the mixture of those of the Raman–Nath diffraction and the Bragg reflection.

In parallel with the theoretical work previously mentioned, a number of investigations have been made using the diffraction phenomena for several purposes: measurements of the elastic moduli, the acoustic absorption coefficients, and the ratio of photoelastic constants; visualization of the acoustic field in media; and determination of the acoustic waveform [17, 18]. On the other hand, before the laser was invented in 1961, no significant applications of the light–sound interaction to practical devices had been made except to television [19, 20].

Development of the laser and the high-frequency acoustic technique in association with the discovery of new piezoelectric materials with a high coupling factor made it possible to apply acousto-optics to devices such as light deflectors, modulators, and signal processors. The excellent light-deflection systems were constructed by Korpel et al. [21, 22] for laser television and by Anderson [23–25] for a readout deflector of the page-organized holographic memory. The number of resolvable spots strikingly increased even in solid-state deflectors [26, 27] and the use of these devices for display, laser printer, and laser facsimile may be realized in the near future.

3.3 Principle of Acousto-optic Effect

The acousto-optic effect is a specific case of photoelasticity, where there is a change in a material's permittivity, ε, due to a mechanical strain ε. Photoelasticity is the variation in the optical indicatrix coefficients B_i caused by the strain ε_j given by [28]

$$\Delta B_i = p_{ij}\varepsilon_j \quad i,j = 1-6 \tag{3.1}$$

where p_{ij} is a second-rank tensor, called the strain-optical (elasto-optic) coefficients.

Specifically, in the acousto-optic effect, the strains ε_j are the result of the acoustic wave which has been excited within a transparent medium. This then gives rise to the variation in the refractive index. For a plane acoustic wave propagating along the z-axis, the change in the refractive index can be expressed as [28]

$$n(z,t) = n_0 + \Delta n \cos(\Omega t - Kz) \tag{3.2}$$

where n_0 is the undisturbed refractive index, Ω is the angular frequency, K is the wave number of the acoustic wave, and Δn is the amplitude of variation in the refractive index generated by the acoustic wave and is given as [28]

$$\Delta n = -\frac{1}{2}\sum_j n_0^3 p_{zj}\varepsilon_j \tag{3.3}$$

The generated refractive index $n(z,t)$, described by Eq. (3.2), gives a diffraction grating moving with the velocity given by the speed of the sound wave in the medium. Light which then passes through the transparent material is diffracted due to this generated refraction index grating, forming a prominent diffraction pattern. This diffracted beam emerges at an angle (θ) that depends on the wavelength (λ) of the light relative to the wavelength (Λ) of the sound wave [29]

$$2\Lambda\sin\theta = m\lambda \tag{3.4}$$

where "$m = \ldots -2, -1, 0, 1, 2, \ldots$" is the order of diffraction.

Light diffracted by an acoustic wave of a single frequency produces two distinct diffraction types. These are Raman–Nath diffraction and Bragg diffraction. Raman–Nath diffraction is observed with relatively low acoustic frequencies, typically less than 10 MHz, and with a small acousto-optic interaction length, l, which is typically less than 1 cm. This type of diffraction occurs at an arbitrary angle of incidence, θ_0 (Figure 3.1).

In contrast, Bragg diffraction occurs at higher acoustic frequencies, usually exceeding 100 MHz. The observed diffraction pattern generally consists of two diffraction maxima; these are the zeroth and the first orders. However, even these two maxima only appear at definite incidence angles close to the Bragg angle, θ_B. The first order maximum or the Bragg maximum is formed due to a selective reflection of the light from the wave fronts of the ultrasonic wave (Figure 3.2).

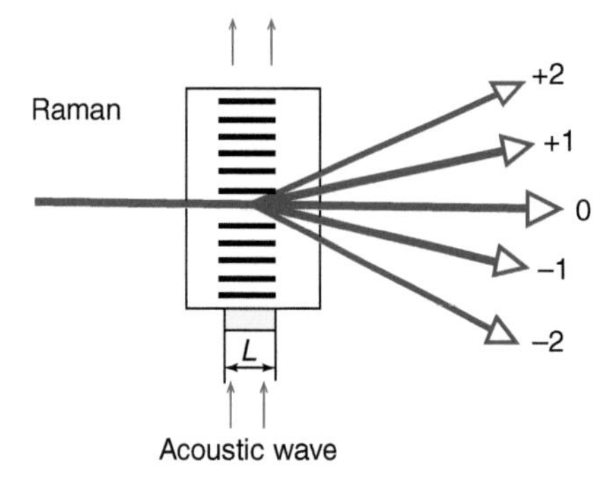

Figure 3.1 Raman–Nath diffraction ($Q \ll 1$). The laser beam is incident roughly normal to the acoustic beam and there are several diffraction orders (−2, −1, 0, 1, 2, …) with intensities given by Bessel functions.

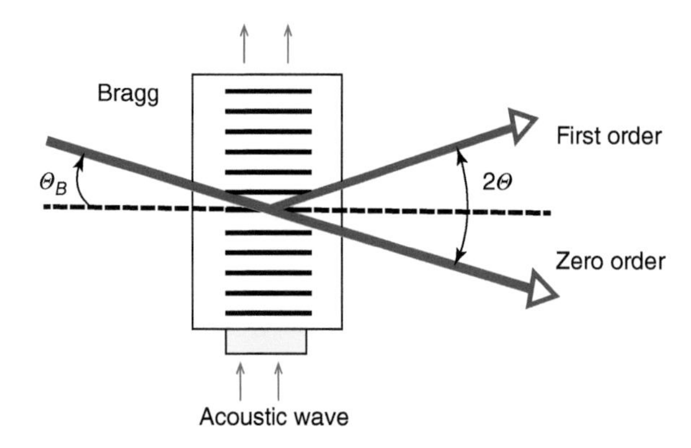

Figure 3.2 Bragg diffraction ($Q \gg 1$). At one particular incidence angle θ_B, only one diffraction order is produced – the others are annihilated by destructive interference.

The Bragg angle is given by the expression [28]

$$\sin\theta_B = -\frac{\lambda f_a}{2n_i v_a}\left[1 + \frac{v_a^2}{\lambda^2 f_a^2}(n_i^2 - n_d^2)\right] \tag{3.5}$$

where λ is the wavelength of the incident light wave (in a vacuum), f_a is the acoustic frequency, v_a is the velocity of the acoustic wave, and n_i and n_d are the refractive indices for the incident and the diffracted optical waves, respectively.

In general, there is no point at which Bragg diffraction takes over from Raman–Nath diffraction. It is simply a fact that as the acoustic frequency increases, the number of observed maxima is gradually reduced due to the angular selectivity of the acousto-optic interaction. Traditionally, the type of diffraction, Bragg or Raman–Nath, is determined by the conditions $Q \gg 1$ and $Q \ll 1$, where Q is given by [28]

$$Q = \frac{2\pi\lambda l f_a^2}{n v_a^2} \tag{3.6}$$

which is known as the Klein–Cook parameter. Since, in general, only the first-order diffraction maximum is used in acousto-optic devices, Bragg diffraction is preferable due to the lower optical losses. However, the acousto-optic requirements for Bragg diffraction limit the frequency range of acousto-optic interaction. As a consequence, the speed of operation of acousto-optic devices is also limited.

3.4 Acousto-optic Devices

The development of optoelectronic technology in the past two decades has intensified research on devices that control and manipulate optical radiation. Acousto-optic devices are based on acousto-optic effects in which the optical medium is altered by the presence of ultrasound. Examples of acousto-optic devices include optical modulators, deflectors, scanners, *Q*-switches, isolators, and frequency shifters. These devices have many applications in high-speed laser printers, laser lithography, optical communications and computing, large-screen laser projectors, frequency shifting, particle inspection, optical spectrum analysis, signal processing, radar, range finder, and target designation.

Progress in acousto-optics has been stimulated by the development of growth methods for acousto-optic crystals and by new methods for fabricating piezoelectric transducers that can efficiently convert electrical energy into acoustic energy at frequencies ranging from a few tens of megahertz up to several gigahertz. One promising development in acousto-optic devices is the recent commercial availability of acousto-optic tunable filters (AOTFs). In these filters, the interaction between an ultrasonic wave and light in an acousto-optic crystal is used to spectrally filter the light. In operation, tunable filters resemble interference filters and can be used in applications that require a filter wheel, grating, or prism.

3.4.1 Acousto-optic Modulator

An AOM is a device which can be used for controlling the power, frequency, or spatial direction of a laser beam with an electrical drive signal. It is based on the acousto-optic effect, i.e. the modification of the refractive index by the oscillating mechanical pressure of a sound wave.

The key element of an AOM is a transparent crystal (or a piece of glass) through which the light propagates. A piezoelectric transducer attached to the crystal is used to excite a sound wave with a frequency on the order of 100 MHz. Light can then experience Bragg diffraction at the traveling periodic refractive index grating generated by the sound wave; therefore, AOMs are sometimes called Bragg cells. The optical frequency of the scattered beam is increased or decreased by the frequency of the sound wave (depending on the propagation direction of the acoustic wave relative to the beam) and propagates in a slightly different direction. The frequency and direction of the scattered beam can be controlled via the frequency of the sound wave, whereas the acoustic power is the control for the optical powers.

A diagram of the relationship between the acoustic wave and the laser beam is shown in Figure 3.3. The acoustic wave may be absorbed at the other end of the crystal. Such a *traveling wave geometry* makes it possible to achieve a broad modulation bandwidth of many megahertz. Other devices are resonant for the sound wave, exploiting the strong reflection of the acoustic wave at the other end of the crystal. The resonant enhancement can greatly increase the modulation strength (or decrease the required acoustic power), but reduces the modulation bandwidth.

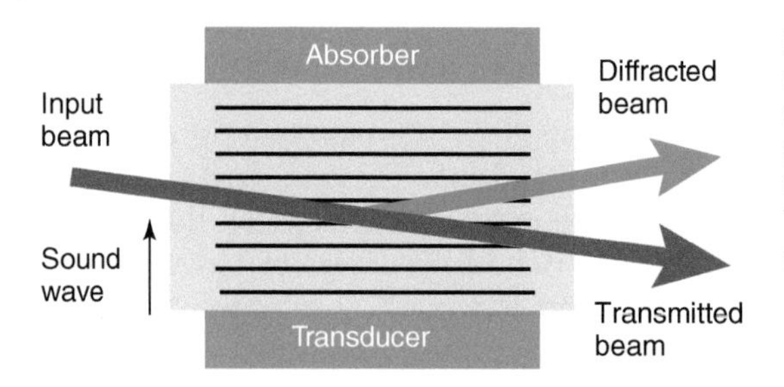

Figure 3.3 Schematic setup of a nonresonant acousto-optic modulator. A transducer generates a sound wave at which a light beam is partially diffracted. The diffraction angle is exaggerated.

With appropriate design of the modulator and proper adjustment of the incident angle between the laser light and the axis of acoustic propagation in the optical material (Bragg angle), the first-order beam can be made to have the highest efficiency. The angle (θ), the light diffracted is defined by the equation:

$$\theta = \frac{\lambda f_a}{v_a} = 2\theta_B \tag{3.7}$$

where λ is the wavelength of the incident light wave in air, v_a is the velocity of the acoustic wave, f_a is the acoustic frequency, and θ_B is the Bragg angle. This is the angle between the incident laser beam and the diffracted laser beam with the acoustic wave direction propagating at the base of the triangle formed by three vectors.

The intensity of the light diffracted is proportional to the acoustic power (P_{ac}), the figure of merit (M_2) of the optical material, geometry factors (L/H, L is the interaction length and H is the transducer height), and inversely proportional to the square of the wavelength. This is seen in the following equation:

$$\text{DE} = \eta = \sin^2\left[\frac{\pi}{\lambda}\left(\frac{M_2 P_{ac} L}{2H}\right)^{\frac{1}{2}}\right] \tag{3.8}$$

A variety of materials are used for AOMs depending on the laser parameters such as wavelength, polarization, and power density. For the visible (VIS) region and near-infrared (NIR) region, the most common modulators are made from dense flint glass, fused silica, crystal quartz, tellurium dioxide (TeO_2), or chalcogenide glass. At the infrared (IR) region, germanium is the most common material with a relative high figure of merit. Lithium niobate (LN) and gallium phosphate are used for high-frequency signal processing devices.

3.4.1.1 Acousto-optic Modulator Construction

A schematic of the focused modulator setup is shown in Figure 3.4. Once the acousto-optic material is selected, it is optically polished. The surfaces of the material that are to be optical windows are optically anti-reflection (AR) coated to reduce optical reflections. Typical losses are from a few percent for external cavity devices to 0.2% for intracavity devices. The side of the material that the acoustic energy is to originate from has an LN transducer metal vacuum bonded to the modulator medium. The transducer converts radio frequency (RF) energy applied to it into acoustic energy. Then the transducer is lapped to the fundamental resonant frequency such as 80 MHz. The top surface of the transducer is then

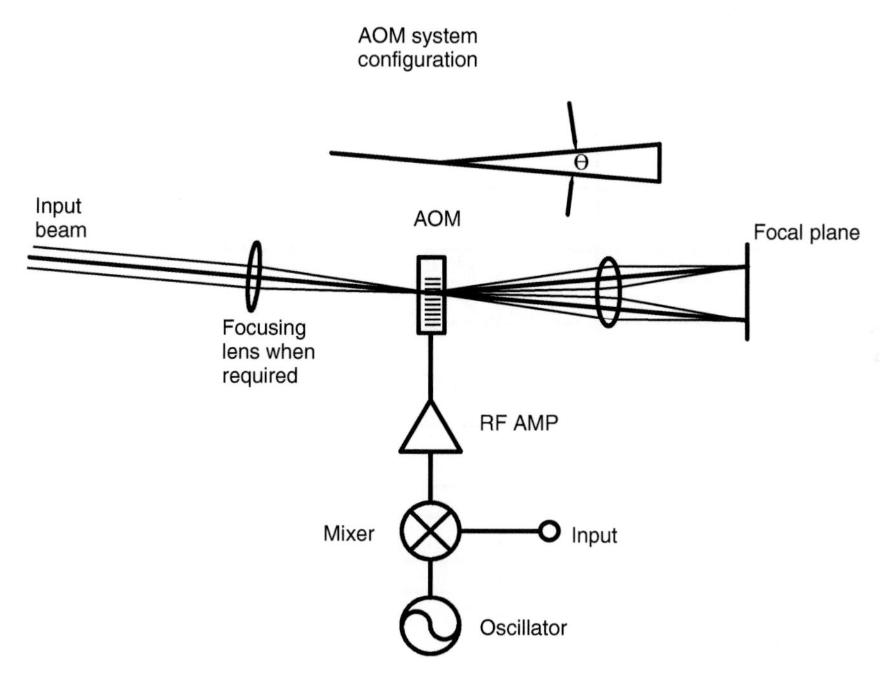

Figure 3.4 Schematics of the focused modulator setup (www.optoscience.com).

metalized with the transducer shape and size defined in this process. The modulator is then tuned to match the electrical impedance of the RF driver which will supply the RF energy at the frequency of the transducer's resonant frequency.

The RF driver is typically a fixed frequency oscillator and usually consists of an RF oscillator and amplitude modulator with an interface, which accepts input modulation, and an RF amplifier which supplies the AO modulator with the level of RF power needed to achieve the highest diffraction efficiency (DE).

3.4.1.2 Digital Modulation

An AOM can be used to shutter a laser beam "on" and "off." By applying a digital TTL (transistor–transistor logic) signal to the modulator's driver digital modulation input, the RF energy applied to the modulator is modulated "on" and "off." To support the on–off signal, the rise time of the modulator system has to follow the digital waveform transition. The limit of the AOM rise and fall time is the transit time of the acoustic wave propagation across the optical beam. The rise time is given by

$$t_r = \frac{\mathrm{DIA}}{1.5v_a} \tag{3.9}$$

where DIA is the laser beam diameter and v_a is the velocity of the acoustic wave. A typical rise time for a 1-mm-diameter laser beam is around 150 ns. To achieve faster rise times, it is necessary to focus the laser beam through the modulator and decrease the acoustic transit time. Since the incident beam is a convergent instead of a collimated beam, the DE decreases as the ratio of the optical beam

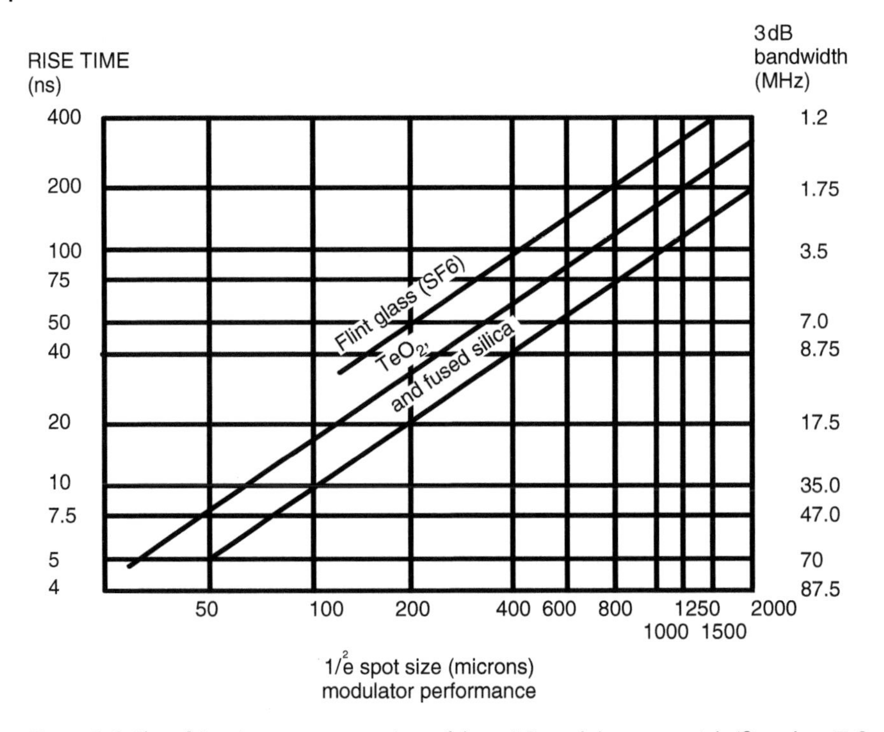

Figure 3.5 Plot of rise time versus spot time of three AO modulator materials (flint glass, TeO_2, and fused silica) (www.optoscience.com).

convergence and acoustic beam convergence angle increases [158]. A plot of rise time versus spot size for three common AOM materials is given in Figure 3.5.

3.4.1.3 Analog Modulation

The AOM has a nonlinear transfer function; therefore, care must be exercised when using it as an analog modulation system. For simple gray level control, the best approach is to characterize the transfer function and apply the appropriate voltage levels into the driver's analog modulation input port. For sinusoidal modulation, a bias is required to move the operating point to the linear region of the transfer function and focusing may be necessary to ensure that the rise time is adequate. The modulation transfer function (MTF) model is given by the following equation:

$$\mathrm{MTF} = \exp\left[-\left(\frac{f_m}{1.2f_0}\right)^2\right]; \quad f_0 = \frac{0.35}{t_r} \tag{3.10}$$

where f_m is the modulating frequency. A typical MTF function is shown in Figure 3.6. The modulation contrast ratio at any f_m can also be obtained from experimental measurements:

$$\mathrm{MT} = \frac{(I_{\max} - I_{\min})}{(I_{\max} + I_{\min})} \tag{3.11}$$

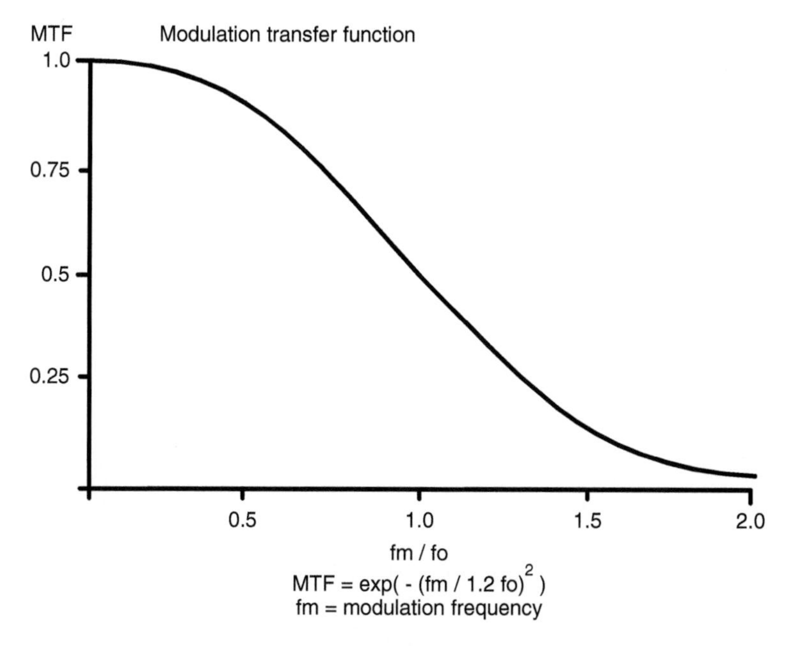

Figure 3.6 A typical modulation transfer function (www.optoscience.com).

where I_{max} = maximum laser intensity measured and I_{min} = minimum laser intensity measured.

3.4.1.4 Dynamic Contrast Ratio

The dynamic contrast (DC) ratio decreases as the modulation frequency increases and the modulator frequency responses degrade in performance. The DC contrast ratio is defined as

$$\mathrm{CR} = \frac{I_{max}}{I_{min}}, \text{ for the first order diffracted beam} \tag{3.12}$$

In the DC case, the I_{min} consists of contributions of the scattered light and of light leakage due to the extinct RF power driving the modulator. For optimum contrast ratio, I_{max} must be optimized. This is done by maximizing the DE of the AOM through careful adjustment of the Bragg angle and optimizing the RF drive power. Application of too much RF drive power causes the DE to be reduced. Light leakage (I_{min}) due to the extinct RF power driving the modulator can be reduced by changing the driver's operating frequency to an idle frequency when the light is desired to be off; thereby, the residual light can be directed away from the optical path. The DC scattered light can be reduced by use of a beam block. An optimized DC contrast ratio is between 500 : 1 and 1000 : 1.

For higher modulation rates, the contrast ratio is reduced due to the loss in DE from the application of the required lens to focus laser light to the needed smaller spot size to achieve the needed rise and fall time in the AOM. Also, I_{min} increases and I_{max} decreases as the modulation frequency increases as both the modulator's frequency response and the RF driver's frequency response degrades in performance due to rise and fall time (Eq. 3.10).

3.4.1.5 Applications of Acousto-optic Modulators

The AOM can perform other tasks in modulating the laser beam in addition to digital and analog modulation. By careful design, a special class of AOM that modulates more than one wavelength at a time can be made. By coupling the laser light into and out of the modulators with a fiber optical cable, the modulators can be used as a switch in the communication industry. By careful broadband design of the transducer and by varying the frequency of the drive signal, the angle that the laser beam is deflected will change. This can be an acousto-optic beam deflector (AOBD).

When the laser beam passes through the acoustic wave in the acousto-optic material, the interaction causes the frequency of the light (wavelength λ) to be shifted by an amount equal to the acoustic frequency. This frequency shift can be used for heterodyne detection applications, where precise phase information is measured and can be used to measure distance and velocity accurately. An AO modulator, called a *Q*-switch, typically operating internal to the cavity of a continuous wave (CW)-pumped Nd:YAG laser, produces greater than 10 kW power pulses with pulse widths of 40–200 ns wide and repetition rates of up to 100 kHz. An AO modulator, called a cavity dumper, typically operating internal to an Ar+ laser cavity, produces peak power around 100 W and a pulse width of 15 ns and has a repetition rate of up to 1 MHz. An AO modulator, called a mode locker, typically operating internal to a titanium sapphire laser cavity, modulates the laser at the resonance frequency of the laser cavity and causes the longitudinal modes of the laser to be in phase. This produces very narrow laser pulses having less than 100 fs pulse width and typically with peak power of around 150 kW.

A two-channel AOM (AOM) makes simultaneous assignment of the frequency and phase of radio signals possible. AOM is a fundamental element of a spectrum analyzer based on an acousto-optic receiver. An analysis of the influence of structural elements on characteristics of a two-channel modulator is shown in Figure 3.7 and the procedure of optimization of its construction described [30]. Based on this procedure, the modulator for spectrum analysis at global system for mobile communications (GSM) band was designed, performed, and investigated. The two-channel AOMs give the possibility of detection and defining direction of radiation with accuracy of 0.5°.

An arrangement scheme for analysis of radio-signal phase is shown in Figure 3.7. Separated blocks represent light source (laser), beam-forming optics, two-channel AOM, analyzing optics, and elements of image registration and

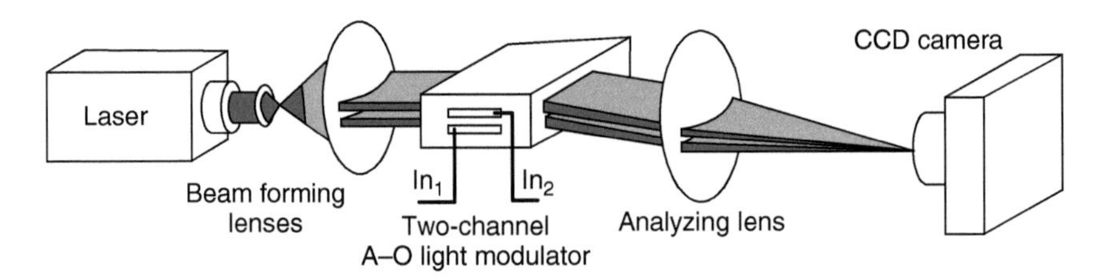

Figure 3.7 Scheme of a system for analysis of radio-signal phase with use of two-channel A–O modulator. In1 and In2 are the signal and reference inputs, respectively. Source: Jodłowski 2003 [30]. Reproduced with permission of Opto-Electron. Rev.

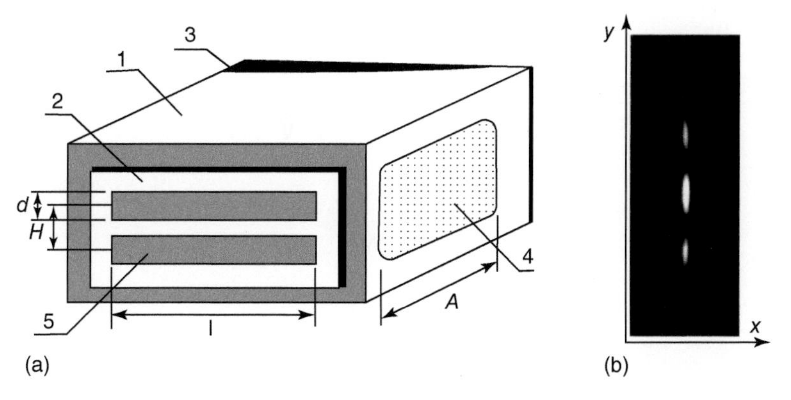

Figure 3.8 Details of construction of A–O modulator (a) and a CCD picture (b) taken in a laboratory setup. Source: Jodłowski 2003 [30]. Reproduced with permission of Opto-Electron. Rev.

analysis (charge-coupled device [CCD] camera, personal computer [PC] with frame-grabber card). Collimating optics is used to obtain uniform intensity distribution of laser beam in the area of AOM's part overtaken by acoustic beam. As a result of acousto-optic effect, part of the laser beam is deflected with an angle proportional to the frequency of the electric signal. Analyzing optics (lens) separates and focuses deflected and main beams in different places; therefore, in an image plane of the CCD camera, only the deflected beam is visible.

A structural scheme of the two-channel AOM is presented in Figure 3.8a. Interaction between light beam and ultrasonic wave takes place in an acoustic buffer region (1). Ultrasonic beam is generated and formed by the piezoelectric transducer (2) with the electrodes (5) of dimensions $d \times l$. The length of ultrasound–light interaction equals A (the width of the optical window (4)). A distance between parallel electrodes is H. The absorber (3) is intended for attenuation of the acoustic wave after having passed the whole buffer. An interference picture recorded by CCD camera is shown in Figure 3.8b. Projection of interference picture on the x-axis gives information about frequency of signal utilized in a spectrum analyzer. Distribution of light intensity in the y-direction depends on differences between phases of analyzed and reference signals.

The method for calculation of two-channel AO deflectors completes existing work devoted to acousto-optic deflectors. A two-channel modulator is a key element in any acousto-optic receiver used in spectrum analyses. The present work shows construction of a device for spectrum and signal phase measurements at a frequency bandwidth of 80 MHz with resolution of 0.2 MHz. Such a device can be based on an inexpensive acousto-optic deflector made of SF4 glass. This modulator can be used for analysis and recognition of signals in mobile phone networks (digital GSM, analog NMT [nordic mobile telephone]) or ultrahigh frequency (UHF) band.

3.4.2 Acousto-Optic Beam Deflector

An AOBD spatially controls the optical beam. In the operation of an acousto-optic deflector, the power driving the acoustic transducer is kept on, at a constant

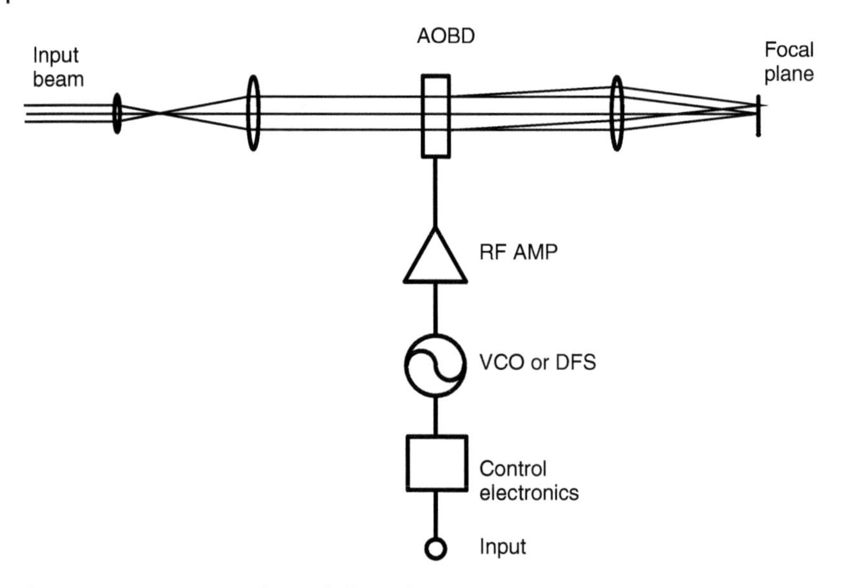

Figure 3.9 Acousto-optic beam deflector (www.optoscience.com).

level, while the acoustic frequency is varied to deflect the beam to different angular positions. The AOBD typically deflects the laser beam over a fraction of a degree to a couple of degrees with a resolution of a few hundred spots to an upper limit of about 2000 spots. Typical diffraction efficiencies are 40–70%.

A schematic setup of the AOBD and drive electronics is shown in Figure 3.9. One of the AOD's physical characteristics of concern in an optical system design is its optical aperture dimension optical height (H) and the width (D). Usually, the optical width is much larger than the height because of performance and design constraints. As a result, the input and output optical laser beam will require cylindrical optics to transform the incident laser beam from a circular beam to a truncated profile rectangular beam and then back to a circular beam after the deflector. The output optics usually focuses the deflected circular beam to a line of focused spots in the output plane.

3.4.2.1 Definition of Optical Deflector Resolution

Optical deflectors, whether they are mechanical or solid state in nature, obey the same fundamental equations for resolution. Assuming the deflector aperture is D, the natural divergence of a collimated laser beam of width D is given by

$$\Delta\Theta = \frac{\lambda}{D} \tag{3.13}$$

If the total scan angle of the deflector is defined as $\Delta\theta$, then the total number of resolvable spots is

$$N = \frac{\Delta\theta}{\Delta\Theta} = \Delta\theta\frac{D}{\lambda} \tag{3.14}$$

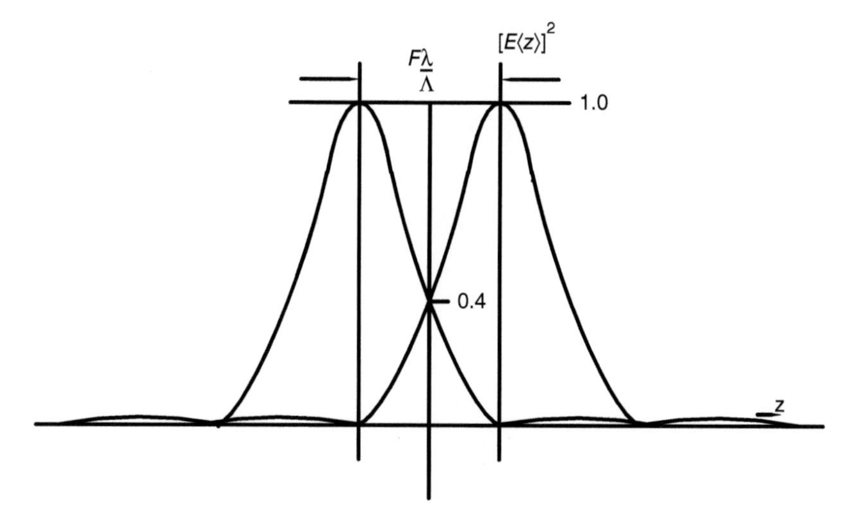

Figure 3.10 Spot profiles of AOBD (www.optoscience.com).

Equation 3.14 holds for all deflectors. Now, this equation is applied to the AOBD. The total angular sweep of the AOBD is

$$\Delta\theta = \frac{\lambda \Delta f_a}{v_a} \tag{3.15}$$

where λ is the optical wavelength, Δf_a is the acousto-optic bandwidth, and v_a is the acoustic velocity. Now substituting $\Delta\theta$ into the resolution Eq. (3.14), we get

$$N = \frac{\Delta f_a D}{v_a} = \Delta f_a \times \Delta T \tag{3.16}$$

Therefore, the number of resolution elements N is equal to the aperture time ΔT of the AOBD multiplied by the acousto-optic bandwidth Δf_a (commonly known as time bandwidth product). The value N is obtained with uniform illumination of the aperture D. When the output of the deflector is focused to a spot, the neighboring spots are such that the peak of one intensity spot is on the first zero intensity of the neighbor. The two spots cross over at the 40% intensity points and the spot profiles are shown in Figure 3.10. There are several factors that will degrade the total number of resolution elements and these are discussed subsequently.

3.4.2.2 Modulation Transfer Function

When dealing with laser deflection or scanning of an entire line or frame, it is necessary to consider the MTF or the contrast ratio. A parameter p, the truncation ratio of the laser beam illuminating the AOBD, is defined as

$$p = \frac{D}{W} \tag{3.17}$$

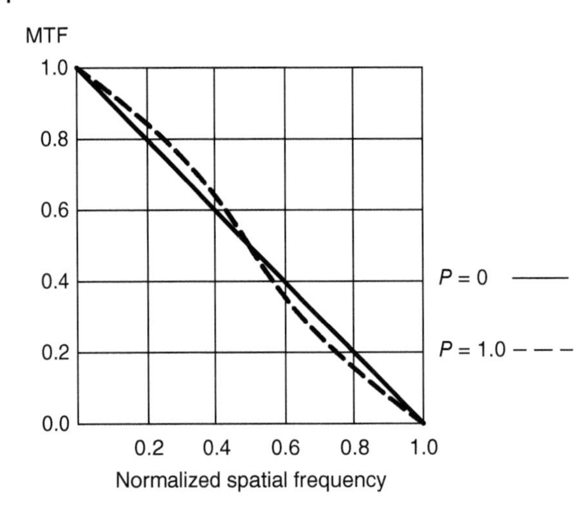

Figure 3.11 Plot of modulation transfer function (www.optoscience.com).

where W is the diameter of the laser beam at the $1/e^2$ intensity points. A plot of MTF is shown in Figure 3.11. For example, with $p = 0$ (uniform illumination) and MTF of 0.5, the maximum number cycles per line is equal to $\Delta f_a \times \Delta T/2$. With $p = 1$, the intensity drops to $1/e^2$ at the ends of the aperture. The resolution in cycles per line is about $\Delta f_a \times \Delta T/2.1$.

3.4.2.3 Scan Flyback Time

Since it takes a finite time for the acoustic energy to fill the AOBD, the total number of resolvable spots is reduced to

$$N = \left(1 - \frac{\Delta T}{T - \Delta T}\right)\left(\frac{\Delta T \Delta f_a}{a}\right) \tag{3.18}$$

where T is the total linear frequency modulated (FM) scan time, "a" is the parameter for uniformity of illumination ($a = 1$ for uniform beam illumination and $a = 1.34$ for Gaussian beam illumination).

3.4.2.4 Cylinder Lensing Effect

The linear FM modulation in the AOBD produces a lensing effect in addition to deflection. The focal length (FL) of the acoustic lens is given by

$$\mathrm{FL}_a = \frac{v_a^2}{\lambda \frac{df_a}{dt}} \tag{3.19}$$

where df_a/dt is the FM slope. This lensing effect must be taken into the design of any optical system using an AOBD. This lensing effect can also be useful in some applications.

3.4.2.5 Applications of AOBD

A variety of operations can be performed with AOBD devices: They include single-axis (1D) and two-axis (2D) laser beam deflection and optical signal processing. The electronics for the deflector are arranged in one of three ways depending on the application. First, for continuous laser beam deflection, the deflection angle is directly proportional to the RF. Therefore, a linear voltage-controlled oscillator (VCO) or a digital frequency synthesizer (DFS) is used to drive the RF amplifier for the AOBD. For a continuous line scan, a linear sawtooth waveform drives the VCO, outputting a linear FM signal. Since the frequency linearity is extremely important, it is necessary to have additional digital electronics to correct for small nonlinearity of the VCO. This signal will drive the AOBD to output a line scan. The scan rate is limited by the scan flyback time (Eq. 3.18) and the lensing effect (Eq. 3.19).

In the second application where vector (random) scanning is needed, the electronic input is usually a digital word, which causes a different frequency to be the output for each word. The location of the AOBD output beam is represented by the digital word. A D/A circuit converts the digital signal to an analog signal and the analog signal in turn drives the linear VCO. With this electrical input, the AOBD deflects the laser beam to a specific point in the output plane. To address the next location, consideration must be given to the minimum access time, which is equal to the sum of the AOBD aperture time (D/v_a) plus the electronics retrace time.

In the third application, for signal processing, the AOBD or Bragg cell is driven by an input RF signal from an amplifier which brings the signal of interest to the appropriate RF power level for the best performance of the AOBD. Typical signal processing involves spectral analysis of the input signal for frequency information or presence or detection of a specific signal; correlation to the presence of the specific signal; tempest testing, where the original signal cannot be present in the encoded signal, and radar signal analysis for ambiguity.

3.4.3 Acousto-optic Frequency Shifter

The diffracted beam of the AOM and AOD is also shifted in frequency (wavelength), by the acoustic beam. This is called the *Doppler shift*. If the incident acoustic wave is introduced in the direction opposite to that of the incident optical wave, the laser shifts toward the lower frequency side (Figure 3.12a). If the incident acoustic wave is introduced in the direction of the incident optical wave, the laser frequency shifts toward the higher side (Figure 3.12b).

3.4.3.1 Principles of Operation

An RF signal applied to a piezoelectric transducer, bonded to a suitable crystal, will generate an acoustic wave. This acts like a "phase grating," traveling through the crystal at the acoustic velocity of the material and with an acoustic wavelength dependent on the frequency of the RF signal. Any incident laser beam will be diffracted by this grating, generally giving a number of diffracted beams.

A parameter called the "quality factor, Q," determines the interaction regime. Q is given by

$$Q = \frac{2\pi\lambda_0 L}{n\Lambda^2} \tag{3.20}$$

where λ_0 is the wavelength of the laser beam, n is the refractive index of the crystal, L is the distance the laser beam travels through the acoustic wave, and Λ is the acoustic wavelength.

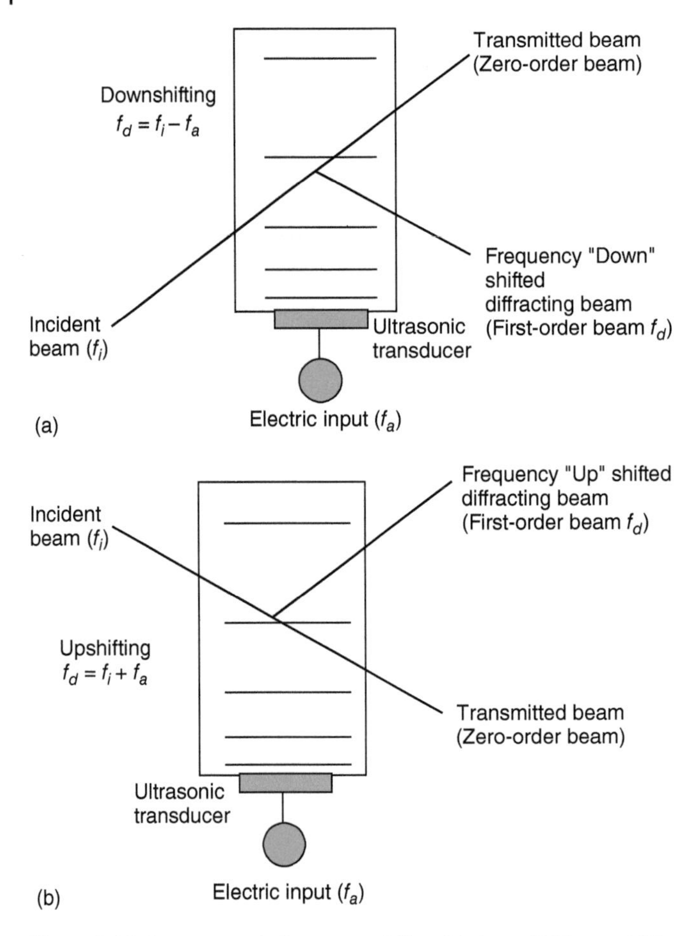

Figure 3.12 Acousto-optic frequency shifter, (a) downshifting and (b) upshifting (http://www.brimrose.com/pdfandwordfiles/aofshift.pdf).

Traditionally, the type of diffraction is determined by the conditions $Q \ll 1$ and $Q \gg 1$. If $Q \ll 1$, this is the Raman–Nath regime. The laser beam is incident roughly normal to the acoustic beam and there are several diffraction orders (…−2 −1 0 1 2 3…) with intensities given by Bessel functions (Figure 3.1). If $Q \gg 1$, this is the Bragg regime. At one particular incidence angle θ_B, only one diffraction order is produced – the others are annihilated by destructive interference (Figure 3.2).

In the intermediate situation, an analytical treatment is not possible and a numerical analysis would need to be performed by a computer. Most acousto-optic devices operate in the Bragg regime, the common exception being acousto-optic mode lockers and Q-switches.

An acousto-optic interaction can be described as a transfer of energy and momentum. However, conservation of the energy and momentum must be maintained. The equation for conservation of momentum can be written as [34, 35]

$$k_d = k_i \pm k_a \tag{3.21}$$

where $k_i = 2\pi n_i/\lambda_0$ – wave vector of the incident beam, $k_d = 2\pi n_d/\lambda_d$ – wave vector of the diffracted beam, and $k = 2\pi f_a/v_a$ – wave vector of the acoustic wave.

Here, f_a is the frequency of the acoustic wave traveling at velocity v_a. n_i and n_d are the refractive indexes experienced by the incident and diffracted beams ($n_i \neq n_d$). Energy conservation leads to

$$f_d = f_i \pm f_a \tag{3.22}$$

So, the optical frequency of the diffracted beam (f_d) is by an amount equal to the frequency of the acoustic wave (f_a). This "Doppler shift" can generally be neglected. Since $f_a \ll f_d$ or f_i, it can be of great interest in other applications such as heterodyning, Doppler, or optical time-domain reflectometer (OTDR) applications. It is important to notice that the frequency shift can be positive or negative.

Double Pass The double pass inside the same AOS (acousto-optic shifter) allows to double the frequency shift linked to the interaction. With this method, we can create high shift values over 500 MHz.

Low Frequency Shifts The cascade of two frequency shifters, one with a positive shift and the second with a negative shift, allows creating small values of frequency shift as low as zero. This method is commonly used for low-frequency shifters below 35 MHz.

3.4.3.2 Laser Doppler Vibrometer (LDV)

A laser Doppler vibrometer (LDV) is a scientific instrument that is used to make noncontact vibration measurements of a surface. The laser beam from the LDV is directed at the surface of interest and the vibration amplitude and frequency are extracted from the Doppler shift of the laser beam frequency due to the motion of the surface. The output of an LDV is generally a continuous analog voltage that is directly proportional to the target velocity component along the direction of the laser beam.

A vibrometer is generally a two-beam laser interferometer that measures the frequency (or phase) difference between an internal reference beam and a test beam. The most common type of laser in an LDV is the helium–neon laser, although laser diodes, fiber lasers, and Nd:YAG lasers are also used. The test beam is directed to the target and scattered light from the target is collected and interfered with the reference beam on a photodetector, typically a photodiode. Most commercial vibrometers work in a heterodyne regime by adding a known frequency shift (typically 30–40 MHz) to one of the beams. This frequency shift is usually generated by a Bragg cell or AOM.

A schematic of a typical laser vibrometer is shown in Figure 3.13. The beam from the laser, which has a frequency f_o, is divided into a reference beam and a test beam with a beam splitter (BS). The test beam then passes through the Bragg cell, which adds a frequency shift f_b. This frequency shifted beam is then directed to the target. The motion of the target adds a Doppler shift to the beam given by $f_d = 2v(t)\cos(\alpha)/\lambda$, where $v(t)$ is the velocity of the target as a function of time, α is the angle between the laser beam and the velocity vector, and λ is the wavelength of the light.

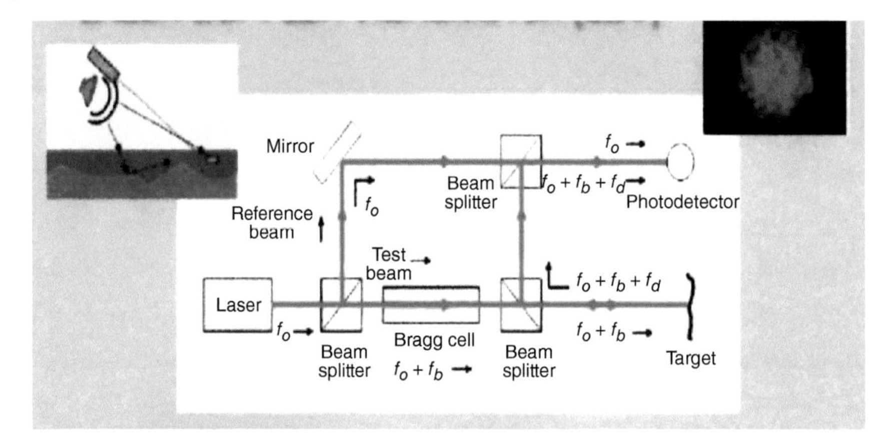

Figure 3.13 A schematic of a typical LDV (http://www.pegasus-optik.de/PDF/VAR-SHIFTER-FIO.pdf).

Light scatters from the target in all directions, but some portion of the light is collected by the LDV and reflected by the BS to the photodetector. This light has a frequency equal to $f_o + f_b + f_d$. This scattered light is combined with the reference beam at the photodetector. The initial frequency of the laser is very high (>10^{14} Hz), which is higher than the response of the detector. The detector does respond, however, to the beat frequency between the two beams, which is at $f_b + f_d$ (typically in the tens of megahertz range).

The output of the photodetector is a standard FM signal with the Bragg cell frequency as the carrier frequency and the Doppler shift as the modulation frequency. This signal can be demodulated to derive the velocity v_s time of the vibrating target.

3.4.4 Acousto-optical *Q*-Switch

The acousto-optical *Q*-switch often used in laser marking makes use of the mutual interaction between an ultrasonic wave and a light beam in a scattering medium. The light beam that enters in a direction forming a Bragg angle to the wave surface of the acoustic wave in the scattering medium is diffracted in accordance with periodic changes in the diffraction rate produced by the acoustic wave.

First of all, an RF signal is impressed to the transducer adhered to the molten quartz and thickness extensional vibration is produced. Ultrasonic shear waves are caused to advance in the molten quartz by this vibration and phase grating formed by acoustic waves is produced. The laser beam is diffracted when it satisfies the Bragg angle with respect to this phase grating and is separated in space from the incident light.

If the laser optical resonator is constructed against 0-dimensional diffracted light (undiffracted light), the diffracted light deviates from the laser optical resonator axis when an RF signal is impressed. As a result, loss occurs in the laser optical resonator and laser oscillation is suppressed. To make use of this

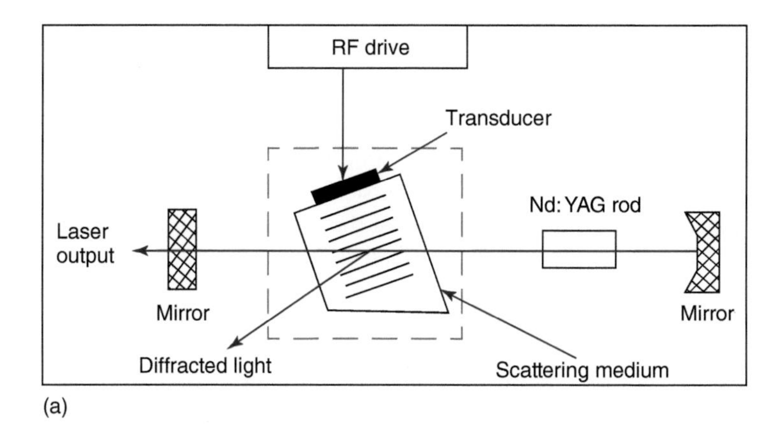

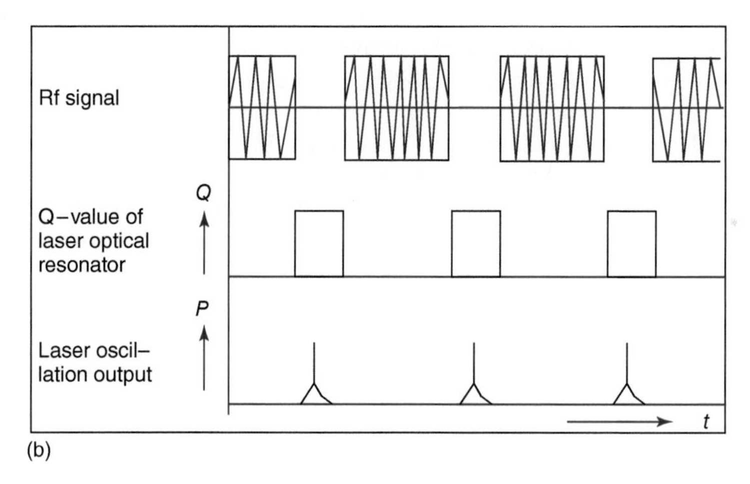

Figure 3.14 (a) Schematic of acousto-optical *Q*-switch, (b) *Q*-switch output pulses (http://www.sintec.sg/catalog/qswitch.pdf).

phenomenon, an RF signal is impressed for a certain length of time only (status of low *Q*-value) to suspend laser oscillation. In the meantime, the population inversion of the Nd:YAG rod is accumulated by continuous pumping. When the RF signal is reduced to zero (status of high Q-value) and the loss to the laser optical resonator is removed, the accumulated energy is activated as laser oscillation in a pulse form within an extremely short length of time. They are *Q*-switch pulses. The situation is briefly shown in Figure 3.14a,b.

When an RF signal is subjected to pulse modulation, it is possible to periodically take out a *Q*-switch pulse. When the period of *Q*-switch pulses becomes shorter than the life (about 200 μs) of the higher order of the Nd:YAG rod, however, the population inversion decreases and the peak value of *Q*-switch pulses decreases.

3.4.4.1 Applications of Acousto-optical *Q*-Switches

Acousto-optic devices have long been used in a variety of laser intracavity applications. These applications can be divided into two categories: zero-order beam applications and diffracted beam applications. One of the zero-order beam applications is acousto-optic Q-switching. A Q-switched laser is actually a variable cavity loss laser. Most of the time, the laser is placed in a state where the cavity loss is larger than the gain. As a result, there is no lasing and the pumping source produces a very large population inversion. After the cavity loss is suddenly decreased, a very intense pulse of stimulated radiation is obtained. Using this operation principle, some of the important applications of acousto-optic Q-switched fiber laser are described as follows.

Q-switching by Intermodal Acousto-optic Modulation in an Optical Fiber The Q-switched fiber lasers have attractive applications in different fields, such as in remote sensing and material processing. The mechanism of short pulse emission is based on the modulation of the Q-factor of the cavity, which can be done either passively or actively [33]. In the former, the setups are simpler, but the repetition rate only varies with the pump power of the medium gain. Further, they usually show long-term instability, and frequently the amplitude of the pulses is randomly modulated in time [34]. On the other hand, active Q-switching is independently and accurately controlled by an electrical signal, which triggers the modulator. The bulk approach, such as electro-optic [35] or AOMs [36], is not adapted to compact fiber laser systems required nowadays; further, they have large optical coupling losses and stringent alignment requirements. For these reasons, the all-fiber approach is of permanent interest, being advantageous in terms of cost, loss, packaging, robustness, and simplicity. One solution widely investigated has been enabled by the use of fiber Bragg gratings (FBGs) as cavity mirrors. In this way, the tuning of the wavelength of one of the FBGs has been used to achieve active Q-switching [37].

C. Cuadrado-Laborde et al. [38] described an actively Q-switched ytterbium-doped strictly all-fiber laser. The setup used for the Q-switched fiber laser is schematically illustrated in Figure 3.15. The gain was provided by 0.65 m of a heavily doped ytterbium-doped single-mode fiber-Nufern SM-YSF-HI, cutoff wavelength of 860 ± 70 nm, numerical aperture of 0.11, and fiber absorption of 250 dB m^{-1} at 975 nm. Q-switching modulation was achieved by intermodal modulation induced by flexural acoustic waves traveling in a tapered optical fiber.

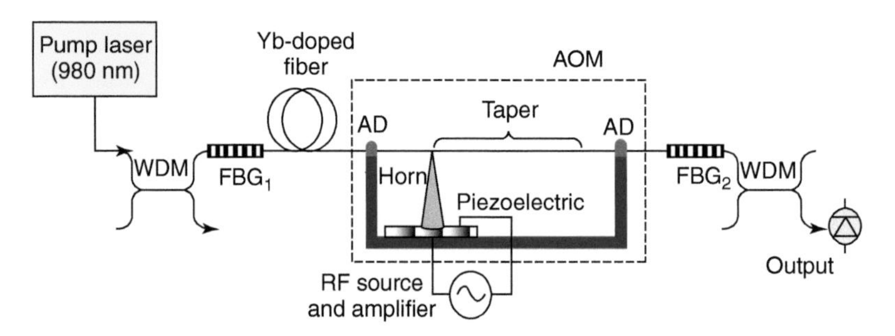

Figure 3.15 Q-switched fiber-laser setup; the acousto-optic modulator is defined by the elements inside the dashed line; AD stands for acoustic dumper. Source: Cuadrado-Laborde et al. 2011 [38]. Reproduced with permission of InTech.

Q-switched light pulses at 1064.1 nm were successfully obtained at repetition rates in the range 1–10 kHz, with pump powers between 59 and 88 mW. Best results were for laser pulses of 118 mW peak power, 1.8 μs of time width with a pump power of 79 mW at 7 kHz repetition rate.

Q-switching of an All-fiber Laser by Acousto-optic Interaction in a FBG in the Long-wavelength Regime When a traveling axially propagating longitudinal acoustic wave is launched through an FBG, the periodic strain field of the acoustic wave perturbs the grating in two different ways. First, the average index changes in response to the stress-optical effect and, second, the otherwise uniform Bragg grating pitch changes being modulated by the acoustical signal [39]. As a consequence of both effects, the reflectivity changes; the main features of these changes depend on the ratio between the acoustical wavelength and the grating length [37]. Thus, we can distinguish two very different situations: the long-wavelength and the short-wavelength regimes. In the first, the Bragg grating pitch is homogeneously perturbed along its length. The successive cycles of compression and expansion generated by the longitudinal wave will shift periodically in time the spectral response of the grating as a whole to longer and shorter wavelengths [37, 40]. On the contrary, in the short-wavelength regime, the acoustic wave generates many compressed and expanded sections, which gives rise to a superstructure within the grating. In this case, the spectral response of the original FBG shows new and narrow reflection bands symmetrically at both sides of the original Bragg wavelength [41, 42]. The position and strength of these sidebands can be controlled by varying the frequency and voltage applied to the piezoelectric, respectively.

As an example of the use of the long-wavelength regime in FBGs, C. Cuadrado-Laborde et al. [37] presented an in-fiber resonant AOM suitable for *Q*-switching applications. The modulator consists of a short-length FBG modulated by a long-wavelength standing longitudinal acoustic wave. Figure 3.16 shows a diagram of the laser setup, together with the details of the AOM. The acoustic wave shifts periodically in time the reflection band of the grating along a given wavelength range; using this modulator, they demonstrated an actively *Q*-switched all-fiber laser. Output light pulses of 1.6 W peak power and 172 ns time width were obtained at 18 kHz repetition rate.

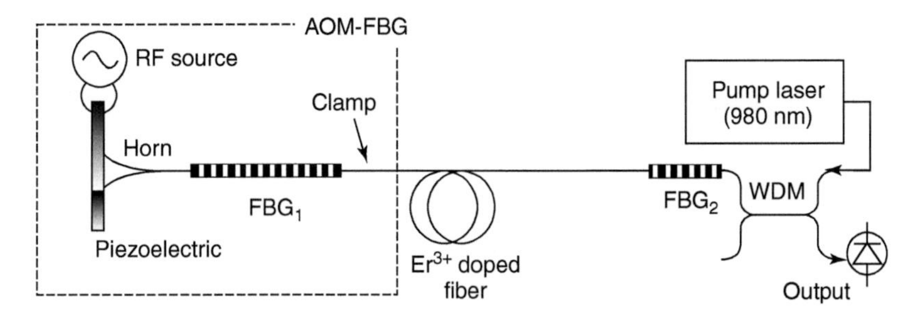

Figure 3.16 Laser setup, the dashed line defines the acousto-optic modulator. Source: Andres et al. 2008 [37]. Reproduced with permission of IOP.

Mode Locking and Q-switching Mode Locking by Acousto-optic Interaction in an FBG in the Short-wavelength Regime The spectral response of the original FBG in this acoustical regime shows new and narrow reflection bands symmetrically at both sides of the original Bragg wavelength [41, 42]. The position of these sidebands can be controlled by varying the frequency at a slope of 0.15 and 0.30 nm/MHz for the first- and second-order sideband, respectively. The strength of the reflection bands, on the other hand, can be controlled independently by varying the voltage applied to the piezoelectric. Since these sidebands can be regarded as weak ghosts of the strong permanent Bragg grating, its FWHM bandwidth is that of a weak Bragg grating of the same length [41].

Figure 3.17 shows the setup for a typical reflectivity measurement on an acousto-optic super-lattice modulation (AOSLM). The AOSLM in turn is composed of an RF source, an electrical RF amplifier, a piezoelectric disk, a silica horn, and an FBG. The tip of the silica horn was reduced by chemical etching to the same diameter of FBG – 125 μm and subsequently fusion spliced to the fiber. The uniform and non-apodized grating was written in photosensitive fiber using a double argon laser and a uniform period mask; the FBG was 120 mm long. The reflection properties of the AOSLM were investigated by illuminating the FBG through an optical circulator with a broadband light source and detecting the reflected light with an optical spectrum analyzer (OSA).

In summary, an AOSLM can be driven in two different regimes, either using traveling or standing acoustic waves. In both cases, it can be used as an amplitude modulator. However, in the first case – traveling acoustic waves – the light reflected on the sidebands is modulated at the same frequency of the acoustical signal, whereas in the second case – standing acoustic waves – it is modulated at two times this frequency, respectively.

Cuadrado-Laborde et al. showed first the construction of a mode-locked laser driven by standing acoustic waves [43, 44]. The setup proposed for the standard mode-locking laser is schematically illustrated in Figure 3.18. The gain was provided by an erbium-doped fiber (EDF) containing 300 ppm of Er^{3+} with a cutoff

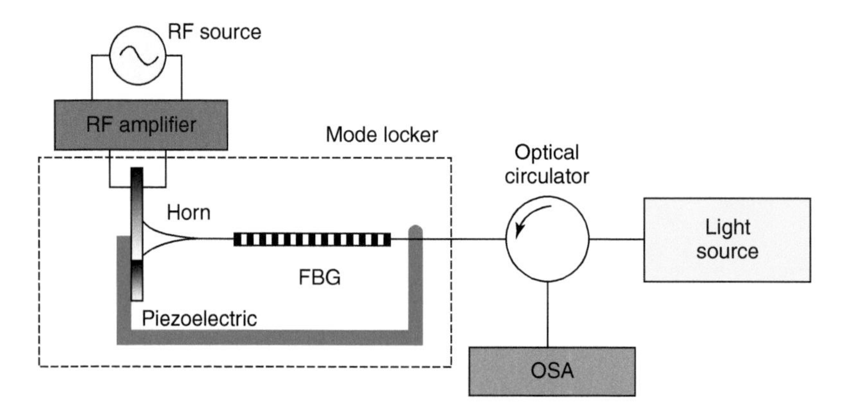

Figure 3.17 Setup for the characterization of the acousto-optic superlattice modulator; OSA, optical spectrum analyzer; FBG, fiber Bragg grating; and RF, radiofrequency. Source: Liu et al. 1997 [41]. Reproduced with permission of OSA Publishing.

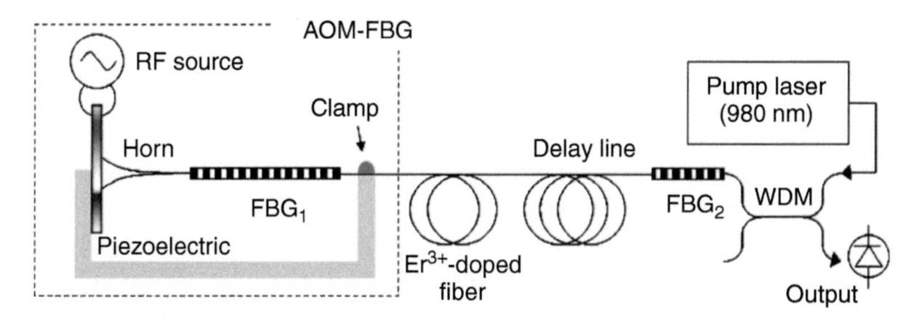

Figure 3.18 Setup of the mode-locked all-fiber laser by using an acousto-optic super-lattice modulator (AOSLM) as mode locker [43, 44].

wavelength of 939 nm and a numerical aperture of 0.24. In this configuration, they obtained transform-limited optical light pulses of up to 120 mW peak power and 780 ps pulse width generated at a fixed repetition rate of 9 MHz, with an emission line width of 2.8 pm at 1530.5 nm. They also studied the influence of different parameters on the mode-locking process such as frequency detuning, EDF length, amplitude modulation (AM), and dispersion. In this case, narrower pulses were obtained at higher modulation depths, normal dispersion and shorter lengths of active fiber. Best results were reached when the laser was optimized according to these variables (160 mW peak power and 630 ps pulse width).

On the other hand, Cuadrado-Laborde et al. showed that by slightly modifying the cavity, it is possible to operate the laser in a double-active *Q*-switching mode-locking regime [45]. It is worth mentioning that this approach was unique, this double-active all-fiber laser being the first of its kind. The setup of our double-active *Q*-switching mode-locking all-fiber laser is shown in Figure 3.19. It is basically the same as it was described before for the standard mode locking as shown in Figure 3.18, with the same length of EDF and delay line, except by the magnetostrictive device controlling the FBG_2. It is composed of a 15-mm-long (1 mm^2 cross section) magnetostrictive rod of Terfenol-D bonded to the FBG_2. The rod and the fiber were placed inside a small coil driven by an electronic circuit designed to drive square current pulses with amplitudes up to 260 mA of any required duty cycle. When the magnetic pulses are generated in the solenoid stretch FBG_2, its central wavelength is brought to match the short-wavelength

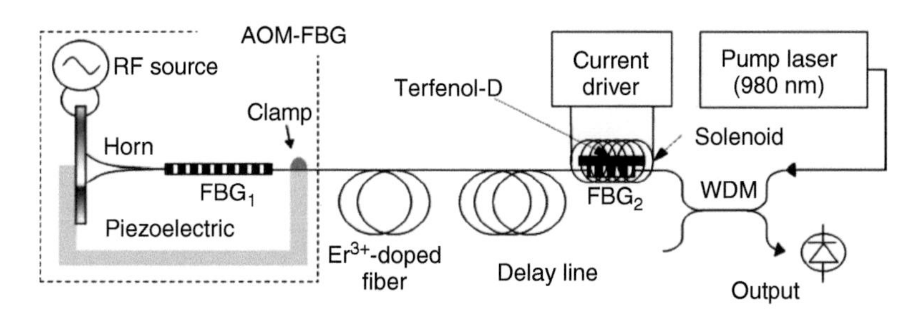

Figure 3.19 Setup of the doubly-active *Q*-switching mode-locking all-fiber laser when the AOSLM is driven by standing acoustic waves. Source: Cuadrado-Laborde et al. 2009 [45]. Reproduced with permission of OSA Publishing.

sideband of FBG_1 for a short period of time, which results in an increased Q-value. In this way, by modulating the coil current with a given Q-switched frequency, the Q-factor is actively modulated at the same frequency. The magnetostriction has a low-pass frequency response and, consequently, presents the advantage of permitting a continuous tuning of both the Q-switched repetition rate and the duty cycle of the modulation pulses.

Fully modulated Q-switched mode-locked trains of optical pulses were obtained for a wide range of pump powers and repetition rates. For a Q-switched repetition rate of 500 Hz and a pump power of 100 mW, the laser generates trains of 12–14 mode-locked pulses of about 1 ns each, within an envelope of 550 ns, an overall energy of 0.65 μJ, and a peak power higher than 250 W for the central pulses of the train.

Cuadrado-Laborde et al. discussed the construction of a mode-locked laser when the mode locker is driven by traveling acoustic waves. In this case, the modulation frequency is half the frequency obtained when standing acoustic waves are used. Optical pulses were obtained of 530 mW peak power, 700 ps pulse width, at a repetition rate of 4.1 MHz. The variation in the pulse parameters under frequency detuning and applied voltage was also studied. Finally, they demonstrated that it is not necessary to modify the setup in order to reach double-active Q-switching and mode locking, when traveling acoustic waves were used to drive the mode locker. In this case, the commutation between mode locking and Q-switching mode locking is remarkably simple; it just needs use of a different electrical signal to drive the piezoelectric of the mode locker, i.e. from a sinusoidal to a burst-sinusoidal electrical signal. In this case, fully modulated 10–25 mode-locked pulses around 700 ps each within a Q-switching envelope around 1 μs and a maximum overall energy of 0.68 μJ were obtained.

Q-switching in a Distributed Feedback Fiber Laser When the acoustic perturbation is not a harmonic wave but a single pulse, its passage through the FBG creates a defect which can be used to control the Q-factor in a distributed feedback (DFB) all-fiber laser. Figure 3.20 shows the scheme of the proposed DFB all-fiber laser. The FBG was 100 mm long and was written in a 1500 ppm erbium hydrogen-loaded fiber (codoped with germanium and aluminum) of the same length using a doubled argon laser and a uniform period mask. The FBG shows more than

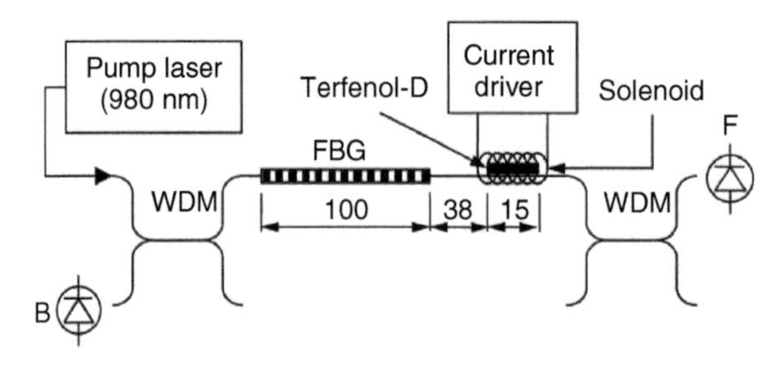

Figure 3.20 Q-switched distributed feedback-fiber laser setup. Lengths are in millimeters, whereas B and F stand for backward and forward outputs of the laser, taking as reference the pump direction. Source: Cuadrado-Laborde et al. 2008 [46]. Reproduced with permission of OSA Publishing.

30 dB attenuation at the Bragg wavelength of 1532.45 nm and a 3 dB bandwidth of 88 pm. The FBG was pumped through a 980-/1550-nm wavelength division multiplexer (WDM) with a 980-nm semiconductor laser, providing a maximum pump power of 130 mW. A square-shaped rod of a magnetostrictive material (Terfenol-D, 15 mm long and 1 mm^2 section) was bonded outside the FBG to a free section of fiber at 88 mm from the center of the grating and placed inside a small coil, as shown in Figure 3.20.

Cuadrado-Laborde et al. showed a single-mode, transform-limited, actively *Q*-switched DFF fiber laser [46]. Optical pulses of 800 mW peak power, 32 ns temporal width, and up to 20 kHz repetition rates were obtained. The measured linewidth demonstrates that these pulses are transform limited: 6 MHz for a train of pulses of 10 kHz repetition rate, 80 ns temporal width, and 60 mW peak power. Efficient excitation of spontaneous Brillouin scattering was demonstrated.

In summary, photonic devices can benefit highly from a strictly all-fiber configuration which provides them a series of attractive advantages. Among all the proposed in-fiber solutions, devices controlled by acoustic waves have been by far the most employed, especially in mode-locked lasers, providing a broad range of alternatives. The recent advances in acoustically controlled photonic systems positioned them as a promising candidate for commercially available systems in the near future.

3.4.5 Acousto-optic Tunable Filter

AOTFs are solid-state devices that act as electronically tunable spectral bandpass filters. An AOTF consists of a properly oriented birefringent uniaxial crystal to which a piezoelectric transducer is bonded [47]. The application of an RF signal to the transducer produces an acoustic wave that propagates inside the crystal. The traveling acoustic wave modulates the refraction index of the material periodically, due to the elasto-optic effect. The process acts like a volume phase grating, leading to the diffraction of a particular wavelength that satisfies a specific momentum-matching condition inside the birefringent medium. The wavelength filtered by the crystal can be rapidly tuned across a wide spectral range by changing the applied RF signal. The acousto-optic interaction also changes the polarization state of a single wavelength. The structure of an anisotropic uniaxial medium permits two normal modes (ordinary and extraordinary) to propagate in any direction with mutually orthogonal polarizations and different velocities. If the AOTF input light beam at the tuned wavelength is extraordinarily polarized, then it emerges from the AOTF ordinarily polarized, and vice versa.

The acousto-optic filter design uses two basic geometric configurations, collinear and noncollinear. In the noncollinear configuration [49], depicted in Figure 3.21, the directions of the optical and acoustic waves are different, causing diffracted and nondiffracted rays to be angularly separated at the filter output.

3.4.5.1 Principles of AOTF Operation

AOTF is an all-solid-state, electronic dispersive device which is based on the diffraction of light in a crystal [31, 32]. Light is diffracted by an acoustic wave because an acoustic wave when propagated in a transparent material will produce

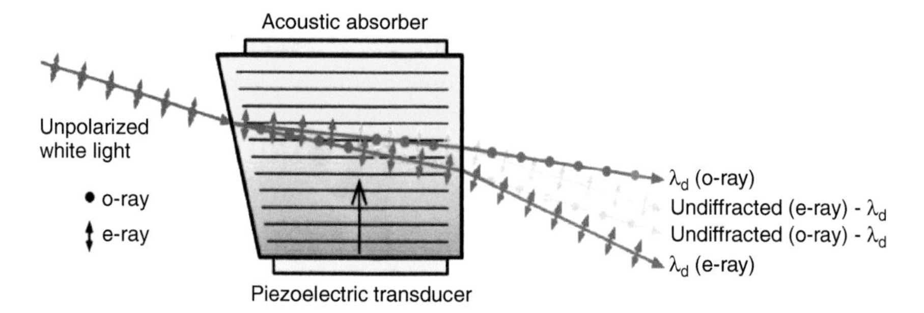

Figure 3.21 Noncollinear AOTF design. The zero-order (undiffracted) and first-order (diffracted) outputs are separated by a certain angle. Source: Vila et al. 2005 [48]. Reproduced with permission of ACM Digi Library.

a periodic modulation of the index of refraction (via the elasto-optical effect). This, in turn, will create a moving grating which diffracts portions of an incident light beam. The diffraction process can, therefore, be considered as a transfer of energy and momentum. However, conservation of the energy and momentum must be maintained. The equation for conservation of momentum can be written as [31, 32]

$$\boldsymbol{k}_d = \boldsymbol{k}_i \pm \boldsymbol{k}_a \tag{3.23}$$

where $\boldsymbol{k}_i$, $\boldsymbol{k}_d$, and $\boldsymbol{k}_a$ are the wave vector of the incident and diffracted light and of the phonon.

In the case of the AOTF, the acousto-optic interaction occurs in an anisotropic medium and the polarization of the diffracted beam is orthogonal to that of the incident beam. The momenta of incident and diffracted photons are

$$|\boldsymbol{k}_i| = \frac{2\pi n_i}{\lambda} \tag{3.24}$$

$$|\boldsymbol{k}_d| = \frac{2\pi n_d}{\lambda} \tag{3.25}$$

They are not equal since one is an ordinary ray and the other is an extraordinary ray (i.e. $n_i \neq n_d$). In the case of the collinear AOTF, the incident and diffracted light beams and the acoustic beam are all collinear. If the incident light is an extraordinary ray and the diffracted light is an ordinary ray, the momentum matching condition becomes

$$\boldsymbol{k}_d = \boldsymbol{k}_i - \boldsymbol{k}_a \tag{3.26}$$

$$f_a = \frac{v_a(n_e - n_o)}{\lambda} \tag{3.27}$$

where the acoustic wave vector $|\boldsymbol{k}_a| = 2\pi f_a/v_a$, f_a and v_a are the frequency and velocity of the acoustic wave. Equation (3.27) can be generalized for all types of AOTFs including the collinear and the noncollinear AOTFs:

$$f_a = [v_a(n_e - n_o)]\frac{1}{\lambda}(\sin^4\theta_i + \sin^2 2\theta_i)^{1/2} \tag{3.28}$$

where θ_i is the incident angle. When $\theta_i = 90°$, Eq. (3.28) is reduced to Eq. (3.27), i.e. the case of collinear.

It is thus evidently clear that in an anisotropic crystal where the phase matching requirement is satisfied, diffraction occurs only under optimal conditions. These conditions are defined by the frequency of the acoustic waves and the wavelength of a particular diffracted light. For a given acoustic frequency, only light whose wavelength satisfies either Eq. (3.27) or (3.28) is diffracted from the crystal. The filter can, therefore, be spectrally tuned by changing the frequency of the acoustic waves (f_a).

Generally, the AOTF is fabricated from an anisotropic TeO_2 crystal onto it an array of $LiNbO_3$ piezoelectric transducers are bonded. An RF signal is applied to the transducers which, in turn generates an acoustic wave propagating through the TeO_2 crystal. These propagating acoustic waves produce a periodic moving grating which will diffract portions of an incident light beam. Figure 3.22a–c shows collinear- and noncollinear-type AOTFs. As illustrated,

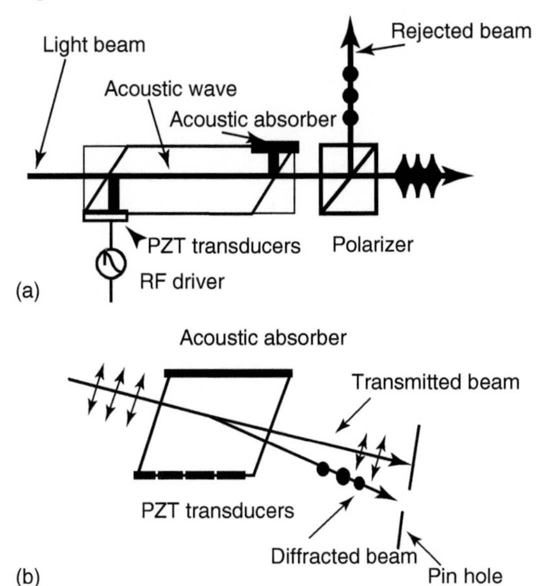

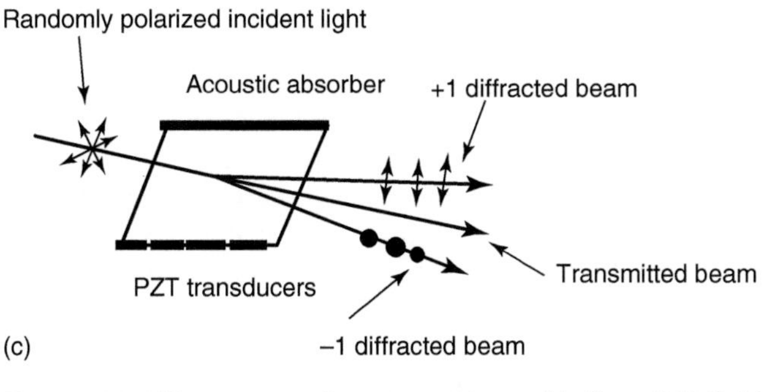

Figure 3.22 Different types of acousto-optic tunable filters (AOTF): (a) collinear AOTF; (b) noncollinear AOTF with polarized light incident light; and (c) noncollinear AOTF with randomly polarized light incident light. Source: Tran 2003 [50]. Reproduced with permission of T&F.

a light beam propagating as an e-ray is converted into an o-ray by interaction with, and diffraction from, an acoustic wave propagating in the same medium. In a collinear-type AOTF (Figure 3.22a), the diffracted o-beam is separated from the transmitted beam (e-beam) by means of a polarizer (analyzer). Since the diffracted beam is spatially separated from the transmitted beam in a noncollinear AOTF (Figure 3.22b,c), it is not necessary to use an analyzer.

The scanning speed of the filter is defined by the speed of the acoustic wave in the crystal, which is on the order of microseconds. As a consequence, compared to conventional gratings, AOTFs offer such advantages as being all-solid state (containing no moving parts), having rapid scanning ability (μs), wide spectral tuning range, allowing high-speed random or sequential wavelength access, and giving high resolution.

3.4.5.2 Infrared Multispectral Imaging

IR multispectral imaging, referred to in literature as spectroscopic imaging, hyperspectral imaging, or chemical imaging, is a relatively new technique that combines spectroscopy and imaging, namely, it records the chemically rich information available from spectroscopy in a spatially resolved manner [51]. In this instrument, the recorded images contain signals that are generated by molecules or units in a sample, plotted as a function of spectral and spatial distribution [51]. Recent advances in material sciences, electronics, and computer sciences have led to the development of novel types of electronically tunable filters. AOTFs belong to these types of electronic tuning devices. They are particularly suited for use in multispectral imaging instruments, as they contain no moving parts and can be spectrally tuned over a wide spectral range (from NIR to middle-IR) with much faster speeds (from micro- to milliseconds).

As depicted in Eq. (6.28), for a fixed acoustic frequency and sufficiently long interaction length, only a very narrow band of optical frequencies can approximately satisfy the phase matching condition and be diffracted. The wavelength of the diffracted light can, therefore, be tuned over large spectral regions by simply changing the frequency of the applied RF signal. Figure 3.23a,b shows the spectral tuning course of a noncollinear AOTF fabricated from TeO_2 for the VIS and NIR region. As illustrated, light can be spectrally tuned from 400 to 700 nm and from 1.1 to 2.4 mm by simply scanning the frequency of the applied RF signal from 222 to 100 MHz and from 60 to 30 MHz, respectively [32].

Depending on the spectral region, different birefringent crystals are used to fabricate AOTFs. Three most widely used crystals for AOTFs are shown in Table 3.1. As illustrated, quartz is often used for AOTF in the ultraviolet (UV) and VIS regions [52]. Quartz is known to have relatively good optical transparency for UV and VIS regions. However, because it has a relatively low acoustic figure of merit (ability to couple acoustic wave to the crystal), AOTFs based on quartz often require, in addition to collinear configuration (to enhance acousto-optic interaction length l), relatively high power of applied RF signal. The high RF power consumption somewhat limits applications of the quartz-based AOTFs [52]. In addition, the collinear configuration of this type of AOTF renders it difficult to use this type of AOTF in a multispectral imaging instrument. Different from quartz, TeO_2 crystal is good for AOTF from the

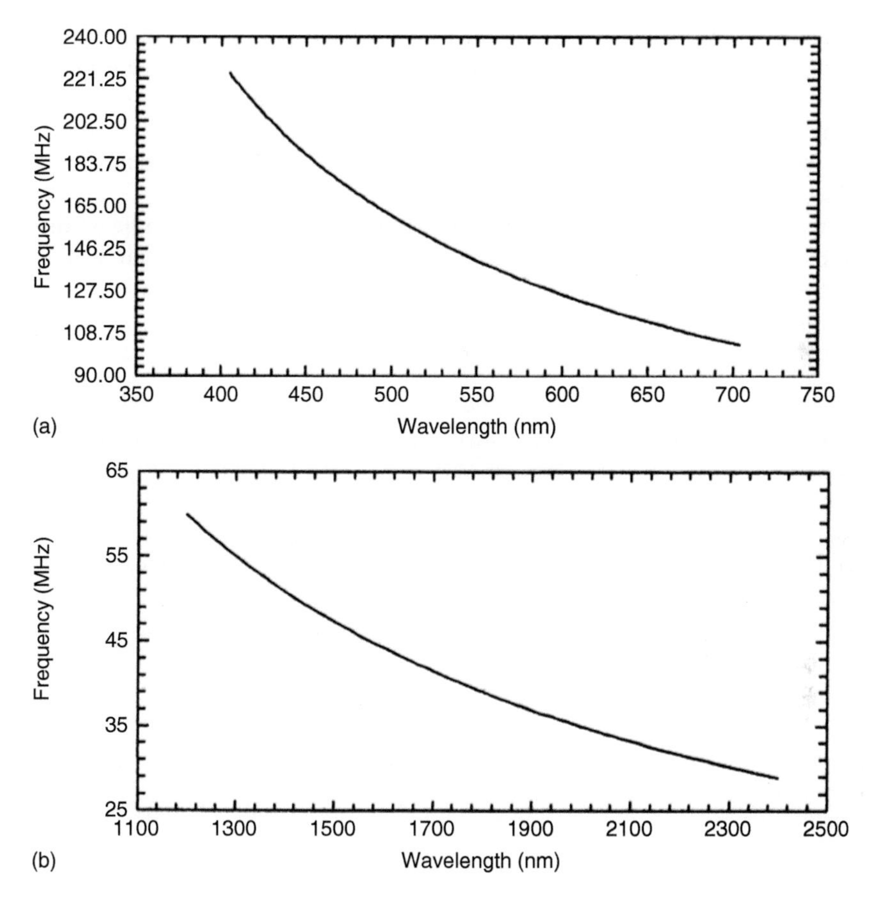

Figure 3.23 Spectral tuning curves of noncollinear TeO_2 AOTF for (a) the visible region and (b) for the near-infrared region. Source: Tran 2003 [50]. Reproduced with permission of T&F.

Table 3.1 Properties of different AOTF crystals.

Materials	Transparency range (μm)	Acoustic figure of merit
Quartz	0.2–4.5	1
TeO_2	0.35–5.0	795
Tl_3AsSe_3	1.0–16	900

Source: Tran 2003 [50]. Reproduced with permission of T&F.

VIS region up about 5 μm where it becomes opaque. Tl_3AsSe_3 (often known as TAS) can be used for AOTF from 1 to 16 μm. Because these two crystals have relatively higher acoustic figures of merit (795-fold and 900-fold higher than quartz), AOTFs based on these crystals require relatively lower power of applied RF and signal. In addition, they can be constructed using noncollinear configuration [32]. As a consequence, TeO_2- and Tl_3AsSe_3-based AOTFs are particularly suited for use in multispectral imaging instruments.

Of particular importance to the multispectral imaging technique is the scanning speed of the AOTF. The scanning speed of the filter is defined by the speed of the acoustic wave in the crystal, which is on the order of microseconds [32]. As a consequence, AOTFs have the fastest scanning speed compared to other dispersive devices including conventional gratings, liquid crystal tunable filters (LCTFs), and interferometers.

Spectral resolution ($\Delta\lambda$) of the AOTF is related to the wavelength of the diffracted light λ, acousto-optic interaction length l, and incident angle θ_i and Δn ($\Delta n = n_e - n_o$) by [32]:

$$\Delta\lambda = \frac{\lambda^2}{2l\Delta\sin^2\theta_i} \tag{3.29}$$

As depicted in the equation, the spectral resolution is proportional to the wavelength of diffracted light and inversely proportional to the interaction length. As a consequence, quartz collinear AOTFs for the UV often have high spectral resolution compared to noncollinear TeO_2 AOTFs for the VIS, NIR, and middle-IR. Figure 3.24a–c shows the spectral resolution of a quartz collinear AOTF for the UV and VIS region (Figure 3.23a), the TeO_2 noncollinear AOTF for the VIS (Figure 3.23b) and NIR. As illustrated, spectral resolution of a few angstroms can be readily achieved with the quartz collinear AOTF for the UV region. However, the resolution deteriorates somewhat in an AOTF with a noncollinear configuration fabricated from TeO_2 up to a few nanometers (Figure 3.24b). As wavelength becomes longer in the NIR, the spectral resolution increases to tens of nanometers (Figure 3.24c).

Taken together, the AOTFs offer such advantages as being compact, all-solid state (contains no moving parts), having rapid scanning ability (microseconds), wide spectral tuning range (from VIS to NIR and IR) and high throughput (>90% DE for incident polarized light), allowing high-speed random or sequential wavelength access and giving high resolution (a few angstroms). As a consequence, AOTF is particularly suited for use as a dispersive device in a multispectral imaging instrument. In fact, AOTF-based multispectral imaging spectrometers have been developed for the NIR and middle-IR. A schematic diagram of such an instrument is shown in Figure 3.25 [53–62]. As illustrated, advantages of the AOTF provide not only the simplicity in construction of the instrument but also ease of operation. Features such as high sensitivity and fast scanning speed made it possible for this imaging spectrometer to be used for studies which are not possible with other ([laser communication test system] LCTS- and interferometer-based) imaging instruments [53–62]. These include the determination of inhomogeneity in the kinetics of curing of epoxy resins and kinetics and chemical inhomogeneity in the sol–gel formation [53–62]. Furthermore, because the AOTF-based imaging spectrometer is compact, contains no moving parts, and requires relatively low electrical power, it can be powered by a battery. As a consequence, it can be deployed directly in the fields of agricultural and environmental measurements.

3.4.5.3 Analytical Applications of AOTF

The AOTF has offered unique means to develop novel analytical instruments which are not feasible otherwise. The instrumentation development and unique

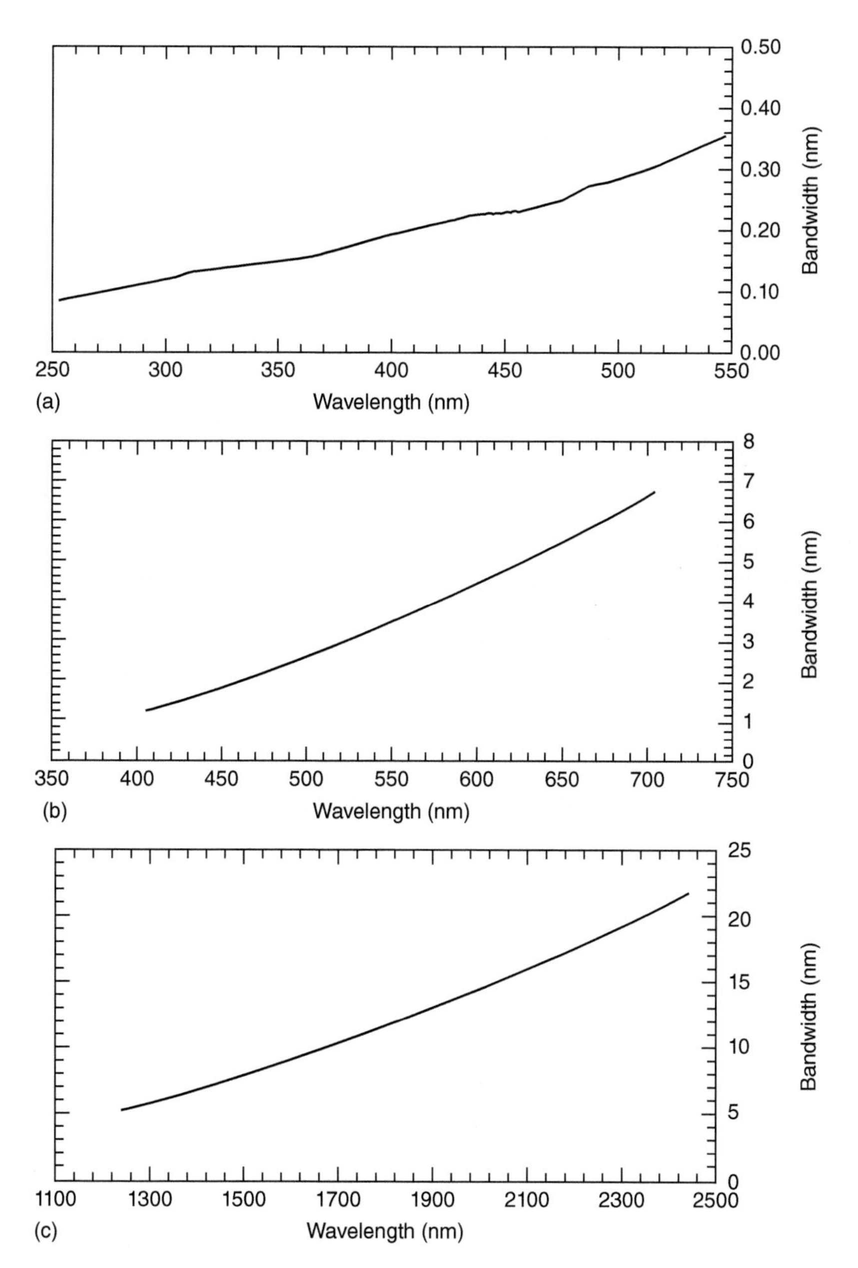

Figure 3.24 Spectral resolution of (a) collinear quartz AOTF for the visible and UV region (b) noncollinear TeO_2 AOTF for the visible region; and (c) noncollinear TeO_2 AOTF for the near-IR region. Source: Tran 2003 [50]. Reproduced with permission of T&F.

features of such AOTF-based instruments including the multidimensional fluorimeter, multiwavelength thermal lens spectrometer, NIR spectrometer based on erbium-doped fiber amplifier (EDFA), and detectors for high-performance liquid chromatography (HPLC) and flow injection analysis (FIA) are described here.

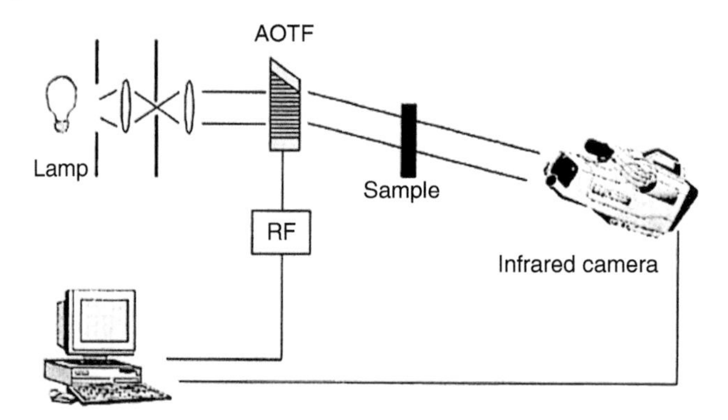

Figure 3.25 Schematic diagram of an infrared multispectral imaging instrument based on an acousto-optic tunable filter: AOTF (acousto-optic tunable filter); RF (radio frequency signal generator) [53–62].

AOTF-Based Fluorimeter The AOTF can be incorporated with an incoherent light source or multiline white-light laser to produce a fast tunable light source. It can be switched between any randomly selected wavelength in microseconds with resolution from several nanometers up to 0.1 nm. One of the applications for this source is fluorescence spectroscopy [63]. Fluorescent probes can indicate diverse properties as ion concentration, pH, electronic potential in live cells and tissues. In acquiring kinetic data from fluorescent probes, it is often necessary to monitor the ratio of two (or several) excitation or emission wavelengths to cancel out the intensity of excitation and dye concentration, giving an accurate estimation of target ion concentration. It is also essential to alternate wavelengths as rapidly as possible. Conventionally, a spinning filter is used; but it is slow, hard to synchronize with opto-electronic data collection, and also causes mechanical vibration that can be troublesome in a microscopy setup.

With an AOTF, these problems can be circumvented, because it is a solid-state device with no moving parts. Also, a single AOTF can be used as a multiwavelength modulator. By coupling the AOTF to a broadband light source (or a multiline white laser) and driving the AOTF with two RF frequencies, two excitation wavelengths can be generated simultaneously. Furthermore, each wavelength can be modulated electronically at different frequencies, and lock-in amplifiers can demodulate the fluorescent emission into its two components.

The schematic diagram of the AOTF-based multidimensional fluorimeter is shown in Figure 3.26 [64, 65]. Two AOTFs were used in this instrument: one for excitation and the other for emission. The first AOTF was used to specifically diffract white incident light into a specific wavelength(s) for excitation. Depending on the needs, the second AOTF (i.e. the emission AOTF) can be used as either a very fast dispersive device or a polychromator. In the first configuration (i.e. the rapid scanning fluorimeter), the sample was excited by a single excitation wavelength; the emitted light was analyzed by the emission AOTF which was scanned very fast. A speed of 4.8 Å/μs was found to be the fastest speed at which the AOTF can be scanned with a reasonable signal-to-noise ratio (S/N) and resolution. With this speed, a spectrum of 150 nm can be measured in 312 μs.

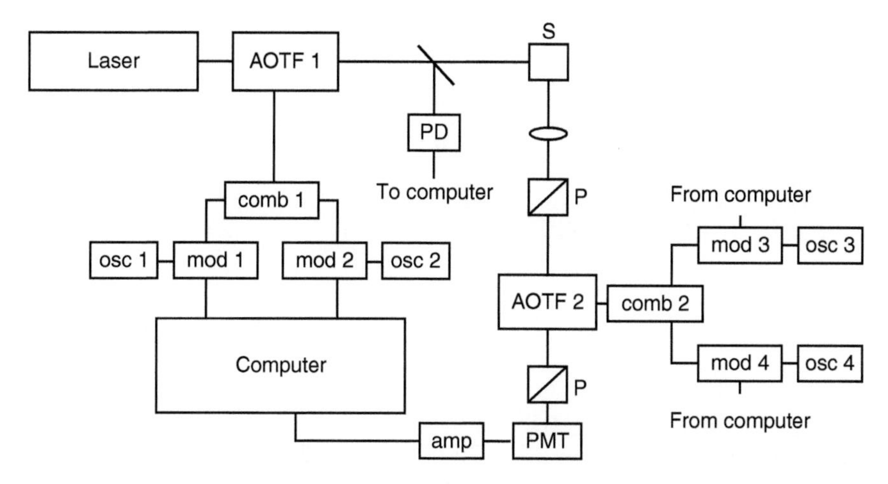

Figure 3.26 Schematic diagram of the AOTF-based fluorimeter: PD, photodiode; S, sample; P, polarizer; PMT, photomultiplier tube; amp, amplifier; osc, oscillator; mod, modulator; comb, combiner. Source: Tran 1997 [66]. Reproduced with permission of Elsevier.

Faster scanning is possible, but because of the limitation due to the speed of the acoustic wave, it may undesirable lead to the degradation in the S/N and spectral resolution [64, 65]. In the second configuration (i.e. multidimensional fluorimeter), both AOTFs were used as a polychromator. Several different RF signals were simultaneously applied into the first AOTF to provide multiple excitation wavelengths. The emission was simultaneously analyzed at several wavelengths by the emission AOTF. With this configuration, the fluorimeter can be used for the analysis of multicomponent samples and the maximum number of components it can analyze is, in principle, $a \times b$, where a and b are the number of excitation and emission wavelengths, respectively [65].

An AOTF can be coupled with an argon-krypton white laser that simultaneously emits 12 wavelengths. By simultaneously adding several RF frequencies with properly adjusted power levels, artificial composite laser color that may find applications in confocal microscopy, holograph, and laser entertainment can be generated.

Scientists at the National Institutes of Health (Bethesda, MD) are pioneering innovative biomedical and chemical applications of AOTFs. In one experiment, an argon laser equipped with an AOTF was used as a light source for fluorescent microscopy [67]. The narrow bandwidth, rapid wavelength selection (microseconds), and the intensity control enable a variety of measurements to be made in an extremely short time. Recently, this system was used to bleach a small region of a sample periodically at 514 nm to determine rates of flow and diffusion in two fibroblast cells stained with a fluorescent dye. The sharp details and high S/N ratio result from the narrow bandwidth of the exciting light produced by the laser-AOTF combination.

In another experiment, an AOTF coupled with a tungsten lamp provided a fast-tuning NIR light source for spectroscopic imaging of water (Figure 3.27) [68]. The microscope images of water were collected at 960 and 850 nm; the darkening

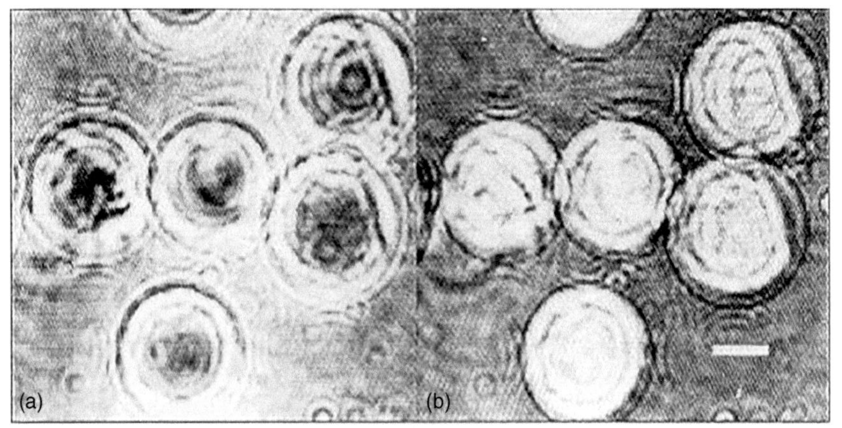

Figure 3.27 Near-IR light source illuminates water droplets at 960 nm (a) and 850 nm (b); wavelengths are selected by AOTF. Darkening of spots results from water absorption; reference bar corresponds to 7 μm (http://www.brimrose.com/pdfandwordfiles/tunable_filter.pdf).

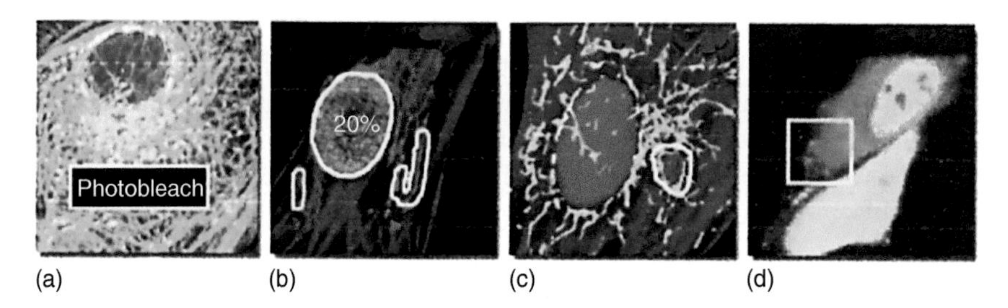

Figure 3.28 AOTF section of specific regions for excitation in confocal microscopy. (a) FRAP (b) ROI intensity (c) FRET (d) Photoactivation. (http://www.olympusconfocal.com/theory/aotfintro.html).

of the droplets at 960 nm corresponds to the vibrational absorption of water (specifically, the second overtone O—H stretch). For presentation, the images are corrected for instrument response and background contributions.

Figure 3.28 illustrates several examples of user-defined regions of interest (ROIs) that were created for advanced fluorescence applications in laser scanning confocal microscopy. In each image, the ROI is outlined with a black and white border. The rat kangaroo cell presented in Figure 3.28a has a rectangular area in the central portion of the cytoplasm that has been designated for photobleaching experiments. Fluorophores residing in this region can be selectively destroyed by high-power laser intensity and the subsequent recovery of fluorescence monitored for determination of diffusion coefficients. Several freehand ROIs are illustrated in Figure 3.28b, which can be targets for selected illumination intensities or photobleaching experiments. Fluorescence emission ratios in resonance energy transfer (FRET) can be readily determined using selected regions in confocal microscopy by observing the effect of bleaching the acceptor fluorescence in these areas (Figure 3.28c; Indian Muntjac cells with yellow fluorescent protein). AOTF control of laser excitation in selected regions with confocal microscopy is also useful for investigations of protein diffusion in photoactivation studies using fluorescent proteins, as illustrated in Figure 3.28d. This image frame presents the fluorescence emission peak of the Kaede protein as it shifts from green to red in HeLa (human cervical carcinoma) cells using selected illumination (yellow box) with a 405-nm violet-blue diode laser.

The development of the AOTF has provided substantial additional versatility to techniques such as fluorescence recovery after photobleaching (FRAP), fluorescence loss in photobleaching (FLIP), as well as in localized photoactivated fluorescence (uncaging) studies (Figure 3.28). The AOTF allows near-instantaneous switching of light intensity, but it also can be utilized to selectively bleach randomly specified regions of irregular shape, lines, or specific cellular organelles and to determine the dynamics of molecular transfer into the region.

By enabling precise control of illuminating beam geometry and rapid switching of wavelength and intensity, the AOTF is a significant enhancement to the application of the FLIP technique in measuring the diffusional mobility of certain cellular proteins. This technique monitors the loss of fluorescence from continuously illuminated localized regions and the redistribution of fluorophore from distant locations into the sites of depletion. The data obtained can aid in the determination of the dynamic interrelationships between intracellular and intercellular components in living tissue and such fluorescence loss studies are greatly facilitated by the capabilities of the AOTF in controlling the microscope illumination.

Because the AOTF functions without use of moving mechanical components to electronically control the wavelength and intensity of multiple lasers, great versatility is provided for external control and synchronization of laser illumination with other aspects of microscopy experiments. When the confocal instrument is equipped with a controller module having input and output trigger terminals, laser intensity levels can be continuously monitored and recorded. The operation of all laser functions can also be controlled to coordinate with other experimental specimen measurements, automated microscope stage movements, sequential time-lapse recording, and any number of other operations.

AOTF-Based Thermal Lens Spectrometer The fluorescence technique is very sensitive and can be used for the determination of trace chemical species at very low concentrations. However, the technique is not applicable to all compounds because only a few molecules are fluorescent. Therefore, it is important that a novel technique which has the same sensitivity as the fluorescence technique but is applicable to nonfluorescent compounds be developed. The thermal lens technique is one such possibility.

The technique is based on the measurement of the temperature rise that is produced in an illuminated sample by nonradioactive relaxation of the energy absorbed from a laser [69–74]. Because the absorbed energy is directly measured in this case, the sensitivity of the technique is similar to that of the fluorescence technique and is relatively higher than the conventional absorption measurements. In fact, it has been calculated and experimentally verified that the sensitivity of the thermal lens technique is 237 times higher than that by conventional absorption techniques, when a laser of only 50 mW power was used for excitation [69–74]. Absorptivities as low as 10–7 have been measured using this ultrasensitive technique [69–74]. Potentially, the technique should serve as an excellent method for trace chemical analysis because it has high sensitivity in situ and nondestructive ability and requires a minimum amount of sample. Unfortunately, its applications to the area of general trace chemical analysis are not so widespread in comparison to other spectrochemical methods. A variety of reasons might account for its limited use, but the most likely one is probably due to the low selectivity. The majority of reported thermal lens spectrometers employs only a single excitation wavelength [69–74] and, as a consequence, can only be used for the analysis of one component samples. A multiwavelength thermal lens spectrometer is needed to analyze, in real-time, multicomponent samples without any pretreatment.

The schematic diagram of the AOTF-based multiwavelength thermal lens spectrometer is shown in Figure 3.29 [76, 77]. This instrument is based on the use of the AOTF as a polychromator. An argon ion laser operated in the multiline mode was used as the light source. This multiwavelength laser beam was converted into the excitation beam with appropriate number of wavelengths by means of a TeO_2 AOTF. In this case, four different wavelengths were used. They were obtained by simultaneously applying four different RF signals to the AOTF. These RF signals were provided by four different oscillators (oscl, osc2, osc3, and osc4). To differentiate each wavelength from the others, the applied RFs were sinusoidally amplitude modulated at four different frequencies by four modulators (modl, mod2, mod3, and mod4). The four AM-modulated RF signals were then combined by means of a combiner and amplified by an RF power amplifier (amp).

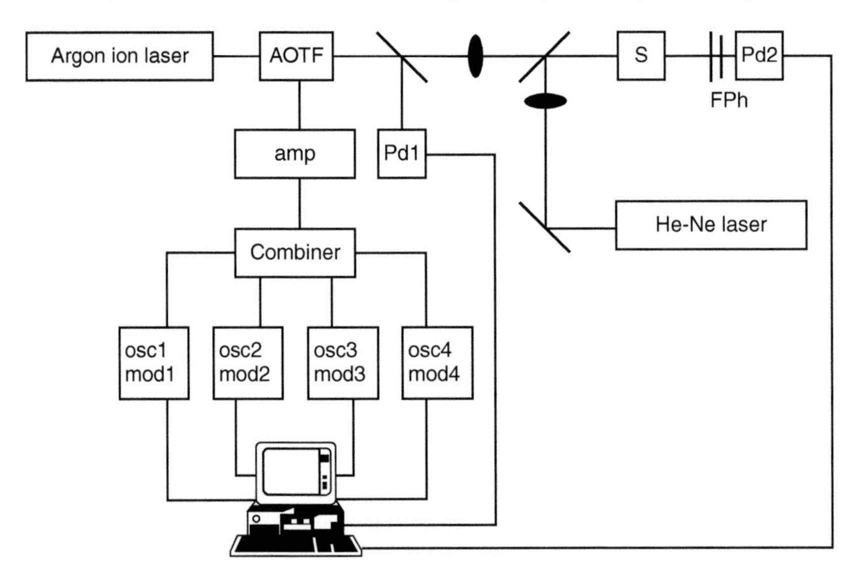

Figure 3.29 Schematic diagram of the multiwavelength thermal lens spectrometer based on an AOTF as a polychromator: amp, RF power amplifier; osc, oscillator; mod, sinusoidal modulator; PIN, photodiode; S, interference filter; Ph, pinhole. Source: Tran 1997 [75]. Reproduced with permission of Elsevier.

The amplified, AM-modulated signal was then applied onto the AOTF to enable it to diffract the incident multiline laser beam into a beam which has the four different wavelengths. The probe beam, provided by a He—Ne laser, was aligned to overlap with the pump beam at the sample cell by means of a dichroic filter (DF). The heat generated by the sample absorption of the pump beams changes the intensity of the probe beam. The intensity fluctuation of the probe beam was measured by a photodiode (Pd2) placed behind a 632.8-nm interference filter (F) and a slit (S). A lens was used to focus the probe beam and the distance between this lens and the sample was adjusted to give maximum thermal lens signals. The signal intensity, measured as the relative change in the probe beam center intensity, was recorded by a microcomputer through an AD interface board [76, 77].

Compared with other multiwavelength thermal lens instruments, this all-solid-state thermal lens spectrophotometer has advantages which include its ability to simultaneously analyze multicomponent samples in microsecond time scale, without the need for any prior sample preparation. In fact, with this apparatus and with the use of only 12 mW multiwavelength excitation beam, multicomponent samples including mixtures of lanthanide ions (Er^{3+}, Nd^{3+}, Pr^{3+}, and Sm^{3+}) can be simultaneously determined with a limit of detection (LOD) of 10^{-6} cm^{-1} [77].

NIR Spectrophotometer Based on AOTF and EDFA The use of the NIR spectrometry in chemical analysis has increased significantly in the past few years [72–74, 78–84]. The popularity stems from the advantages of the technique, namely, its wide applicability and noninvasive and online characteristics. The NIR region covers the overtone and combination transitions of the C—H, O—H, and N—H groups, and since all organic compounds possess at least one or more of these groups, the technique is applicable to all compounds [78–84]. There is no need for pretreatment of the sample; and since NIR radiation can penetrate a variety of samples, the technique is noninvasive and has proved to be potentially useful for noninvasive, online measurements. In fact, the NIR technique has been used in industries for online measurements for quality control and assurance. However, its applications are not as wide as expected. This may be due to a variety of reasons, but the most likely ones are probably due to the limitations on the speed, stability, and light throughput of the currently available instruments. To be effectively used as a detector for online measurements, the NIR instrument needs to have a high and stable light throughput, so as to not suffer a drift in the baseline and can be rapidly scanned. These impose severe limitations on conventional NIR spectrophotometers because such instruments generally have a relatively low and unstable light throughput, suffer some degree of baseline drift, and can only be scanned very slowly. AOTF, with its ability to rapidly scan as well as to control and maintain the intensity of light at a constant level, offers a means to alleviate some of these limitations.

It has been shown recently that stimulated emission can be achieved in a fiber when the fiber is doped with rare earth ions such as Er^{3+} and optically pumped by an ion laser or a YAG laser [85, 86]. Now, for the first time, a lasing medium can be confined in a material as flexible and as small (<10 μm) as a single-mode fiber [78, 79]. Because of such features, the length of the doped fiber can be adjusted to as long as a few kilometers to enable the fiber to have optical output in the range of kilowatts. Furthermore, the doped fiber can be pumped by a high-power diode laser fusion spliced directly into the doped fiber. It is thus evidently clear that this all-fiber, compact EDFA can provide NIR light with highest intensity and widest spectral bandwidth compared to other (cw) NIR light sources currently available. A novel, compact, all-solid-state, fast-scanning NIR spectrophotometer which has no moving part, high and very stable light throughput can, therefore, be developed by use of this EDFA as a light source and AOTF as a dispersive element to scan and control the intensity of the diffracted light.

The schematic diagram of such a spectrophotometer is shown in Figure 3.30 (with the inset showing the construction of the EDFA). As illustrated, the

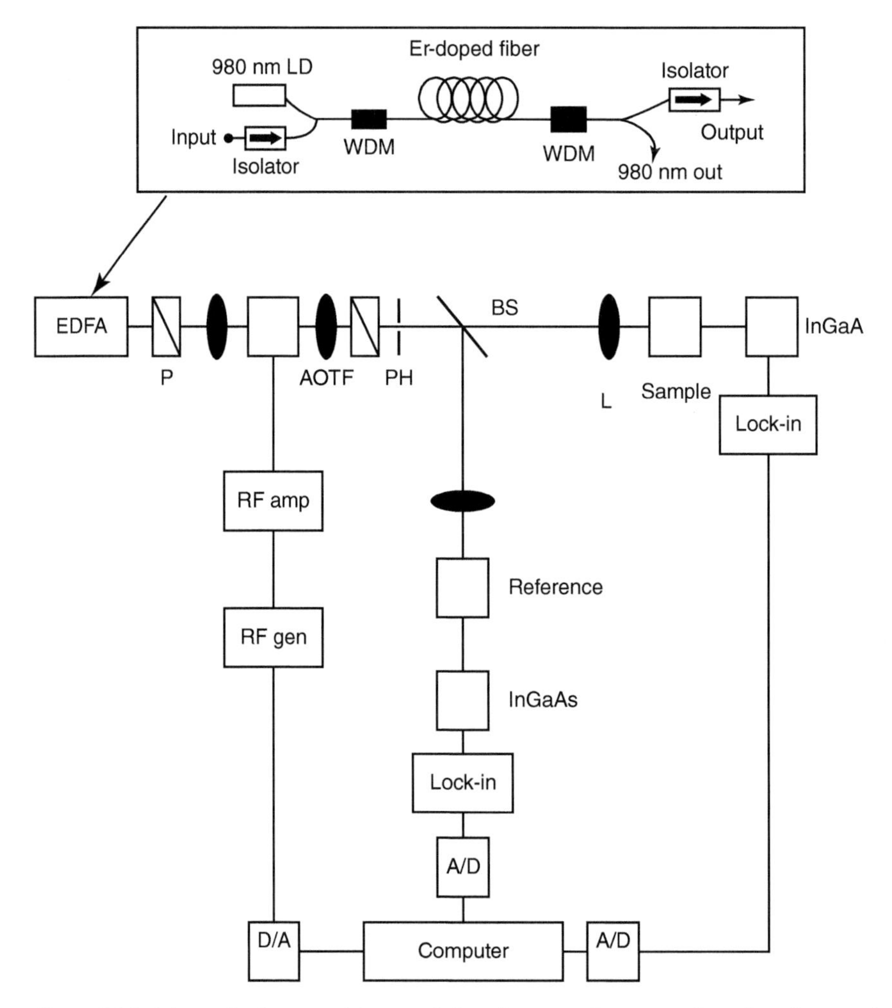

Figure 3.30 Schematic diagram of the near infrared spectrophotometer based on the use of the EDFA as a light source and AOTF as a dispersive element: EDFA, erbium-doped fiber amplifier; P, polarizer; AOTF, acousto-optic tunable filter; PH, pinhole; BS, beam splitter; L, lens; RF amp, RF power amplifier; RF gen, RF generator; InGaAs, detector. Inset: Schematic diagram of the erbium-doped fiber amplifier system: LD, laser diode; WDM, wavelength division multiplexer. Source: Tran 1997 [66]. Reproduced with permission of Elsevier.

NIR light from the output of the EDFA was dispersed to monochromatic light and spectrally scanned by an AOTF. The RF signal, provided by a driver, was amplitude modulated at 50 kHz by a home-built modulator and amplified by an RF power amplifier prior to being applied to the AOTF. The light diffracted from the AOTF was split into two beams (i.e. sample and reference beams) by means of a BS. Intensity of the light in the sample and reference beams was detected by thermoelectrically cooled InGaAs detectors. The output signals (reference and sample) from the detectors which were amplitude modulated at 50 kHz were connected to lock-in amplifiers (Stanford Research Systems Model SR 810) for demodulating and amplifying. The signal from the reference beam can be used either as a reference signal for a double-beam spectrophotometer or as a reference signal for the fed-back loop to stabilize the intensity of the light in the sample beam in a manner similar to those used previously [76, 87]. The signals from the lock-in amplifiers were then connected to a microcomputer through a 16-bit A/D interface board.

As expected, the sensitivity of this EDFA- and AOTF-based spectrophotometer is comparable with those of the halogen tungsten lamp–AOTF-based instruments. In fact, this spectrophotometer can be used to detect water in ethanol at an LOD of 10 ppm. More important is its high light throughput. The intensity of this EDFA light source was found to be about 20 times higher than that of a 250 W halogen tungsten lamp. As a consequence, it can be used for measurements which are not possible with lamp-based instruments. Two measurements were performed to demonstrate this advantage.

Figure 3.31 is the intensity of the light transmitted through six sheets of photocopy paper (Cascade X-9000 white paper). As illustrated, because of the high

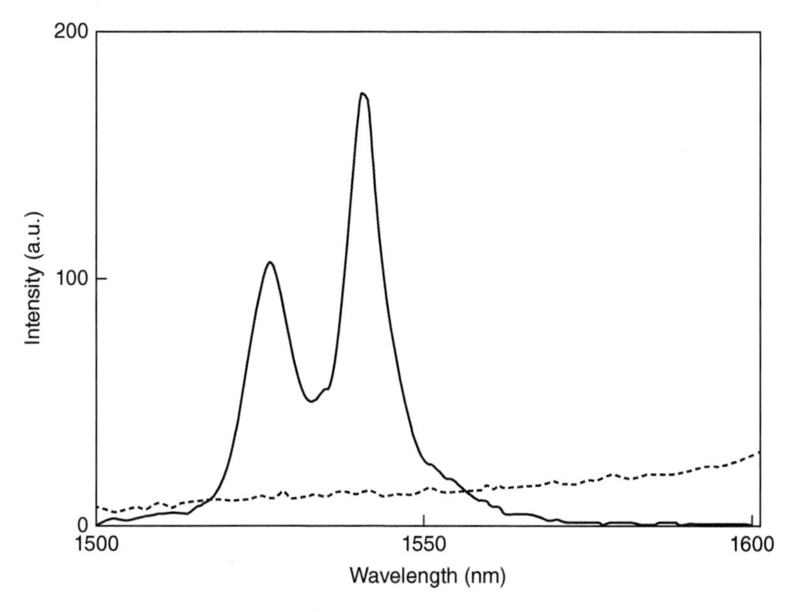

Figure 3.31 Relative intensity of the light transmitted through six sheets of paper, measured with a 250 W halogen tungsten lamp based spectrophotometer (---), and with the EDFA-based spectrophotometer (__). Source: Tran 1997 [66]. Reproduced with permission of Elsevier.

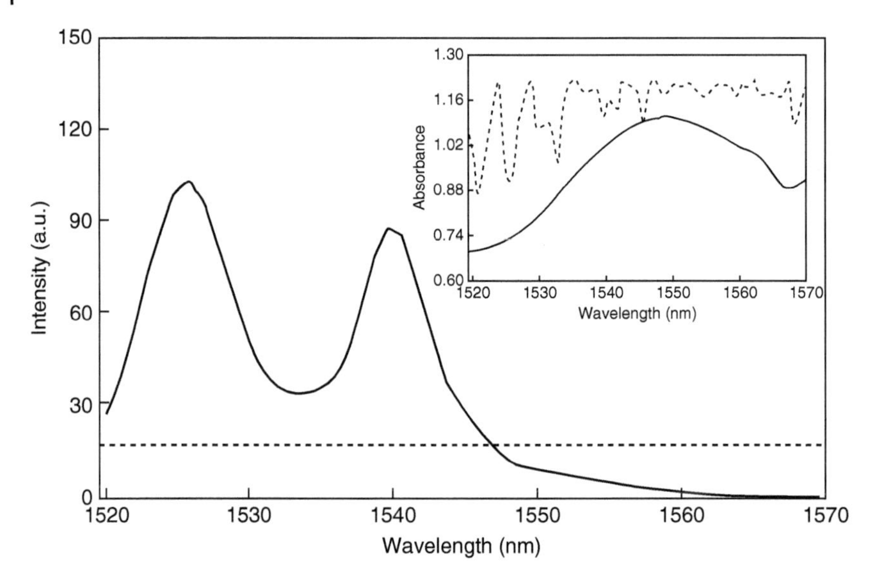

Figure 3.32 Relative intensity of the light transmitted through four sheets of paper and 1.0 M solution of Pr^{3+} in D_20 (in a 2-mm path length cell), measured with a 250 W halogen tungsten lamp–based spectrophotometer (---), and with the EDFA-based spectrophotometer (__). Inset: Spectra of the same sample plotted as absorption spectra. Source: Tran 1997 [66]. Reproduced with permission of Elsevier.

absorption (of the papers) and low intensity (of the halogen tungsten lamp), no light was transmitted when the halogen tungsten lamp–based spectrophotometer was used. Because of its high intensity, a substantial amount of light was transmitted through the papers. In the second measurement, which is shown in Figure 3.32, it was not possible to use a lamp-based spectrometer to measure absorption spectrum of 1.0 M solution of Pr^{3+} in D_20 (in a 2-mm cell) placed after four sheets of photocopy paper (dashed line). Because of its high intensity, a substantial amount of EDFA light was transmitted through the papers and Pr^{3+} solution, and, as a consequence, absorption spectrum of the latter can be recorded (Figure 3.32, inset). This high-throughput advantage is of particular importance in NIR measurements because NIR techniques are often used for measurements in which the signal of interest is very small and riding on top of a very large background signal. As such, it is very difficult, inaccurate, and sometimes impossible to perform such measurements with low light-throughput spectrophotometers.

NIR Spectropolarimeter Based on AOTF In 2008, C. F. Pereira et al. described the construction of a new NIR spectropolarimeter based on an AOTF [88]. An AOTF is an electro-optical device that functions as an electronically tunable excitation filter to simultaneously modulate the intensity and wavelength from a radiation source. These devices are manufactured of a specialized birefringent crystal whose optical properties vary upon interaction with an acoustic wave. Changes in the acoustic frequency alter the diffraction properties of the crystal, enabling very rapid wavelength tuning, limited only by the acoustic transit time across the crystal [87, 89].

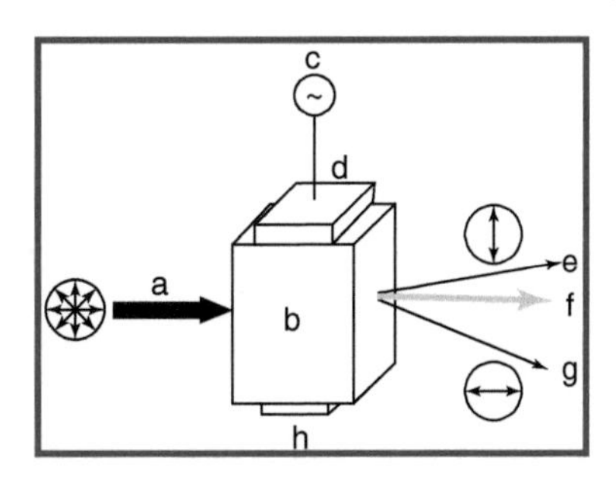

Figure 3.33 Acousto-optical tunable filter based on an anisotropic crystal. (a) Nonpolarized polychromatic input beam; (b) TeO_2 crystal; (c) radio frequency signal; (d) piezoelectric transducer; (e) ordinary polarized diffracted narrow wavelength band beam; (f) nondiffracted, zero-order beam; (g) extraordinary polarized diffracted narrow wavelength band beam; (h) acoustic absorber. Source: Pereira et al. 2008 [88]. Reproduced with permission of ACS.

An AOTF used for the NIR region, such as shown in Figure 3.33, frequently consists of a tellurium dioxide (TeO_2) crystal to which a piezoelectric transducer is bonded. In response to the application of an oscillating RF electrical signal, the transducer generates a high-frequency mechanical (acoustic) wave that propagates through the crystal. The alternating ultrasonic acoustic wave induces a periodic redistribution of the refractive index through the crystal that acts as a transmission diffraction prism to deviate a portion of incident light into a first-order beam. Changing the frequency of the transducer signal applied to the crystal alters the period of the refractive index variation and, therefore, the wavelength of light that is diffracted. The relative intensity of the diffracted beam is determined by the amplitude (power) of the RF signal applied to the crystal [90].

When a nonpolarized incident light beam is employed in the noncollinear configuration (Figure 3.33) with an anisotropic crystal, the diffracted portion of the beam comprises two spatially separated first-order beams, which are orthogonally polarized. If the input beam to the AOTF is linearly polarized at 90°, only one diffracted beam exits the device, with its polarization state rotated 90° relative to the input polarization axis. However, if the input beam to the AOTF is linearly polarized at 45° in relation to the AOTF crystal, as employed in this work, two diffracted beams (with ideally equal intensities), still orthogonally polarized, exit the device. The two beams are typically spatially separated by a few degrees (5–12°), which is a function of the device design.

The property of the anisotropic AOTF in producing two monochromatic and orthogonally polarized beams forms the basic principle utilized to develop a new approach to spectropolarimetry, based on a new use of the acousto-optic effect from birefringent crystals. The prototype of an instrument that operates under this new approach has been assembled to demonstrate its feasibility, advantages, and limitations.

In order to demonstrate the feasibility of the proposed approach to spectropolarimetry and optical rotation measurements, the instrument prototype depicted in Figure 3.34 has been constructed and evaluated. The instrument employs a tungsten/halogen filament source (30 W, 12 V) whose radiation is collimated by a quartz lens and passed through a mechanical chopper set at 1 kHz. The modulated beam passes through a polarizer sheet (ColorPol IR 1300 BC5, Codixx AG) and is directed through a cylindrical sample cell with a nominal 10-cm optical path and 6 mm of inner diameter. This polarizer has recently become available and ensures a high transmittance (~85%) and high contrast ($> 5 \times 10^5$ up to 1×10^7) in the spectral range from 800 to 1600 nm. Although the AOTF could

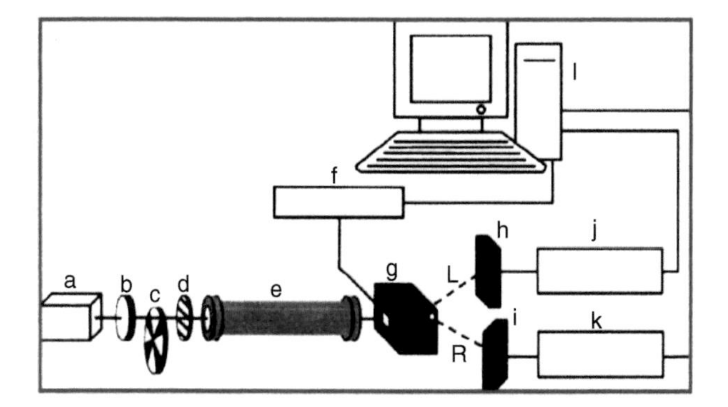

Figure 3.34 Schematic diagram of the spectropolarimeter based on AOTF. (a) Tungsten/halogen radiation source; (b) collimating lens; (c) mechanical chopper; (d) polarizer sheet; (e) sample cell; (f) RF generator; (g) acousto-optical tunable filter; (h, i) InGaAs detectors; (j, k) miniature lock-in amplifiers, and (l) controlling microcomputer. R and L, right and left beams, considering the intensity increase caused by a dextro- and levorotatory optically active substance, respectively. Source: Pereira et al. 2008 [88]. Reproduced with permission of ACS.

also be used to modulate the radiation (by on–off switching the RF signal), the prototype employs a mechanical chopper in order to certify that only the intensity of the polarized radiation passing the sampling cell is monitored by the detectors connected to the lock-in amplifiers. This allows the instrument to operate without a cover, under minimal influence of ambient radiation. If modulation was provided by the AOTF, spurious nonpolarized, although modulated radiation, might be detected.

The polarized beam impinges onto an AOTF (Brimrose TEAF 0.8–1.6, spectral range 800–1600 nm, nominal resolution between 2 and 6 nm) made of TeO_2. The RF signal necessary to drive the AOTF was produced by an RF signal generator (TGR 1040, Turlby Thandar Instruments) whose output signal was amplified by an RF amplifier (RFGA 0101-05, RF Gain). The RF signal frequency and amplitude are computer controlled through an RS-232 serial interface. The intensities of the two diffracted beams produced by the AOTF were monitored by twin InGaAs detectors (Hamamatsu, G8605-22), and the preamplified signals were sent to two miniature lock-in amplifiers (LIA-MV- 150, Femto). From the lock-in amplifiers, the signals are directed to two of the analog inputs of a parallel interface (PCL-711S) to be digitalized with 12-bit resolution. A program written in Visual Basic 5.0 was employed to control the RF generator, for signal acquisition and storage and for data treatment.

The proposed scheme for optical rotation measurement of optically active substances made by employing two diffracted beams of an anisotropic AOTF has been demonstrated. The proposed system is simple and shows the feasibility of optical rotatory dispersion (ORD) measurement in the NIR spectral region, which can contribute to studies of many interesting compounds. The instrument allows for simultaneous acquisition of the absorption spectra of optically active and inactive species. For optically active species, the intensities of the R and L beams must be added for the reference (background) and for the sample scans. Thus, the usual calculations of transmittance can be employed. This facility of attaining complementary information for optically active substances (absorbance and ORD spectra) can be useful to increase the analytical information for these types of substances and can provide, in the near future, a way to improve the selectivity and sensitivity of analytical methods based on the NIR spectral region using chemometric multivariate data treatments.

The proposed instrument is simple, robust, and versatile and makes use of the high-energy throughput of the AOTF in order to improve the S/N ratio of spectropolarimetric measurements. Because the prototype employs a double-beam system, the optical rotation measurement is less prone to fluctuations in the radiation source. For instance, it has been observed that a 15% change in source intensity (deliberately induced by lowering or increasing its supply voltage) results in alterations in the measured rotation angle that are within the precision of the present instrument (0.11°).

Autonomous Tunable Filter System – Based on AOTF In certain application fields where image processing is involved, the use of spectral information in the VIS and NIR range is critical. For example, vegetation characteristics such as fractional vegetation cover, wet biomass, leaf area index, or chlorophyll content can be easily estimated from a set of multispectral reflectance measures [91, 92]. In the fruit processing industry, the VIS range is necessary to detect defects on the skin. However, two key parameters for the ripeness of a fruit, the total soluble solids, and the skin chlorophyll are better measured in the NIR domain [93], since NIR energy is absorbed by certain chemical groups and not by others.

Systems for spectral measuring can be classified according to their spectral resolution or their spatial resolution. Narrowband hyperspectral systems are usually based on point spectrometers or linear spectrometers. These systems can measure many precise narrow spectral bands of a point light source or a line source in the VIS and NIR ranges and beyond. On the other hand, red, green, and blue (RGB) cameras provide a very high 2D spatial resolution, while the spectral information is badly separated into three overlapping bands in the VIS range. Some systems can acquire narrowband spectral information with high 2D spatial resolution but at the cost of a very long acquisition time [94].

Smart-Spectral is a smart multispectral system that fits somewhere between these two approaches. The Smart-Spectra camera provides six bands with fully configurable spectral shape from snapshot to snapshot, in the range 400–1000 nm. Each band can be as narrow as 5 nm or as wide as an RGB band. At the same time, the limited number of bands reduces the acquisition time of the system, yet maintaining the relevant spectral information. The small number of bands also reduces the subsequent processing time, encouraging the use of the system in real-time applications. As a variety of classification and quantification problems demonstrate, six bands is a good trade-off between resolving the power and response time of the system [95, 96].

For some applications, each band can be configured to acquire a linear combination of narrow bands. This functionality is useful in order to estimate convenient parameters such as the optimum multiple narrowband reflectance (OMNR) index [91] with a single Smart-Spectra band. The Smart-Spectra camera acts like an RGB camera in terms of spatial resolution and integration time. Furthermore,

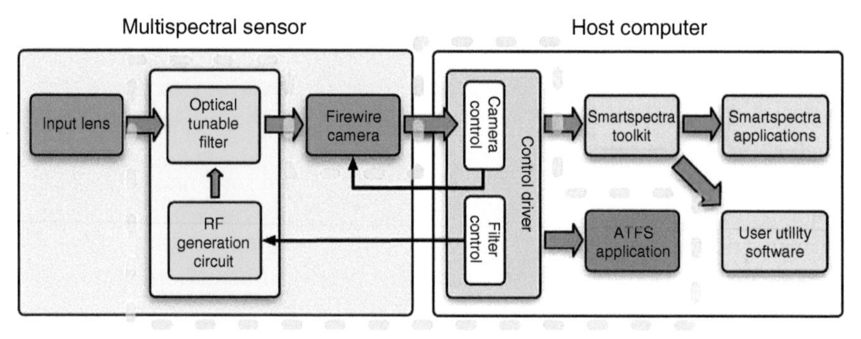

Figure 3.35 A block scheme diagram of the Smart-Spectra system and ATFS. Source: Vila et al. 2005 [48]. Reproduced with permission of Elsevier.

the flexibility of band configuration allows the system to adapt to a wide variety of problems and changing situations in real time. Smart-Spectra technology is affordable while assuring robustness. The system is intended to make multispectral techniques accessible to industrial, environmental, and commercial applications. Although the system is not fully operative yet, a first prototype called the autonomous tunable filter system (ATFS) has been completed and is functional.

The system can be divided into two main blocks, the sensor and the host computer, as depicted in Figure 3.35. The sensor part mainly involves optics and electronics, while the host computer comprises the driver for sensor configuration and image acquisition, image processing software, and the development of specific algorithms for the application fields of fruit quality assessment, agriculture, and environmental monitoring. The camera acquires six bands that may be located in the VIS and NIR spectra and simulates two common Firewire RGB cameras. The ATFS prototype is a reduced version of the Smart-Spectra system that uses an off-the-shelf Firewire camera and does not include the Toolkit software.

The Smart-Spectra system acquires multispectral images by means of an optical tunable filter in front of a monochrome camera. The tunable filter is configured to select which light wavelengths reach the sensor. The selected spectral region can be a single passband with a bandwidth between 5 and 100 nm or a linear combination of up to four passbands summing a total bandwidth of up to 100 nm.

The selected filter technology is the AOTF, which diffracts a wavelength of the incoming light by means of an acoustic wave in the RF range. The frequency, power, and spectral distribution of the RF signal define the intensity and bandwidth of the diffracted light [97]. When a pure sinusoidal RF signal is applied to the AOTF, the crystal diffracts a single wavelength with a very narrow bandwidth. The DE is directly proportional to the RF power. Moreover, if a linear combination of pure sinusoids is applied to the crystal, the corresponding combination of wavelengths is diffracted [98]. This characteristic can be used to increase the bandwidth of the filtered passband (using a set of very close sine signals) [99] or to generate a complex comb-shaped output (using a set of separate sinusoids) [100].

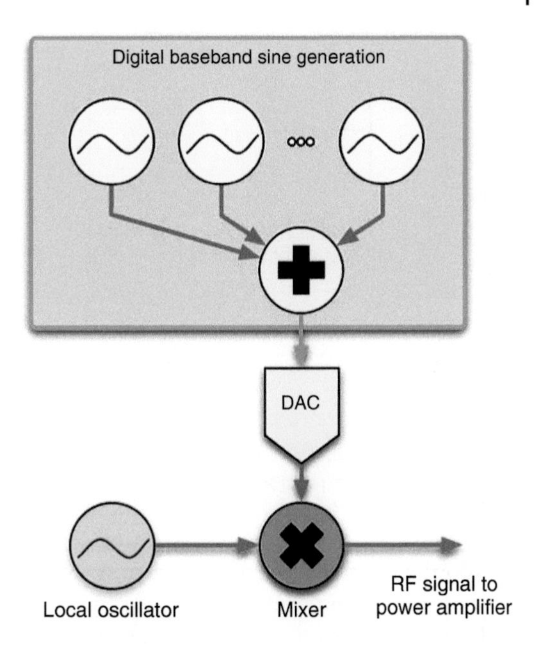

Figure 3.36 Multi-sine AOTF driver. Source: Vila et al. 2005 [48]. Reproduced with permission of Elsevier.

The first prototype of the Smart-Spectra optical tunable filter used this multi-sine approach to excite the AOTF, implemented as a custom RF driver design [101]. The concept of this RF driver is shown in Figure 3.36. A digital signal processing (DSP) is used to generate a multi-sine signal in baseband which is upconverted to the proper frequency range with an RF mixer circuit. This driver was tested on a Brimrose AOTF showing a moderate performance. The response was not very good because the mixer added some intermodulation products to the RF signal. These products spread the power density and therefore reduced the DE in the selected band and increased the out-of-band diffracted light.

Another approach takes advantage of the fast dynamic response of the acousto-optic devices and the fast switching capabilities of modern digital direct synthesizers (DDSs). This technique drives the AOTF with a sweeping acoustic frequency, i.e. the frequency applied to the crystal varies cyclically from an initial frequency to a final frequency in constant steps. By taking an integration time that is a multiple of the complete sweeping time, the acquired image is proportional to the intensity of the spectral range diffracted by the different frequencies.

A new RF driver was developed using this approach. Figure 3.37 shows the block diagram of the driver, which was implemented in the ATFS prototype. The core of the driver is the AD9858, a 1 Giga sample per second DDS, capable of generating a frequency-agile analog output sine wave at up to 400 MHz. The DDS is clocked by a 1 GHz signal generated by the AD4360-2 integrated synthesizer and VCO. Both integrated circuits are controlled by an ADuC832 MicroConverter. The RF signal amplitude is controlled by an AD8367 variable gain amplifier. All integrated circuits are from Analog Devices Inc. (ADI; www.analog.com). The low-level RF signal output is amplified by the AP5300-2 power amplifier from radio frequency power amplifier (RFPA) (www.rfpa.com).

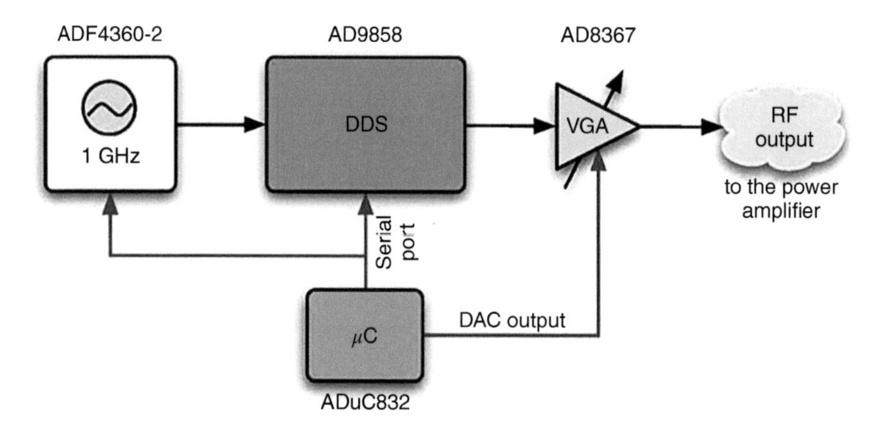

Figure 3.37 Block diagram of the sweeping frequency driver. Source: Vila et al. 2005 [48]. Reproduced with permission of Elsevier.

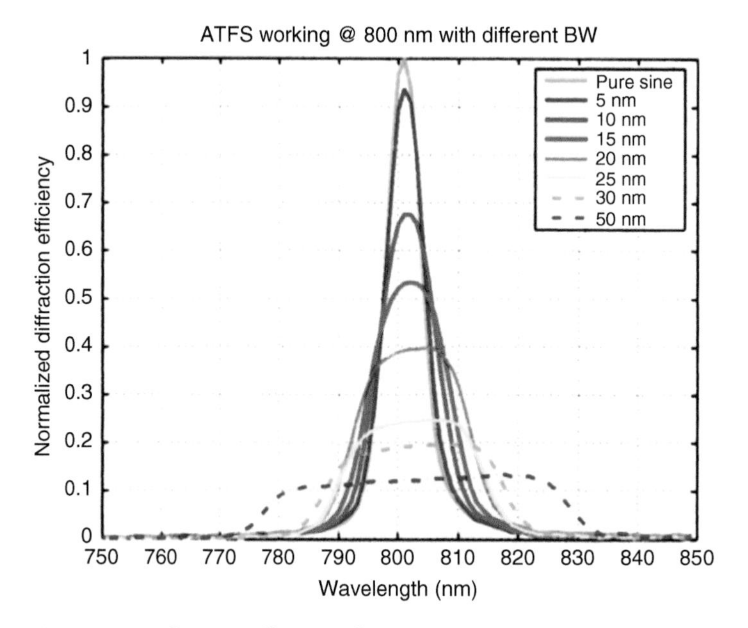

Figure 3.38 Diffraction efficiency of the sweeping frequency driver. Source: Vila et al. 2005 [48]. Reproduced with permission of Elsevier.

The DDS has been programmed to generate a complete sweep within a 5-ms period. The initial and final frequencies are calculated in real time from the bandwidth requirements, as well as from the frequency step. Figure 3.38 shows the performance of the system, measured over the Brimrose AOTF. The figure shows the DE of the system when the RF generator was configured to filter a passband centered at 800 nm with a variable bandwidth. Although the DE decreases as the bandwidth increases, the total energy contained in the passband is kept almost constant, i.e. the efficiency of the system is not altered. The DDS can be switched instantly among four frequency profiles. This feature allows the generation of an arbitrary four-sine signal in which each sine is used one-fourth of the time. For an integration time that is a multiple of the signal period, the result is equivalent to the multi-sine approach with four sines.

The performance of the ATFS prototype has been evaluated in the estimation of chlorophyll content in plant leaves. The Smart-Spectra system provides spectral and spatial information simultaneously and allows regular repetition of the measurements in order to detect variations in the plant status as a whole.

Figure 3.39 shows the application of this index to the multispectral image of a sunflower (*Helianthus annus*) leaf. The leaf was treated with herbicide and thus presented high variations in chlorophyll content. Figure 3.39a represents a color image of the leaf showing the damage caused by the herbicide in the areas

Figure 3.39 Study of the chlorophyll content of a sunflower leaf treated with herbicide. (a) RGB image of the leaf, where the gray rectangle represents the selected study area. (b) Reflectance of the leaf in the 680-nm band. (c) Reflectance of the leaf in the 800-nm band. (d) Image of the chlorophyll distribution calculated with the $NDVI_{800, 680}$ index. The figure is shown in false color (light gray) so that details can be clearly visualized. The dark gray color corresponds to the lowest concentration and gray to the highest. Source: Vila et al. 2005 [48]. Reproduced with permission of Elsevier.

surrounding the veins. Figure 3.39b shows the image of the leaf in the 680-nm band, while Figure 3.39c shows the 800-nm band. Finally, Figure 3.39d shows the chlorophyll distribution calculated with the index. A lower chlorophyll concentration was observed close to the central nerve, where the herbicide effect was clearly stronger.

The Smart-Spectra system proposes a new concept for multispectral imaging. It is a hybrid of common color cameras (three broad overlapping bands) and spectrometers (many very narrow bands). It uses six bands that are configurable in central wavelength and band width from snapshot to snapshot. Stress is placed on robustness, flexibility, and affordable cost in order to make it accessible to final users. A software platform is built to simplify the use of this camera. The use of a reduced number of quickly configurable bands makes the system especially suited for real-time applications.

AOTF-Based Detectors for HPLC HPLC has increasingly become the technique of choice for chemical separation. As the technique becomes more prevalent, the demand for detectors that can provide quantitative as well as qualitative information on the analyte increases. Variable wavelength absorption detectors are the most widely used detectors for HPLC. However, this type of detector can only be used as a quantitative technique because the qualitative information obtained from this detector is rather limited, namely, it relies on the use of the retention time as the only tool for identification. The development of diode array detectors (DADs) in the early 1980s made it possible to obtain information on peak purity and identity [102]. Specifically, with a DAD-based detector, the spectrum obtained for each peak in the chromatogram can be stored and the subsequent comparison with standard spectra will facilitate the identification of peaks [102]. The optimum wavelength for single-wavelength detection can easily be found [102]. Wavelength changes can be programmed to occur at different points in the chromatogram, either to provide maximum sensitivity for peaks, or to edit out unwanted peaks, or both [102]. Unfortunately, in spite of their advantages, DADs still suffer from limitations including their relatively high cost and their low sensitivity (compared to variable and fixed wavelength detectors). It is, therefore, of particular importance that a novel detector which has higher sensitivity, low cost, and possesses all of the DADs' advantages be developed.

The AOTF with its unique features is particularly suited for the development of such a detector. Specifically, with its microsecond scanning speed, the AOTF-based detector can rapidly record absorption spectrum of a compound as it elutes from the column. The random access to wavelength(s) makes it possible to change and/or to program the detector to any wavelength(s) to obtain optimal detection. However, different from the DADs, the AOTF-based detector is a single-channel detection technique, i.e. it is based on a photomultiplier tube. Its sensitivity is therefore higher and its cost is lower than the multichannel detectors (i.e. DADs).

The schematic diagram of the AOTF-based detector is shown in Figure 3.40. A 150 W xenon arc lamp was used as the light source. Its output radiation which contains UV and VIS light was focused onto the AOTF by a combination of a reflector, collimator, and lens. Acoustic waves will be generated in the AOTF

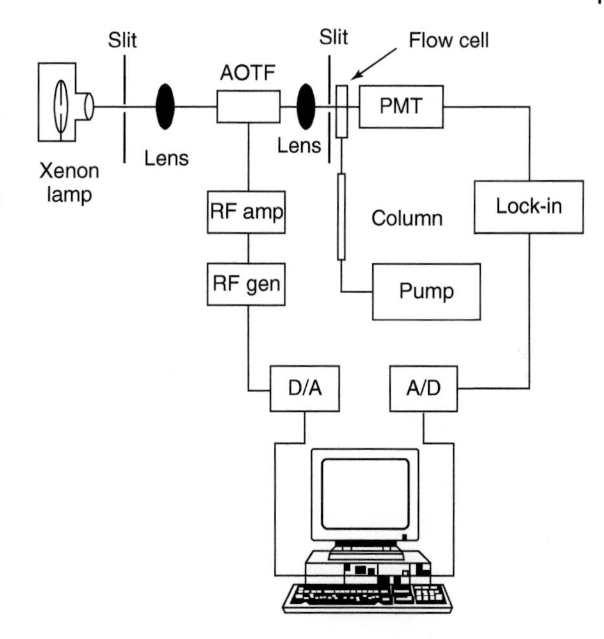

Figure 3.40 Schematic diagram of the AOTF-based detector for HPLC: PMT (photomultiplier tube); lock-in (lock-in amplifier); RF (RF power amplifier); RF gen (RF signal generator). Source: Tran 1997 [75]. Reproduced with permission of Elsevier.

when the RF signal is applied into the filter by a signal generator. To reduce noise and to facilitate the phase-sensitive detection, the RF signal was sinusoidally modulated at 50 kHz by the microcomputer through the D/A. Prior to being connected to the AOTF, the AM-modulated RF signal was amplified to 5 W power by an RF power amplifier. The intensity of the light diffracted from the AOTF was detected by a photomultiplier tube and demodulated (at 50 kHz) by a lock-in amplifier prior to being recorded by a microcomputer.

The chromatographic system consists of an isocratic pump and a sample injector valve equipped with a 40-μl loop. A 250 mm × 4.6 mm I.D. stainless-steel column packed with Nucleosil 5 silica was used. The chromatographic microflow cell used in this study, which has a path length of 8 mm and volume of 8 μL, is similar to that used previously [73]. Shown in Figure 3.41 is the three-dimensional graph plotting the chromatogram as a function of time and wavelength of a sample which was a mixture of three phenol derivatives, i.e. 4-chloro-, trichloro-, and pentachlorophenol. The chromatogram as a function of time was obtained by setting the AOTF at a single wavelength of 292 nm. Absorption spectrum of each compound was measured as it eluted out of the column by rapidly scanning the AOTF. Each single spectrum was obtained by scanning the AOTF for 100 nm (from 250 to 350 nm) and recording 100 points (i.e. 1 point for each nanometer). The setting was selected so that it required 2 ms to record each point. Therefore, the time required to record a single spectrum is 200 ms, and it took four seconds to obtain the spectrum, which is the average of 20 spectra for each compound. However, as evident from the figures, only 60 nm (i.e. from 260 to 320 nm) is required to record the whole spectrum for all three compounds. Therefore, with the optimal setting of 300 μs time constant, 12 dB roll-off, and 2 ms pt^{-1} (on the lock-in amplifier), it requires only 900 ms to obtain an average of five spectra which has relatively good S/N.

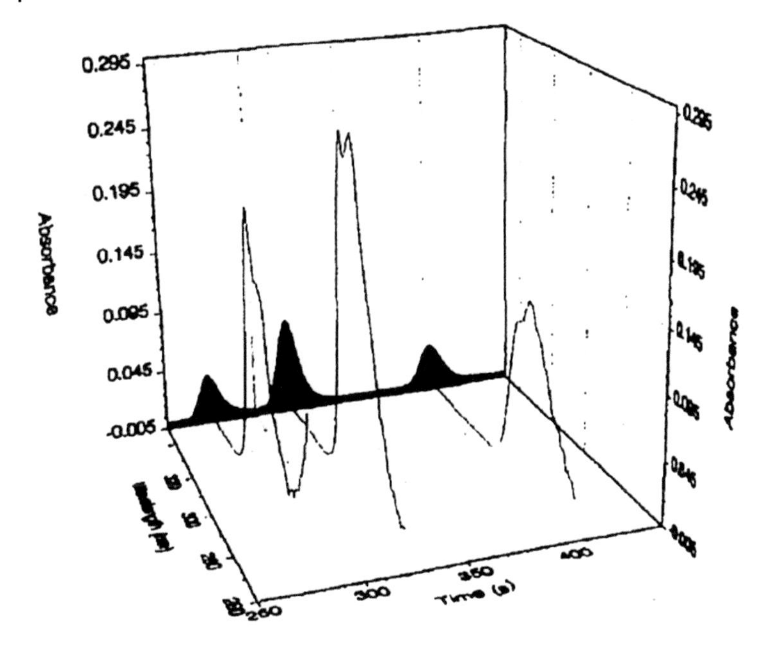

Figure 3.41 Three-dimensional graph plotting the chromatogram of mixture of pentachloro-, trichloro-, and 4-chlorophenol as a function of time and wavelength. Source: Tran 1997 [75]. Reproduced with permission of Elsevier.

The calibration curve was constructed for each compound over a concentration range of 1.0×10^{-4}–2.0×10^{-3} M using the data obtained when the AOTF was fixed at a single wavelength corresponding to the peak of the absorption spectrum of each compound, i.e. 285, 300, and 306 nm for 4-chloro-, trichloro-, and pentachlorophenol, respectively. As expected, good linear relationship was obtained for all three compounds (the correlation coefficients for all three compounds were larger than 0.999). The LODs defined as twice the peak-to-peak noise of the baseline divided by the slope of the calibration graph, are estimated to be 1.0×10^{-5}, 1.1×10^{-5}, and 1.7×10^{-5} M for trichloro-, pentachloro-, and 4-chlorophenol, respectively. These LOD values correspond to the mass defectivity of 59, 88, and 65 ng, respectively, and to the absorbance unit of 4.0×10^{-4}. These detection limits are comparable with those found on commercially available (grating or filter based) single-wavelength absorption detectors. Particularly, its detectability of 4.0×10^{-4} absorption unit is similar to the value of 3.9×10^{-4} absorbance unit, which was previously determined for 4-chlorphenol using the Shimadzu model SPD-6AV VU-VIS absorption detector [73]. The LOD value of 4.0×10^{-4} AU is much smaller than those obtained using commercially available DADs [102]. This is as expected because the present AOTF-based detector is a single-channel detection technique, which is more sensitive than the multichannel detection employed in the DADs. Furthermore, the light source used in this AOTF-based detector is modulated at 50 kHz (through modulating the applied RF signal) which facilitates the phase lock detection. The S/N is further enhanced by this phase-sensitive detection. The other feature which makes this AOTF-based detector more desirable than DADs is its high spectral resolution. Specifically, the resolution of this AOTF-based detector is 0.82 Å at 253 nm [103]. This spectral resolution is much smaller than those of the DAD, which are generally on the order of several nanometers [69].

AOTF-Based NIR Detectors for Flow Injection Analysis FIA is among the most widely used methods for automated analysis. Its applications to several fields of chemistry have been demonstrated [104, 105]. Several operational modes of FIA have been realized by appropriately modifying traditional wet chemical methods (dilution, extraction, titration, fast kinetic reactions) into automated flow devices [70, 71]. Different types of detectors, including electrochemical (UV and VIS), spectrophotometric, and luminescent, have been applied to the FIA [104, 105]. However, there has not been an FIA detector which is truly universal.

Spectrochemical applications of the NIR absorption technique have increased significantly in recent years [72]. The popularity stems from advantages of the technique including the wide applicability (all compounds that have C—H, O—H, and/or N—H groups have absorption in this spectral region), the possibility of in situ applications (no need for sample pretreatment), and the availability of powerful and effective multivariate statistical methods for data analysis. These features enable the NIR to serve as a universal detector for FIA. However, the detection of FIA by NIR has not been fully exploited. In fact, there are only two reports describing the utilization of NIR for continuous-flow FIA detection [106, 107]. Unfortunately, in these studies, the potentials of the NIR technique have not been fully exploited because they were based on the use of only a single wavelength [106, 107]. As a consequence, the multivariate calibration methods cannot be used to analyze data in these studies. This drawback can be overcome using an AOTF-based NIR detector.

The construction of the AOTF-based NIR detector for FIA, shown in Figure 3.42, is essentially the same as that of the AOTF-based UV–VIS detector for HPLC. The only differences were those of the light source (a 100 W, 12 V halogen tungsten lamp), the AOTF for the NIR region, and the detector (InGaAs photodiodes). This AOTF-based detector covers an NIR region from 1000 to 1600 nm. Because the combination and overtone absorption bands of O—H and C—H groups are in this region, this FIA-AOTF detector can be used for such determinations as water in chloroform, and water and benzene in ethanol.

Figure 3.43 shows the FIA absorption peak profile (i.e. absorption spectrum as a function of time and wavelength), obtained when a solution of 0.10% (v/v) water was injected into chloroform. It is evident that as the concentration of water in the flow cell increases, there is an increase in the absorption in the 1300- to 1500-nm region. This can be attributed to the first overtone transition and the combination of stretching and bending of the O—H group at 1450 and 1180 nm, respectively [74]. The absorption reaches its maximum 12 seconds after the injection and then starts to decrease. Using the spectra measured 12 seconds after the injection for different concentrations of water, a calibration model based on the partial least square (PLS) method was developed for the determination of water in chloroform. Good correlation was obtained between the concentration of water injected and the concentration of water calculated by the model ($r = 0.99$).

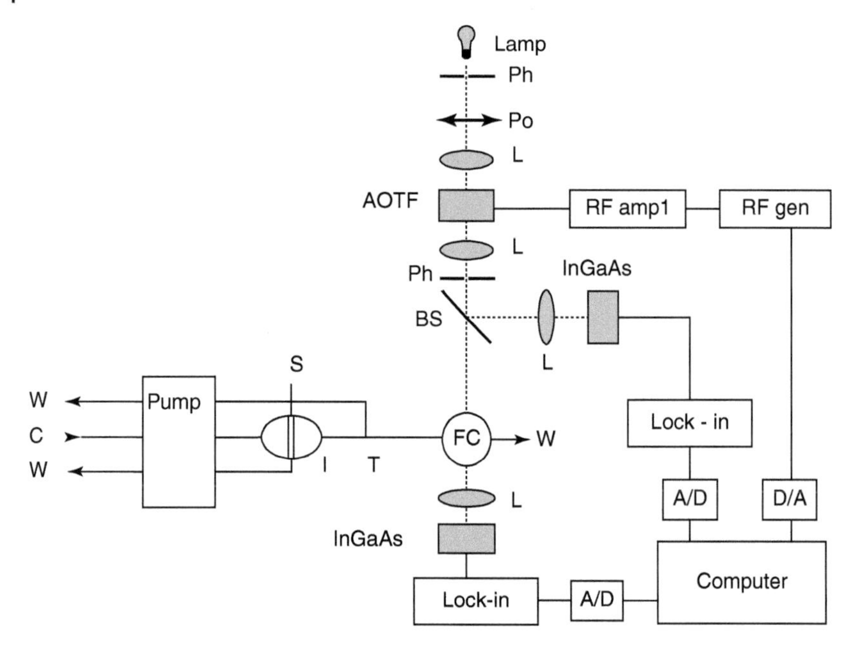

Figure 3.42 Near-IR-FIA instrument: W, waste; C, carrier; Pump, peristaltic pump; S, sample; I, injector; T, T connector; FC, flow cell; Ph, pinhole; Po, polarizer; L, lens; BS, beam splitter. Source: Tran 1997 [75]. Reproduced with permission of Elsevier.

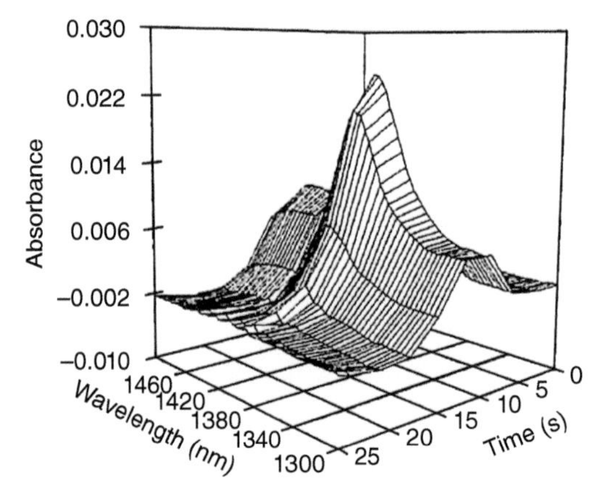

Figure 3.43 Absorption as a function of time and wavelength after the injection of 0.10% (v/v) water solution in chloroform (absorption spectra measured with a 1.7-cm flow cell). Source: Tran 1997 [75]. Reproduced with permission of Elsevier.

The root-mean-square deviation (RMSD) for this determination was calculated to be 0.002%. The LOD at 1400 nm was found to be 15 ppm of water.

It is possible to sensitively and simultaneously determine the concentrations of water and benzene in ethanol by NIR spectrometry. Figure 3.44 shows the absorption measured as a function of time and wavelength after the injection of the sample containing 0.6% and 1.5% of water and benzene, respectively. As can be seen from the figure, when the sample is passing through the detector between 8 and 26 seconds, there is an increase in the absorption in the 1650- to 1700-nm and 1390- to 1500-nm regions (due to absorption of benzene and

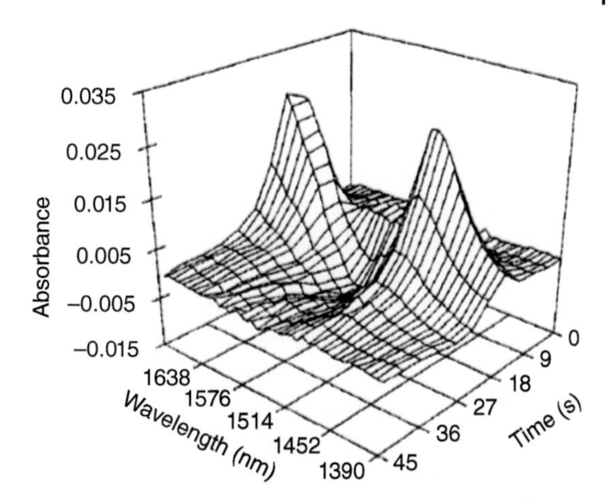

Figure 3.44 Absorption (0.75-cm flow cell using ethanol as blank) as a function of time and wavelength measured after the injection of ethanol solution containing 0.6% and 1.5% (v/v) of water and benzene, respectively. Source: Tran 1997 [75]. Reproduced with permission of Elsevier.

water, respectively) and a decrease in the 1500- to 1650-nm region (due to benzene). These spectral profiles are the same as those observed in the previous measurements using the non-flowing 1-cm cuvette (figure not shown). In order to produce a calibration model for the simultaneous determination of benzene and water in ethanol using the NIR-FIA technique, 21 samples containing different concentrations of water and benzene were prepared. Each sample was injected four times. Good correlation was obtained between the concentration of water and benzene injected and the concentration of both components calculated by the PLS model (SR = 1450–1496 nm, 1672–1700 nm, 1583–1602 nm; NF = 4 and 3 for water and benzene, respectively). The statistical parameters obtained were $r = 0.997$ and RMSD = 0.015% and $r = 0.997$ and RMSD = 0.033% for water and benzene, respectively.

In summary, it has been demonstrated that due to its high scanning velocity and wavelength accuracy, the AOTF-based detector for FIA can measure whole NIR spectra of flowing samples within the time frame required for flow analysis. This allowed the utilization of multivariate statistical methods of analysis, which, in turn, increase the sensitivity, accuracy, and applicability of the technique. In fact, it was possible to perform not only a simple analysis, such as the determination of dryness of organic solvent (i.e. the concentration of water in chloroform) but also a more complex analysis including the simultaneous determination of two component systems (i.e. the concentration of water and benzene in ethanol). For simple one-component systems, the LOD of this AOTF-NIR-FIA technique is comparable to that of the FIA spectrophotometry as well as to that of the single-wavelength NIR-FIA technique. In this case, the advantages of the AOTF-NIR-FIA instrumentation are its sensitivity, automation, and wide applicability.

For relatively more complex systems (e.g. two-component systems) the advantages become more prevalent. In fact, there is not a method currently available that can be easily coupled with the FIA instrument for the sensitive and simultaneous determination of the concentrations of two or more components without any sample pretreatment. The present automated and real-time determination of water and benzene in ethanol is important because ethanol is increasingly being used as a substitute and/or additive to gasoline (i.e. it is important to know the concentrations of water and benzene impurities in such systems).

3.4.5.4 Satellite- and Space-Based Applications of AOTF

Many AO-based devices, such as the AOTF spectrometer and the AOM are well suited for space environments. AOTF spectrometers can be powerful tools for the in situ surface and subsurface chemical analysis of soils, rocks, and ices [108], as well as for astrobiology-related experiments. AOTF is an electronically tunable optical filter in which optical beams passing through the AO crystal can be manipulated by frequency-generated acoustic waves in an anisotropic crystal. Devices based on AOTF technology have been successfully used in space-based instruments where compact AOTF-based spectrometers and cameras are used for research and process control. JPL has developed an AOTF-based spectrometer for aerospace applications in which the AOTF device was reported to have undergone radiation hardness testing [109]. One of the first spectroscopic applications of an AOTF-based spectrometer used on a civilian spacecraft is in the SPICAM Light optical package [110, 111]. This system is presently acquiring data on the ESA Mars Express mission. A standoff, AOTF-based Raman imaging system that can be used for planetary measurements has also been reported [112].

AOTF-based IR – spectrometers Spectrometers employing AOTFs have rapidly gained popularity in space, and in particular on interplanetary missions. They allow for reducing volume, mass, and complexity of the instrumentation. To date, they are used for analyzing ocean color, greenhouse gases, atmospheres of Mars and Venus, and for lunar mineralogy. More instruments for the Moon, Mars, and asteroid mineralogy are in flight, awaiting launch, or in the state of advanced development. The AOTFs are used in point (pencil-beam) spectrometers for selecting echelle diffraction orders, or in hyperspectral imagers and microscopes.

The ISEM (Infrared Spectrometer for ExoMars) is a pencil-beam IR spectrometer that will measure reflected solar radiation in the NIR range for context assessment of the surface mineralogy in the vicinity of the ExoMars rover. The instrument will be accommodated on the mast of the rover and will be operated together with the panoramic camera (PanCam), high-resolution camera (HRC). ISEM will study the mineralogical and petrographic composition of the Martian surface in the vicinity of the rover; and in combination with the other remote sensing instruments, it will aid in the selection of potential targets for close-up investigations and drilling sites. Of particular scientific interest are water-bearing minerals, such as phyllosilicates, sulfates, carbonates, and minerals indicative of astrobiological potential, such as borates, nitrates, and ammonium-bearing minerals. The instrument has an $\sim 1^0$ field of view (FOV) and covers the spectral range between 1.15 and 3.30 μm with a spectral resolution varying from 3.3 nm at 1.15 μm to 28 nm at 3.30 μm. The ISEM optical head is mounted on the mast, and its electronics box is located inside the rover's body. The spectrometer uses an AOTF and a Peltier-cooled InAs detector. The mass of ISEM is 1.74 kg, including the electronics and harness.

The ExoMars rover is a mobile laboratory equipped with a drill designed to sample the surface of Mars to a maximum depth of 2 m, and a suite of instruments to analyze the samples. The drilling device is the only means to access near-subsurface materials and introduce them to the internal analytical laboratory. As the number of samples obtained with the drill will be limited, the selection of high-value sites for drilling will be crucial. The rover's mast is therefore equipped with a set of remote sensing instruments to assist the selection process by characterizing the geological and compositional properties of the surrounding terrains. It includes several cameras – a pair of navigation cameras (NavCam), and a PanCam. PanCam consists of a stereo multispectral wide-angle camera pair (the WACs) and an HRC [142], and will provide the context images used to plan traveling and sampling. To complement and enhance the capabilities of the remote sensing suite, an IR spectrometer with the capability to unambiguously distinguish many rocks and minerals from their spectral reflectance will allow remote characterization and selection of potential astrobiological targets. This mast-mounted IR spectrometer was proposed during an early discussion of the new ESA-Roscosmos ExoMars configuration as a useful addition to the rover science and to help operations by characterizing from afar the mineralogical interest of targets that the rover could visit.

ISEM is a derivative of the lunar infrared spectrometer (LIS) [143] being developed at the Space Research Institute (IKI) in Moscow for the Luna-25 and Luna-27 Russian landers planned for flight in 2019 and 2021, respectively [144]. Both the ISEM and LIS instruments have been conceived with similar spectral capabilities. The ISEM design is improved with respect to that of LIS, and modifications were also necessary to comply with the more stringent environmental conditions on the ExoMars rover.

The main goal of ISEM is to establish the mineral composition of Mars' surface materials remotely. ISEM, together with PanCam (Figure 3.45), offers high

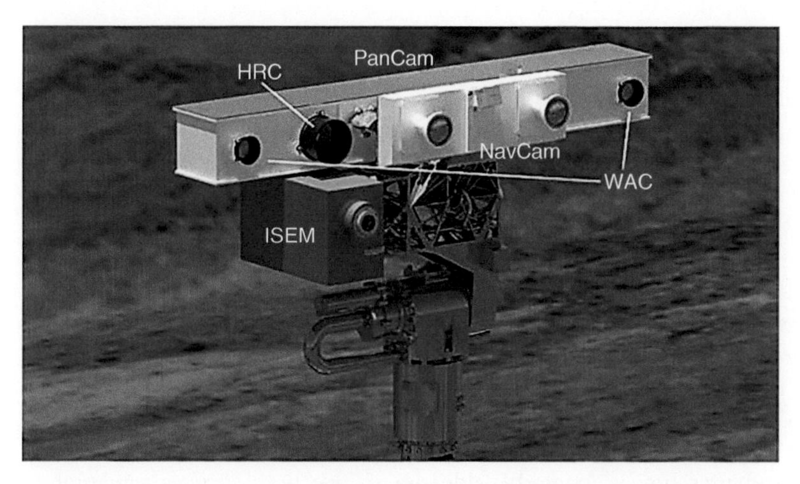

Figure 3.45 Schematic view of the ExoMars rover mast instruments: PanCam, navigation cameras, and ISEM. ISEM (infrared spectrometer for ExoMars); PanCam (panoramic camera). Source: Korablev et al. 2017 [145]. Reproduced with permission of Astrobiology.

potential for the remote identification and characterization of any scientifically high-value target in the vicinity of the rover, including proximal and distant rocks, outcrops, and other geological formations. ISEM will help establish the geological context of each site along the rover traverse, discriminating between various classes of minerals and rocks. ISEM will also be important in the selection of promising sites for subsurface sampling.

IR reflectance spectroscopy allows the study of the composition in the uppermost few millimeters of a rock's surface. It allows discriminating between various classes of silicate (hydr-)oxides, hydrated/hydroxylated salts, and carbonates. As shown in Figure 3.46, the 1.3 FOV of ISEM lies within the 5^0 FOV of the color PanCam HRC, and they are both within a much wider FOV (38.6^0) of the WACs with multispectral capabilities [142]. The multispectral data are produced using a filter wheel with 11 filter positions for each of the two WACs. Out of the 22 filters, 6 are devoted to red, green, and blue broadband color and are duplicated in both cameras. Twelve are optimized for mineralogy in the 400- to 1000-nm range, and four "solar" filters are dedicated to atmospheric studies. By extending the wavelength range beyond PanCam, ISEM will enable many more spectral features diagnostic of specific mineralogy to be detected. Together, PanCam and ISEM provide spectrally resolved information from 0.4 to 3.3 μm. The identification and mapping of the distribution of aqueous alteration products in the upper surface layer, combined with subsurface data from the neutron detector ADRON and the ground-penetrating radar WISDOM, will help understand the subsurface structure and the exobiology potential at each prospective drilling site.

The collected drill samples will be analyzed in the rover's Analytical Laboratory Drawer by several instruments. The first is an IR hyperspectral microscope MicrOmega [146]. The principle of MicrOmega is very similar to that of ISEM. The reflectance spectroscopy is performed in the NIR, but the analysis is done at the microscopic scale. The sample is illuminated by a monochromatic light

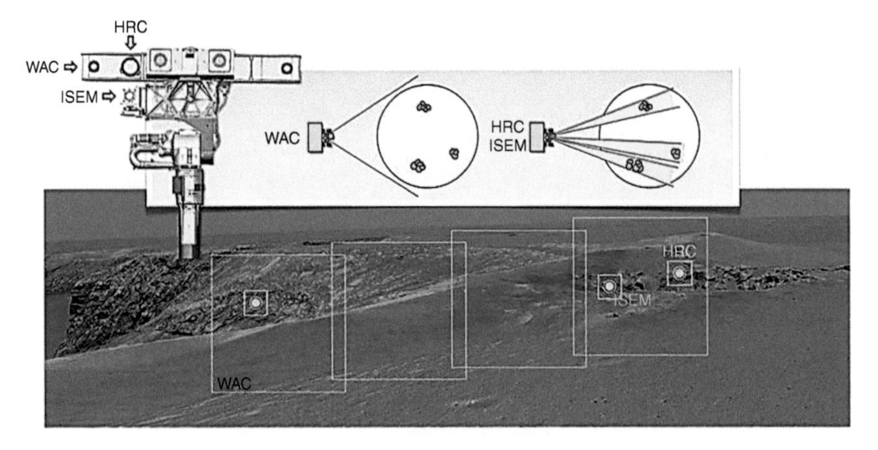

Figure 3.46 Schematic representation of possible ISEM and PanCam joint observation scenario showing a sequence of ISEM measurements acquired together with WAC and HRC PanCam frames. HRC, high-resolution camera; WAC, wide-angle camera. Source: Korablev et al. 2017 [145]. Reproduced with permission of Astrobiology.

source. In contrast to MicrOmega, ISEM has a much wider FOV and range of detection. The distance to a target is not significantly limited, and practically may reach hundreds of meters. Thus, ISEM is better suited for accommodation on the rover mast, where it can be used for target identification of faraway objects and also for investigating outcrop, rock, and soil mineralogy at close range.

There is no specific instrument dedicated to environmental characterization on the rover. Although hampered by the limited number of observation cycles, ISEM, jointly with PanCam, will deliver information regarding atmospheric aerosol opacity and the atmospheric gaseous composition. The data on water vapor content and aerosol will be retrieved as a by-product of reflectance spectra, from inflight calibration, from the direct Sun imaging by PanCam, and from sky observations by PanCam and ISEM.

A brief description of ISEM is given here for better understanding of the device that includes (a) instrumentation concept, (b) optical box (OB), (c) electronics box (EB), (d) calibration target, and (e) environmental requirements and characterization.

(a) *Instrument concept:* The measurement principle of ISEM is based on the use of an AOTF. The core element of an AOTF is a birefringent crystal (typically of paratellurite, TeO_2, due to the combination of acoustic and optical properties) with a welded piezoelectric transducer. The RF applied to the transducer generates an acoustic field in the crystal, implementing acousto-optic interactions in Bragg's regime. The spectral selectivity of the acousto-optic diffraction allows the filtering of light. The diffraction occurs for a single wavelength, and there are no diffraction orders. The applied RF controls the tuning of the AOTF. AOTFs are technologically mature and widely used for spectral analysis. The robust design, small dimensions, and mass, coupled with the absence of moving parts in an AOTF-based spectrometer, make them popular for space applications. So far, AOTF-based spectrometers have been used in space science in the following ways: (i) to study the atmospheric composition of Mars and Venus [113, 114]; (ii) on the Moon within the Chang'e-3 VNIS spectrometer (0.45–2.4 μm) mounted on the Yutu rover [147]; (iii) for isolation of echelle-spectrometer diffraction orders in high-resolution instruments [148–151]; and (iv) to illuminate the sample of an IR microscope with monochromatic light [152, 153]. Wider application of the AOTFs in remote sensing is hampered by the inherent requirement to sequentially scan the spectrum. On an orbital mission with a short dwell time, this scanning interferes with the spacecraft or line of sight motion, complicating the analysis. Even for a pencil-beam device, different parts of the acquired spectrum would correspond to different observed areas. Conversely, observations from a static point, such as a planetary lander, or a rover, which remains immobile during the measurement, are well suited for AOTF-based instruments.

To reduce the influence of the extreme temperature conditions at the rover's mast on the electronics, the instrument is implemented as two separate boxes, a mast-mounted OB and the EB (Figure 3.47). The thermally stabilized EB is

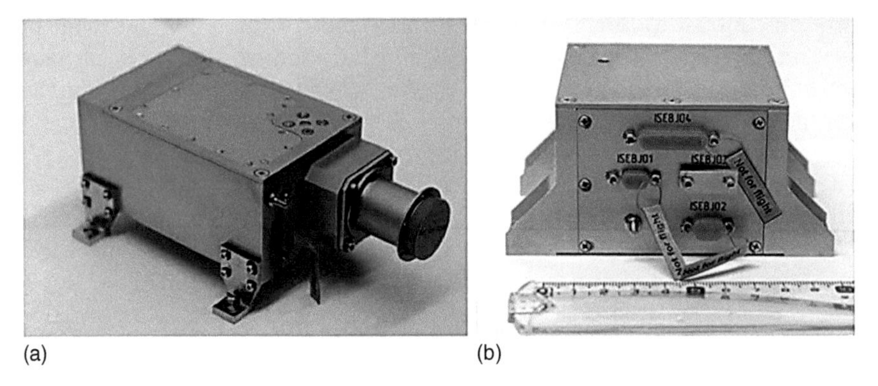

(a) (b)

Figure 3.47 The ISEM instrument, the photographs of the optical (a) and electronics box (b) structural and thermal models. Source: Korablev et al. 2017 [145]. Reproduced with permission of Astrobiology.

mounted within the rover's body. The parameters of ISEM are summarized in Table 3.2.

(b) *Optical Box:* Although most of the electronics are in the EB, some electronics such as the detector's preamplifier and the RF conditioning electronics remain in the OB. Neither the weak photodiode current nor the power RF can be transmitted via the 5-m harness. The OB contains all the optical elements, the AOTF with associated electronics (ultrasound frequency synthesizer and amplifier boards), the photodetector, and the photodetector board (Figure 3.48). The spectrometer is built following a standard layout for an AOTF spectrometer, with the AOTF in the path of a quasi-parallel beam. The optical scheme is based on a Galileo system with a remote pupil built using CaF2 and ZnSe lenses, and it is presented in Figure 3.49. The image is transferred through the optical system, and the FOV of 1.3° is formed on the detector's sensitive area. The achromatic lens entry telescope (1) has an aperture of 25 mm. The AOTF crystal (5) is placed in a quasi-parallel beam between collimating lenses (3, 7) and a pair of polarizers (4, 6); the output collimating lens (8) serves also as a focusing optic for the detector (8).

A wide-angle AOTF from NII Micropribor in Zelenograd, Russia, is manufactured on the base of a tellurium dioxide crystal. The ultrasound frequency range of 23–82 MHz provides a spectral range from 1.15 to 3.3 μm with two piezoelectric transducers. The transducers operate in the sub-bands of 23–42 and 42–82 MHz. The crystal cutoff angle is 12.5° in the (110) crystallographic plane. The anisotropic Bragg diffraction regime is used. The incident optical radiation has ordinary polarization, and the diffracted optical beam has the extraordinary polarization. The angle between the passed and diffracted optical beams is 6° at the output of the AO crystal. A pair of polarizers with crossed polarizing planes is used to filter out the non-desired zero diffraction order. The AOTF and its electronics are assembled in a single functional unit, which includes the acousto-optic cell, the polarizers, a proximity RF matching board, an RF synthesizer, the driver of the AO crystal (the RF power amplifier), and

Table 3.2 Infrared spectrometer for ExoMars main characteristics and resources.

Parameter	Value
Spectral range	1.15–3.3 μm
Spectral resolution	Better than 25 cm^{-1}, 3.3 nm at 1.15 μm, 16 nm at 2.5 μm, 28 nm at 3.3 μm
Field of view	1.3^0
Temperature range, operational	−45 °C, …, +30 °C (OB)
	−40 °C, …, +50 °C (EB)
	−130 °C, …,+60 °C for OB pending final confirmation
AOTF	
Material	TeO_2
Spectral range	1.15–3.39 μm
Effectiveness	>50% (in polarized light)
Aperture	Ø 5 mm, $5^0 \times 5^0$;
Mean RF power	5 W
RF range	23–82 MHz
Detector	InAs photodiode, Ø1 mm, 1–3.45 μm
	Teledyne Judson Technologies J12TE3-66D-R01M, 3-stage Peltier cooler
ADC	16-bit
Number of points per spectral range	Variable, by default 1024 for one observation
Data volume	Variable, by default 20 Kbit for one observation
Data/command interface	RS-422
Power supply voltage	28 V
Power consumption (W)	
Peak	14
Average	11.5
Standby	9
Dimensions	160 × 80 × 96 mm^3 OB
	116 × 84 × 55 mm^3 EB
Mass, overall	1.740 kg
OB	0.690 kg
EB	0.560 kg
Calibration target (ISEM part)	0.014 kg
Harness	0.476 kg including 20% margin (TAS-I data)

ADC, analog-to-digital converter; AOTF, acousto-optic tunable filter; EB, electronics box; OB, optical box; RF, radio frequency; TAS-I, Thales Alenia Space-Italy.
Source: Korablev et al. 2017 [145]. Reproduced with permission of Astrobiology.

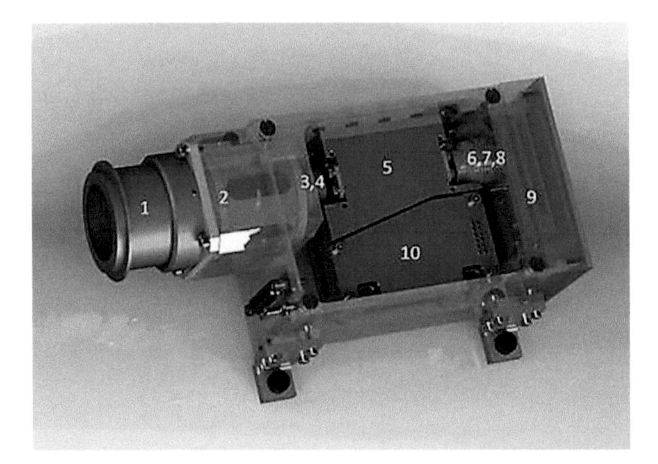

Figure 3.48 The three-dimensional model open view of the ISEM optical box. The main elements are visible: 1-baffle, 2-entry optics; 3, 7-AOTF collimating optics; 4, 6-polarizers; 5-AOTF crystal; 8-detector; 9-detector's preamplifier; 10-AOTF RF proximity electronics. AOTF, acousto-optic tunable filter; RF, radio frequency. Source: Korablev et al. 2017 [145]. Reproduced with permission of Astrobiology.

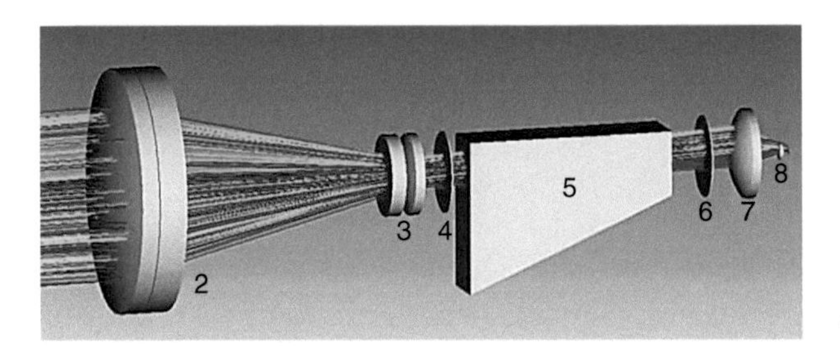

Figure 3.49 ISEM optical scheme. The numbering is the same as in Figure 3.48: 2-foreoptics; 3, 7-AOTF collimating optics; 4, 6-polarizers; 5-AOTF crystal; 8-sensitive area of the detector. Source: Korablev et al. 2017 [145]. Reproduced with permission of Astrobiology.

a dedicated internal microcontroller, which communicates with the main controller (MC) in the EB and controls all the functions of the acousto-optic module (AOM).

Commands define magnitude and frequency of the RF signal applied, as well as the RF driver "ON"/"OFF" states. Within the RF range of 23–82 MHz, the minimum step of frequency sweeping is 10 kHz (5900 frequency points). The RF amplitude may be set at 1 of 16 even levels. The AOM transmits back to the MC a few housekeeping parameters, such as measured RF voltage and AO crystal temperature. The MC and the AOM are connected via an RS-485 interface running at a speed of 115.2 Kbit/s.

During the measurement, the RF level is alternated between "ON" and "OFF" states, with the cadence being defined by integration time. The measured signal is then processed as the AC allows to remove offsets caused by the detector's dark current and stray light, and improving the dynamic range of the instrument. The detector is a single-pixel InAs thermoelectrically cooled photodiode.

A detector module J12TE3-66D-R01M from Teledyne Judson Technologies is used. The built-in three-stage thermoelectric Peltier cooler maintains a detector temperature of about 90 °C below that of the hot side. For the ISEM OB operating in the range from −10 to +30 °C, it results in the detector's temperature ranging between −100 and −60 °C; the corresponding detector's shunt resistance is therefore 60–400 kΩ. The temperature of the sensitive area is monitored by a built-in thermistor. The detector's photocurrent is amplified with a two-stage circuit. The first stage is a transimpedance amplifier, AC coupled (one-second time constant) to the second stage. The second stage has a gain of 83 and a time constant of 80 μs and is based on an ADA4610 (Analog Devices) operational amplifier characterized by low current and voltage noise (50 fA $Hz^{-1/2}$, 7.5 nV $Hz^{-1/2}$). Its output signal is transmitted via the harness to the EB.

(c) *Electronics box:* The EB is mounted inside the rover at the rear balcony, and it is thermally stabilized. It includes the ADC, the main and auxiliary controllers, power conditioning (power supply unit, PSU), and the interface and bridge boards, which support RS-422 communication between the ISEM and the rover data and command system. The two ISEM blocks are connected via a 5-m harness running from the inside of the rover to the mast. It is fabricated by Thales Alenia Space-Italy (TAS-I). The cable includes the analog signals from the detector and the thermal sensor, the power supply lines, and a digital RS-485 connection to control the AOFT RF synthesizer and the power amplifier.

The MC located in the EB commands the operation of all modules of the instrument. It uses an MSP430FR5739 Texas Instruments circuit running at 19.68 MHz. The controller is equipped with 1 kB of RAM and 16 kB of ferroelectric memory (FRAM) used for program and data storage. Compared to commonly used FLASH memory, the FRAM has better radiation immunity. A 10-bit internal ADC of the MC is used to digitize the signal from the detector's thermistor. The main ADC serves to digitize the signal from the detector, amplified in the OB. A 16-bit ADC (ADS8320; Texas Instruments) is used. The full-scale range of the ADC is 2.9 V and the peak-to-peak noise is 3 LSB (22 μV RMS). The MC receives the ADC output data via a serial interface operating at 2 Mbit/s. The preamplifier and the ADC contribute little to the total noise in the signal path, and thus the Johnson noise of the detector controls the limit of signal detection.

The auxiliary controller, based on C8051F121 Silicon Laboratories circuit operating at 12.25 MHz, incorporates a 12-bit ADC and serves to monitor power supply voltages and dispatch the operation of different memory devices. In all, ISEM uses two built-in FLASH memories within the microcontroller and two external chips of FRAM and magnetoresistive memory (MRAM). The two microcontrollers hold four identical copies of their firmware in program memories. At the start of operation, the controllers check the copies, repair damaged ones, and run a validation program. To increase its reliability, ISEM contains two redundant MC units. After turning on, first the MC1 is powered. Its sequence, if performed correctly, commands the PSU to keep power at MC1. If not commanded in five seconds, the PSU automatically powers on the MC2, and so on. The rover onboard computer (OBC) controls ISEM operations via the RS-422 interface. There are two (nominal and redundant) RS-422 links. Communication via only one of the links is available at a given time. The baud rate is 112.179% ± 1% Kbit/s.

(d) *Calibration target:* To determine the incident solar illumination spectrum for deriving I/F data and to verify the in-flight performance and stability of the instrument, ISEM will observe the radiometric calibration target before each measurement. The target is used for the in-flight radiometric calibration of both PanCam and ISEM and will be located on the front deck of the rover, as shown in Figure 3.50. At this location, it will be interrogated by ISEM from a distance of 1.1 m at an emittance angle of 24°. The calibration target occupies an area of $67 \times 76\,\text{mm}^2$ and has a mass of 40 g.

The target includes eight stained glass diffuse reflectance calibration patches with different spectral reflectance properties. Two of these calibration patches will be used by ISEM – the "white," which has a reflectance near 100% in the 0.4–3.0 μm spectral range, and the multiband patch, which has distinct spectral features. The white patch is manufactured from Pyroceram provided by the Vavilov State Optical Institute in St. Petersburg, and the multiband patch is manufactured from WCT-2065, a rare earth-doped glass manufactured by Schott and supplied by Avian Technologies in the United States. The calibration patches

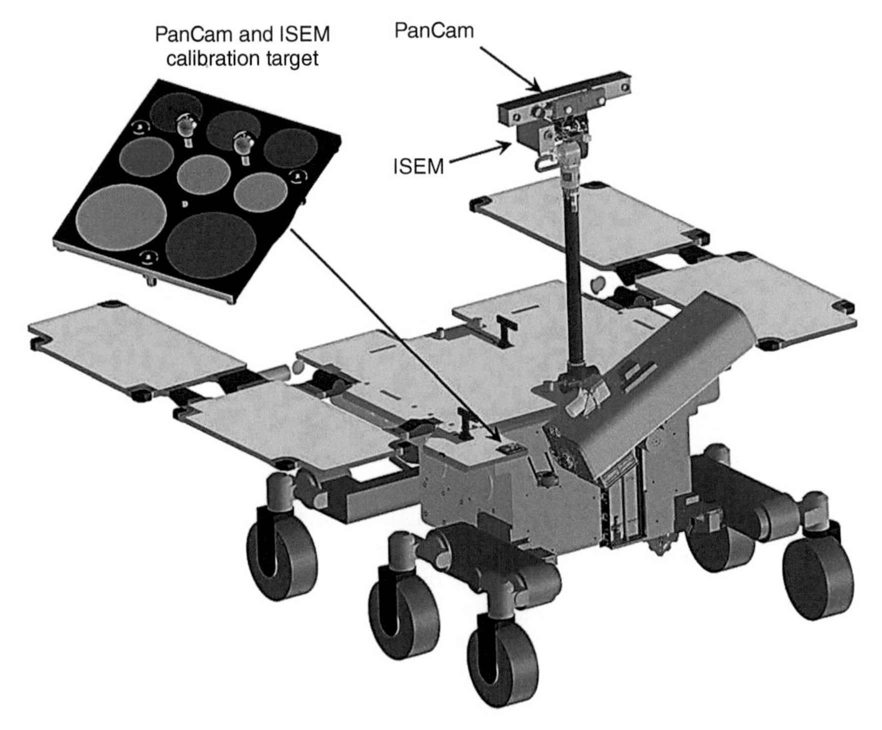

Figure 3.50 PanCam and ISEM calibration target and its location on the rover. Larger circles (Ø30 mm) are for both ISEM and PanCam. The smaller circles (Ø18 mm) and the shadow posts will be used by PanCam only. Source: Korablev et al. 2017 [145]. Reproduced with permission of Astrobiology.

will be calibrated for absolute total hemispherical reflectance and bidirectional reflectance distribution function, and all measurements will be traceable to photometric standards. Dust deposition on the radiometric calibration target during the ExoMars mission will be accounted for in the data processing by developing a model of the calibration target and dust system, which will build on the results of previous missions, measurements of settling rates on the rover panels, solar arrays, and so on [154, 155], and from PanCam calibration results.

(e) *Environmental requirements and characterization:* The placement of the ISEM optical head on the rover mast with no possibility of thermal control imposes stringent requirements on survival temperature. The AOTF is a critical device involving bonding of highly anisotropic materials, such as TeO_2 and $LiNbO_3$. Anisotropic thermal expansion makes the AOTF devices vulnerable to large temperature excursion, in particular at low temperatures. AOTF technology has demonstrated survival and even operation down to −130 °C and below [156, 157]. Several nonoperational thermal cycles attaining −130 °C lower bound were also performed on two dedicated ISEM acousto-optic components; more testing to fully validate the technology is to be completed. An important aspect of the AOTF spectrometer characterization is the thermal calibration, because the optical heads are intended for operations across a relatively wide temperature range. Some results of thermal characterization of the AOM were described by Mantsevich et al. [157]. The operation of the AO filters was tested in the range of −50 to +40 °C. The temperature affects the elastic properties of TeO_2, changing the ultrasound velocity and birefringence. The variation of the TeO_2 refraction coefficient as a function of temperature is negligible. In turn, the change of the slow acoustic wave velocity causes a noticeable shift of the dispersion curve, comparable with the AO filter pass band in the operating temperature range. The characterization of the flight instrument will therefore include calibration of the dispersion curve within the operational temperature range.

Sampling from beneath the Martian surface with the intent of reaching and analyzing material unaltered or minimally affected by cosmic radiation is a significant feature and perhaps the strongest advantage of the ExoMars rover. The selection of the sampling sites and, in more general terms, the remote reconnaissance and studies of the landing site area are a major part of the rover's mission. ISEM offers a method of mineralogical characterization using IR reflectance spectroscopy, a technique well proved in orbital studies and expected to be even more valuable and precise at the local scale.

AOTF-Based SPICAM-IR Spectrometer The SPICAM-IR spectrometer on Mars Express mission (1.0–1.7 μm, spectral resolution 0.5–1.2 nm) is dedicated primarily to nadir measurements of H_2O abundance. It is one of the two channels of SPICAM UV-IR instrument. In this spectrometer, O. Korablev et al. [113] applied for the first time in planetary research the technology of an AOTF that allowed unprecedented mass reduction for such an instrument: 0.75 kg. The electrically commanded acousto-optic filter scans sequentially at a desired sampling with random access over the entire spectral range.

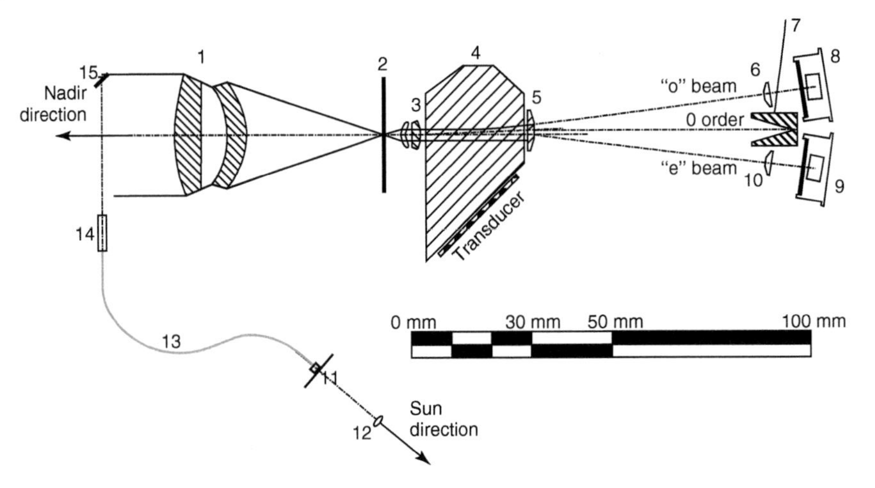

Figure 3.51 Optical scheme of the IR spectrometer. 1-front-end telescope objective; 2-FOV diaphragm; 3 and 5-collimator; 4-AOTF crystal with actuator; 6 and 10-detector proximity lenses; 7-zero-order light trap; 8-detector "1" (ordinary); 9-detector "2" (extraordinary); 11-Sun aperture entry lens; 12-Sun FOV diaphragm; 13-optical fiber; 14-collimating gradient lens; 15-Sun entry folding mirror. Source: Korablev et al. 2017 [113]. Reproduced with permission of Astrobiology.

The AOTF IR spectrometer is included in the UV package, but it is assembled on a dedicated base plate. The optical scheme of the AOTF IR spectrometer is shown in Figure 3.51. The upwelling radiation from Mars is collected by a telescope with its optical axis parallel to the optical axis of the UV spectrometer and other nadir-looking instruments on Mars Express. The three-lens telescope has a diameter of 30 mm and a focal ratio of 1 : 1.4. A circular diaphragm of 0.7 mm diameter placed in the focal plane of the telescope forms the FOV of 1°. This FOV corresponds to 4.5 km on the surface of Mars when observed from the Mars Express pericenter altitude of about 250 km. The AOTF cell and the associated electronics are placed in a shielded unit. The telescope, the FOV diaphragm, and the collimator are assembled in a tubelike structure attached to the AOTF unit. The dimensions of this block, including the AOTF with its electronics and the telescope are 104 × 60 × 40 mm and its mass is 330 g.

The AOTF is made of tellurium dioxide (TeO_2). In the case of the SPICAM spectrometer, the AOTF is employed in a specific noncollinear configuration [115], which provides high spectral resolution concomitantly a relatively large acceptance angle. Such a crystallographic configuration, in which both polarizations of the diffracted light are conserved, is unique for TeO_2. The acoustic wave in the crystal is excited by a piezoelectric transducer with a voltage varying at high frequency (85–140 MHz). The RF drive signal is applied to the transducer placed at the oblique side of the crystal. The length of the active zone (in which the light–sound interaction occurs) is 23 mm.

A collimator forms a quasi-parallel beam in the crystal; the divergence inside the crystal is limited at ±5.5°. The pupil is minimal in the center of the crystal and does not exceed 3.5 mm at the edges. The incoming light beam is transformed in the birefringent crystal into two partly overlapping output beams. As soon as the acoustic wave is exited in the crystal (RF is turned on), two diffracted beams are observed. The diffracted beams are monochromatic but tunable with RF and in orthogonal polarizations. They are deflected to both sides at 7.5°, the divergence being 5°. Using an AOTF exit lens ($f = 59$ mm), the output beams are spatially separated from the relatively bright undiffracted orders which are captured by a central light trap. The angular walkoff (i.e. the drift in the diffracted beam angle as the device is tuned) is insignificant and there is no need for its compensation in a point spectrometer with relatively large single-pixel detectors.

Two identical detectors with proximity lenses are used to analyze simultaneously the diffracted beams. The detectors are InGaAs photodiodes 1 mm (Hamamatsu) providing unstabilized cooling (incorporated Peltier elements) with $\Delta T = 25°$. In the present design, the noise of the electronic preamplifier is larger than the detector noise at temperature around 0 °C.

A dedicated side port lying in the XY plane of the spacecraft and directed at 90^0 from the spacecraft axis $+Z$ (away from the all nadir-looking instruments) is used for solar occultation. The entry optics (a small lens with useful diameter of 3 mm and a circular diaphragm) provides an angular FOV of about 4 arc minutes when observing the Sun. An optical fiber delivers the solar light to the IR spectrometer objective. A gradient collimator lens at the output of the fiber and a 45° flat mirror mounted at the baffle of the NIR objective completes the design of the solar entry (Figure 3.51). There is no Sun detector. The pointing to the Sun is provided by the Mars Express spacecraft, oriented to the Sun direction with an accuracy of generally much better than 6 arc minutes (spacecraft specification). The side port for Sun observations is normally closed by a shutter, controlled by the data processing unit (DPU), which is opened only during the observation.

The main characteristics of the SPICAM IR spectrometer are summarized in Table 3.3. The complete IR AOTF spectrometer is shown in Figure 3.52. This instrument has fully demonstrated the capability of mapping the H_2O contents in the atmosphere [116]. The use of the tunable filter implies sequential registration of a spectrum and a relatively large measurement time, during which the spectrum could change because of the orbital motion. This drawback is fully compensated by such advantages as flexible sampling within the spectral range, and the modulation capability of the AOTF, minimizing the stray light. The AOTF spectrometer is a rugged and stable device and it contains no moving parts. Besides H_2O measurements, the exited state of O_2 due to photodissociation of ozone is routinely measured [116].

In addition, O. Korablev et al. detected the signatures of H_2O and CO_2 ices, characterizing the conditions on the surface and in the atmosphere. They believe that the concept of the instrument capable of measuring a number of important atmospheric and environmental characteristics from the orbit on Mars or from a lander for less than 1 kg with a potential to upgrade to an imaging spectrometer has a great future. With a mass of only 0.75 kg, this instrument should fly as a low-cost passenger on all future missions to Mars.

Table 3.3 Characteristic of the SPICAM – IR spectrometer.

Spectral range	**1.0–1.7 μm**
Spectral resolution	0.5 nm at 1.1 μm, 0.95 nm at 1.5 μm
FOV	1°
Telescope	Lens type, Ø30 mm
AOTF	TeO_2 aperture $3.6 \times 3.6\,mm^2$, ±3.5°
AOTF RF	85–140 MHz, 0.5–2.5 W
Detector	Two InGaAs photodiodes Ø1 mm (Hamamatsu G5832)
Transmission of optics	20%
NER	$\sim 5.\,10^{-5}\,W/m^2/sr$
Gain control	4 preset gain values
Number of spectral points	2 spectra with different polarizations, 332 to 3984 points each
Dynamic range	2^{16}, rounded to 2^{12}
Power consumption	5 W average
Dimensions	$220 \times 85 \times 65\,nm^3$
Mass	≤700 g

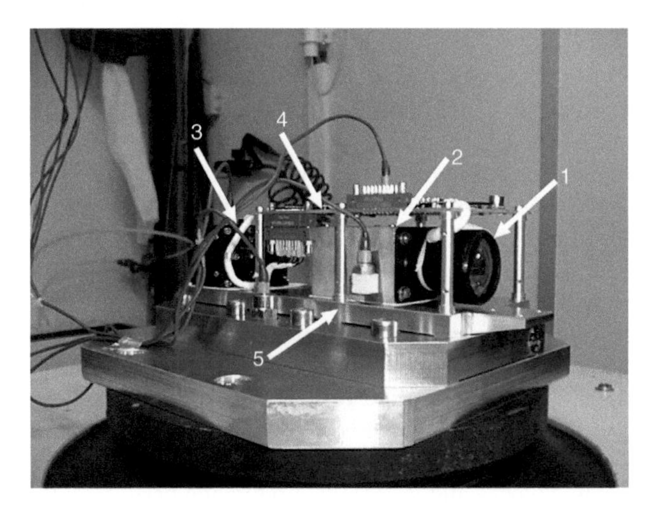

Figure 3.52 Complete IR channel of SPICAM at the vibration test bench. 1-objective; 2-AOTF unit; 3-detectors unit; 4-electronics; 5-mechanical structure. The overall mass is 750 g. Source: Korablev et al. 2017 [113]. Reproduced with permission of Astrobiology.

3.5 Acousto-optic Materials and Their Applications

To design acousto-optic devices with high DE, the first choice is an acousto-optic medium with high figures of merit (M_2) as described in Refs. [24, 117, 118]. In terms of physical properties of the medium, this is equivalent to selecting a material with a large refractive index, a large photoelastic constant, and low acoustic velocity. For the light deflector, low acoustic velocity also favors the large angle of deflection and large number of resolvable spots, but it is adverse to the attainment of a high-speed operation and also prevents use in the high-frequency region because such a medium has high acoustic attenuation in general. Thus, in selecting the acousto-optic materials, a certain compromise has to be made between their figures of merit and acoustic attenuation, depending on the purpose of application. In addition, the following requirements have to be satisfied for the medium to be important practically: (i) high optical transparency at the wavelength to be used, (ii) chemical stability and mechanical durability, (iii) established technology of the crystal growth to obtain large and high-quality boules, and (iv) low temperature coefficient of physical constants, especially of the acoustic velocity.

Some descriptions of typical acousto-optic crystals are given here.

3.5.1 Lead Molybdate ($PbMoO_4$)

Excellent acousto-optic properties of this crystal were first reported in 1969 [119], and a detailed study was made in 1971 [120], including its elastic, optical, and thermal properties. This crystal has been known as wulfenite and has the scheelite-type structure belonging to the point group $C_{4h}(4/m)$. The growth conditions were studied by Bonner and Zydzik [121] and large-size single crystals up to 4 cm diameter × 12 cm long and with high optical quality were grown reproducibly. The characteristic feature observed in acousto-optic performance is that the longitudinal acoustic wave propagating along the c-axis scatters light with the same efficiency for both ordinary and extraordinary rays. The large value of the figure of merit, $M_2 = 36 \times 10^{-18}$ s^3/g, as well as low acoustic loss for this acoustic mode, makes the crystal very useful for practical devices operable at optical wavelengths between 0.45 and 5 μm and acoustic frequencies up to about 500 MHz.

3.5.2 Tellurium Dioxide (TeO_2)

Laser quality tellurium dioxide (TeO_2 or paratellurite) is an excellent piezo-optical material. It is extensively used because of its high acousto-optical figure of merit in making acousto-optical modulators, imaging devices, splitters, deflectors, tunable polarization filters, RF spectral analyzers, and other acousto-optoelectronic equipment for laser radiation control. TeO_2 crystals are grown by the Czochralsky method. It has higher damage threshold and better optical quality compared with TeO_2 crystals grown by some other methods.

The first single crystal was grown by Liebertz [122], and its elastic, piezoelectric, and optical properties were reported by Arlt and Schweppe [123]. Two notable features of TeO_2 observed by them are that a slow shear wave propagating along the [96] direction with the displacement along [96] has an exceptionally low acoustic velocity of 0.616×10^5 cm/s, and that the crystal has large optical activity. Combined with high refractive indices of this crystal, the shear mode previously mentioned should have a high figure of merit, and this feature was confirmed by Uchida and Ohmachi [124]. Detailed study of the photoelastic properties [124] revealed that the longitudinal mode along the *c*-axis has an acousto-optic performance comparable to that of $PbMoO_4$. Based on the discovery of these performances that were preferable for the acousto-optic application, they investigated temperature characteristics [125], acoustic-wave propagation [126], acoustic attenuation [127], and optical properties [128], all required for the deflector design. Technology of crystal growth was refined by Miyazawa and Iwasaki [129], and high-quality crystal boules have been obtained starting with high purity source material, by choosing the [96] direction as the pulling axis and by controlling the pulling rate, the rotation speed, and the shape of the liquid solid interface. The importance of the use of high-purity source material was also discussed recently by Bonner et al. [130]. During the years, tellurium dioxide found wide application in all the areas of engineering devices. Some of the devices are described here.

(a) *AOTF:* In 2003, A. K. Zaitsev et al. [131] designed and investigated the properties of AOTF based on a tellurium dioxide (TeO_2) single crystal. They used the so-called subcollinear geometry of anisotropic acousto-optic interaction that occurs when the wave vector of incident light is collinear to acoustic energy vector. They obtained the analytical formula of spectrum spread function of proposed AOTF and found the condition determining the shape of that characteristic. Also, they designed and fabricated the experimental prototype of subcollinear AOTF. In their design, they used the reflection of acoustic wave that allowed use of the crystal's media more effectively and reduced the AOTF size.

In order to minimize the cell's size, they chose a variant of acoustic reflection (Figure 3.53). The piezoelectric transducer irradiates a shear acoustic wave with wave normal $\boldsymbol{K}_{a1}$, which shifting vector is parallel to the surface of reflecting facet. The vector $\boldsymbol{v}_{g1}$ shows the direction of energy transport of the irradiated wave. The acoustic reflection happens without wave polarization transformation and is characterized by a high-energy transfer coefficient. The reflected acoustic wave has the wave normal $\boldsymbol{K}_{a2}$ and the energy transport $\boldsymbol{v}_{g2}$. It should be mentioned that due to strong acoustic anisotropy of TeO_2, the reflection on the facet occurs when the incident angle is larger than 90°. The geometry of acoustic reflection is shown in Figure 3.54.

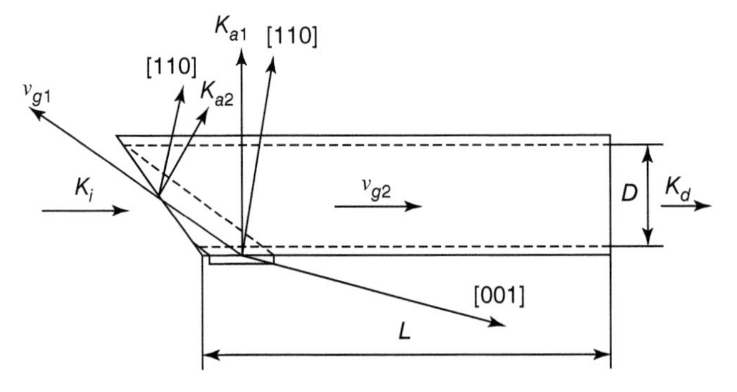

Figure 3.53 The protopype's layout of AOTF. Source: Zaitsev and Kludzin 2003 [131]. Reproduced with permission of Elsevier.

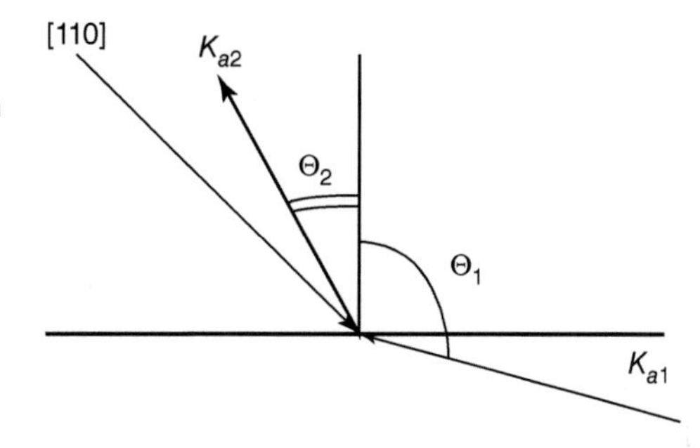

Figure 3.54 The geometry of acoustic reflection on the facet. Source: Zaitsev and Kludzin 2003 [131]. Reproduced with permission of Elsevier.

Snell's law is still correct in this case and can be written as

$$\frac{\sin \Theta_1}{\nu_1} = \frac{\sin \Theta_2}{\nu_2}$$

where Θ_1, Θ_2, and ν_1, ν_2 are angles and velocities of incident and reflected acoustic waves, respectively. An incident optical wave corresponding to ordinary polarization is diffracted on reflected acoustic wave. The vector of wave normal of optical wave $\boldsymbol{K}_i$ is coincided with the energy transport of acoustic wave $\boldsymbol{v}_{g2}$. The diffracted optical wave with wave normal $\boldsymbol{K}_d$ corresponds to extraordinary polarization (Figure 3.53). It is important that the designed prototype is very compact. The main advantage of the filter based on proposed geometry is the extremely low value of electrical driving power (20 mW) that makes it a strong candidate in many applications.

The acousto-optic properties of tellurium crystals have been examined in order to develop an AOTF operating in the long-wave infrared (LWIR) region [132]. The AOTF design was based on the wide-angle regime of light diffraction in the YZ plane of the birefringent crystal operating from 8.4 to 13.6 μm (Figure 3.55).

Device characteristics were obtained from both theoretical and experimental investigations. Experiments were carried out using both a 10.6-μm pulsed CO_2 laser and a tunable CO_2 laser operating in a continuous wave mode from 9.2 to 10.7 μm. The AOTF was tuned over the acoustic frequency range of 81.5–94.7 MHz. The filtering performance in the tellurium device was provided by a pure shear elastic wave propagating at a 95.8° angle with respect to the positive direction of the optic axis, while an ordinary polarized optical beam was incident at the Bragg angle of 6.0° relative to the acoustic wavefront. At 10.6 μm, the measured spectral bandwidth of the filter was 127 nm and the optical transmission coefficient was around 8.8% with 2.0 W drive power.

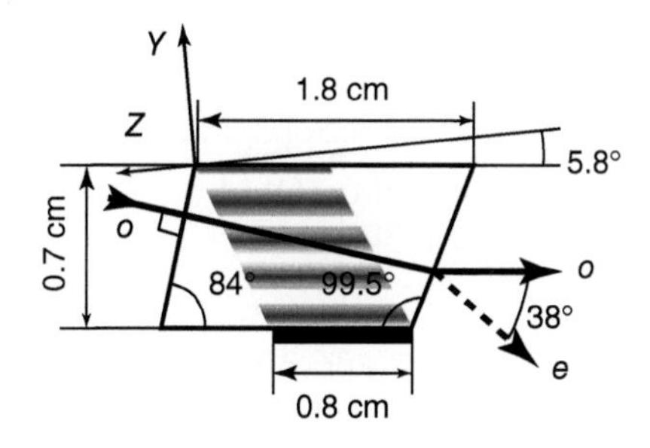

Figure 3.55 Configuration of the acousto-optic cell of the tunable filter. Source: Gupta et al. 2012 [132]. Reproduced with permission of IOP.

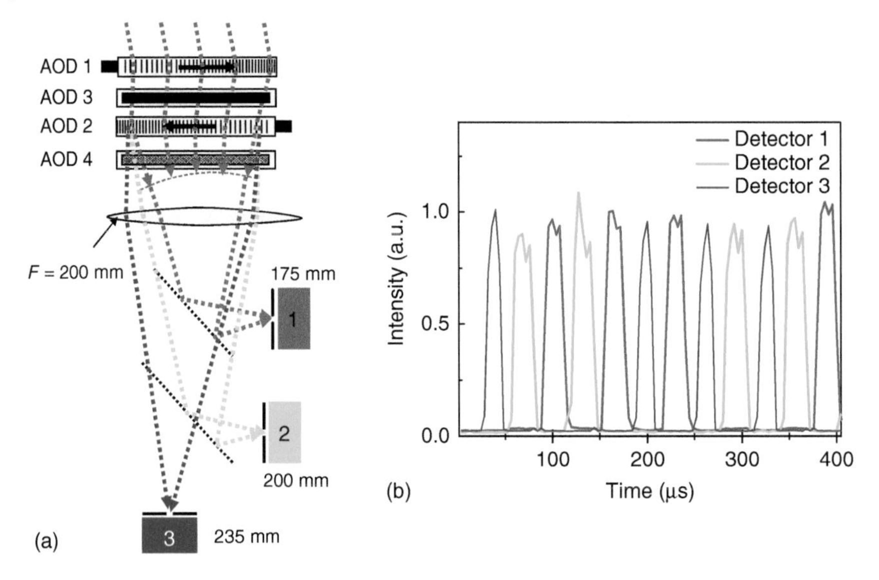

Figure 3.56 Focusing of a laser beam to arbitrarily chosen focal points at 30 kHz. (a) Simplified diagram of four AOD AOL, 200-mm focal length lens, two partial reflectors, and three silicon detectors with 20-μm pinholes. Detectors were placed at arbitrary XY positions and axial distances from the lens of 175 mm (gray), 200 mm (light gray) and 235 mm (dark gray). (b) The signal from each detector, plotted with the appropriate color code for its detector, during random access pointing where the laser beam is focused sequentially into each of the pinholes.

The results obtained during this research can be used in future to determine the optimal AO interaction geometries applicable to various AO devices such as modulators, deflectors, and tunable filters operating in the LWIR region of the spectrum.

(b) *Acousto-Optic Lens Microscope:* In 2010, P.A. Kirkby et al. [133] described a high-speed 3D acousto-optic lens microscope (AOLM) for femtosecond 2-photon imaging. By optimizing the design of the four acousto-optic deflectors (AODs) and by deriving new control algorithms, they developed a compact spherical acousto-optic lens (AOL) with a low temporal dispersion that enables 2-photon imaging at 10-fold lower power than previously reported [134, 135], as shown in Figure 3.56. The high-efficiency AODs are made of TeO_2.

They showed that the AOLM can perform high-speed 2D raster-scan imaging (>150 Hz) without scan-rate-dependent astigmatism (Figure 3.57). It can deflect and focus a laser beam in a 3D random access sequence at 30 kHz and has an extended focusing range (>137 μm; 40× 0.8NA objective). These features are likely to make the AOLM a useful tool for studying fast physiological processes distributed in 3D space.

(c) *Acousto-optic deflector:* AODs are promising ultrafast scanners for nonlinear microscopy. Their use has been limited until now by their small scanning range and by the spatial and temporal dispersions of the laser beam

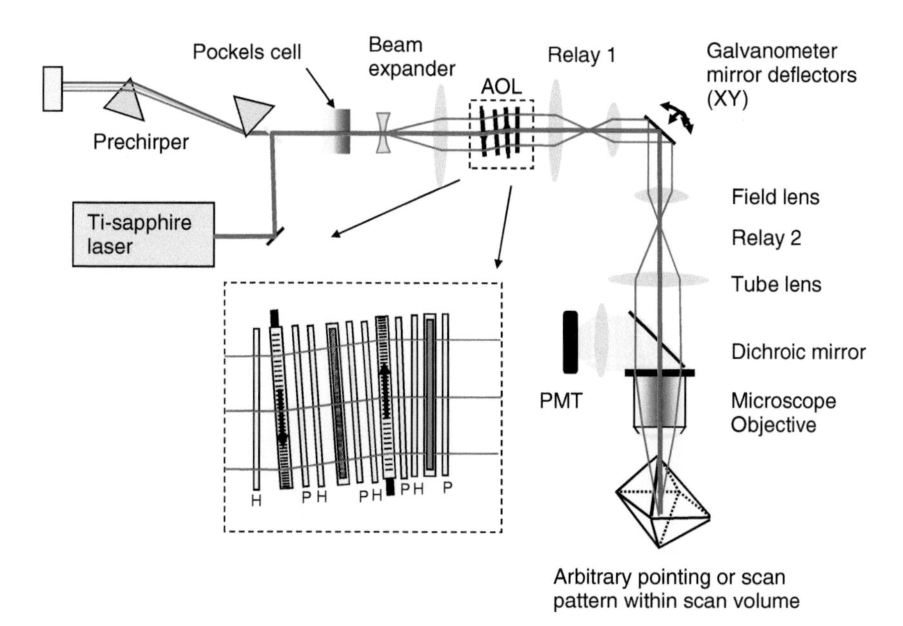

Figure 3.57 Schematic diagram of complete acousto-optic lens (AOL) 2-photon microscope. Inset shows details of AOL. The four AODs are placed as close to one another as possible (40–60 mm center to center). Before each AOD is a half waveplate (H). After each AOD is a polarizer (P) to absorb the residual undiffracted zero order light. By fixing the galvanometers, AOL deflection and focusing can be used to focus anywhere within the octahedral-shaped scan volume below the objective. Fluorescence was either collected through the objective onto a photomultiplier tube (PMT), or additionally through the condenser (not shown).

going through the deflectors. In 2008, Y. Kremer et al. [136] showed that the use of AOD of large aperture (13 mm) compared to standard deflectors allows accessing much larger FOV while minimizing spatiotemporal distortions. They fully characterized a TeO_2-made AOD scanner from the company A&A having exceptional physical dimensions and performance: 13-mm aperture, 50-mrad scanning range, a tuning range of 720–880 nm allowing access to about 850 points per line (7.10^5 points in the FOV). The spatial and temporal compensations are obtained using an AOM placed at a distance of the AOD. Using a GRENOUILLE, an easy to align frequency-resolved optical gating (FROG), i.e. a spectrally resolved auto-correlator, they showed that temporal broadening and spatial dispersion can be minimized using this setup. Moreover, they characterized the spatiotemporal distortions including spatial chirp (SC) and pulse front tilt (PFT) which arise from the single pass in the AOM-AOD setup. They demonstrated that the fine tuning of the AOM-AOD setup and the use of large aperture AOD allow minimization of PFT and SC.

The setup (Figure 3.58) consisted of a Tsunami Ti:sapphire femtosecond laser source (Spectra Physics, Irvine, CA) which provided ~ 90 fs pulses having a spectrum $\delta\lambda$ ~11+/−0.5 nm (full width at half maximum [FWHM]). The beam was first magnified by an afocal telescope (magnification ~8) to a beam diameter D_0 of about 11 mm (measured at $1/e^2$), which slightly undercovered the AOM

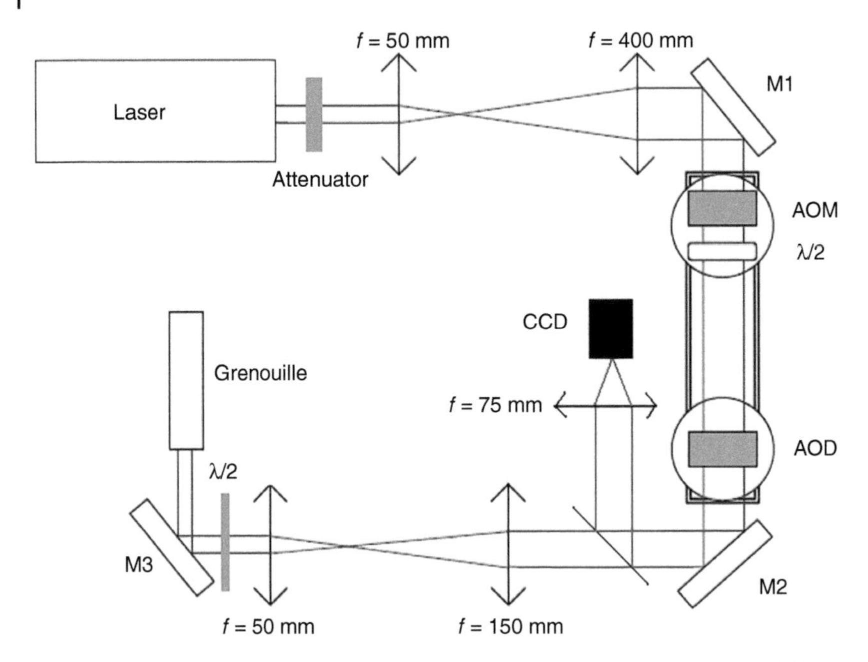

Figure 3.58 The optical setup. Laser: Tsunami Ti:Sa laser. AOM: acousto-optic modulator and AOD: acousto-optic deflectors. Both were mounted on goniometric plates for alignment and placed on a rail to adjust their distance. $\lambda/2$: achromatic half waveplates. GRENOUILLE: single-shot second harmonic generation frequency-resolved optical gating devices. M1, M2, M3: dielectric mirrors. Attenuator: variable intensity attenuator.

aperture (13 mm, ref. AA.MTS/A15@720–920 nm, A&A, Orsay, France). The AOD pair (A&A, ref. AA.DTS.XY/A15@720–920 nm) was placed at a distance D of the AOM, where D was about 60 cm and was carefully adjusted for optimal temporal compensation. The AOD aperture (13 mm) was chosen so that it did not diaphragm the laser beam, despite its spectral divergence after the AOM. A $\lambda/2$ achromatic waveplate was placed between the AOM and the AOD to adjust the pulse polarization. The AOD exit aperture was imaged by a telescope of magnification 1/3 onto a spectrally resolved auto-correlator (GRENOUILLE 10-50, Swamp Optics, Atlanta). A second $\lambda/2$ achromatic waveplate was used for convenience to adjust the light polarization. A QuickFrog software (version 3.1, Swamp optics) was used to reconstruct the electric field from the measurement. A glass slide placed after the AOD was used to collect a small fraction of the beam which was focalized onto a digital CDD camera (IMI-141- FT, IMI tech, Seoul, Korea) using a $f = 75$ mm lens to measure the spatial dispersion of the pulses and to optimize its compensation. The ultrasonic wave frequency and amplitude were controlled in the AOM using a AA.MOD-nC driver (A&A) with a remote control and in the AOD using a DDS-XY-70-140-D8b-3W driver (A&A) and a program written with Labview using a fast digital I/O board (NI PCI DIO 32HS, National Instruments). This large AOM-AOD device is referred to as *scanner* 1 in the following.

For comparison, a second scanning system was used. It consists of an AOM (AA.MTS.141/B20/A5@840 nm, A&A) and two crossed AODs (AA.DTS.XY .250@840 nm, A&A) optimized at a wavelength of 840 nm, having a smaller aperture of 4.2 mm and driven by AA.MOD-nC (A&A) and AA.DDS.1-250.2V (A&A) drivers respectively. In that case, a beam diameter D_0~3 mm was used at the AOM entrance pupil. This small AOM-AOD setup is referred to as *scanner* 2 in the following. It was used with a Mai-Tai Ti:Sapphire laser (Spectra Physics), with a spectrum of $\delta\lambda \sim 14+/- 0.5$ nm (FWHM).

Y. Kremer et al. has shown that a 13-mm aperture AOD offering potentially a FOV of typically 300 μm using a 20× objective can be used over a bandwidth of 160 nm. Spatial compensation at all wavelengths is achieved using an AOM of identical aperture, whereas temporal dispersion compensation is obtained at an AOM–AOD distance of 63 cm. This distance is in agreement with the expected value of the GDD (group delay dispersion) introduced by AOD knowing their physical dimension and the GVD (group velocity dispersion) of TeO_2. They obtained an excellent temporal compensation which allows using an almost fully compensated pulse in a two-photon microscope.

Finally, these large AODs limit the SC introduced by the single-pass configuration. SC is due to the spatial separation of the pulse wavelengths after the AOD relative to the beam diameter. Since AOD scanners, whatever their size, introduces the same angular dispersion and since temporal compensation is obtained at distances of same order of magnitude, the use of larger AOD minimizes SC. This might be a great advantage compared to smaller systems where SC will eventually lead to temporal pulse broadening in scattering or aberrant media. For example, in nonlinear microscopy, high numerical aperture objectives are used to image deep in highly scattering medium. In the presence of SC, the different frequencies of the pulse travel different path lengths in the tissue. As a result of scattering, the pulse at the objective focus might contain fewer frequencies as expected, thus increasing the pulse width. Finally, the use of the GRENOUILLE allows also to finely minimize the PFT so that it almost vanishes at the distance which compensates for the temporal dispersion.

3.5.3 Lithium Niobate ($LiNbO_3$)

This well-known ferroelectric material has various usages for electro-optic, nonlinear-optic, piezoelectric, and acoustic devices. The utilization of the crystal for high-frequency acousto-optics has been indicated from the photoelastic measurement made by Dixon and Cohen [137] and acoustic loss measurements made by Wen and Mayo [138], Grace et al. [139] and Spencer and Lenzo [140]. The large value of M_2/Γ makes $LiNbO_3$ one of the future candidates for a high-frequency acousto-optic medium.

An acousto-optic device requires a material with good acoustic and optical properties and high optical transmission. An acousto-optic material can modulate the light which passes through it when influenced by an acoustic field. LN is an example of an AOM. LN continues to be a material of interest for various optical and surface acoustic wave (SAW) applications due to its unique combination of piezoelectric and optical properties.

Since 1967, Crystal Technology has been supplying high-quality, leading edge LN for a variety of applications including SAW devices, integrated optic devices, refractive elements, and passive detectors. An intersecting waveguide polarization beam splitter (PBS), a film-loaded surface acoustic waveguide (SAWG) and an o-gap directional coupler-type reflector have been developed using Ti:$LiNbO_3$ materials in order to design a new integrate AOTF in a single chip [140] and are described in the following section.

(d) **Polarization Beam Splitter (PBS):** The proton-exchange-type PBS is the most popular PBS used in AOTFs. It has several excellent characteristics; namely, a high extinction ratio and a small wavelength dependency. However, it also has a large excess loss caused by mode field mismatch between the Ti in diffused waveguide and the proton-exchange waveguide. It also has the demerit of being a long device (~10 mm).

A new intersecting waveguide PBS has been developed in which L_c and W_c are the length and center width in the intersecting region (Figure 3.59). It functions as a directional coupler whose operation depends on the beat between the odd-mode light and even-mode light for transverse electric (TE)-mode light and transverse-magnetic (TM)-mode light, respectively. The PBS waveguide propagates the TE-mode light to the cross output port and the TM-mode light to the parallel output port. This polarization selectivity is due to the birefringence of $LiNbO_3$ thin film materials. The extinction ratios of both polarizations are less than −30 dB. This PBS also has the merit of a low excess loss. The excess loss is minimized by the shape of the intersecting waveguide and has a minimum value of 0.15 dB.

(e) **Film-loaded Surface Acoustic Waveguide (SAWG):** The performance of an AOTF greatly depends on the design of its SAWG. A Ti-deep diffused SAWG (Figure 3.60a) is the most popular, but it is quite wide because of the need for sufficient distance between the Ti-SAWG and the Ti-optical waveguide. This makes it difficult to integrate multiple AOTFs in a single chip and also decreases the design tolerances.

A new film-loaded SAWG has been designed with the transparent materials of SiO_2 or In_2O_3-doped SiO_2 for the film, as shown in Figure 3.60b. The confinement in this new type of SAWG has an exact relation with the device's width, thickness,

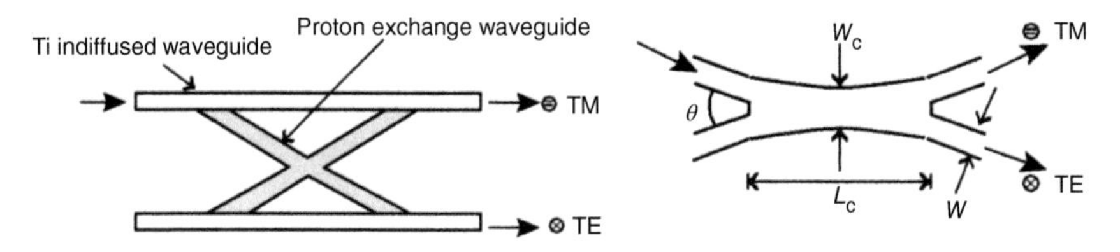

Figure 3.59 (a) Proton exchange waveguide and (b) Ti indiffused intersecting waveguide. Source: Nakazawa et al. 1999 [141]. Reproduced with permission of Fujitsu Sci Tech J.

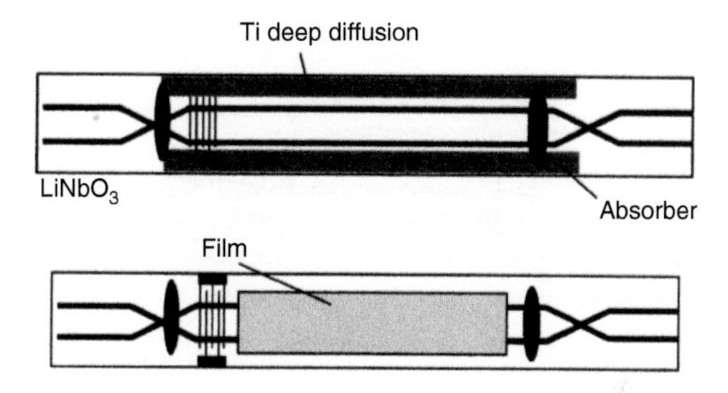

Figure 3.60 (a) Ti-deep diffusion SAWG and (b) film-loaded-type SAWG. Source: Nakazawa et al. 1999 [141]. Reproduced with permission of Fujitsu Sci Tech J.

and film material but does not have a distinct relation with the SAW propagation speed on the film. The new film-loaded SAWG has an advantage when it comes to integrating multiple AOTFs in a single chip.

(f) **Waveguide Reflector:** The conventional waveguide reflector reflects the light geometrically over the large reflecting angle θ, which is usually more than 4°. But from the manufacturing point of view, the excess loss of this reflector is too high.

To solve this problem, a new type of waveguide reflector has been developed which uses a 0-gap directional coupler waveguide that directs both the TE-mode and TM-mode light to the cross output port (Figure 3.61). The waveguide is cut at the center of the intersecting region and a metal mirror is set at the cut end-face. In this design, the intersecting region is three times longer compared with the case in a PBS because the beat period of the TM-mode is three times longer than that of the TE-mode. This reflector has more than 10 times the manufacturing tolerance of the conventional design.

Using these techniques, a new integrated AOTF has been designed as shown in Figure 3.62 [141]. The device is a single chip containing five AOTFs and four reflectors that interconnect them. The input light is divided into the through light and drop light by the first-stage AOTF (center AOTF), which consists of a PBS, the new film-loaded SAWG, and an interdigital transducer (IDT). First, the input light propagates through the middle SAWG and then back in the opposite direction through the two adjacent SAWGs after being reflected by the newly designed

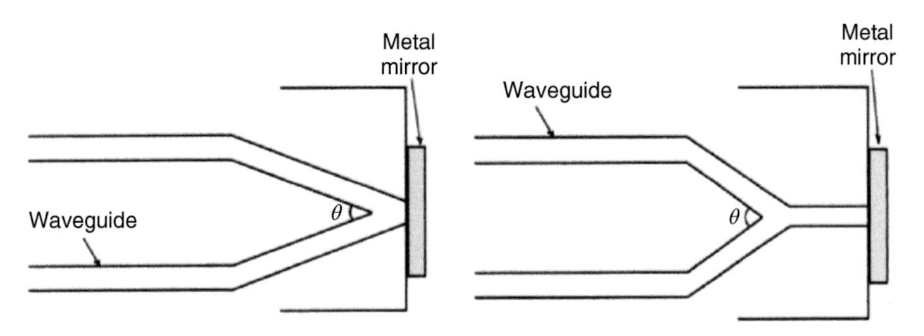

Figure 3.61 (a) Conventional waveguide reflector and (b) 0-gap directional coupler type reflector. Source: Nakazawa et al. 1999 [141]. Reproduced with permission of Fujitsu Sci Tech J.

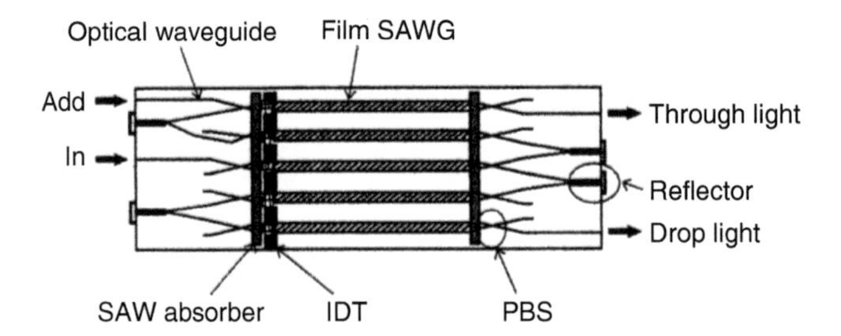

Figure 3.62 Integrated AOTF. Source: Nakazawa et al. 1999 [141]. Reproduced with permission of Fujitsu Sci Tech J.

reflectors. The light is again reflected by two more of the new reflectors, propagates through the top and bottom SAWGs, and emerges as the through light and the drop light from top and bottom, respectively. The through light and drop light, therefore, is filtered once by the same SAWG and then twice more by two different SAWGs.

References

1 Brillouin, L. (1922). Diffusion de la lumière et des rayons x par un corps transparent homogène, infuence de l'agitation thermique. *Annales de physique paris* 17: 88–122.

2 Debye, P. and Sears, F.W. (1932). On the scattering of light by supersonic waves. *Proc. Natl. Acad. Sci. U.S.A.* 18 (6): 409–414.

3 Lucas, R. and Biquard, P. (1932). Nouvelles propriétés optiques des liquides soumis ǎ des ondes ultra-sonores. *C. R. Acad. Sci.* 194: 2132–2134.

4 Raman, C.V. and Nath, N.S.N. (1936). The diffraction of light by high frequency sound waves. *Proc. Indian Acad. Sci.-A* 2: 406–412.

5 Sanders, F.H. (1936). Intensity measurements in the diffraction of light by ultrasonic waves. *Can. J. Res.* 14A: 158–171.

6 Nomoto, O. (1940). Intensitätsverteilung des Lichtes in den an fertschreitenden Ultraschallwellen in Flüssigkeiten erzeugten Beugungsspektren. *Proc. Phys. Math. Soc. Jpn.* 22: 314–320.

7 Nomoto, O. (1942). Studien über die Beugung von Licht an Ultraschallwellen. *Proc. Phys. Math. Soc. Jpn.* 24: 380–400.

8 Rytow, S.M. (1935). Lightbeugung an Ultraschallwellen. *Phys. Z. Sowjetunion* 8: 626–649.

9 Bhagavantam, S. and Rao, B.R. (1948). Diffraction of light by very high frequency ultrasonic waves; effect of tilting the wave front. *Proc. Indian Acad. Sci.* 28A: 54–62.

10 Phariseau, P. (1956). On the diffraction of light by progressive supersonic waves. *Proc. Indian Acad. Sci.* 44A: 165–170.

11 Klein, W.R. (1966). Theoretical efficiency of Bragg devices. *Proc. IEEE* 54: 803–804.

12 Exterman, R. and Wannier, G. (1936). Theorie de la diffraction de la lumiére par les ultrasons. *Helv. Phys. Acta* 9: 520–532.
13 Bhatia, A.B., Noble, W.J., and Born, M. (1953). Diffraction of light by ultrasonic waves I. General theory. *Proc. R. Soc.* 220 (1142): 356–368, 369–385.
14 Berry, M.V. (1966). *The Diffraction of Light by Ultrasound*. New York: Academic Press.
15 Nomoto, O. (1971). Diffraction of light by ultrasound: extension of the Brillouin theory. *Jpn. J. Appl. Phys.* 10: 611–622.
16 Klein, W.R. and Cook, B.D. (1967). Unified approach to ultrasonic light diffraction. *IEEE Trans. Sonics Ultrason.* SU-14: 123–134.
17 Bergmann, L. (1954). *Der Ultraschall*. Stuttgart, Germany: Hirzel.
18 Hargrove, L.E. and Achyuthan, K. (1965). Use of light diffraction in measuring the parameter of nonlinearity of liquids and the photoelastic constants of solids. In: *Physical Acoustics*, vol. 2, Pt. B (ed. W.P. Mason). New York: Academic Press.
19 Okolicsanyi, F. (1937). The wave-slot, an optical television system. *Wireless Eng.* 14: 527–536.
20 Robinson, D.M. (1939). The supersonic light control and its application to television with special reference to the Scophony television receiver. *Proc. IRE* 27: 483–486.
21 Korpel, A., Adler, R., Desmares, P., and Watson, W. (1966). A television display using acoustic deflection and modulation of coherent light. *Proc. IEEE* 54: 1429–1437.
22 Korpel, A., Lotsoff, S.N., and Whitman, R.L. (1969). The interchange of time and frequency in television displays. *Proc. IEEE* 57: 160–170.
23 Anderson, L.K. (1968). Holographic optical memory fòr bulk data storage. *Bell Lab. Rec.* 46: 318–325.
24 LaMacchia, J.T. (1970). Optical memories: a progress report. *Laser Focus,* vol. 6, pp. 35–39.
25 Anderson, L.K. (1970). An experimental read-only holographic optical memory. Presented at the 6th International Quantum Electronics Conference, Kyoto, Japan.
26 Hrbek, G. and Watson, W. (1971). A high speed laser alphanumeric generator. *Proceedings of Electro-Optical Systems Design Conference*, pp. 271–275 (New York, September).
27 Warner, A.W., White, D.L., and Bonner, W.A. (1972). Acousto-optic light deflectors using optical activity in paratellurite. *J. Appl. Phys.* 43: 4489–4495.
28 "Acousto-optic effect". Retrieved 2007-08-07.
29 Scruby, C.B. and Drain, L.E. (1990). *Laser Ultrasonics: Techniques and Applications*. Taylor & Francis.
30 Jodłowski, L. (2003). Optimisation of construction of two-channel acousto-optic modulator for radio-signal detection. *Opto-Electron. Rev.* 11 (1): 55–63.
31 Tran, C.D. and Bartelt, M. (1992). Performance characteristics of acousto-optic tunable filter for optical spectrometry. *Rev. Sci. Instrum.* 63: 2932–2939.

32 Tran, C.D. (2000). Acousto-optic tunable filter: a new generation monochromator and more. *Anal. Lett.* 33: 1711–1732.
33 Siegman, E. (1986). *Lasers*. Mill Valley, CA: University Science.
34 Vicente, S.G.C., Gámez, M.A.M., Kiryanov, A.V. et al. (2004). Diode-pumped self-Q-switched erbium-doped all-fibre laser. *Quantum Electron.* 34 (4): 310–314.
35 Kee, H.H., Lees, G.P., and Newson, T.P. (1998). Narrow linewidth CW and Q-switched erbiumdoped fibre loop laser. *Electron. Lett.* 34 (13): 1318–1319.
36 Álvarez-Chávez, J.A., Offerhaus, H.L., Nilsson, J. et al. (2000). High-energy, high-power ytterbium-doped Q-switched fiber laser. *Opt. Lett.* 25 (1): 37–39.
37 Andres, M.V., Cruz, J.L., Díez, A. et al. (2008). Actively Q-switched all-fiber lasers. *Laser Phys. Lett.* 5 (2): 93–99.
38 Cuadrado-Laborde, C., Díez, A., Andrés, M.V. et al. (2011). Applications of in–fiber acousto–optic devices. In: *Acoustic Waves – From Microdevices to Helioseismology* (ed. M.G. Beghi), 595–636. InTech. ISBN: 978-953-307-572-3.
39 Russell, P.S.J. and Liu, W.F. (2000). Acousto-optic superlattice modulation in fiber Bragg gratings. *J. Opt. Soc. Am. A:* 17 (8): 1421–1429.
40 Cuadrado-Laborde, C., Delgado-Pinar, M., Torres-Peiró, S. et al. (2007). Q-switched all-fibre laser using a fibre-optic resonant acousto-optic modulator. *Opt. Commun.* 274 (2): 407–411.
41 Liu, W.F., Russell, P.S.J., and Dong, L. (1997). Acousto-optic super lattice modulator using a fiber Bragg grating. *Opt. Lett.* 22 (19): 1515–1517.
42 Liu, W.F., Russell, P.S.J., and Dong, L. (1998). 100% efficient narrow-band acousto-optic tunable reflector using fiber Bragg grating. *J. Lightwave Technol.* 16 (11): 2006–2009.
43 Cuadrado-Laborde, C., Díez, A., Delgado-Pinar, M. et al. (2009). Mode locking of an all-fiber laser by acousto-optic superlattice modulation. *Opt. Lett.* 34 (7): 1111–1113.
44 Cuadrado-Laborde, C., Diez, A., Cruz, J.L., and Andres, M.V. (2010). Experimental study of an all-fiber laser actively mode-locked by standing-wave acousto-optic modulation. *Appl. Phys. B: Lasers Opt.* 99 (1–2): 95–99.
45 Cuadrado-Laborde, C., Diez, A., Cruz, J.L., and Andres, M.V. (2009). Doubly active Q-switching and mode locking of an all-fiber laser. *Opt. Lett.* 34 (18): 2709–2711.
46 Cuadrado-Laborde, C., Perez-Millán, P., Andres, M.V. et al. (2008). Transform-limited pulses generated by an actively Q-switched distributed fiber laser. *Opt. Lett.* 33 (22): 2590–2592.
47 Xu, J. and Stroud, R. (1992). *Acousto-optic devices: principles, design and applications*. New York: Wiley-Interscience.
48 Vila, J., Calpe, J., Pla, F. et al. (2005). Smart spectra: applying multispectral imaging to industrial environments. *Real-Time Imaging* 11: 85–98.
49 Chang, I.C. (1974). Noncollinear acousto-optic filter with large angular aperture. *Appl. Phys. Lett.* 25 (7): 370–372.
50 Chieu, D. (2003). Tran, infrared multispectral imaging: principles and instrumentation. *Appl. Spectrosc. Rev.* 38 (2): 133–153.

51 Morris, M.D. (1993). *Microscopic and Spectroscopic Imaging of the Chemical State*. New York: Marcel Dekker.

52 Tran, C.D. and Lu, J. (1995). Characterization of the acousto-optic tunable filter for the ultraviolet and visible regions and development of an AOTF-based rapid scanning detector for HPLC. *Anal. Chim. Acta* 314: 57–66.

53 Tran, C.D., Cui, Y., and Smirnov, S. (1998). Simultaneous multispectral imaging in the visible and near infrared region: applications in document authentication and determination of chemical inhomogeneity of copolymers. *Anal. Chem.* 70: 4701–4708.

54 Fischer, M. and Tran, C.D. (1999). Evidence for kinetic inhomogeneity in the curing of epoxy using the near-infrared multispectral imaging technique. *Anal. Chem.* 71: 953–959.

55 Fischer, M. and Tran, C.D. (1999). Investigation of solid phase peptide synthesis by the near infrared multispectral imaging technique: a novel detection method of combinatorial chemistry. *Anal. Chem.* 71: 2255–2261.

56 Tran, C.D. (2000). Visualizing chemical composition and reaction kinetics by near infrared multispectral imaging technique. *J. Near Infrared Spectrosc.* 8: 87–99.

57 Khait, O., Smirnov, S., and Tran, C.D. (2000). Time resolved multispectral imaging spectrometer. *Appl. Spectrosc.* 54: 1734–1742.

58 Khait, O., Smirnov, S., and Tran, C.D. (2001). Multispectral imaging microscope with millisecond time resolution. *Anal. Chem.* 73: 732–739.

59 Alexander, T. and Tran, C.D. (2001). Near infrared spectrometer determination of di- and tripeptides synthesized by combinatorial solid phase method. *Anal. Chem.* 73: 1062–1067.

60 Tran, C.D. (2001). Development and analytical applications of multispectral imaging techniques. *Fresenius J. Anal. Chem.* 369: 313–319.

61 Alexander, T. and Tran, C.D. (2001). Near-infrared multispectral imaging technique for visualizing sequences of di- and tripeptides synthesized by solid phase combinatorial method. *Appl. Spectrosc.* 55: 939–945.

62 Politi, M. and Tran, C.D. (2002). Visualizing chemical compositions and kinetics of sol-gel by near-infrared multispectral imaging technique. *Anal. Chem.* 74: 1604–1610.

63 Kurtz, I., Dwelle, R., and Katzkaq, P. (1987). *Rev. Sci. Instrum.* 58 (11): 1996.

64 Tran, C.D. and Furlan, R.J. (1992). *Anal. Chem.* 64: 2775–2782.

65 Tran, C.D. and Furlan, R.J. (1993). *Anal. Chem.* 65: 1675–1681.

66 Tran, C.D. (1997). Principles and analytical applications of acousto-optic tunable filters, an overview. *Talanta* 45: 237–248.

67 Flamion, B., Bungay, P.M., Gibson, C.C., and Spring, K.R. (1991). *Biophys. J.* 60: 1229.

68 Treado, P.J., Levin, I.W., and Lewis, E.N. (1992). *Appl. Spectrosc.* 46 (3).

69 Tran, C.D. (1992). Analytical thermal lens spectrometry: past, present and future prospects. In: *Photoacoustic and Photothermal Phenomena Ill* (ed. D. Bicanic), 463–473. Berlin: Springer Verlag.

70 Franko, M. and Tran, C.D. (1996). *Rev. Sci. Instrum.* 67 (1): 18.

71 Tran, C.D. and Grishko, V.I. (1994). *Appl. Spectrosc.* 48: 96–100.

72 Tran, C.D. and Grishko, V.I. (1994). *Anal. Biochem.* 218: 197–203.
73 Tran, C.D., Huang, G., and Grishko, V.I. (1994). *Anal. Chim. Acta* 299: 361–369.
74 Tran, C.D., Grishko, V.I., and Baptista, M.S. (1994). *Appl. Spectrosc.* 48: 833–842.
75 Chieu, D. (1997). Tran, Principles and analytical applications of acousto-optic tunable filters, an overview. *Talanta* 45: 237–248.
76 Tran, C.D. and Simianu, V. (1992). *Anal. Chem.* 64: 1419–1425.
77 Tran, C.D., Furlan, R.J., and Lu, J. (1994). *Appl. Spectrosc.* 48: 101–106.
78 Murray, L. and Cowe, L.A. (1992). *Making Light Work: Advances in Near Infra-red Spectroscopy*. New York: VCH Publishing.
79 Hildrum, K.L., lsaksson, T., Naes, T., and Tandberg, A. (1992). *Near Infra-Red Spectroscopy, Bridging the Gap between Data Analysis and NIR Applications*. Chichester: Ellis Horwood chapter 23.
80 Burns, D.A. and Ciurczak, E.W. (1992). *Handbook of Near-Infrared Analysis*. New York: Marcel Dekker chapter 3.
81 Patonay, G. (1993). *Advances in Near-lnfrared Measurements*. Greenwich, Connecticut: JAI Press.
82 Weyer, L.G. (1985). *Appl. Spectrosc. Rev.* 21: 1–43.
83 Patonay, G. and Antoine, M.D. (1991). *Anal. Chem.* 63: 321A–327A.
84 Imasaka, T. and Ishibashi, N. (1990). *Anal. Chem.* 62: 363A–371A.
85 Bjarklev, A. (1993). *Optical Fiber Amplifiers: Design and System Applications*. Boston: Artech House.
86 Desurvire, E. (1994). *Erbium-Doped Fiber Amplifiers Principles and Applications*. New York: John Wiley.
87 Tran, C.D. (1992). *Anal. Chem.* 64: 971A–981A.
88 Pereira, C.F., Gonzaga, F.B., and Pasquini, C. (2008). Near-Infrared Spectropolarimetry Based on Acousto-Optical Tunable Filters. *Anal. Chem.* 80: 3175–3181.
89 Bei, L., Dennis, G.I., Miller, H.M. et al. (2004). *Prog.Quantum Electron.* 28: 67–87.
90 Gonzaga, F.B. and Pasquini, C. (2005). *Anal. Chem.* 77: 1046–1054.
91 Thenkabail, P.S., Smith, R.B., and Pauw, E.D. (2000). Hyperspectral vegetation indices and their relationships with agricultural crop characteristics. *Remote Sens. Environ.* 71: 158–182.
92 Purevdorj, T., Tateishi, R., Ishiyama, T., and Honda, Y. (1998). Relationships between percent vegetation cover and vegetation indices. *Int. J. Remote Sens.* 19 (18): 3519–3535.
93 Ozanich, R.M. (1999). Near-infrared spectroscopy: background and application to tree fruit quality measurements. *Tree Fruit Postharvest Journal* 10 (1): 18–19.
94 Gat, N. (2000). Imaging spectroscopy using tunable filters: a review. *Proceedings of the SPIE* 4056: 50–64.
95 Sotoca, J.M., Pla, F., and Klaren, A.C. (2004). Unsupervised band selection for multispectral images using information theory. In: *Proceedings of the 17th international conference on pattern recognition*, Cambridge, UK. IEEE.

96 Gómez-Chova, L., Calpe-Maravilla, J., Soria, E. et al. (2003). CART-based feature selection of hyperspectral images for crop cover classification. In: *Proceedings of the IEEE international conference on image processing*, Barcelona, Spain, vol. 2, 11–24. IEEE.
97 Denes, L.J., Gottlieb, M.S., and Kaminsky, B. (1998). Acousto-optic tunable filters in imaging applications. *Opt. Eng.* 37 (4): 1262–1267.
98 Gupta, N., Dahmani, R., and Choy, S. (2002). Acousto-optic tunable filter based visible-to near-infrared spectropolarimetric imager. *Opt. Eng.* 41 (5): 1033–1038.
99 Wachman, E.S., Niu, W., and Frakas, D.L. (1996). Imaging acousto-optic tunable filter with 0.35-micrometer spatial resolution. *Appl. Opt.* 35 (25): 5220–5226.
100 Shnitser, P.I. and Agurok, I.P. (1997). Spectrally adaptive light filtering. In: *Proceedings of SPIE Conference on Photometric Engineering of Sources and Systems*, vol. 3140, 117–127. SPIE.
101 Vila-Frances, J. (2003). Design of the filtering and sensing part of the Smart-spectra camera. MSc thesis. University of Valencia.
102 Huber, L. and George, S.A. (1993). *Diode Array Detection in HPLC*. New York: Marcel Dekker, Inc.
103 Pasquini, C., Lu, J., Tran, C.D., and Smirnov, S. (1996). *Anal. Chim. Acta* 319: 315–324.
104 Fang, Z. (1993). *Flow Injection Separation and Preconcentration*. New York: VCH Press.
105 Ruzicka, J. and Hansen, E.H. (1988). *Flow Injection Analysis*, 2e. New York: Wiley.
106 Anreus, E.L., Garrigues, S., and Guardia, M. (1995). *Fresenius J. Anal. Chem.* 351: 724–728.
107 Garrigues, S., Gallignani, M., and Guardia, M. (1993). *Anal. Chim. Acta* 351: 259–264.
108 McCord, T.B., Hansen, G.B., and Fanale, F.P. (1998). Salt on Europa's Surface Detected by Galileo's Near Infrared Mapping Spectrometer. *Science* 280, 1242.
109 Zhang, H., Wang, X.L., Soos, J.I., and Crisp, J.A. (1995). Design of a miniature solid state NIR spectrometer. *Proc. SPIE* 2475: 376–383.
110 Korablev, O., Bertaux, J.L., Dimarellis, E. et al. (2002). An AOTF-based spectrometer for MARS atmosphere sounding. *Proc. SPIE* 4818: 261–271.
111 Korablev, O., Bertaux, J.L., Grigoriev, A. et al. (2002). An AOTF-based spectrometer for the studies of Mars atmosphere for MARS express ESA mission. *Adv. Space Res.* 29 (2): 143–150.
112 Chance Carter, J., Scaffidi, J., Burnett, S. et al. (2005). Stand-off Raman Detection using dispersive and tunable filter based systems. *Spectrochim. Acta A* 61: 2288–2298.
113 Korablev, O., Bertaux, J.-L., Fedorova, A. et al. (2006). SPICAM IR acousto-optic spectrometer experiment on Mars Express. *J. Geophys. Res.* 111, E09S03, 17 pp.

114 Korablev, O., Fedorova, A., Bertaux, J.-L. et al. (2012). SPICAV IR acousto-optic spectrometer experiment on Venus Express. *Planet. Space Sci.* 65: 38–57.

115 Epikhin, V. M., F. L. Vizen, and V. I. Pustovoit, Acousto-optic filter, Patent of Russia 1247816, 22 pp., *Byull. Otkrytiya, Izobreteniya, Moscow*, 22 Oct. (1984).

116 Fedorova, A., Korablev, O., Bertaux, J.-L. et al. (2006). Mars water vapor abundance from SPICAM IR spectrometer: Seasonal and geographic distributions. *J. Geophys. Res.* 111 (E9): 1–18.

117 Fowler, V.J. and Schlafer, J. (1966). A survey of laser beam deflection techniques. *Proc. IEEE* 54: 1437–1444.

118 Nelson, T.J. (1964). Digital light deflection. *Bell Syst. Tech. J.* 43: 821–845.

119 Pinnow, D.A., Van Uitert, L.G., Warner, A.W., and Bonner, W.A. (1969). Lead molybdate: A melt-grown crystal with a high figure of merit for acousto-optic device applications. *Appl. Phys. Lett.* 15: 83–86.

120 Coquin, G.A., Pinnow, D.A., and Warner, A.W. (1971). Physical properties of lead molybdate relevant to acousto-optic device applications. *J. Appl. Phys.* 42: 2162–2168.

121 Bonner, W.A. and Zydzik, G.J. (1970). Growth of single crystal lead molybdate for acousto-optic applications. *J. Cryst. Growth* 7: 65–68.

122 Liebertz, J. (1969). Einkristallziichtung von Paratellurit (TeO_2). *Krist. Tech.* 4: 221–225.

123 Arlt, G. and Schweppe, H. (1968). Paratellurite, a new piezoelectric material. *Solid State Commun.* 6: 783–784.

124 Uchida, N. (1969). Elastic and photoelastic properties of TeO_2 single crystal. *J. Appl. Phys.* 40: 4692–4695.

125 Ohmachi, Y. and Uchida, N. (1970). Temperature dependence of elastic, dielectric and piezoelectric constants in TeO_2 single crystals. *J. Appl. Phys.* 41: 2307–2311.

126 Ohmachi, Y., Uchida, N., and Niizeki, N. (1972). Acoustic wave propagation in TeO_2 single crystal. *J. Acoust. Soc. Am.* 51: 164–168.

127 Uchida, N. (1972). Acoustic attenuation in TeO_2. *J. Appl. Phys. (Commun.)* 43: 2915–2917.

128 (1971). Optical properties of single-crystal paratellurite (TeO_2). *Phys. Rev.* B4: 3736–3745.

129 Miyazawa, S. and Iwasaki, H. (1970). Single crystal growth of paratellurite TeO_2. *Jpn. J. Appl. Phys.* 9: 441–445.

130 Bonner, W.A., Singh, S., Van Uitert, L.G., and Warner, A.W. (1972). High quality tellurium dioxide for acousto-optic and non-linear application. *J. Electron. Mater.* 1: 155–165.

131 Zaitsev, A.K. and Kludzin, V.V. (2003). Subcollinear acousto-optic tunable filter based on the medium with a strong acoustic anisotropy. *Opt. Commun.* 219: 277–283.

132 Gupta, N., Voloshinov, V.B., Knyazev, G.A., and Kulakova, L.A. (2012). Tunable wide-angle acousto-optic filter in single-crystal tellurium. *J. Opt.* 14: 035502 (9pp).

133 Kirkby, P.A., Naga Srinivas Nadella, K.M., and Silver, R.A. (2010). A compact acousto-optic lens for 2D and 3D femtosecond based 2-photon microscopy. *Opt. Express* 18 (13): 13721–13745.

134 Reddy, G.D., Kelleher, K., Fink, R., and Saggau, P. (2008). Three-dimensional random access multiphoton microscopy for functional imaging of neuronal activity. *Nat. Neurosci.* 11 (6): 713–720.

135 Reddy, D. and Saggau, P. (2007). Fast three-dimensional random access multi-photon microscopy for functional recording of neuronal activity. *Proc. SPIE* 6630, 66301A 66301–66308.

136 Kremer, Y., Léger, J.-F., Lapole, R. et al. (2008). *Opt. Express* 16 (14): 10066–10076.

137 Dixon, R.W. and Cohen, M.G. (1966). A new technique for measuring magnitudes of photoelastic tensors and its application to lithium niobate. *Appl. Phys. Lett.* 8: 205–201.

138 Wen, C.P. and Mayo, R.F. (1966). Acoustic attenuation of a single domain lithium niobate crystal at microwave frequencies. *Appl. Phys. Lett.* 9: 135–136.

139 Grace, M.I., Kedzie, R.W., Kestigian, M., and Smith, A.B. (1966). Elastic attenuation in lithium niobate. *Appl. Phys. Lett.* 9: 155–156.

140 Spencer, E.G. and Lenzo, P.V. (1967). Temperature dependence of microwave elastic losses in $LiNbO_3$ and $LiTaO_3$. *J. Appl. Phys.* 38: 423–424.

141 Nakazawa, T., Taniguchi, S., and Seino, M. (1999). Ti:$LiNbO_3$ Acousto-optic Tunable Filter (AOTF). *FUJITSU Sci. Tech. J.* 35 (1): 107–112.

142 Coates, A.J., Jaumann, R., Griffiths, A.D. et al. (2017). The PanCam Instrument for the ExoMars Rover. *Astrobiology* 17 (6-7): 511–541.

143 Korablev, O., Ivanov, A., Fedorova, A. et al. (2015). Development of a mast or robotic arm-mounted infrared AOTF spectrometer for surface Moon and Mars probes. *Proc SPIE* 9608: 07–10.

144 Zelenyi, L., Mitrofanov, I., Petrukovich, A., Khartov, V., Martynov, M., and Lukianchikov, A. (2014) Russian plans for lunar investigations. Stage 1. In: European Planet Science Congress 2014, vol. 9 EPSC2014-702 [abstract].

145 Korablev, O.I., Dobrolensky, Y., Evdokimova, N. et al. (2017). Infrared spectrometer for ExoMars: a mast-mounted instrument for the rover. *Astrobiology* 17 (6–7): 542–564.

146 Bibring, J.-P., Hamm, V., Pilorget, C. et al. (2017). The MicrOmega investigation onboard ExoMars. *Astrobiology* 17: 621–626.

147 He, Z.P., Wang, B.Y., Lu, G. et al. (2014). Visible and near-infrared imaging spectrometer and its preliminary results from the Chang'E 3 project. *Rev. Sci. Instrum.* 85: 083104.

148 Nevejans, D., Neefs, E., Van Ransbeeck, E. et al. (2006). Compact high-resolution spaceborne echelle grating spectrometer with acousto-optical tunable filter based order sorting for the infrared domain from 2.2 to 4.3 μm. *Applied Optics* 45: 5191–5206.

149 Korablev, O.I., Kalinnikov, Y.K., Titov, A.Y. et al. (2011). The RUSALKA device for measuring the carbon dioxide and methane concentration in the atmosphere from on board the International Space Station. *J. Optical Technol.* 78: 317–327.

150 Korablev, O., Trokhimovsky, A., Grigoriev, A.V. et al. (2014). Three infrared spectrometers, an atmospheric chemistry suite for the ExoMars 2016 trace gas orbiter. *J. Appl. Remote Sensing* 8: 4983.

151 Neefs, E., Vandaele, A.C., Drummond, R. et al. (2015). NOMAD spectrometer on the ExoMarstracegasorbitermission:part 1 – design, manufacturing and testing of the infrared channels. *Applied Optics* 54: 8494.

152 Pilorget, C. and Bibring, J.P. (2013). NIR reflectance hyperspectral microscopy for planetary science: application to the Micr Omega instrument. *Planet Space Sci* 76: 42–52.

153 Bibring, J.-P., Hamm, V., Pilorget, C. et al. (2017). The Micr Omega investigation onboard ExoMars. *Astrobiology* 17: 621–626.

154 Kinch, K.M., Sohl-Dickstein, J., Bell, J.F. et al. (2007). Dust deposition on the Mars Exploration Rover Panoramic Camera (Pancam) calibration targets. In: *J Geophys Res 112:E06S03.*

155 Kinch, K.M., Bell, J.F., Goetz, W. et al. (2015). Dust deposition on the decks of the Mars Exploration Rovers: 10 years of dust dynamics on the Panoramic Camera calibration targets. *Earth Space Science* 2: 144–172.

156 Leroi, V., Bibring, J.P., and Berthe, M. (2009). Micromega/IR: design and status of a near-infrared spectral microscope for in situ analysis of Mars samples. *Planet Space Sci* 57: 1068–1075.

157 Mantsevich, S.N., Korablev, O.I., Kalinnikov, Y.K. et al. (2015). Wide-aperture TeO_2 AOTF at low temperatures: operation and survival. *Ultrasonics* 59: 50–58.

158 Young, E.H. and Yao, S.K. (1981). Design considerations for acousto-optic devices. In: *Proc. IEEE,* 54–64.

4

Magneto-optics

4.1 Introduction

A magneto-optic (MO) effect is any one of a number of phenomena in which an electromagnetic wave propagates through a medium that has been altered by the presence of a quasistatic magnetic field. In such a material, which is also called gyrotropic or gyromagnetic, left- and right-rotating elliptical polarizations can propagate at different speeds, leading to a number of important phenomena. When light is transmitted through a layer of magneto-optic material, the result is called the Faraday effect (FE): the plane of polarization can be rotated, forming a Faraday rotator. The results of reflection from a magneto-optic material are known as the magneto-optic Kerr effect (MOKE).

In general, magneto-optic effects break time reversal symmetry locally (i.e. when only the propagation of light and not the source of the magnetic field is considered) as well as Lorentz reciprocity, which is a necessary condition to construct devices such as optical isolators (through which light passes in one direction but not the other).

4.1.1 Gyrotropic Permittivity

In particular, in a magneto-optic material the presence of a magnetic field (either externally applied or because the material itself is ferromagnetic) can cause a change in the permittivity tensor ε of the material. The ε becomes anisotropic, a 3×3 matrix, with complex off-diagonal components, depending of course on the frequency ω of incident light. If the absorption losses can be neglected, ε is a Hermitian matrix. The resulting principal axes become complex as well, corresponding to elliptically polarized light where left- and right-rotating polarizations can travel at different speeds (analogous to birefringence).

More specifically, for the case where absorption losses can be neglected, the most general form of Hermitian ε is

$$\varepsilon = \begin{pmatrix} \varepsilon'_{xx} & \varepsilon'_{xy} + ig_z & \varepsilon'_{xz} - ig_y \\ \varepsilon'_{xy} - ig_z & \varepsilon'_{yy} & \varepsilon'_{yz} + ig_x \\ \varepsilon'_{xz} + ig_y & \varepsilon'_{yz} - ig_x & \varepsilon'_{zz} \end{pmatrix} \tag{4.1}$$

Crystal Optics: Properties and Applications, First Edition. Ashim Kumar Bain.

Or, equivalently, the relationship between the displacement field $\boldsymbol{D}$ and the electric field $\boldsymbol{E}$ is

$$\boldsymbol{D} = \varepsilon \boldsymbol{E} = \varepsilon' \boldsymbol{E} + i\boldsymbol{E} \times \boldsymbol{g} \tag{4.2}$$

where ε' is a real symmetric matrix and $\boldsymbol{g} = (g_x, g_y, g_z)$ is a real pseudo vector called the gyration vector, whose magnitude is generally small compared to the eigenvalues of ε'. The direction of $\boldsymbol{g}$ is called the axis of gyration of the material. To first order, $\boldsymbol{g}$ is proportional to the applied magnetic field:

$$\boldsymbol{g} = \varepsilon_0 \chi^{(m)} \boldsymbol{H} \tag{4.3}$$

where $\chi^{(m)}$ is the magneto-optical susceptibility (a scalar in isotropic media, but more generally a tensor). If this susceptibility itself depends upon the electric field, one can obtain a nonlinear optical effect of magneto-optical parametric generation (somewhat analogous to a Pockels effect whose strength is controlled by the applied magnetic field).

The simplest case to analyze is the one in which g is a principal axis (eigenvector) of ε', and the other two eigenvalues of ε' are identical. Then, if we let $\boldsymbol{g}$ lie in the z- direction for simplicity, the ε tensor simplifies to the form:

$$\varepsilon = \begin{pmatrix} \varepsilon_1 & +ig_z & 0 \\ -ig_z & \varepsilon_1 & 0 \\ 0 & 0 & \varepsilon_2 \end{pmatrix} \tag{4.4}$$

Most commonly, one considers light propagating in the z-direction (parallel to $\boldsymbol{g}$). In this case, the solutions are elliptically polarized electromagnetic waves with phase velocities $1/\sqrt{\mu(\varepsilon_1 \pm g_z)}$(where μ is the magnetic permeability). This difference in phase velocities leads to the Faraday effect.

For light propagating purely perpendicular to the axis of gyration, the properties are known as the Cotton–Mouton effect and used for a circulator.

4.1.2 Kerr Rotation and Kerr Ellipticity

Kerr rotation and Kerr ellipticity are changes in the polarization of the incident light that comes in contact with a gyromagnetic material. Kerr rotation is a rotation in the angle of transmitted light, and Kerr ellipticity is the ratio of the major to minor axis of the ellipse traced out by elliptically polarized light on the plane through which it propagates. Changes in the orientation of polarized incident light can be quantified using these two properties.

According to classical physics, the speed of light varies with the permittivity of a material:

$$v_p = \frac{1}{\sqrt{\varepsilon\mu}} \tag{4.5}$$

where v_p is the velocity of light through the material, ε is the material permittivity, and μ is the material permeability. Because the permittivity is anisotropic, polarized light of different orientations will travel at different speeds.

This can be better understood if we consider a wave of light that is circularly polarized (Figure 4.1). If this wave interacts with a material at which the horizontal component (light grey sinusoid) travels at a different speed than the vertical

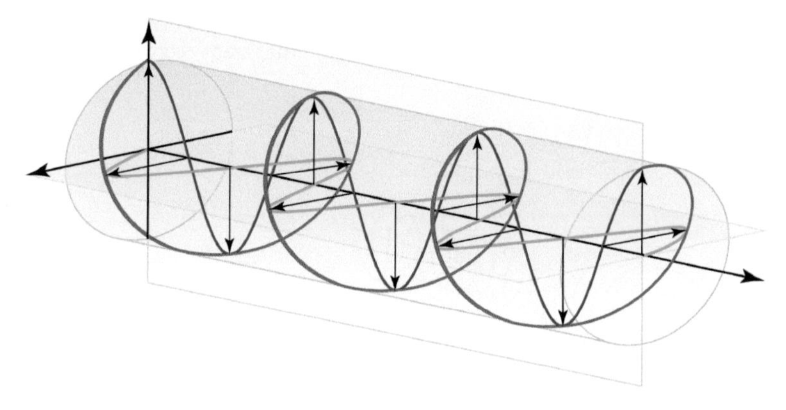

Figure 4.1 Circular polarized light.

component (grey sinusoid), the two components will fall out of the 90° phase difference (required for circular polarization) changing the Kerr ellipticity.

A change in Kerr rotation is most easily recognized in linearly polarized light, which can be separated into two circularly polarized components: left-handed circular polarized (LCP) light and right-handed circular polarized (RCP) light. The anisotropy of the magneto-optic material permittivity causes a difference in the speed of LCP and RCP light, which will cause a change in the angle of polarized light. Materials that exhibit this property are said to be birefringent.

From this rotation, we can calculate the difference in orthogonal velocity components, find the anisotropic permittivity, find the gyration vector, and calculate the applied magnetic field H.

4.2 Mode of Interaction

Magneto-optic effects are divided into three main categories related to transmission, reflection, and absorption of light by magnetic materials.

4.2.1 Transmission Mode

The first category deals with the interaction of the internal magnetization of matter with the electromagnetic wave propagating through it. When linearly polarized light travels through a magnetized sample, the plane of polarization is rotated.

The magnetization in the sample might be due to the presence of an external magnetic field, parallel or perpendicular to the optical path. The first of the configurations is called the Faraday effect and is shown in Figure 4.2. Phenomenologically, this effect is related to the normal Zeeman effect. The magnetic field causes a splitting of the energy levels, thus shifting the original resonant frequency ν_0 of an absorption line to slightly higher and lower frequencies ν_1 and ν_2. A double dispersion effect is generated. LCP light is dispersed at ν_1 and RCP light is dispersed at ν_2 as shown in Figure 4.3 [1].

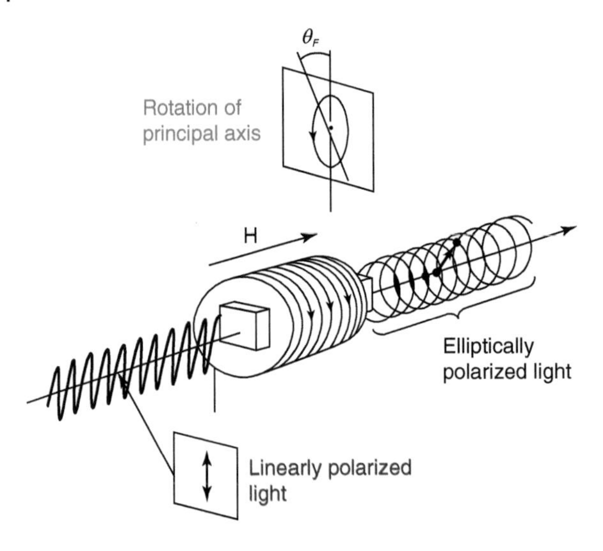

Figure 4.2 Polarized light in the Faraday configuration.

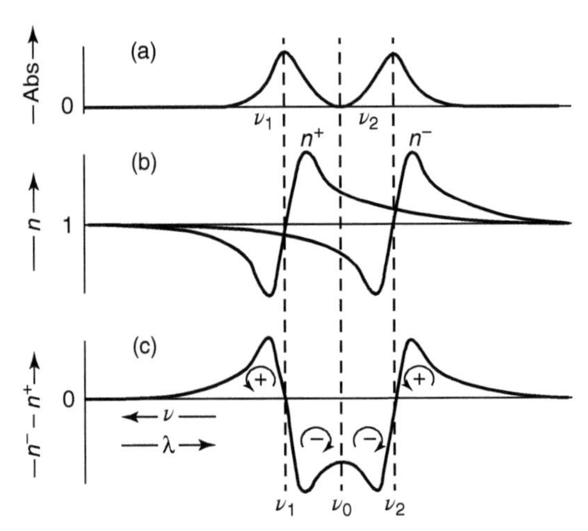

Figure 4.3 Magnetic splitting of an absorption line causing Faraday effect [1]. (a) Absorption, (b) dispersion, and (c) rotation.

This causes the LCP and RCP components to have different indices of refraction. Thus, a retardation of one component relative to the other gives rise to a phase difference between them. When emerging, these components will combine to give an elliptic polarization whose major axis is rotated relative to the original direction of polarization. This angle is called Faraday's rotation θ_F.

The Faraday rotation is directly proportional to the magnetization M and to the length of the light path through the material l. The proportionality constant is called the Verdet constant V.

$$\theta_F = VMl \tag{4.6}$$

Because of the linear dependence of θ_F on the internal magnetization, or in most cases on H (the external magnetic field), the Faraday effect is called an "odd" effect. The Faraday rotation θ_F varies from a few degrees per centimeter to 10^6 deg/cm. Sometimes, the Faraday effect is referred to as "circular magnetic birefringence" (CMB).

The birefringence Δn is related to the angle of rotation by the equation

$$\theta_F = \frac{(n^+ - n^-)\pi d}{\lambda} = \frac{(\Delta n)\pi d}{\lambda} \tag{4.7}$$

where n^+ and n^- are two refraction indices, d is the thickness of the sample, and λ is the wavelength.

The intensity variation associated with the birefringence is

$$I = I_o \sin^2\theta_F \tag{4.8}$$

When the magnetic field H is directed perpendicular to the optical path and the incident light is linearly polarized, the interaction with the material is different. This configuration is called the Voigt effect. It is a magneto-optical phenomenon that rotates and elliptizes a linearly polarized light transmitted into an optically active medium. Instead of viewing the linearly polarized light as LCP or RCP, the electric field vector of the light is now parallel or perpendicular to the magnetic field as shown in Figure 4.4.

The parallel component is in no way affected by the field, while the two perpendicular components are oppositely interacting with it. This interaction implies, again, a difference in the refracting index ($\pm\Delta n$) and thus a birefringence for the two perpendicular components. This is a linear magnetic birefringence (LMB) and is independent of the field's orientation (axial). It is proportional to the square of the magnetic field H^2 (or the magnetization M^2) and is called an "even" effect. Although misleading, it is often referred to as "Cotton–Mouton" effect. The Voigt effect is very small in para- and diamagnetic materials. This is not necessarily the case in ferromagnets. The angle of rotation θ_V is related to the birefringence by

$$\theta_V = \frac{(\Delta n)2\pi d}{\lambda} \tag{4.9}$$

The phenomenological theory of the Faraday and Voigt effects is given by Freiser [2]. These two magneto-optic effects related to the transmission mode have been used extensively to get insight on the magnetic state of materials. They are also used in a large variety of optical applications. But, evidently, their use is limited by the absorption of radiation by the media in the spectral range of interest. The significant factor is the ratio between θ_F, the specific Faraday rotation, and α, the absorption coefficient. These two combine to give the "figure

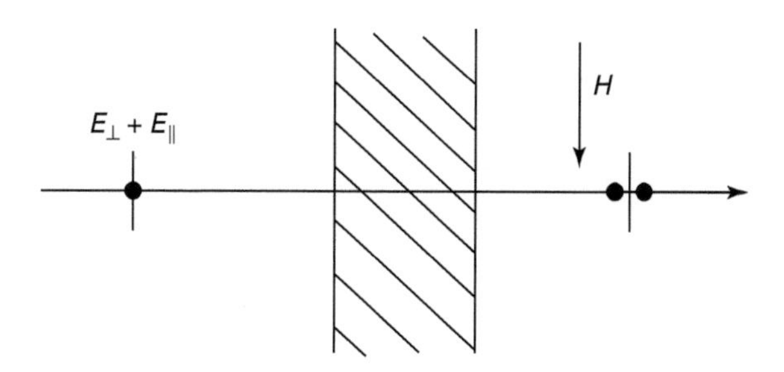

Figure 4.4 Polarized light in the Voigt configuration.

of merit" of the material defined as

$$FOM\left(\frac{\text{deg}}{\text{dB}}\right)=\frac{\theta_F\left(\frac{\text{deg}}{\text{cm}}\right)}{\alpha\left(\frac{\text{dB}}{\text{cm}}\right)} \tag{4.10}$$

A third transmission mode magnetic effect, one that is not related to the Zeeman effect, i.e. to an electronic dispersion, is the genuine Cotton–Mouton effect. When a linearly polarized wave propagates perpendicular to the magnetic field (e.g. in a magnetized plasma), it can become elliptized. Because a linearly polarized wave is some combination of in-phase extraordinary and ordinary modes, and because extraordinary and ordinary waves propagate with different phase velocities, this causes elliptization of the emerging beam. As the waves propagate, the phase difference (δ) between E_X and E_O increases. The configuration is similar to the Voigt configuration and thus displays an even effect.

$$\theta_{CM} \propto M^2 \tag{4.11}$$

where θ_{CM} is the rotation of the main axis of the elliptically polarized transmitted light. This effect is usually found in colloidal solutions of magnetic particles, and as M might be large θ_{CM} is too. It is the equivalent of the Kerr electro-optic effect.

4.2.2 Reflection Mode

The second category of magneto-optic effects deals with reflected light from the surface of a magnetized material. It is labeled in all its different configurations as the Kerr effect. There are three different configurations of the light with respect to the magnetization in the sample: the polar, longitudinal, and equatorial modes (Figure 4.5).

For thin (ferromagnetic) films, the polar configuration usually requires strong magnetic fields to overcome the large demagnetization factor or some other kind of induced anisotropy.

The analysis of the reflection mode is complicated because of the oblique incidence involved. Usually, one would expect a mixture of all the possible polarizations in each direction. However, a simplified computation based on the electromagnetic boundary conditions at the reflecting surface for polar configurations at normal incidence gives some insight into the effect. The Fresnel coefficient for this case, representing the reflected amplitude is

$$r=\frac{n-1}{n+1} \tag{4.12}$$

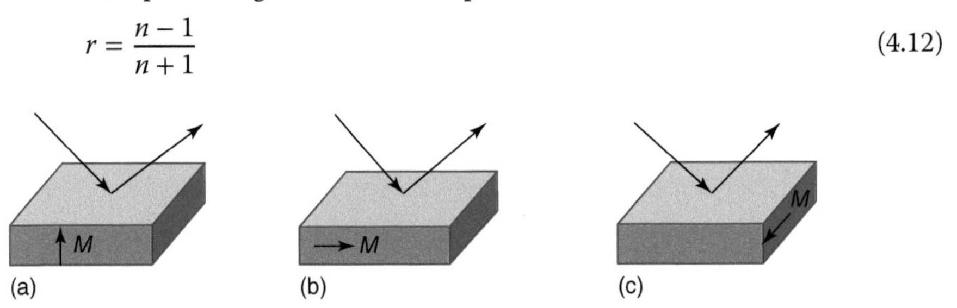

Figure 4.5 (a) Polar, (b) longitudinal, and (c) equatorial configuration of light with respect to magnetization of the sample.

The dependence on n, the refraction index, hints that as in the transmission mode, the difference in n for RCP or LCP light will create a difference in amplitude and a phase shift between the two.

If the incident light is linearly polarized, the reflected light will be elliptically polarized, because the two circular components, composing the incident beam, will no longer be equal in amplitude or in phase as happens in the Faraday effect. The main axis of the ellipse will be rotated by the angle θ_K relative to the original polarization. This angle, θ_K, is very small in order of magnitude of minutes.

A special situation, however, exists for one case. In the equatorial configuration, the calculated components of the reflected lights show that only the component parallel to the field is affected by the magnetization and its amplitude is linearly proportional to it ($R_{\|} \propto M$), unlike the quadratic Voigt effect, which is similar in configuration. As a result of this uniaxiality, switching the magnetization by 180° affects directly the intensity of the reflected light. This change can be detected without the use of an analyzer, which indicates considerable increase in efficiency. A typical value for θ_K in the visible range is 10 minutes. This has been reported for nickel films at saturation. The smallness of the Kerr rotation θ_K has been overcome to a certain extent by different kinds of multi-reflection methods [3, 4], thus increasing the signal by a factor of 20.

It is also worthwhile to note that ferromagnetic alloys such as permalloy or GdFe have a greater Kerr rotation than the ferromagnetic transition metals. The equatorial Kerr effect is a principal tool in the investigation of the magnetic domain structure and the related magnetic properties of ferromagnetic materials. It also is used for computer memory read-out applications [5].

The angle of rotation θ_F, θ_V, or θ_K can be measured by a static method with a Babinet compensator or by a dynamic method in which the magnetization is switched periodically and the intensity variations are measured by means of a lock-in amplifier [6, 7].

4.2.3 The Absorption Mode

The last magneto-optic effect to be mentioned is the circular magnetic dichroism (CMD). This effect is the difference in the absorption coefficient for RCP or LCP light. This difference changes slightly with the spectrum of absorption of a sample magnetized in the beam direction. As in the transmission configuration, there exists also a linear magnetic dichroism (LMD), equivalent to the Voigt effect configuration where the magnetic field is transverse to the light.

The CMD and LMD can give an insight on band structure of crystals and magnetically induced transitions [8, 9]. It has not found any other applications.

4.3 Magneto-optic Materials Classification

Materials used in magneto-optic devices can be grouped according to their magnetic ordering. There are three spontaneous kinds of magnetic order: ferromagnetic, anti-ferromagnetic, and ferrimagnetic. All three are characterized

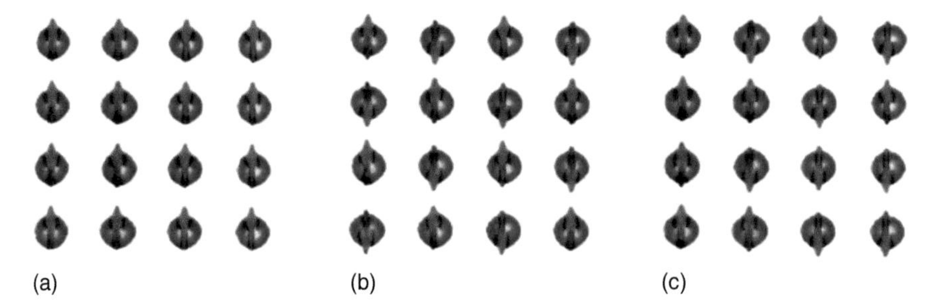

Figure 4.6 (a) Parallel alignment of magnetic moments of atoms in ferromagnetic materials. (b) Antiparallel alignment of magnetic moments of atoms in antiferromagnetic materials. (c) Mixture of parallel and antiparallel alignments of magnetic moments of atoms in ferrimagnetic materials.

by a nearly regular magnetic domain pattern, not directly related to the crystal structure. The size of the domains varies from 20 to 0.02 μm.

In the ferromagnetic state, each atom possesses an unbalanced magnetic moment (related to the spin of the unfilled 3d band). Owing to the strong exchange interaction between adjacent atoms, these magnetic moments are all parallel giving rise to a strong resultant magnetization (Figure 4.6a). This, in turn, generates a strong demagnetization effect that favors energetically a multidomain structure. Within one domain, all the magnetic dipoles are parallel. When applying an external magnetic field, the domains whose magnetization is parallel to the field direction grow on account of the others, thus gradually forming an overall single domain [10]. As ferromagnetic materials are heated thermal agitation of the atoms occurs and the degree of alignment of the atomic magnetic moments decreases and hence the saturation magnetization also decreases. Eventually, the thermal agitation becomes so pronounced that the material becomes paramagnetic; the temperature of this transition is the Curie temperature, T_c (Fe: $T_c = 770\,°C$, Co: $T_c = 1131\,°C$, and Ni: $T_c = 358\,°C$). All the magnetization-related properties, such as the Faraday rotation, vanish above T_c. The 3d transition group metals such as iron, cobalt, nickel, manganese, chromium, and most of their alloys are typical ferromagnets.

In antiferromagnets, each individual atom or ion also possesses an unbalanced magnetic moment; but these moments are in an antiparallel arrangement and cancel the net magnetization completely (Figure 4.6b). However, there is no demagnetization effect, and the antiferromagnetic order is also broken into a domain pattern [11]. This pattern is not controllable by an external magnetic field but can be affected by stresses or other induced anisotropies [12]. The antiferromagnetic ordering has slight temperature dependence but undergoes an abrupt change at the Neel temperature (T_N), above which the magnetic moments become randomly oriented. Typical antiferromagnetic substances are the oxides of transition metals such as FeO, NiO, and CoO as well as the fluorides and chlorides such as FeF_2, NiF_2, CoF_2 (the rutile crystals), $FeCl_2$, $CoCl_2$, and $NiCl_2$.

The third spontaneous magnetic ordering is one typical of the ferrites. Within these materials, the exchange interactions lead to parallel alignment of atoms in some of the crystal sites and antiparallel alignment of others (Figure 4.6c). The material breaks down into magnetic domains, similarly to a ferromagnetic material and the magnetic behavior is also very similar, although ferrimagnetic materials usually have lower saturation magnetizations. The number of ions and the magnitudes of their magnetic moment are different and thus a net magnetization occurs. For example, in barium ferrite ($BaO{\cdot}6Fe_2O_3$), the unit cell contains 64 ions of which the barium and oxygen ions have no magnetic moment, $16Fe^{3+}$ ions have moments aligned parallel, and $8Fe^{3+}$ aligned antiparallel giving a net magnetization parallel to the applied field, but with a relatively low magnitude as only 1/8 of the ions contribute to the magnetization of the material.

As in the ferromagnets, a field controlled domain pattern is detected; again, at a critical temperature the whole magnetic order is destroyed giving rise to a sharp transition in all the properties related to the magnetization. Typical ferrimagnets are $MnFe_2O_4$, $CoFe_2O_4$, F_2O_3 (spinels) and $Y_3Fe_5O_{12}$ (YIG), and $Gd_3Fe_5O_{12}$ (garnets).

4.3.1 Ferromagnetic Metals and Alloys

The transition group metals and some of their binary alloys have the largest known Faraday rotation, but at the same time they are highly absorbent. In the bulk, these materials are suitable only for Kerr effect measurements. As thin films, they can be transmitting in the visible and near visible regions and are now widely being used in magneto-optic devices.

Recently, Osada et al. [13] have demonstrated a fabrication procedure for spin-electronic devices by using two-dimensional titania nanosheets. Magneto-optical Kerr measurements demonstrate that Co-substituted titania nanosheets ($Ti_{0.8}Co_{0.2}O_2$) act as nanoscale ferromagnetic layers at room temperature, and their multilayer assemblies exhibit robust magnetic circular dichroism (MCD) (10^4–10^5 deg/cm) near the absorption edge at 260 nm, being the shortest operating wavelength attained so far. The availability of ferromagnetic nanostructures and their molecule-level assembly allow the rational design and construction of high-efficiency magneto-optical devices by forming superlattices.

Jia et al. [14] has successfully fabricated a hetero structured multilayer film of two different nanocrystals by layer-by-layer stacking of $Ti_{0.8}Co_{0.2}O_2$ nanosheet and Fe_3O_4 nanoparticle films. UV–vis spectroscopy and atomic force microscope (AFM) observation confirmed the successful alternating deposition in the multilayer buildup process. The average thickness of both $Ti_{0.8}Co_{0.2}O_2$ nanosheet and Fe_3O_4 nanoparticle layers was determined to be about 1.4–1.7 and 5 nm, which was in good agreement with transverse electro magnetic (TEM) results. MOKE measurements demonstrated that the hetero assemblies exhibit gigantic MCD (2×10^4 deg/cm) at 320–360 nm, deriving from strong interlayer $[Co^{2+}]t_{2g}$–$[Fe^{3+}]e_g$ d–d charge transfer, which was confirmed by X-ray photoelectron spectroscopy. Their structure-dependent MCD showed high potential in the rational design and construction of high-efficiency magneto-optical devices.

With the aim of optimizing ferromagnetic metals for use in semiconductor optical isolators, Shimizu et al. [15] characterized the transverse MOKE in the ferromagnetic metals Fe, Co, and $Fe_{50}Co_{50}$ at the telecommunication wavelength of 1550 nm. $Fe_{50}Co_{50}$ showed the largest transverse Kerr effect. They compared the experimental results with theoretical calculations. From this comparison, $Fe_{50}Co_{50}$ is the most suitable ferromagnetic metal among the three materials for semiconductor optical isolators operating at 1550 nm.

A 1.5-μm nonreciprocal-loss waveguide optical isolator having improved transverse-magnetic-mode (TM-mode) isolation ratio was developed by Amemiya et al. [16]. The device consisted of an InGaAlAs/InP semiconductor optical amplifier (SOA) waveguide covered with a ferromagnetic epitaxial MnSb layer. Because of the high Curie temperature ($T_c = 314\,°C$) and strong magneto-optical effect of MnSb, a nonreciprocal propagation of 11–12 dB/mm has been obtained at least up to 70 °C.

Pirnstill et al. [17] have developed a method to modulate the plane of polarized light through the use of a high permeability ferrite core design. A proof-of-principle, optical Faraday effect device has been constructed and tested. Magnetic fields were generated to provide up to 1 degree of rotation at frequencies of direct current up to 10 kHz using a terbium gallium garnet crystal rod.

Recently, ferromagnetic semiconductors and magnetic fluids (MFs) have been the major focus of magneto-optic research. Some important research works on these materials are described here.

4.3.1.1 Ferromagnetic Semiconductor

The discovery of ferromagnetism in Mn-doped III–V semiconductors [18, 19], such as InMnAs and GaMnAs, has paved a path to new device capabilities. These include the ability to change refractive properties using magnetic fields, in addition to temperature and/or electric fields commonly used in advanced optical modules such as distributed feedback (DFB) semiconductor tunable lasers [20, 21]. Another possible advantage is that they may enable the monolithic integration of a laser and an isolator [22]. Furthermore, the possibility of manipulating magnetic properties in these materials [22], together with the recent demonstration of demagnetization on the time scale of femtoseconds [23], makes them attractive material systems for optical communication applications. Relatively low ferromagnetic phase transition temperatures limit their practical applications today, although a worldwide search for ferromagnetic semiconductors with higher Curie temperatures is underway [24].

In 2009, Umnov et al. [25] have discussed the potential of (III, Mn)V ferromagnetic semiconductors as novel optoelectronic materials based on the MOKE. These materials have novel properties that make them attractive candidates for such applications. They performed measurements of magneto-optical Kerr angles for very thin (about 10 nm) films of these materials. Two $In_{1-x}Mn_xAs$ samples with different magnetization easy axis directions were investigated, which showed distinctly different magnetic field dependences. The Kerr angle value was strongly dependent on both the wavelength and temperature, reflecting the fact that Kerr spectroscopy probes not only the magnetization but also the band structure of the system. Additionally, they proposed a scheme for a polarization-base switch with a method for increasing the spatial separation between the two polarization beams. Finally, Umnov et al. considered the possible use of ultrafast optical modifications of ferromagnetism in these materials for ultrafast switching.

Xua et al. [26] have successfully grown single-crystal Fe films on InAs(100) and InAs(graded)/GaAs(100) substrates using molecular beam epitaxy. In situ MOKE and ex situ alternating gradient field magnetometry (AGFM) measurements show that the films have well-defined magnetic properties, and I–V measurements in the temperature range 2.5–304 K show that Fe forms an ohmic contact on InAs. This demonstrates that Fe/InAs is a very promising system for use in future magneto-electronic devices as it has both favorable magnetic and electrical properties. They also show that with careful substrate preparation and suitable growth conditions Fe/GaAs films do not exhibit a magnetically "dead" layer at the interface. A spin-polarized field effect transistor based on Fe/InAs/GaAs has been proposed, which could operate using either an external electric field or an external magnetic field.

The magneto-optical response of the ferromagnetic semiconductor $HgCdCr_2Se_4$ at terahertz (THz) frequencies is studied by Huisman et al. [27] using polarization-sensitive THz time-domain spectroscopy. It is shown that the polarization state of broadband terahertz pulses, with a spectrum spanning from 0.2 to 2.2 THz, changes as an even function of the magnetization of the medium. Analyzing the ellipticity and the rotation of the polarization of the THz radiation, Huisman et al. [27] showed that these effects originate from linear birefringence and dichroism, respectively, induced by the magnetic ordering. These effects are rather strong and reach 102 rad/m at an applied field of 1 kG, which saturates the magnetization of the sample. Their observation serves as a proof of principle showing strong effects of the magnetic order on the response of a medium to electric fields at terahertz frequencies. These experiments also suggest the feasibility of spin-dependent transport measurements on a sub-picosecond timescale.

In 2016, Tanaka and coworkers [28] showed high-temperature ferromagnetism in heavily Fe-doped ferromagnetic semiconductor $(Ga_{1-x}Fe_x)Sb$ ($x = 23\%$ and 25%) thin films grown by low-temperature molecular beam epitaxy. MCD spectroscopy and anomalous Hall effect measurements indicate intrinsic ferromagnetism of these samples. The Curie temperature reaches 300 and 340 K for $x = 23\%$ and 25%, respectively, which are the highest values reported so far in intrinsic III–V ferromagnetic semiconductors. Their results open a way to realize semiconductor spintronic devices operating at room temperature.

4.3.1.2 Magnetic Fluid

The MF, which is named as ferrofluid, is a highly stable magnetic colloidal suspension consisting of well-dispersed single-domain ferromagnetic nanoparticles in polar or nonpolar liquid carriers with the assistance of suitable surfactants [29]. A grinding process for ferrofluid was invented in 1963 by NASA's Steve Papell as a liquid rocket fuel that could be drawn toward a pump inlet in a weightless environment by applying a magnetic field [30]. Since then, methods for fabricating MF had rapid development, which led to a great improvement in the stability and homogeneity of MF. The Brownian motion of magnetic nanoparticles prevents them from sticking to each other due to van der Waals attraction. The typical size of the magnetic nanoparticles, such as Fe_3O_4, $MnFe_2O_4$, and $CoFe_2O_4$, is around 3–15 nm. The carriers are usually water, kerosene, and heptane.

MF is a promising function material that possesses diverse attractive magneto-optical effects, including refractive index tunability, birefringence, dichroism, field-dependent transmission property, Faraday rotation, magneto-chromatics, thermal–optical effects, and so on [31–36]. Taking advantage of these properties, a lot of MF-based photonics devices, especially optical fiber devices, were developed such as switches, modulators, magnetic field sensors, tunable filters, and wavelength division multiplexer (WDM) [37–49]. Among these applications, magneto-optical modulators or switches were most intensively investigated and heavily reported [50–52], which can be further classified into two classes according to the modulation methods: intensity-modulation type and phase-modulation type. In intensity-modulation-type modulators, the light intensity is absorbed by the MF or attenuated by the MF-assisted fiber directly. The field-dependent transmission property of the MF is responsible for the absorption of the light, which was applied to realize light modulation with an extinction ratio (ER) of 0.8 dB [44]. ER is one of the most important parameters of the modulator, which is defined as the ratio of maximum to minimum of the optical transmission power when it is operated as the "ON"- and "OFF"-state, respectively. Higher ER up to 10 dB can be achieved by increasing the thickness of the MF film at the cost of reducing modulation speed to hundreds of seconds [33]. Besides, MF-assisted fiber modulators were carried out by utilizing the MF as the refractive-index-tunable cladding of an etched multimode fiber (MMF) [43] or a tampered single-mode fiber (SMF) [46] with ERs of 0.5–0.9 dB. The refractive index of the MF, which is susceptible to temperature, should be carefully selected and controlled. In phase-modulation-type modulators, various optical interferometers are employed to convert the phase shift, which is induced by the magneto-optical effects of MF, to light intensity modulation indirectly. A Fabry–Perot interferometer (FPI) by utilizing the refractive index tunability of MF was reported as a tunable wavelength filter [39]. A fiber Mach–Zehnder interferometer (MZI) was demonstrated for optical signal processing with an ER of 0.55 dB [53]. A Sagnac interferometer (SI) with the MF film inserted into the fiber loop was also carried out to implement light intensity modulation with a greatly improved ER of 19.5 dB [42]. Most of current MF-based modulator applications are mainly based on its field-dependent transmission property or tunable refractive index. Birefringence is another important magneto-optical effect; however, few applications were reported.

A novel magneto-optical fiber modulator with a high ER based on a polarization interference structure is proposed [54]. A MF film and a section of a polarization maintaining fiber (PMF) are inserted into the structure to generate a suitable sinusoidal interference spectrum for light modulation. The MF film leads to a spectrum shift under external magnetic field due to its magnetically controllable birefringence, whereas the PMF is used to control the period of the interference spectrum. The light intensity was modulated into square wave and sine wave with high ERs of 38 and 29 dB, respectively. The PMF-assisted structure is very simple, which can be developed into an integrated fiber device, potentially.

A ferrofluid (portmanteau of ferromagnetic and fluid) is a liquid that becomes strongly magnetized in the presence of a magnetic field. The effective dielectric constant or refractive index of the MF is dependent on the dispersion state of the magnetic nanoparticles. Under zero magnetic field, the ferromagnetic nanoparticles disperse uniformly in the liquid carrier, and the MF as a whole exhibits homogeneity. If the MF is subjected to the external magnetic field, the homogeneous state is broken, and the situation changes. Agglomeration occurs to the ferromagnetic nanoparticles in the liquid carrier, which leads to phase separation between the nanoparticles and liquid carrier. Subsequently, the dielectric constant or refractive index of the MF is changed accordingly [51]. It has been shown that the higher the external magnetic field strength, the more significant the phase separation and then the greater the variation of the magnetic field. Accompanied by the phase separation, the nanoparticles agglomerate and tend to form magnetic columns or magnetic chains along the direction of magnetic field to some extent; hence, the MF becomes an anisotropic structure under the external magnetic field. In this instance, the MF can be treated like a traditional optical crystal with optical axes. Since the magnetic columns or chains of ferromagnetic nanoparticles are aligned along the magnetic field direction, the slow axis of the MF is also aligned along the magnetic field direction, and, in turn, the fast axis of the MF is perpendicular to the magnetic field direction. The MF exhibits different refractive indices on the fast axis and slow axis. This phenomenon is referred to as the birefringence effect of the MF, which follows a Langevin function as the magnetic field strength is increased [32, 50]. If the light is launched into the MF perpendicularly to the magnetic field direction, then it will fall into two orthogonal polarized light beams, namely, the extraordinary light (e-ray, whose electric vector orients parallel to the slow axis) and the ordinary light (o-ray, whose electric vector orients parallel to the fast axis). The two orthogonal light beams travel with different propagation constants due to the birefringence effect, and a phase difference is accumulated over the thickness of the MF. Meanwhile, the e-ray and o-ray also experience different absorption coefficients in the MF due to the anisotropic structure, and this phenomenon is referred to as dichroism of the MF. Usually, the birefringence effect and the dichroism effect coexist in the MF under external magnetic field.

Recently, the optical and MO properties of the MF in terahertz regime have been reported rarely. Shalaby et al. [55, 56] have demonstrated the Faraday rotation effect in a ferrofluid under the weak external magnetic field and its absorption loss is low in the terahertz regime. Fan et al. [57] investigated terahertz (THz) magneto-optical properties of a ferrofluid and a ferrofluid-filled photonic crystal (FFPC) by using the THz time-domain spectroscopy. A magneto-plasmon resonance splitting and an induced THz transparency phenomenon were demonstrated in the FFPC. Further investigation reveals that the induced transparency originates from the interference between magneto-plasmon modes in the hybrid magneto-optical system of FFPC, and THz modulation with a 40% intensity modulation depth can be realized in this induced transparency frequency band. This device structure and its tunability scheme will have great potential applications in THz filtering, modulation, and sensing.

The dielectric property and magneto-optical effects of ferrofluids have been investigated in the terahertz (THz) regime by using THz time-domain spectroscopy [58]. The experiment results show that the refractive index and absorption coefficient of ferrofluid for THz waves rise with the increase of nanoparticle concentration in the ferrofluid. Moreover, two different THz magneto-optical effects have been found with different external magnetic fields, the mechanisms of which have been theoretically explained well by microscopic structure induced refractive index change in the magnetization process and the transverse magneto-optical effect after the saturation magnetization, respectively. This work suggests that ferrofluid is a promising magneto-optical material in the THz regime, which has wide potential applications in THz functional devices for THz sensing, modulation, phase retardation, and polarization control.

4.3.2 Ferrimagnetic Compounds

A very important group of materials that display high quality of magneto-optic properties are the rare earth garnets. These are ferrimagnetic oxides have a quite complex cubic structure, similar to the natural mineral by the name: $Mg_3Al_2(SiO_4)_3$. In the basic and most investigated garnet, known by the name "yttrium iron garnet (YIG)," the silicon ions are replaced by three iron ions on one kind of lattice site (fourfold oxygen coordination) and two iron ions on a different kind of lattice site (sixfold oxygen coordination). The other metal ions are replaced by a nonmagnetic trivalent ion such as yttrium. A crystal of the composition $Y_3Fe_5O_{12}$ is thus synthesized [59]. This is a body-centered cubic lattice of complex structure, and contains 160 atoms in its unit cell, which is 12.4 Å in size (Figure 4.7).

Over 50 years ago, the discovery of yttrium iron garnet [60, 61] ($Y_3Fe_5O_{12}$ or YIG) led to remarkable advances in microwave technology. The combination of low magnetic damping, soft magnetization behavior, and a bandgap of 2.66 eV, making it a good insulator, qualify this material for microwave applications such as filters [62, 63] or sensors [64, 65]. Its low absorption in the optical and near-infrared wavelength region, combined with a magneto-optic Faraday

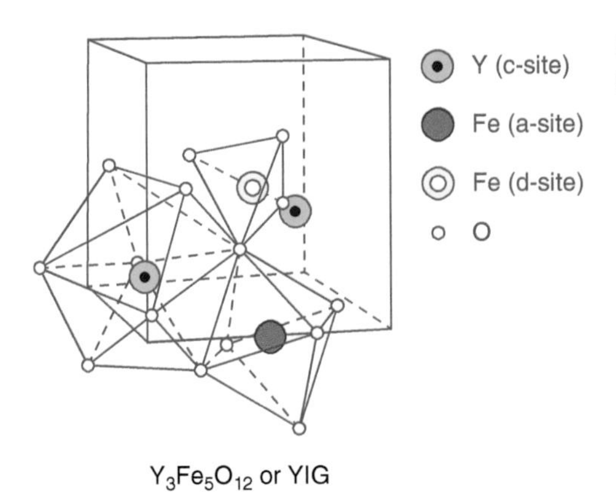

Figure 4.7 Structure of yttrium iron garnet crystal.

effect, render this material interesting for telecommunication devices such as magneto-optic isolators [66, 67]. For such applications, a higher FE is desirable, which allows for the miniaturization of the device [66, 68].

For miniaturized devices with low power requirements, integration of garnet with semiconductors is preferred. Although the silicon-on-insulator (SOI) platform does not lend itself readily to optical sources, it has been demonstrated that low optical losses and the current fabrication technologies allow for complex and flexible photonic integrated circuits (PICs), potentially compatible with CMOS (complementary metal–oxide–semiconductor). The first Si-core nonreciprocal device was proposed by Yokoi et al. where direct wafer bonding of Ce:YIG was used to integrate a garnet cladding on SOI guides [69]. The key to this structure was the low-index (SiO_2) lower cladding, which provided an asymmetry [70] that enhanced the interaction of the propagating mode with the top garnet cladding, leading to remarkable enhancement in the nonreciprocal phase shift (NRPS) [70, 71]. The first silicon-core isolator, using NRPS in a MZI configuration, had 21 dB optical isolation with 8 dB insertion loss at 1559 nm [66]. Recently, a similar Ce:YIG bonded MZI optical isolator employing a GaInAsP core was also demonstrated with an isolation ratio of 28.3 dB with 15 dB insertion loss at 1558 nm [72]. Optical circulators [73] can also be achieved using NRPS in Ce:YIG cladded semiconductor-core MZIs that have 3 dB 2×2 couplers on either side as opposed to the Y-couplers or 3×2 couplers used in optical isolators. These structures magnetize the branches of the MZI in opposite directions to obtain a "push/pull" effect, which allows shorter devices than those with phase shift in only one branch (Figure 4.8). One concern with MZI designs is the small bandwidth for wavelengths at which high isolation can be obtained. This can be mitigated if the reciprocal phase shifter can cancel the wavelength dependence of the NRPS [75].

Recently, Shoji et al. [76] have fabricated and demonstrated silicon-based optical isolator and circulator using direct bonding technology. A magneto-optic garnet Ce:YIG is directly bonded on a silicon waveguide without any adhesive. A magneto-optic effect induces NRPS in an MZI, which enables nonreciprocal propagations. So far, isolation of 30 dB is the highest record among waveguide optical isolators ever reported. Also, a four ports circulator with silicon waveguides has been demonstrated with maximum isolations of 15.3 and 9.3 dB in cross and bar ports, respectively, at a wavelength of 1531 nm [77].

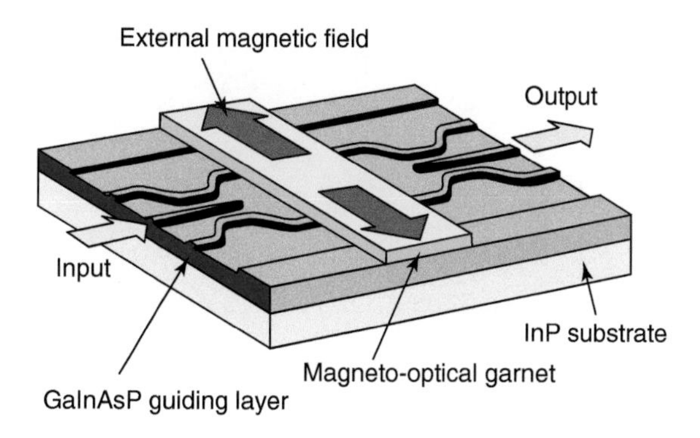

Figure 4.8 Schematic of semiconductor core isolator. Source: Yokoi and Mizumoto 1997 [74]. Reproduced with permission of IEEE.

Most of these NRPS structures using semiconductor cores were simulated and realized with direct wafer bonding. In addition, changes in the sign of Faraday rotation were obtained by inhomogeneously applied magnetic fields. However, Ghosh et al. [78, 79] demonstrated a silicon MZI optical isolator with an isolation of 25 dB. The advantage of this device is that it operates in a unidirectional magnetostatic field. They bonded a Ce:YIG die on a silicon MZI waveguide through an adhesive benzocyclobutene (BCB) bonding technique. A Ce:YIG layer was grown on an SGGG (substituted $Gd_3Ga_5O_{12}$) substrate through a sputter epitaxial growth technique. It should be noted that since the evanescent field penetrating the Ce:YIG layer is responsible for the magneto-optical phase shift, the thickness of the intermediate layer between the silicon and garnet must be stringently controlled in order to obtain the desired magneto-optical phase shift.

To continue miniaturization of isolators, Tien et al. [68] demonstrated a ring isolator for the first time by directly bonding a cerium-substituted yttrium iron garnet (Ce:YIG) onto a silicon ring resonator using oxygen plasma enhanced bonding. The silicon waveguide is 600 nm wide and 295 nm thick with 500 nm thick Ce:YIG on the top to have reasonable nonreciprocal effect and low optical loss. With a radial magnetic field applied to the ring isolator, it exhibits 9 dB isolation at resonance in the 1550 nm wavelength regime.

Finally, garnets with negative and positive gyrotropy can be easily patterned by today's lithographical methods that would allow "push/pull" devices with total Faraday rotations that are closer to simulated values than those obtained using traditional laser-anneal patterning of Ga-garnets [67]. Figure 4.9 shows an example of alternating YIG/Ce:YIG, which has a beat length of 7.5 μm according to finite-difference time-domain (FDTD) simulations. If the waveguide had been solely comprised of Ce:YIG, the Faraday rotation would only have oscillated between 1.5° and 0° due to birefringence. By alternating YIG and Ce:YIG as shown, 45° will be obtained using a 350 μm long device. This push/pull (Ce:YIG/YIG) simulation used realistic values of $\theta_F = -3000$ deg/cm (typical for Ce garnets that are monolithically integrated) and $\theta_F = +250$ deg/cm for YIG. This device could be fabricated using a 0.8 μm thick YIG film that is patterned in 3.75 μm wide strips where the areas between the strips would not be fully etched down to the substrate. After rapid thermal anneal (RTA), Ce:YIG could

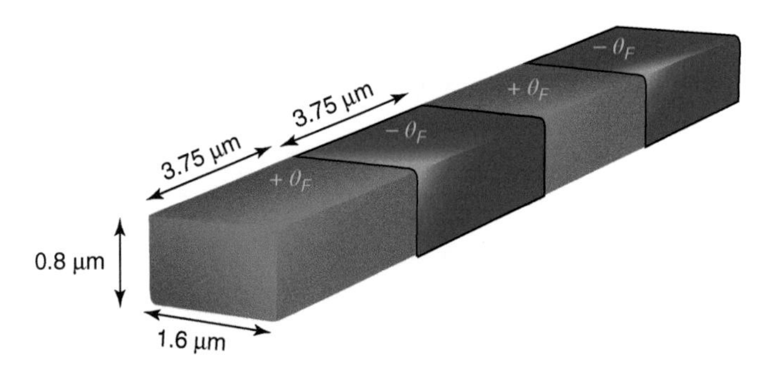

Figure 4.9 Waveguide design with alternating segments of positive and negative Faraday rotation. Source: Stadler and Mizumoto 2014 [67]. Reproduced with permission of IEEE.

then be deposited on top of the thin regions, which would act as seed layers for the Ce:YIG. Finally, ridges would be etched perpendicular to the strips to obtain the periodic waveguides shown in Figure 4.9. Device sizes of 350 μm are comparable to other published isolator designs but without narrow-band ring or interferometer structures. This quasi phase matching (QPM) method can also be used with garnet cladded rectangular Si waveguides that experience birefringence [80].

4.3.3 Antiferromagnetic Compounds

The transition metal fluorides such as MnF_2, FeF_2, CoF_2, NiF_2, and $RbFeF_2$ have a noticeable magnetic birefringence, i.e. a large Faraday rotation [81]. They have a distinct magnetic ordering transition, which is connected by temperature as well as by magnetic field [82]. The fluorides limitations for use in magneto-optical devices rise from the fact that this controllable magnetic birefringence is masked by a much larger natural birefringence. This natural birefringence might be minimized by using polycrystalline bulk or films.

MOKE, normally found in magnetic materials with nonzero magnetization such as ferromagnets and ferrimagnets, has been known for more than a century. Using first-principles density functional theory, Feng et al. [83] demonstrated large MOKE in high temperature noncollinear antiferromagnets Mn_3X (X = Rh, Ir, or Pt), in contrast to usual wisdom. The calculated Kerr rotation angles are large, being comparable to that of transition metal magnets such as bcc Fe. The large Kerr rotation angles and ellipticities are found to originate from the lifting of the band double degeneracy due to the absence of spatial symmetry in the Mn_3X noncollinear antiferromagnets, which together with the time-reversal symmetry, would preserve Kramer's theorem. Their results indicate that Mn_3X would provide a rare material platform for exploration of subtle magneto-optical phenomena in noncollinear magnetic materials without net magnetization.

Using symmetry arguments and a tight-binding model, Sivadas et al. [84] showed that for layered collinear antiferromagnets, magneto-optic effects can be generated and manipulated by controlling crystal symmetries through a gate voltage. This provides a promising route for electric field manipulation of the magneto-optic effects without modifying the underlying magnetic structure. They further demonstrated the gate control of MOKE in bilayer $MnPSe_3$ using first-principles calculations. The field-induced inversion symmetry breaking effect leads to gate-controllable MOKE whose direction of rotation can be switched by the reversal of the gate voltage.

Transition metal oxides such as MnO, CoO, and NiO have also been investigated [85–87] for their magneto-optic properties. They do not possess a crystalline birefringence. Usually, the oxides are antiferromagnets at low temperatures and display a sharp magnetic disordering transition at the Neel temperature. This transition is accompanied by a more gradual decrease in the birefringence.

$FeBO_3$ has been of interest because of its high figure of merit in the visible region, but like the fluorides possesses a natural birefringence [88]. A considerable amount of work has been done on the ferromagnetic oxide Fe_2O_3 [89, 90].

This oxide has a Morin transition, i.e. a transition from an antiferromagnetic state to a ferromagnetic state. The Faraday rotation has a very distinct change in sign, jumping from 640 deg/cm to −2000 deg/cm. Near 265 K, the exact temperature of the jump is controlled by a magnetic field.

Antiferromagnetic spintronics is a new rapidly developing field whose focus is the manipulation of antiferromagnets with electrical current. Antiferromagnets are promising materials for spintronic applications as they are fast, nonvolatile, and robust with respect to external fields. They can be manipulated by spin and charge currents. Their complex structures, strong magnetoelastic coupling, and compatibility with technologically important semiconductors make antiferromagnets interesting materials and open a way for new functionalities in comparison with ferromagnetic materials.

Almost all modern electronic devices require memory devices for large-scale data storage with the ability to write, store, and access information. There are strong commercial drives for increased speed of operation, energy efficiency, storage density, and robustness of such memories. Most, large scale data storage devices, including hard drives, rely on the principle that two different magnetization orientations in a ferromagnet represent the "zeros" and "ones." By applying a magnetic field to a ferromagnet, one can reversibly switch the direction of its magnetization between different stable directions and read out these states/bits from the magnetic fields they produce. This is the basis of ferromagnetic media used from the nineteenth century to current hard drives.

Today's magnetic memory chips (MRAMs) do not use magnetic fields to manipulate magnetization with the writing process done by current pulses, which can reverse magnetization directions due to the spin-torque effect. In the conventional version of the effect, switching is achieved by electrically transferring spins from a fixed reference permanent magnet. More recently, it was discovered that the spin torque can be triggered without a reference magnet, by a relativistic effect in which the motion of electrons results in effective internal magnetic fields. Furthermore, the magnetization state is read electrically in such MRAMs. Therefore, the sensitivity of ferromagnets to external magnetic fields and the magnetic fields they produce is not utilized. In fact, they become problems since data can be accidentally wiped by magnetic fields, and can be read by the fields produced making data insecure. Also, the fields produced limit how closely data elements can be packed.

Recently, Marti et al. [91] have shown that antiferromagnetic materials FeRh can be used to perform all the functions required of a magnetic memory element. Antiferromagnets have the north poles of half of the atomic moments pointing in one direction and the other half in the opposite direction, leading to no net magnetization and no external magnetic field. For antiferromagnets with specific crystal structures, they predicted and verified that current pulses produce an effective field that can rotate the two types of moments in the same directions. Marti et al. [91] were able to reverse the moment orientation in antiferromagnets by a current-induced torque and to read out the magnetization state electrically.

Since antiferromagnets do not produce a net magnetic field they do not have all the associated problems discussed above. The dynamics of the magnetization in antiferromagnets occur on timescales orders of magnitude faster than in ferromagnets, which could lead to much faster and more efficient operations. Finally, the antiferromagnetic state is readily compatible with metal, semiconductor, or insulator electronic structures and so their use greatly expands the materials basis for such applications.

4.4 Magneto-optic Devices

Magneto-optic materials have unique physical properties that offer the opportunity of constructing devices with many special functions not possible from other photonic devices. The most significant of these properties are that the linear magneto-optic effect can produce circular birefringence and that, unlike other optical effects in dielectric media, it is nonreciprocal. All practical magneto-optic devices exploit one or both of these two properties. Recently, research works on magneto-optical microstructure devices in terahertz (THz) regime has attracted special interest from the scientific community. Important applications of these devices include isolators, modulators, sensors, directional beam scanners, etc. Some of the magneto-optic devices are described in more detail here.

4.4.1 Magneto-optic Modulator

The magneto-optic modulator is based on the rotation of optical polarization of light that propagates along the magnetic field in a material by the Faraday effect. In recent years, many scientists have designed magneto-optic modulators, and some of them are described here.

4.4.1.1 Magneto-optic Spatial Light Modulator

The magneto-optic spatial light modulator (MOSLM) is a category of optical device that spatially transforms an input signal (either electrical or optical) as encoded information in the intensity and/or phase of an optical beam. The spatial light modulator (SLM) is integral to a myriad of applications. Displays (LCD and projection), optical data processing, optical fiber switching, and adaptive optics (correcting imagery for atmospheric distortion) are just some of the many tasks routinely accomplished using these devices [92].

Common to all MOSLM devices is the Faraday rotator. Faraday rotators alter the polarization vector E of input light using an applied magnetic field (Figure 4.10). Faraday effect, a magneto-optic effect, is thus employed to modify intensity and/or phase of a propagating optical beam. These devices have no moving parts and consist of solenoids with magneto-optical (MO) material cores, usually bismuth-substituted yttrium iron garnet (Bi:YIG) or similar crystals. The solenoid can be either a permanent magnet or an electromagnet (Figure 4.11). The magnetic flux density B is a function of applied voltage.

In terms of operation, monochromatic light is input through a linear polarizer either parallel or antiparallel to an applied magnetic field H (Figure 4.12). When the magnetization direction coincides with the light propagation direction, the light polarization vector is rotated by a Faraday rotation angle β in a right-hand

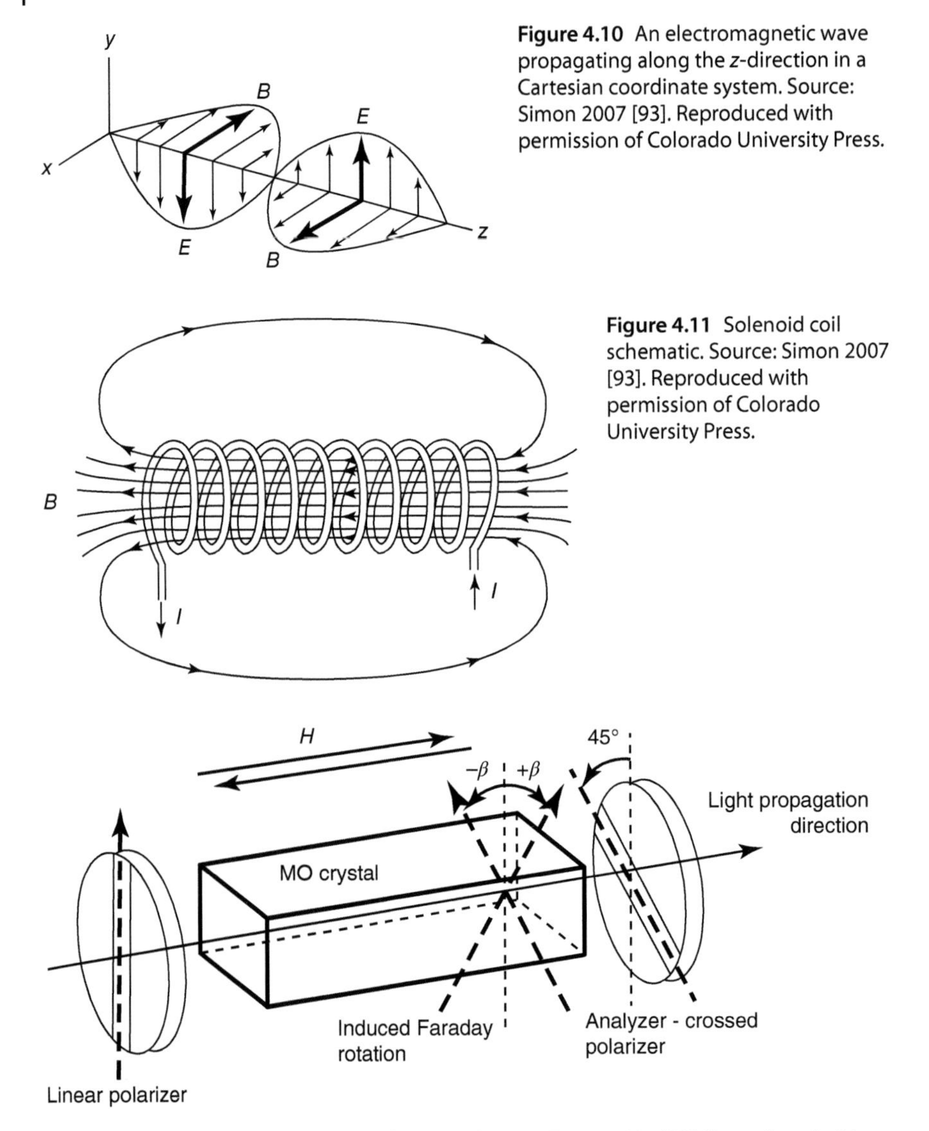

Figure 4.10 An electromagnetic wave propagating along the *z*-direction in a Cartesian coordinate system. Source: Simon 2007 [93]. Reproduced with permission of Colorado University Press.

Figure 4.11 Solenoid coil schematic. Source: Simon 2007 [93]. Reproduced with permission of Colorado University Press.

Figure 4.12 Typical Faraday rotator schematic. Source: Simon 2007 [93]. Reproduced with permission of Colorado University Press.

screw sense, i.e. clockwise is positive. Reversing the field direction so that it is opposite to the light propagation direction reverses the induced rotation to $-\beta$. The analyzer is a linear polarizer rotated 45° from the linear polarizer axis. By adjusting the magnetic field strength and direction, the polarization vector of the propagated light can be changed such that it crosses the analyzer, allowing the device to serve as an intensity modulator [94, 95].

In 1982, Ross et al. designed the first MOSLM device employing a bismuth substituted iron garnet thin film with a rectangle-shaped pixel array as the active element [96–98]. To construct an MEMS (microelectromechanical systems)

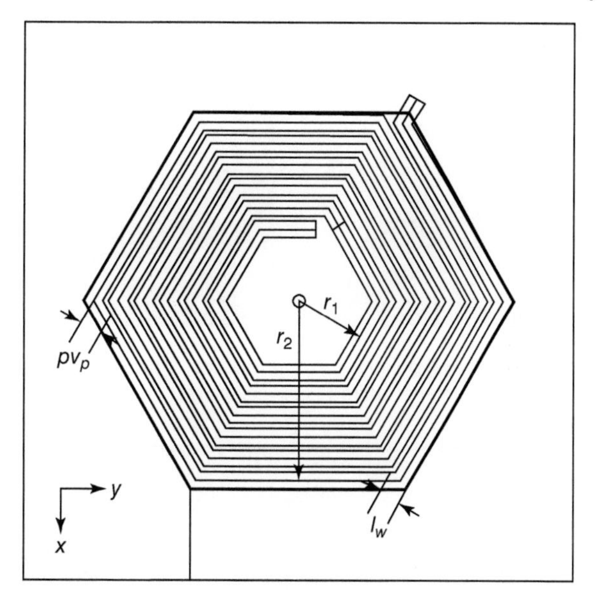

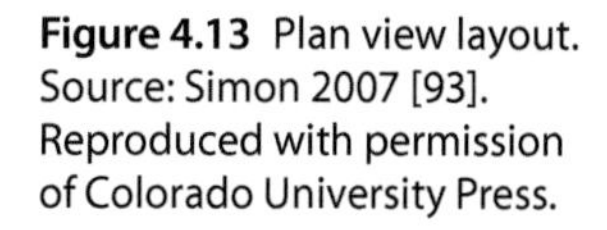
Figure 4.13 Plan view layout. Source: Simon 2007 [93]. Reproduced with permission of Colorado University Press.

Faraday rotator, it is required to fabricate a solenoid and integrate a magneto-optic crystal. There are potentially many ways to accomplish this task; however, the challenge and the constraint are in the solenoid construction. In 2007, M. Simon [93] designed an MEMS Faraday rotator layout in terms of a right-handed coordinate frame with z as the up axis. The actual plan view coil geometry (Figure 4.13) was chosen to be a hexagon to achieve the highest possible fill factor for 2D array configurations.

The layout was accomplished using Coventor Ware 2005 software. Several MUMPS (multiuser MEMS) process modifications were required. The substrate thickness was increased to 20 μm and due to its conductivity, copper was chosen for the metal layer. Other slight deposition method modifications were required to mesh the solid model with a minimal number of elements prior to finite element simulation.

The innermost turn radius r_1 was 9 μm and the outermost turn radius r_2 was 22.5 μm. The plan view pitch pv_p between turns was 2.25 μm. The coil line width l_w was 2.078 μm. The total substrate thickness d including the silicon nitride layer was 20.75 μm. The coil was designed to have six turns. Extending the layout to accomplish a full polarization controller (Figure 4.14) was considered. Also, a system schematic was determined for 2D MOSLM array composed of individual Faraday rotators (Figures 4.15 and 4.16.) It is envisioned that the MOSLM would be aligned and flip chip bonded to a VCSEL (vertical cavity surface emitting lasers) array, with one VCSEL per pixel. Inductor leads would penetrate the VCSEL array and be solder bonded to matched inductor pads on the MOSLM substrate underside. A control ASIC (application-specific integrated circuit) would also be flip chip bonded to the VCSEL array.

An estimate of Faraday rotation for the developed Faraday rotator geometry as a function of input voltage was conducted to determine approximately the input voltage required to obtain a 45° Faraday rotation. The Bi:YIG crystals are known

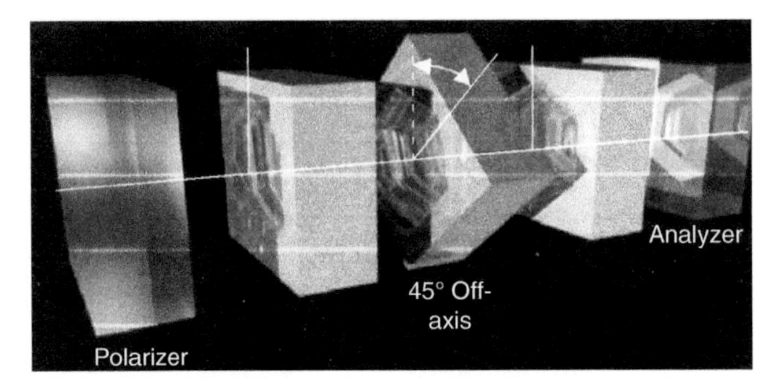

Figure 4.14 Full polarization controller configuration. Source: Simon 2007 [93]. Reproduced with permission of Colorado University Press.

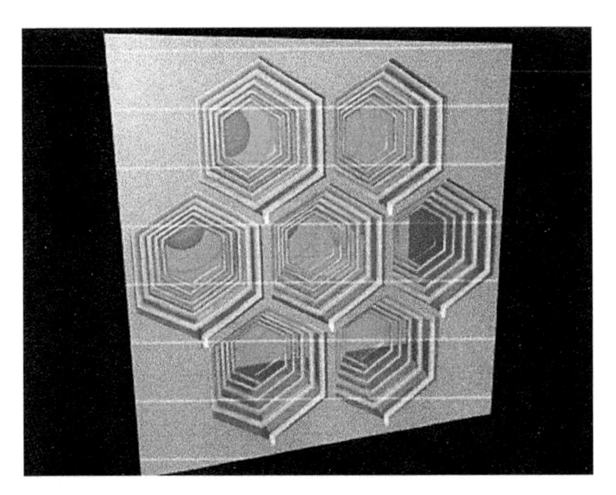

Figure 4.15 Rendering of 2D MOSLM array composed of individual MEMS Faraday rotators. Source: Simon 2007 [93]. Reproduced with permission of Colorado University Press.

to exhibit very large Verdet constants, on the order of −806 (ν) [99], indicating potentially substantial Faraday rotations. For an ideal solenoid, the magnetic flux density B at the solenoid center (T) is given by [100].

$$B = \mu_e nI = \mu_0 \mu_r nI \tag{4.13}$$

$$\mu_0 = 4\pi \times 10^{-7} \left(\frac{\mathrm{T\,m}}{\mathrm{A}}\right) \tag{4.14}$$

where μ_0 is the free space permeability, μ_r is the relative permeability of the solenoid core, μ_c is the permeability of the solenoid core $\left(\frac{\mathrm{T\,m}}{\mathrm{A}}\right)$, n is the number of coil turns, and I is the current (A). The relative permeability of a material is given by

$$\mu_r = \frac{\mu}{\mu_0} \tag{4.15}$$

While a specific permeability value for Bi:YIG crystal could not be located, a conservative relative permeability value for ferrites (Bi:YIG is a ferrite) is on the order of 1000. In general, the permeability for ferrites is a nonlinear function of B and H. However, for this analysis a linear assumption is sufficient since at optical and near-optical wavelengths, μ is close to unity [101].

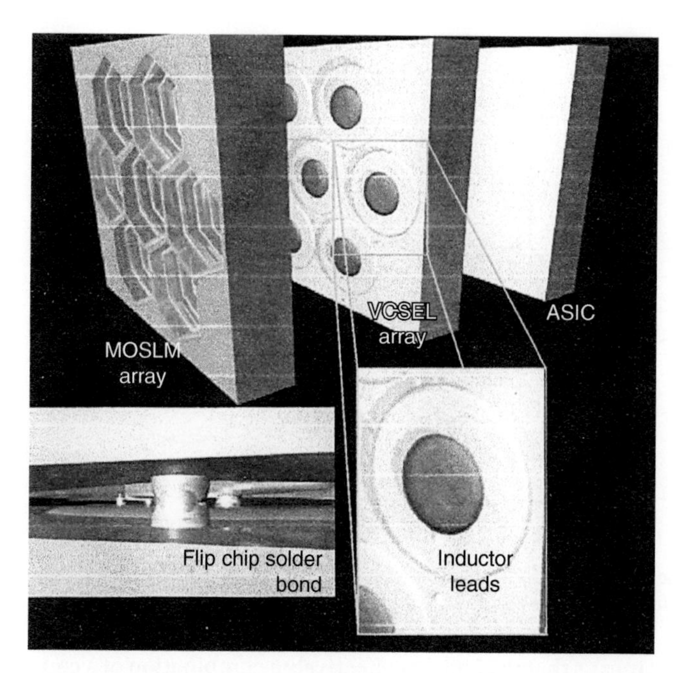

Figure 4.16 2D MOSLM array assembly schematic. Source: Simon 2007 [93]. Reproduced with permission of Colorado University Press.

To compute B, it was necessary to compute the current I. However, to do so, the resistance $R(\Omega)$ of the coil for an input voltage V was also required using

$$I = \frac{V}{R} \tag{4.16}$$

R is given by

$$R = \rho \frac{L_W}{A_W} \tag{4.17}$$

where ρ is the resistivity of the coil material (Ω m), L_w is the length of the coil wire (m), and A_w is the cross-sectional area of the coil wire (m^2). The resistivity value for copper $\rho = 1.72 \times 10^{-8}$ Ω m was used. L_w and A_w both obviously depended on specific coil geometry. A_w was simply L_w multiplied by the metal layer thickness.

The equation for Faraday rotation is given by [94]

$$\beta = vBd \tag{4.18}$$

where β is the Faraday rotation angle in arc minutes, v is the Verdet constant $\left(\frac{\text{arc min}}{\text{cm G}}\right)$, B is specified in Gauss, and d is the optical distance that light travels through the medium (cm), in this case the solenoid length.

Critical to this discussion was the procedure for obtaining an accurate value for L_W. This required adequate knowledge of extended coil geometry. When planar, the coil had a constant pitch. When extended, the coil possessed variable pitch in the z direction with the pitch varying exponentially downward from the outer (r_2)

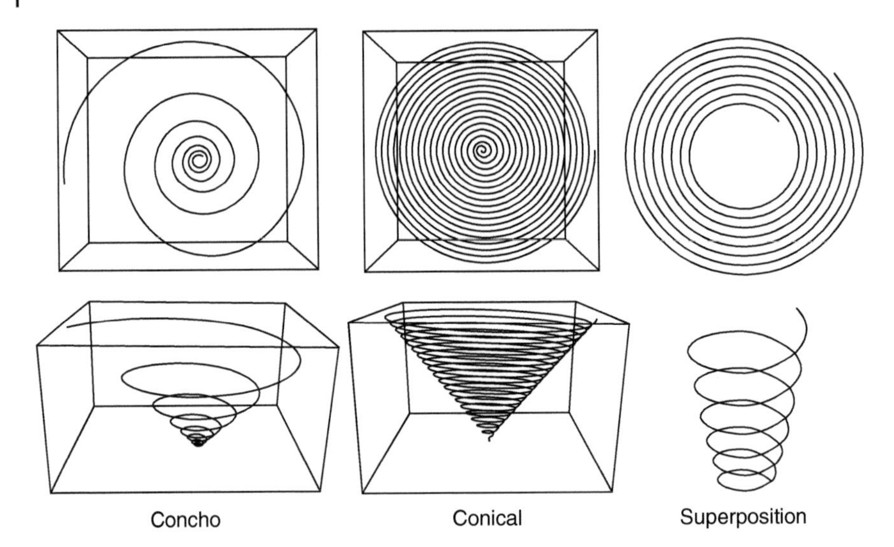

Figure 4.17 Spiral geometry combination. Source: Simon 2007 [93]. Reproduced with permission of Colorado University Press.

to inner (r_1) turns. This means that the coil was effectively a combination of a conical spiral and a concho-spiral. This superposition effect is shown in Figure 4.17.

Considering this geometry combination, the computed length of the coil wire was found to be 594.08 μm (details calculation in Ref. [93]). Thus, it is concluded that 0.002 V is sufficient to achieve a Faraday rotation of ~52° using this MEMS Faraday rotator design. The Bi:YIG modulator has shown thermal stability between 10 and 300 K [102, 103]. The MEMS waveguide modulator exploiting the Faraday effect has been realized with modulation rates up to 1.8 GHz [104]. Also, intriguing is that this coil design can be used to implement MEMS variable inductors, which are important in RF (radio frequency) circuits for filtering and reactive impedance matching [105].

4.4.1.2 Mach–Zehnder Magneto-optic Modulator

The technology involves a new type of compact device based on magneto-optics in a fiber micro modulator. This type of modulator allows the user to inter manipulate and control the propagation of the incoming light. Ultrafast superconducting optoelectronics is an acknowledged field of technological importance in which much research has been performed in recent years [106]. The fastest digital electronic circuits are superconducting single-flux-quantum (SFQ) logic systems, based on resistively shunted Josephson tunnel junctions [107]. Implementation of the predicted speed of SFQ circuitry in any full-performance system, or even in a demonstration prototype, requires a new paradigm regarding digital input/output (I/O) communication between the SFQ processor and the outside world.

In 2001, Sobolewski and Park [108] designed an ultrafast superconducting Mach–Zehnder magneto-optic (MO) modulator for the SFQ-to-optical digital interface. The modulator is based on the Faraday effect and consists of a

Table 4.1 Properties of selected magneto-optic materials.

Compound	Verdet constant V (K/Oe μm)	Critical temperature T_c (K)	Wavelength λ (μm)
EuS	4.8	17	0.546
EuSe	9.6	7	0.546
EuS/EuF_2	2.0	17	0.546
YIG	0.02	585	>1.3
(Tb)YIG	0.03	~520	>1.15
(SmLuCa)YIG	0.08	>600	>0.83
Glass	0.000 07	~800	0.85

Source: Sobolewski and Park 2001 [108]. Reproduced with permission of IEEE.

microwave microstrip line (MSL) with a polarization-sensitive magneto-optic active medium and fiber-optic continuous wave (CW) light delivery.

Europium-based magnetic and diluted magnetic semiconductors, such as europium oxides (EuO) and europium monochalcogenides (Eus, EuTe, and EuSe), show an interesting range of behavior in the region of their ordering temperature T_c (Table 4.1). Both EuO and EuS order ferromagnetically, while EuTe and EuSe order antiferromagnetically, but they can be driven ferromagnetic with the application of an external H field. Below T_c, they also act as polarization rotors when placed in a magnetic field – a property known as the Faraday effect. The angle of Faraday rotation β is given by the simple relationship:

$$\beta = VlH \tag{4.19}$$

where V, H, and l denote the Verdet constant, magnetic-field intensity, and the light path length in the magneto-optic (MO) material, respectively. Other things being equal, the high V value is desirable because it allows the use of low external H field modulation and a thinner, less absorbing medium.

Europium chalcogenides are characterized by some of the highest known values of V, but the MO effect itself has been demonstrated in a very large range of solid materials and liquids, including even glass. Recently, rare earth iron garnets, such as yttrium iron garnet (YIG) and its derivatives, have been demonstrated as effective MO materials. Their major advantages are excellent transmissivity in the near-infrared radiation and T_c well above the room temperature. Thus, garnets have been proposed as an active medium for modulators for optical communication [109]. The properties of selected MO materials, such as V, T_c, and optical operating wavelength, are summarized in Table 4.1.

Garnets have been most recently pursued as the MO material of choice for ultrafast room temperature optical communication; europium chalcogenides are clearly ideal for low-temperature superconducting optoelectronic applications: they exhibit the highest values of the Verdet constant and single picosecond response times and can be easily deposited in a thin film form, using either vacuum evaporation [110] or laser ablation [111]. Thus, Sobolewski and Park

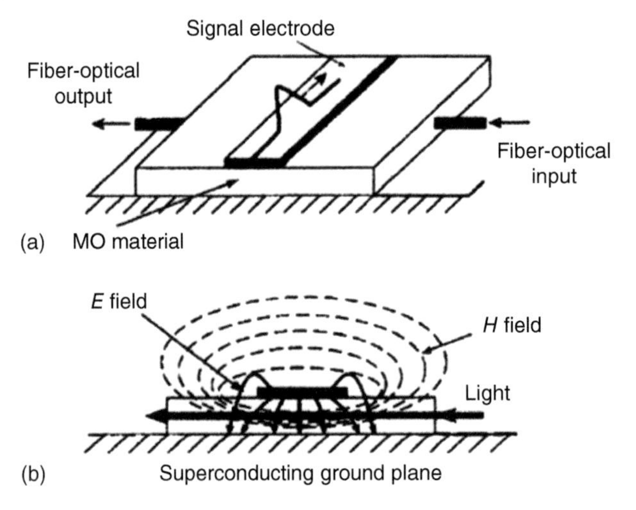

Figure 4.18 Structure of the superconducting magneto-optic modulator. (a) MSL geometry and (b) cross section schematically showing the spatial distribution of the *E* and *H* fields. Source: Sobolewski and Park 2001 [108]. Reproduced with permission of IEEE.

[108] have chosen EuSe as the material for proposing a superconducting MO modulator for optical output of the Nb-based SFQ electronics.

Figure 4.18 shows the structure of a superconducting MO modulator based on a microwave MSL with a polarization-sensitive MO active medium and fiber-optic CW light delivery (Figure 4.18a). Implementation of the MSL can obtain a very long interaction distance l and the low characteristic impedance of the line, which ensures that the H field component of the electromagnetic signal is uniform along the modulator length (Figure 4.18b). The MO modulator operation is based on the Faraday effect; thus, light modulation direction occurs in parallel to the H field and perpendicular to the signal propagation (shown as the fiber-optic light input and output in Figure 4.18a). The uniformity of the H field inside the modulator was extensively simulated numerically in order to optimize the device design (primarily the length-to-height aspect ratio) and obtain the actual value of H along the optical path.

Figure 4.19 graphically shows the H field distribution along the length of the MO crystal. In simulations, the quasi-static approximation method was used with E and H fields treated as quasi-TEM (transverse electromagnetic). The MO material EuSe was characterized at high frequencies by $\mu_r \approx 1$ and $\varepsilon_r \approx 15$. The amplitude of the input current pulse was assumed to be 1 mA, which corresponds to 10 kA/cm^2 critical current density for a nominal 10 μm^2 Josephson tunnel junction – the specification easily achievable within the current Nb-trilayer technology. For the 100 μm wide top electrode and 5 μm high MSL, filled with the EuSe MO material, a better than 98% H uniformity was obtained along the optical pass with the H_{max} at the device center equal to 2.51 Oe, which, according to Eq. (4.19), corresponds to $\beta = 4.52°$. The device characteristic impedance at high frequencies was 4.4 Ω.

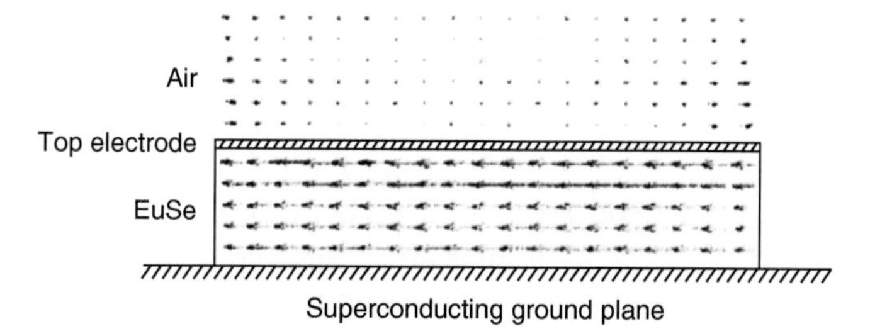

Figure 4.19 Simulation of the magnetic field distribution inside the MO modulator. Source: Sobolewski and Park 2001 [108]. Reproduced with permission of IEEE.

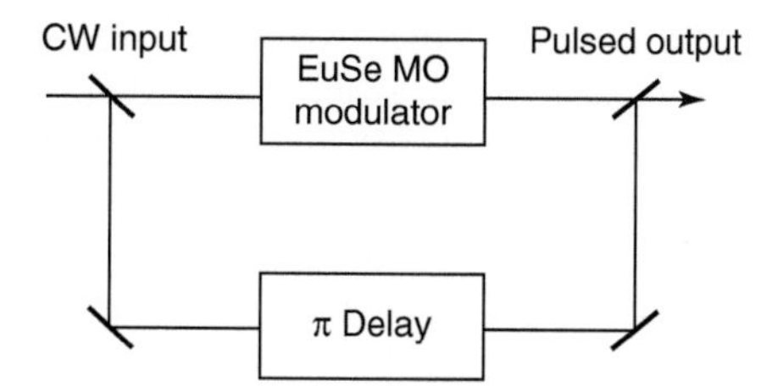

Figure 4.20 Mach–Zehnder design of a EuSe MO modulator. Source: Sobolewski and Park 2001 [108]. Reproduced with permission of IEEE.

A much higher value of the Faraday rotation was obtained for a macroscopic device, consisting of a 150 μm wide top electrode MSL and 20 Ω high-frequency characteristic impedance, but containing the 0.5 mm high and 5 mm long EuSe crystal. In this case, the total β was given by V times the integral of H over the optical interaction distance $l = 5$ mm, leading to a very respectable 37° of polarization rotation; such a device, based on the EuSe single crystal, could be implemented in the superconducting circuit as a hybrid element.

To operate as an efficient electrical-to-optical transducer, the MO modulator (shown schematically in Figure 4.20) is incorporated into the Mach–Zehnder configuration and works as an intensity modulator. As a light source, a commercial green light CW laser diode was used, operating at 0.546 μm – the optimal wavelength for EuSe. To obtain the highest contrast aspect between the SFQ logical "1" and "0" (the highest signal-to-noise (SN) ratio at the modulator optical output), the phase difference between the interferometer's arms must be equal to π, so that with no signal applied to the MO material, light at the output will be completely extinguished. In this situation, the light signal is present at the output of the device only when the SFQ pulse is applied to the modulator. The transmittance T of the Mach–Zehnder modulator in such a configuration is given as

$$T = \cos^2\left[\frac{\pi + \beta}{2}\right] \tag{4.20}$$

It is noticed that even for the modulator with $l = 100$ μm, Eq. (4.20) gives the easily detectable $T = 0.2\%$ modulation depth, while the single crystal device should provide at the output with approximately 10% of the input light intensity, making the EuSe modulator very attractive for direct, SFQ-to-optical digital interface.

4.4.1.3 Magnetic Fluid-Based Magneto-optic Modulator

MF is a highly stable colloidal suspension of single domain ferromagnetic nanoparticles in a suitable carrier liquid. It has attracted a great deal of interest from researchers due to its remarkable and versatile magneto-optic properties such as tunable refractive index [31, 112], birefringence [29, 113], Faraday effect [114], magneto-chromatics [36, 115], and field-dependent transmission [40, 44, 114]. Additionally, it is a superparamagnetic function material that exhibits no hysteresis effect in photonics devices and sensing applications.

Many MF-based photonics devices for characterization of magnetic modulation, especially optical fiber devices, have been developed. H. E. Hrong demonstrated a magneto-optic modulator by employing MF as a cladding layer with tunable refractive index [43, 112]. The ER achieved was 4%, which is quite small. The MMF employed limited the applications in SMF systems. Configuration of a free space MZI, which made use of the tunable refractive index property of MF films, was proposed to realize modulation and improve the ER slightly [116]. Optical fiber-based MZI and cascaded modulators utilizing MF were also demonstrated to realize optical logical operations with a further improved ER of 20% [53]. Additionally, a magneto-optic modulator, which acted as an optical switch, was implemented by means of the field-dependent transmission property of MFs [40].

The structure of the MF-based modulator is shown in Figure 4.21. The plane of the *MF* film is normal to the light propagating direction and parallel to the external magnetic field (H) direction. The collimated light, which is linearly polarized after passing through the polarizer (P_1), is decomposed into two orthogonal linearly polarized beams when traveling in the *MF* film as well as the PMF. The e-ray and o-ray propagate at different speeds due to the birefringence effects of the *MF* film as well as the *PMF* and experience different absorption losses due to the dichroism effect of the *MF* film. Subsequently, the components of e-ray and o-ray in the direction of the analyzer (P_2) interfere, and a wavelength-dependent polarization interference spectrum is produced at the analyzer (P_2).

In the configuration, the slow axis of the *MF* film (parallel to the magnetic field direction) is configured to be parallel to the fast axis of the *PMF*. So, the total phase difference is given as $\varphi = \varphi_{PM} - \varphi_{MF}$, where $\varphi_{PM} = 2\pi B_{PM} L/\lambda$ is the phase difference caused by the birefringence of *PMF* B_{PM} over the fiber length L, $\varphi_{MF} = 2\pi B_{MF} d/\lambda$ is the phase difference caused by the birefringence of *MF* film

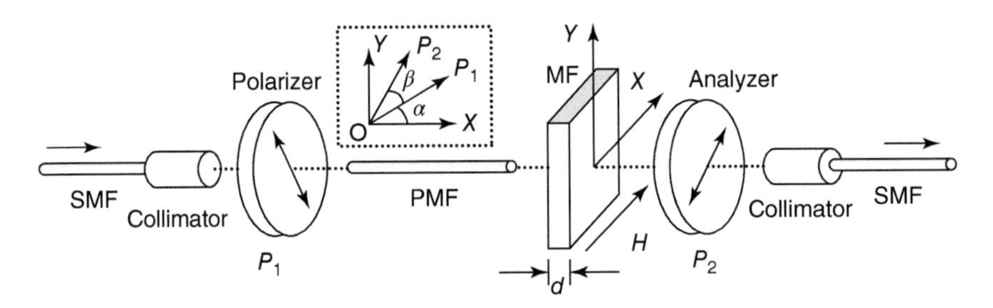

Figure 4.21 Schematic diagram of the structure of the MF-based modulator. Source: Zu et al. 2012 [54]. Reproduced with permission of IEEE.

B_{MF} over the thickness of d, and λ is an operating wavelength. φ_{PM} is a constant, whereas φ_{MF} is a variable with the external magnetic field strength.

The linearly polarized light after P_1 is decomposed into two components along the slow axis and fast axis of the *MF* film, which can be given as $E_0 \cos\alpha$ and $E_0 \sin\alpha$ respectively. The E_0 is the amplitude of the electrical vector of the linear polarized light after P_1; α is the angle between P_1 direction and magnetic field direction X. After the *MF* film and the *PMF*, the two components become $E_0 \cos\alpha \cdot e^{-\alpha_e d}$ and $E_0 \sin\alpha \cdot e^{-\alpha_0 d} \cdot e^{-i\varphi}$ for the X and Y direction, respectively, where α_e and α_0 are the corresponding absorption coefficients for the two components, respectively. Subsequently, the two light components project on the P_2 direction again and become

$$E_1 = E_0 \cos\alpha \cdot e^{-\alpha_e d} \cdot \cos(\alpha + \beta) \tag{4.21}$$

$$E_2 = E_0 \sin\alpha \cdot e^{-\alpha_0 d} \cdot e^{-i\varphi} \cdot \sin(\alpha + \beta) \tag{4.22}$$

where β is the angle between P_1 direction and P_2 direction. Then, the transmission function of the modulator $T = (E_1 + E_2) \cdot (E_1 + E_2)^*$ is deduced as

$$T = E_0^2 e^{-(\alpha_0+\alpha_e)d} \begin{bmatrix} \cos^2\alpha \cos^2(\alpha+\beta) e^{-\Delta\alpha d} + \sin^2\alpha \sin^2(\alpha+\beta) e^{\Delta\alpha d} \\ +2\sin\alpha\cos\alpha\sin(\alpha+\beta)\cos(\alpha+\beta)\cos\varphi \end{bmatrix} \tag{4.23}$$

where $\Delta\alpha = \alpha_e - \alpha_0$ describes the dichroism property of *MF*. If the analyzer P_2 is adjusted to be perpendicular to the polarizer P_1 (i.e. $\beta = 90^\circ$), Eq. (4.23) is simplified as

$$T = E_0^2 e^{-(\alpha_0+\alpha_e)d} \cdot \sin^2 2\alpha \cdot [\cosh(\Delta\alpha) - \cos\varphi]/2 \tag{4.24}$$

The transmission spectrum of the modulator is sinusoidal approximately (Figure 4.22). It is shown in Figure 4.22 that *ER* of the modulator is equal to ΔT. If the condition $\varphi = 2m\pi$ (m is an integer) is fulfilled, the transmission spectrum T is the minimum, and the transmission dips appear at the corresponding wavelengths on the transmission spectrum. The wavelength space between the two adjacent transmission dips S is given by

$$S = \lambda^2/(B_{PM}L - B_{MF}d) \tag{4.25}$$

A tunable laser with a narrow bandwidth is used to realize light modulation, and its wavelength is selected to be coincident with the dip wavelength of the interference spectrum under zero magnetic field (e.g. 1570.2 nm). It is shown in Figure 4.22 that without the external magnetic field, the transmission power reaches its minimum. As the external magnetic field is applied, the transmission spectrum shifts to the left, and the transmission power reaches its maximum. In order to obtain the light modulation with the best *ER*, the spectrum shift is expected to cover at least half period S of the spectrum.

Compared with the length L of *PMF*, the thickness d (tens of micrometers) of the *MF* film is so small that $B_{MF}\, d$ can be neglected, so S is mainly dependent on L. It is shown in Eq. (4.24) that the modulating *ER* is also affected by the visibility of the interference spectrum, which is determined by $e^{-(\alpha_0+\alpha_e)d}$ and $\sin^2 2\alpha$. In order to obtain the best visibility, α is set to be 45°. Finally, highly modulating *ER* can be realized by selecting a suitable length of *PMF*.

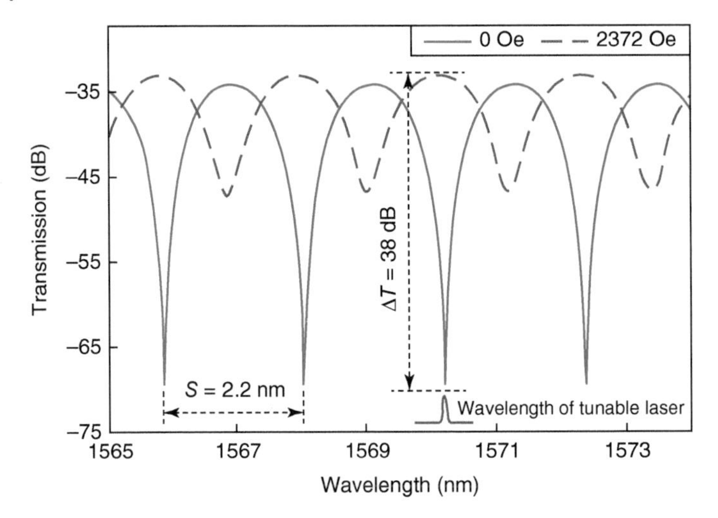

Figure 4.22 Operating principle of the magneto-optical modulator based on the magnetically controllable transmission spectrum. Source: Zu et al. 2012 [54]. Reproduced with permission of IEEE.

The water-based *MF* (EMG605, Ferrotec), containing 10 nm diameter ferromagnetic nanoparticles, Fe_3O_4, was sealed between two 14×14 mm^2 optical glass plates to form a 15 μm thick *MF* film. The concentration of the magnetic nanoparticles of the *MF* sample was 3.9%, and the *MF* film was light-brown and translucent. The saturated magnetization and initial magnetic susceptibility of the *MF* used were 220 Oe and 2.96 Gs/Oe, respectively. The length of the *PMF* (PM-1550-HP) used in the experiment was 3 m, and its polarization extinction ratio (PER) was higher than 40 dB. The *PERs* of the polarizer and analyzer (Thorlabs) were both 40 dB. The central wavelength of the collimators was 1550 nm, and the beam diameter of the collimated light was about 0.5 mm. The *MF* film was placed amidst a uniform magnetic field, which was generated with an electromagnet (LakeShore) and calibrated with a gauss meter (LakeShore). The dimension of the uniform region was about 65×25 mm^2, and the deviation of the uniformity was better than 0.05%. The magnetic field strength can be controlled by a personal computer. The experiment was conducted at room temperature, 23.7 °C.

The transmission spectra of the magneto-optical modulator under different magnetic field strengths were measured with a broadband light source (1520–1610 nm) and an optical spectrum analyzer (OSA, AQ6370), which are shown in Figures 4.22 and 4.23. The visibility of the transmission spectrum under zero magnetic field is about 36 dB, and the period is about 2.2 nm. As the magnetic field strength was increased from 0 to 2372 Oe, the transmission spectrum shifted about 1.2 nm to the red side, as well as the visibility decreased from 36 to 13 dB, gradually. In order to investigate the variation trend of the transmission spectrum, the relationship between the dip wavelength (the third dip is chosen as an example) and the magnetic field strength is shown in Figure 4.24a. During the initial segment of the progression of magnetic field

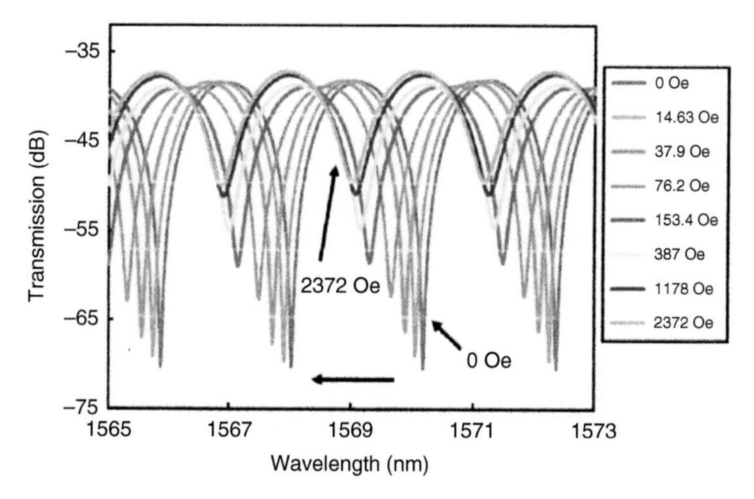

Figure 4.23 Transmission spectra of the magneto-optical modulator under different applied magnetic field strengths. Source: Zu et al. 2012 [54]. Reproduced with permission of IEEE.

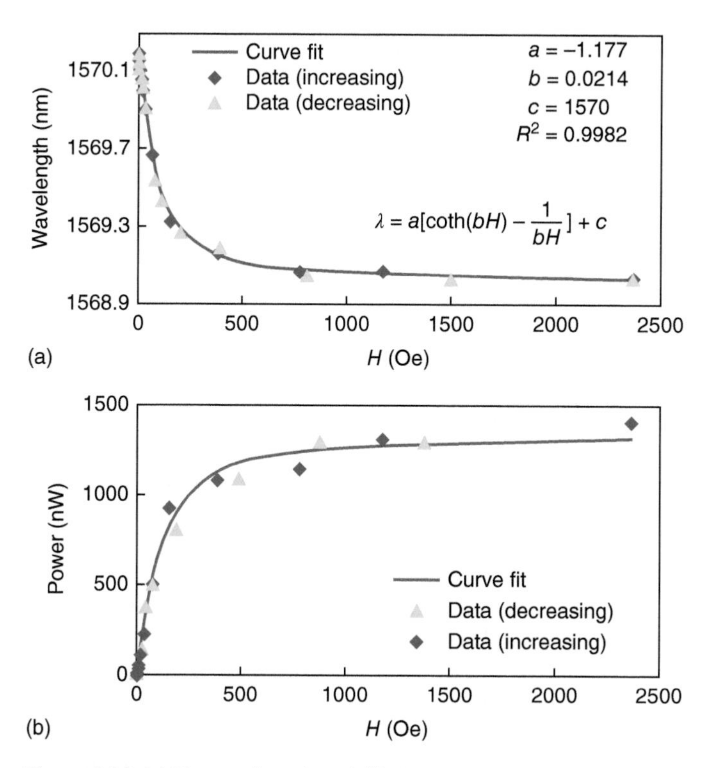

Figure 4.24 (a) Dip wavelength and (b) transmission power versus the magnetic field strength. Source: Zu et al. 2012 [54]. Reproduced with permission of IEEE.

strength (e.g. $H < 200$ Oe), the dip wavelength varied quickly and linearly. As the magnetic field strength increased further, the shift of the dip wavelength became slow and nonlinear. As the magnetic field strength increased over a certain value (e.g. $H > 750$ Oe), the shift of the dip wavelength tended to be saturated. The curve for the decreasing process of the magnetic field strength was almost coincident with the one for the increasing process, which can be fitted well by modified Langevin function with a high goodness of fit R^2 value of 0.9982.

In order to verify the light intensity modulation property of the proposed modulator, a tunable laser (AQ8203) and a photodiode were applied to test the modulator. The wavelength of the tunable laser was tuned to 1570.213 nm, which was coincident with the wavelength of the third dip on the transmission spectrum. The results are shown in Figure 4.24b. Similar to the variation trend of transmission spectrum, the light intensity varied quickly under small magnetic field strength and tended to be saturated under large magnetic field strength. The results in the processes of increasing and decreasing of the magnetic field strength are almost coincident, which show a good repeatability and indicate that hysteretic effect is absent in the *MF*. The curve is fitted with modified Langevin function with an R^2 value of 0.9912, which can be used to calibrate the nonlinearity between the light intensity and the magnetic field strength during the modulation.

The modulation characteristics of the proposed magneto-optical modulator are demonstrated by applying a square-wave and a sine-wave modulating magnetic fields on the *MF* film, and the corresponding light intensity modulation results are shown in Figure 4.25. The magnetic field strengths for square-wave and sine-wave modulation were 1200 and 100 Oe, whereas the corresponding *ERs* were about 38 and 29 dB, respectively. Limited by the speed of the magnetic field generation device, waveform distortion appears. The frequency response of the modulator is dependent on the response time of *MF*, which ranged from 10 to 600 ms [41, 48, 53]. Hence, the highest frequency of such modulator is about 100 Hz. The response time can be further improved by reducing the concentration of the *MF* or by using other types of the magneto-optical material such as YIG [117, 118]. Owing to the response time of *MF*, the modulator cannot work in high modulation frequency. However, it can be used to generate low-frequency control signal in the optical communication system. Also, it is good enough for applications such as switch, routers, and display.

4.4.1.4 Terahertz Magneto-optic Modulator

Magneto-optical (MO) material introduced into the artificial microstructure, such as magnetic photonic crystal [119, 120] or magneto-plasmonics [121], has become a research hotspot in recent years [122, 123]. Through a reasonable design of device structure, MO effect can be significantly enhanced by plasmonic resonance or bandgap effect [124, 125]; conversely, MO effect leads to some new physical mechanism and phenomena such as the splitting of plasmonic resonance, nonreciprocal transmission, and giant Faraday rotation effect [126–129]. For example, giant Faraday rotations were observed in some high electron mobility semiconductors, such as InSb [130], HgTe [131], and graphene [132]. A magnetically induced THz transparency (that is a sudden

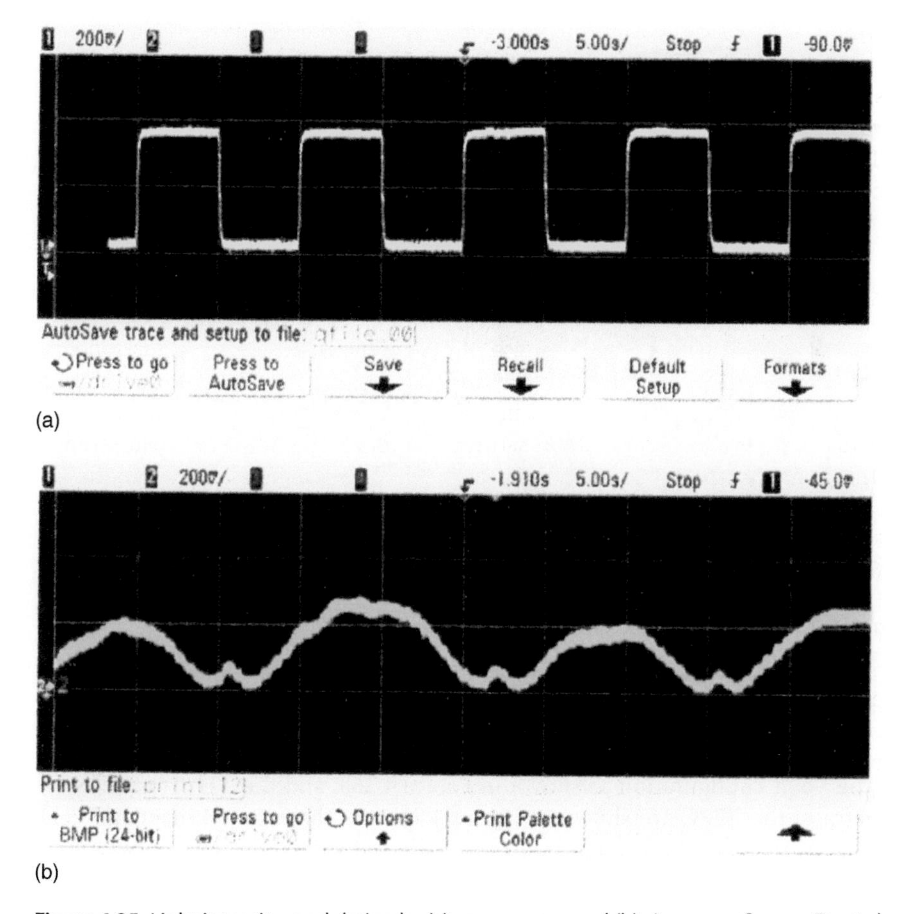

Figure 4.25 Light intensity modulation by (a) square wave and (b) sine wave. Source: Zu et al. 2012 [54]. Reproduced with permission of IEEE.

appearance and disappearance of THz wave transmission) in the n-doped InSb was demonstrated, owing to the interference between left and right circularly polarized magneto-plasmon eigenmodes [133]. Tunable THz magneto-plasmon and MO splitting were observed in graphene [134] and ferrofluid [135]. Moreover, the properties of MO devices can be controlled by an external magnetic field. The unique nonreciprocal effect and magnetic tunability of the MO device make it play an irreplaceable role in the high-performance isolator, phase shifter, MO modulator, and magnetic field sensor. However, due to the lack of high-performance THz MO materials and the limitation of device fabrication, improvement of THz MO devices is still challenging.

In 2013, Fan et al. [135] designed a schematic diagram of THz MO modulator based on the MO microstructure. Since ferrofluid has MO effects in the THz regime and it can be tuned by the external magnetic field, filling the ferrofluid into an artificial microstructure may enhance the MO response and have some new effects in the THz regime. Fan et al. [135] filled ferrofluid on the surface of Si photonic crystals to form a FFPC. They applied the external magnetic field

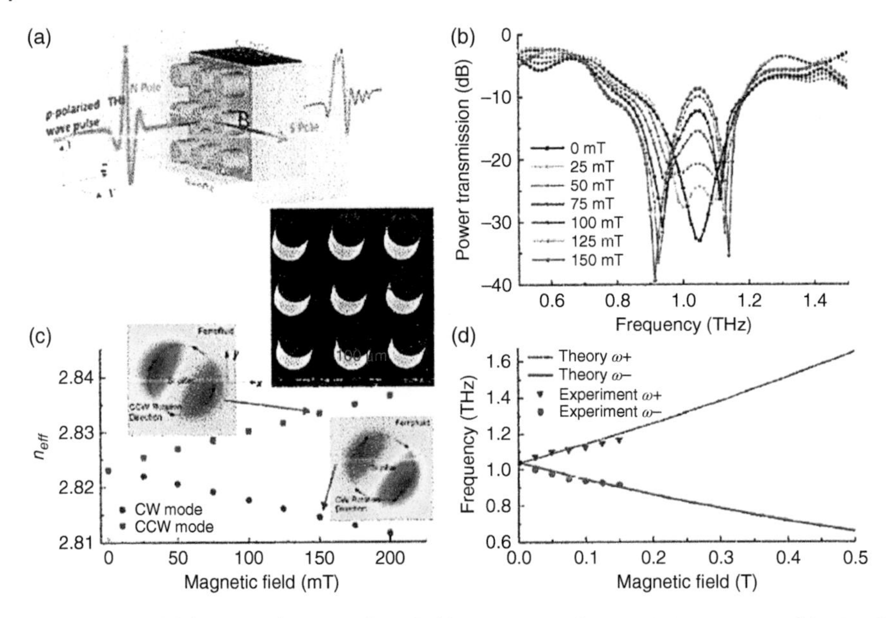

Figure 4.26 (a) Schematic diagram of FFPC. (b) Experimental transmission spectra of the FFPC under different external magnetic fields. (c) Dependence of the effective refractive indexes of the CCW and CW modes at 150 mT. (d) The dependence of the magneto plasmonic resonance frequency ω_- and ω_+ on the external magnetic field both from experimental data and theory. Source: Fan et al. 2013 [135]. Reproduced with permission of AIP.

as the Voigt configuration as shown in Figure 4.26a, and the THz transmission spectra of the FFPC are shown in Figure 4.26b. As the external magnetic field increases, the original resonance dip gradually splits into two resonances. Their resonance frequencies gradually move to lower and higher frequency compared to the original resonance, respectively, and their resonance intensities gradually increase with the external magnetic field increasing. In this process, the transmittance at 1.04 THz gradually rises up from 1.1% at 0 mT to 40% at 150 mT, which experiences a transition from a resonance dip to a transmission peak. Therefore, this FFPC realizes a magnetically induced THz wave transparency at the central frequency of 1.04 THz with 150 GHz bandwidth. THz wave modulation can be realized in this frequency band with an intensity modulation depth of 40%, and other remarkable modulations occur at 0.9 and 1.14 THz. The mode splitting, resonance strength, and modulation depth increase with increasing magnetic field.

It is necessary to analyze the mechanism of the resonance splitting and induced transparency in the FFPC. Fan et al. have pointed out that the p-polarized wave in the ferrofluid becomes elliptically polarized in the Voigt configuration with a dielectric tensor [135], while the wave in the Si column is still linear polarized, so this periodically MO-dielectric hybrid model leads the magneto-plasmon mode splitting [126, 134]. They used the finite element method (FEM) to simulate this model, and confirmed the existence of the clockwise (CW) and counter-clockwise (CCW) rotating magneto-plasmon modes shown in Figure 4.26c. Both magneto-plasmon modes have the dipole patterns and rotate around the ferrofluid-Si column interface in the CCW and CW directions in the x–y

plane, respectively, not oscillating in the z-direction. Thus, a single original guided-mode resonance turns to two different magneto-plasmon resonances. Meanwhile, these two modes have different effective refractive indexes, i.e. n_- for the CCW mode and n_+ for the CW mode, and they are calculated by the FEM shown in Figure 4.26c. It can be seen that $n_+ = n_-$ without the external magnetic field, and n_+ becomes smaller while n_- becomes larger with the external magnetic field increasing. $\Delta n = (n_- - n_+)$ reaches 0.023 at 150 mT. The refractive index changes lead to the frequency movement of the magneto-plasmon resonances, which is as follows [126, 133, 134]:

$$\omega_{\pm} = \sqrt{\frac{\omega_c^2}{4} + \omega_0^2} \pm \frac{\omega_c}{2} \tag{4.26}$$

where ω_0 is the original resonance frequency, and ω_c is proportional to the external magnetic field, so the $\omega_{\pm}$ are mainly dependent on the external magnetic field. The calculation results are shown in Figure 4.26d and well agree with the experiment data. Similar phenomena of magneto-plasmon and MO mode splitting were also observed in the monolayer graphene and patterned graphene disk arrays [126, 134]. The mechanism of induced transparency can be well explained by the interference between the CCW and CW magneto-plasmon mode. This THz microstructure MO device and its tunability scheme will have great potential applications in THz filtering, modulation, and sensing.

4.4.2 Magneto-optical Circulator

An optical circulator, in which a cubic prism for a signal coupler, a lens-like magneto-optic structure, and a reflecting structure are utilized as important constituents, acts a role of circulation in optical region. Basic features of the optical circulator are such that a signal light cumulatively affects the Faraday rotation on repeated passages through the magneto-optic structure and therefore a shortened magneto-optic structure will provide sufficient roles in performing a circulator action, and furthermore, such multiple passage of the signal light will introduce useful resonant modes in the magneto-optic structure in order to produce operating modes that can achieve various multiple circulation frequency operations in the optical region. And also, the optical circulator has low loss characteristics and broad availability, especially in the field of optical communication. Over the years, several kinds of magneto-optical (MO) circulators have been designed, and some of them are described here.

4.4.2.1 T-shaped Magneto-optical Circulator

Magneto-optical (MO) circulators have attracted much attention for their special applications in photonic crystal integrated optical circuits, which are believed to be one of the most potential candidates to realize future all-optical integrated microchips [136, 137]. With the increasing scale of integrated optical circuits, interferences among waves from different elements or devices on a microchip become more and more serious, and it would destroy the functions of the designed systems [138–141]. Therefore, optimizing optical circuits becomes a key technical issue in developing integrated optical microchips.

Magneto-photonic crystal (MPC) circulators, a kind of nonreciprocal devices [119, 142–145] that make light circulating along a single direction, have useful roles in modulating optical circuits. Optical circulators can be used to isolate light reflected from nearby optical elements or devices, which are important for reducing unwanted light interferences and improving stabilization of large-scale integrated optical circuits. To date, several kinds of MPC circulators have been designed based on the structures with a triangle lattice of air holes embedded in dielectric material [146–150]. In microwave frequencies, the Y-typed optical circulators are also studied [151, 152]. They are achieved by coupling three waveguides to a single magneto-optical material cavity.

In 2012, Wang et al. [153] designed a compact MPC circulator with a T-shaped structure based on a square-lattice structure with dielectric rods embedded in air. In the square-lattice photonic crystals, a wide photonic bandgap for TE (transverse electric) mode, whose electric vector is transverse to the propagation vector of the operating wave, can be obtained. More importantly, it is easy to combine multiple optical devices into an integrated system with a cross structure distribution that is compact in size [154]. Here, through the coupling of magneto-optical rods, nonreciprocal light transmission in two 90° angle waveguides, or a path of single-direction 90° bend of light, is achieved. With the aid of a side-coupled cavity, two paths of the above single-direction 90° bend of light are effectively cascaded, which implements nonreciprocal transmission in two waveguides arranged and linked on a straight line. These solve the key problems for realizing T-shaped optical circulators in square-lattice photonic crystals. The properties of MPC circulators are investigated theoretically by the FEM.

Several kinds of three-port optical circulators have been proposed in triangle-lattice photonic crystals by coupling three photonic crystal waveguides to a magneto-optical material cavity [146, 147, 150], such as a windmill-shaped structure shown in Figure 4.27a and a Y-shaped structure shown in Figure 4.27b. The solid circles labeled M denote the magneto-optical material cavities and the rectangles marked with W_1, W_2, and W_3 denote the waveguides. The circulators have the functions of single-direction light circulation among the three ports, from ports 1 to 2, 2 to 3, and 3 to 1.

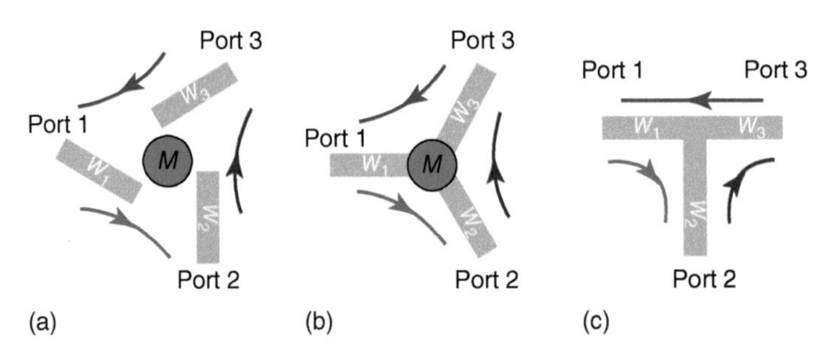

Figure 4.27 Three kinds of circulators. (a) A windmill-shaped circulator. (b) A Y-shaped circulator based on triangle-lattice photonic crystals. (c) A T-shaped circulator based on a square-lattice photonic crystal. Source: Wang et al. 2012 [153]. Reproduced with permission of Elsevier.

The common feature of such structures is that the three waveguides exhibit a 120° rotational symmetry, or triangle symmetry, with respect to the center cavity. This kind of symmetry provides an obvious advantage for designing three-port circulators based on triangle-lattice photonic crystals. However, the symmetry in square-lattice photonic crystals is different from that in triangle-lattice photonic crystals. For devices made in square-lattice photonic crystals, circulators with a T-shaped structure as shown in Figure 4.27c, is needed. In this figure, the bend of light is 90° from port 1 to 2 or port 2 to 3, but no bend from port 3 to 1. So, with only one magneto-optical material cavity located at the intersection of the three waveguides such as that in the triangle-lattice photonic crystals, one could not obtain the three-port circulator shown in Figure 4.27c.

Here, four coupled magneto-optical rods and a side-coupled cavity are used to realize a T-shaped circulator in a square-lattice photonic crystal. As shown in Figure 4.28a, a square-lattice photonic crystal is composed of dielectric rods embedded in air. The refractive index and the radius of the rods are $n_o = 3.4$ and $r_0 = 0.16a$, respectively, where a is the lattice constant. The rods can be chosen to be low-loss ceramic material [155]. In the T-shaped structure, four uniform magneto-optical rods labeled A, B, C, and D are introduced and located near to the waveguide cross center. The magneto-optical rods have the same radius of r_1 to keep a high symmetry in structure. In order to couple neighboring magneto-optical rods efficiently, a dielectric rod labeled H is added at the center. The refractive index of the rod H is the same as that of the other dielectric rods and its radius is set to be $r_2 = 1.25r_1$. Furthermore, a side-coupled cavity is introduced and the details are illustrated in Figure 4.28b. The radius of the dielectric rod in the center of the side-coupled cavity is decreased to be r_3. The three dielectric rods near the cavity center are also reduced and pulled away from the cavity

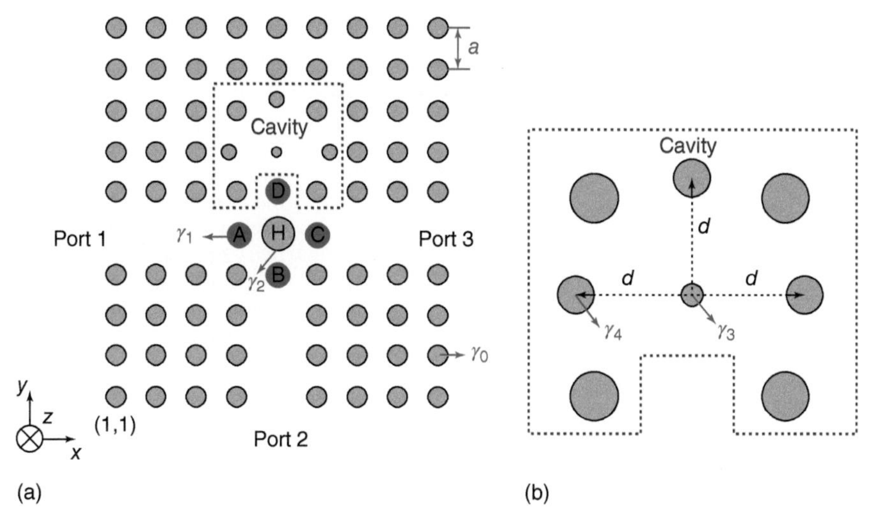

Figure 4.28 (a) Schematic diagram of a T-shaped circulator in a square-lattice photonic crystal with four uniform ferrite rods and a side-coupled cavity. The four ferrite rods are labeled A, B, C, and D respectively; (b) details of the side-coupled cavity. Source: Wang et al. 2012 [153]. Reproduced with permission of Elsevier.

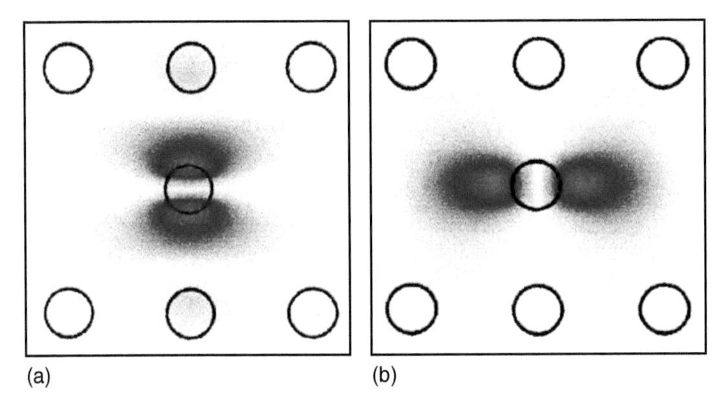

Figure 4.29 E_z-field distributions of two degenerate modes. Source: Wang et al. 2012 [153]. Reproduced with permission of Elsevier.

center. And r_4 and d denote the radius and the distance away from the cavity center, respectively.

Without an external magnetic field, the ferrite rod can support a pair of degenerate modes, as shown in Figure 4.29. When the external magnetic field is introduced, coupling between the two modes takes place, which leads to a rotation of E_z-field distribution or a change of wave front in the area of the ferrite rod. Thus, the direction of light transmission is rotated accordingly. At a certain frequency, a 45° rotation of light can be obtained under the magneto-optical effect with proper operating parameters.

The detailed operating principle of the circulator is shown in Figure 4.30a–c. Figure 4.30a demonstrates the path of light from port 1 to 2. The light is first rotated by an angle of 45° through the magneto-optical effect of ferrite rod A and then another angle of 45° through the magneto-optical effect of ferrite rod B, and finally travels out from port 2, isolating port 3. As a result, a 90° bend of light is realized by the coupling of the two ferrite rods. Similarly, Figure 4.30b illustrates

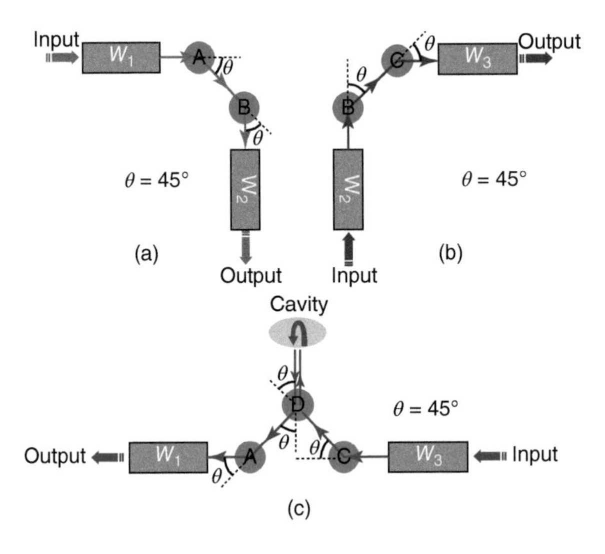

Figure 4.30 Schematic diagram of the light circuit in the T-shaped circulator. (a) From ports 1 to 2, (b) from ports 2 to 3, and (c) from ports 3 to 1. Source: Wang et al. 2012 [153]. Reproduced with permission of Elsevier.

a 90° rotation of light from port 2 to 3 by the cooperation of coupled ferrite rods B and C. Figure 4.30c demonstrates the traveling path of light from port 3 to 1. The light in waveguide W_3 is first downloaded to the side-coupled cavity, with a 90° rotation of light by coupling of ferrite rods C and D, and then is uploaded from the side-coupled cavity to waveguide W_1, with another 90° rotation of light by coupling of ferrite rods D and A. So, the single-direction transmission along the straight line is achieved by bending the light twice by 90° in a single direction through the aid of the side-coupled cavity. For the three cases, each ferrite rod provides a 45° bend of light based on the magneto-optical effect.

To investigate the T-shaped MPC circulator, the FEM was used to analyze the system. It demonstrates that the insertion losses are below 0.24 dB and the isolations are better than 20 dB at the optimized operating frequency. In order to check the feasibility of the circulator, the distributions of electric field E_z for different cases are simulated at the operating frequency of $\omega = 0.4132(2\pi c/a)$, as shown in Figure 4.31a–c. Here c is the light velocity. Figure 4.31a shows that the light launched from port 1 is almost totally transmitted to port 2 (the output port), isolating port 3 (the isolated port). Figure 4.31b demonstrates that the light launched from port 2 is transmitted to port 3 (the output port), isolating

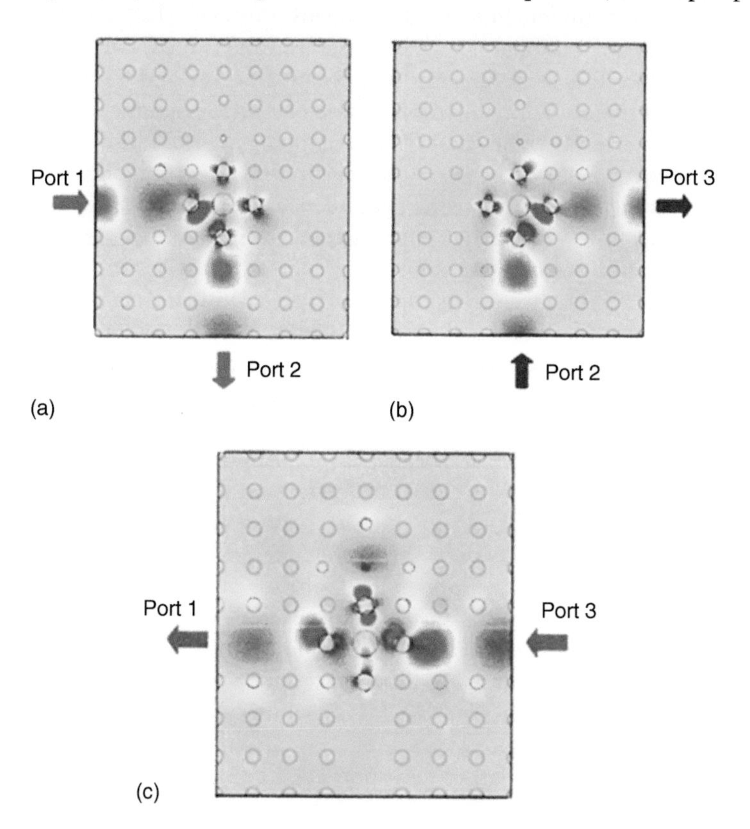

Figure 4.31 Distributions of E_z-field in the T-shaped *MPC* circulator operating at the frequency of $\omega = 0.4132(2\pi c/a)$ for input light launched (a) from port 1, (b) from port 2, and (c) from port 3. Source: Wang et al. 2012 [153]. Reproduced with permission of Elsevier.

port 1 (the isolated port). Figure 4.31c indicates that the light launched from port 3 is transmitted to port 1 (the output port), isolating port 2 (the isolating port). These phenomena demonstrate that single-direction optical circulation is indeed realized in the T-shaped structure.

The circulator shown above can be matched and used with other square-lattice photonic crystal devices to construct optical circuits [156, 157]. Also, it can be cascaded to build multi-port nonreciprocal devices, e.g. isolators between source and load components, echo signal extractors, optical bridges, and optical add-drop multiplexers. These can find applications in microwave sensing and communication. On the other hand, according to the scaling property of photonic crystals, the structure designed above can be applied to that operating at other frequency bands by enlarging or reducing the lattice constant and the rod radii with the same scaling factor and by using materials working well at the corresponding frequencies.

In 2015, Dmitriev et al. [158] proposed a T-shaped circulator with a very simple and compact structure. Through a series of adjustments in the crystalline geometry and using the Nelder–Mead optimization method [159], they achieved a high level of isolation and low insertion losses. The circulator consists of a square lattice of dielectric cylinders immersed in air. Dmitriev et al. [158] considered a junction consisting of a resonator with MO material and three waveguides coupled to the resonator. This structure can operate in sub-THz and THz frequency range and perform nonreciprocal transmission of electromagnetic waves. The proposed device based on photonic crystal technology can be built with reduced dimensions, favoring an increase in the component integration density in communication systems. Owing to strong dependence of parameters of photonic crystals with respect to geometry, adjustments were made in the crystal structure by using of an optimization technique.

Considering excitation in the three ports of the circulator, Dmitriev et al. [158] looked for its good transmission and isolation for a particular frequency band. The objective function was defined as S parameters of the circulator. Thus, optimized values for the radius and the position of the ferrite and dielectric cylinders comprising the resonant were obtained.

Figure 4.32a shows the structure of the crystal before the optimization process. The ferrite cylinder is positioned in the center of the axis between the connecting waveguides. It is noticed that for this geometrical configuration, there is no nonreciprocal transmission, which leads to seeking change in the parameters of the cylinders near the resonator. There are also high losses of the structure for this geometrical arrangement, as can be seen in Figure 4.32b. Then one realizes that optimization needs to be applied to this problem. From the values obtained using the optimization module, the final optimal design with the changes made in the crystal structure can be seen in Figure 4.33. The white cylinders are related to the periodic structure of the employed photonic crystal and each of them has radius equal to $0.2a$, where a is the lattice constant. For frequency $f = 100$ GHz, $a = 1.065$ mm.

After the changes made in the central structure, the radius of the A cylinder remained equal to $0.30562a$, but was displaced in relation to the axis of the waveguide between ports 2 and 3 (Δy_1) of $0.69086a$. The C cylinder has a reduced radius

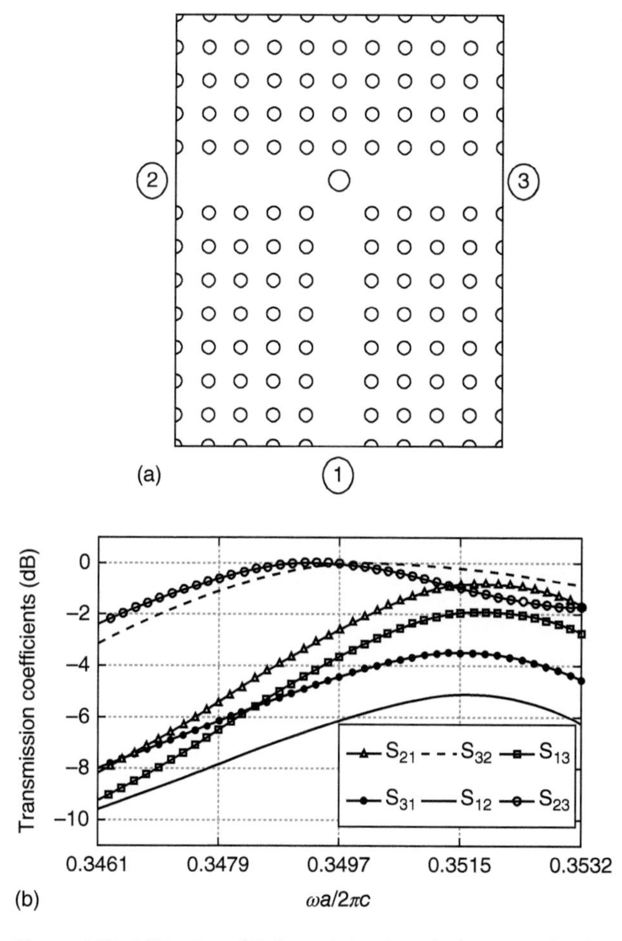

Figure 4.32 (a) Design of T-shaped circulator before optimization process. (b) Frequency responses of the transmission coefficients. Source: Dmitriev et al. 2015 [158]. Reproduced with permission of IARIA.

of $0.01249a$ and was moved vertically to the axis of the upper cylinders (Δy_2) in $0.2563a$. The radii of the B cylinders were increased to $0.07439a$.

The frequency splitting of the rotating modes ω^+ and ω^- versus k/μ is shown in Figure 4.34. It is apparent that the circulator works with low parameter $k/\mu = 0.17$, i. e. it can be projected for THz region. The resonant cavity is based on a nickel-zinc based ferrite rod inserted in the center of the device and the dipole modes are excited in this rod. The ferrite used is produced by TransTech [160] and its product code is TT2-111.

The device frequency response is shown in Figure 4.35. In the normalized central frequency $\omega a/2\pi c = 0.3499$, the insertion losses are smaller than −0.05 dB, where ω is the angular frequency (in radians per second); c is the speed of light in free space. In the frequency band located around 100 GHz, the bandwidth defined at the level of −15 dB of isolation is equal to 620 MHz for excitation at port 1, 680 MHz for excitation at port 2, and 730 MHz for excitation at the port 3.

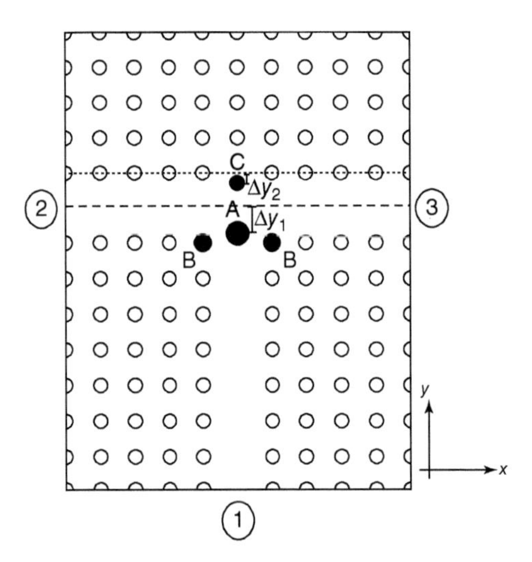

Figure 4.33 Optimal design of the T-shaped circulator. Source: Dmitriev et al. 2015 [158]. Reproduced with permission of IARIA.

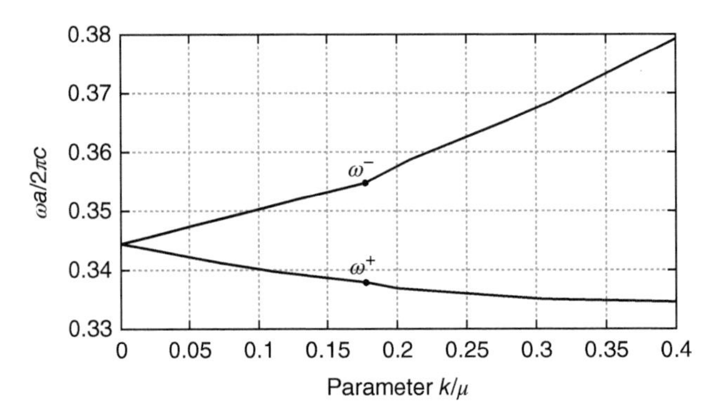

Figure 4.34 Frequency splitting of dipole modes excited in an MO resonator. Source: Dmitriev et al. 2015 [158]. Reproduced with permission of IARIA.

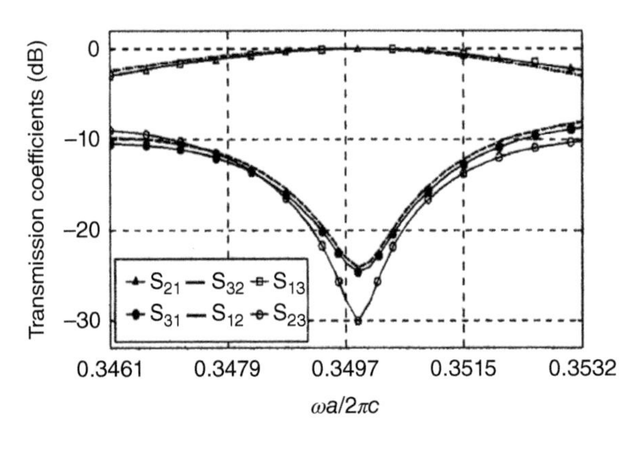

Figure 4.35 Frequency responses of T-circulator for excitation at ports 1, 2, and 3. Source: Dmitriev et al. 2015 [158]. Reproduced with permission of IARIA.

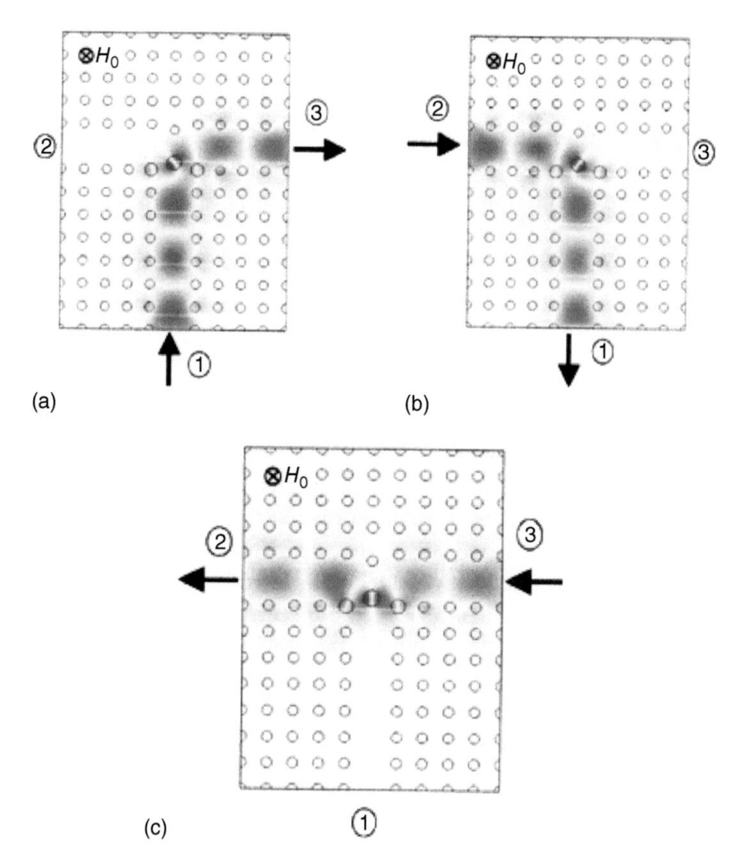

Figure 4.36 Distribution of E_z-component of the electromagnetic field for T- circulator at central frequency $f = 98.55$ GHz. (a) For excitation at port 1. (b) For excitation at port 2, and (c) For excitation at port 3. Source: Dmitriev et al. 2015 [158]. Reproduced with permission of IARIA.

The propagation of electromagnetic waves is given as follows: when the excitation is applied at port 1, there is signal transmission from this port to port 3, with isolation of port 2 due to the special alignment of the dipole mode, as can be seen in Figure 4.36a. Similarly, when the input signal is applied at port 2 (Figure 4.36b), this is transferred to port 1, with isolation of port 3, and when applied at port 3 (Figure 4.36), this is transferred to port 2, with isolation of port 1. This case corresponds to the propagation in a counterclockwise direction. If the signal of the external dc magnetic field H_0 is reversed, the propagation of signals is clockwise (1 to 2, 2 to 3, and 3 to 1).

In the cases illustrated in Figure 4.36a,b, it can be seen that the stationary dipole mode excited in the resonant cavity is rotated by an angle of 45°, which provides isolation of ports 2 and 3, respectively. On the other hand, in the case illustrated in Figure 4.36c, it is shown that the stationary dipole mode suffers no rotation, transferring the input signal applied in port 3 to port 2 with port 1 isolated. Analyzing the intensity of the electric field in the T-junction, one can see that the resonant cavity is formed by a central ferrite cylinder and two dielectric cylinders with increased diameters compared to other cylinders that comprise the photonic crystal.

4.4.2.2 Multiple-Port Integrated Optical Circulators

The circulator is a multiple-port device that routes light such that light entering any port exits from the next, much like a roundabout for photons. Many applications in data centers, telecommunications, and sensors would benefit from bidirectional operation, which requires an optical circulator. Owing to optical nonreciprocity, circulators often operate based on the magneto-optic Faraday effect. However, the transition from discrete to integrated optical circulators has been hindered by lattice and thermal mismatch between commonly used magneto-optic materials and silicon or III–V substrates.

Huang et al. [161] demonstrated for the first time heterogeneously integrated optical circulators on silicon operating in the TM mode with up to 14.4 dB of isolation ratio. They exploit a micro ring based structure with a significantly reduced footprint (20 μm radius). Furthermore, they do not use a permanent magnet to generate the magnetic field. Instead, a localized magnetic field is generated from an integrated microstrip that serves as an electromagnet and can be tailored to match various device geometries. This current-induced magnetic field can be switched with a rise and fall time of 400 ps, which flips the direction of the magnetic field and reroutes all the optical pathways in the device. Thus, the circulator can be dynamically reconfigured on a sub-nanosecond timescale.

The proposed design of the device is shown in Figure 4.37. It consists of two identical ring resonators and three bus waveguides fabricated using deep ultraviolet lithography on a 230 nm thick SOI wafer with a bonded 400 nm Ce:YIG upper cladding. The waveguides are 600 nm wide, so the largest NRPS can be achieved [162]. The two rings share a common bus waveguide, but are not coupled together. The ring radius is set to 20 μm and the ring–waveguide gap is 200 nm.

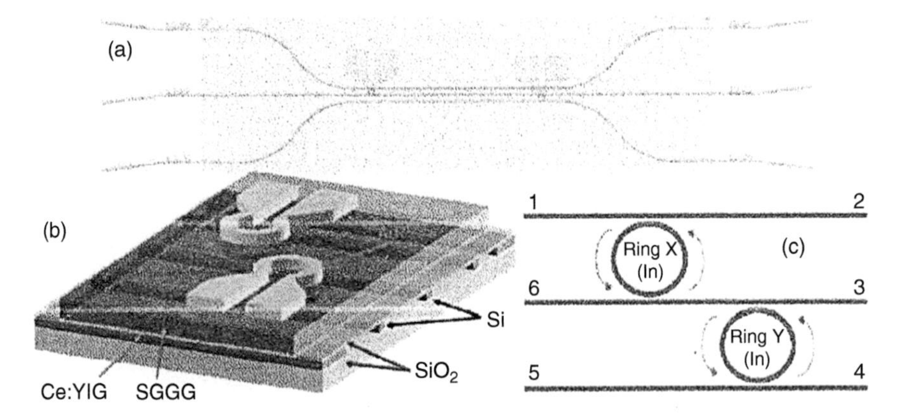

Figure 4.37 (a) Mask layout of the six-port circulator. (b) Schematic of the device. (c) One possible configuration of the circulator with the fields radially outward in both rings. The circulation direction is given by the arrows. Source: Huang et al. 2017 [161]. Reproduced with permission of OSA Publishing.

Following an O_2 plasma assisted direct bonding procedure, Huang et al. [161] reduced the thickness of the substrate of the bonded Ce:YIG die to roughly 10 μm, and then patterned a gold microstrip on top of each ring using a lift-off process. When current is applied through the two microstrips, they induce a radially outward or inward magnetic field with respect to the corresponding ring resonator. This causes a resonance wavelength split (RWS) in the ring between the clockwise (CW) and counterclockwise (CCW) propagating TM modes. Thus, light propagating between two ports will experience a different transfer function according to the direction.

In the configuration shown in Figure 4.37c, the magnetic field is orientated outward for both rings. When the operating wavelength is aligned to the CCW resonating mode, light going from the left to right will observe the ring out of resonance. Vice versa, for the light going from right to left, the rings are on-resonance and the light is dropped to the adjacent waveguide (i.e. port). As a result, the circulation among the ports is $1 \rightarrow 2 \rightarrow 3 \rightarrow 4 \rightarrow 5 \rightarrow 6 \rightarrow 1$. In this configuration, the proposed device can be seen as a cascade of two identical optical circulators presented in Refs. [163, 164].

Furthermore, the magnetic field can be easily switched between inward and outward directions by reversing the current. Therefore, it is possible to dynamically reconfigure the circulation path. For the six-port circulator, Huang et al. [161] have fabricated four possible configurations; each microstrip can be independently controlled. These are shown in Figure 4.38. In general, the reconfigurable multi-port architecture presented here can be expanded to obtain an arbitrary number of ports by adding additional rings and bus waveguides in this manner. Using $(N-1)$ rings with N bus waveguides can result in the realization of a $2N$-port circulator with $2^{(N-1)}$ possible configurations. Circulators with an odd number of ports can be fabricated by adding a loop mirror at one port.

The six-port circulator is characterized by injecting TM polarized light into one of the ports using a tunable laser that is swept near 1550 nm. Owing to the

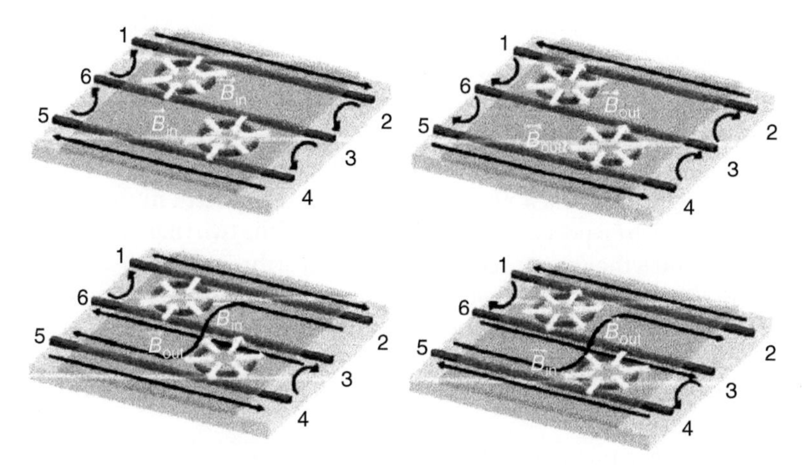

Figure 4.38 The four different configurations of the six-port circulator. Source: Huang et al. 2017 [161]. Reproduced with permission of OSA Publishing.

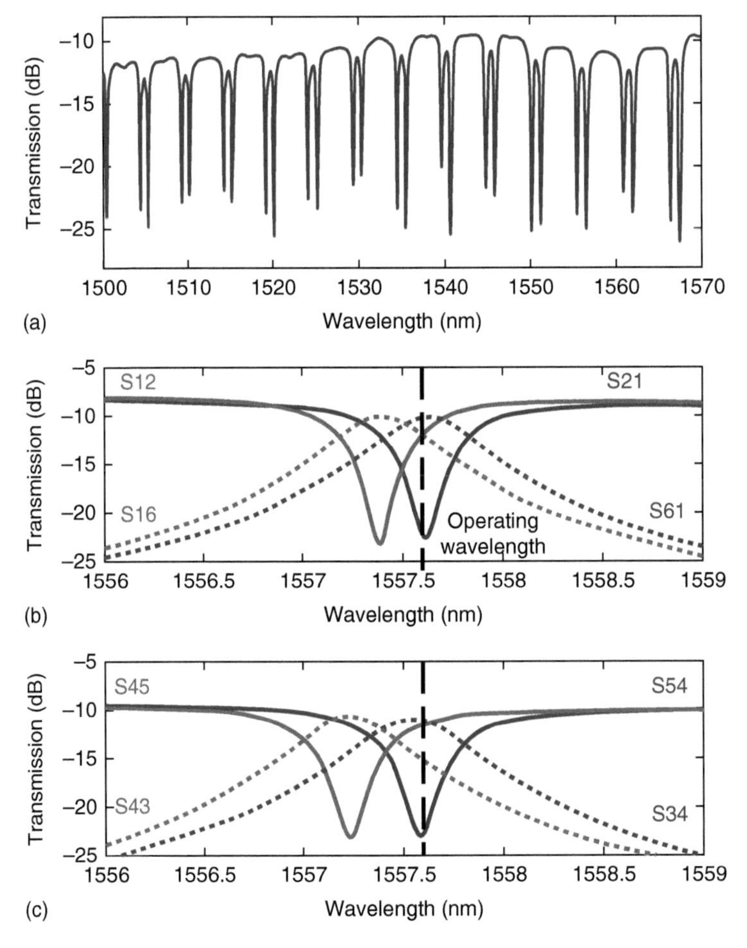

Figure 4.39 (a) Intrinsic spectrum of the two rings with no magnetic field. Transmission through the top ring (b) and the bottom ring (c) are shown, with clear evidence of *RWS* and optical nonreciprocity. Source: Huang et al. 2017 [161]. Reproduced with permission of OSA Publishing.

angled facets designed to reduce the reflections, Huang et al. [161] were unable to simultaneously measure the transmittance through all ports. Instead, they successively injected light from each port and measured the spectra in the corresponding through and drop ports. The intrinsic spectra of the two rings are shown in Figure 4.39a in which they measured the transmittance through the middle bus waveguide (Ports 3 and 6 in Figure 4.38), capturing the resonance wavelength of both rings. Although the rings were designed to be identical, fabrication error and wafer nonuniformity cause more than 1 nm difference between the two resonances. Therefore, the two rings must also be aligned to achieve circulation at a common wavelength.

Huang et al. [161] has previously shown that the microstrip can also serve as a thermal tuner for the ring [165]. They utilize this and apply different amounts of current to the two rings in order to align them in addition to generating the necessary magnetic field. They find the optimal conditions to be 185 mA for the top ring and 262 mA for the bottom ring. Under these conditions, the two resonances align, and there is sufficient magnetic field to establish a clear RWS between the CW and CCW modes. This is depicted in Figure 4.39b for the top ring and Figure 4.39c for the bottom ring. Since the magnetic field strength is different for the two rings, the RWS is around 0.25 nm for the top ring and 0.35 nm for the bottom ring.

This is reasonable considering that the applied current was larger for the bottom ring in order to compensate for the fabrication variation. In this configuration, the operating wavelength near 1557.6 nm is on resonance with the CCW modes of both rings, causing a circulation path of $1 \rightarrow 2 \rightarrow 3 \rightarrow 4 \rightarrow 5 \rightarrow 6 \rightarrow 1$. From the measurements, the scattering parameters of the device can be extracted at the working wavelength $\lambda = 1557.6$ nm. The largest isolation ratio was found to be 14.4 dB between ports 2 and 3, while the smallest isolation ratio was 2.5 dB between ports 1 and 6. The insertion losses along the forward circulating path range from 10.1 to 14.3 dB, which is similar to what was measured in the four-port device, and can be reduced by shortening the length of the Ce:YIG cladding above the bus waveguides. Overall, a better device performance can be achieved by aligning the resonances with better fabrication accuracy or a separate thermal tuner.

4.4.2.3 Terahertz Magneto-optical Circulator

In 2012, Fan et al. [166] investigated the gyromagnetic properties of ferrite materials and the nonreciprocal property of a silicon-ferrite photonic crystal cavity in the terahertz region. They designed a magnetically tunable circulator, the central operating frequency of which can be tuned in the sub-THz region (180–205 GHz) and the maximum isolation is 65.2 dB.

The structure of a three-port junction circulator is shown in Figure 4.40a, with three 120° rotational symmetry branches of waveguides coupled to a ferrite rod cavity at the center. This 120° rotational symmetry breaks the space symmetry along the propagation direction. In the photonic crystal cavity, two orthogonal doubly degenerate dipole modes exist without an external field, while the external magnetic field makes them couple and forms two linear combinations corresponding to left- and right-rotating modes with frequencies ω_L and ω_R [146, 167]. Figure 4.40b shows one of the rotating modes. Only rotating modes can well support the function of the circulator, and the frequency $(\omega_L + \omega_R)/2$ is just the central operating frequency of the circulator. When $\Delta\omega = \omega_L - \omega_R$ becomes larger, the intensity of MO coupling is stronger. The MO coupling strength is not only determined by the gyrotropy κ/μ, but also closely related to the spatial overlap between the ferrite domain and the defect modes field.

In the analysis, the photonic crystal cavity affects the mode confinement and spatial overlap of the magnetic material domain and defect modes, while the gyrotropy and ferromagnetic loss of the ferrite, induced and affected by magnetic field and THz frequency, have a great influence on the transmission and isolation property of this device. By structure optimization and analysis of defect mode coupling, a tunable circulator is designed, the central operating frequency of which can be tuned from 180 to 205 GHz and the maximum isolation is 65.2 dB

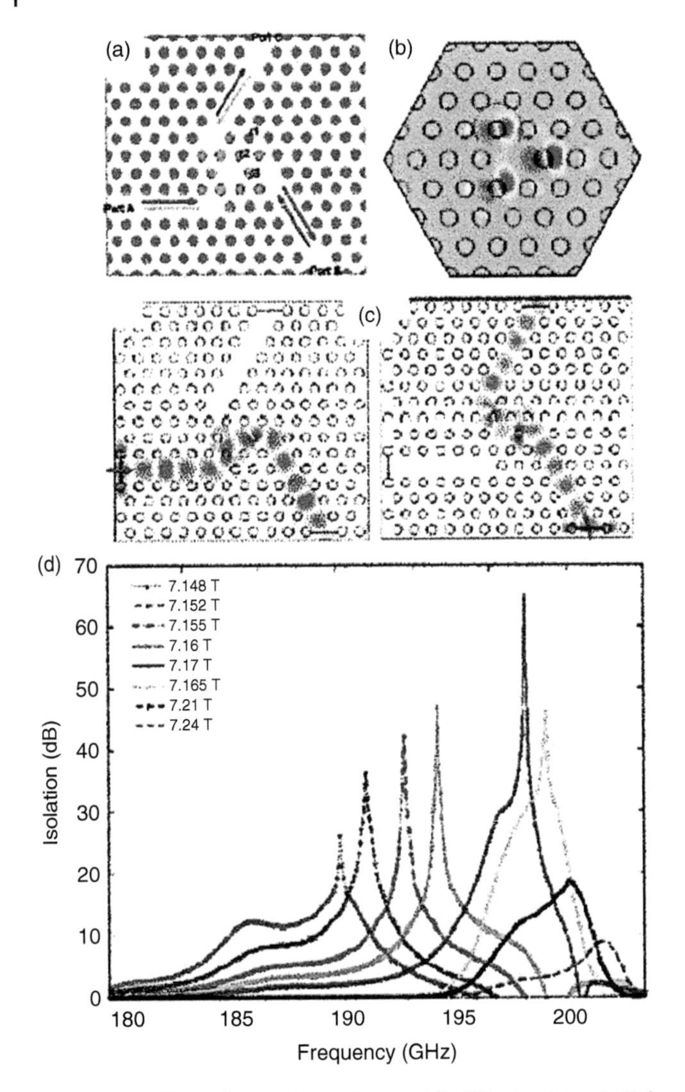

Figure 4.40 Silicon-ferrite photonic crystal for THz circulator. (a) Schematic diagram. (b) Electric field pattern of defected rotating dipole mode. (c) Electric field patterns of the circulator. (d) Isolation spectra under different external magnetic fields. Source: Fan et al. 2012 [166]. Reproduced with permission of Elsevier.

as shown in Figure 4.40d. The field patterns at operating frequency are shown in Figure 4.40b,c. This circulator can realize the controllable splitting, routing, and isolating for THz applications.

4.4.3 Magneto-optical Isolator

Magneto-optical isolators are one of the most important passive components in optical communication systems. The function of an optical isolator is to let a light beam pass through in one direction, that is, the forward direction only, like

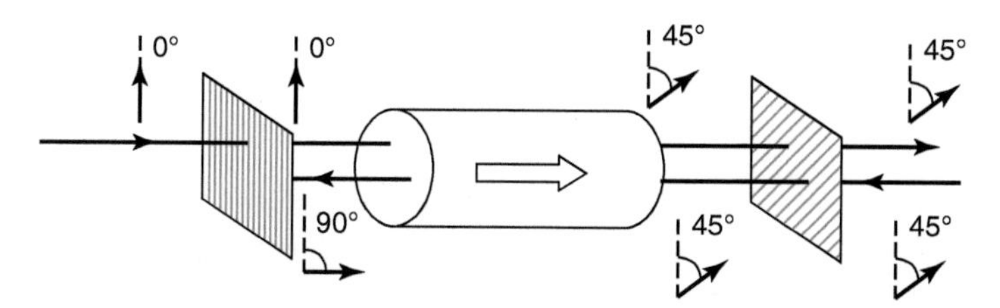

Figure 4.41 Design of free-space optical isolator. The Faraday rotator is placed between the entrance polarizer (left side) and the exit polarizer (right side). Upper diagrams show polarization in the forward direction. Lower diagrams show polarization in the backward direction. Source: Zayets and Ando 2010 [168]. Reproduced with permission of InTech.

one-way traffic. Optical isolators are used to prevent destabilizing feedback of light that causes undesirable effects such as frequency instability in laser sources and parasitic oscillation in optical amplifiers.

The conventional bulk-type optical isolator consists of a 45° Faraday rotator placed between two polarizers (Figure 4.41). The angle between the axes of the entrance polarizer and exit polarizer is 45°. In forward direction, the polarization of light is 45° rotated by the Faraday rotator to be along the axis of the exit polarizer. Therefore, the light can pass through the isolator in the forward direction. In backward direction, the direction of polarization rotation is opposite to that in forward direction due the nonreciprocal nature of the magneto-optical effect. At the entrance polarizer, the polarization is 90° to the polarizer axis and the light is fully blocked.

In present optical networks, ferrimagnetic garnet oxide crystals such as $Y_3Fe_5O_{12}$ (YIG) and $(GdBi)_3Fe_5O_{12}$ are used as magneto-optical materials for discrete optical isolators. The ordinary optical isolator is bulky (therefore called a bulk isolator) and incompatible with waveguide-based optical devices, so it cannot be used in PICs. It has, however, superior optical characteristics (low forward loss and high backward loss). Such good performance is a target in developing waveguide optical isolators.

There are several strategies to develop waveguide optical isolators that can be integrated monolithically with waveguide-based semiconductor optical devices on an InP substrate. The strategies can be classified into two types. One is to use the Faraday effect as in conventional bulk isolators. Transferring the principle of bulk isolators to a planer waveguide geometry raises a number of inherent difficulties such as the incoherence of polarization rotation induced by structural birefringence. Therefore, newer ideas are needed to use the Faraday effect in waveguide structure. Sophisticated examples are the Cotton–Mouton isolator [169, 170] and the QPM Faraday rotation isolator [171, 172]. The latter, in particular, has attracted attention in recent years because of its compact techniques for producing the device. The other strategy to make waveguide isolators is to use asymmetric magneto-optic effects that occur in semiconductor waveguides combined with magnetic material. Leading examples are the nonreciprocal-phase-shift isolator [69, 173–175] and the nonreciprocal-loss isolator [176–181]. The nonreciprocal-loss isolator does not use rare earth garnet, so it is compatible with standard semiconductor manufacturing processes.

Outlines of the QPM Faraday rotation isolator and the nonreciprocal-phase-shift isolator are given in the following sections.

4.4.3.1 Quasi-Phase-Matching Faraday Rotation Isolator

Figure 4.42 shows a schematic of the QPM Faraday rotation isolator. The device consists of a Faraday rotator (non-reciprocal) section and a polarization rotator (reciprocal) section integrated with a semiconductor laser diode that provides a TE-polarized output. The Faraday rotator section consists of an AlGaAs/GaAs waveguide combined with a sputter coated film of magnetic rare earth garnet $CeY_2Fe_5O_{12}$(Ce:YIG). To obtain appropriate polarization rotation, this device uses the QPM Faraday effect in an upper-cladding that periodically alternates between magneto-optic (MO) and non-MO media. Incident light of TE mode traveling in the forward direction will first pass through the Faraday rotator section to be rotated by +45°. The light then passes through the reciprocal polarization rotator section and is rotated by −45°. Consequently, the light keeps its TE-mode and passes through the output edge. In contrast, backward traveling light of TE-mode from the output filter is first rotated by +45° in the reciprocal polarization rotator and then nonreciprocally rotated by +45° in the Faraday rotator section. Consequently, backward light is transformed into a TM mode and therefore has no influence on the stability of the laser because the TE-mode laser diode is insensitive to TM-polarized light. The point of this device is TE–TM mode conversion in the waveguide. At present, efficient mode conversion cannot be achieved, so practical devices have yet to be developed.

Using magneto-optical waveguides made of $Cd_{1-x}Mn_xTe$ is effective in achieving efficient mode conversion [183, 184]. Diluted magnetic semiconductor $Cd_{1-x}Mn_xTe$ has the zinc blende crystal structure, the same as that of ordinary electro-optical semiconductors such as GaAs and InP. Therefore, a single-crystalline $Cd_{1-x}Mn_xTe$ film can be grown epitaxially on GaAs and InP substrates. In addition, $Cd_{1-x}Mn_xTe$ exhibits a large Faraday effect near its absorption edge because of the anomalously strong exchange interaction between the sp-band electrons and localized d-electrons of Mn^{2+}. Almost

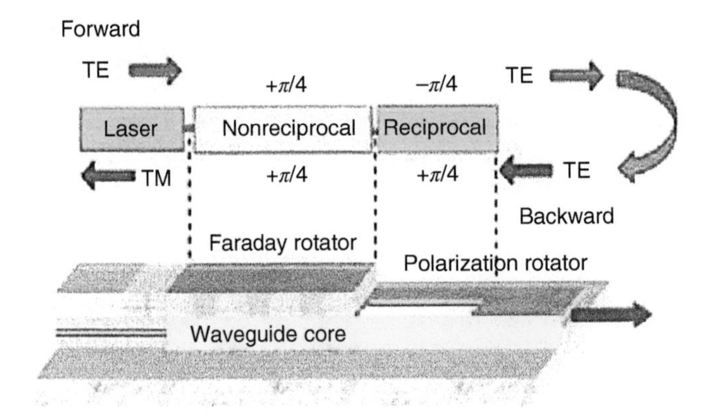

Figure 4.42 Schematic of QPM Faraday rotation isolator. Source: https://creativecommons.org/licenses/by/3.0/, [182].

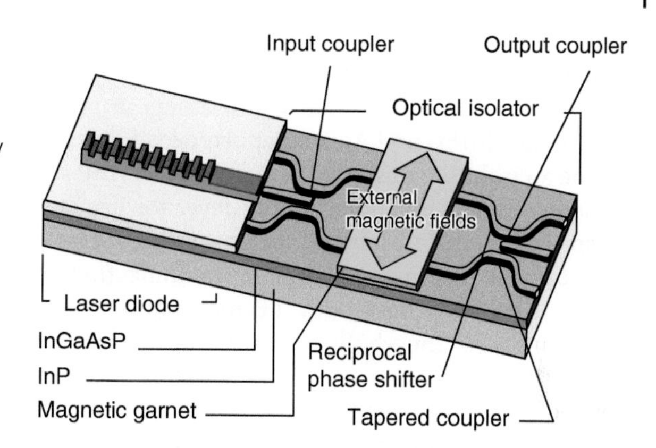

Figure 4.43 Nonreciprocal-phase-shift isolator uses modified Mach–Zehnder interferometer. Source: https://creativecommons.org/licenses/by/3.0/, [182].

complete TE–TM mode conversion (98%±2% conversion) was observed in a $Cd_{1-x}Mn_xTe$ waveguide layer on a GaAs substrate [183, 184].

4.4.3.2 Nonreciprocal Phase-Shift Isolator

The nonreciprocal-phase-shift isolator uses a modified MZI that is designed so that light waves traveling in two arms will be in phase for forward propagation and out of phase for backward propagation. Figure 4.43 shows the structure of the isolator combined with a laser. The InGaAsP MZI consists of a pair of three-guide tapered couplers, and an ordinary reciprocal 90° shifter on one of the arms. Reciprocal phase shifting is achieved simply by setting a difference in dimensions or a refractive index between the optical paths along the two arms. A magnetic rare earth garnet YIG:Ce layer is placed on the arms to form a nonreciprocal 90° phase shifter on each arm. The garnet layer was pasted on the interferometer by means of a direct-bonding technique. Two external magnetic fields are applied to the magnetic layer on the two arms in an antiparallel direction, as shown in Figure 4.43; this produces a NRPS in the interferometer in a push–pull manner.

The isolator operates as follows. A forward-traveling light wave from the laser enters the central waveguide of the input coupler and is divided between the two arms. During the light wave's travel in the arms, a −90° nonreciprocal phase difference is produced, but it is canceled by a +90° reciprocal phase difference. The two waves that are divided recouple at the output coupler, and output light will appear in the central waveguide. In contrast, for a backward-traveling wave from the output coupler, the nonreciprocal phase difference changes its sign to +90°, and it is added to the reciprocal phase difference to produce a total difference of 180°. Consequently, output light will appear in the two waveguides on both sides of the input coupler and not appear in the central waveguide.

4.4.3.3 TM-Mode Waveguide Isolator

One of the promising ways to create waveguide optical isolators is by making use of the phenomenon of nonreciprocal loss. This phenomenon is a nonreciprocal magneto-optic phenomenon where the propagation loss of light is larger in backward propagation than in the forward direction. Use of this phenomenon can provide new waveguide isolators that use neither Faraday rotator nor polarizer and, therefore, are suitable for monolithic integration with other optical devices on an InP substrate. The theory of the nonreciprocal loss phenomenon was first proposed by Takenaka, Zaets, and others in 1999 [185, 186]. After that, Ghent University-IMEC and Alcatel reported leading experimental results in 2004; they made an isolator consisting of an InGaAlAs/InP semiconductor waveguide combined with a ferromagnetic CoFe layer for use at 1.3 μm wavelength [176, 177]. Inspired by this result, aiming to create polarization insensitive waveguide isolators for 1.5 μm band optical communication systems, Amemiya and Nakano [182] have developed both TE-mode and TM-mode isolators based on this phenomenon. They built prototype devices and obtained a nonreciprocity of 14.7 dB/mm for TE-mode devices and 12.0 dB/mm for TM-mode devices, the largest values ever reported for 1.5 μm band waveguide isolators. The TE-mode device consisted of an InGaAsP/InP waveguide with a ferromagnetic Fe layer attached on a side of the waveguide [179]. For the TM-mode device, instead of ordinary ferromagnetic metals, Amemiya and Nakano [182] used ferromagnetic intermetallic compounds MnAs and MnSb, which are compatible with semiconductor manufacturing processes.

Figure 4.44 illustrates the TM-mode waveguide isolators with a cross section perpendicular to the direction of light propagation. Two kinds of structure are shown. The device consists of a magneto-optical planar waveguide that is composed of a TM-mode SOA waveguide on an InP substrate and a ferromagnetic layer attached on top of the waveguide. To operate the SOA, a metal electrode is placed on the surface of the ferromagnetic layer (a driving current for the SOA flows from the electrode to the substrate). Incident light passes through the SOA waveguide perpendicular to the figure (z-direction). To operate the device, an external magnetic field is applied in the x-direction so that the ferromagnetic layer is magnetized perpendicular to the propagation of light. Light traveling along the waveguide interacts with the ferromagnetic layer.

Nonreciprocal propagation loss is caused by the magneto-optic transverse Kerr effect in the magneto-optical planar waveguide. To put it plainly for TM-mode light, the nonreciprocity is produced when light is reflected at the interface between the magnetized ferromagnetic layer and the SOA waveguide.

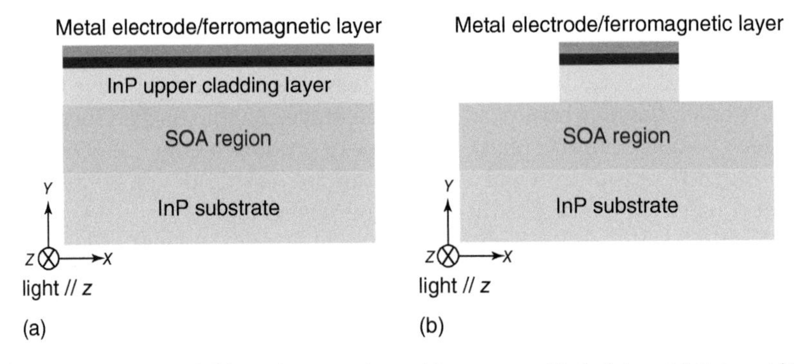

Figure 4.44 Typical TM-mode nonreciprocal-loss waveguide isolators. (a) Gain guiding structure and (b) ridge waveguide structure. Source: https://creativecommons.org/licenses/by/3.0/, [182].

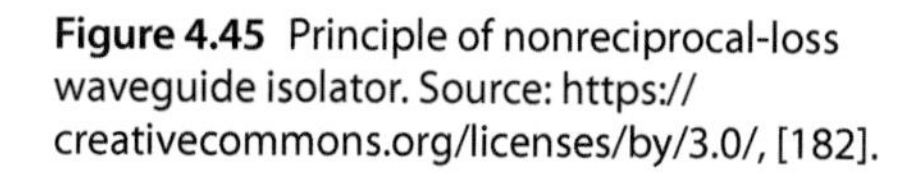

Figure 4.45 Principle of nonreciprocal-loss waveguide isolator. Source: https://creativecommons.org/licenses/by/3.0/, [182].

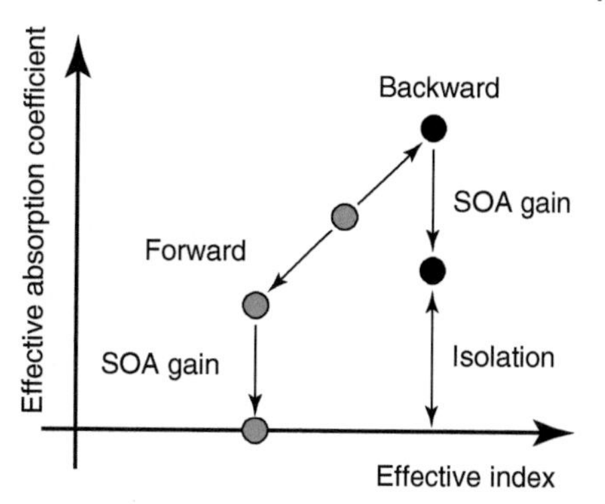

The light reduces its intensity when reflected from the ferromagnetic layer, which absorbs light strongly, and the reduction is larger for backward propagating light than forward propagating light because of the transverse Kerr effect. As a result, the propagation loss is larger for backward propagation (z-direction) than for forward propagation (z-direction). Figure 4.45 illustrates the operation of the isolator on the propagation constant plane of the waveguide. The backward light is attenuated more strongly than forward light. Since forward light is also attenuated, the SOA is used to compensate for the forward loss; the SOA is operated so that the net loss for forward propagation will be zero. Under these conditions, the waveguide can act as an optical isolator.

Figure 4.46a is a cross-sectional diagram of the TM-mode waveguide isolator with a ferromagnetic MnAs layer. Figure 4.46b is simply the cross section of SEM device. The MnAs layer covers the SOA surface, and two interface layers (a highly doped p-type InGaAs contact layer and a p-type InP cladding layer) are inserted between the two. The InGaAs contact layer has to be thin so that 1.5 μm light traveling in the SOA will extend into the MnAs layer (the absorption edge of the contact layer is about 1550 nm). An Au/Ti double metal layer covers the MnAs layer, forming an electrode for current injection into the SOA. Light passes through the SOA waveguide in a direction perpendicular to the

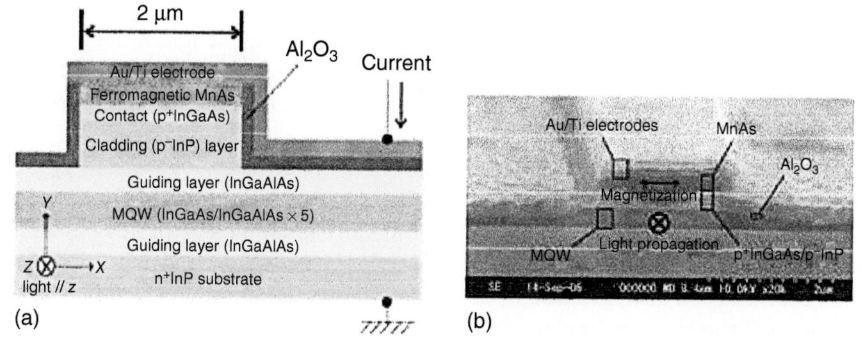

Figure 4.46 (a) Schematic cross section of the waveguide isolator for 1.5 μm TM mode, consisting of a ridge-shaped optical amplifying waveguide covered with a MnAs layer magnetized in the x-direction. Light propagates along the z-direction. (b) SEM cross section of device. Source: https://creativecommons.org/licenses/by/3.0/, [182].

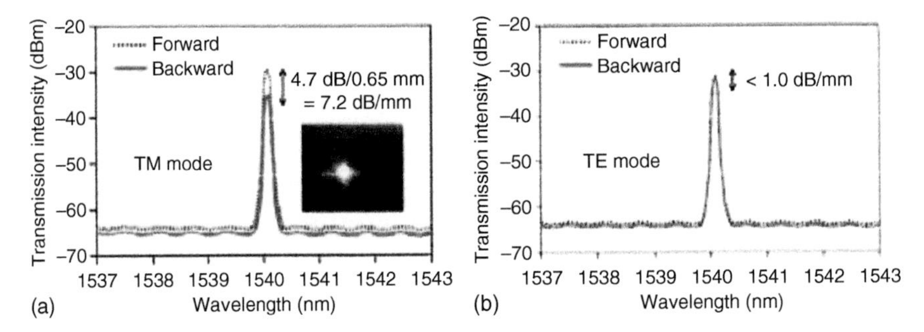

Figure 4.47 Transmission spectra of device for forward transmission (dashed line) and backward transmission (solid line), measured for (a) TM-mode and (b) TE-mode at 1.54 μm wavelength, 100 mA driving current, and 0.1 T magnetic field. Device is 0.65 mm long. Data on transmission intensity include loss caused by measurement system. Inset is the near-field pattern of TM-mode forward propagating light. Source: https://creativecommons.org/licenses/by/3.0/, [182].

figure (z-direction). An Al_2O_3 insulating layer separates the SOA surface from the Au–Ti electrode except on the contact region. Incident light passes through the SOA waveguide perpendicular to the figure (z-direction).

The intensity of light transmitted in the device (or the output light from the device) was measured using the OSA and the power meter. Figure 4.47 shows the transmission spectra of the device with a length of 0.65 mm. The intensity of the output light from the device is plotted as a function of wavelength for forward (dashed line) and backward (solid line) propagation of (a) TM-polarized and (b) TE-polarized light. The wavelength of incident light was fixed at 1.54 μm, which was the gain peak wavelength of the SOA. For TM-mode light, the output intensity changed by 4.7 dB by switching the direction of light propagation. The device operated efficiently as a TM-mode isolator with an isolation ratio of 7.2 dB/mm (=4.7 dB/0.65 mm). In contrast, the output intensity for TE-mode light was not dependent on the direction of light propagation. Small periodic ripples in amplified spontaneous emission spectra are shown in Figure 4.47. They are caused by Fabry–Perot (FP) interference due to reflection from cleaved facets; the period was consistent with the value predicted from the length and effective refractive index of the device. The inset in Figure 4.47a is the near-field pattern of the TM-mode forward propagating light and shows that the device operated successfully in a single mode.

The data of transmission intensity in Figure 4.47 includes the loss caused by the measurement system. To examine the intrinsic transmission loss of the device, Amemiya and Nakano [182] measured the transmission intensity for devices with different lengths. Figure 4.48 shows the output intensity for forward and backward transmission as a function of device length (isolation ratio is also plotted). The slope of the forward line gives the intrinsic transmission loss (or absorption loss) per unit length. Amemiya and Nakano [182] estimated that forward loss in the device was 10.6 dB/mm – still large for practical use. This is so because the gain of the SOA was lower than they had expected, and therefore, insufficient to compensate for the intrinsic transmission loss in the device. The loss

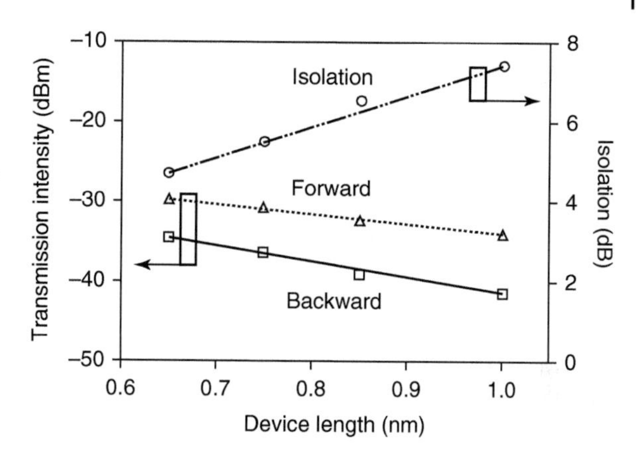

Figure 4.48 Transmission intensity as a function of device length, measured for 1.54 μm TM mode, with 100 mA driving current and 0.1 T magnetic field. Isolation ratio is also plotted. Source: https://creativecommons.org/licenses/by/3.0/, [182].

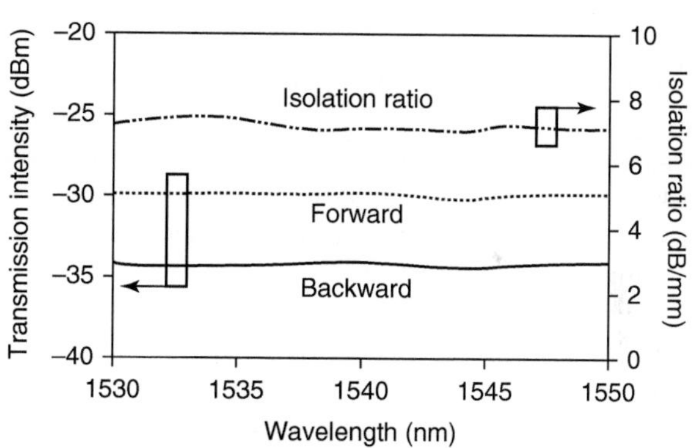

Figure 4.49 Isolation ratio as a function of a wavelength from 1.53 to 1.55 μm for a 0.65-mm long device. Transmission intensity is also plotted for forward and backward propagation (including measurement system loss). Source: https://creativecommons.org/licenses/by/3.0/, [182].

caused by the measurement system can also be calculated using the vertical-axis intercept of the forward line and the output intensity of the tunable laser. It was estimated to be 28 dB – output coupler loss 3 dB plus lensed-fiber coupling loss 12.5 dB/facet × 2 between the measurement system and the device.

Figure 4.49 is a plot of the isolation ratio, as a function of wavelength from 1.53 to 1.55 μm. The device was 0.65 mm long. The output intensities for forward and backward propagations are also plotted (including the measurement system loss). In this range of wavelength, the isolation ratio was almost constant. The isolation ratio 7.2 dB/mm of this waveguide isolator was still small for practical use. In addition, the device was unable to operate at temperatures higher than room temperature because the Curie temperature of MnAs is only 40 °C. Although MnAs is not common material at present for integrated optics, it will soon bring technical innovation in functional magneto-optic devices for large-scale PICs.

4.4.3.4 Silicon-Based MO Isolator and Circulator

Nonreciprocal devices are indispensable for highly functional optical communication systems. Such device with semiconductor platform partially bonded magneto-optic garnet can be integrated with other devices. Using the NRPS, which is the first-order magneto-optic effect in the Voigt configurations, is more suitable for waveguide devices than using the Faraday rotation, which requires strict phase matching between TE and TM modes.

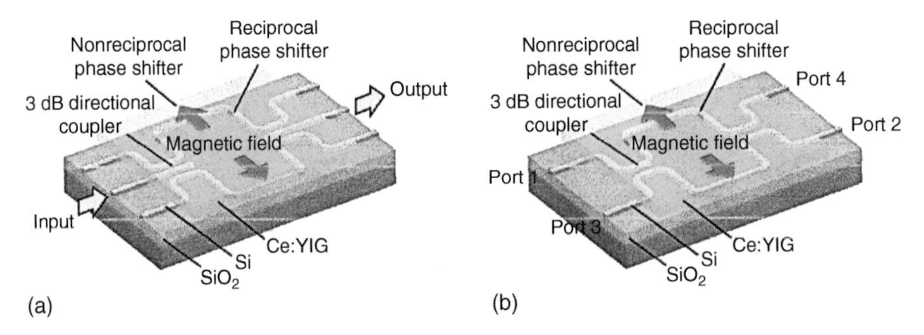

Figure 4.50 Schematics of (a) MZI optical isolator and (b) MZI optical circulator with silicon waveguides. Source: Shoji et al. 2013 [187]. Reproduced with permission of PIERS.

Shoji et al. [187] have designed a magneto-optical isolator and circulator on silicon photonics circuits using the direct bonding of silicon and a magneto-optic garnet. The devices are based on a MZI composed of silicon waveguides. Bonded magnet-optic garnet as an upper cladding layer induces NRPS in a push–pull manner by applying magnetic fields in the Voigt configuration. Depending on the number of input and output ports, an MZI becomes an isolator or a circulator. In this development, an optical isolator with 30 dB isolation and a four-port circulator are demonstrated.

Figure 4.50 shows the schematic of silicon-based MZI optical isolator and circulator. The center ports of 3 dB coupler are the input and output ports. A magneto-optic garnet Ce-substituted yttrium iron garnet (Ce:YIG) is bonded on the MZI as an upper cladding layer. A NRPS is induced by the magneto-optic effect in Ce:YIG by applying magnetic fields to the arms in antiparallel directions. Also, a reciprocal phase shift is induced by an optical path difference between the MZI arms. The NRPS provides a $\pm\pi/2$ phase difference depending on the propagation direction. This is canceled for the forward direction with a reciprocal phase difference of $\pi/2$, but is added for the backward direction. The forward light exhibits in-phase interference, while the backward light exhibits anti-phase interference. The light exhibiting in-phase interference couples to the center port. The light exhibiting anti-phase interference is radiated to the side ports. Then, the MZI functions as an optical isolator.

In the same way, when the 3 dB directional coupler is composed of 2×2 ports, in- and anti-phase interferences select the transmission ports of the MZI. That is, the light transmits from port 1 to port 2 and from port 3 to port 4 for in-phase interference and, on the other hand, from port 2 to port 3 and from port 4 to port 1 for anti-phase interference, respectively. Then, the MZI functions as an optical circulator.

The NRPS is a directionally dependent propagation constant and can be simulated by solving optical mode field and using perturbation theory [188, 189]. This shift is brought about by the directionally dependent dielectric permittivity of the magneto-optic garnet, where the off-diagonal elements change its sign due to the first-order magneto-optic effect. The length of the NRPS is designed to be 400 μm for a silicon waveguide dimension of $450 \times 220\ \text{nm}^2$ and the Faraday rotation coefficient of −4500 deg/cm of Ce:YIG at a wavelength of 1550 nm.

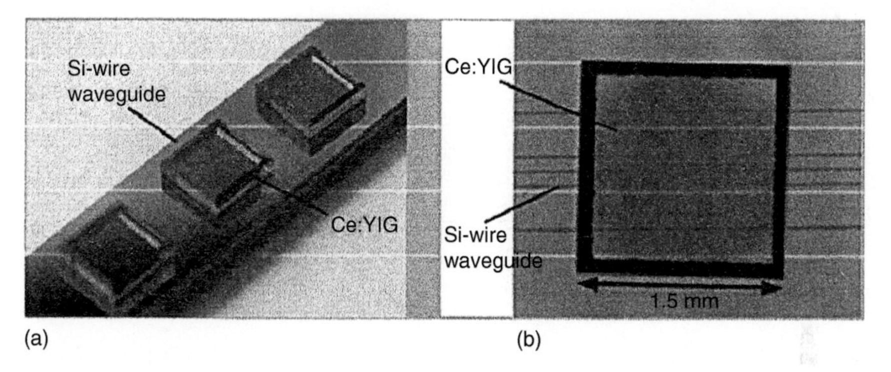

Figure 4.51 Photographs of fabricated silicon waveguide optical isolators. Source: Shoji et al. 2013 [187]. Reproduced with permission of PIERS.

Silicon waveguides were fabricated by electron-beam lithography and reactive ion etching technique. A single crystalline Ce:YIG layer was grown on a gadolinium gallium garnet substrate by RF sputtering deposition technique. A $1.5 \times 1.5\,\text{mm}^2$ Ce:YIG die was bonded on the silicon waveguide using a plasma assisted direct bonding technique [190]. The surface of a patterned silicon layer and Ce:YIG die were exposed to N_2 plasma for 10 seconds. Then, they were brought into contact and pressed at 6 MPa at 200 °C in a vacuum chamber. Figure 4.51 shows the photos of fabricated silicon waveguide optical isolators. Although the bonded region should be only on the NRPS, it is difficult to manipulate such a small die to place on the precise position. In this case, all the MZI was covered by the Ce:YIG die.

The fabricated isolator was characterized by measuring the transmission spectra in a 1550 nm wavelength range as shown in Figure 4.52 [191]. The resonant spectra are due to the optical path difference of the reciprocal phase shifter, that was designed much longer than that for minimum length to be $\pi/2$ so that the wavelength shift can be easily observed. By using a compact permanent magnet with three reversed poles, an external magnetic field was applied to the MZI arms in antiparallel directions. The dashed line indicates the transmittance without

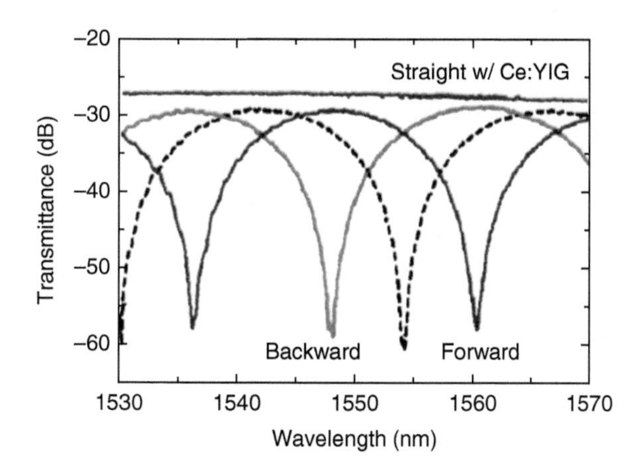

Figure 4.52 Measured transmittance of silicon waveguide isolator [187].

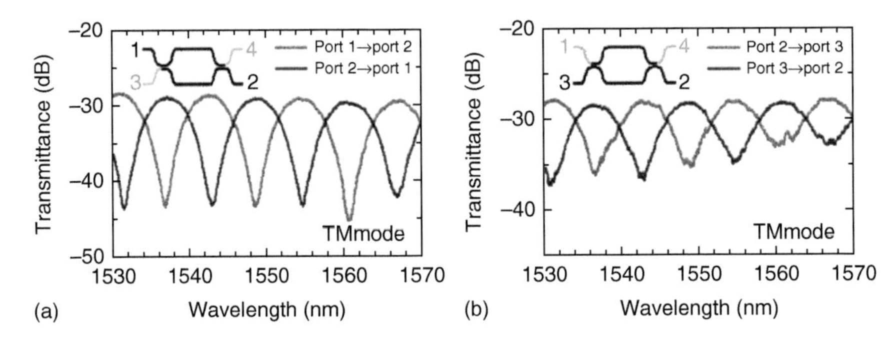

Figure 4.53 Measured transmittance spectra of a circulator with port pairs of (a) port 1–2 and (b) ports 2–3. Source: Shoji et al. 2013 [187]. Reproduced with permission of PIERS.

Table 4.2 Measured transmittance for input/output port combinations at a wavelength of 1531 nm.

	Transmittance measured at output (dB)			
Input	**Port 1**	**Port 2**	**Port 3**	**Port 4**
Port 1	—	−28.4	—	−34.5
Port 2	−43.7	—	−28.0	—
Port 3	—	−37.3	—	−29.0
Port 4	−27.8	—	−42.1	—

Source: Shoji et al. 2013 [187]. Reproduced with permission of OSA Publishing.

applying a magnetic field. When the magnetic field was applied, the transmission spectra exhibited wavelength shifts with different signs depending on the light propagation direction as shown by the light gray (backward) and dark gray (forward) solid lines. A maximum isolation ratio of 30 dB, which was defined by the difference in transmittance between two propagation directions, was obtained at a wavelength of 1548 nm. This is the highest isolation among silicon waveguide optical isolators ever reported.

Figure 4.53 shows the measured transmittance spectra of silicon waveguide circulator [192]. The transmittance measured at 1531 nm is summarized in Table 4.2 for four input/output combinations. Port 2, 3, 4, and 1 are conducting ports for the input from port 1, 2, 3, and 4, respectively, at this wavelength. On the other hand, port 4, 1, 2, and 3 are isolating ports for the input from port 1, 2, 3, and 4, respectively. Isolations defined as the difference in transmittance between two ports are better for cross port pairs 1–2 and 3–4 than bar port pairs 2–3 and 1–4. This is because an exact 3 dB dividing characteristic was not provided in the directional couplers. Since the deviations in two couplers are almost identical in actual fabrication, power imbalance due to the first coupler is nullified by the second coupler for the cross port. On the other hand, power imbalances are added for the bar port. Therefore, better isolations are obtained for the cross port pairs. A maximum isolation of 15.3 dB was obtained for the pair of ports 1–2.

The measured transmittance included a coupling loss of 8 dB/facet between fiber and silicon waveguide. The propagation loss of an air clad silicon waveguide is estimated to be 2.7 dB/cm at a wavelength of 1550 nm. The measured insertion loss of a straight waveguide adjacent to the device, shown by dark gray (straight) line in Figure 4.52, is estimated from optical simulations to include the junction losses at the boundary between air and Ce:YIG cladding regions of 7.6 dB and the optical absorption of Ce:YIG of 4.1 dB. The junction loss can be reduced to be less than 2 dB by covering the air cladding region with SiO_2. The optical absorption of Ce:YIG can be reduced to be about 1 dB by annealing the Ce:YIG after the growth. The difference between the transmittance of straight waveguide and that of maximum peak of the MZI was 3 dB, which is due to excess loss in directional couplers by a fabrication error.

4.4.3.5 THz Isolators Based on Plasmonics

In 2012, Fan et al. [193] proposed a tunable THz isolator based on a metal/MO plasmonic waveguide (MMOPW). As shown in Figure 4.54a, the structure is composed of a metal wall and a single line InSb pillar array with a square pillar in the periodic structure. The air gap of the waveguide between the edge of the semiconductor pillar and metal wall is 50 μm. An external magnetic field is applied along the pillar axis, and the TM polarized THz waves propagate in the air gap of the waveguide between the edge of the semiconductor pillar and metal wall. It is an asymmetric structure of metal/air/InSb periodic array, and because InSb has MO effects under a magnetic field at THz frequency range, the two necessary conditions for nonreciprocal one-way transmission can be satisfied.

Owing to the introduction of periodic structure, the MMOPW shows a unique photonic bandgap characteristic that greatly enhances the one-way transmission, and obtains a lower insertion loss and higher isolation. Because the time reversal symmetry of the waveguide is broken, the dispersion curves in the positive and negative propagation directions are different, indicating an asymmetric photonic bandgap structure. Figure 4.54b shows the dispersion relation of the MMOPW at $T = 195$ K and $B = 1$ T. The dispersion curves show a significant photonic band structure due to the periodic structure. For the forward wave, a passband with high transmittances locates in 1.16–1.28 THz, and the maximum transmittance is 95%. For the backward wave, no backward waves can propagate through the MMOPW in this frequency band. As can be seen from the isolation spectra in Figure 4.54c, the device obtains its 30 dB isolation spectral width of larger than 80 GHz, the maximum isolation of higher than 90 dB at 1.18 THz, and the corresponding insertion loss of less than 5%. Figure 4.54d shows the electric field distributions of the MMOPW when $T = 195$ K and $B = 1$ T of 1.2 THz forward wave and backward wave. Further discussions show that the operating frequency band of this MMOPW isolator can be controlled by magnetic and thermal methods.

Another isolator based on plasmonics is a nonreciprocal plasmonic lens. Using surface plasmonic polaritons (SPPs), the plasmonic lens can focus light beyond the diffraction limit. Different phase retardations can be achieved by structuring the plasmonic lens to manipulate the spatial distribution of the output beam [194–196]. Based on this, Fan et al. [197] proposed a THz isolator composed of

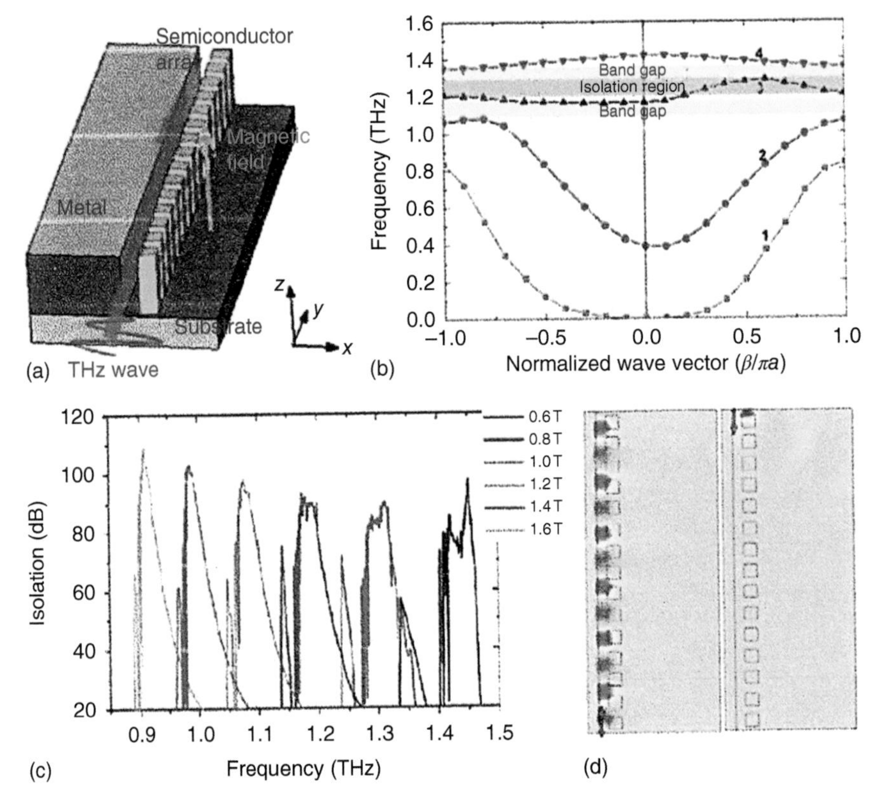

Figure 4.54 (a) Schematic structure of the proposed MMOPW. (b) Band structure of the MMOPW when $T = 195$ K, $B = 1$ T. (c) Isolation spectra of the SMPW (surface magneto-plasmonic waveguide) under different external magnetic fields at $T = 195$ K. (d) Electric field distributions of the MMOPW when $T = 195$ K and $B = 1$ T of 1.2 THz forward wave and backward wave. Source: Fan et al. 2012 [193]. Reproduced with permission of IEEE.

a slit array with a periodic arrangement of the metal and InSb grating in turn as shown in Figure 4.55a called metal/magneto-optic plasmonic lens (MMOPL) [197]. Owing to the MO material and asymmetric periodic structure, forward wave is focused, while backward wave is reflected, forming nonreciprocal transmission. Compared with the previous single waveguide isolators, this MMOPL structure is an area array, which is much easier coupled in the space transmission system. Since this structure is larger than the area of THz beam, the THz beam does not have to be coupled by any waveguides. Therefore, the MMOPL has a smaller insertion loss.

The asymmetric dispersion relations of the proposed MMOPL can be obtained by both analytical and numerical methods as shown in Figure 4.55b. When there is no external magnetic field applied, the dispersion curve is symmetric for the forward and backward propagations. When an external magnetic field is applied, due to the gyrotropy of the InSb under the external magnetic field and the non-symmetry structure, the SPPs split as two different magneto-plasmons. Thus, the dispersion curve becomes asymmetric, the branch of the forward

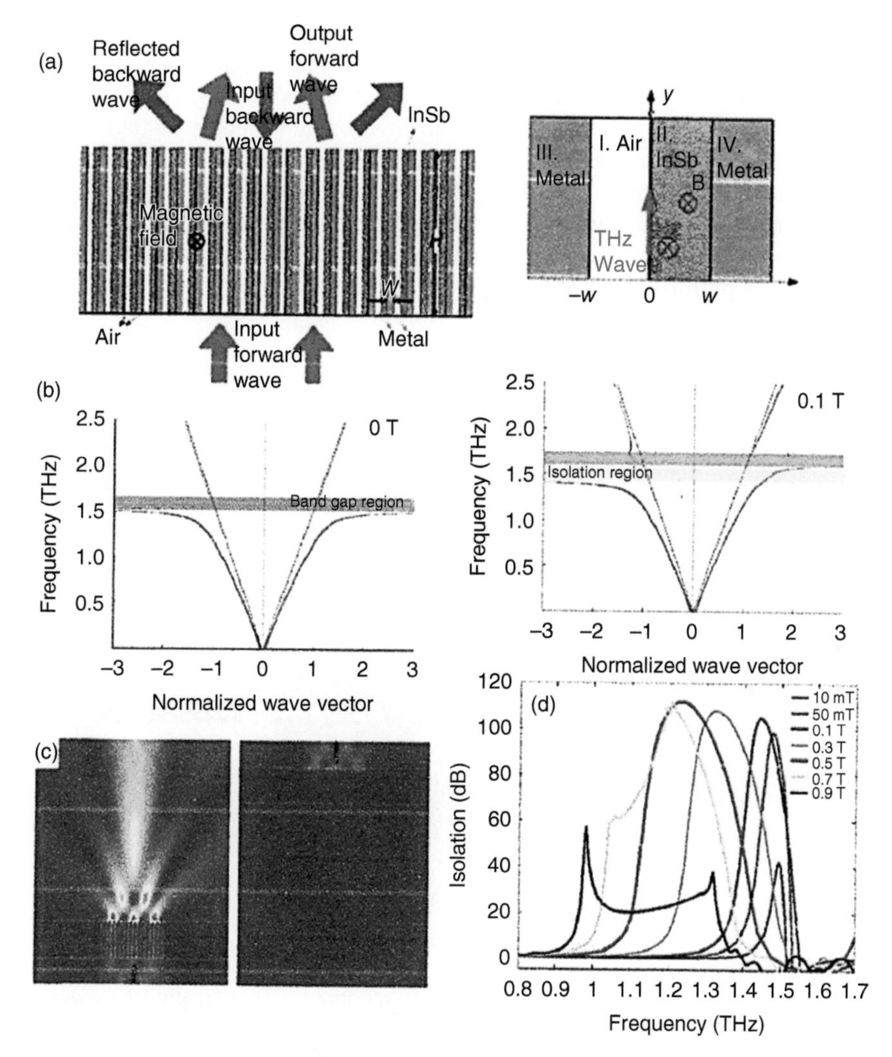

Figure 4.55 (a) Schematic structure of the proposed MMOPL and the model of metal/air/InSb waveguide. (b) Dispersion relations of MMOPL at 185 K, 0 T and 0.1 T. (c) Forward and backward power flow distributions through the MMOPL at 1.25 THz. (d) Isolation spectra of the MMOPL under different external magnetic fields at 185 K. Source: Fan et al. 2013 [197]. Reproduced with permission of OSA Publishing.

propagation moves to a higher frequency, and the backward propagation moves to a lower frequency. For the backward wave, its bandgap is located at 1.45–1.55 THz (light grey region in Figure 4.55c), while the forward wave can still transmit at this frequency range. Therefore, this is an isolation region that permits the forward wave but forbids the backward wave. In the blue region, the kink of the backward wave curve is the second-order plasmonic mode with a strong damping, which can only transmit a very short length in the waveguide, and the forward wave is just located in the forbidden band between the first and second plasmonic modes. The numerical results show that the proposed isolator has an isolation bandwidth of larger than 0.4 THz and a maximum isolation of higher than 110 dB at the temperature of 185 K as shown in Figure 4.55d. The operating frequency of this device also can be broadly tuned by changing the external magnetic field or temperature. Further discussions show that transmission enhancement through this MMOPL is about 30 times larger than that of the ordinary plasmonic lens. This low-loss, high-isolation, broadband tunable nonreciprocal THz transmission mechanism has a great potential application in promoting the performances of the THz application systems.

4.4.3.6 THz Isolators Based on Metasurfaces

For easy coupling with the free spatial THz waves, Chen et al. [198] proposed a kind of a nonreciprocal metasurface for THz isolator. The metasurfaces are flat, ultrathin optical components that produce abrupt changes over the scale of the free space wavelength in the phase, amplitude, and/or polarization of a light beam [199, 200]. MO material is applied into the metasurface structure to form a magneto-metasurface, as shown in Figure 4.56a,b [198]. A periodically patterned InSb layer with the thickness $D = 100\,\mu m$ is coated on the silica substrate layer with the thickness of 50 μm. The unit cell of the patterned InSb is axial symmetry

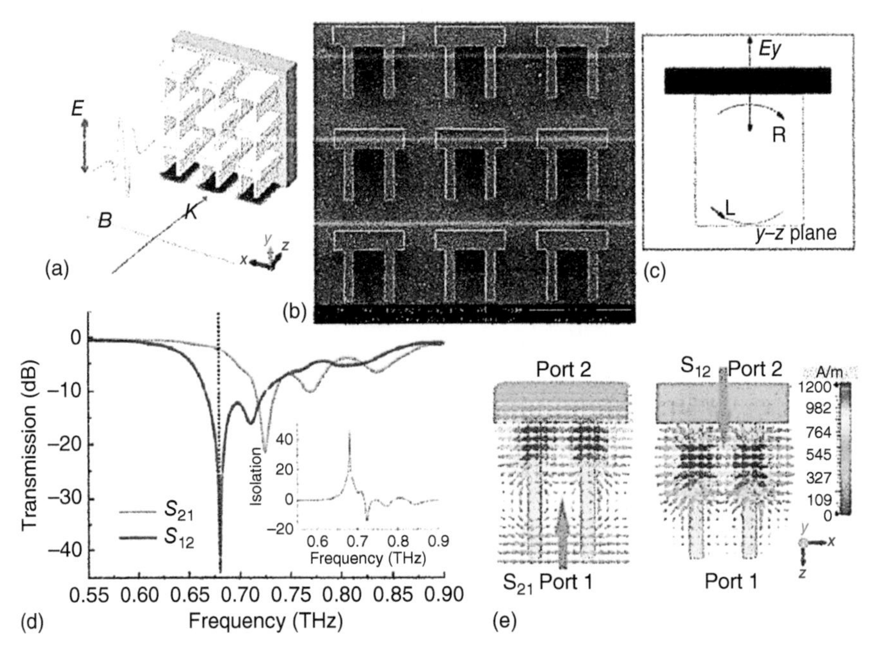

Figure 4.56 (a) 3D view of the device including the directions of wave polarization and external magnetic field and the coordinate system. (b) SEM photograph of the metasurface. (c) Asymmetric structure makes the effective refractive indexes of the left and right rotating magneto-plasmon modes become different. (d) The transmission spectrum of the forward waves $|S21|^2$ and backward waves $|S12|^2$, when $T = 195$ K, $B = 0.3$ T; the inset picture is the isolation spectra of the isolator. (e) Steady magnetic field of the isolator at 0.68 THz when $T = 195$ K, $B = 0.3$ T in the x–z cut plane. Source: Chen et al. 2015 [198]. Reproduced with permission of OSA Publishing.

for the y-axis and asymmetry for the x-axis. THz waves are perpendicularly incident into the periodic plane along the x-axis, the polarization direction of the THz waves is along the y-axis, and an external magnetic field is applied along the x-axis. In this case, the vectors of K, E, and B are orthogonal to each other, and along the E direction the geometry of the device is asymmetric. As shown in Figure 4.56d, the resonances of forward and backward waves happen at not same frequency. When the backward wave has resonance at 0.678 THz, the forward wave at the same frequency band can transmit through this device.

If a circularly polarized light has left-handed rotation in the y–z plane for the forward transmission, there should be a right-handed rotation for the backward transmission. If the device structure is symmetric, the forward and backward waves (left and right rotating waves) are totally equivalent, so it is a reciprocal transmission. When the structure is asymmetric along the polarization direction of incident wave as shown in Figure 4.56c (the asymmetric geometry along the y-direction), the forward and backward waves are no longer equivalent. The effective refractive indexes of the left and right rotating magneto-plasmon modes become different, so the resonance frequencies of the forward and backward waves move to a higher and lower frequency respectively. This is a pair of the nonreciprocal resonance modes in the asymmetric magneto-metasurface, which lead to a high isolation in the THz transmission.

The discussions above can be concluded as follows: (i) the generation and frequency of the resonances are related to the unit structure and polarization state in the metasurface, indirectly related to the external magnetic field. (ii) The MO interaction takes place between the incident waves and external magnetic field, so the metasurface should have MO material in the THz regime and the THz waves polarization direction should be orthogonal to the external magnetic field in the Voigt configuration. (iii)The metasurface structure is asymmetric along the polarization direction of normal incident wave, which satisfies the first condition. Chen et al. [198] used the structural asymmetry in this work, but other methods can also break the symmetry of the transmission system and realize nonreciprocal transmission in the MO systems. For examples, symmetric gratings covered on a MO material substrate can realize the nonreciprocal transmission by an oblique incidence at the optical frequency [201, 202]. Their maximum isolation only reaches 17 dB and the nonreciprocal performance is highly sensitive to the incidence angle. Therefore, this nonreciprocal device based on oblique incidence is not appropriate for an isolator. Instead of oblique incidence, asymmetric geometry in the magneto-metasurface is a new way to realize the nonreciprocity, which is more suitable for the THz isolator, due to its normal incidence and a high isolation of over 40 dB.

4.4.4 Magneto-optical Sensor

Reliable use of magnetic materials requires accurate information about the distribution, intensity, and orientation of magnetic fields in manufacturing, quality control, and research and development. The principles of established magnetic field measurement systems are based on different physical effects. A common feature of all these systems is the analysis of changes in electrical parameters such as voltage and current. Parameter measurement capabilities depend on sensor design and are altered depending on the properties of the applied magnetic field. Measured electrical values and specific material constants make it possible for sensors to determine flux density and strength of the magnetic field. For example, with Hall sensors, the Hall effect in conductive materials (e.g. semiconductor materials) causes an electrical voltage (known as Hall voltage) that directly depends on thc magnetic flux density.

Magnetoresistive field sensors are also widely used. The principle is based on the change in resistance of the sensor material as a function of an applied magnetic field. Magnetoresistive sensors use the change in resistance (measured by electrical voltage) to determine the magnetic field intensity. In contrast, magneto-optical sensors (MO-sensors) are based on the Faraday effect instead of electrical effects to analyze magnetic fields. MO-sensors have the technical benefit of immediately obtaining measurement data directly above the surface of the magnetic material depending on the sensor size. Thus, real-time investigations of the magnetic field distribution can be performed without the need for time-consuming, point-to-point scans, such as that required using Hall sensors, for example. MO-sensor principles, classifications, and typical applications are presented here.

Magneto-optical sensors are based on the Faraday effect discovered in 1845 by Michael Faraday, who recognized that light passing through a transparent medium with an external applied magnet field alters the light wave depending on the magnetic field. This discovery was the first indication of interaction between light and magnetism and later led to the establishment of Maxwell's equations, which among other things, describes light as electromagnetic waves. The fundamentals of electromagnetic interactions in classical physics were created through these discoveries.

The Faraday effect describes the rotation of a polarization plane (plane of vibration) of linear polarized light passing through a magneto-optical medium under the influence of an external magnetic field parallel to propagation direction of the light wave (Figure 4.57). The rotation angle of the polarization plane is defined by the empirical equation

$$\theta_F = V \cdot d \cdot B \tag{4.27}$$

where (referring to MO-sensors) θ_F is proportional to the static magnetic flux density of the external magnetic field B, d is the distance through which the light passes within the MO medium, and V is the material-specific Verdet constant used to express the material-specific rotation intensity and differs from material to material. Therefore, the Verdet constant is dependent on the wavelength of light and the MO material-specific refraction index. Rearranging the equation allows determining the external applied magnetic field for constant boundary conditions depending only on the rotation angle. To minimize errors, it is recommended to use materials having the highest possible Verdet constants to maximize the resulting relative plane rotation, thereby maximizing the accuracy of measurement of θ_F.

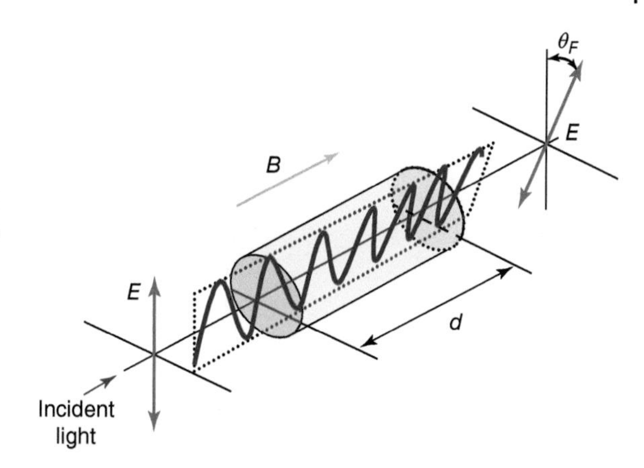

Figure 4.57 Interaction between light and magnetic fields within a magneto-optical (MO) medium; the difference in rotation of the light's polarization plane before and after passing the MO medium is sketched for comparison. Source: Koschny and Lindner 2012 [203]. Reproduced with permission of Advanced Materials & Processes.

Nowadays, there is a great diversity of MO-sensors. Taking into account the sensing mechanism employed and the materials used, MO-sensors can be organized in three main groups: (i) all-fiber sensor, (ii) bulk optic sensor, and (iii) magnetic force sensor.

4.4.4.1 All-Fiber Sensors

All-fiber sensors use very simple configurations because the fiber can be simply coiled around the electric conductor to be measured. Also, the sensor sensibility can be changed by simply changing the number of turns of the optical fiber around the conductor [204]. Since such devices usually use several meters of fiber, the sensors are more vulnerable to pressure and temperature gradients, mechanical vibrations, and other environmental noises than smaller devices (such as bulk-optic).

When operating at high currents, the magnetic crosstalk between different conductors can also be a relevant problem. Schemes to reduce the magnetic crosstalk in three-phase electric systems have been proposed [205]. However, if the detection scheme forms a closed circuit around the electrical conductor, as most of them do, it will not be sensitive to external magnetic fields.

Typically, standard silica fibers have a low Verdet constant when compared with bulk-optic glasses. Many studies have been conducted in order to increase the Verdet constant of the fibers. Materials such as flint glass have been tested and showed a Verdet constant six times higher than fused silica and a much lower photoelastic coefficient (780 times) [204, 206–208]. The study provides specific information on the sensors (setup configurations, temperature dependence, accuracy, etc.) and the fibers used (Verdet constants, photoelastic constants, wavelength dependence, etc.) as well as results of field tests performed. In Figure 4.58, a typical current sensor with flint fiber is represented. A comparison between the performances of flint glass fiber and twisted SMF (usually used before the flint glass) as a Faraday element was presented [209], and showed that the system with the flint glass fiber is more stable than the one with twisted fiber.

In 2009, a new optical glass for optical fibers with high refractive index for applications in MO-sensors was reported. The results showed that these fiber glasses could have 10 times higher Verdet constant and small internal mechanical stress

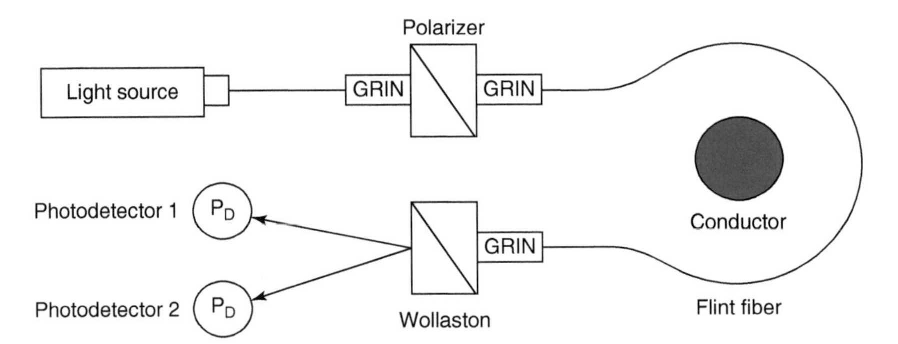

Figure 4.58 Typical configuration of an optical current sensor with flint fiber. Source: Hotate et al. 1998 [209]. Reproduced with permission of SPIE.

[210]. Recently, a terbium-doped core phosphate optical fiber with a Verdet constant six times larger than standard optical fibers was presented [211]. In 2010, the same authors also presented a high concentration of terbium-doped fiber with a record Verdet constant of −32 rad/(T m), which is 27 times larger than standard optical fibers and corresponds to 83% of the Verdet constant of commercially available crystals used in bulk optics–based isolators [212].

Another problem of using the fiber as the transducer is the effect of the linear birefringence that is induced by mechanical stress (when the fiber is bent for example), thermal stress, manufacture imperfections, and other effects. The presence of linear birefringence significantly reduces the sensor sensitivity due to the polarization state degeneration. The linear birefringence can be neglected if circular birefringence is high enough. The use of a twisted SMF to impose a circular birefringence in the fiber has been demonstrated [213]. A similar approach is to use spun high birefringence fibers [214]. Linear birefringence can also be reduced by fiber annealing process [215, 216]. Theoretical models and experimental measurements for the Verdet constant dispersion in annealed fibers have been presented [217].

Some methods for linear birefringence compensation using reflected light propagation have been presented, such as Faraday rotating mirrors [218] and fiber polarization rotators [219], which also doubles the sensitivity for the same fiber length. In Figure 4.59, a Faraday rotator mirror is shown. This component leads to a polarization shift of 90° and light that propagates in one axis is coupled to the other axis, and vice versa. Since the linear birefringence is a reciprocal effect, the phase difference introduced by the linear birefringence will be compensated [218].

A sensor using reflected light propagation for compensation of linear birefringence, made of low-birefringent flint fiber with a very low photoelastic constant, achieved the accuracy required for the 0.1% class of current metering transformers in the range of 1 kA [220].

Commercial fiber-optic dc current sensors have also been demonstrated, [221] being able to measure up to 500 kA with an error of 0.1%. More recently, a signal processing methodology based on artificial neural networks (ANN) was developed, which processes signals from a typical MO-sensor and a

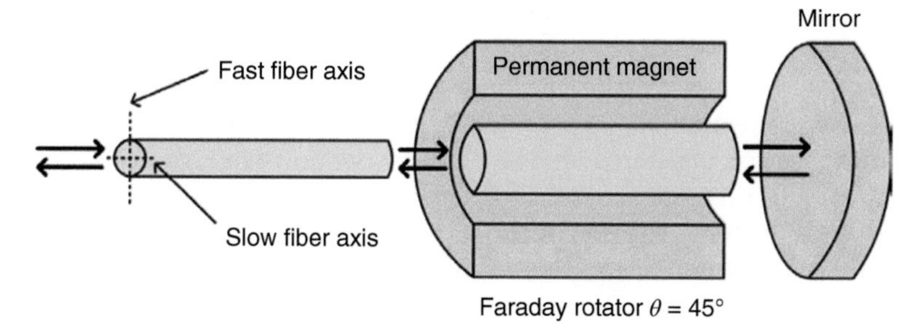

Figure 4.59 Faraday rotator mirror. Source: Drexler and Fiala 2008 [218]. Reproduced with permission of RADIOENGINEERING.

thermometer, achieving higher accuracy with temperature and nonlinearity compensation [222]. The great benefit of ANN is to get a transfer function for the measurement system taking into account all variables, even those from unwanted and unknown effects, providing a compensated output, after the ANN training session.

Some MO-sensors are already commercially available. Considering all-fiber MO-sensors, for instance, ABB and NXTphase are two companies that have available sensors with working principle identical to the one shown in Figure 4.62. The first company assures a range of operation up to 500 kA and 0.1% accuracy [224]. The second company permits dc and ac measurements with a range of operation from 1 to 63 kA_{RMS}, being also in the 0.1% class category [225, 226].

4.4.4.2 Bulk-optic Sensors

As seen before, all-fiber MO-sensors are easy to implement and to increase the sensitivity, but suffer from linear birefringence and have relatively low Verdet constants. In this context, bulk optic sensors are rapidly imposed and are the most industrially implemented technology [227]. When compared to all-fiber devices, bulk optic sensor presents some obvious advantages. These are usually smaller devices and mechanically stiffer. Therefore, mechanical and thermal gradients, vibrations, and other external noises in the bulk optic material are very small. Also, their Verdet constants are typically two times higher than the ones found in optical fibers, and due to their low photoelastic coefficients, the intrinsic linear birefringence is also very small, which allows for sensors with high sensitivities. Another big advantage is that the sensor around the conductor does not have to be a single piece of material. This allows an easy installation without the need of interrupting the current in the conductor.

The main disadvantages of these sensors are related to the fact that reflections need to occur inside the bulk optic material for light to go around the conductor. Since the reflection angle will be dependent on the refractive index of the crystal and outside surrounding medium (such as air), the sensor will be affected by a number of external factors such as humidity. This can be solved by isolating the sensor from external factors but will add higher cost to installation and maintenance operation of the sensor [228]. Also, every internal reflection of the light that occurs will introduce an optical phase difference between the two polarizations.

In a polarimetric or interferometric scheme, this leads to an unavoidable error source [229]. The problem can be solved with the use of double reflections in each corner of the bulk optic material, where the optical phase difference introduced in the first reflection is compensated by the second reflection. A patent describing this idea was registered in 1986 [230] and is shown in Figure 4.60.

However, this idea presents another problem: in the optical path between the first and second reflections, the polarization state is elliptical and the signal will be nonlinearly affected by the magnetic field to be measured. This will cause the system to be affected by external magnetic fields. This problem takes special relevance when measuring current in three-phase systems, since the three conductors are usually close to each other. Studies have also been developed to minimize the crosstalk between different conductors [205]. Another possible solution is using a triangular shaped bulk optic material in which the light is always reflected at the critical angle [231], as shown in Figure 4.61.

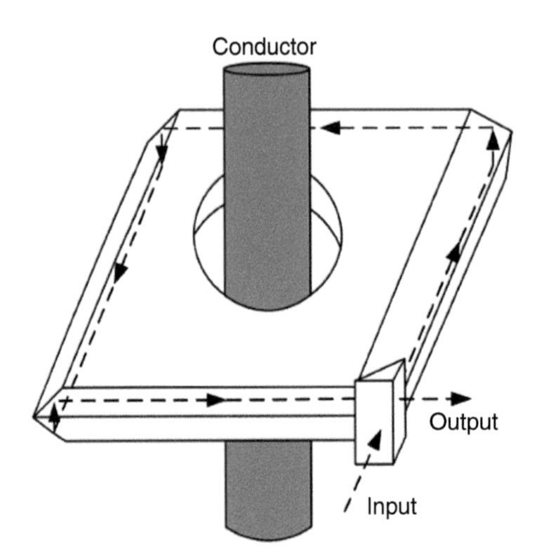

Figure 4.60 Bulk optic current sensor with double reflection. Source: Sato et al. 1986 [230]. Reproduced with permission of Hitachi.

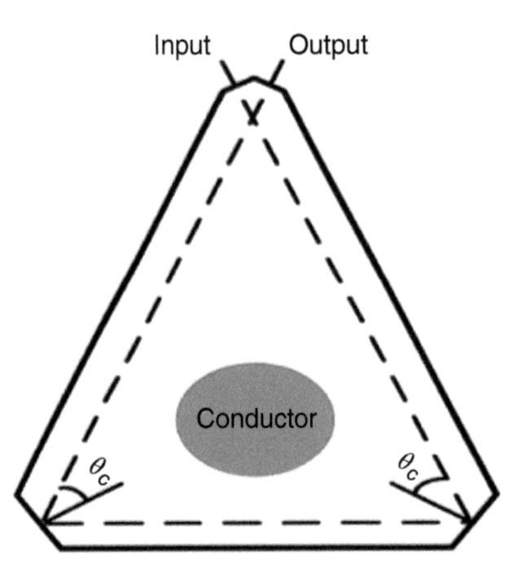

Figure 4.61 Bulk optic triangular shaped current sensor. Source: Fisher and Jackson 1996 [231]. Reproduced with permission of IOP.

This configuration is simpler than the previous one. However, this requires a precise reflection angle cut and a near perfect bulk optic material well isolated from external noise sources. Several schemes have been proposed to further increase the sensitivities of these sensors, including field concentrators (which consist in an air-gap core surrounding the conductor) and extending the light optical path inside the crystal (by using total internal reflections before the light exits from the sensing element) [232]. For this purpose, a circular sensor head was developed, where light is injected through a prism and at specific angles, light travels five times before exiting the prism. This scheme enhances sensitivity and it is not sensitive to external fields [233]. Other sensors are available in a square configuration where light passes around the conductor several times [234, 235].

Also, in order to increase the sensitivity of the sensor, the compensation of the linear birefringence is very important. It was shown that the use of a Faraday rotator mirror can reduce the angle of rotation of the polarization plane due to the linear birefringence from 20° to 5° [236]. The dependence of the Verdet constant with source wavelength is also important, because it changes the sensitivity of the sensor [237]. Also, it was shown that the errors introduced by a small bandwidth optical source are very small and therefore it is reasonable to use models that assume a monochromatic light source [238]. As in all-fiber MO-sensors, bulk-optic sensors also exhibit temperature dependency of the Verdet constant. One solution is to measure the temperature and compensate its effect by introducing a correction factor that depends on the temperature [239].

Many interesting ideas for MO-sensors may be found in literature, but they are not designed for high voltage and/or high current sensing operations. However, specific applications for high current [240] and high voltage [241] can be found. In the case of high currents, the sensors are usually limited to measurements between 0° and 360°, in order to avoid rotation angle ambiguities. This problem was solved [240] by using a technique that counted the number of times that the phase crossed 0° and allowed measurements up to 720 kA.

Also, this type of sensor has been successfully introduced to the market. PowerSense [242] is a company that has a commercial product based on the sensing scheme such as the one shown in Figure 4.58, which permits ac current measurement from 20 to 20 000 A, with maximum errors of 2%.

4.4.4.3 Magnetic Force Sensors

Optical fiber magnetic field sensors exploit the magnetostrictive effect that produces a mechanical strain in a ferromagnetic material when it is subjected to an external magnetic field. For that an optical fiber is bonded to a sample, or thin-film deposited on a magnetostrictive material. The most commonly used magnetostrictive sensing elements were optical fibers coated with nickel [243], metallic glasses such as Metglass 2605S2 or Vitrovac 40–60 [223, 244–246], and ceramic thin films (Fe_2O_4, $NiFe_2O_4$, or $Ni_xCo_{1-x}Fe_2O_3$) [247, 248].

Since 2000, an intensive investigation on an alloy material with high magnetostriction coefficient (Terfenol-D) has been reported in the literature. The most studied optical configuration is the rod or ribbon shape of Terfenol-D attached to a fiber Bragg grating (FBG) [249]. The advantages of this type of sensors are the small size and possibility to have various sensing elements in the same optical fiber, thus allowing optical multiplexing. Alloys using Terfenol-D have also been tested in microstructured fibers [250]. Recently, another kind of sensors to detect magnetic fields have been proposed, where the sensor is based on MF.

Magnetostrictive Sensors During the 1980s a great effort was taken for detecting magnetic fields and/or electric current using optical fibers in combination with materials that exhibit magnetostriction effects. In 1980, Yariv and Winsor [243] studied the possibility of detecting weak magnetic fields by using magnetostrictive perturbation in optical fibers. A low-loss optical fiber of length L was wrapped around a nickel jacket that suffers a longitudinal mechanical strain when immersed in a magnetic field.

The first experimental evidence of the magnetic field sensing characteristics of optical fibers jacketed with either nickel or metallic glass magnetostrictive materials was reported by Dandridge et al. [223]. Both bulk magnetic stretchers as well as thin films directly deposited on a SMF were analyzed. An all-fiber MZI was used to detect the magnetically induced mechanical changes in the optical path length that contained the magnetostrictive jacket. Figure 4.62 shows the schematic diagram of the magnetometer proposed by Dandridge et al. [223]. The sensor sensitivity achieved with the bulk nickel material was 6.37×10^{-6} A/m^2.

A SMF wound under tension around a magnetically sensitive nickel cylindrical piece was proposed by Rashleigh [251]. The magnetic field changed the state of polarization (SOP) of light in the fiber. An optical phase sensitivity of 1.76×10^{-2} rad/m Oe was achieved, allowing a detection of magnetic fields as small as 3.5×10^{-4} A/m^2 of fiber.

The magneto-optic coupling coefficient as a function of frequency for a MZI incorporating a nickel toroid was measured by Cole et al. [252]. They observed that the frequency response of the coupling coefficient was found to be essentially smooth in the range between 15 and 600 Hz. Hartman et al. [253] reported the fabrication and characterization of optical fibers coated with nickel thin film. The thickest coating was greater than 90 μm and was fabricated by chemical vapor deposition (CVD). The optical phase sensitivity obtained by the fiber MZI was 1.9×10^{-8} rad m/A.

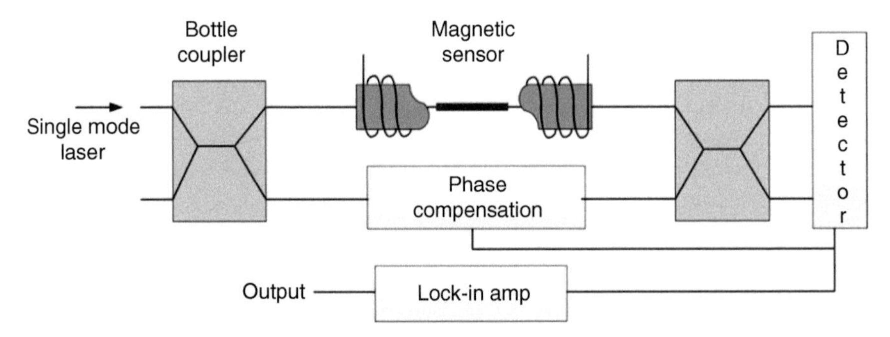

Figure 4.62 Experimental scheme of fiber-optic magnetic sensor. Source: Dandridge et al. 1980 [223]. Reproduced with permission of IEEE.

The use of a metallic glass as sensing element for detection of very low magnetic fields was described by Koo and Sigel [244]. In their experiment, different metallic glasses were used on an all-fiber MZI in order to measure their magnetic field detection sensitivities. A minimum detectable magnetic field of 3.98×10^{-7} A/m^2 was reported.

A configuration for measuring dc magnetic fields using an optical fiber coated with amorphous $Fe_{0.8}B_{0.2}$ alloy in one arm of a modified magnetometer was developed by Willson and Jones [254]. In order to obtain a maximum sensitivity, the system was analyzed in quadrature allowing the elimination of polarization fluctuations. A linear region was analyzed and a sensitivity of 102 rad/T m was obtained.

A technique for the measurement of weak dc and low frequency ac magnetic fields using an all-fiber SMF magnetometer was described by Kersey et al. [255]. For the first time, the experimental scheme used consisted of a Michelson interferometer where the two cleaved fiber ends were coated with silver film to form the mirrors. Detection sensitivities of $\approx 1 \times 10^{-7}$ T/m at 20 Hz were achieved by the bulk nickel magnetostrictive element. Figure 4.63 represents the experimental configuration used.

A fiber MZI passively stabilized using a 3×3 fiber coupler was used by Koo et al. [256]. The system response was examined by subjecting the metallic glass to dc and ac magnetic fields. The sensor sensitivity was in the order of 7.96×10^{-5} A/m^2 for a fiber length of 1 m at 1 Hz bandwidth.

The first demonstration of a closed-loop fiber optic magnetometer with dynamic magnetostrictive response was reported by Kersey et al. [257]. The sensor was capable of detecting low frequency and dc changes of magnetic field with a sensitivity of ≈ 2 nT at frequencies below 2 Hz. One year later, the same group proposed a similar configuration, where the main difference in the

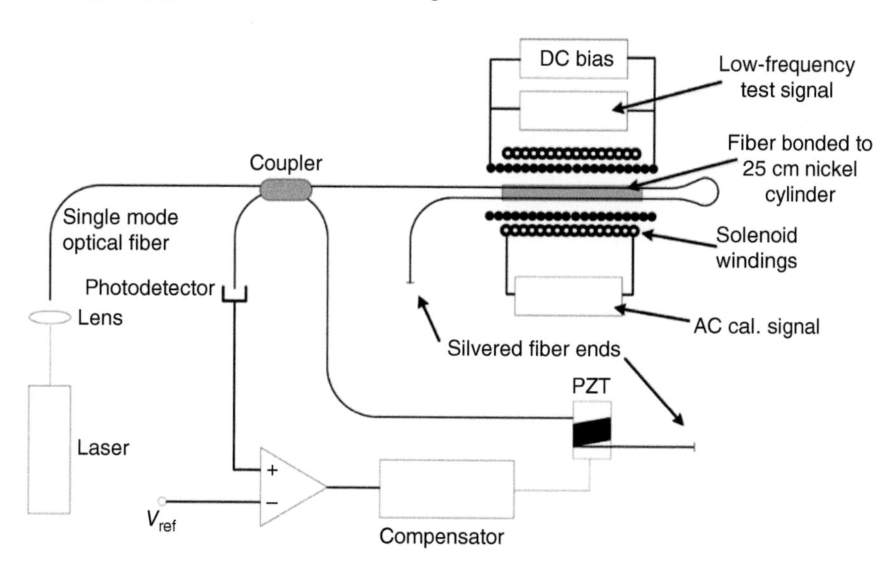

Figure 4.63 Scheme of the all-fiber magnetometer. Source: Kersey et al. 1983 [255]. Reproduced with permission of IEEE.

magnetometer was that they kept at constant value the total local field (bias + ambient) [245]. By active-bias field stabilization they ensured that the magnetostrictive sensing element was maintained in zero state zero magnetization. This method eliminated any problems arising due to magnetic hysteresis of the metallic glass materials used (Metglass and Vitrovac). A minimum detectable magnetic field of 8×10^{-5} A/m was observed over the frequency range from dc to 20 Hz using a small length of optical fiber ($\approx$0.5 m).

The effects of external perturbations, such as mechanical stress and vibrations on the operation of fiber-optic magnetometers, were investigated by Bucholtz et al. [246]. These perturbations were controlled by maintaining the dc magnetic field at prescribed levels. The analysis performed has shown that the method used by the authors could be applicable in all interferometric fiber sensors employing nonlinear transducing mechanisms.

In 1989, the first demonstration of the detection of magnetic fields at frequencies above 50 kHz using a fiber-optic magnetometer was reported by Bucholtz et al. [258]. A MZI with a thick cylindrical metallic glass transducer in one of the fiber arms was used. Mixing high frequency signals in the magnetostrictive sensor allowed a heterodyne detection scheme, where the transducer acted as both the receiving and nonlinear mixing element. The same group reported a minimum detectable ac magnetic field of 70 $fT/Hz^{1/2}$ at 34.2 kHz [259].

Optical fibers coated with magnetostrictive ceramic films were tested by Sedlar et al. [248] by using a modified MZI operating in an open-loop configuration. The sensors exhibited excellent linearity and good sensitivity. The materials used were magnetite, Fe_2O_3, nickel ferrite, and cobalt-doped nickel ferrite (NCF_2) jackets, being the last one with the best response to magnetic field. They achieved a minimum detectable magnetic field of 3.2×10^{-3} A/m for optical fibers jacketed with 2 μm thick and 1 m long NCF_2 material.

A fiber-optic sensor for measuring dc magnetic fields based on the extrinsic Fabry–Perot interferometer (EFPI) was proposed by Oh et al. [260]. An SMF and a Metglass wire magnetostrictive transducer constituted the input-and-output and reflector arms of the EFPI sensor. Owing to the sensor optical geometry design, it showed a low vibration sensitivity and high thermal induced compensation (better than 99%).

Pérez-Millán et al. [261] demonstrated a new approach for measuring electric current on high-voltage systems. The fiber-optic sensor employed the intrinsic magnetostriction of the ferromagnetic core of a standard current transformer and was integrated with a MZI. The sensor avoided certain defects of Faraday based optical current sensors reported, such as optical power source fluctuations, phase drifts, and polarization dependency. Figure 4.64 represents the experimental configuration used by Pérez-Millán et al. [261].

More recently, Djinovic et al. [262] presented results of measurement of ac and dc magnetic fields by using a fiber-optic interferometric sensor for structural health monitoring. The principle of operation was based on changes in the optical path length of the cavity between a magnetostrictive wire and a fiber-optic tip. The sensing configuration consisted of a Michelson fiber interferometer using a 3×3 SMF coupler. They were able to detect the instant separation between the

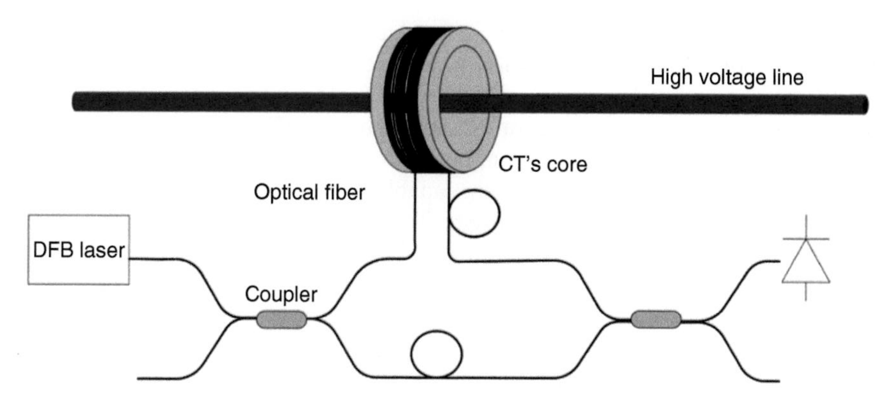

Figure 4.64 Experimental arrangement used for measuring electric current on high voltage systems. Source: Perez-Millan et al. 2002 [261]. Reproduced with permission of IEEE.

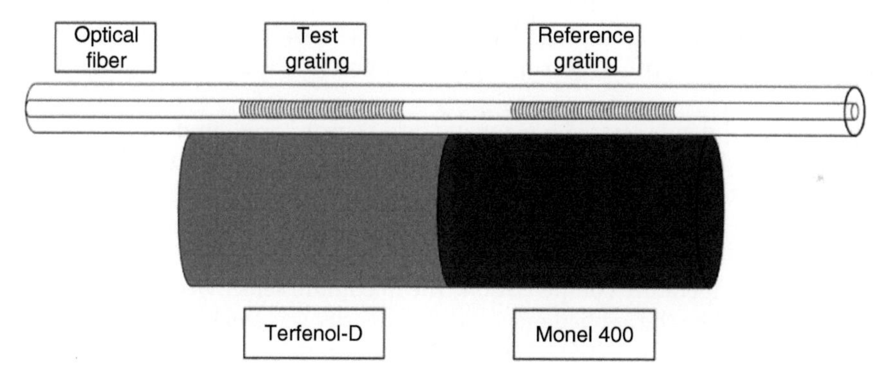

Figure 4.65 Schematic diagram of the fiber sensor. Source: Mora et al. 2000 [249]. Reproduced with permission of IEEE.

wire and fiber end with an accuracy of 50 nm corresponding to a magnetic field in a range of 50 to 800 μT.

Terfenol-D Sensors The first work combining Terfenol-D alloy with optical fiber sensors to measure static magnetic fields was reported by Mora et al. [249] as seen in Figure 4.65. The magnetostrictive sensor with temperature compensation scheme was composed by two different alloys with similar thermal expansion coefficient, one of them being Terfenol-D and the other Monel 400. The mechanical expansion of both materials due to temperature and magnetic field variations was detected by the two FBGs attached. The spectral difference between the two Bragg wavelengths was proportional to the amplitude of the magnetostriction, and the wavelength shift produced by the grating bonded to the nonmagnetic alloy (Monel 400) was proportional to the temperature variation.

For applied magnetic fields smaller than 4.7×10^{-4} A/m ($B < 60$ mT) a linear response of the fiber sensor was verified. The spectral sensitivity in this range was independent of temperature, having a value of $(2.31 \pm 0.05) \times 10^{-5}$ nm/(A^2/m^2). In order to compensate the temperature effects in the FBG based on magnetostrictive materials, two simple techniques were demonstrated by Yi et al. [263].

The first technique consisted of two FBGs placed perpendicular to each other and bounded onto a single Terfenol-D layer material. In the second technique, two FBGs were stacked onto two different magnetostrictive bars (Terfenol-D and nickel) and were physically parallel with each other. The materials used had similar thermal expansion coefficients but with magnetostrictive coefficients of opposite signs. The two techniques were capable of measuring the magnetostrictive effects with temperature insensitivity. The sensitivities due to the magnetic field were 2.44×10^{-4} and 1.8×10^{-4} nm/mT, for the first and second techniques, respectively.

An electric current optical sensor based on a passive prototype magnetostriction device for high-power applications was developed and demonstrated by Satpathi et al. [264]. The optical current sensor concept was achieved by attaching a piece of Terfenol-D to a FBG, thus producing mechanical strain that was proportional to the magnetic field amplitude. In order to obtain a linear response from the magnetostrictive alloy, the material was subjected to mechanical pre-stress and dc magnetic field bias tuning. The sensor had a linear electrical current range from 100 to 1000 A, with a measured phase shift of 30° for a frequency signal at 60 Hz.

Li et al. [265] demonstrated a magnetic field sensor based on dual FBG configuration consisting essentially of a rod of Terfenol-D attached to the optical fiber. One of the gratings was fixed on both ends of the magnetostrictive alloy, while the other was only attached on one point, the other end being free to move. The configuration of dual FBGs was employed for point measurement reference and for easy temperature compensation. The maximum sensitivity achieved was 0.018 nm/mT when the applied magnetic flux density was smaller than 70 mT.

In 2006, Mora et al. [266] presented a fiber-optic ac current sensor for high-voltage lines based on a uniform FBG mounted on a Terfenol-D piece. An innovative way of processing the optical signal from the sensing head allowed a simultaneous measurement of temperature and the ac current. The first physical parameter was coded on wavelength shift, while the second one was coded on the amplitude of the signal. The sensor could also operate at long distances with a capacity to be multiplexed.

Also, a sensor for simultaneous measurement of ac current and temperature was reported by Reilly et al. [267]. The device was composed of a magnetically biased Terfenol-D piece fixed to one FBG. The sensor was capable of measuring ac current at temperatures from 18 to 90 °C. The nonlinear effects originated by the magnetostrictive alloy were identified and methods for compensating these effects were proposed.

A hysteresis compensation technique for a magnetic field sensor, containing a magnetostrictive alloy device, was presented by Davino et al. [268]. The sensing head integrated a rod of Terfenol-D and a FBG. Owing to the nonlinear effects taking place in such materials the magnetoelastic material was accurately modeled in order to compensate for hysteresis. The algorithm allowed improvement in the sensor performance.

In 2009 and for the first time, a fiber-optic magnetic field sensor with a thin film of Terfenol-D instead of the bulk magnetostrictive materials was reported by Yang et al. [269] as shown in Figure 4.66. By magnetron sputtering deposition

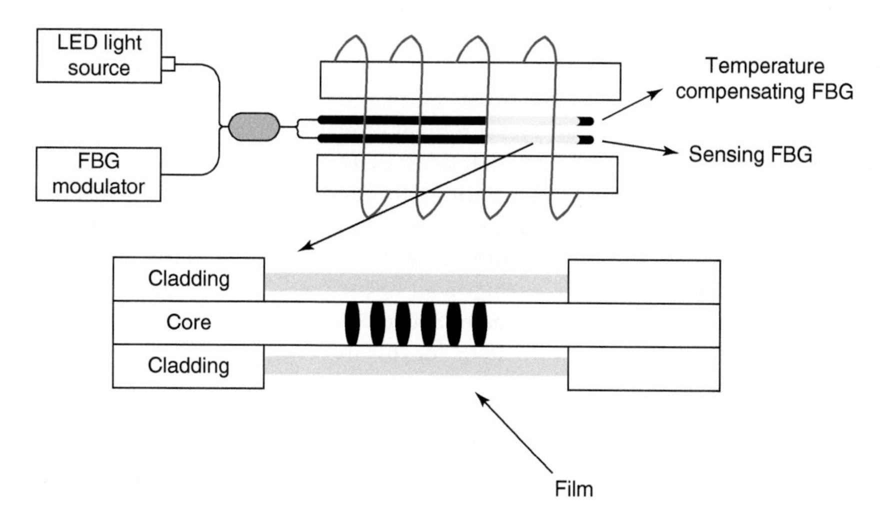

Figure 4.66 Configuration of optical fiber magnetic sensor. Source: Yang et al. [269]. Reproduced with permission of OSA Publishing.

process, Terfenol-D thin films were deposited on etched FBGs. Two methods to improve the sensitivity were demonstrated. In one, the magnetostrictive alloy was deposited over different diameters of cladding-etched FBGs. The maximum sensitivity response to magnetic field was 0.95 pm/mT for an 85 μm diameter sensor. The other method was coating FBGs with a layer of magnetostrictive alloys (FeNi and Terfenol-D) and a multilayer with both materials. The multilayer has the highest sensitivity of 1.08 pm/mT. In this experiment, the authors not only reduced the dimensions of the sensor but also improved the sensitivity to magnetic field.

The dependence of shape anisotropy on the strain response of magnetic field sensors based on magnetostrictive alloys and FBGs was studied by Pacheco et al. [270]. The maximum sensor sensitivity was 18 με/mT when a biased uniform field was applied for a 20 mm long Terfenol-D cuboid shape. For gradient magnetic fields, they were able to detect a significant change in the magnetostrictive response at different positions, attaching FBGs along the cuboid containing the magnetic field direction.

Quintero et al. [271] presented a sensor for dc and ac magnetic field measurements. The grating was coated with a layer of a composite mixed with particles of Terfenol-D. Different compositions were tested and the best magnetostrictive response was achieved using a 30% volume fraction of Terfenol-D mixed in a matrix of epoxy resin as blinder. The effect of a compressive prestress in the sensor was also investigated. They achieved a resolution of 0.4 mT without and 0.3 mT with prestress. The sensitivity obtained for the magnetic field was $2.2 \times 10^{-6}\ \text{mT}^{-1}$, using an FBG interrogation system.

Smith et al. [272] presented a novel sensor for detecting static magnetic fields using an inscribed FBG and a micromachined slot sputtered with Terfenol-D. The grating and the slot were created with a femtosecond laser inscribing. The sensitivity to magnetic field in transmission was 0.3 pm/mT. They also measured in-reflection with different polarization states, achieving sensitivities between 0.2 and 0.1 pm/mT. The use of a femtosecond laser to inscribe a FBG and create a micromachined slot simplifies greatly the fabrication process.

The first fiber-optic sensor for measurement of magnetic field by integrating a highly birefringent photonic crystal fiber (HiBi PCF) combined with a Terfenol-D epoxy resin composite material was reported by Quintero et al. [250]. An in-fiber modal interferometer was assembled by evenly exciting both eigenmodes of the HiBi PCF. Changes in the cavity length as well as the effective refractive index were induced by exposing the sensing head to magnetic fields. The sensitivity obtained was 0.006 nm/mT to a range of 300 mT. Also, an FBG coated with the same composite material was fabricated. For this sensor, the sensitivity achieved was 0.003 nm/mT.

Magnetic Fluid Sensor MF is a promising functional material that attracts much research interest and finds a variety of applications in photonic devices due to its diverse magneto-optical effects such as refractive index tunability, birefringence effect, dichroism effect, Faraday effect, field-dependent transmission, and so on [29, 31, 112, 114, 273]. Generally, in optical devices, MF is employed either in the form of thin films in which light is transmitted and modulated, or in the form of reflective interface where the evanescent field of light interacts at the surface layer of MF [42, 50, 51, 54, 112]. For example, several magnetic field sensors were demonstrated by using the MF films in optical fiber interferometers such as the SI or FPI [39, 274].

A fiber-optic current sensor based on MF was developed by Hu et al. [39] and achieved a sensitivity of 1.25 pm/Oe. MF was used as the medium in a FP resonant cavity. The refractive index characteristic of the MF suffers variations due to the external magnetic field and the current was measured by the output wavelength of the FP fiber sensor. A signal demodulation method using an FBG wavelength measurement system was proposed. The results indicated good linearity, and the thickness and initial concentration of the MF affected the performance of the sensor.

Dai et al. [275] investigated a novel fiber-optic sensor based on MF and etched gratings with a sensitivity of 0.344 pm/Oe. Nanoparticles of Fe_3O_4 (the MF used) were injected into capillaries containing etched FBGs acting as sensing elements. The grating with smaller diameter exhibited a wavelength shift of 86 pm, being the most sensitive. Experimental results showed a reversible response to magnetic field under 16 mT.

A magnetic field sensor consisting of a PCF and small amount of Fe_3O_4 magnetic nano-fluid trapped in the cladding holes of a polarization maintaining PCF were reported by Thakur et al. [276]. They demonstrated that a magnetic field of a few millitesla can be easily and very well detected with higher sensitivity of 242 pm/mT. Later on, a magneto-optic modulator with a MF film inserted on a Sagnac fiber interferometer was proposed by Zu et al. [42]. When the magnetic field was applied, it exhibited a variable birefringence, and this Faraday effect led to phase and polarization state variations. Zhao et al. [277] reported the use of hollow core photonic crystal fiber (HC PCF) in the fabrication of the cavity of the FP sensor, as shown in Figure 4.67, which exhibits low transmission losses and

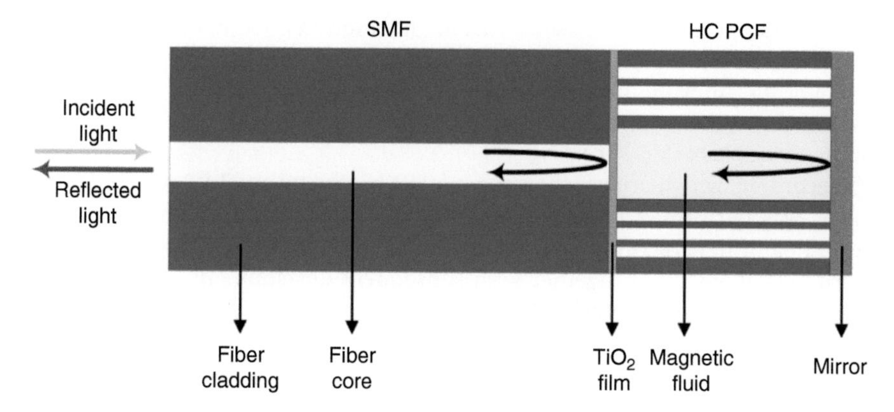

Figure 4.67 Structure diagram of magnetic filled HC PCF FP sensor. Source: Zhao et al. 2012 [277]. Reproduced with permission of Elsevier.

insensitivity to temperature variations. The core of the HC PCF was filled with MF, allowing this way its use as a magnetic sensor. The sensitivity achieved was 0.41 pm/(A m).

Zu et al. [50] proposed a new magnetic field fiber sensor design based on MFs. The sensor was configured as a SI with a MF film and a section of PMF inserted in a fiber loop. The wavelength shift sensitivity to magnetic field was 0.21 pm/(A/m) with a resolution of 47.7 A/m. The optical power also varied with the magnetic field strength, obtaining a sensitivity of 5.02×10^{-3} dB/(A/m).

4.4.5 Magneto-optical Recording

Nowadays, various kinds of optical storage techniques are in practical use, such as CD (compact disk), CD-ROM (compact disk read-only memory), CD-RW (compact disk rewritable), DVD (digital versatile disk), and MO (magneto-optical) disk. These optical recording techniques have many advantages: removability is the most important, which means that storage media can be easily removed from the storage system and be replaced by other media. Optical recording media are usually disk shaped with a hole for the spindle at the center. Because of the removability we can access infinitely large amounts of information by using a single storage system. The second advantage is the nonvolatility. The data stored in the media will not disappear even if the electric power supply is switched off. A further advantage common in optical storage systems is the bit cost, which is far less than those in semiconductor memories.

Among various types of optical recording media, both the MO disk and the CD-RW are rewritable, but their physical principles are completely different. The MO disk is a magnetic medium such as magnetic tape or hard disk drive systems, while the CD-RW is not magnetic but utilizes the phase change between crystalline and amorphous states. Among various optical recording media including read-only media, only the MO recording does not accompany the movement of atoms or electrons for information storage. In the case of the MO disk, such as magnetic recording, only the reversal of the angular momentum, in other words, time reversal, occurs in the media. This might result in good rewritability of the magnetic media compared with other optical storage media utilizing the pit formation or the amorphous–crystalline phase transition.

4.4.5.1 Principles of MO Recording

MO recording, which utilizes thermomagnetic recording and MO read out techniques, was performed first by Chen et al. [278] using MnBi films. Owing to the finding of new recording materials of rare earth–transition metal (RE-TM) amorphous alloys [279] together with subsequent development of optical disk technology, the first-generation MO drive was put to practical use in 1988. The technology of MO recording is a combined technique of optical recording and magnetic recording. The MO recording disks are already widely used as removable memory in various computer systems and in consumer audio and video recording systems.

In the writing process, the MO recording technique utilizes laser energy as a heat source but not as photon energy. MO recording is one of the so-called heat mode optical recording techniques, and is often referred to as thermomagnetic recording, which was performed first by Williams et al. [280] in 1957. The principle of thermomagnetic recording is shown in Figure 4.68. The direction of the magnetization is changed by the simultaneous application of the thermal energy and the magnetic field. As shown in the figure, when a small region

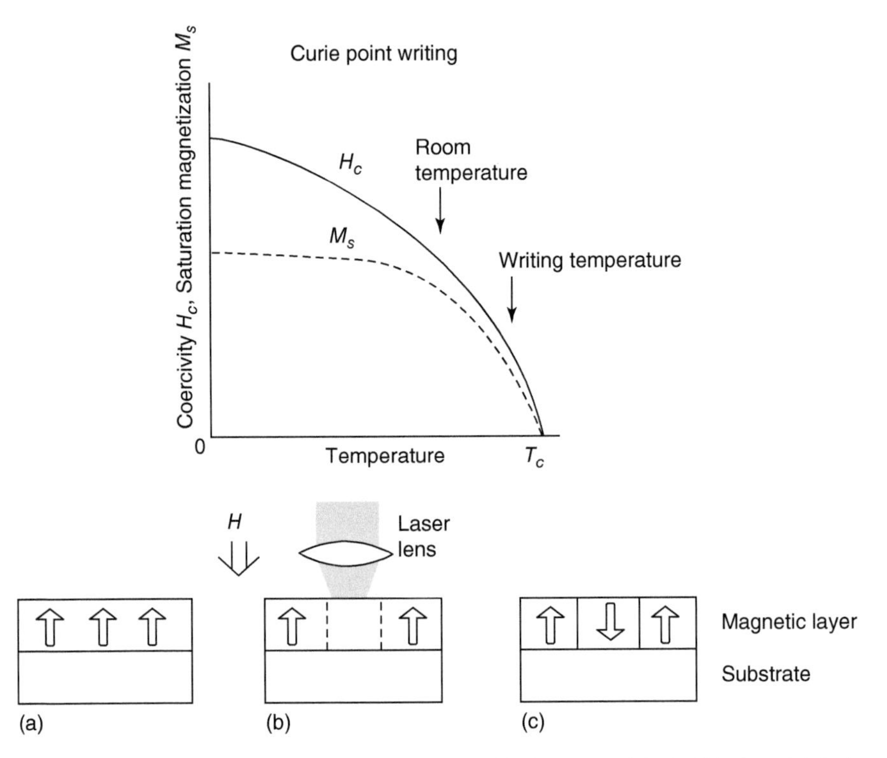

Figure 4.68 Principle of thermomagnetic recording (Curie point writing): (a) before, (b) during, and (c) after the writing. Source: Tsunashima 2001 [281]. Reproduced with permission of IOP.

of magnetic medium is heated up to near the Curie temperature, the spontaneous magnetization M_s of this part becomes much smaller and the coercivity H_c also decreases dramatically. If a magnetic field is applied during the heating, the medium in the local region is easily magnetized along the field direction. This is phenomenologically the same as when volcanic rocks are magnetized by the earth's field after cooling down to ambient temperature. The field strength required to change the magnetization direction is much smaller in the thermomagnetic recording process than that required in pure magnetic recording where only the magnetic field is applied to the media. This is one of the advantages of MO recording.

The materials of MO recording have strong perpendicular magnetic anisotropy. This type of anisotropy favors the "up and down" directions of magnetization over all other orientations. The disk is initialized in one of these two directions, say "up," and the recording takes place when small regions are selectively reverse-magnetized by the thermomagnetic process. The resulting magnetization distribution then represents the pattern of recorded information. For instance, binary sequences may be represented by a mapping of zeros to up-magnetized regions and ones to down-magnetized regions.

The thermomagnetic recording process is expressed most simply by Huth's equation [282], in which a circular magnetic domain is assumed to appear with a cylindrical domain wall at the domain boundary. Huth's equation, which takes into account all the pressures acting on the wall of circular magnetic domains in thin films, shows the stability condition of the "bubble" domains. If the effective field corresponding to the total pressure is lower than the wall coercivity, as shown in Eq. (4.28), the bubble domain is stable

$$\frac{\sigma_W}{2rM_s} + \frac{1}{2M_s}\frac{\partial \sigma_W}{\partial r} - H_d - H_{ext} < H_c \tag{4.28}$$

Here, r is the domain radius, H_d, the demagnetizing field, H_{ext}, the applied external field, and σ_{w}, the wall energy density of the magnetic medium. All parameters are those at the position of the wall. The first and second terms, originating from the surface energy of the cylindrical domain wall, tend to collapse the domain, while the third term, resulting from the magnetostatic energy, tends to enlarge the domain. It is noted that the second term is due to the temperature gradient of the media, which is irradiated by the focused laser beam. When the temperature profile has cylindrical symmetry, all the parameters in the above equation can be easily evaluated. If the absolute value of the effective field is higher than H_c, the bubble domain will become either smaller or larger. During the cooling process, H_c usually becomes larger and will become equal to the effective field and then the domain shape will be fixed.

The time needed for recording is an important parameter for present storage systems. The speed of recording is determined by the time needed for increasing the temperature of the media, and the speed; that is, the "data bit rate" is limited by the power of the semiconductor laser diodes irradiating the medium. If there is no limit to the laser power, the recording time will be determined by the relaxation time of the spin system near the Curie temperature. The relaxation time seems to be in the order of several picoseconds [283], which is much shorter than the time for the magnetization reversal in a typical magnetic field. This fact shows the potential of thermomagnetic recording for very fast operation of the storage system.

The density of data storage is the most important parameter. As long as a two-dimensional storage medium is considered, the areal density of the optical storage is determined by the spot size of the laser beam. In the case of a typical objective lens, the beam spot size d defined as the distance between half power points of the focused beam is approximately given by

$$d = \frac{\lambda}{2NA} \tag{4.29}$$

Here, λ is the wavelength of the laser, NA the numerical aperture of the objective lens, which is defined as $NA = n \sin\theta$ where n is the refractive index of the image space, and θ the angle between the outermost rays and the optic axis. Since in usual lens systems NA is smaller than 1, the minimum beam spot diameter is roughly equal to or slightly smaller than the wavelength. Thus, the area of the heated region is of the same order as the wavelength of the laser. More precisely speaking, the temperature profile of the medium that is irradiated by the laser beam spot depends on the thermal properties of the medium as well as the velocity with which it is moved. Figure 4.69a shows an example of the temperature profile calculated for a moving medium irradiated continuously. The maximum recording density depends on the temperature profile and the resulting thermal variation of the magnetic properties of the medium.

4.4.5.2 MO Recording Process

In a practical MO system, two kinds of recording methods are used. One is laser power modulation (LPM) and the other is magnetic field modulation (MFM) process. Details description of LPM and MFM are given here.

LPM Process In this traditional approach to thermomagnetic recording, the electromagnet produces a constant field, while the information signal is used to modulate the power of the laser beam. As the disk rotates under the focused spot, the pulsed laser beam creates a sequence of up/down domains along the track. The Lorentz electron micrograph in Figure 4.70b shows a number of domains recorded by LPM. The domains are highly stable and may be read over and over again without significant degradation. If, however, the user decides to discard a recorded block and to use the space for new data, the LPM scheme does not allow direct overwrite; the system must erase the old data during one revolution and record the new data in a subsequent revolution cycle.

During erasure, the direction of the external field is reversed, so that up-magnetized domains in Figure 4.70a now become the favored ones. While writing is achieved with a modulated laser beam, in erasure the laser stays on for a relatively long period of time, erasing an entire sector. Selective erasure of individual domains is not practical, nor is it desired, since mass data storage systems generally deal with data at the level of blocks, which are recorded onto and read from individual sectors. Note that at least one revolution cycle elapses between the erasure of an old block and its replacement by a new block. The

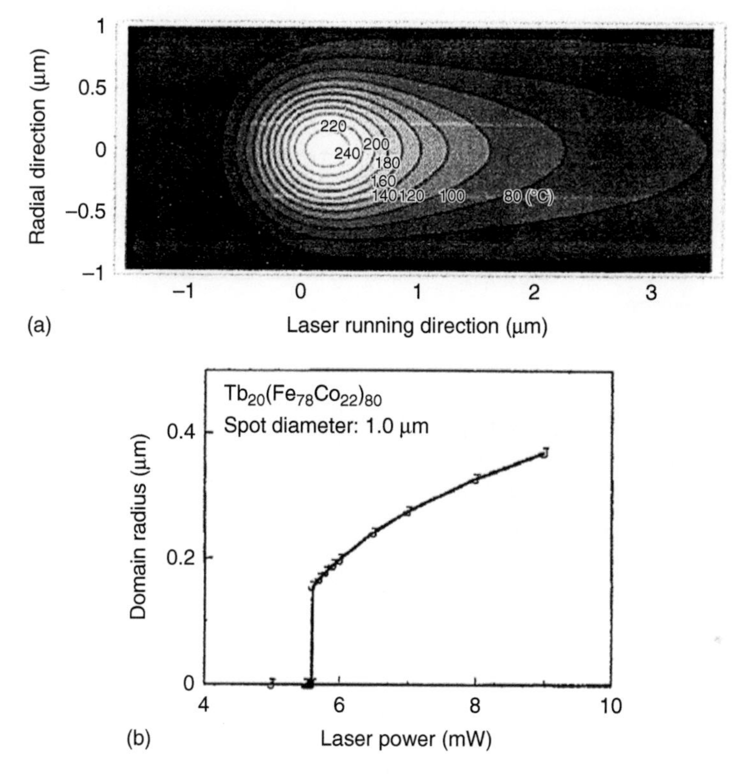

Figure 4.69 Example of (a) the temperature profile calculated for a TbFeCo amorphous alloy films on a polymethyl methacrylate (PMMA) disk moving at 10 m/s irradiated continuously, where a beam spot radius and laser power are 1 μm in diameter and 7 mW, respectively; and (b) recorded mark radius calculated using Eq. (4.28). Source: Tsunashima 2001 [281]. Reproduced with permission of IOP.

electromagnet therefore need not be capable of rapid switching. (When the disk rotates at 3600 rpm, for example, there is a period of 16 ms or so between successive switching events.) This kind of slow reversal allows the magnet to be large enough to cover all the tracks simultaneously, thereby eliminating the need for a moving magnet and an actuator. It also affords a relatively large gap between the disk and the magnet tip, which enables the use of double-sided disks and relaxes the mechanical tolerances of the system without overburdening the magnet's power supply.

The obvious disadvantage of LPM is its lack of direct overwrite capability. A subtler concern is that it is perhaps unsuitable for the PWM (pulse width modulation) scheme of representing binary waveforms. Owing to fluctuations in the laser power, spatial variations of material properties, lack of perfect focusing and track-following, etc., the length of a recorded domain along the track may fluctuate in small but unpredictable ways. If the information is to be encoded in the distance between adjacent domain walls (i.e. PWM), then the LPM scheme of thermomagnetic writing may suffer from excessive domain-wall jitter. LPM works well, however, when the information is encoded in the position of domain

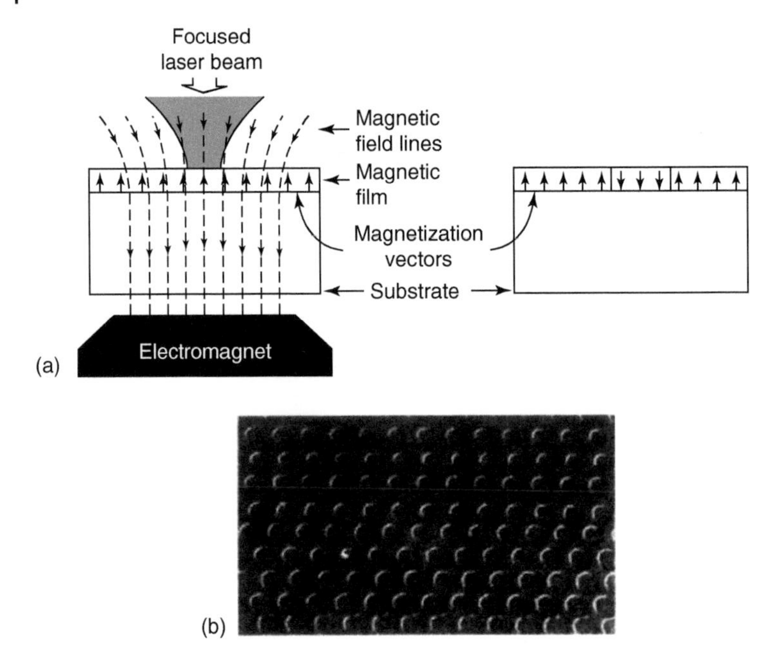

Figure 4.70 (a) Thermomagnetic recording process. The field of the electromagnet helps reverse the direction of magnetization in the area heated by the focused laser beam. (b) Lorentz micrograph of domains written thermos-magnetically. The various tracks shown here were written at different laser powers, with power level decreasing from top to bottom. Source: Mansuripur 1995 [284]. Reproduced with permission of Cambridge University Press.

centers (i.e. pulse position modulation [PPM]). In general, PWM is superior to PPM in terms of the recording density, and methods that allow PWM are therefore preferred.

MFM Process Another method of thermomagnetic recording is based on MFM, and is depicted schematically in Figure 4.71a. Here, the laser power may be kept constant while the information signal is used to modulate the direction of the magnetic field. Photomicrographs of typical domain patterns recorded in the MFM scheme are shown in Figure 4.71b. Crescent-shaped domains are the hallmark of the field modulation technique. If one assumes (using a simplified model) that the magnetization aligns itself with the applied field within a region whose temperature has passed a certain critical value, T_{crit}, then one can explain the crescent shape of these domains in the following way: with the laser operating in the CW (constant wave) mode and the disk moving at constant velocity, temperature distribution in the magnetic medium assumes a steady-state profile, such as that in Figure 4.71c. Of course, relative to the laser beam, the temperature profile is stationary, but in the frame of reference of the disk the profile moves along the track with the linear track velocity. The isotherm corresponding to T_{crit} is identified as such in the figure; within this isotherm the magnetization always aligns itself with the applied field. A succession of critical isotherms along the track, each obtained at the particular instant of time when the magnetic field switches

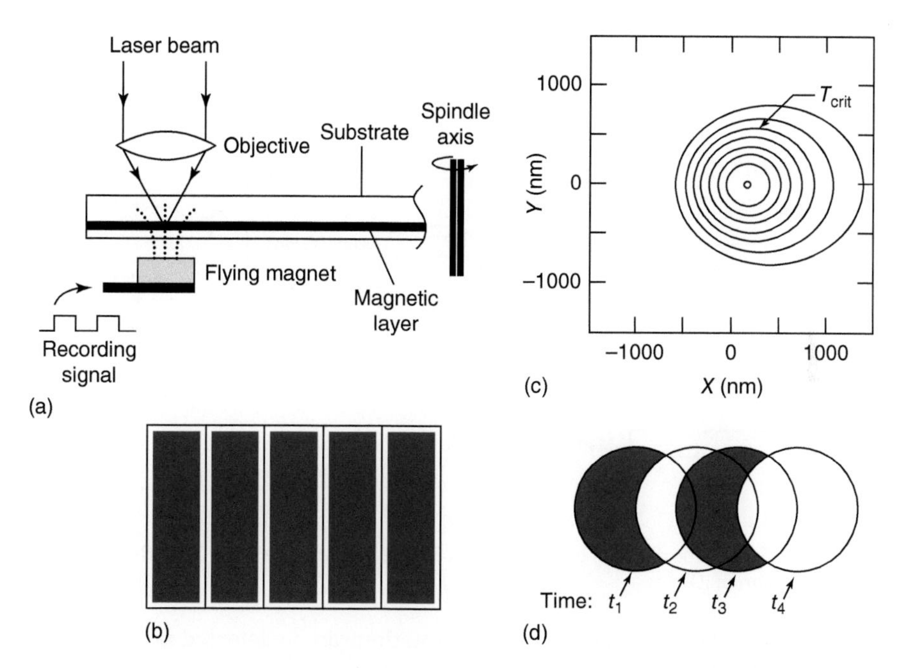

Figure 4.71 (a) Thermomagnetic recording by magnetic field modulation. (b) Polarized-light microphotograph of recorded domains. (c) Computed isotherms produced by a CW laser beam, focused on the magnetic layer of a disk. (d) Magnetization within the heated region (above T_{crit}) follows the direction of the applied magnetic field, whose switching occurs at times t_n. The resulting domains are crescent shaped. Source: Mansuripur 1995 [284]. Reproduced with permission of Cambridge University Press.

direction, is shown in Figure 4.71d. From this picture, it is not difficult to see how the crescent-shaped domains form and also to understand the relation between the waveform that controls the magnet and the resulting domain pattern.

The advantages of MFM recording are that (i) direct overwriting is possible, and (ii) domain wall positions along the track, being rather insensitive to defocus and laser power fluctuations, are fairly accurately controlled by the timing of the magnetic field switching. On the negative side, the magnet must now be small and fly close to the disk surface if it is to produce rapidly switched fields with a magnitude of a few hundred gausses. Systems that utilize MFM often fly a small electromagnet on the opposite side of the disk from the optical stylus. Since mechanical tolerances are tight, this might compromise the removability of the disk in such systems. Moreover, the requirement of close proximity between the magnet and the storage medium dictates the use of single-sided disks in practice.

4.4.5.3 Magneto-optical Readout

For the readout, MO recording systems utilize the MOKE. Figure 4.72 shows schematically the recording system, where the optical head is used for both writing and reading. Usually the incident light is linearly polarized, and the polarization direction of the light reflected from the surface of the magnetic media is rotated somewhat, usually around 1° or less due to the Kerr effect of MO media.

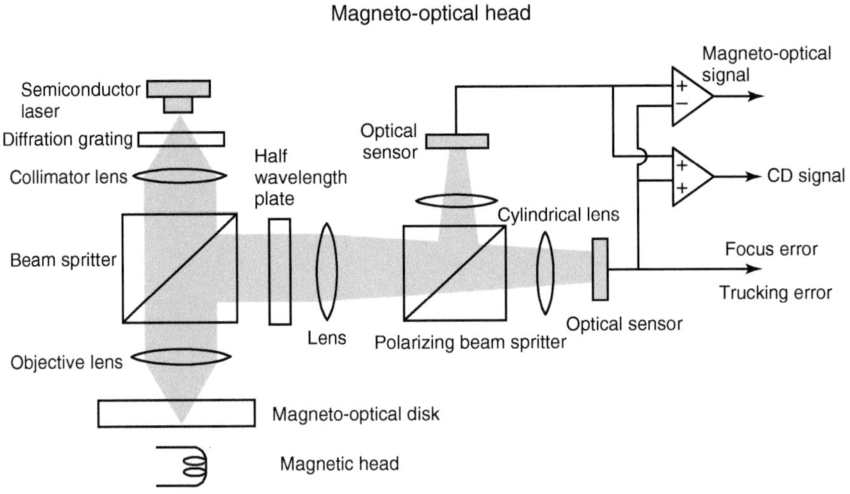

Figure 4.72 Schematic diagram of an MO head. Source: Tsunashima 2001 [281]. Reproduced with permission of IOP.

By using the polarized beam splitter, the rotation can be detected as a change in light intensity that can be detected by photodiodes.

The MOKE involves transfer of angular momentum from electron to photon, occurring with optical transition of electrons. The effect appears through the spin–orbit interaction in spin-polarized substances. Since the Kerr effect is most efficiently detected when the electron spin direction is parallel to that of the photon spin, which is parallel to the light propagation direction, the magnetization of the media is desired to be normal to the film plane. Such magnetic media can be obtained in magnetic films having a large perpendicular (out-of-plane) magnetic anisotropy. The SN ratio of the readout signal is in principle determined by the magnitude of the Kerr rotation angle. When shot noise is dominant in the readout signal, the theoretical signal-to-noise ratio is given as [285]

$$\frac{S}{N} = \left(\frac{\eta}{2Bh\nu}\right)^{1/2} (RI_0)^{1/2} \sin 2\theta_k \tag{4.30}$$

where $h\nu$ is the photon energy, η is the quantum efficiency of the detector, B is the bandwidth of detection, R is the reflectivity of the media, and I_0 is the power of detecting light. As can be seen from this equation, the *SN* ratio is proportional to the Kerr rotation angle and the square root of the readout power, namely, the number of photons for readout.

4.4.5.4 MO Recording Materials

To realize the requirement, MO recording media must satisfy many kinds of physical conditions. In conventional disk storage systems, the media are disk-shaped with a magnetic film deposited on a flat substrate. The following properties are required for the magnetic film.

(1) High coercivity that drops sharply when it is illuminated with a focused laser beam.

(2) MO effect large enough to read out the recorded magnetic marks.
(3) Magnetic anisotropy large enough to make the magnetization perpendicular to the film plane and to attain rectangular hysteresis loops.
(4) Appropriate optical absorption and reflectivity.
(5) No optical and magnetic inhomogeneity that contributes to the readout noise.
(6) Thermal stability against the heating process of writing and reading.
(7) Chemical stability and further for practical application and wide use.
(8) Efficient, low-cost production.

Table 4.3 [281] shows various kinds of materials so far considered as MO recording media. The materials shown here satisfy conditions (1)–(7) and have been examined for practical disk storage systems. The materials can be categorized into metals and oxides, or into crystalline and amorphous materials. Roughly speaking, crystalline films consist of metals or oxides exhibiting large MO effect but often requiring a high-temperature process for crystallization. This requires thermally stable substrates and often results in grain noise due to grain growth during the heat treatment. On the other hand, amorphous films and metallic multilayers can be made on substrates kept at room temperature, which suppresses the grain growth and results in low grain noise.

Uniaxial magnetic anisotropy with the easy axis perpendicular to the film plane, the so-called perpendicular magnetic anisotropy, is necessary for MO media. In

Table 4.3 Characteristics of typical magneto-optical recording materials.

Material	Crystal structure	Deposition method	Annealing temperature (°C)	Curie temperature (°C)	Squareness	Coercivity 10^5 (A m^{-1})$^{-1}$	Kerr rotation (°)
Metal							
MnBi	Hexagonal	Vacuum evaporation	300	360	1	2	0.6 (633 nm)
MnCuBi	Cubic	Vacuum evaporation	400	180	1	1–2	0 43 (830 nm)
PtCo	Tetragonal	Sputtering	600	400	0.8		
TbFeCo	Amorphous	Sputtering	RT	150–250	1	4	0.2 (830 nm)
Pt/Co	Cubic	Vacuum evaporation	RT	250–300	1	1	0.36 (410 nm)
Oxide							
BiDy1G	Garnet	Thermal decomposition	500–600	250	1	2	1[a](633 nm)
Co-ferrite	Spinel	Thermal decomposition	300–600	0.9	2		

a) Faraday rotation (deg/mm).
Source: Tsunashima 2001 [281]. Reproduced with permission of IOP.

the early stage of research and development in the 1960s, polycrystalline MnBi films with preferred orientation were examined [278]. Although MnBi films have perpendicular magnetization together with a large MO effect, MO disks did not enter practical use since many technologies related to the optical storage, such as semiconductor laser, had not yet been developed. In 1973, a new material, amorphous RE–TM, was discovered [279]. In spite of the amorphous structure, the RE–TM films have perpendicular magnetic anisotropy larger than 10^6 erg/cm^3 so that they exhibit perpendicular magnetization and are considered as a candidate for magnetic bubbles and MO recording media. The performance of amorphous Gd–Co films was examined as MO recording media together with various crystalline materials [286]. The amorphous Gd–Co films showed very little media noise due to optical inhomogeneity. This was a big advantage over polycrystalline magnetic films considered formerly.

RE–TM Amorphous Alloy Films RE–TM films are made usually on water-cooled substrates by sputtering. The films, which contain 20–30 atm% RE, have amorphous structure as long as they are confirmed by X-ray diffraction. For practical media, ternary or quaternary alloy systems based on Tb–Fe are used. Here, magnetic properties of typical amorphous alloy systems are explained to understand the fundamental properties of the RE–TM films.

In RE–TM films, the RE spin and the TM spin are coupled antiferromagnetically with each other, while within the RE element the orbital moment is coupled with the spin moment according to Hund's rules. As a result, RE–TM alloys made of LRE (light rare earth) atoms, whose orbital moment is larger than the spin moment and aligned antiparallel to the spin moment, behave like a ferromagnet while alloys consisting of HRE (heavy rare earth), where the orbital moment is parallel to the spin moment, behave as a ferrimagnet. Ferromagnetic LRE–TM films with a large saturation magnetization support perpendicular magnetization with difficulty due to the large demagnetizing energy and are very hard to make with a completely square hysteresis loop. The amorphous films applied to MO recording media usually consist of HRE and TM, and the composition is close to the so-called compensation composition, where the magnetizations of RE and TM subnetworks are nearly compensated and the net magnetization is small. In such compositions, perpendicularly magnetized films with the remanence ratio of one can be obtained even when the perpendicular anisotropy energy is relatively small. Figure 4.73 schematically shows the configuration of the atomic moments in ferrimagnetic RE–TM films.

The temperature profile of the spontaneous magnetization is the most important parameter for MO recording media. Figure 4.74 shows the temperature dependence of the magnetization for a GdCoMo amorphous alloy film, where the spontaneous magnetization disappears at a compensation temperature far below the Curie point. This type of temperature profile is called N-type ferrimagnetism. The temperature dependence is usually analyzed by using mean-field approximation introducing exchange energies J_{TM-TM}, J_{RE-TM}, and J_{RE-RE} [286].

The magnetism of Fe in amorphous alloys is much different from that in *bcc* crystals. Amorphous Fe alloys such as FeY and FeZr do not exhibit strong

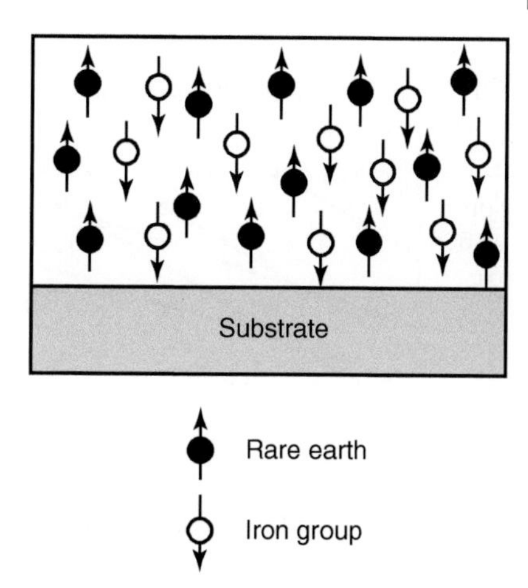

Figure 4.73 Schematic configuration of the magnetic moments in ferrimagnetic RE–TM films. Source: Tsunashima 2001 [281]. Reproduced with permission of IOP.

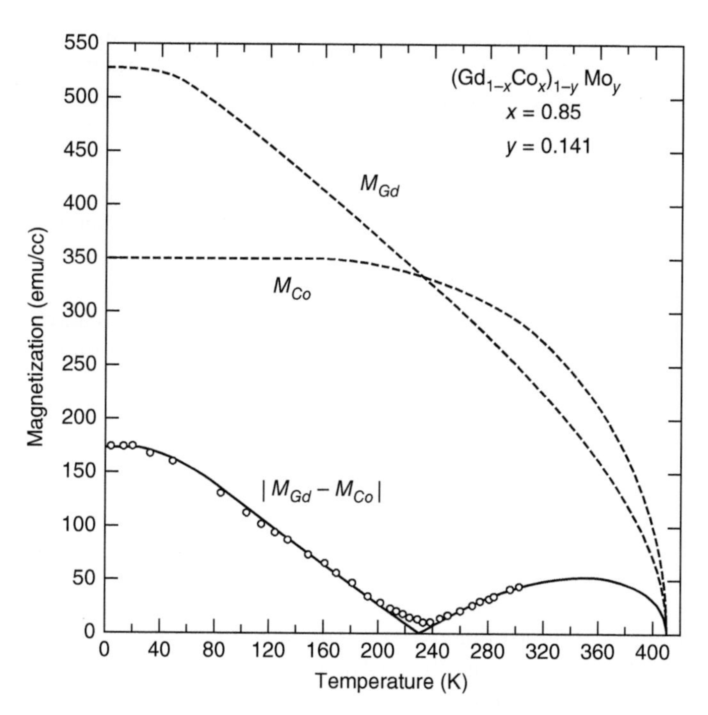

Figure 4.74 Temperature dependence of the magnetization for a GdCoMo amorphous alloy film and the sublattice magnetization calculated with mean field approximation. Source: Hasegawa et al. 1975 [285]. Reproduced with permission of AIP.

ferromagnetism but exhibit spin-glass-like behavior at low temperatures [287, 288]. When some Fe atoms are replaced with Co, the amorphous alloy becomes a strong ferromagnet and the Curie temperature becomes higher with increasing Co content [289]. For this reason, RE–Fe amorphous alloys, having lower Curie temperatures than the corresponding crystalline alloys, are suitable for thermomagnetic recording. In contrast, RE–Co amorphous

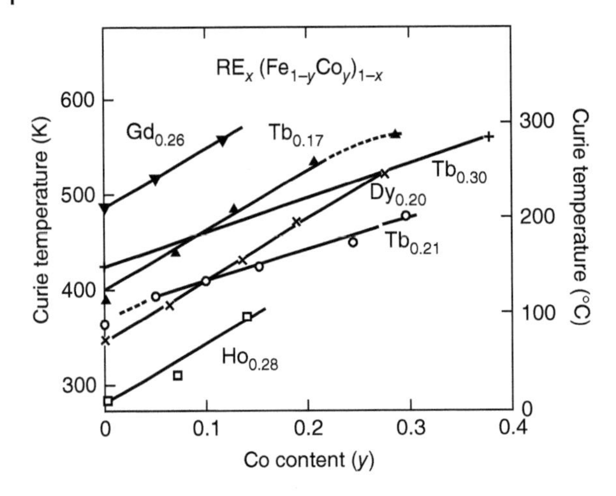

Figure 4.75 Curie temperature as a function of Co content for RE–FeCo amorphous alloy systems. Source: Tanaka et al. 1987 [290]. Reproduced with permission of IOP.

alloys have higher Curie temperatures, which are too high for them to be used as thermomagnetic recording media. Figure 4.75 [290–292] shows the Curie temperature for the RE–FeCo amorphous alloy system. In RE–Fe binary alloys, the Curie temperature is much influenced by the kind of RE elements through the interaction between RE and TM atoms, which is proportional to the magnitude of RE spin. The Curie temperature is the highest for GdFe with the largest RE spin, and it decreases in the order of TbFe and DyFe. As shown in Figure 4.75, the Curie temperature is quite easily controlled by adding Co, which dramatically enhances the TM–TM interaction.

Figure 4.76 [293] shows the temperature dependence of the saturation magnetization and the coercivity where curves are drawn for two kinds of composition. One composition is a TM dominated alloy, and the other is an RE dominated alloy. Even when the Curie temperatures of both types of film are almost the same, the temperature profile of M_s is much different. In MO media, a large MO effect is desirable but large magnetization is not required since the readout process does not utilize the magnetic field from the magnetic moments but utilizes the MO effect of polarized sublattice moments. A large magnetization and resulting large magnetostatic energy rather make it difficult for the magnetization vector to be normal to the film plane. For these reasons, RE–TM films with nearly compensated magnetization are used for MO media. It should be noted here that a wide variety of temperature profiles can be easily obtained in the amorphous RE–TM system by adjusting Co and/or RE contents, which resulted in various unique techniques to improve the recording properties of MO media.

Perpendicular magnetic anisotropy K_u is thought to originate from the film preparation process. RE–TM films are prepared by the sputter deposition or vacuum deposition technique. The anisotropy depends on the deposition method and conditions, such as sputtering gas pressures. Anisotropic structures are thought to be introduced by incident atoms on the film surface. In the case of sputtering, the kinetic energy of incident atoms seems to be of the order of 10 eV [294]. This high energy will introduce anisotropy into the near neighbor atomic arrangement in the film and will induce perpendicular magnetic anisotropy. In the case of thermal evaporation, the energy of the incident atom is relatively

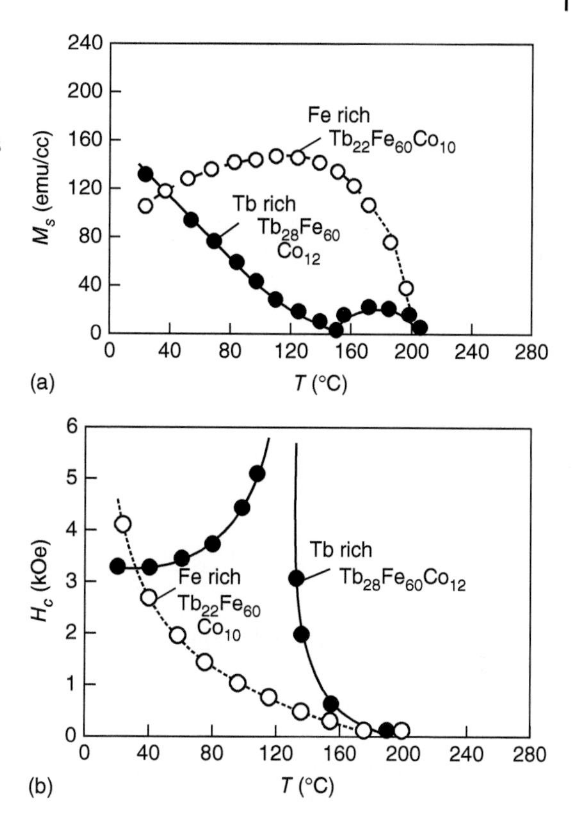

Figure 4.76 Temperature dependence of (a) the saturation magnetization and (b) the coercivity of TbFeCo amorphous films. Source: Takahashi et al. 1988 [293]. Reproduced with permission of Elsevier.

low but the film preparation process involves unidirectional anisotropy and may introduce magnetic anisotropy.

The magnitude of perpendicular magnetic anisotropy is in the order of 10^5 erg/cc in Gd–Co and Gd–Fe films while it is 10^6 or larger for other RE–TM films such as Tb– Fe and Dy–Co films. This shows that RE atoms with a 4f shell of non-S state enhances the anisotropy.

Figure 4.77 shows the perpendicular anisotropy energy per each RE atom that is substituted for Gd atoms of $Gd_{19}Co_{81}$ amorphous films prepared by RF sputtering [295]. The results can be explained by a single ion anisotropy model [296], where the anisotropy is given by the following equation:

$$K_u = 2\alpha_j J\left(J - \frac{1}{2}\right)A_2\langle r^2\rangle \tag{4.31}$$

where A_2 is the uniaxial anisotropy of the crystal field around 4f electrons, α_j the Steven's factor, J the total angular momentum quantum number, and $\langle r^2\rangle$ the average of the square of the orbital radius of 4f electrons. The 4f electron cloud of pancake (positive α_j) or cigar (negative α_j)-like shape exhibits magnetic anisotropy when it exists in anisotropic crystal fields. The line in the figure is drawn using Eq. (4.31) with appropriate magnitude of the anisotropic crystal field strength. As can be seen in the figure, the anisotropy is positive for Tb, Dy, and Ho while it is negative for Tm and Yb, which agrees well with the single ion model. In Gd–TM alloys, such as Gd–Fe and Gd–Co, the origin of the anisotropy is not

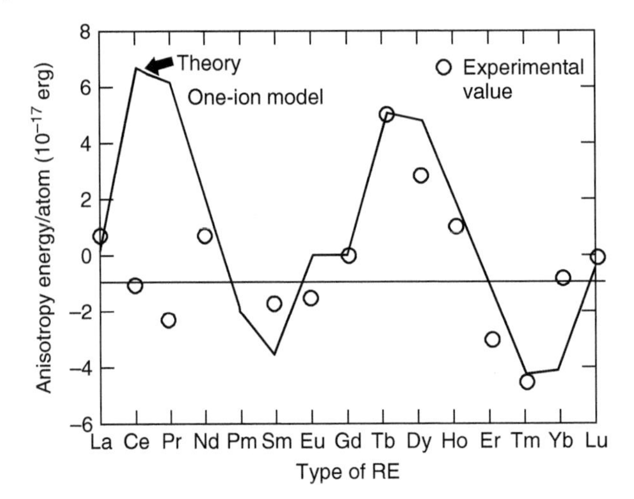

Figure 4.77 Perpendicular anisotropy energy per one RE atom substitution in $Gd_{19}Co_{81}$ amorphous films prepared by RF sputtering. Source: Suzuki et al. 1987 [295]. Reproduced with permission of IEEE.

clearly understood, although single-ion and two-ion anisotropies are conceivable origins. Stress-induced anisotropy due to the substrate constraint also possibly contributes to the anisotropy.

From an application point of view, it is important to enhance the perpendicular anisotropy not only to obtain the remanent magnetization perpendicular to the film plane but also to stabilize very fine domains for high-density recording. To enhance the anisotropy, it is effective to introduce an artificial uniaxial structure by simply stacking very thin atomic layers. This so-called artificial superlattice will be described later, but here it is noted that the anisotropy of amorphous RE–TM films can be enhanced by introducing the multilayer structure.

The MO effect of RE–TM films is also important from an application point of view. In RE–TM films, the MO effect seems to originate from the polarization of both RE and TM atoms. Roughly speaking, the contribution of TM is dominant in the near infrared region, while RE is dominant in the ultraviolet region. This tendency has been confirmed from the spectra of MO Kerr rotation in related amorphous alloys. Figures 4.78 and 4.79 [297] show the MO Kerr spectra of Gd–Ni and Y–Fe, Y–Co amorphous films, respectively, where Ni and Y are thought to be almost nonmagnetic. Both Gd–Ni and Y–Fe exhibit single peak spectra compared with the corresponding crystalline Gd and Fe with double peak spectra. This might be due to the character of d electrons in the amorphous structure.

In the short-wavelength region, the Kerr spectrum of RE–TM films changes significantly depending on the kind of RE, which reflects the difference in the electron occupancy of the 4f shell. Figure 4.80 [297] shows the contribution of RE to the Kerr spectra of various RE–Co amorphous films. As can be seen from the figure, LRE elements with less than half-filled 4f electron shells make a positive contribution to Co while HRE elements with more than half-filled shells become negative.

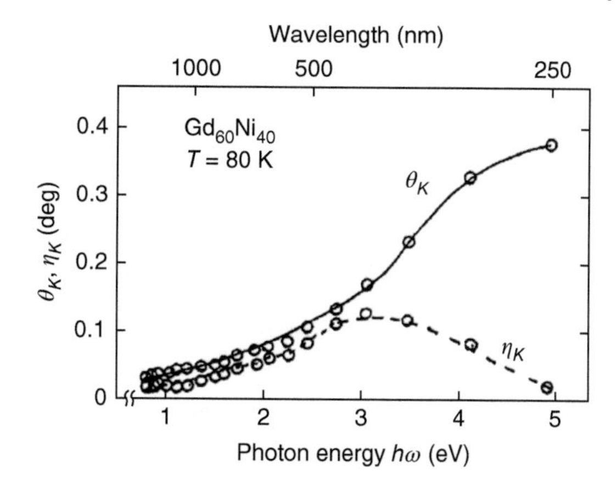

Figure 4.78 MO Kerr rotation θ_K and ellipticity η_K of a Gd–Ni amorphous film. Source: Courtesy of Choe [297].

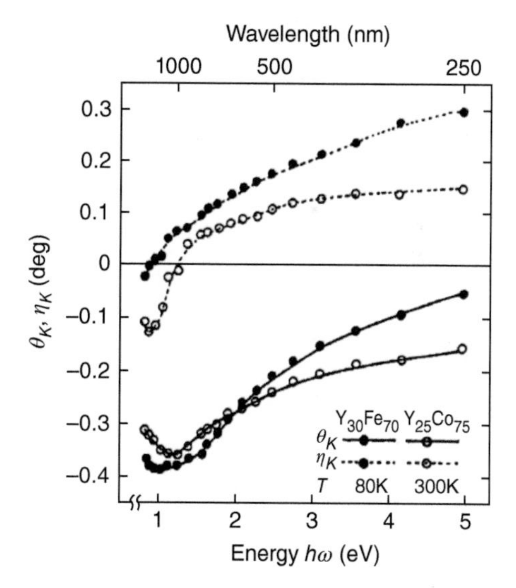

Figure 4.79 MO Kerr rotation θ_K and ellipticity η_K of Y–Fe and Y–Co amorphous films. Source: Courtesy of Choe [297].

Figure 4.81 shows the spectra of typical MO recording materials including RE–TM films and a metallic multilayer, where $(\theta_k^2 + \eta_k^2)^{1/2}$ shows the magnitude of the MOKE [298]. Among the materials, TbFeCo, which is most often used for MO media, shows a reasonable value at long wavelength, but at short wavelength the value becomes smaller. In contrast, GdFeCo, which is magnetically soft and not appropriate for recording media, has a larger effect at short wavelength. Nevertheless, GdFeCo is useful for MO readout when it is coupled with a recording layer that is magnetically hard.

Exchange Coupled Films The coercivity and Curie temperatures of RE–TM films can be controlled very easily by changing the material composition while maintaining the amorphous microstructure. This makes such films very suitable for exchange coupling between materials with different thermomagnetic properties. MO recording media must provide two different functions: thermomagnetic recording of data and readout of the data through MO effects. By delegating

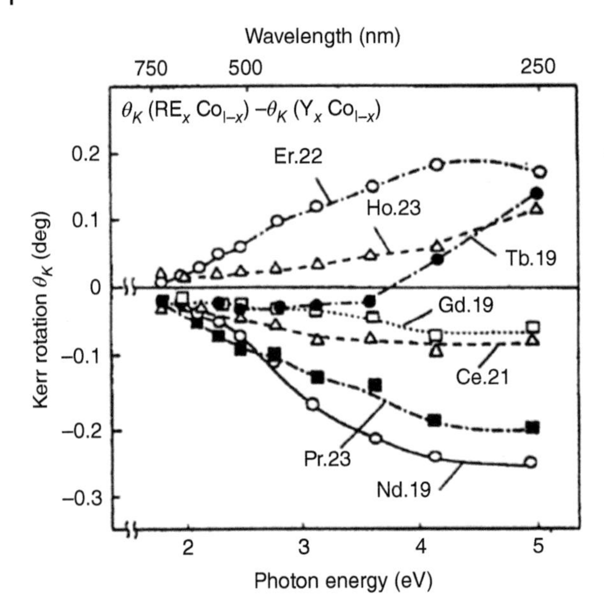

Figure 4.80 Contribution of RE to the Kerr spectra in various RE–Co amorphous films. Source: Courtesy of Choe [297].

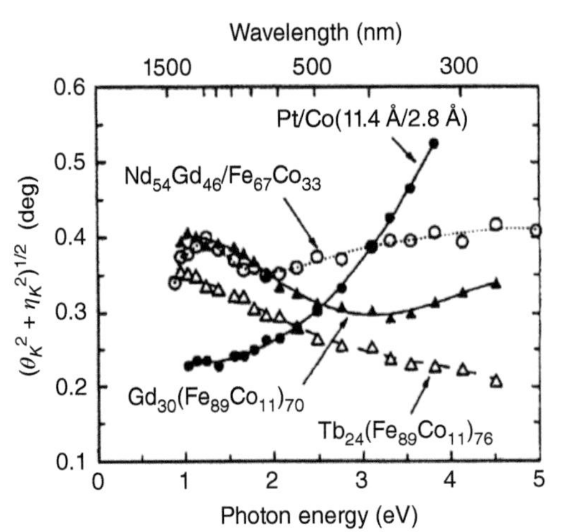

Figure 4.81 Spectra of MO Kerr effect for typical MO recording materials, where $(\theta_k^2 + \eta_k^2)^{1/2}$ shows the effective magnitude of the Kerr effect. Source: Tsunashima et al. 1997 [298]. Reproduced with permission of IEEE.

these two functions to two different types of magnetic layer and appropriately coupling the two layers, various high-performance media have been developed [299]. Currently, exchange coupled films are used for magnetic super resolution (MSR) MO media, which enables 1.3 GB recording for a 3.5 in. size disk.

Here we consider two magnetic layers (I and II) in contact on an atomic scale. The electron spins of magnetic atoms A and B, in the respective layers I and II, are coupled through an exchange interaction at the interface. This exchange interaction is extremely strong: between atoms in proximity its magnetic field equivalent reaches 10^4 kOe, orders of magnitude higher than magnetostatic coupling forces, the upper limit of which is limited by the saturation flux density. But while magnetostatic forces act over comparatively long distances, powerful direct exchange forces mainly act between the nearest neighbor atoms.

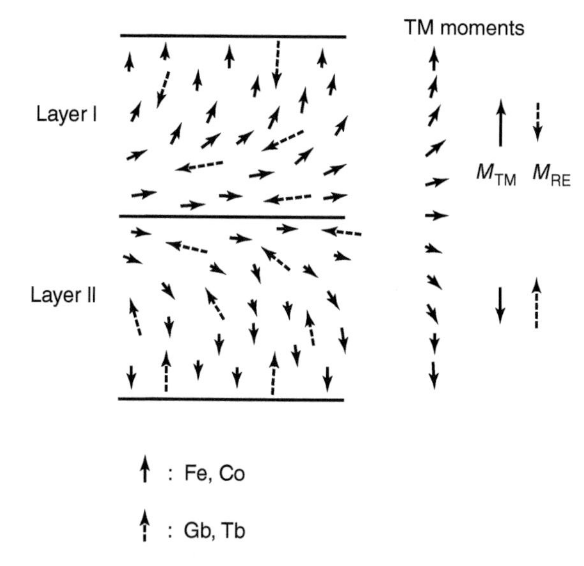

Figure 4.82 Interface wall appearing between two exchange-coupled layers with out-of-plane magnetization. Source: Tsunashima 1999 [300]. Reproduced with permission of IEEE.

We further consider heavy RE–TM amorphous films [299, 300]. Then the net magnetization of the film is given by the difference in magnetization between the RE and TM sublattices. If in one layer the RE magnetization is dominant and in the other layer the TM magnetization dominates, when a sufficiently strong magnetic field is applied, the net magnetization of both layers will align parallel to the magnetic field. Here the atomic magnetic moments (spins) of the first and second layer are oppositely directed, so that between the layers a region will appear in which the spin gradually changes in direction. As shown in Figure 4.82, a domain wall similar to a Bloch wall (in the case of perpendicular magnetization, the wall differs from the usual Bloch wall in the sense that the magnetization on both sides of the wall is perpendicular to the plane of the wall) will appear at the interface between the two layers [300]. This is called an interface wall, where similar to the Bloch wall, exchange and anisotropy energies are stored. In the case of perpendicularly magnetized film, demagnetization energy is also stored there.

In the case of HRE–TM double layer films, the TM magnetic moments of the two layers tend to align parallel because of the ferromagnetic exchange interaction between TM moments and of the antiferromagnetic exchange between TM and HRE moments. As a result, the magnetization states (a) or (c) on the left-hand side in Figure 4.83 are stable under zero magnetic field. Here the double-layer film (a) is called the A (antiparallel)-type where one layer has the composition opposite to that of the other layer relative to the compensation composition; if TM magnetization is dominant in one layer, RE magnetization is dominant in the other. Double-layer film (c) is called the P (parallel)-type; both layers have their compositions on the same side of the compensation composition. In A-type films the magnetizations of the two layers tend to be antiparallel, while in P-type films they tend to be parallel. When a strong magnetic field is applied in an upward direction to the A-type film shown in Figure 4.83a, the magnetization of the first layer is also directed upward as in Figure 4.83b, and the magnetizations of the two layers tend to assume a parallel orientation. Since atomic magnetic moments

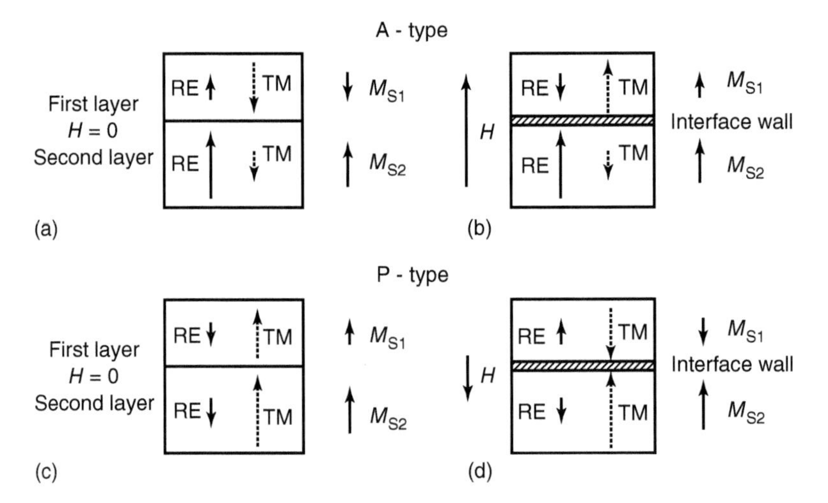

Figure 4.83 Magnetization states in exchange-coupled RE–TM films with out-of-plane magnetization [299, 300].

(spin) in the first and second layers are oppositely directed, an interface wall should form between the two layers. Even in a P-type film, under certain conditions an interface wall will form as shown in Figure 4.83d.

The magnetization process of both A-type and P-type films can be deduced by introducing the interface wall energy. Magnetization curves thus obtained are summarized in Figure 4.84 together with the Kerr hysteresis loops to be observed from the surfaces of the first and the second layers. Here σ_w is the interface wall energy density and M_{si}, H_{ci}, h_i $(i = 1, 2)$ are the saturation magnetization, the coercivity, and the thickness of the layer 1 or 2, respectively. Figure 4.85 is an example of measured magnetization curves for A-type double-layer films. Here H_1 and H_2 are the magnetic fields at which the magnetization is reversed in the GdFe and TbFe layers, respectively. The value of H_1 observed is of the order of several kilo-Oe, and the coercive force of the Gd–Fe film is known to be of the order of 100 Oe or less, except when the film temperature is very close to the compensation point, indicating that the coercivity of the GdFe layer is apparently much increased by the exchange coupling.

The reversing field of the GdFe layer H_1 is expressed as

$$H_1 = \frac{\sigma_w}{2M_{s1}h_1} - H_{c1} \tag{4.32}$$

Here the quantity given by $\sigma_w/2M_{s1}h_1$ is observed as the shift of the center of the minor loop corresponding to the magnetization reversal of the GdFe layer. This loop shift, called "exchange bias," can be used to control the apparent coercivity.

Metallic Multilayers As shown in Table 4.3, many kinds of materials have been considered as candidates of MO recording media, but only a few metallic crystalline materials are described. Perpendicular magnetic anisotropy is a necessary condition for MO media, but very few materials satisfy this condition. In crystalline materials, large perpendicular magnetic anisotropy appears mostly when

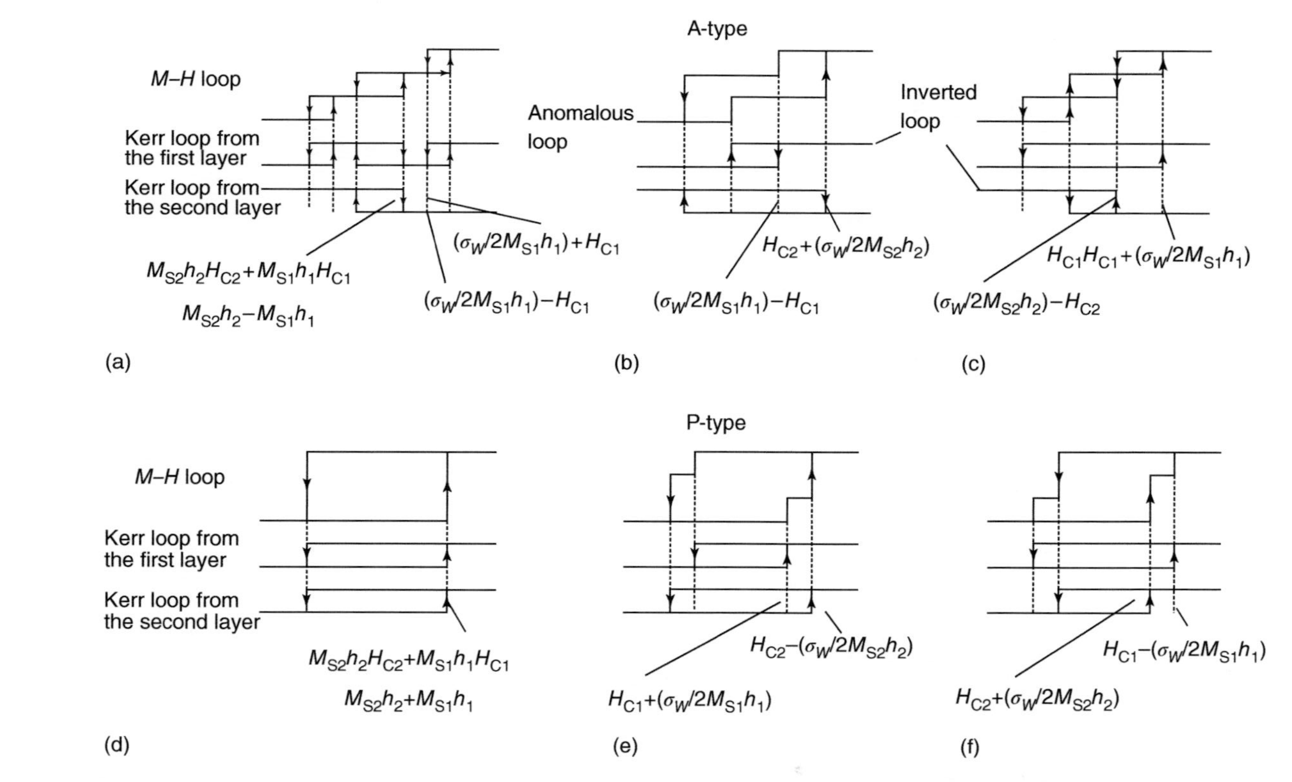

Figure 4.84 Kerr hysteresis loops to be observed in exchange-coupled RE–TM films with out-of-plane magnetization [299, 300].

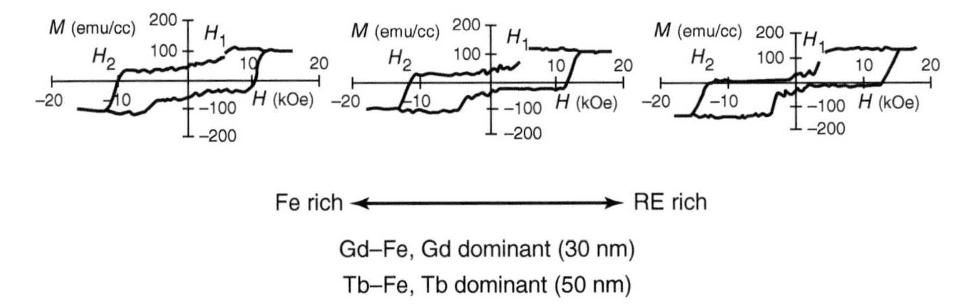

Figure 4.85 An example of measured magnetization curves for A-type.

the crystalline anisotropy is very large and its easy axis is perpendicular to the film plane. One example is a hexagonal crystal structure with the *c*-axis perpendicular to the film plane. MnBi typified this and was used in the early stage of development. A difficulty of polycrystalline films often appears in the growth process. To obtain good crystallinity and preferred orientation, usually high-temperature annealing is necessary, but such a heat treatment often results in the growth of large crystallites, which degrades the optical media through an increase in the media noise due to large grains.

Metallic multilayers are one solution of the above problem. In 1985, Carcia et al. found that Pd/Co multilayers show very large anisotropy, which is enough to attain perpendicular magnetization [301]. They further found that Pt/Co multilayers also show perpendicular magnetization and exhibit large MOKE at short wavelength, as shown in Figure 4.81 [302]. The large perpendicular magnetic anisotropy is thought to originate from the interface between the Co and noble metal layers. This can be phenomenologically understood from the fact that the effective anisotropy per unit area, $K_{\mathrm{eff}}\, t_{Co}$, linearly increases with decreasing Co layer thickness, t_{Co}, as shown in Figure 4.86 [303], where the extrapolated intercept with the vertical axis is considered to be the surface anisotropy of the interface. In these multilayers, perpendicular anisotropy appears when films are made on substrates kept at room temperature so that the size of the crystallite is in the

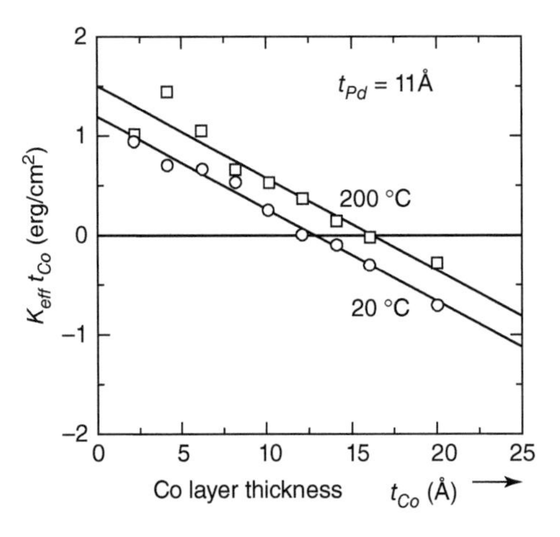

Figure 4.86 Effective anisotropy per unit area $K_{eff}\, t_{Co}$ as a function of Co layer thickness for epitaxial [109] Co/Pd multilayers deposited at 20 and 200 °C. Source: den Broeder et al. 1991 [303]. Reproduced with permission of Elsevier.

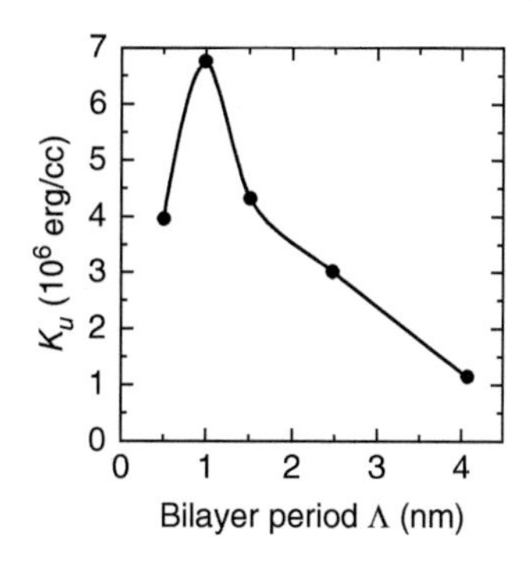

Figure 4.87 Perpendicular magnetic anisotropy as a function of bilayer thickness for Tb/Fe multilayers prepared by RF sputtering, where the thickness of each Tb layer is equal to that of the Fe layer. Source: Fujiwara et al. 1997 [304]. Reproduced with permission of IOP.

range of 10 nm, which may be small enough for the present recording media. The Pt/Co multilayers are candidates for MO media operating with blue or ultraviolet laser.

In RE–TM films described above, anisotropy is also enhanced by introducing the multilayer structure. Figure 4.87 shows the perpendicular magnetic anisotropy as a function of bilayer thickness for Tb/Fe multilayers prepared by RF sputtering, where the thickness of each Tb layer is equal to that of the Fe layer. The anisotropy shows a maximum at a bilayer thickness of 1 nm, where it was confirmed by X-ray absorption fine structure (EXAFS) experiments that the anisotropic alignment of neighboring atoms becomes most significant [304].

4.4.5.5 High-Density MO Recording

The recording density of conventional optical disks is limited by the resolution of the readout optical system. To overcome this limitation, several methods have been proposed. With these methods, the resolution is improved by using the restricted temperature "area" on the medium narrower than the light spot. Some of the methods are described as follows.

Resolution of Optical Readout Generally speaking, the higher the recording density the better the performance of the storage system. In optical recording systems, the readout process limits the density of the storage since the minimum beam waist size of the focused laser beam determines the minimum size of the marks to be read out. In the case of conventional optical systems, the minimum beam waist size is roughly equal to the wavelength of the light.

Alternatively, the recording process does not always limit the density. The recorded mark size can be made much smaller than that of the laser beam waist. This is because the thermomagnetic writing is not done in the whole beam spot area but only in the area where the temperature is raised above or a certain point around the Curie temperature. If we use LPM recording (Figure 4.70b), this area can be made much smaller than the laser beam spot when the laser power is made sufficiently low. This is called "writing by the tip of brush." The recorded mark area can be estimated by using Huth's equation for a given temperature profile, as shown in Figure 4.69b. If we use MFM recording, the marks are recorded according to the temperature profile and the field strength during the cooling process. In both recording methods, marks (magnetic domains) much

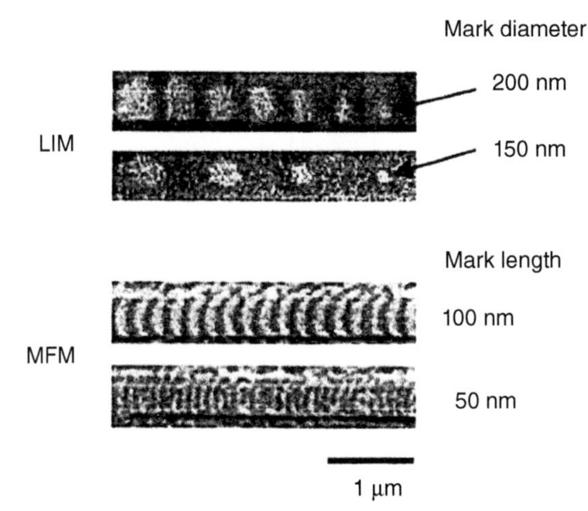

Figure 4.88 Images of marks recorded by LPM and MFM methods in TbFeCo amorphous media recorded with a magnetic X-ray microscope. Source: Fischer et al. 2001 [305]. Reproduced with permission of J-Stage.

smaller than the laser beam spot size can be recorded as seen in the images of Figure 4.88 taken with a magnetic X-ray microscope [305]. In MFM recording, a mark length as small as 50 nm can be recorded by using the laser wavelength of 635 nm and a conventional objective lens of $NA = 0.60$.

The upper limit of the recording density is actually determined by the readout signal quality. If the mark size and the mark separation are made too small, the readout beam spot will cover the area containing many neighboring marks as shown in the upper right side of Figure 4.89, and the signal from individual marks cannot be distinguished. To overcome this difficulty, new techniques have been developed that utilize multilayered magnetic films. The multilayers consist of at least two magnetic layers. One is a recording layer to store the information, and the other is a readout layer to detect the information. Both layers are coupled with each other by either exchange interaction or magnetic dipole interaction. The techniques utilize the temperature profile of the media within and around the beam spot together with the unique temperature dependence of the magnetic properties of RE–TM films.

One technique to overcome the difficulty is called magnetic super resolution (MSR) [307], where only one mark stored in the readout layer is transferred to the reading layer but other marks are masked even when they are within the beam spot. By using this technique, the interference of neighboring marks is removed during the readout. Although many kinds of magnetic super-resolution techniques have been developed so far, here only a technique called center aperture detection (CAD-MSR in Figure 4.89) is explained [306]. As shown in the figure CAD utilizes magnetostatically coupled double layers. In the CAD technique, the readout layer is usually made of GdFeCo film with smaller perpendicular magnetic anisotropy and the film composition is adjusted to have a compensation temperature above room temperature. At room temperature, the magnetization of the readout layer lies in the film plane and the in-plane magnetization masks marks in the recording layer. When the medium comes near the center of the beam spot and its temperature rises, the magnetization of the layer becomes very small and as a result the demagnetization energy drops very sharply. This brings

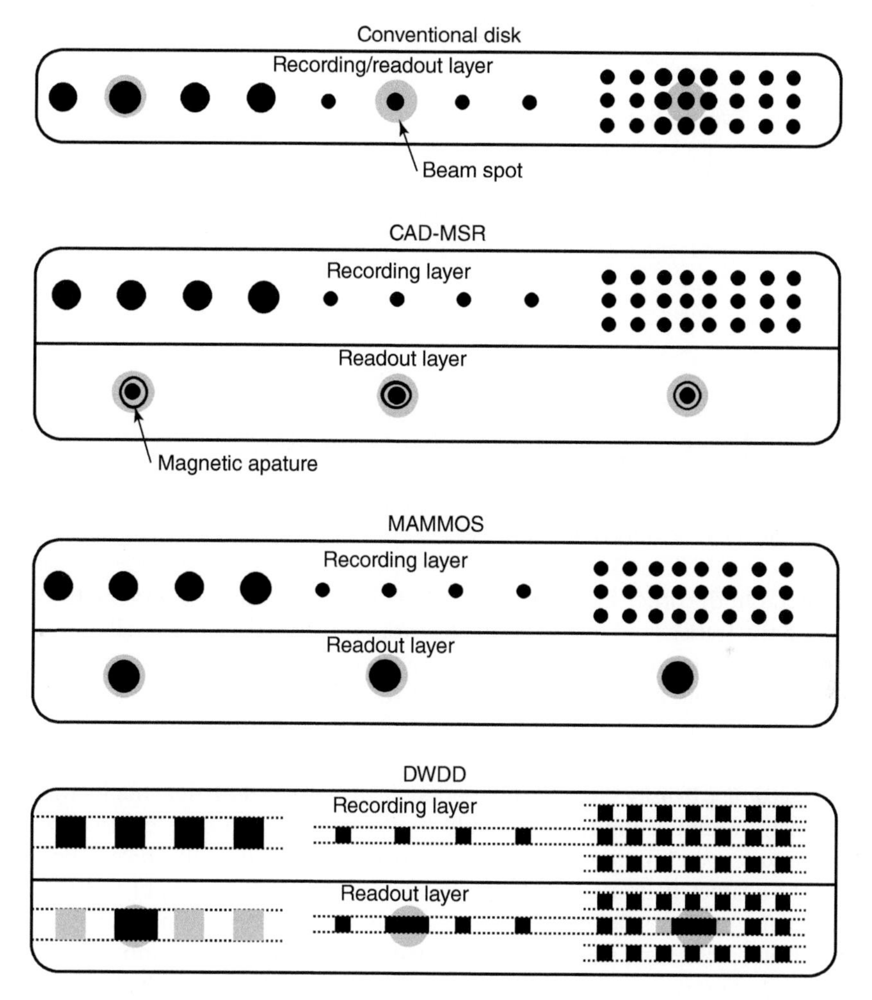

Figure 4.89 Readout techniques for recorded marks smaller than the laser beam spot. Source: Takahashi et al. 1994 [306]. Reproduced with permission of IEEE.

about a transition of the magnetization direction from in-plane to out-of-plane. When the readout layer has out-of-plane magnetization, as shown in Figure 4.90, the readout layer acts as an aperture since its magnetization direction reflects the mark in the recording layer due to the magnetic dipole coupling force.

Other techniques are of the domain wall expansion type called MAMMOS (magnetic amplifying magneto optical system) [308] and DWDD (domain wall displacement detection) [309]. In both techniques (Figure 4.89), the mark transferred to the readout layer is expanded to cover the whole or most of the area of the laser beam spot so that the signal magnitude is kept constant even when the recorded mark size is much smaller than the beam spot size. MAMMOS utilizes the magnetostatic coupling force between the readout and recording layers to transfer the mark. To expand the magnetic domain in the readout layer, an external field is applied synchronously at the time when the mark enters into the

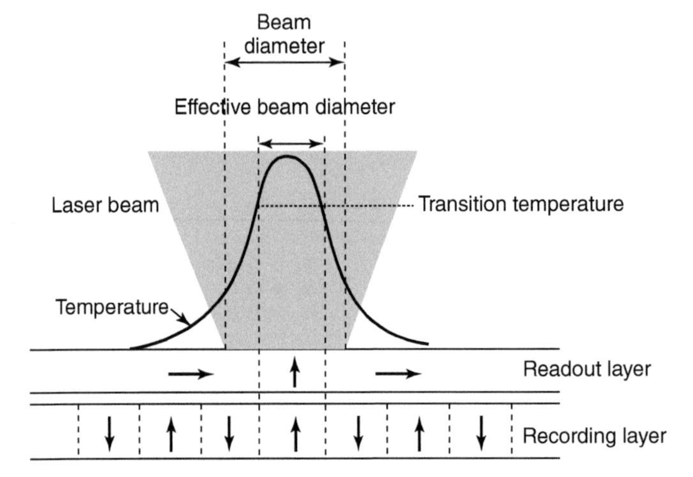

Figure 4.90 Schematic illustration of CAD magnetic super resolution. Source: Takahashi et al. 1994 [306]. Reproduced with permission of IEEE.

beam spot. On the other hand, DWDD utilizes exchange coupling and the force to expand the transferred marks originates from the gradient of the wall energy due to the temperature gradient around the beam center. By using these techniques, MO readout has become possible for recorded marks as small as 100 nm and even smaller in the track length direction.

Ultimate Recording Density If it becomes possible to read out domains smaller than the optical diffraction limit, what would limit the recording density? In some sense, this question is common to the usual magnetic recording. Recording density will be ultimately limited by the appearance of super paramagnetism. When the size of magnetic particles becomes extremely small and as a result their magnetic anisotropy energy, proportional to their volume, becomes comparable to the thermal energy, the ferromagnetic ordering of the spin system is not stable any more. The factor that characterizes the stability of the magnetization of the fine particle is given by

$$N = \frac{K_u V}{kT} \tag{4.33}$$

where K_u is the uniaxial magnetic anisotropy constant and V is the volume of the particle. Simulations for longitudinal magnetic storage media have shown that N is required to be larger than 60 [310]. This will hold for magnetic recording media that consist of magnetic fine particles. If the particle model can be applied to the MO media, the minimum mark (domain) size will be determined in the same manner. For MO media of TbFeCo, if we assume $K_u = 10^0$ erg/cc, and $T = 300$ K, $K_u V/kT = 100$ gives a cylinder of 10 nm height and 20 nm diameter, which is enough to attain recording density in future MO systems. In the case of MO recording, however, the media are continuous films where atomic spins are exchange-coupled with each other throughout the film. Thus, the situation is a little different from conventional magnetic recording media. We must take into account the effect of the domain wall around the recorded domains. When we take into account both effects of the wall energy and the thermal agitation on the uniform precession of the domain magnetization, the factor for the thermal stability given by Eq. (4.33) will be modified to

$$N = \left(K_u - \frac{\sigma_w}{2r}\right)/kT \tag{4.34}$$

As a result, thermal stability will become worse compared with the case of Eq. (4.33), as pointed out experimentally [311].

Apart from the appearance of super paramagnetism, the minimum mark size is determined by the stability of bubble domains. By introducing a phenomenological parameter, the coercivity H_c, the stability of the cylindrical domain is described by Eq. (4.28). When the temperature of the media is uniform, the second term of Eq. (4.28) is negligible and if the third term, stray fields, is also neglected, the minimum domain radius is determined by the first term and the coercivity as given by $r_{min} = \sigma_w/2M_sH_c$. If the domain radius is smaller than r_{min}, the domain will collapse due to the wall energy, which acts as the surface energy of the domain. The coercivity is difficult to derive theoretically for practical media, but if we simply apply single domain theory $H_c = 2K_u/M_s$ and consider the relation $\sigma_w = 4(AK_u)^{1/2}$, the minimum domain radius becomes

$$r_{\min} = \left(\frac{A}{K_u}\right)^{1/2} \tag{4.35}$$

In the case of $A = 10^{-6}$ erg/cm and $K_u = 10^6$ erg/cm^3, the minimum domain radius $2r_{min}$ becomes 20 nm. Besides the above, we must take into account the width of the domain wall. Here the width was not taken into account but was assumed to be infinitely thin. The minimum domain diameter would be larger than twice the 180° wall width, which is expressed as

$$\delta_w = \pi\left(\frac{A}{K_u}\right)^{1/2} \tag{4.36}$$

If we consider the minimum domain radius to be the same as δ_w, the minimum domain diameter becomes a factor of π larger than that of Eq. (4.35).

For high-density recording, it is very important where and how the domain walls are pinned. If the density of the pinning sites is very low, the domain wall or recorded marks cannot form and keep their shape as determined by conditions such as in Eq. (4.28) but they might form a shape much influenced by the distribution of the pinning sites. Alternatively, if the density of pinning sites is high enough, the domain wall will be pinned according to Eq. (4.28) or similar conditions. The distribution of pinning sites depends on the mechanism of the coercivity, which is still not clearly understood in the case of amorphous RE–TM films. It was proposed that in amorphous magnetic materials random anisotropy exists for a magnetic atom on each site. This anisotropy, which originates from the random distribution of nearest neighbor atoms around each magnetic atom, will not appear as bulk magnetic anisotropy when averaged over a macroscopic size, but will appear only as coercivity of the bulk magnetization [312]. For 4f elements other than Gd, the random anisotropy is expected to be very large because of their nonspherical 4f electron cloud. If random anisotropy is the real origin of the wall coercivity, the wall might be pinned anywhere in a scale of wall width, resulting in a coercivity which is extremely uniform throughout the medium. However, if another nonuniformity on a scale much larger than the wall width was the origin of wall coercivity, the wall would move from one pinning site to another and would not be stabilized between the sites. As a result, recording density would be limited by the density of pinning sites.

Near-Field Recording The density of optical recording is governed by the size of the focused laser beam spot, which is limited to be about $\lambda/2NA$ by diffraction. This limits the density of optical recording systems, especially in the readout process. To circumvent the diffraction limit, near-field optical techniques have been recently developed. One way to overcome the limit is imaging through a pinhole that is smaller than the diffraction limited spot size. If the pinhole is placed very close to the recording media, the spot size becomes equal to the size of the pinhole. Betzig et al. used a near-field scanning optical microscope based on this principle to image and record magnetic domains [313]. They used an optical fiber probe, the tip of which was coated with an Al layer, obtaining a resolution of 30–50 nm in the imaging mode and 60 nm in the recording mode. When using this approach, low optical efficiency and low data rate are problems to be solved in the application of the optical storage system.

Another way is to use a solid immersion lens (SIL) that was demonstrated by Kino and coworkers [314, 315]. The SIL is a trenched sphere of high index glass, which is placed between the objective lens and the sample of interest. Owing to the high index inside the SIL together with increased θ_{max} by the refraction at the inner surface of the sphere, the spot size is much reduced at the bottom surface of the SIL. By placing the media within the evanescent decay length from the bottom of the SIL, the small sized spot can be transmitted across the air gap. Terris et al. achieved a 317 nm diameter spot size by illuminating with a 780 nm laser. They have written and read 350 nm diameter magnetic domains [315]. By using an air-bearing slider, Terris et al. further made a realistic demonstration of near-field MO data storage as shown in Figure 4.91. They achieved reading and writing at a density of 3.8×10^6 bits/mm^2 with a data rate of 3.3×10^6 bits/s.

Hybrid Recording A new method called "hybrid recording" combines thermomagnetic recording and magnetic flux readout using a magnetoresistance (MR) head [316, 317]. As described above, optical readout encounters difficulty in the resolution, which is governed by diffraction, while conventional magnetic recording encounters the problem of thermal stability of the magnetic particle, which must become finer and finer with increasing recording density. To improve the stability, it is necessary to make the recording media magnetically harder, but such media require a large recording magnetic field, which cannot be generated by usual magnetic heads. In contrast, a high magnetic field is not necessary in the case of thermomagnetic recording. So, the concept of hybrid recording is thermally assisted recording. This kind of recording has become realistic since giant magnetoresistance (GMR) heads have been developed and

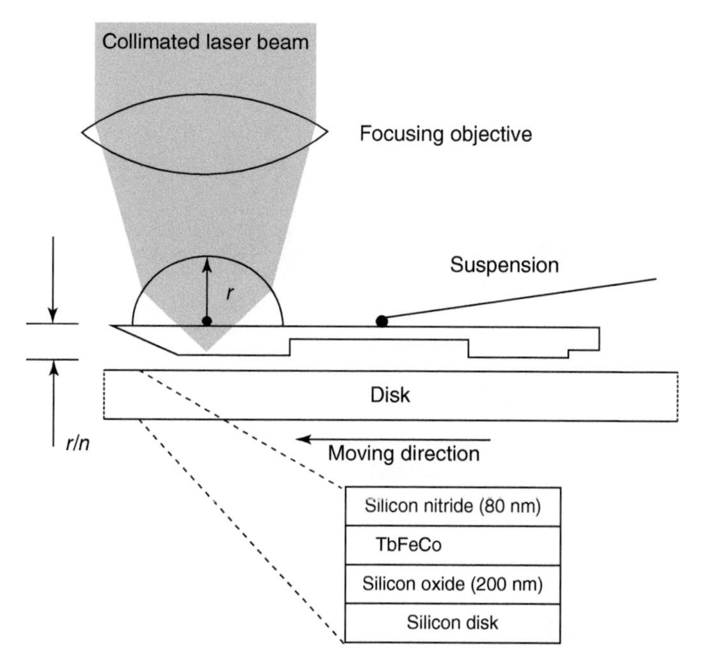

Figure 4.91 Schematic of the flying MO head using an SIL. Source: Terris et al. 1994 [315]. Reproduced with permission of AIP.

applied to the readout head of hard disk drives. The GMR head has not only a high sensitivity but also high lateral resolution down to below 100 nm [318].

Figure 4.92 shows the read and write methods of the hybrid recording technique proposed. The method is essentially the same as MFM MO recording as described above. The recording area and resulting density are determined by the temperature profile, especially by the temperature gradient in the area. The resolution of the readout in the track length direction is determined by the construction of GMR heads. In the approach of Katayama et al. [316], the readout area of the media was heated by a focused laser beam also in the readout process. On heating, the magnetization of the RE–TM films, the compensation point of which is around room temperature, significantly increases. As a result, a high readout signal level can be obtained. The laser heating is also effective in increasing density in the track width direction if the width of the magnetic head is larger than the laser spot size. Saga et al. also examined hybrid recording using RE–TM films for the recording media [319]. They did not apply laser heating for readout, but instead they used exchange-coupled double-layer films. Large magnetization for magnetic flux readout was achieved by using a readout layer with high magnetization at room temperature. The readout layer is exchange coupled with a recording layer with higher Curie temperature and higher coercivity at room temperature.

4.4.5.6 Ultrahigh-Density MO Recording

Femtosecond (fs) laser-induced ultrafast magnetization reversal dynamics is one of the most attractive topics in magnetics in recent years and related to the development of ultrafast magneto-optical (MO) storage devices with ultrahigh

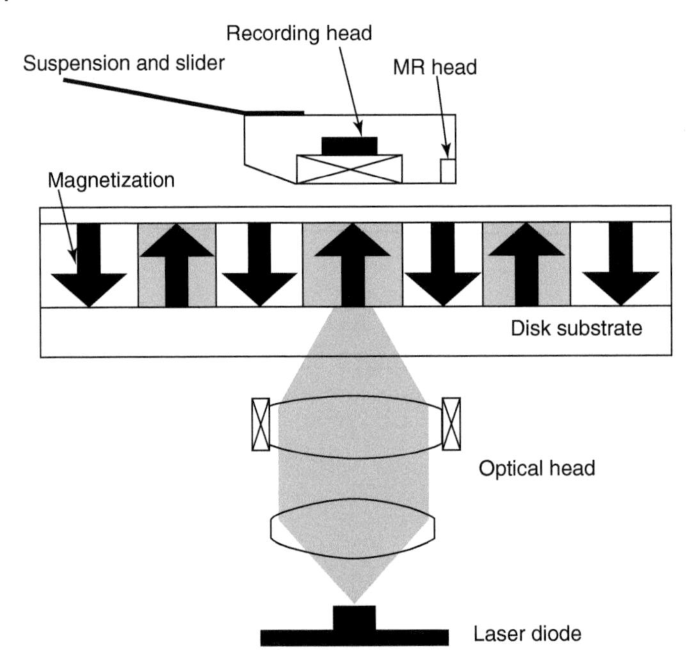

Figure 4.92 Schematic of hybrid recording with optical recording and GMR readout heads. Source: Tsunashima 2001 [281]. Reproduced with permission of IOP.

areal density recording. It is well known that RE-TM ferrimagnetic amorphous alloy films, such as TbFeCo and GdFeCo, are excellent ultrahigh-areal-density recording mediums [320, 321]; thus their ultrafast dynamics of magnetization reversal induced by femtosecond (fs) laser pulses has been studied intensively [322–326]. A sub-picosecond magnetization reversal of MO recordings [323] and an all-optical magnetization reversal within 10 ps (picosecond) have been reported in GdFeCo films [325, 326]. Such ultrafast reversals were explained by the extreme speedup of domain wall motion as fs-laser-induced transient temperature rose up over the angular momentum compensation point of GdFeCo films. Such an explanation implied that the magnetization reversal mechanism of the nucleation and growth of oppositely directed domains was adopted [323, 325].

Recently, ultrafast dynamics of fs-laser-induced MO recording across T_M (magnetization compensation temperature) was studied by Chen et al. [327], using the time-resolved pump probe polar Kerr spectroscopy combined with a laser synchronized SAMF (sinusoidal alternating magnetic field). The magnetization reversal dynamics of genuine MO recording was demonstrated and the external field dependence of the magnetization reversal dynamics across T_M shows that the reversal rate increases with the external fields. Quantitative analysis reveals a good linear dependence of the reversal rate on the external fields, which supported that the magnetization reversal mechanism of MO recording of GeFeCo films is related to the formation and expansion of oppositely directed domains driven by external fields. The nucleation field is revealed to be the minimum external field that enables MO recording. It is shown by the ultrafast dynamic results that the rate of MO recording across T_M is indeed faster than that across T_C, hence ultrafast MO recording is feasible. This research is closing in on the next-generation of terabit-rate ultrahigh-density magneto-optical storage devices that could store more than 6000 Terabits (6 petabits) of data, more than 70 times the contents of the entire US Library of Congress, on a single 5-in. disk (http://www.newswise.com/articles/progress-toward-terabit-rate-high-density-recording).

In 2015, Yamamoto et al. measured the magnetization reversal of Fe in GdFeCo alloy within several 100 fs using a seeded-FEL (free electron laser) tuned to the Fe M-absorption edge [328]. The results are compared with the calculations based on the resonant scattering theory. The results have revealed the magnetization switching due to ultrafast heating effect and demonstrate that studies combining FEL and TR-RMOKE (time resolved- resonant magneto-optical Kerr effect) experiments are opening new possibilities for investigating ultrafast magnetism.

References

1 Wolf, A., Awschalom, D.D., Buhrman, R.A. et al. (2001). *Science* 294: 1488.
2 Dietl, T., Ohno, H., Matsukura, F. et al. (2000). *Science* 287: 1019.
3 Han, S.-J., Jang, T.-H., Kim, Y.B. et al. (2003). *Appl. Phys. Lett.* 83: 920.
4 Sharma, P., Rao, K.V., Owens, F.J. et al. (2005). *Nat. Mater.* 2: 673.
5 Lim, S.W., Jeong, M.C., Ham, M.H., and Myoung, J.M. (2004). *Jpn. J. Appl. Phys., Part 2* 43: L280.
6 Rao, C.N.R. and Deepak, R.L. (2005). *J. Mater. Chem.* 15: 573.
7 Lawes, G., Risbud, A.S., Ramirez, A.P., and Seshadri, R. (2005). *Phys. Rev. B: Condens. Matter* 71: 045201.
8 Park, J.H., Kim, M.G., Jang, H.M. et al. (2004). *Appl. Phys. Lett.* 84: 1338.
9 Baik, J.M. and Lee, J.-L. (2004). *Met. Mater. Int.* 10: 555.
10 Chang, Y.Q., Wang, D.B., Luo, X.H. et al. (2003). *Appl. Phys. Lett.* 83: 4020.
11 Heo, W., Ivill, M.P., Ip, K. et al. (2004). *Appl. Phys. Lett.* 84: 2292.
12 Ip, K., Frazier, R.M., Heo, Y.W. et al. (2003). *J. Vac. Sci. Technol., B* 21: 1476.
13 Osada, M., Ebina, Y., Takada, K., and Sasaki, T. (2007). *Key Eng. Mater.* 350 (4): 15–18.
14 Jia, B., Zhang, W., Liu, H. et al. (2016). *J. Nanopart. Res.* 18 (9): 9.
15 Shimizu, H., Umetsu, S., and Kaihara, T. (2013). *Jpn. J. Appl. Phys.* 52: 028006.
16 Amemiya, T., Ogawa, Y., Shimizu, H. et al. (2008). *Appl. Phys. Express* 1: 022002.
17 Pirnstill, C.W., Malik, B.H., Thomas, E., and Coté, G.L. (2013). Design and characterization of a ferromagnetic, air gapped magneto-optic Faraday rotator. *Opt. Lett.* 38 (8): 1298–1300.
18 Munekata, H., Ohno, H., von Molnár, S. et al. (1989). Diluted magnetic III–V semiconductors. *Phys. Rev. Lett.* 63: 1849.

19 Ohno, H., Munekata, H., Penney, P. et al. (1992). Magnetotransport properties of p-type (In, Mn)As diluted magnetic III–V semiconductors. *Phys. Rev. Lett.* 68: 2664.
20 Svelto, O. (1998). *Principles of Lasers*, 4e, 394–415. New York: Plenum Press.
21 Sanders, S., Mattison, D., Jeffries, J.B., and Hanson, R.K. (2001). Rapid temperature tuning of a 1.4-mcm diode laser with application to high-pressure H_2O absorption spectroscopy. *Opt. Lett.* 26 (20): 1568–1570.
22 Matsukura, F., Ohno, H., and Dietl, T. (2002). Ferromagnetic semiconductors. In: *Handbook of Magnetic Materials*, vol. 14 (ed. K.H.J. Buschow), 1. Amsterdam: Elsevier.
23 (a) Wang, J., Sun, C., Kono, J. et al. (2005). Ultrafast quenching of ferromagnetism in InMnAs induced by intense laser irradiation. *Phys. Rev. Lett.* 95: 167401. (b) Wang, J., Sun, C., Hashimoto, Y. et al. (2006). Ultrafast magneto-optics in ferromagnetic III–V semiconductors. *J. Phys. Condens. Matter* 18: R501–R530. (c) Wang, J., Cywiński, Ł., Sun, C. et al. (2008). Femtosecond demagnetization and hot-hole relaxation in ferromagnetic GaMnAs. *Phys. Rev. B* 77: 235308.
24 Jungwirth, T., Wang, K.Y., Mašek, J. et al. (2005). Prospects for high temperature ferromagnetism in (Ga,Mn)As semiconductors. *Phys. Rev. B* 72: 165204.
25 Umnov, A.G., Naito, T., Munekata, H. et al. (2009). Possible applications of (III,Mn)V ferromagnetic semiconductors for optoelectronic devices. *Proc. SPIE* 7212: 72120X–72121X.
26 Xua, Y.B., Kernohana, E.T.M., Freelanda, D.J. et al. (2000). *Sens. Actuators, A* 81 (1–3): 258–262.
27 Huisman, T.J., Mikhaylovskiy, R.V., Telegin, A.V. et al. (2015). *Appl. Phys. Lett.* 106 (13): 132411.
28 Tu, N.T., Hai, P.N., Anh, L.D., and Tanaka, M. (2016). *Appl. Phys. Lett.* 108: 192401.
29 Di, Z.Y., Chen, X.F., Pu, S.L. et al. (2006). Magnetic-field-induced birefringence and particle agglomeration in magnetic fluids. *Appl. Phys. Lett.* 89 (21): 211106-1–211106-3.
30 Stephen, P.S. Low viscosity magnetic fluid obtained by the colloidal suspension of magnetic particles. US Patent 3215572, filed 9 October 1963 and issued 02 November 1965.
31 Horng, H.E., Hong, C.Y., Yang, S.Y., and Yang, H.C. (2003). Designing the refractive indices by using magnetic fluids. *Appl. Phys. Lett.* 82 (15): 2434–2436.
32 Xu, M. and Ridler, P.J. (1997). Linear dichroism and birefringence effects in magnetic fluids. *J. Appl. Phys.* 82 (1): 326–332.
33 Li, J., Liu, X.D., Lin, Y.Q. et al. (2007). Field modulation of light transmission through ferrofluid film. *Appl. Phys. Lett.* 91 (25): 253108-1–253108-3.
34 Pan, Y.T., Du, C.W., Liu, X.D. et al. (1993). Wavelength dependence of the Faraday-effect and magneto-birefringence in ferrofluid thin-films. *J. Appl. Phys.* 73 (10): 6139–6141.
35 Pu, S.L., Chen, X.F., Liao, W.J. et al. (2004). Laser self-induced thermo-optical effects in a magnetic fluid. *J. Appl. Phys.* 96 (10): 5930–5932.

36 Horng, H.E., Yang, S.Y., Lee, S.L. et al. (2001). Magneto-chromatics of the magnetic fluid film under a dynamic magnetic field. *Appl. Phys. Lett.* 79 (3): 350–352.

37 Liu, T., Chen, X., Di, Z. et al. (2007). Tunable magneto-optical wavelength filter of long-period fiber grating with magnetic fluids. *Appl. Phys. Lett.* 91 (12): 121116-1–121116-3.

38 Huang, Y.W., Hu, S.T., Yang, S.Y., and Horng, H.E. (2004). Tunable diffraction of magnetic fluid films and its potential application in coarse wavelength-division multiplexing. *Opt. Lett.* 29 (16): 1867–1869.

39 Hu, T., Zhao, Y., Li, X. et al. (2010). Novel optical fiber current sensor based on magnetic fluid. *Chin. Opt. Lett.* 8 (4): 392–394.

40 Horng, H.E., Chen, C.S., Fang, K.L. et al. (2004). Tunable optical switch using magnetic fluids. *Appl. Phys. Lett.* 85 (23): 5592–5594.

41 Pu, S.L., Chen, X.F., Di, Z.Y., and Xia, Y.X. (2007). Relaxation property of the magnetic-fluid-based fiber-optic evanescent field modulator. *J. Appl. Phys.* 101 (5): 053532-1–053532-5.

42 Zu, P., Chan, C.C., Siang, L.W. et al. (2011). Magneto-optic fiber Sagnac modulator based on magnetic fluids. *Opt. Lett.* 36 (8): 1425–1427.

43 Chieh, J.J., Yang, S.Y., Horng, H.E. et al. (2007). Magnetic-fluid optical-fiber modulators via magnetic modulation. *Appl. Phys. Lett.* 90 (13): 133505-1–133505-3.

44 Horng, H., Yang, S., Tse, W. et al. (2002). Magnetically modulated optical transmission of magnetic fluid films. *J. Magn. Magn. Mater.* 252: 104–106.

45 Park, S.Y., Handa, H., and Sandhu, A. (2009). High speed magneto-optical valve: rapid control of the optical transmittance of aqueous solutions by magnetically induced self-assembly of superparamagnetic particle chains. *J. Appl. Phys.* 105 (7): 07B526.

46 Pu, S.L., Chen, X.F., Chen, Y.P. et al. (2006). Fiber-optic evanescent field modulator using a magnetic fluid as the cladding. *J. Appl. Phys.* 99 (9): 093516-1–093516-4.

47 Deng, H.D., Liu, J., Zhao, W.R. et al. (2008). Enhancement of switching speed by laser-induced clustering of nanoparticles in magnetic fluids. *Appl. Phys. Lett.* 92 (23): 233103-1–233103-3.

48 Chieh, J.J., Yang, S.Y., Chao, Y.H. et al. (2005). Dynamic response of optical-fiber modulator by using magnetic fluid as a cladding layer. *J. Appl. Phys.* 97 (4): 043104-1–043104-4.

49 Seo, J.W., Kim, H., and Sung, S. (2004). Design and fabrication of a magnetic microfluidic light modulator using magnetic fluid. *J. Magn. Magn. Mater.* 272: E1787–E1789.

50 Zu, P., Chan, C.C., Lew, W.S. et al. (2012). Magneto-optical fiber sensor based on magnetic fluid. *Opt. Lett.* 37 (3): 398–400.

51 Zu, P., Chan, C.C., Lew, W.S. et al. (2012). Temperature-insensitive magnetic field sensor based on nanoparticle magnetic fluid and photonic crystal fiber. *IEEE Photonics J.* 4 (2): 491–498.

52 O'Faolain, L., Beggs, D.M., White, T.P. et al. (2010). Compact optical switches and modulators based on dispersion engineered photonic crystals. *IEEE Photonics J.* 2 (3): 404–414.

53 Chieh, J., Hong, C., Yang, S. et al. (2010). Study on magnetic fluid optical fiber devices for optical logic operations by characteristics of superparamagnetic nanoparticles and magnetic fluids. *J. Nanopart. Res.* 12 (1): 293–300.
54 Peng, Z., Chan, C.C., Lew, W.S. et al. (2012). High extinction ratio magneto-optical fiber modulator based on nanoparticle magnetic fluids. *IEEE Photonics J.* 4 (4): 1140–1146.
55 Shalaby, M., Peccianti, M., Ozturk, Y. et al. (2012). Terahertz faraday rotation in a magnetic liquid: High magneto-optical figure of merit and broad band operation in a ferrofluid. *Appl. Phys. Lett.* 100: 241107.
56 Shalaby, M., Peccianti, M., Ozturk, Y. et al. (2014). Terahertz magnetic modulator based on magnetically clustered nanoparticles. *Appl. Phys. Lett.* 105: 151108.
57 Fan, F., Chen, S., Lin, W. et al. (2013). Magnetically tunable terahertz magneto-plasmons in ferrofluid-filled photonic crystals. *Appl. Phys. Lett.* 103: 161115.
58 Chen, S., Fan, F., Chang, S. et al. (2014). Tunable optical and magneto-optical properties of ferrofluid in the terahertz regime. *Opt. Express* 22 (6): 6313–6321.
59 Morrish, A.H. (1965). *The Physical Principles of Magnetism*. Wiley.
60 Bertaut, F. and Forrat, F. (1956). Structure of ferrimagnetic rare-earth ferrites. *C.R. Acad. Sci.* 242 (382).
61 Geller, S. and Gilleo, M.A. (1957). Structure and ferrimagnetism of yttrium and rare-earth-iron garnets. *Acta Crystallogr.* 10: 239.
62 Röschmann, P. (1973). Compact YIG bandpass filter with finitepole frequencies for applications in microwave integrated circuits. *IEEE Trans. Microwave Theory Tech.* 21: 52.
63 Murakami, Y., Ohgihara, T., and Okamoto, T. (1987). A0.5–4.0-GHz tunable bandpass filter using YIG film grownby LPE. *IEEE Trans. Microwave Theory Tech.* 35: 1192.
64 Deeter, M.N., Rose, A.H., and Day, G.W. (1990). Fast sensitive magnetic-field sensors based on the Faraday effect in YIG. *J. Lightwave Technol.* 8: 1838.
65 Higuchi, S., Ueda, K., Yahiro, F. et al. (2001). Fabrications of cerium-substituted YIG thin films for magnetic field sensor by pulsed-laser deposition. *IEEE Trans. Magn.* 37: 2451.
66 Shoji, Y., Mizumoto, T., Yokoi, H. et al. (2008). Magneto-optical isolator with silicon waveguides fabricated by direct bonding. *Appl. Phys. Lett.* 92 (7): 071117-1–071117-3.
67 Stadler, B.J.H. and Mizumoto, T. (2014). Integrated magneto-optical materials and isolators: A review. *IEEE Photonics J.* 6: 1.
68 Tien, M.-C., Mizumoto, T., Pintus, P. et al. (2011). Silicon ring isolators with bonded nonreciprocal magneto-optic garnets. *Opt. Express* 19: 11740.
69 Yokoi, H., Mizumoto, T., and Shoji, Y. (2003). Optical nonreciprocal devices with a silicon guiding layer fabricated by wafer bonding. *Appl. Opt.* 42 (33): 6605–6612.
70 Mizumoto, T., Takei, R., and Shoji, Y. (2012). Waveguide optical isolators for integrated optics. *IEEE J. Quantum Electron.* 48 (2): 252–260.

71 Espinola, R.L., Izuhara, T., Tsai, M. et al. (2004). Magneto-optical nonreciprocal phase shift in garnet/silicon-on-insulator waveguides. *Opt. Lett.* 29 (9): 941–943.

72 Sobu, Y., Shoji, Y., Sakurai, K., and Mizumoto, T. (2013). GaInAsP/InP MZI waveguide optical isolator integrated with spot size converter. *Opt. Express* 21 (13): 15 373–15 381.

73 Takei, R. and Mizumoto, T. (2010). Design and simulation of silicon waveguide optical circulator employing nonreciprocal phase shift. *Jpn. J. Appl. Phys.* 49 (5): 052203-1–052203-6.

74 Yokoi, H. and Mizumoto, T. (1997). Proposed configuration of integrated optical isolator employing wafer-direct bonding technique. *Electron. Lett.* 33 (21): 1787–1788.

75 Shoji, Y. and Mizumoto, T. (2007). Ultra-wideband design of waveguide magneto-optical isolator operating in 1.31 μm and 1.55 μm band. *Opt. Express* 15 (2): 639–645.

76 Shoji, Y., Mitsuya, K., and Mizumoto, T. (2013). Silicon-based magneto-optical isolator and circulator fabricated by direct Bonding technology. *Progress In Electromagnetics Research Symposium Proceedings*, Stockholm, Sweden (21–24 August 2013).

77 Shoji, Y. and Mizumoto, T. (2014). Magneto-optical non-reciprocal devices in silicon photonics. *Sci. Technol. Adv. Mater.* 15: 1–10.

78 Ghosh, S., Keyvaninia, S., Van Roy, W. et al. (2012). Ce:YIG/silicon-on-insulator waveguide optical isolator realized by adhesive bonding. *Opt. Express* 20 (2): 1839–1848.

79 Ghosh, S., Keyvaninia, S., Shoji, Y. et al. (2012). Compact Mach–Zehnder interferometer Ce:YIG/SOI optical isolators. *IEEE Photonics Technol. Lett.* 24 (18): 1653–1656.

80 Hutchings, D.C. and Holmes, B.M. (2011). A waveguide polarization toolset design based on mode beating. *IEEE Photonics J.* 3 (3): 450–461.

81 Jahn, I.R. (1973). Linear magnetic birefringence in the antiferromagnetic iron group difluorides. *Phys. Status Solidi B* 57 (2): 681–692.

82 Borovik-Romanov, A.S., Kreines, N.M., Pankov, A.A., and Talalaev, M.A. (1973). *Sov. Phys. JEPT* 37 (5): 890.

83 Feng, W., Guo, G.-Y., Zhou, J. et al. (2015). Large magneto-optical Kerr effect in noncollinear antiferromagnets Mn_3X (X = Rh, Ir or Pt). *Phys. Rev. B* 92: 144426.

84 Sivadas, N., Okamoto, S., and Xiao, D. (2016). Gate-controllable magneto-optic Kerr effect in layered collinear antiferromagnets. *Phys. Rev. Lett.* 117: 267203.

85 Germann, K.H., Maier, K., and Straus, E. (1974). *Phys. Status Solidi B* 61: 449.

86 Treindl, A. and Germann, K.H. (1977). *Phys. Status Solidi B* 80: 159.

87 Germann, K.H., Maier, K., and Straus, E. (1974). *Solid State Commun.* 14: 1309.

88 Andlauer, B., Schneider, J., and Wettling, W. (1976). *Appl. Phys.* 10: 139.

89 Leycuras, C., Legall, H., Minella, D. et al. (1977). *Physica* 89B: 43.

90 Legall, H., Leycuras, C., Minella, D. et al. (1977). *Physica* 86-88b: 1223.

91 Martí, X., Fina, I., and Jungwirth, T. (2015). Prospect for antiferromagnetic spintronics. *IEEE Trans. Magn.* 51 (4): 1–4.
92 Efron, U. (ed.) (1995). *Spatial Light Modulator Technology*. Marcel Dekker, Inc.
93 Simon, M. (2007). A MEMS Faraday Rotator for MEMS Polarization Controllers, MEMS Magneto Optic Spatial Light Modulators and Other Enabled Technologies. http://www.colorado.edu/MCEN/MEMSII/IndProj/MEMS%20Magneto_Simon.pdf (accessed 18 March 2007).
94 Hecht, E. (2002). *Optics*, 4e. Addison Wesley.
95 Scott, G.B. and Lacklinson, D.E. (1976). Magneto-optic properties and applications of bismuth substituted iron garnets. *IEEE Trans. Magn.* 12 (4).
96 Ross, W.E., Psaltis, D., and Anderson, R.H. (1982). Two-dimensional magneto-optic spatial light modulator for signal processing, Proc. SPIE 0341, Real-Time Signal Processing V, 191 (December 28, 1982).
97 Ross, W.E., Psaltis, D., and Anderson, R.H. (1983). Two-dimensional magneto-optic spatial light modulator for signal processing. *Opt. Eng.* 22: 495–490.
98 Ross, W.E., Snapp, K.M., and Anderson, R.H. (1983). Fundamental characteristics of the Litton iron garnet magneto-optic spatial light. *SPIE Proc.* 388: 55–64.
99 Kovacs, G.T.A. (1998). *Micromachined Transducers Sourcebook*. The McGraw-Hill Companies, Inc.
100 Cutnell, J.D. and Johnson, K.W. (1995). *Physics*, 3e. Wiley.
101 Kahl, S. (2004). Bismuth iron garnet films for magneto-optical photonic crystals. Doctoral Dissertation. Condensed Matter Physics, Laboratory of Solid State Devices, IMIT, Royal Institute of Technology, Stockholm.
102 Guillot, M., Le Gall, H., Desvignes, J.M., and Artinana, M. (1994). Faraday rotation of bismuth substituted terbium iron garnets. *IEEE Trans. Magn.* 30 (6).
103 Kucera, M., Visnovsky, S., and Prosser, V. (1984). Low temperature Faraday rotation spectra of $Y_{3-x}Bi_xFe_5O_{12}$. *IEEE Trans. Magn.* 20 (5).
104 Irvine, S.E. and Elezzabi, A.Y. (2003). A miniature broadband bismuth-substituted yttrium iron garnet magneto-optic modulator. *J. Phys. D: Appl. Phys.* 36: 2218–2221.
105 Lubecke, V.M., Barber, B., Chan, E. et al. (2001). Self-assembling MEMS variable and fixed RF inductors. *IEEE Trans. Microwave Theory Tech.* 49 (11).
106 Sobolewski, R. and Butler, D.P. (2000). *Optical Sensors, to Appear in Handbook of Superconducting Materials*. Bristol: IOP Publishing Ltd.
107 Likharev, K.K. and Semenov, V.K. (1991). RSFQ logic/memory family: a new Josephson-junction technology for sub-terahertz-clock-frequency digital systems. *IEEE Trans. Appl. Supercond.* 1: 3–28.
108 Sobolewski, R. and Park, J.R. (2001). Magneto-optical modulator for superconducting digital output interface. *IEEE Trans. Appl. Supercond.* 11 (1): 727–730.
109 Tien, P.K. and Martin, R.J. (1972). Switching and modulation of light in magneto-optic waveguides of garnet films. *Appl. Phys. Lett.* 21: 394–396.

110 Schuster, T., Koblischka, M.R., Ludescher, B. et al. (1991). EuSe as magneto-optical active coating for use with the high resolution Faraday effect. *Cryogenics* 31: 811–816.

111 Mulloy, M.P., Blau, W.J., and Lunney, J.G. (1993). Pulsed laser deposition of magnetic semiconductor EuS, EuSe and EuTe thin films. *J. Appl. Phys.* 73: 4104–4106.

112 Horng, H.E., Chieh, J.J., Chao, Y.H. et al. (2005). *Opt. Lett.* 30 (5): 543.

113 Scholten, P.C. (1980). *IEEE Trans. Magn.* 16: 221.

114 Martinez, L., Cecelja, F., and Rakowski, R. (2005). *Sens. Actuators, A* 123–124: 438.

115 Horng, H., Hong, C., Yeung, W., and Yang, H. (1998). *Appl. Opt.* 37: 2674.

116 Hong, C.Y., Yang, S.Y., Fang, K.L. et al. (2006). *J. Magn. Magn. Mater.* 297: 71.

117 Hutchings, D. and Holmes, B. (2011). A waveguide polarization tool set design based on mode beating. *IEEE Photonics J.* 3 (3): 450–461.

118 Attygalle, M., Gupta, K., and Priest, T. (2011). Broad band extended dynamic range analogue signal transmission through switched dual photonic link architecture. *IEEE Photonics J.* 3 (1): 100–111.

119 Wang, Z., Chong, Y., Joannopoulos, J., and Soljačić, M. (2009). Observation of unidirectional backscattering-immune topological electromagnetic states. *Nature* 461: 772–775.

120 Fan, F., Guo, Z., Bai, J.J. et al. (2011). Magnetic photonic crystals for terahertz tunable filter and multifunctional polarization controller. *J. Opt. Soc. Am. B* 28: 697–702.

121 Temnov, V.V., Armelles, G., Woggon, U. et al. (2010). Active magneto-plasmonics in hybrid metal–ferromagnet structures. *Nat. Photonics* 4: 107–111.

122 Lodewijks, K., Maccaferri, N., Pakizeh, T. et al. (2014). Magneto-plasmonic design rules for active magneto-optics. *Nano Lett.* 14: 7207–7214.

123 Armelles, G., Cebollada, A., García-Martín, A., and González, M.U. (2013). Magneto-plasmonics: combining magnetic and plasmonic functionalities. *Adv. Opt. Mater.* 1: 10–35.

124 Belotelov, V., Akimov, I.A., Pohl, M. et al. (2011). Enhanced magneto-optical effects in magneto-plasmonic crystals. *Nat. Nanotechnol.* 6: 370–376.

125 Kreilkamp, L.E., Belotelov, V.I., Chin, J.Y. et al. (2013). Waveguide-plasmon polaritons enhance transverse magneto-optical Kerr effect. *Phys. Rev. X* 3: 041019.

126 Yan, H.G., Li, Z., Li, X. et al. (2012). Infrared spectroscopy of tunable Dirac terahertz magneto-plasmons in graphene. *Nano Lett.* 12: 3766–3771.

127 Belotelov, V., Kreilkamp, L.E., Akimov, I.A. et al. (2013). Plasmon-mediated magneto-optical transparency. *Nat. Commun.* 4: 2128.

128 Floess, D., Chin, J.Y., Kawatani, A. et al. (2015). Tunable and switchable polarization rotation with nonreciprocal plasmonic thin films at designated wavelengths. *Light Sci. Appl.* 4: e284.

129 Chinetal, J.Y. (2013). Nonreciprocal plasmonics enables giant enhancement of thin-film Faraday rotation. *Nat. Commun.* 4: 1599.

130 Arikawa, T., Wang, X., Belyanin, A.A., and Kono, J. (2012). Giant tunable Faraday effect in a semiconductor magneto-plasma for broadband terahertz polarization optics. *Opt. Express* 20: 19484–19492.

131 Shuvaev, A.M., Astakhov, G.V., Pimenov, A. et al. (2011). Giant magneto-optical Faraday effect in HgTe thin films in the terahertz spectral range. *Phys. Rev. Lett.* 106: 107404.

132 Shimano, R., Yumoto, G., Yoo, J.Y. et al. (2013). Quantum Faraday and Kerr rotations in graphene. *Nat. Commun.* 4: 1841.

133 Wang, X., Belyanin, A.A., Crooker, S.A. et al. (2010). Interference-induced terahertz transparency in a semiconductor magneto-plasma. *Nat. Phys.* 6: 126–130.

134 Crassee, I., Orlita, M., Potemski, M. et al. (2012). Intrinsic terahertz plasmons and magneto-plasmons in large scale monolayer graphene. *Nano Lett.* 12: 2470–2474.

135 Fan, F. et al. (2013). Magnetic ally tunable terahertz magneto-plasmons in ferrofluid-filled photonic crystals. *Appl. Phys. Lett.* 103: 161115.

136 Johnson, S.G. and Joannopoulos, J.D. (2002). *Photonic Crystals: The Road from Theory to Practice*. Kluwer.

137 Joannopoulos, J.D., Johnson, S.G., Winn, J.N., and Meade, R.D. (2008). *Photonic Crystals: Moulding the flow of light*. Princeton University Press.

138 Asakawa, K., Sugimoto, Y., Watanabe, Y. et al. (2006). *J. Phys.* 8: 208.

139 Shinya, A., Mitsugi, S., Tanabe, T. et al. (2006). *Opt. Express* 14: 1230.

140 Zhu, Z.H., Ye, W.M., Ji, J.R. et al. (2006). *Opt. Express* 14: 1783.

141 Andalib, P. and Granpayeh, N. (2011). *J. Opt. Soc. Am. B: Opt. Phys.* 26: 10.

142 Zhu, H.B. and Jiang, C. (2011). *IEEE J. Lightwave Technol.* 29: 708.

143 Liu, S.Y., Lu, W.I., Lin, Z.F., and Chui, S.T. (2010). *Appl. Phys. Lett.* 97: 201113.

144 Inoue, M., Baryshev, A.V., Khanikaev, A.B. et al. (2008). *IEICE Trans. Electron.* E91-C: 1630.

145 Ao, X.Y., Lin, Z.F., and Chan, C.T. (2009). *Phys. Rev. B* 80: 033105.

146 Wang, Z. and Fan, S. (2005). Optical circulators in two-dimensional magneto-optical photonic crystals. *Opt. Lett.* 30: 1989–1991.

147 Smigaj, W., Romero-Vivas, J., Gralak, B. et al. (2010). Magneto-optical circulator designed for operation in a uniform external magnetic field. *Opt. Lett.* 35: 568.

148 Fan, S. and Wang, Z. (2006). *J. Magn. Soc. Jpn.* 30: 641.

149 Fan, S. (2007). *Physica B* 394: 221.

150 Wang, Q., Ouyang, Z.B., and Liu, Q. (2011). *J. Opt. Soc. Am. B: Opt. Phys.* 28: 703.

151 Ivanov, S.A. (1995). *IEEE Trans. Microwave Theory Tech.* 43: 1253.

152 Adams, R.S., O'Neil, B., and Young, J.L. (2009). *IEEE Antennas Wirel. Propag. Lett.* 8: 165.

153 Wang, Q., Ouyang, Z., Tao, K. et al. (2012). T-shaped optical circulator based on coupled magneto-optical rods and a side-coupled cavity in a square-lattice photonic crystal. *Phys. Lett. A* 376: 646–649.

154 Liu, Q., Ouyang, Z.B., Wu, C.J. et al. (2008). *Opt. Express* 16: 18992.

155 Sebastian, M.T. and Jantunen, H. (2008). *Int. Mater. Rev.* 53: 57.

156 Chen, S.W., Du, J.I., Liu, S.Y. et al. (2008). *Opt. Lett.* 33: 2476.

157 Fu, J.X., Lian, J., Liu, R.J. et al. (2011). Unidirectional channel-drop filter by one-way gyromagnetic photonic crystal waveguides. *Appl. Phys. Lett.* 98 (21): 211104.

158 Dmitriev, V., Portela, G., and Martins, L. Design and optimization of T-shaped circulator based on magneto-optical resonator in 2D-photonic crystals. In: *ADVCOMP 2015: The Ninth International Conference on Advanced Engineering Computing and Applications in Sciences*, 85–87.

159 COMSOL Multiphysics. www.comsol.com (accessed 18 May 2015).

160 TransTech. www.trans-techinc.com (accessed 18 May 2015).

161 Huang, D., Pintus, P., Zhang, C. et al. Multiple-port integrated optical circulators. https://optoelectronics.ece.ucsb.edu/sites/default/files/2017-05/multiport_IPC2016_FINAL.pdf

162 Pintus, P., Tien, M.-C., Bowers, J.E. et al. (2011). Design of magneto-optical ring isolator on SOI based on the finite-element method. *IEEE Photonics Technol. Lett.* 23: 1670–1672.

163 Pintus, P., Di Pasquale, F., Bowers, J.E. et al. (2013). Integrated TE and TM optical circulators on ultra-low loss silicon nitride platform. *Opt. Express* 21 (4): 5041–5052.

164 Huang, D., Pintus, P., Zhang, C. et al. (2016). Reconfigurable integrated optical circulator. In: *Conference on Lasers and Electro-Optics*. Optical Society of America, paper SM3E.1.

165 Huang, D., Pintus, P., Zhang, C. et al. (2016). Silicon micro ring isolator with large optical isolation and low loss. In: *Optical Fiber Communication Conference*. Optical Society of America, paper Th1K.2.

166 Fan, F., Chang, S.-J., Niu, C. et al. (2012). Magnetically tunable silicon-ferrite photonic crystals for terahertz circulator. *Opt. Commun.* 285: 3763–3769.

167 Wang, Z. and Fan, S. (2005). Magneto-optical defects in two-dimensional photonic crystals. *Appl. Phys. B* 81: 369–375.

168 Zayets, V. and Ando, K. (2010). Magneto-optical devices for optical integrated circuits. In: *Frontiers in Guided Wave Optics and Optoelectronics* (ed. B. Pal). InTech. ISBN: 978-953-7619-82-4 http://www.intechopen.com/books/frontiers-in-guided-wave-optics-and-optoelectronics/magneto-opticaldevices-for-optical-integrated-circuits.

169 Ando, K., Takeda, N., Koshizuka, N., and Okuda, T. (1985). *J. Appl. Phys.* 57: 1277–1281.

170 Dammann, H., Pross, E., Rabe, G. et al. (1986). *Phys. Lett.* 49: 1755–1757.

171 Holmes, B.M. and Hutchings, D.C. (2006). *Appl. Phys. Lett.* 88: 061116.

172 Holmes, B.M. and Hutchings, D.C. (2006). Towards the monolithically integrated optical isolator on a semiconductor laser chip. In: *Proceedings of IEEE Lasers and Electro-Optics Society*, 897–898. IEEE.

173 Yokoi, H., Mizumoto, T., Shinjo, N. et al. (2000). *Appl. Opt.* 39: 6158–6164.

174 Sakurai, K., Yokoi, H., Mizumoto, T. et al. (2004). *Jpn. J. Appl. Phys.* 43: 1388–1392.

175 Shoji, Y. and Mizumoto, T. (2006). *Appl. Opt.* 45: 7144–7150.

176 Vanwolleghem, M., Van Parys, W., Van Thourhout, D. et al. (2004). *Appl. Phys. Lett.* 85: 3980–3982.

177 Van. Parys, W., Moeyersoon, B., Van Thourhout, D. et al. (2006). *Appl. Phys. Lett.* 88: 071115.
178 Shimizu, H. and Nakano, Y. (2004). *Jpn. J. Appl. Phys.* 43: L1561–L1563.
179 Shimizu, H. and Nakano, Y. (2006). *IEEE J. Lightwave Technol.* 24: 38–43.
180 Amemiya, T., Shimizu, H., Nakano, Y. et al. (2006). *Appl. Phys. Lett.* 89: 021104.
181 Amemiya, T., Shimizu, H., Hai, P.N. et al. (2007). *Appl. Opt.* 46: 5784–5791.
182 Amemiya, T. and Nakano, Y. (2010). Single mode operation of 1.5-μm waveguide optical isolators based on the nonreciprocal-loss phenomenon. In: *Advances in Optical and Photonic Devices* (ed. K.Y. Kim). InTech. ISBN: 978-953-7619-76-3 http://www.intechopen.com/books/advances-in-optical-andphotonic-devices/single-mode-operation-of-1-5-micro-m-waveguide-optical-isolators-based-on-thenonreciprocal-loss-phe.
183 Zaets, W. and Ando, K. (2000). *Appl. Phys. Lett.* 77: 1593–1595.
184 Zayets, V., Debnath, M.C., and Ando, K. (2004). *Appl. Phys. Lett.* 84: 565–567.
185 Takenaka, M. and Nakano, Y. (1999). Proposal of a novel semiconductor optical waveguide isolator. In: *Proceedings of the Eleventh International Conference on Indium Phosphide and Related Materials*, 289–292. IEEE.
186 Zaets, W. and Ando, K. (1999). *IEEE Photonics Technol. Lett.* 11: 1012–1014.
187 Shoji, Y., Mitsuya, K., and Mizumoto, T. (2013). Silicon-based Magneto-optical Isolator and Circulator Fabricated by Direct Bonding Technology. *Progress In Electromagnetics Research Symposium Proceedings*, Stockholm, Sweden (12–15 August 2013). pp. 21–24.
188 Dötsch, H., Bahlmann, N., Zhuromskyy, O. et al. (2005). Applications of magneto-optical waveguides in integrated optics: review. *J. Opt. Soc. Am. B: Opt. Phys.* 22 (1): 240–253.
189 Okamura, Y., Inuzuka, H., Kikuchi, T., and Yamamoto, S. (1986). Nonreciprocal propagation in magneto-optic YIG rib waveguides. *J. Lightwave Technol.* 4 (7): 711–714.
190 Mizumoto, T., Shoji, Y., and Takei, R. (2012). Direct wafer bonding and its application to waveguide optical isolators. *Materials* 5 (5): 985–1004.
191 Shirato, Y., Shoji, Y., and Mizumoto, T. (2013). High isolation in silicon waveguide optical isolator employing nonreciprocal phase shift. In: *Optical Fiber Communication Conference and Exposition and the National Fiber Optic Engineers Conference (OFC/NFOEC)*, (17–21 March 2013).
192 Mitsuya, K., Shoji, Y., and Mizumoto, T. (2013). Demonstration of a silicon waveguide optical circulator. *IEEE Photonics Technol. Lett.* 25 (8): 721–723.
193 Fan, F., Chang, S.-J., Gu, W.-H. et al. (2012). Magnetically tunable terahertz isolator based on structured semiconductor magneto plasmonics. *IEEE Photonics Technol. Lett.* 24 (22): 2080–2083.
194 Fang, N., Lee, H., Sun, C., and Zhang, X. (2005). Sub-diffraction-limited optical imaging with a silver superlens. *Science* 308: 534–537.
195 Liu, Z., Steele, J.M., Srituravanich, W. et al. (2005). Focusing surface plasmons with a plasmonic lens. *Nano Lett.* 5: 1726–1729.
196 Srituravanich, W., Pan, L., Wang, Y. et al. (2008). Flying plasmonic lens in the near field for high-speed nanolithography. *Nat. Nanotechnol.* 3: 733–737.

197 Fan, F., Chen, S., Wang, X.-H., and Chang, S.-J. (2013). Tunable nonreciprocal terahertz transmission and enhancement based on metal/magneto-optic plasmonic lens. *Opt. Express* 21: 8614–8621.
198 Chen, S., Fan, F., Wang, X. et al. (2015). Terahertz isolator based on nonreciprocal magneto-metasurface. *Opt. Express* 23: 1015–1024.
199 Kildishev, A.V., Boltasseva, A., and Shalaev, V.M. (2013). Planar photonics with metasurfaces. *Science* 339: 1232009.
200 Yu, N. and Capasso, F. (2014). Flat optics with designer metasurfaces. *Nat. Mater.* 13: 139–150.
201 Khanikaev, A.B., Mousavi, S.H., Shvets, G., and Kivshar, Y.S. (2010). One-way extraordinary optical transmission and nonreciprocal spoof plasmons. *Phys. Rev. Lett.* 105: 126804.
202 Zhu, H. and Jiang, C. (2011). Nonreciprocal extraordinary optical transmission through subwavelength slits in metallic film. *Opt. Lett.* 36: 1308–1310.
203 Koschny, M. and Lindner, M. (2012). Magneto-optical sensors: accurately analyse magnetic field distribution of magnetic materials. *Adv. Mater. Processes* 13–16.
204 Jorge, P. (2001). Sensores Ópticos para a Medição de Corrente Elétrica em Alta-tensão. Master Thesis. Faculdade de Ciências da Universidade do Porto, Porto, Portugal.
205 Perciante, C.D. and Ferrari, J.A. (2008). Magnetic crosstalk minimization in optical current sensors. *IEEE Trans. Instrum. Meas.* 57: 2304–2308.
206 Kurosawa, K., Sakamoto, K., and Yoshida, S. (1994). Polarization-maintaining properties of the flint glass-fiber for the Faraday sensor element. *Tenth International Conference on Optical Fiber Sensors,* Glasgow, Scotland (11 October 1994). pp. 28–35.
207 Yamashita, T., Watabe, A., Masuda, I. et al. (1996). Extremely small stress-optic coefficient glass single mode fibers for current sensor. *Optical Fiber Sensors 11,* Japan (21 May 1996). pp. 168–171.
208 Kurosawa, K. (1997). Optical current transducers using flint glass fiber as the Faraday sensor element. *Opt. Rev.* 4: 38–44.
209 Hotate, K., Thai, B.T., and Saida, T. (1998). Comparison between flint glass fiber and twisted/bent single-mode fiber as a Faraday element in an interferometric fiber optic current sensor. *European Workshop on Optical Fibre Sensors,* Scotland (8 July 1998). pp. 233–237.
210 Barczak, K., Pustelny, T., Dorosz, D., and Dorosz, J. (2009). New optical glasses with high refractive indices for applications in optical current sensors. *Acta Phys. Pol. A* 116: 247–249.
211 Sun, L., Jiang, S., Zuegel, J.D., and Marciante, J.R. (2009). Effective verdet constant in terbium-doped-core. *Opt. Lett.* 34: 1699–1701.
212 Sun, L., Jiang, S., and Marciante, J.R. (2010). Compact all-fiber optical Faraday components using 65-wt%-terbium-doped fiber with a record Verdet constant of −32 rad/(Tm). *Opt. Express* 18: 12191–12196.
213 Rose, A.H., Ren, Z.B., and Day, G.W. (1996). Twisting and annealing optical fiber for current sensors. *J. Lightwave Technol.* 14: 2492–2498.
214 Laming, R.I. and Payne, D.N. (1989). Electric-current sensors employing spun highly birefringent optical fibers. *J. Lightwave Technol.* 7: 2084–2094.

215 Bohnert, K., Gabus, P., and Brandle, H. (2000). Towards commercial use of optical fiber current sensors. *Conference on Lasers and Electro-Optics (CLEO 2000)*, San Francisco, CA, USA (7–12 May 2000). pp. 303–304.
216 Tang, D., Rose, A.H., Day, G.W., and Etzel, S.M. (1991). Annealing of linear birefringence in single-mode fiber coils – application to optical fiber current sensors. *J. Lightwave Technol.* 9: 1031–1037.
217 Rose, A.H., Etzel, S.M., and Wang, C.M. (1997). Verdet constant dispersion in annealed optical fiber current sensors. *J. Lightwave Technol.* 15: 803–807.
218 Drexler, P. and Fiala, P. (2008). Utilization of Faraday mirror in fiber optic current sensors. *Radioengineering* 17: 101–107.
219 Zhou, S. and Zhang, X. (2007). Simulation of linear birefringence reduction in fiberoptical current sensor. *IEEE Photonics Technol. Lett.* 19: 1568–1570.
220 Kurosawa, K., Yamashita, K., Sowa, T., and Yamada, Y. (2000). Flexible fiber faraday effect current sensor using flint glass fiber and reflection scheme. *IEICE Trans. Electron.* 83: 326–330.
221 Bohnert, K., Philippe, G., Hubert, B., and Guggenbach, P. (2005). Highly accurate fiber-optic DC current sensor for the electrowinning industry. *IEEE Trans. Ind. Appl.* 43: 180–187.
222 Zimmermann, A.C., Besen, M., Encinas, L.S., and Nicolodi, R. (2011). Improving optical fiber current sensor accuracy using artificial neural networks to compensate temperature and minor non-ideal effects. *The 21st International Conference on Optical Fiber Sensors*, Ottawa, Canada (21 May 2011). pp. 77535–77448.
223 Dandridge, A., Tveten, A.B., Sigel, G.H. et al. (1980). Optical fiber magnetic-field sensors. *Electron. Lett.* 16: 408–409.
224 The ABB Group. Automation and Power Technologies. Available online: http://www.abb.com/ (accessed 14 December 2011).
225 Rahmatian, F. and Blake, J.N. (2006). Applications of high-voltage fiber optic current sensors. *IEEE Power Engineering Society General Meeting*, Montreal, Canada (18 June 2006). pp. 1–6.
226 Alstom Grid. Available online: http://www.nxtphase.com/ (accessed 12 December 2011).
227 Ripka, P. (2010). Electric current sensors: a review. *Meas. Sci. Technol.* 21: 1–23.
228 Walsey, G.A. and Fisher, N.E. (1997). Control of the critical angle of reflection in an optical current sensor. *Optical Fiber Sensors 12*, Williamburg, VA, USA (28 October 1997). pp. 237–240.
229 Bush, S.P. and Jackson, D.A. (1992). Numerical investigations of the effects of birefringence and total internal reflection on Faraday effect current sensors. *Appl. Opt.* 31: 5366–5374.
230 Sato, T., Takahashi, G.T., and Inui, Y. (1986). Method and apparatus for optically measuring a current. US patent 4564754.
231 Fisher, N.E. and Jackson, D.A. (1996). Vibration immunity and Ampere's circuital law for a near perfect triangular Faraday current sensor. *Meas. Sci. Technol.* 7: 1099–1102.
232 Yi, B., Chu, B., and Chiang, K.S. (2002). Magneto-optical electric-current sensor with enhanced sensitivity. *Meas. Sci. Technol.* 13: N61–N63.

233 Ning, Y.N., Chu, B., and Jackson, D.A. (1991). Miniature Faraday current sensor based on multiple critical angle reflections in a bulk-optic ring. *Opt. Lett.* 16: 1996–1998.

234 Ning, Y.N., Wang, Z.P., Palmer, A.W., and Gratan, K. (1995). A Faraday current sensor using a novel multi-optical-loop sensing element. *Meas. Sci. Technol.* 6: 1339–1342.

235 Benshun, Y., Andrew, C., Madden, I. et al. (1998). A Novel bulk-glass optical current transducer having an adjustable multiring closed-optical-path. *IEEE Trans. Instrum. Meas.* 47: 240–243.

236 Wang, Z.P., Wang, H., Jiang, H., and Liu, X. (2007). A magnetic field sensor based on orthoconjugate reflection used for current sensing. *Opt. Laser Technol.* 39: 1231–1233.

237 Wang, Z.P., Qing, B., Yi, Q. et al. (2006). Wavelength dependence of the sensitivity of a bulk-glass optical current transformer. *Opt. Laser Technol.* 38: 87–93.

238 Wang, Z.P., Xiaozhong, W., Liu, X. et al. (2005). Effect of the spectral width of optical sources upon the output of an optical current sensor. *Meas. Sci. Technol.* 16: 1588–1592.

239 Madden, W.I., Michie, W.C., Cruden, A. et al. (1999). Temperature compensation for optical current sensors. *Opt. Eng.* 38: 1699–1707.

240 Deng, X.Y., Li, Z., Qixian, P. et al. (2008). Research on the magneto-optic current sensor for highcurrent pulses. *Rev. Sci. Instrum.* 79: 1–4.

241 Cruden, A., Michie, C., Madden, I. et al. (1998). Optical current measurement system for high-voltage applications. *Measurement* 24: 97–102.

242 PowerSense A/S. DISCOS System. Available online: http://www.powersense.dk/ (accessed 15 December 2011).

243 Yariv, A. and Winsor, H.V. (1980). Proposal for detection of magnetic-fields through magnetostrictive perturbation of optical fibers. *Opt. Lett.* 5: 87–89.

244 Koo, K.P. and Sigel, G.H. (1982). Characteristics of fiberoptic magnetic-field sensors employing metallic glasses. *Opt. Lett.* 7: 334–336.

245 Kersey, A.D., Jackson, D.A., and Corke, M. (1985). Single-mode fibre-optic magnetometer with DC bias field stabilization. *J. Lightwave Technol.* 3: 836–840.

246 Bucholtz, F., Koo, K.P., and Dandridge, A. (1988). Effect of external perturbations on fiber-optic magnetic sensors. *J. Lightwave Technol.* 6: 507–512.

247 Jarzynski, J., Cole, J.H., Bucaro, J.A., and Davis, C.M. (1980). Magnetic-field sensitivity of an optical fiber with magnetostrictive jacket. *Appl. Opt.* 19: 3746–3748.

248 Sedlar, M., Paulicka, I., and Sayer, M. (1996). Optical fiber magnetic field sensors with ceramic magnetostrictive jackets. *Appl. Opt.* 35: 5340–5344.

249 Mora, J., Diez, A., Cruz, J.L., and Andres, M.V. (2000). A magnetostrictive sensor interrogated by fiber gratings for DC-Current and temperature discrimination. *IEEE Photonics Technol. Lett.* 12: 1680–1682.

250 Quintero, S.M.M., Martelli, C., Braga, A.M.B. et al. (2011). Magnetic field measurements based on terfenol coated photonic crystal fibers. *Sensors* 11: 11103–11111.

251 Rashleigh, S.C. (1981). Magnetic-field sensing with a single-mode fiber. *Opt. Lett.* 6: 19–21.

252 Cole, J.H., Lagakos, N., Jarzynski, J., and Bucaro, J.A. (1981). Magneto-optic coupling coefficient for fiber interferometric sensors. *Opt. Lett.* 6: 216–218.

253 Hartman, N., Vahey, D., Kidd, R., and Browning, M. (1982). Fabrication and testing of a nickel-coated single-mode fiber magnetometer. *Electron. Lett.* 18: 224–226.

254 Willson, J.P. and Jones, R.E. (1983). Magnetostrictive fiber-optic sensor system for detecting DC magnetic-fields. *Opt. Lett.* 8: 333–335.

255 Kersey, A.D., Corke, M., Jackson, D.A., and Jones, J.D.C. (1983). Detection of DC and low-frequency AC magnetic-fields using an all single-mode fiber magnetometer. *Electron. Lett.* 19: 469–471.

256 Koo, K., Dandridge, A., Tveten, A., and Sigel, G. Jr. (1983). A fiber-optic DC magnetometer. *J. Lightwave Technol.* 1: 524–525.

257 Kersey, A.D., Corke, M., and Jackson, D.A. (1984). Phase-shift nulling dc-field fibre-optic magnetometer. *Electron. Lett.* 20: 573–574.

258 Bucholtz, F., Dagenais, D.M., and Koo, K.P. (1989). Mixing and detection of RF signals in fibre-optic magnetostrictive sensor. *Electron. Lett.* 25: 1285–1286.

259 Bucholtz, F., Dagenais, D.M., and Koo, K.P. (1989). High-frequency fiberoptic magnetometer with 70 Ft/square-root (Hz) resolution. *Electron. Lett.* 25: 1719–1721.

260 Oh, K.D., Ranade, J., Arya, V. et al. (1997). Optical fiber Fabry-Perot interferometric sensor for magnetic field measurement. *IEEE Photonics Technol. Lett.* 9: 797–799.

261 Perez-Millan, P., Martinez-Leon, L., Diez, A. et al. (2002). A fiber-optic current sensor with frequency-codified output for high-voltage systems. *IEEE Photonics Technol. Lett.* 14: 1339–1341.

262 Djinovic, Z., Tomic, M., and Gamauf, C. (2010). Fiber-optic interferometric sensor of magnetic field for structural health monitoring. Vol. 5. *Eurosensors XXIV Conference*, Linz, Austria (5–8 September 2010). pp. 1103–1106.

263 Yi, B., Chu, B.C.B., and Chiang, K.S. (2003). Temperature compensation for a fiber-Bragg-grating-based magnetostrictive sensor. *Microwave Opt. Technol. Lett.* 36: 211–213.

264 Satpathi, D., Moore, J.A., and Ennis, M.G. (2005). Design of a Terfenol-D based fiber-optic current transducer. *IEEE Sens. J.* 5: 1057–1065.

265 Li, M.F., Zhou, J.F., Xiang, Z.Q., and Lv, F.Z. (2005). Giant magnetostrictive magnetic fields sensor based on dual fiber Bragg gratings. *2005 IEEE Networking, Sensing and Control Proceedings*, Arizona, AZ, USA (19–22 March 2005). pp. 490–495.

266 Mora, J., Martinez-Leon, L., Diez, A. et al. (2006). Simultaneous temperature and ac-current measurements for high voltagelines using fiber Bragg gratings. *Sens. Actuators, A* 125: 313–316.

267 Reilly, D., Willshire, A.J., Fusiek, G. et al. (2006). A fiber-Bragg-grating-based sensor for simultaneous AC current and temperature measurement. *IEEE Sens. J.* 6: 1539–1542.

268 Davino, D., Visone, C., Ambrosino, C. et al. (2008). Compensation of hysteresis in magnetic field sensors employing fiber Bragg grating and magneto-elastic materials. *Sens. Actuators A* 147: 127–136.
269 Yang, M.H., Dai, J.X., Zhou, C.M., and Jiang, D.S. (2009). Optical fiber magnetic field sensors with TbDyFe magnetostrictive thin films as sensing materials. *Opt. Express* 17: 20777–20782.
270 Pacheco, C.J. and Bruno, A.C. (2010). The effect of shape anisotropy in giant magnetostrictive fiber Bragg grating sensors. *Meas. Sci. Technol.* 21: 065205–065209.
271 Quintero, S.M.M., Braga, A.M.B., Weber, H.I. et al. (2010). A magnetostrictive composite-fiber Bragg grating sensor. *Sensors* 10: 8119–8128.
272 Smith, G.N., Allsop, T., Kalli, K. et al. (2011). Characterisation and performance of a Terfenol-D coated femtosecond laser inscribed optical fibre Bragg sensor with a laser ablated microslot for the detection of static magnetic fields. *Opt. Express* 19: 363–370.
273 Yang, S.Y., Chieh, J.J., Horng, H.E. et al. (2004). *Appl. Phys. Lett.* 84 (25): 5204.
274 Zu, P., Chan, C.C., Jin, Y.X. et al. (2011). *Proc. SPIE* 7753: 77531G.
275 Dai, J.X., Yang, M.H., Li, X.B. et al. (2011). Magnetic field sensor based on magnetic fluid clad etched fiber Bragg grating. *Opt. Fiber Technol.* 17: 210–213.
276 Thakur, H.V., Nalawade, S.M., Gupta, S. et al. (2011). Photonic crystal fiber injected with Fe_3O_4 nanofluid for magnetic field detection. *Appl. Phys. Lett.* 99: 161101:1–161101:3.
277 Zhao, Y., Lv, R.Q., Ying, Y., and Wang, Q. (2012). Hollow-core photonic crystal fiber Fabry-Perot sensor for magnetic field measurement based on magnetic fluid. *Opt. Laser Technol.* 44: 899–902.
278 Chen, D., Ready, J.F., and Bernal, E. (1968). *J. Appl. Phys. Lett.* 39: 3916.
279 Chaudhari, P., Cuomo, J.J., and Gamino, R.J. (1973). *Appl. Phys. Lett.* 42: 202.
280 Williams, H.J., Sherwood, R.C., Forester, F.G., and Kelly, E.M. (1957). *J. Appl. Phys.* 28: 1181.
281 Tsunashima, S. (2001). Magneto-optical recording. *J. Phys. D: Appl. Phys.* 34: R87–R102.
282 Huth, B.G. (1974). *IBM J. Res. Dev.* 10: 100–109.
283 Beaurepaire, E., Merle, J.-C., Daunois, A., and Bigot, J. (1996). *Phys. Rev. Lett.* 76: 4250.
284 Mansuripur, M. (1995). *The Physical Principles of Magneto-optical Recording*, 652. Cambridge University Press.
285 Hasegawa, R., Argyle, B.E., and Tao, L.-J. (1975). *AIP Conf. Proc.* 24: 110.
286 Brown, B.R. (1973). *Appl. Opt.* 13: 761.
287 Hiroyoshi, H. and Fukamichi, K. (1982). *J. Appl. Phys.* 53: 2226.
288 Ryan, D.H., Coey, J.M.D., Batalla, E. et al. (1987). *Phys. Rev. B* 35: 8630.
289 Ohnuma, S., Shirakawa, K., Nose, M., and Masumoto, T. (1980). *IEEE Trans. Magn.* 15: 1129.
290 Tanaka, F., Tanaka, S., and Imamura, N. (1987). Magneto-optical recording. *J. Phys. D: Appl. Phys.* 26: 231.
291 Heiman, N., Lee, K., and Potter, R.I. (1976). *J. Appl. Phys.* 47: 2634.

292 Kusiro, Y., Imamura, N., Kobayashi, T. et al. (1978). *J. Appl. Phys.* 49: 1208.
293 Takahashi, M., Niihara, T., and Ohta, N. (1988). *J. Appl. Phys.* 64262.
294 Oechsner, H. (1970). *Z. Phys.* 238: 433.
295 Suzuki, Y., Takayama, S., Kirino, F., and Ohta, N. (1987). *IEEE Trans. Magn.* 23: 2275.
296 Rhyne, J.J. (1972). *Magnetic Properties Rare Earth Metals* (ed. R.J. Elliott), 156. London: Plenum Press.
297 Choe Y J 1989 Doctoral Thesis Nagoya University ch. 4.
298 Tsunashima S, Iwata S and Yu X Y 1997 Research Report No 98. Magnetics Society of Japan (in Japanese). p. 61.
299 Kobayashi, T., Tsuji, H., Tsunashima, S., and Uchiyama, S. (1981). *Jpn. J. Appl. Phys.* 20: 2089.
300 Tsunashima, S. (1999). Exchange-coupled films Chapter 7. In: *Magneto-Optical Recording Materials* (ed. R.J. Gambino and T. Suzuki). Piscataway, NJ: IEEE.
301 Carcia, P.F., Meinhaldt, A.D., and Suna, A. (1985). *Appl. Phys. Lett.* 47: 178.
302 Zeper, W.B., van Kesteren, H.W., Jacobs, B.A.J. et al. (1991). *J. Appl. Phys.* 70: 2264.
303 den Broeder, E.J.A., Hoving, W., and Bloemen, P.J.H. (1991). *J. Magn. Magn. Mater.* 93: 562.
304 Fujiwara, Y., Masaki, T., Yu, X. et al. (1997). *Jpn. J. Appl. Phys.* 36: 5097.
305 Fischer, P., Eimülle, T., Glück, S. et al. (2001). *J. Magn. Soc. Jpn.* 25: 186.
306 Takahashi, A., Nakayama, J., Murakami, Y. et al. (1994). *IEEE Trans. Magn.* 30: 232.
307 Aratani, K., Fukumoto, A., Ohta, M. et al. (1991). *Proc. SPIE* 1499: 209.
308 Awano, H., Ohnuki, S., Shrai, H., and Ohta, N. (1996). *Appl. Phys. Lett.* 69: 4257.
309 Shiratori, T., Fujii, E., Miyako, Y., and Hozumi, Y. (1998). *J. Magn. Soc. Jpn.* 22: 47.
310 Charap, S.H., Pu-Ling, L., and He, Y. (1997). *IEEE Trans. Magn.* 33: 978.
311 Mochida, M., Birukawa, M., Sbiaa, R., and Suzuki, T. 2000. Technical Digest of Joint MORIS/APDSC 2000 (Nagoya). p. 46.
312 Alben, R., Becker, J.J., and Chi, M.C. (1978). *J. Appl. Phys.* 49: 1653.
313 Betzig, E., Trautman, J.K., Wolfe, R. et al. (1992). *Appl. Phys. Lett.* 61: 142.
314 Mansfield, S.M. and Kino, G.S. (1990). *Appl. Phys. Lett.* 57: 2615.
315 Terris, B.D., Mamin, H.J., Ruger, D. et al. (1994). *Appl. Phys. Lett.* 65: 388.
316 Katayama, H., Sawamura, S., Ogimoto, Y. et al. (1999). *J. Magn. Soc. Jpn.* 23: 233.
317 Nemoto, H., Saga, H., Sukeda, H., and Takahashi, M. (1999). *J. Magn. Soc. Jpn.* 23: 229.
318 Gangopadhyay, S., Subramanian, K., Ryan, P. et al. (2000). 23.8 Gb/in.^{2} areal density demonstration. *J. Appl. Phys.* 87: 5407.
319 Saga, H., Nemoto, H., Sukeda, H., and Takahashi, M. (1999). *J. Magn. Soc. Jpn.* 23: 225.
320 Hamann, H.F., Martin, Y.C., and Wickramasinghe, H.K. (2004). *Appl. Phys. Lett.* 84: 810.
321 Awano, H. and Ohta, N. (1998). *IEEE J. Sel. Top. Quantum Electron.* 4: 815.

322 Hohlfeld, J., Gerrits, T., Bilderbeek, M. et al. (2001). *Phys. Rev. B: Condens. Matter* 65: 012413.
323 Stanciu, C.D., Tsukamoto, A., Kimel, A.V. et al. (2007). *Phys. Rev. Lett.* 99: 217204.
324 Ogasawara, T., Iwata, N., Murakami, Y. et al. (2009). *Appl. Phys. Lett.* 94: 162507.
325 Vahaplar, K., Kalashnikova, A.M., Kimel, A.V. et al. (2009). *Phys. Rev. Lett.* 103: 117201.
326 Hohlfeld, J., Stanciu, C.D., and Rebei, A. (2009). *Appl. Phys. Lett.* 94: 152504.
327 Chen, Z., Gao, R., Wang, Z. et al. (2010). Field-dependent ultrafast dynamics and mechanism of magnetization reversal across ferrimagnetic compensation points in GdFeCo amorphous alloy films. *J. Appl. Phys.* 108: 023902.
328 Yamamoto, S., Taguchi, M., Someya, T. et al. (2015). Ultrafast spin-switching of a ferrimagnetic alloy at room temperature traced by resonant magneto-optical Kerr effect using a seeded free electron laser. *Rev. Sci. Instrum.* 86: 083901.

5

Electro-optics

5.1 Introduction

Electro-optic effect is the property of some transparent materials such as glass or crystals that change their refractive index when subjected to a dc (direct current) or low-frequency electric field. From the fundamental materials viewpoint, the electro-optic effect is due to the change in optical susceptibility caused by molecular, ionic, and electronic polarization as a result of an applied electric field. The dependence of the refractive index on the electric field takes the following forms:

- The refractive index changes in proportion to the applied electric field, in which case the effect is known as the linear electro-optic effect or the Pockels effect. Only non-centrosymmetric materials (mostly crystals) exhibit the linear electro-optic effect.
- The refractive index changes in proportion to the square of the applied electric field, in which case the effect is known as the quadratic electro-optic QEO effect or the Kerr effect. All centrosymmetric materials exhibit the Kerr electro-optic effect.

Materials whose refractive index can be modified by means of an applied electric field are useful for producing a large number of optical devices, such as tunable narrow band interference polarizing monochromators, light modulators, light beam deflectors, frequency shifters, and second harmonic generators.

5.2 History of Electro-optic Effects

In 1875, John Kerr discovered that a piece of glass became birefringent when subjected to an electric field, this phenomenon being known as the Kerr electro-optic effect [1]. Kerr electro-optic birefringence is proportional to the square of the electric field and is also known as QEO effect. In 1883, the linear electro-optic effect was discovered for the first time in quartz and tourmaline crystals by Röntgen [2, 3] and by Kundt [4]. In 1893 Pockels [5, 6] discovered the electro-optic effect in Rochelle salt and worked out the phenomenological theory of linear electro-optic effect. Mueller [7, 8] and Valasek [9] carried out thorough investigations on the quadratic and linear electro-optic effects in Rochelle salt.

Crystal Optics: Properties and Applications, First Edition. Ashim Kumar Bain.

Anistratov and Aleksandrov [10] studied the electro-optic properties of Rochelle salt. Jacquerod [11] studied the Kerr effect in potassium dihydrogen phosphate (KDP) and potassium dihydrogen arsenate (KDA). Zwicker and Scherrer [12, 13] reported the linear electro-optic effect in KDP and the salt resulting after the replacement of hydrogen by deuterium.

After the World War II, interest in investigating the electro-optic effect has grown tremendously, owing to the fact that the large magnitude of the effect in the artificially grown crystals of ammonium dihydrogen phosphate (ADP) and KDP has found practical applications in a variety of situations. In 1947, Billings indicated the application of ADP and KDP for the modulation of light and as optical shutters [14]. This is the beginning of a new era in the history of electro-optic effect and its variety of applications in science and technology. Since the last few decades, innumerable scientists have been working on electro-optical materials to find the applications of these materials in consumer electronics, industry, and defense sectors. A chronological listing of many notable specific events in the history of electro-optic materials is given in Table 5.1. Materials displaying electro-optic effect include $BaTiO_3$, $LiNbO_3$, $LiTaO_3$, barium strontium titanate (BST), lead lanthanum zirconate titanate (PLZT), lead magnesium niobate–lead titanate (PMN–PT), $KNbO_3$, and potassium tantalate niobate (KTN), and are increasingly finding applications in industry, defense, and electronic devices such as optical modulators, displays, image-storage systems, shutters, and data processing.

In 1965, I. P. Kaminow [16] at Bell laboratory fabricated the first barium titanate light phase modulator. Currently, $BaTiO_3$ crystals are widely used for nonlinear optical applications, which include optical waveguides [33, 35, 62–70], optical switching devices [71, 72], integrated optics [73–75], and electro-optic modulators [38, 42, 76–88]. In 1964, R. C. Miller first studied second harmonic generation (SHG) in $BaTiO_3$ crystals [15]. The strong SHG from poled $BaTiO_3$ thin films, electron beam-induced poling of $BaTiO_3$ thin films, and third-order nonlinear optical studies on doped and un-doped $BaTiO_3$ films is reported in the literature [89–91]. M. J. Dicken et al. have designed a plasmonic interferometer by utilizing electro-optic $BaTiO_3$ as a dielectric layer [42]. P. Tang et al. have designed 40 GHz barium titanate thin-film modulators [38].

In 1967, C. E. Land designed PLZT ceramic electro-optic storage and display devices [17]. The PLZT fast-shutter has been designed [19, 92–95] for flight simulators, radar and sonar displays, medical imaging, electron microscopy, etc. The electro-optic phase retardation device has been designed using the PLZT ceramic element. Ferroelectric PLZT thin-film optical waveguides and memory devices have been designed by S. Miyazawa and N. Uchida [20], A. Baudrant et al. [96], C. E. Land [97], M. Bazylenko and I. K. Mann [98], J. D. Levine and S. Essaian [99], and S. L. Swartz [100]. The nano-speed PLZT optical switching device has been designed by K. Nashimoto [40], which can be one of the key elements for next-generation networks based on burst switching and packet switching systems including grid, switched access networks, 100G LAN, interconnections, and phased array antenna.

In 1970, A. A. Ballman et al. designed the $LiTaO_3$ electro-optic modulator [18]. Presently, the $LaTiO_3$ waveguides are widely used in many electro-optic devices

Table 5.1 Notable events in the history of electro-optics.

Timeline	Events
1875	Discovery of the quadratic electro-optic effect in glasses [1].
1883	Discovery of the linear electro-optic effect in quartz and tourmaline crystals [2–4].
1893	Discovery of the linear electro-optic effect in Rochelle salt and working out of the phenomenological theory of linear electro-optic effect [5, 6].
1947	Application of ADP and KDP crystals for modulation of light and optical shutters [14].
1964	Second harmonic generation (SHG) in $BaTiO_3$ crystals [15].
1965	Barium titanate light phase modulator [16].
1967	Design of PLZT ceramic electro-optic storage and display devices [17].
1970	Design of $LiTaO_3$ electro-optic modulator [18].
1972	Design of PLZT fast shutter for flight simulators, radar and sonar display, medical imaging, electron microscopy, etc. [19].
1975	PLZT thin film optical waveguide and memory device [20].
1978	Design of Ti-diffused $LiNbO_3$ strip waveguides [21].
1979	Second harmonic generation (SHG) in $KNbO_3$ [22, 23].
1980	Design of high-speed operation of $LiNbO_3$ electro-optic interferometric waveguide modulator [24].
1980	Design of $LiNbO_3$ electric field sensor [25].
1982	Design of proton exchange for high-index waveguides in $LiNbO_3$ [26].
1991	Second harmonic generation (SHG) of blue light in $LiTaO_3$ waveguides [27–30].
1992	Design of 50 GHz $LiNbO_3$ modulator [31].
1994	Design of $Ti:LiNbO_3$ polarization-independent photonic switch [32].
1997	Thin-film channel waveguide electro-optic modulator in epitaxial $BaTiO_3$ [33].
1999	Design of second harmonic generator, electro-optic lens, and electro-optic scanner in $LiTaO_3$ [34].
2002	Thin-film $BaTiO_3$ Mach–Zehnder modulator [35].
2004	Design of KTN crystal waveguide-based electro-optic phase modulator [36, 37].
2004	Design of 40 GHz $BaTiO_3$ thin-film modulator [38].
2005	Development of dynamic photonic devices (variable optical attenuator, polarization controller, variable gain tilt filters, dynamic gain flattening filters, Q-switches, tunable optical filters) using PMN–PT ceramics [39].
2006	Nano-speed PLZT optical switching device [40].
2007	SHG response in $KNbO_3$ nanowires and potential application as frequency converters [41].
2008	Design of plasmonic interferometer by utilizing electro-optic $BaTiO_3$ as a dielectric layer [42].
2009	Design of KTN optical scanner [43].
2009	Fast varifocal lenses based on KTN single crystals [44].
2009	

Table 5.1 (Continued)

Timeline	Events
2014	Electro-optic plasmonic modulators [45–53].
2012	SHG response in $KNbO_3$ nanoneedles and potential application in nano-optical devices [54].
2013	High-speed KTN deflector [55].
2013	
2014	Electro-optic plasmonic switches [53, 56–58].
2015	Design of $LiNbO_3$ microresonators [59, 60].
2017	Design of nanophotonic $LiNbO_3$ electro-optic modulators [61].

such as beam deflector [101, 102], scanner [34, 103], integrated lasers and nonlinear frequency converter [104], SHG [27–30, 34, 103, 105–108], waveguides, and spatial light modulator [109].

In 1978, M. Fukuma et al. have designed Ti-diffused $LiNbO_3$ strip waveguides [21] and since then extensive work has been carried out on its possible implementation [110–116]. In 1982, the first proton-exchanged $LiNbO_3$ waveguides were designed by J. J. Jackel et al. [26] and then several workers have worked on its applications [117–123]. Schmidt and Kaminow [124] designed metal-diffused optical waveguides in $LiNbO_3$. Thomson et al. [125] designed the z-cut lithium niobate ($LiNbO_3$) optical waveguide using femtosecond pulses in the low repetition rate regime. P. Rabiei and P. Günter [126] designed submicrometer $LiNbO_3$ slab waveguides by crystal ion slicing and wafer bonding. R. Geiss et al. [127] designed the $LiNbO_3$ photonic crystal waveguide. H. Lu et al. [128] designed the $LiNbO_3$ photonic crystal wire optical waveguide on a smart-cut thin film. Nowadays, $LiNbO_3$ waveguides are widely used in many functional electro-optic devices such as modulators [24, 31, 61, 129–146], Q-switches [32, 129, 147], filter [129], SHG [120, 148–155], periodically poled waveguides, coherent receiver [156], electric field sensors [25, 157–179], and microresonators [59, 60, 180–188].

In 2000, Li et al. [189] first investigated the electro-optic effects in $Ba_{1-x}Sr_xTiO_3$ (BST) thin films deposited on LAO (001) (lanthanum aluminate) substrates. Kim et al. [190, 191] reported an extremely large QEO coefficient ($R_c = 1.0 \times 10^{-15}$ m^2/V^2) in BST thin films grown on MgO (001) substrates, which implies that BST thin films are applicable for electro-optic modulators and optical waveguide devices. Wang et al. [192, 193] observed large linear ($r_c = 125.0 \times 10^{-12}$ m/V) and QEO coefficients ($R_c = 10.0 \times 10^{-18}$ m^2/V^2) in $B_{0.7}S_{0.3}TiO_3$ thin films, which is very promising for rib waveguides and Mach–Zehnder electro-optic modulators. Takeda et al. [194] studied the birefringence and electro-optic effect in epitaxial BST thin films, and the Kerr coefficient of BST film was determined to be 3.44×10^{-17} m^2/V^2.

PMN–PT, one of the most widely investigated relaxor ferroelectric with a perovskite structure, was first synthesized in the late 1950s [195]. Its electro-optic effect is about 2–5 times higher than that of PLZT and nearly 100 times higher than that of $LiNbO_3$ at room temperature. The remarkably good transparency over a wide wavelength range of 500–77 000 nm makes PMN–PT ceramic best suited for almost all the visible to mid-infrared (MIR) optical applications, such as optical limiter [196], polarization retarder [197], optical switch [39, 198, 199], optical waveguides [200, 201], variable optical attenuator (VOA), polarization controller, filters [39, 202], and integrated photonic devices [203].

In 1974, P. Günter studied the electro-optic properties of $KNbO_3$ [204]. Recently, potassium niobate has attracted much attention from the scientific community due to its many applications in several technological fields. Large electro-optic [204–212] and nonlinear optical coefficients [213–229] and excellent photorefractive sensitivity [230–235] make the crystal a particularly attractive candidate for applications such as optical waveguides [221, 223, 227, 236–261], electro-optic modulators [236, 262], SHG waveguides [22, 23, 41, 54, 213, 225, 248, 250, 251, 257, 258, 263–276], optical parametric oscillation [270, 277–279] and holographic storage medium [280–282], and optical switching [257]. These properties rank $KNbO_3$ at the top of Tuttle's figure of merit (FoM) comparison [283] and Holman's device-level analysis [284] of materials in this class including $BaTiO_3$, PLZT, and $LiNbO_3$.

KTN crystal, since its birth in the 1950s, has been well known for its distinctive properties. It has attracted interest for more than four decades due to its large QEO coefficient [285, 286]. In recent years, there has been substantial progress in all aspects of KTN materials such as material growth, device development, and real-world applications. For example, in terms of material growth, high-quality KTN crystals with size $>30\,cm^3$ have been successfully grown, which ensures material supply for device development [287]. In terms of device development, by taking advantage of the large quadratic EO coefficient of the KTN material, a low-driving voltage EO modulator has been introduced [36, 37]. Based on the EO effect of KTN single crystals, a fast varifocal lens has been developed that responds as fast as $1\,\mu s$, which is 1000 times faster than the conventional commercial devices [44, 288]. Furthermore, by using the unique space-charge-controlled EO effect, low-driving voltage beam deflectors [55, 289, 290] and scanners [43, 291–293] have also been designed.

Over the past few years, plasmonics is a rapidly expanding scientific discipline that deals with, among others, the properties of surface plasmon polaritons (SPPs), i.e. light waves propagating at the interface between metals and dielectrics. Such transmission of optical signals can take place at deep sub-wavelength scales, not limited by the traditional diffraction limit. Thus, integrated plasmonic circuits are envisaged as the missing link between electronics and photonics that combines the dense integration of the former with the high bandwidth, lower latency, and reduced power dissipation of the latter [294].

Essential parts of integrated optical architectures are tunable components such as modulators and switches that convert the electrical signal to optical pulses and control their route through the circuit. In this respect, the plasmonics platform offers a significant advantage, namely, the capability to use the metallic parts not only as light guides in highly compact geometries, but also as the electrodes that dynamically control the properties of electro-optically responsive materials. Modulation can rely on various phenomena, including the field-effect charge accumulation in nanometer-thick transparent oxide layers [45, 46], the Pockels effect index modulation in $\chi^{(2)}$ nonlinear polymers [47], and the electro-optic switching of nematic liquid crystals [56]. To date, a wide variety of electro-optical plasmonic modulators [42, 48–53] and switches [53, 57, 58, 295] have been proposed for integrated optical signal processing.

5.3 Principles of Electro-optic Effects

If a rectangular quartz prism with length along x-axis, breadth along z-axis, and thickness along y-axis is subjected to an electric field parallel to the x-axis and a parallel beam of monochromatic plane-polarized light is passed through the prism along the y-axis as shown in Figure 5.1, then the optical path difference δ_o may be described as follows:

- Owing to the applied electric field, the QEO effect (*Kerr effect*) produces a certain amount of optical path difference δ_k.
- Owing to the piezoelectric nature of the quartz prism, it experiences a physical strain, which in its turn causes a photoelastic path difference δ_p.
- The external electric field may alter the dipole moment of the medium and hence produce some path difference δ_l. This effect is called the linear electro-optic effect (*Pockels effect*).

Thus, in general the observed artificial path difference δ_o is given by

$$\delta_o = \delta_l + \delta_k + \delta_p \tag{5.1}$$

It is important to note that the linear electro-optic effect is exhibited by piezoelectric crystals for some specific orientations.

The refractive index of an electro-optic medium is a function of the applied electric field E. This function varies slightly with E so that it can be expanded in a Taylor's series about $E = 0$

$$n(E) = n + a_1 E + \frac{1}{2} a_2 E^2 + \cdots \tag{5.2}$$

It is convenient to write Eq. (5.2) in terms of two new coefficients $r = -2a_1/n^3$ and $K = -a_2/n^3$, known as the electro-optic coefficients, so that

$$n(E) = n - \frac{1}{2} r n^3 E - \frac{1}{2} K n^3 E^2 + \cdots \tag{5.3}$$

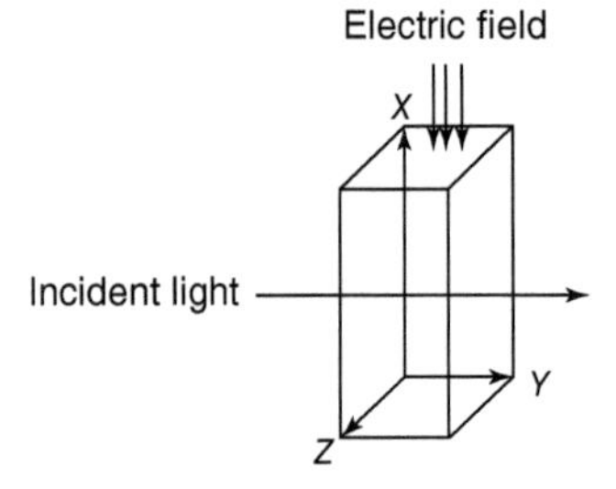

Figure 5.1 A steady electric field is applied to a quartz prism parallel to the x-axis and plane-polarized light is allowed to pass along the y-axis of the prism.

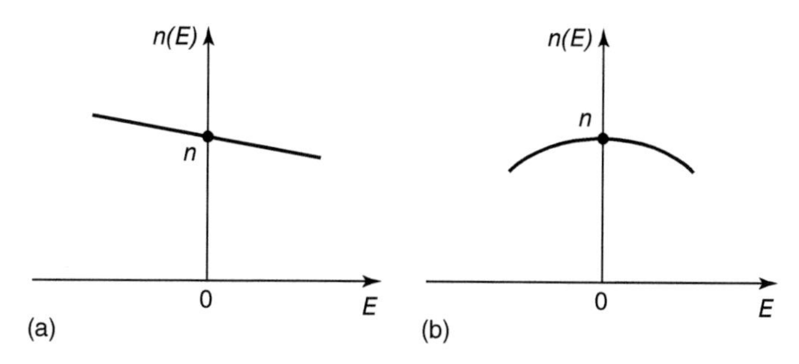

Figure 5.2 Dependence of the refractive index on the electric field: (a) Pockels medium; (b) Kerr medium.

The second- and higher order terms of this series are typically many orders of magnitude smaller than n. Terms higher than the third can be neglected.

In many materials, the third term of Eq. (5.3) is negligible in comparison with the second, whereupon

$$n(E) \approx n - \frac{1}{2} r n^3 E \tag{5.4}$$

This function is illustrated in Figure 5.2a. The medium is then known as a Pockels medium (or a Pockels cell). The coefficient r is called the *Pockels coefficient* or the linear electro-optic coefficient. Typical values of r lie in the range 10^{-12} to 10^{-10} m/V (1 to 100 pm/V).

If the material is centrosymmetric, $n(E)$ must be an even symmetric function as shown in Figure 5.2b, since it must be invariant to the reversal of E. The second term is negligible in comparison with the third in Eq. (5.2), whereupon

$$n(E) \approx n - \frac{1}{2} K n^3 E^2 \tag{5.5}$$

The material is known as a Kerr medium (or a Kerr cell). The coefficient K is called the *Kerr coefficient* or the QEO coefficient. Typical values of K are 10^{-18} to 10^{-14} m^2/V^2 in crystals and 10^{-22}–10^{-19} m^2/V^2 in liquids.

5.4 Phenomenological Theory of Electro-optic Effect

The optical properties of crystals can be expressed in terms of the refractive index ellipsoid, whose semi axes are proportional to the three principal refractive indices. The equation of the refractive index ellipsoid for an electrically unstrained crystal can be expressed in the general form

$$\frac{x^2}{n_{11}^2} + \frac{y^2}{n_{22}^2} + \frac{z^2}{n_{33}^2} + \frac{2yz}{n_{23}^2} + \frac{2zx}{n_{31}^2} + \frac{2xy}{n_{12}^2} = 1 \tag{5.6}$$

The equation of this refractive index ellipsoid or optical index ellipsoid can be written in the form

$$\sum_{i,j} B_{ij} x_i x_j = 1 \quad (i, j = 1, 2, 3) \tag{5.7}$$

where the three optical parameters B_{ij} ($B_{ij} = 1/n_{ij}^2$) are equal to the reciprocals of the squares of the respective refractive indices.

5.4.1 Linear Electro-optic Effect

In order to explain the changes in the optical properties of a crystal under the influence of an applied electric field, Pockels assumed, as in the case of photoelastic behavior of crystals, that the index ellipsoid defining the optical behavior of the crystal gets deformed owing to the applied electric field; the resulting deformation of the index ellipsoid can be described in terms of the changes ΔB_{ij} in the optical parameters of the indicatrix, being linearly related to the components E_k of the applied electric field. Thus, the relation in the three-suffix notation is written in the form

$$\Delta B_{ij} = \sum_k r_{ijk} E_k \quad (i, j = 1, 2, 3; k = 1, 2, 3) \tag{5.8}$$

The coefficient r_{ijk} is a third rank polar tensor, called the linear electro-optic constants (also called the Pockels electro-optic constants). Since the indicatrix components are symmetric ($\Delta B_{ij} = \Delta B_{ji}$), Eq. (5.8) can be expressed in the two-suffix notation

$$\Delta B_i = \sum_j r_{ij} E_j \, (i = 1, \ldots, 6; j = 1, 2, 3) \tag{5.9}$$

or in the expanded form thus

$$\begin{aligned}
\Delta B_1 &= r_{11}E_1 + r_{12}E_2 + r_{13}E_3 \\
\Delta B_2 &= r_{21}E_1 + r_{22}E_2 + r_{23}E_3 \\
\Delta B_3 &= r_{31}E_1 + r_{32}E_2 + r_{33}E_3 \\
\Delta B_4 &= r_{41}E_1 + r_{42}E_2 + r_{43}E_3 \\
\Delta B_5 &= r_{51}E_1 + r_{52}E_2 + r_{53}E_3 \\
\Delta B_6 &= r_{61}E_1 + r_{62}E_2 + r_{63}E_3
\end{aligned} \tag{5.10}$$

where

$r_{111} = r_{11}, r_{112} = r_{12}, r_{113} = r_{13}, r_{222} = r_{22}, r_{121} = r_{61}, r_{122} = r_{62}$, and $r_{123} = r_{63}$, etc.

The changes in optical parameters ΔB_{ij} can be linearly related to the electric polarization P_k in the form

$$\Delta B_{ij} = \sum_k \tau_{ijk} P_k \, (i, j = 1, 2, 3; k = 1, 2, 3) \tag{5.11}$$

Equation (5.11) can be expressed in two-suffix notation of τ_{ij}

$$\Delta B_i = \sum_j \tau_{ij} P_j \, (i = 1, \ldots, 6; j = 1, 2, 3) \tag{5.12}$$

or in the expanded form thus

$$\Delta B_1 = \tau_{11}P_1 + \tau_{12}P_2 + \tau_{13}P_3$$

$$\Delta B_2 = \tau_{21}P_1 + \tau_{22}P_2 + \tau_{23}P_3$$
$$\Delta B_3 = \tau_{31}P_1 + \tau_{32}P_2 + \tau_{33}P_3$$
$$\Delta B_4 = \tau_{41}P_1 + \tau_{42}P_2 + \tau_{43}P_3$$
$$\Delta B_5 = \tau_{51}P_1 + \tau_{52}P_2 + \tau_{53}P_3$$
$$\Delta B_6 = \tau_{61}P_1 + \tau_{62}P_2 + \tau_{63}P_3 \tag{5.13}$$

The coefficients τ_{ij} can be expressed in terms of r_{ij} thus

$$\tau_{ij} = \frac{4\pi r_{ij}}{\varepsilon_i - 1} \tag{5.14}$$

It may be noted that the constants r_{ij} are determined experimentally and thence τ_{ij} are computed because they are of great importance to developing any atomistic or microscopic theory.

In discussing the linear electro-optic effect, mechanical boundary condition is important for large-strain crystals. The crystal is in either free (zero stress) or clamped (zero strain/zero deformation) state. If the crystal is free, a static electric field will cause a strain by the converse piezoelectric effect and this in turn will give a change in refractive index by the photoelastic effect. The electro-optic effect obtained under constant strain is called the *primary effect* and the effect due to piezoelectricity and photoelasticity is then called the *secondary effect*. The observed effect in a free crystal is the sum of the primary and secondary effects. In practice, the primary effect at constant strain may be found by applying an ac (alternating current) of high frequency.

The changes ΔB_i in the coefficients of the index ellipsoid due to both electro-optic and photoelastic effects can be expressed in terms of the applied electric field and either the stress or the strain in the crystal. Thus, Eq. (5.9) can be described in the forms

$$\Delta B_i = \sum_{j=1}^{3} r_{ij}E_j + \sum_{k=1}^{6} q_{ik}\sigma_k \tag{5.15}$$

$$\Delta B_i = \sum_{j=1}^{3} r'_{ij}E_j + \sum_{k=1}^{6} p_{ik}\varepsilon_k \tag{5.16}$$

where σ_k and ε_k are the components of stress and strain respectively in one-suffix notation, q_{ik} and p_{ik} are the stress-optical and strain-optical constants respectively, and r_{ij} and r'_{ij} are the linear electro-optic constants of the crystals at the free and clamped conditions respectively. Owing to the piezoelectric nature of the crystal, the strain components ε_k can be expressed in the form

$$\varepsilon_k = \sum_{l=1}^{6} s_{kl}\sigma_l + \sum_{j=1}^{3} d_{jk}E_j \; (k = 1-6) \tag{5.17}$$

where s_{kl} and d_{jk} are the elastic compliance constants and piezoelectric constants respectively. Substituting Eq. (5.17) in Eq. (5.16), we obtain

$$\Delta B_i = \sum_{j=1}^{3} \left(r'_{ij} + \sum_{k=1}^{6} p_{ik}d_{jk} \right) E_j + \sum_{l=1}^{6} \left(\sum_{k=1}^{6} p_{ik}s_{kl} \right) \sigma_l \tag{5.18}$$

On comparing Eq. (5.18) with Eq. (5.15), it is evident that

$$q_{ij} = \sum_{k=1}^{6} p_{ik} s_{kj} \tag{5.19}$$

and

$$r_{ij} = r'_{ij} + \sum_{k=1}^{6} p_{ik} d_{jk} \tag{5.20}$$

Therefore, the electro-optic coefficients r_{ij} of the crystals measured at zero stress (free crystal) differs from r'_{ij} measured at zero strain (clamped crystal).

5.4.2 Quadratic Electro-optic Effect

A piece of transparent solid becomes birefringent when subjected to an electric field, this phenomenon being known as the Kerr electro-optic effect. The Kerr electro-optic birefringence is proportional to the square of the electric field. The Kerr effect is also called the QEO effect, and is universal in the sense that it is exhibited by all crystals – centrosymmetric and non-centrosymmetric. In fact, in crystals lacking a center of symmetry, the Pockels effect is accompanied by the Kerr effect, although its magnitude may be considerably less than that of the Pockels effect. The QEO effect is the dominant effect in all materials with centrosymmetric structure, such as ferroelectric crystals in the paraelectric state.

The changes in the optical parameters of the indicatrix ΔB_{ij} can be described as linear functions of all the nine terms of $E_k E_l$ of the applied electric fields. Thus, the relation is written in the form

$$\Delta B_{ij} = \sum_{k,l=1,2,3} K_{ijkl} E_k E_l \tag{5.21}$$

where K_{ijkl} is the QEO coefficient, a symmetric tensor of fourth rank. These coefficients can be written in the two-suffix notation in the form

$$\begin{bmatrix} \Delta B_1 \\ \Delta B_2 \\ \Delta B_3 \\ \Delta B_4 \\ \Delta B_5 \\ \Delta B_6 \end{bmatrix} = \begin{bmatrix} K_{11} & K_{12} & K_{13} & K_{14} & K_{15} & K_{16} \\ K_{21} & K_{22} & K_{23} & K_{24} & K_{25} & K_{26} \\ K_{31} & K_{32} & K_{33} & K_{34} & K_{35} & K_{36} \\ K_{41} & K_{42} & K_{43} & K_{44} & K_{45} & K_{46} \\ K_{51} & K_{52} & K_{53} & K_{54} & K_{55} & K_{56} \\ K_{61} & K_{62} & K_{63} & K_{64} & K_{65} & K_{66} \end{bmatrix} \begin{bmatrix} E_1^2 \\ E_2^2 \\ E_3^2 \\ E_2 E_3 \\ E_3 E_1 \\ E_1 E_2 \end{bmatrix} \tag{5.22}$$

Here, $K_{14} = K_{1123}$, $K_{22} = K_{2222}$, etc.

Like the linear electro-optic effect, the QEO effect can be divided into low- and high-frequency conditions. In tensor form

$$K^u_{ijkl} = K^c_{ijkl} + \sum_{m,n} p_{ijmn} M_{mnkl} \tag{5.23}$$

where p_{ijmn} and M_{mnkl} are the elasto-optic (strain-optical) and electrostriction tensors, respectively. The quadratic electro-optic coefficient K^u_{ijkl} is measured with a low-frequency electric field under stress-free condition (unclamped) and K^c_{ijkl} is measured at high frequencies above the mechanical resonance under strain-free condition (clamped). The difference between the two comes from the field-induced electrostriction strain coupled to the elasto-optic effect.

5.5 Electro-optic Devices

Electro-optics is a branch of technology involving components, devices, and systems that operate by modification of the optical properties of a material by an electric field. Thus it concerns the interaction between the electromagnetic (optical) and the electrical (electronic) states of materials. The electro-optic effect relates to a change in the optical properties of the medium, which usually is a change in the birefringence and not simply the refractive index. In a Kerr cell, the change in birefringence is proportional to the square of the electric field, and the material is usually a liquid. In a Pockels cell, the change in birefringence varies linearly with the electric field, and the material is a crystal. Noncrystalline electro-optic materials have attracted interest because of their low cost of production. It is a polymer-based material, and therefore is an organic material. It is called sometimes as organic EO material, plastic EO material, or polymer EO material, which consists of a polymer and a nonlinear optical chromophore. The nonlinear optical chromophore in the molecule has Pockel's effect.

5.5.1 Phase Modulator

When a beam of light propagates through a Pockels cell of length L with applied electric field E, it experiences a phase shift

$$\varphi = \varphi_0 - \pi \left(\frac{rn^3EL}{\lambda_0} \right) \tag{5.24}$$

where $\varphi_0 = 2\pi nL/\lambda_0$ and λ_0 is the free space wavelength. If the electric field E is obtained by applying a voltage V across two faces of the cell separated by distance d, then $E = V/d$, and Eq. (5.24) gives

$$\varphi = \varphi_0 - \pi \left(\frac{V}{V_\pi} \right) \tag{5.25}$$

where

$$V_\pi = \frac{d}{L} \left(\frac{\lambda_0}{rn^3} \right) \tag{5.26}$$

The parameter V_π, known as the half-wave voltage, is the applied voltage at which the phase shift changes by π radian. Equation (5.25) indicates that the phase shift is linearly related to the applied voltage (Figure 5.3). Therefore, one can modulate the phase of an optical wave by varying the voltage V that is applied across a material through which the light passes. The parameter V_π is an important characteristic of the modulator. It depends on the material properties (refractive index and linear electro-optic coefficient), on the wavelength λ_o, and on the aspect ratio d/L.

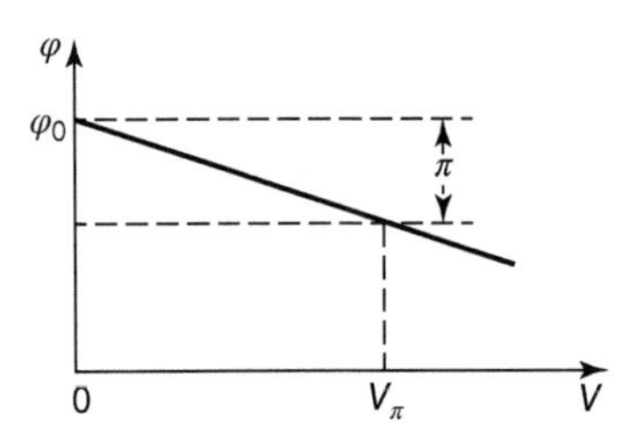

Figure 5.3 Phase shift φ with respect to applied voltage V.

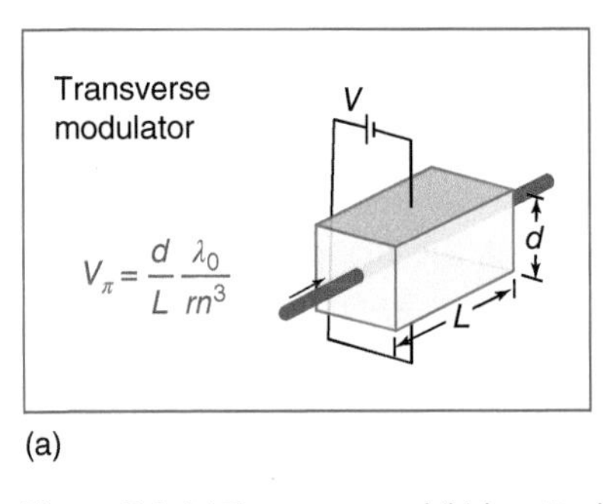

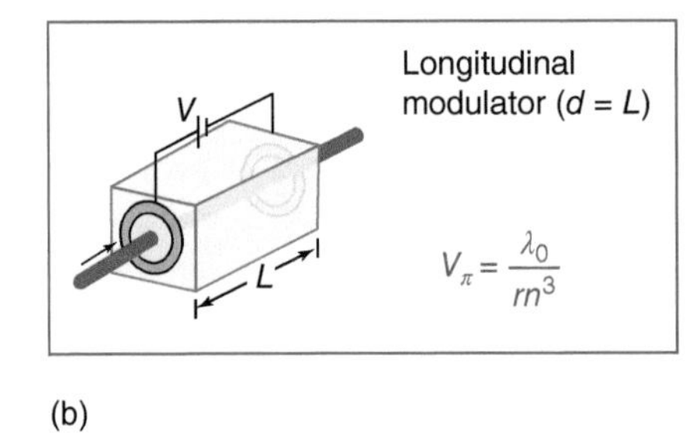

(a) (b)

Figure 5.4 (a) Transverse and (b) longitudinal modulators.

The electric field may be applied in a direction perpendicular to the direction of light propagation (transverse modulators) or parallel to it (longitudinal modulators), in which case $d = L$ (Figure 5.4). The value of the linear electro-optic coefficient r depends on the directions of propagation and the applied electric field since the crystal is anisotropic. Typical values of the half-wave voltage are in the vicinity of one to a few kilovolts for longitudinal modulators and hundreds of volts for transverse modulators.

The speed at which an electro-optic modulator operates is limited by electrical capacitive effects and by the transit time of light through the material. If the electric field $E(t)$ varies significantly within the light transit time T, the traveling optical wave will be subjected to different electric fields as it traverses the crystal. The modulated phase at a given time t will then be proportional to the average electric field $E(t)$ at times from $t - T$ to t. As a result, the transit-time-limited modulation bandwidth is $\approx l/T$. If the velocity of the traveling electrical wave matches that of the optical wave, transit time effects can, in principle, be eliminated. Commercial modulators in the forms shown in Figure 5.4 generally operate at several hundred megahertz, but modulation speeds of several gigahertz are possible.

Electro-optic modulators can also be constructed as integrated-optical devices. These devices operate at higher speeds and lower voltages than do bulk devices. An optical waveguide is fabricated in an electro-optic substrate (often $LiNbO_3$) by diffusing a material such as titanium to increase the refractive index. The electric field is applied to the waveguide using electrodes, as shown in Figure 5.5. Because the configuration is transverse and the width of the waveguide is much smaller than its length ($d \ll L$), the half-wave voltage can be as small as a few volts. These

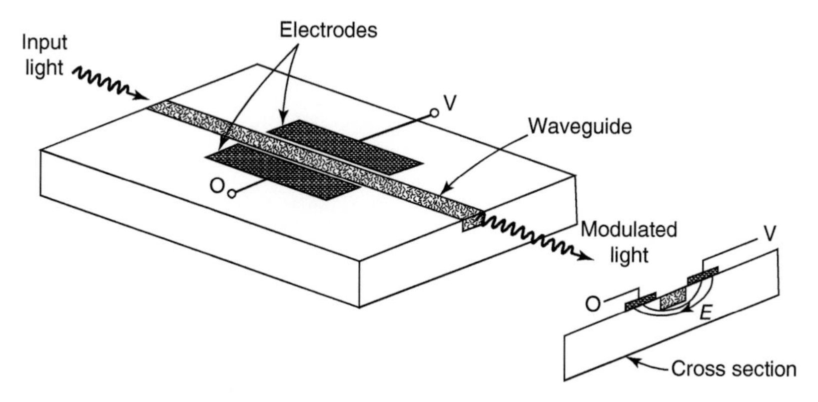

Figure 5.5 An integrated optical phase modulator. Source: Saleh 1991 [296]. Reproduced with permission of John Wiley & Sons.

modulators have been operated at speeds in excess of 100 GHz. Light can be conveniently coupled into, and out of, the modulator by the use of optical fibers.

5.5.2 Dynamic Wave Retarder

In an anisotropic medium, two linearly polarized normal modes propagate with different velocities, i.e. c_0/n_1 and c_0/n_2. If the medium exhibits Pockels effect, then in the presence of a steady electric field E, the two refractive indices can be written as

$$n_1(E) \approx n_1 - \frac{1}{2} r_1 n_1^3 E$$
$$n_2(E) \approx n_2 - \frac{1}{2} r_2 n_2^3 E$$

where r_1 and r_2 are the appropriate Pockels coefficients. After propagation through a distance L, the two modes undergo phase retardation (with respect to each other) given by

$$\Gamma = k_0[n_1(E) - n_2(E)]\,L = k_0(n_1 - n_2)L - \frac{1}{2}k_0(r_1 n_1^3 - r_2 n_2^3)EL \tag{5.27}$$

If the electric field E is obtained by applying a voltage V between two surfaces of the medium separated by a distance d, Eq. (5.27) can be written in the compact form

$$\Gamma = \Gamma_0 - \pi \left(\frac{\mathrm{V}}{V_\pi} \right) \tag{5.28}$$

where $\Gamma_0 = k_0(n_1 - n_2)L$ is the phase retardation in the absence of the electric field and

$$V_\pi = \frac{d}{L} \left(\frac{\lambda_0}{r_1 n_1^3 - r_2 n_2^3} \right) \tag{5.29}$$

where V_π is called the retardation half wave voltage necessary to obtain a phase retardation of π radian. Equation (5.28) indicates that the phase retardation is

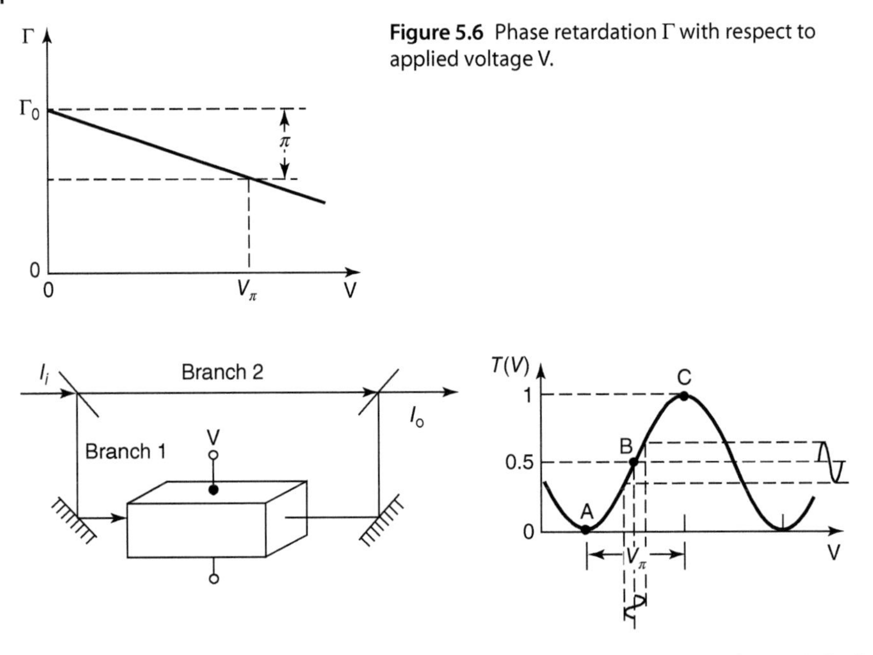

Figure 5.6 Phase retardation Γ with respect to applied voltage V.

Figure 5.7 An intensity modulator operating in a limited region near point B. If V is switched between points A and C, the device serves as an optical switch. Source: Saleh 1991 [296]. Reproduced with permission of John Wiley & Sons.

linearly related to the applied voltage (Figure 5.6). The medium serves as an electrically controllable dynamic wave retarder.

5.5.3 Intensity Modulators (Type 1)

A phase modulator placed in one branch of a Mach–Zehnder interferometer (MZI) can serve as an intensity modulator as illustrated in Figure 5.7. If the beam splitters divide the optical power equally, the transmitted intensity I_0 is related to the incident intensity I_i by

$$I_0 = I_i \cos^2\left(\frac{\varphi}{2}\right)$$

where $\varphi = \varphi_1 - \varphi_2$ is the difference between the phase shifts encountered by light as it travels through the two branches. The transmittance of the interferometer is

$$T = \frac{I_0}{I_i} = \cos^2\left(\frac{\varphi}{2}\right)$$

As the phase modulator is placed in branch 1, the phase shift φ_1 can be expressed as

$$\varphi_1 = \varphi_{10} - \pi\left(\frac{V}{V_\pi}\right)$$

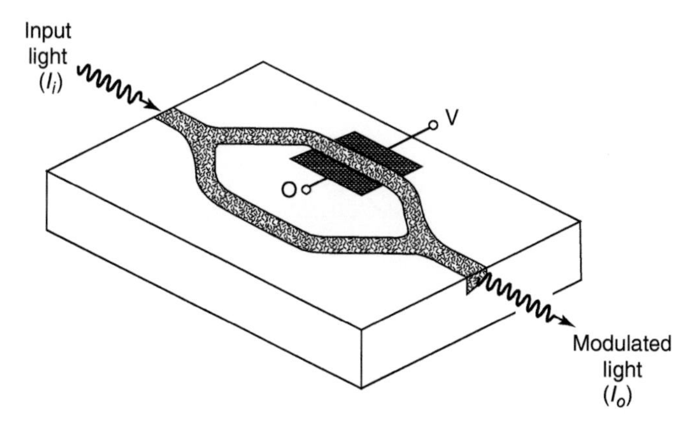

Figure 5.8 An integrated optical intensity modulator (or optical switch). A Mach–Zehnder interferometer and an electro-optic phase modulator are implemented using optical waveguides fabricated from material such as $LiNbO_3$. Source: Saleh 1991 [296]. Reproduced with permission of John Wiley & Sons.

so that φ is controlled by the applied voltage V in accordance with the linear relation

$$\varphi = \varphi_1 - \varphi_2 = \varphi_{10} - \varphi_2 - \pi\left(\frac{\text{V}}{V_\pi}\right) = \varphi_0 - \pi\left(\frac{\text{V}}{V_\pi}\right)$$

where $\varphi_0 = \varphi_{10} - \varphi_2$ depends on the optical path difference. The transmittance of the device is therefore a function of the applied voltage V,

$$T(\text{V}) = \cos^2\left(\frac{\varphi}{2}\right) = \cos^2\left[\frac{\varphi_0}{2} - \frac{\pi}{2}\left(\frac{\text{V}}{\text{V}_\pi}\right)\right] \tag{5.30}$$

This function is plotted in Figure 5.7 for an arbitrary value of φ_0. This device may be operated as a linear intensity modulator by adjusting the optical path difference so that $\varphi_0 = \pi/2$ and operating in the nearly linear region around $T = 0.5$. Alternatively, the optical path difference may be adjusted so that φ_0 is a multiple of 2π. In this case, $T(0) = 1$ and $T(V_\pi) = 0$, so that the modulator switches the light "on" and "off" as V is switched between 0 and V_π.

A Mach–Zehnder intensity modulator may also be constructed in the form of an integrated-optical device. Waveguides are placed on a substrate in the geometry shown in Figure 5.8. The beam splitters are implemented by the use of waveguide Ys. The optical input and output may be carried by optical fibers. Commercially available integrated-optical modulators generally operate at speeds of a few gigahertz but modulation speeds exceeding 25 GHz have been achieved.

5.5.4 Intensity Modulator (Type 2)

A wave retarder is placed between two crossed polarizers, at an angle 45° with respect to the retarder's axes (Figure 5.9), and has an intensity transmittance $T = \sin^2\left(\frac{\Gamma}{2}\right)$. If the retarder is a Pockels cell, then the phase retardation Γ is

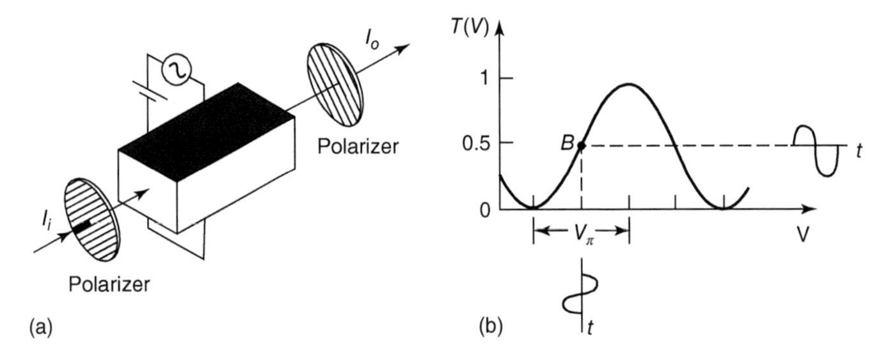

Figure 5.9 (a) An optical intensity modulator using a Pockels cell placed between two crossed polarizers. (b) Optical transmittance versus applied voltage for an arbitrary value of Γ_0; for linear operation the cell is biased near the point B. Source: Saleh 1991 [296]. Reproduced with permission of John Wiley & Sons.

linearly dependent on the applied voltage V as given in Eq. (5.28). The transmittance of the device is then a periodic function of V

$$T(\mathrm{V}) = \sin^2\left[\frac{\Gamma_0}{2} - \frac{\pi}{2}\left(\frac{\mathrm{V}}{\mathrm{V}_\pi}\right)\right] \tag{5.31}$$

This function is plotted in Figure 5.9. The transmittance can be varied between 0 (shutter closed) and 1 (shutter open) by changing the applied voltage V. The device can also be used as a linear modulator if the system is operated in the region near $T(V) = 0.5$. The phase retardation Γ_0 can be adjusted either optically (by assisting the modulator with an additional phase retarder, a compensator) or electrically by adding a constant bias voltage to V.

In practice, the maximum transmittance of the modulator is smaller than unity because of losses caused by reflection, absorption, and scattering. Furthermore, the minimum transmittance is greater than 0 because of misalignments of the direction of propagation and the directions of polarization relative to the crystal axes and the polarizers. The ratio between the maximum and minimum transmittances is called the extinction ratio. Ratios higher than 1000 : 1 are possible.

5.5.5 Scanners

An optical beam can be deflected dynamically by using a prism with an electrically controlled refractive index. The angle of deflection introduced by a prism of small apex angle α and refractive index n is $\theta \approx (n-1)\alpha$. An incremental change in the refractive index Δn caused by an applied electric field E corresponds to an incremental change in the deflection angle

$$\Delta\theta = \alpha\Delta n = -\frac{1}{2}\alpha r n^3 E = -\frac{1}{2}\alpha r n^3 \left(\frac{\mathrm{V}}{d}\right) \tag{5.32}$$

where V is the applied voltage and d is the prism width (Figure 5.10a). By varying the applied voltage V, the angle $\Delta\theta$ varies proportionally, so that the incident light is scanned.

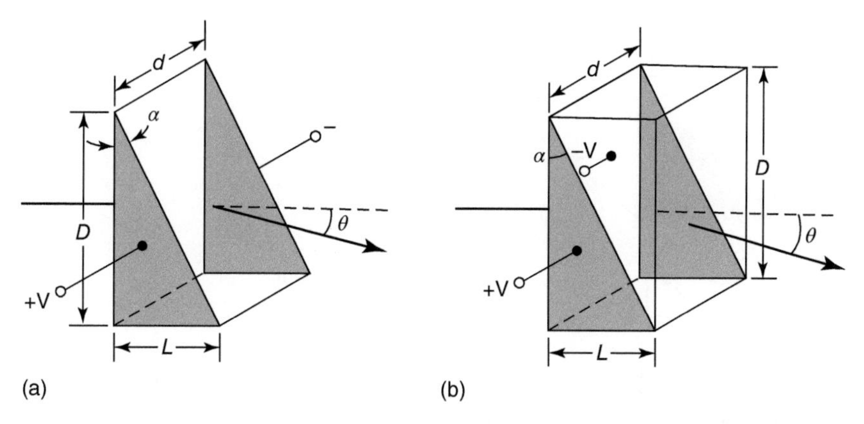

Figure 5.10 (a) An electro-optic prism. The deflection angle θ is controlled by the applied voltage V. (b) An electro-optic double prism. Source: Saleh 1991 [296]. Reproduced with permission of John Wiley & Sons.

It is often more convenient to place triangularly shaped electrodes defining a prism on a rectangular crystal. Two, or several, prisms can be cascaded by alternating the direction of the electric field, as illustrated in Figure 5.10b.

An important parameter that characterizes a scanner is its resolution, i.e. the number of independent spots it can scan. An optical beam of width D and wavelength λ_o has an angular divergence $\delta\theta \approx \lambda_o/D$. To minimize that angle, the beam should be as wide as possible, ideally covering the entire width of the prism itself. For a given maximum voltage V corresponding to a scanned angle $\Delta\theta$, the number of independent spots is given by

$$N \approx \frac{|\Delta\theta|}{\delta\theta} = \frac{\frac{1}{2}\alpha r n^3 \left(\frac{\mathrm{V}}{d}\right)}{\frac{\lambda_0}{D}} \tag{5.33}$$

Substituting $\alpha = L/D$ and

$$\mathrm{V}_\pi = \left(\frac{d}{L}\right)\left(\frac{\lambda_0}{rn^3}\right)$$

we obtain

$$N \approx \frac{\mathrm{V}}{2\mathrm{V}_\pi} \tag{5.34}$$

from which $\mathrm{V} \approx 2N\mathrm{V}_\pi$. This is a discouraging result. To scan N independent spots, a voltage $2N$ times greater than the half-wave voltage is necessary. Since V_π is usually large, making a useful scanner with $N \gg 1$ requires unacceptably high voltages.

The process of double refraction in anisotropic crystals introduces a lateral shift of an incident beam parallel to itself for one polarization and no shift for the other polarization. This effect can be used for switching a beam between two parallel positions by switching the polarization. A linearly polarized optical beam is transmitted first through an electro-optic wave retarder acting as a polarization

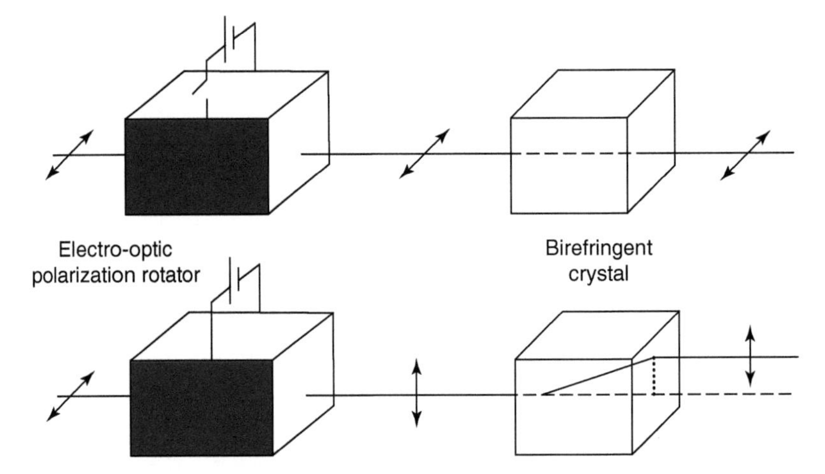

Figure 5.11 A position switch based on electro-optic phase retardation and double refraction. Source: Saleh 1991 [296]. Reproduced with permission of John Wiley & Sons.

rotator and then through the crystal. The rotator controls the polarization electrically, which determines whether the beam is shifted laterally or not, as illustrated in Figure 5.11.

5.5.6 Directional Couplers

The coupling between two parallel planar waveguides in an integrated-optical device can be used to transfer the light from one waveguide to the other, so that the device serves as an electrically controlled directional coupler.

The optical powers carried by the two waveguides, $P_1(z)$ and $P_2(z)$, are exchanged periodically along the direction of propagation z. The parameters governing the strength of this coupling process are the coupling coefficient C (which depends on the dimensions, wavelength, and refractive indices) and the mismatch of the propagation constants $\Delta\beta = \beta_1 - \beta_2 = 2\pi\Delta n/\lambda_0$, where Δn is the difference between the refractive indices of the waveguides. If the waveguides are identical, with $\Delta\beta = 0$ and $P_2(0) = 0$, then at a distance $z = L_0 = \pi/2C$, called the transfer distance or coupling length, the power is transferred completely from waveguide 1 into waveguide 2, i.e. $P_1(L_0) = 0$ and $P_2(L_0) = P_1(0)$, as illustrated in Figure 5.12a.

For a waveguide of length L_0 and $\Delta\beta \neq 0$, the power-transfer ratio $T = P_2(L_0)/P_1(0)$ is a function of the phase mismatch

$$T = \left(\frac{\pi}{2}\right)^2 \sin c^2 \left\{ \frac{1}{2} \left[1 + \left(\frac{\Delta\beta L_0}{\pi} \right)^2 \right]^{1/2} \right\} \tag{5.35}$$

where $\sin c(x) = \sin(\pi x)/(\pi x)$. Figure 5.12b illustrates the dependence of T as a function of $\Delta\beta L_0$. The power-transfer ratio T has its maximum value of unity at $\Delta\beta L_0 = 0$, decreases with increasing $\Delta\beta L_{0,}$ and vanishes when $\Delta\beta L_0 = \sqrt{3}\pi$, at which point the optical power is not transferred to waveguide 2.

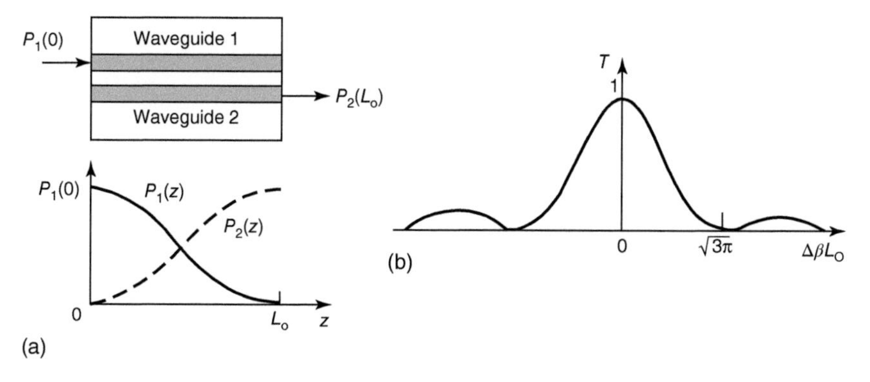

Figure 5.12 (a) Exchange of power between two parallel weakly coupled waveguides that are identical, with the same propagation constant β. At $z = 0$ all of the power is in waveguide 1. At $z = L_0$ all of the power is transferred into waveguide 2. (b) Dependence of the power-transfer ratio $T = P_2(L_0)/P_1(0)$ on the phase mismatch parameter $\Delta\beta L_0$. Source: Saleh 1991 [296]. Reproduced with permission of John Wiley & Sons.

A dependence of the coupled power on the phase mismatch is the key to making electrically activated directional couplers. If the mismatch $\Delta\beta L_0$ is switched from 0 to $\sqrt{3}\pi$, the light remains in waveguide 1. Electrical control of $\Delta\beta$ is achieved by use of the electro-optic effect. An electric field E applied to one of two, otherwise identical, waveguides alters the refractive index by $\Delta n = -\frac{1}{2}n^3rE$, where r is the Pockels coefficient. This results in a phase shift

$$\Delta\beta L_0 = \Delta n\left(\frac{2\pi L_0}{\lambda_0}\right) = -\left(\frac{\pi}{\lambda_0}\right) n^3 r L_0 E$$

A typical electro-optic directional coupler has the geometry shown in Figure 5.13. The electrodes are laid over two waveguides separated by a distance d. An applied voltage V creates an electric field $E = V/d$ in one waveguide and $-V/d$ in the other, where d is an effective distance determined by solving the electrostatics problem (the electric-field lines go downward at one waveguide and upward at the other). The refractive index is incremented in one guide and decremented in the other. The result is a net refractive index difference $2\Delta n = -n^3 r(V/d)$, corresponding to a phase mismatch factor

$$\Delta\beta L_0 = -\left(\frac{2\pi}{\lambda_0}\right) n^3 r \left(\frac{L_0}{d}\right) \mathrm{V}$$

which is proportional to the applied voltage V. The voltage V_0 necessary to switch the optical power is that for which $|\Delta\beta L_0| = \sqrt{3}\pi$, i.e.

$$\mathrm{V}_0 = \sqrt{3}\left(\frac{d}{L_0}\right)\left(\frac{\lambda_0}{2n^3 r}\right) = \left(\frac{\sqrt{3}}{\pi}\right)\left(\frac{c\lambda_0 d}{n^3 r}\right) \tag{5.36}$$

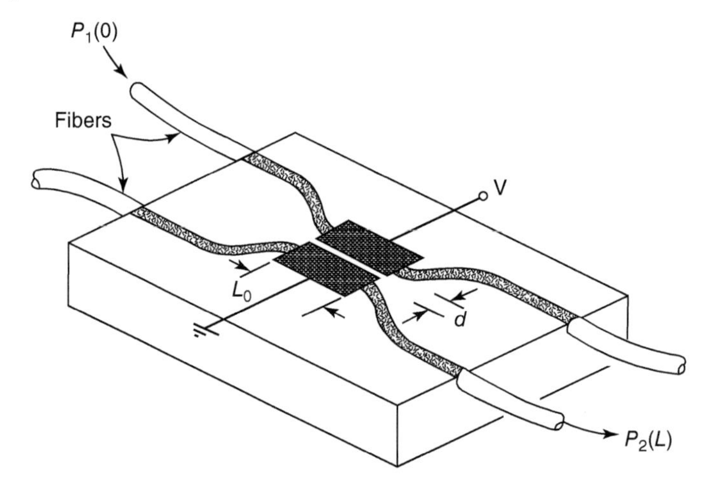

Figure 5.13 An integrated electro-optic directional coupler. Source: Saleh 1991 [296]. Reproduced with permission of John Wiley & Sons.

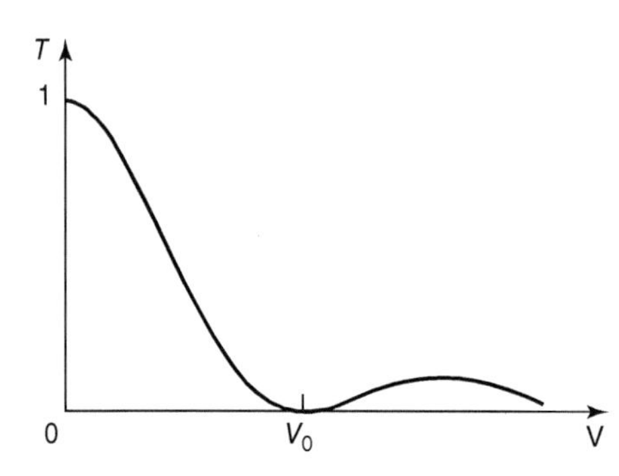

Figure 5.14 Dependence of the coupling efficiency on the applied voltage V. When $V = 0$, all of the optical power is coupled from waveguide 1 into waveguide 2; when $V = V_0$, all of the optical power remains in waveguide 1. Source: Saleh 1991 [296]. Reproduced with permission of John Wiley & Sons.

where $L_0 = \pi/2c$ and V_0 is called the switching voltage. Since $|\Delta\beta L_0| = \sqrt{3}\pi\left(\frac{V}{V_0}\right)$, Eq. (5.35) gives

$$T = \left(\frac{\pi}{2}\right)^2 \sin c^2 \left\{ \frac{1}{2}\left[1 + 3\left(\frac{V}{V_0}\right)^2\right]^{\frac{1}{2}} \right\} \tag{5.37}$$

This equation is plotted in Figure 5.14. It governs the coupling of power as a function of the applied voltage V. An electro-optic directional coupler is characterized by its coupling length L_0, which is inversely proportional to the coupling coefficient C and its switching voltage V_o, which is directly proportional to C. The key parameter is therefore C, which is governed by the geometry and the refractive indices.

Integrated-optic directional couplers may be fabricated by diffusing titanium into high-purity $LiNbO_3$ substrates. The switching voltage V_0 is typically under 10 V, and the operating speeds can exceed 10 GHz. The light beams are focused to spot sizes of a few micrometers. The ends of the waveguide may be permanently

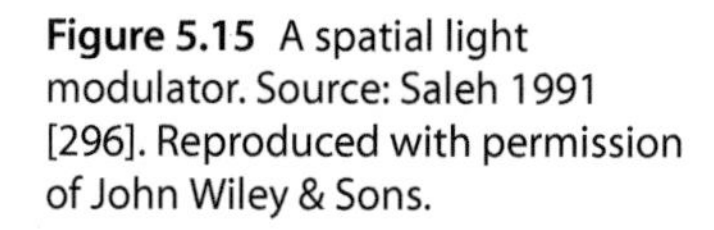

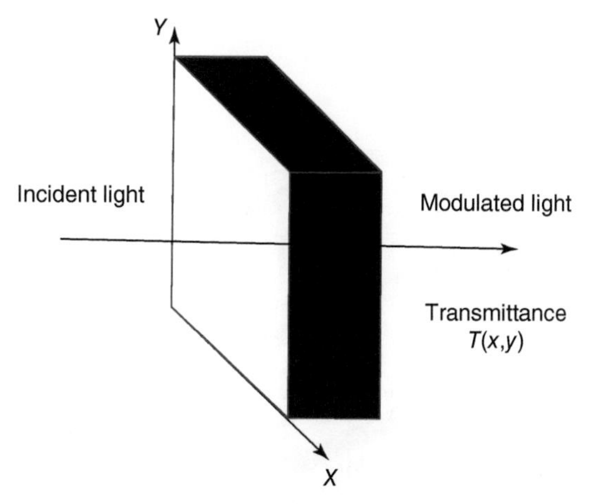

Figure 5.15 A spatial light modulator. Source: Saleh 1991 [296]. Reproduced with permission of John Wiley & Sons.

attached to single-mode polarization maintaining optical fibers. Increased bandwidths can be obtained by using a traveling wave version of this device.

5.5.7 Spatial Light Modulators (Electrically Addressed)

A spatial light modulator is an optical device of controllable intensity transmittance $T(x, y)$ that modulates the intensity of transmitted light at different positions by prescribed factors (Figure 5.15). The transmitted light intensity $I_0(x, y)$ is related to the incident light intensity $I_i(x, y)$ by the product $I_0(x, y) = I_i(x, y)$ $T(x, y)$. If the incident light is uniform (i.e. $I_i(x, y)$ is constant), the transmitted light intensity is proportional to $T(x, y)$. The "image" $T(x, y)$ is then imparted to the transmitted light, much like "reading" the image stored in a transparency by uniformly illuminating it in a slide projector. In a spatial light modulator, $T(x, y)$ is controlled by applying an appropriate electric field.

To construct a spatial light modulator using the electro-optic effect, some mechanism must be devised for creating an electric field $E(x, y)$ proportional to the desired transmittance $T(x, y)$ at each position. One approach is to place an array of transparent electrodes on small plates of electro-optic material placed between crossed polarizers and to apply on each electrode an appropriate voltage (Figure 5.16).

The voltage applied to the electrode centered at position (x_i, y_i) is made proportional to the desired value of $T(x_i, y_i)$. If the number of electrodes is sufficiently large, the transmittance approximates $T(x, y)$. The system is in effect a parallel array of longitudinal electro-optic modulators operated as intensity modulators. It is not practical to address a large number of these electrodes independently; nevertheless, this scheme is used in the liquid crystal spatial light modulators for display, since the required voltages are low.

5.5.8 Spatial Light Modulators (Optically Addressed)

The spatial light modulator (optically addressed) is based on the use of a thin layer of photoconductive material to create the spatial distribution of electric

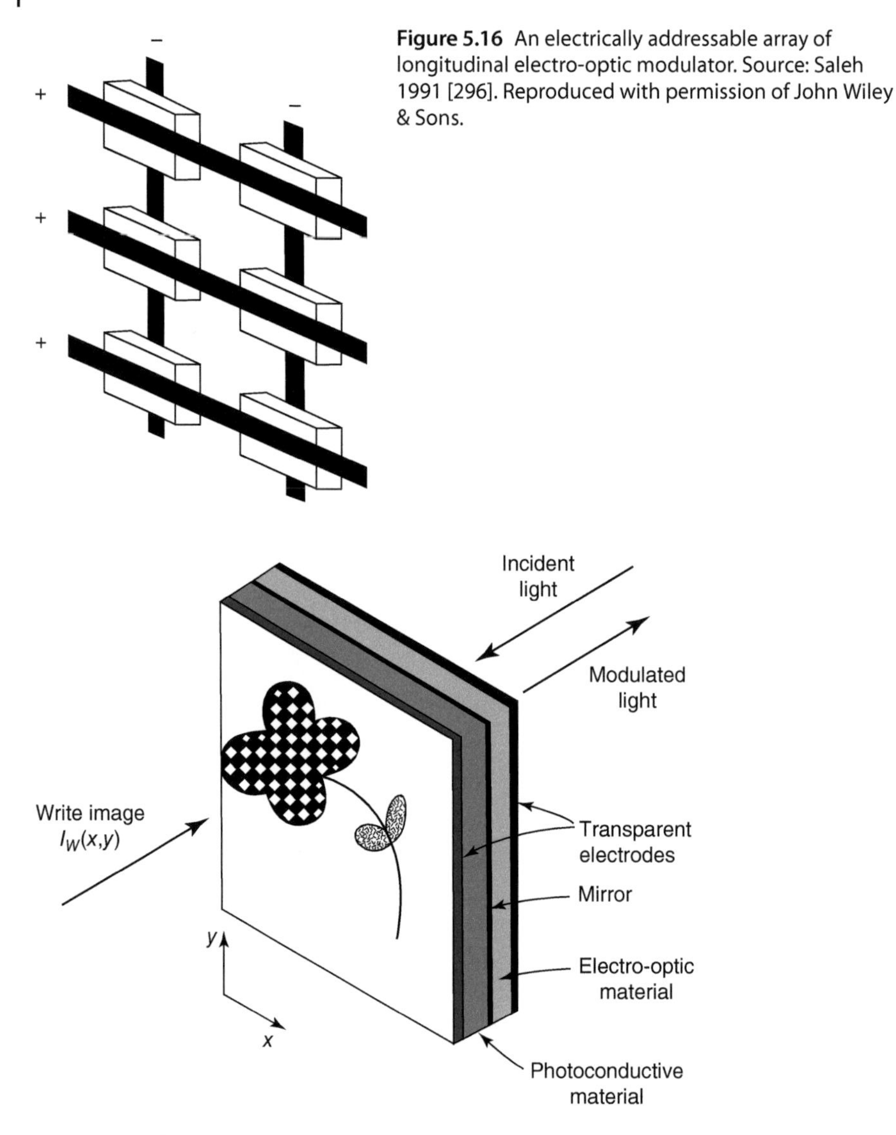

Figure 5.16 An electrically addressable array of longitudinal electro-optic modulator. Source: Saleh 1991 [296]. Reproduced with permission of John Wiley & Sons.

Figure 5.17 The electro-optic spatial light modulator uses a photoconductive material to create a spatial distribution of electric field, which is used to control an electro-optic material. Source: Saleh 1991 [296]. Reproduced with permission of John Wiley & Sons.

field required to operate the modulator (Figure 5.17). When the photoconductive material is illuminated by a light of intensity distribution $I_W(x, y)$, a spatial pattern of conductance $G(x, y)$ is created, which is directly proportional to $I_W(x, y)$, i.e. $G(x, y) \propto I_W(x, y)$. As a result, a spatial distribution of electric field is generated on the layer of the photoconductive material, which controls the transmittance $T(x, y)$ at each position.

To construct a spatial light modulator a layer of photoconductive material is placed between two electrodes that act as a capacitor. The capacitor is initially charged and the electrical charge leakage at position (x, y) is proportional to the local conductance $G(x, y)$. As a result, the charge on the capacitor is reduced in those regions where the conductance is high. The local voltage is therefore proportional to $1/G(x, y)$ and the corresponding electric field $E(x, y) \propto 1/G(x, y) \propto 1/I_W(x, y)$. If the transmittance $T(x, y)$ (or the reflectance $R(x, y)$) of the modulator is proportional to the applied field, it must be inversely proportional to the initial light intensity $I_W(x, y)$.

5.5.9 Pockels Readout Optical Modulator

The Pockels readout optical modulator (PROM) is an ingenious implementation of spatial light modulation technique. The device uses a crystal of bismuth silicon oxide, $Bi_{12}SiO_{20}$ (BSO), which has an unusual combination of optical and electrical properties: (i) it exhibits the electro-optic (Pockels) effect; (ii) it is photoconductive for blue light, but not for red light; and (iii) it is a good insulator in the dark. The PROM (Figure 5.18) is made of a thin wafer of BSO sandwiched between two transparent electrodes. The light that is to be modulated (read light) is transmitted through a polarizer, enters the BSO layer, and is reflected by a dichroic reflector, whereupon it crosses a second polarizer. The reflector reflects red light but is transparent to blue light. The PROM is operated as follows:

- *Priming*: A large potential difference (=4 kV) is applied to the electrodes and the capacitor is charged (with no leakage since the crystal is a good insulator in the dark).
- *Writing*: Intense blue light of intensity distribution $I_W(x, y)$ illuminates the crystal. As a result, a spatial pattern of conductance $G(x, y) \propto I_W(x, y)$ is created, the voltage across the crystal is selectively lowered, and the electric field decreases proportionally at each position, so that $E(x, y) \propto l/G(x, y) \propto 1/I_W(x, y)$. As a result of the electro-optic effect, the refractive indices of the BSO

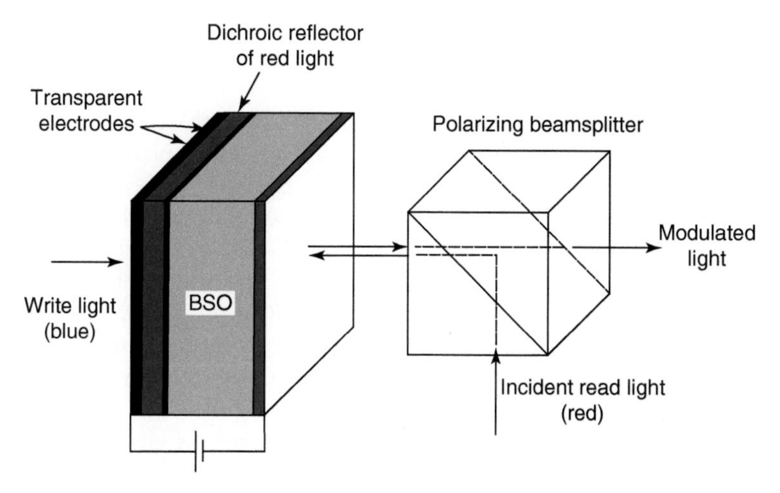

Figure 5.18 The Pockels readout optical modulator (PROM). Source: Saleh 1991 [296]. Reproduced with permission of John Wiley & Sons.

are altered, and a spatial pattern of refractive index change $\Delta n(x, y) \propto 1/I_W(x, y)$ is created and stored in the crystal.

- *Reading*: Uniform red light is used to read $\Delta n(x, y)$ as with the usual electro-optic intensity modulators with the polarizing beam splitter playing the role of the crossed polarizers.
- *Erasing*: The refractive index pattern is erased by the use of a uniform flash of blue light. The crystal is again primed by applying 4 kV, and a new cycle is started.

5.6 Electro-optic Materials and Applications

Technology has made the use of optically transparent material possible. These electro-optic materials find use in consumer electronics, industry, and defense. Electro-optic materials are recently developed materials that help transform electrical information into optical information. Also, their use enables certain optical functions to be carried out through commands in the form of electrical signals.

This new class of material is increasingly finding applications in industry, defense, and electronic devices such as optical shutters, image storage systems, color filters, modulators, displays, and optical data processing. There has been a slow but steady evolutionary growth of electro-optic materials owing to their low cost, high optical transparency, insensitivity to moisture, low power requirement, quick response time, memory capability, high electrical resistivity, dc operation, high electro-optic coefficients, uniform property characteristics of the material, etc. Some examples of electro-optic materials are discussed in detail below.

5.6.1 Barium Titanate ($BaTiO_3$)

Barium titanate ($BaTiO_3$) is one of the most extensively studied ferroelectric materials due to its salient features such as high dielectric constant, large electro-optic coefficients, high transparency in the visible and near infrared range, and its favorable growth characteristics. In 1990, Moretti et al. [62] has fabricated the first planar optical waveguides in $BaTiO_3$ by implantation of 2 MeV He^+ at a dose of 10^{16} cm^{-2}. Currently, $BaTiO_3$ crystals are widely used for nonlinear optical applications, which include optical waveguides [33, 35, 62–70], optical switching devices [71, 72], integrated optics [73–75], and electro-optic modulators [16, 38, 42, 76–88]. In 1963, R. C. Miller et al. first studied the SHG in $BaTiO_3$ and KDP crystals [15, 297]. The strong SHG from poled $BaTiO_3$ thin films, electron beam-induced poling of $BaTiO_3$ thin films, and third-order nonlinear optical studies on doped and un-doped $BaTiO_3$ films are reported in the literature [89–91].

High-transmittance $BaTiO_3$ is a suitable candidate for various electro-optic applications such as dielectric mirrors. Owing to the outstanding properties and capabilities of $BaTiO_3$, recently there have been many attempts to prepare various shapes and products of barium titanate [298–300]. In optics, to meet the miniaturization high-transmittance nanothin films are the most important requirements in many optical applications. In the literature, various deposition techniques such as electrochemical deposition [301], sputtering [302, 303], plasma evaporation [304], hydrothermal [305], solvothermal [306], and sol–gel [307–313] were used to prepare $BaTiO_3$ thin films. Preparing high-transmittance films in this manner can lead to better optical devices based on barium titanate. High-transmittance $BaTiO_3$ nano and ultrathin films have been prepared by Ashiri and Helisaie [314] and Guo et al. [315] for several optical and electro-optical applications.

Several types of electro-optic devices have been developed using $BaTiO_3$ thin films and some are described here.

5.6.1.1 Waveguide Electro-optic Modulator

A simple channel waveguide modulator has been fabricated in epitaxial $BaTiO_3$ on MgO substrate as shown in Figure 5.19 [33]. The films have an effective dc electro-optic coefficient of $r_{eff} \sim 50 \pm 5$ pm/V and $r_{eff} \sim 18 \pm 2$ pm/V at 5 MHz for $\lambda \sim 1.55$ μm light. The electro-optic effect decreases to ~60% of the dc value at 1 Hz, 50% of the dc value at 20 kHz, and ~37% of the dc value at 5 MHz. Epitaxial thin film $BaTiO_3$, therefore, offers the potential for low-voltage, highly confining, guided wave electro-optic modulator structures.

Thin films of $BaTiO_3$ with MgO buffer layers were deposited on patterned GaAs substrates incorporating Al_xO_y for optical confinement [35] as shown in Figure 5.20. The inclusion of Al_xO_y layers provides a means for obtaining thick optical confinement layers as a substitute for MgO cladding layers, which have large thermal expansion mismatch with respect to GaAs and $BaTiO_3$ that typically result in thin-film cracking. Deposition on patterned features was found to reduce thin-film cracking and is attributed to a reduction in thin-film stress resulting from thermal expansion mismatch. A maximum ridge width of 10–20 μm is estimated for 1 μm thick $BaTiO_3$ thin films. Optical waveguiding was observed in $BaTiO_3$/MgO/GaAs/Al_xO_y/GaAs ridges, suggesting the potential application of these structures for integrated optoelectronics.

Epitaxial thin film $BaTiO_3$ Mach–Zehnder modulators with two different film orientations (*c*-axis and *a*-axis) have been designed for optical switching [67] as shown in Figure 5.21. High-quality $BaTiO_3$ epitaxial thin films on MgO substrates have been grown by pulsed-laser deposition. The Mach–Zehnder optical waveguide modulators with a fork angle of 1.7° have been fabricated by ion-beam

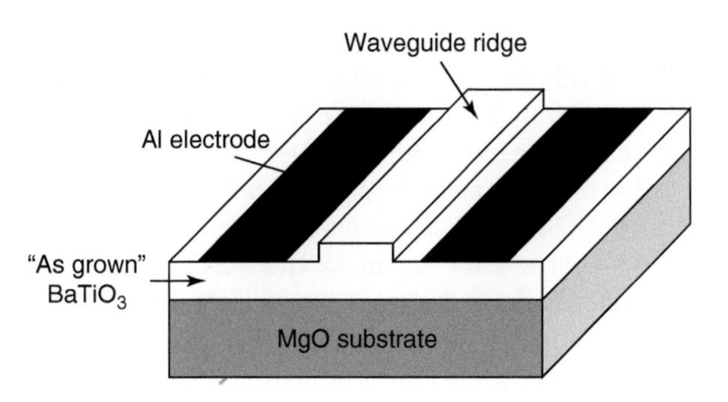

Figure 5.19 Schematic representation of the ridge waveguide modulator structure. Source: Gill et al. 1997 [33]. Reproduced with permission of AIP.

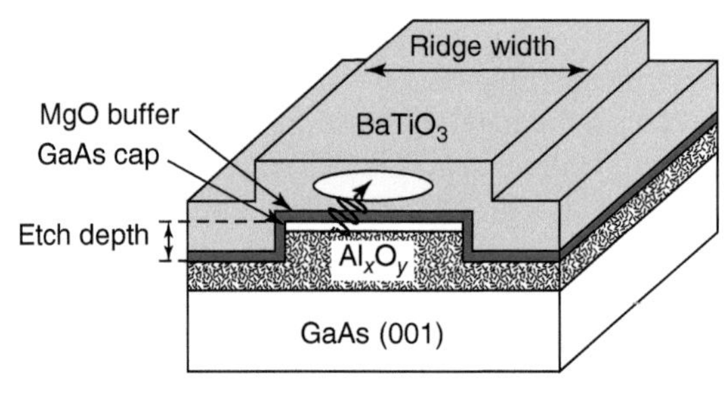

Figure 5.20 Schematic of $BaTiO_3$ ridge waveguide structure. Source: Chen et al. 2004 [35]. Reproduced with permission of AIP.

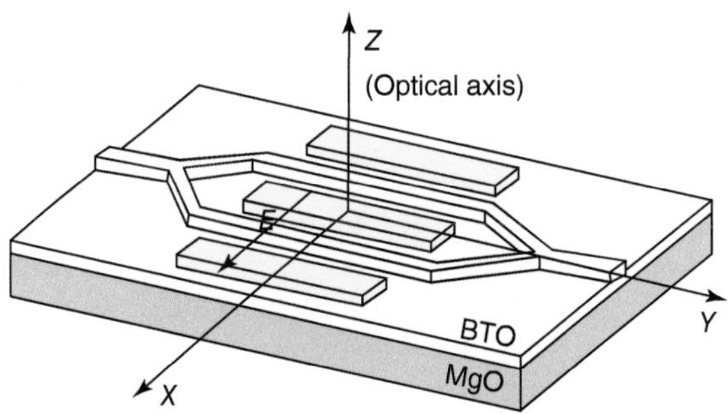

Figure 5.21 The geometry of the *c*-axis optical modulator for optical switching device. Source: Petraru et al. 2002 [67]. Reproduced with permission of Cambridge University Press.

etching. The waveguides are of the ridge type, the $BaTiO_3$ thickness is 1 mm, the ridge step is 50 nm, and the width is 2 mm. Light was coupled into the waveguides from optical fibers. The $BaTiO_3$ waveguide propagation losses are 2–3 dB/cm. The low V_π voltage at 1550 nm wavelength and the short electrodes (~3 mm) make these $BaTiO_3$ thin-film optical modulators attractive candidates for practical application in optics communications (switching device).

The high-frequency operation of a low-voltage electro-optic modulator based on a strip-loaded $BaTiO_3$ thin-film waveguide structure has been designed by Tang et al. [38]. A schematic cross section of the thin film electro-optic waveguide modulator is shown in Figure 5.22a. The waveguide has a strip-loaded structure for low-loss (<1 dB/cm) and single-mode operation [316]. The modulator was fabricated from a 570 nm thick epitaxial $BaTiO_3$ thin film grown on a (100) oriented MgO substrate by a two-step low-pressure metal–organic chemical vapor deposition (MOCVD) process [317]. The Si_3N_4 strip-loaded layer was 4 μm wide and 125 nm thick, fabricated by standard plasma enhanced chemical vapor deposition and reactive ion etching processes [316]. The waveguide was aligned to the <110> crystallographic direction of the MgO substrate. The waveguide pattern was transferred by reactive ion etching of the Si_3N_4 layer using CF_4-based chemistry. A lithographic lift off process with subsequent deposition of 15 nm Cr and 350 nm Au by E-beam evaporation was used to obtain the coplanar electrodes. The electrode length and gap were 3.2 mm and 8 μm, respectively.

The applied electric signal and output modulated signal are shown in Figure 5.22b. The 3.2 mm long modulator exhibits a low half-wave voltage of 3.9 V at a 2.0 V dc bias. The −3 dB electrical bandwidth is 3.7 GHz. The calculated effective electro-optic coefficient is as high as 150 pm/V at 1561 nm wavelength. An effective microwave index as low as 3.3 at 40 GHz was measured for the thin-film modulator. Broadband modulation up to 40 GHz was

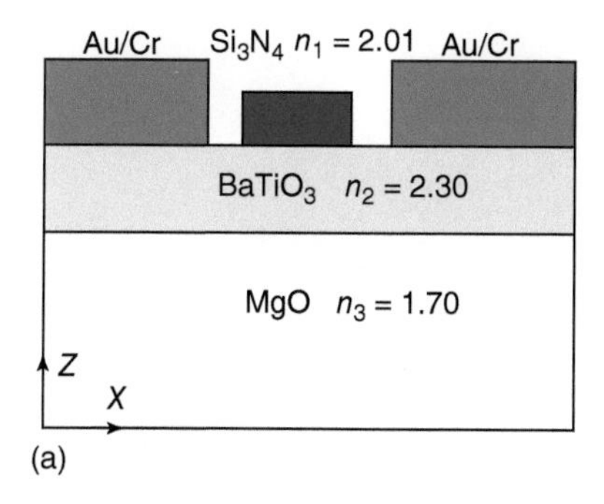

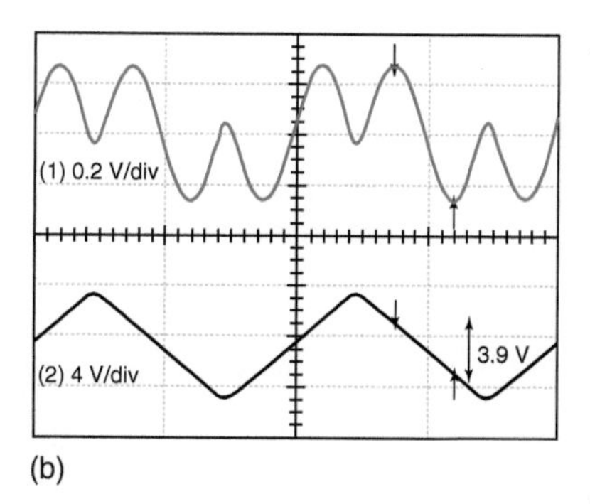

Figure 5.22 (a) Schematic cross section of the electro-optic waveguide modulator. (b) Low-frequency electro-optic modulator performance at 1561 nm wavelength. Applied 1 kHz triangle-driving voltages with 2 V dc bias on 3.2 mm long electrode (bottom trace, 4 V/div) and modulation output signal (top trace, 0.2 V/div). Source: Tang et al. 2004 [38]. Reproduced with permission of OSA Publishing.

observed. Numerical simulation indicates that compact, low-power electro-optic waveguide modulators with a 40 GHz 3 dB bandwidth might be possible using the high EO coefficient of $BaTiO_3$ thin films. The FoM f for $BaTiO_3$ thin film modulator is ~30 GHz/V^2, much larger than that for $LiNbO_3$ modulators, 1 GHz/V^2 [318].

Pernice et al. [319] proposed a novel design concept for compact electro-optic modulators based on a horizontally slotted ridge waveguide using ferroelectric barium titanate ($BaTiO_3$) as the electro-optically active material. A low voltage–length product is achieved by concentrating the propagating electromagnetic fields in the $BaTiO_3$ layer. Thus, high overlap between the modulation electric fields and the guided light is achieved. The proposed waveguide structure features low propagation loss and can be easily fabricated with standard silicon fabrication techniques.

Figure 5.23 shows the layout of the proposed modulator. The device is assumed to be fabricated from standard silicon-on-insulator (SOI) wafers. A thin layer of BTO can then be grown by molecular beam epitaxy onto the top silicon layer [320]. The growth process for such layers has been carefully optimized and single-crystalline layers can be fabricated on a large scale [321]. Subsequently, a layer of amorphous silicon (a-silicon) is deposited onto the BTO layer. By etching the waveguides into the top silicon layers a ridge waveguide with integrated horizontal slot is thus realized. In this case, much of the propagating light will be concentrated in the BTO layer and thus interacts strongly with the driving modulation field. The use of a horizontal slot waveguide does not require patterning of the underlying BTO layer, which is challenging [322]. Instead, only the top a-silicon layer needs to be etched with established silicon photonics fabrication approaches. To excite the modulating fields metal electrodes are fabricated alongside the waveguide. The electrodes are made from gold, which can be conveniently modeled by numerical techniques.

This waveguide structure has the advantage that the optical properties can be controlled precisely by monitoring the growth of the BTO layer. High field concentration within the slot medium leads to good overlap between the

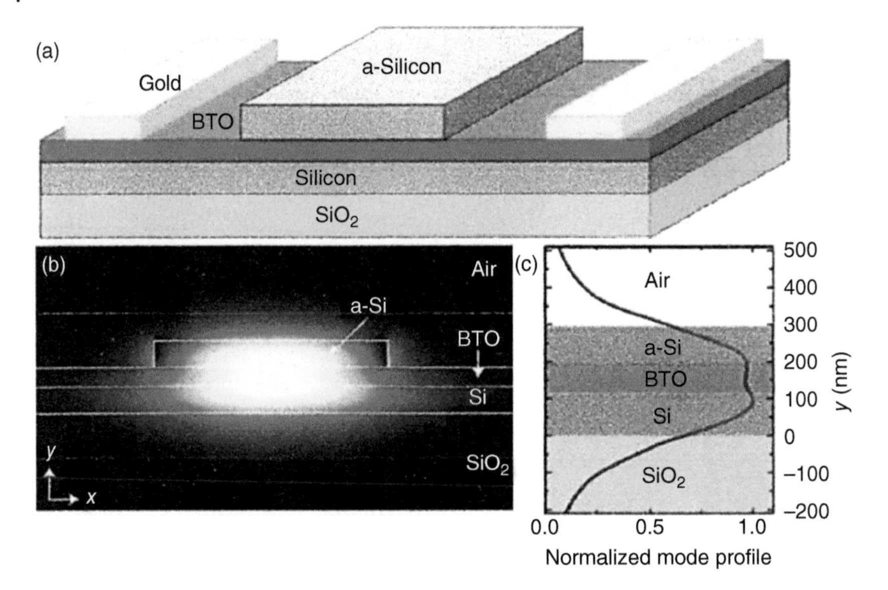

Figure 5.23 (a) Schematic of the electro-optic modulator based on BTO on SOI. Layers of epitaxial BTO and a-silicon are deposited on top of the upper silicon layer. A ridge waveguide with integrated horizontal slot is realized by structuring the top a-silicon layer. Gold electrodes are fabricated alongside the ridge to provide the modulating electric fields. (b) A plot of the E_x (electric field component in the x-direction) component of the optical mode confined in the Si-BTO-a-Si structure. Here, the thickness of both silicon layers is set to 110 nm and the BTO layer is assumed to be 70 nm thick. The waveguide in the above simulation is 900 nm wide. Strong field concentration in the BTO layer is found. (c) A cross-section view of the mode profile in the slotted ridge waveguide quantifies the significant field strength in the BTO layer. Source: Pernice et al. 2014 [319]. Reproduced with permission of IEEE.

modulating electric fields and the propagating electromagnetic fields. By properly engineering the waveguide parameters, modulators shorter than 2 mm are feasible for a switching voltage of 1.0 V. Following the simulation results, the length of the modulator can be reduced by reducing the width of the ridge, increasing the thickness of the BTO layer, and reducing the thickness of the a-Si layer.

5.6.1.2 Plasmonic Interferometer

The control of the SPP wave vector in an active metal-dielectric plasmonic interferometer was demonstrated by utilizing electro-optic $BaTiO_3$ as the dielectric layer. A schematic of the plasmonic interferometer based on double-slit transmission and electro-optic modulation of the SPP wave vector is shown in Figure 5.24 [42].

The structures consist of pairs of parallel slits etched into the metal layer by focused ion beam (FIB) milling (field electron and ion [FEI] Nova 600 dual beam FIB system, Ga+ ions, 30 keV). The slits are 5 μm long by 100 nm wide with slit pitch starting at 500 nm and increasing by 20 nm for each device along a row. Each device is laterally separated by 5 μm. Electrical contacts are made to the strontium ruthenate ($SrRuO_3$) film by mechanically etching the $BaTiO_3$ film and contacting the strontium ruthenate using conventional silver paste. The final device is mounted onto a glass slide and copper tape is used to contact the top chromium/silver layer and bottom strontium ruthenate layer.

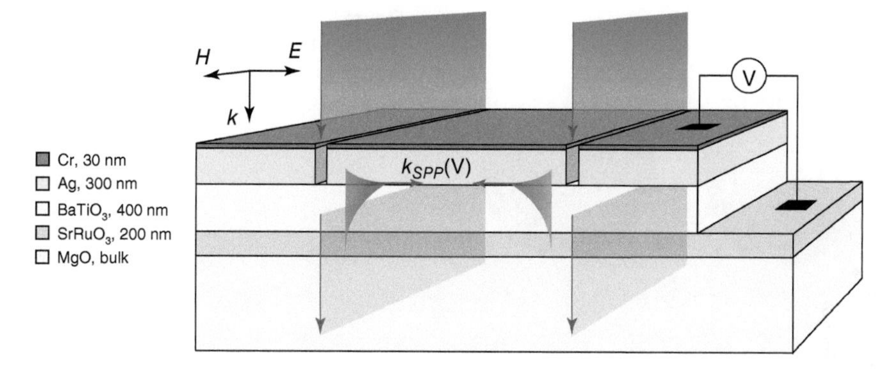

Figure 5.24 Schematic of a plasmonic modulator based on interference of SPPs launched from a set of parallel slits milled into a planar metal film. Source: Dicken et al. 2008 [42]. Reproduced with permission of ACS.

Plasmon-mediated transmission of incident light through the subwavelength slits is modulated by an external voltage applied across the $BaTiO_3$ thin film. Transmitted light modulation is ascribed to two effects, electrically induced domain switching and electro-optic modulation of the barium titanate index. Electrical modulation of the SPP wave vector was achieved by utilizing the electro-optic effect as well as 90° domain switching in barium titanate. The degree of optical switching obtained in these devices is potentially useful for designing new plasmonic and metamaterial structures in which active oxide replaces a static dielectric material. As photonic networks become more prevalent in chip-based microelectronic systems, the need for active nanoscale devices is increasingly apparent. Active plasmonic devices, based on electro-optic modulation, are well suited to fill this nanophotonic niche.

5.6.2 Lead Lanthanum Zirconate Titanate (PLZT)

Since the late 1960s, when transparent PLZT materials were first developed, ferroelectric ceramics have been thoroughly researched, and their characteristics studied to the point that they have now taken their place alongside single crystals as legitimate candidates for electro-optic applications. It was achieved for the first time in 1969 with a precise formulation of 8-atom percent lanthanum (La) in a 65 : 35 ratio of $PbZrO_3$ and $PbTiO_3$.

In 1967, C. E. Land [17] designed ferroelectric PLZT ceramic electro-optic storage and display devices. The utilization of PLZT ceramic elements in electro-optic devices of the fast-shutter type has been described by G. H. Haertling [19, 92–94] and J. T. Cutchen et al. [95]. Such an arrangement is used in eye-protective systems against thermal and flash blindness developed by Sandia National Laboratory for the US Air Force. The PLZT shutters are used in home entertainment, flight simulators, medical imaging, contour mapping, radar and sonar displays, electron microscopy, etc. G. B. Trapani [323] designed a rigidly bonded electro-optic phase retardation device using PLZT ceramic element. Ferroelectric PLZT thin film optical waveguides and memory display devices have been designed by S. Miyazawa and N. Uchida [20], A. Baudrant et al. [96], C. E. Land [97], M. Bazylenko and I. K. Mann [98], J. D. Levine and S. Essaian [99], and S. L. Swartz [100].

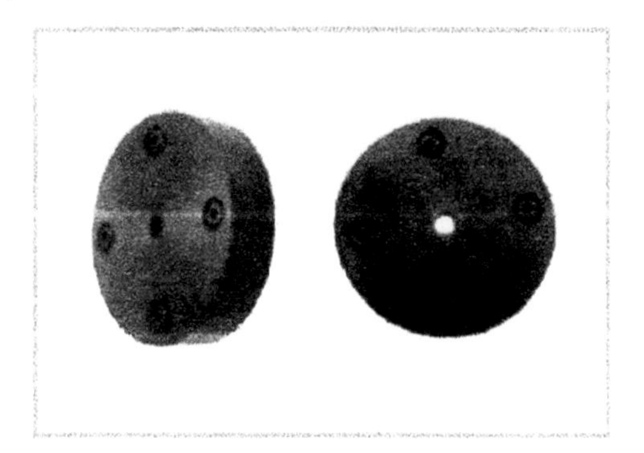

Figure 5.25 TE1000 electro-optic tunable etalon. Source: Reproduced with permission of AC Photonics [324].

A number of electro-optic devices have been developed using PLZT materials and some are described here.

5.6.2.1 Electro-optic Tunable Etalon

AC Photonics TE1000 tunable optical etalon incorporates a novel monolithic PLZT electro-optic chip, as shown in Figure 5.25 [324]. The refractive index of the PLZT chip is controlled and changed electronically. As a result, optical free spectral range (FSR) shifting is fast and accurately controlled. The device has been designed and optimized for a wide range of FSRs with minimum optical loss. It is based on tried and tested PLZT technology with proven reliability. The devices are packaged into compact low profile disks designed to be used in 1″ diameter optical mirror and lens mounts. They are compatible with standard fiber-optic collimators and external free space optics. The TE1000 can be used in interferometers, optical resonators, and lasers.

5.6.2.2 Nanosecond Speed PLZT Optical Switch

The PLZT optical switch module and PLZT optical switch subsystem are powerful tools to switch optical signal at nanosecond speed. The PLZT optical switch subsystems are equipped with PLZT switches and high-speed drivers in 19″ cases as shown in Figure 5.26 [40].

With the efficient electro-optic properties and high refractive index of PLZT crystals, EpiPhotonics' [40] unique PLZT waveguide technology enables a new generation of efficient optical switch systems with potent advantages such as the following:

- Ultrafast switching (10 ns)
- Low driving voltage (5–10 V)
- Low insertion loss
- Low polarization dependence
- Low power consumption
- Miniaturization

In addition, the PLZT products are very reliable and environmentally stable. The crystal structure of the PLZT materials and the fact that the optical switches have absolutely no moving parts guarantee high reliability and stability.

- High reliability (high-power laser admissive)
- Environmentally stable

Figure 5.26 Nanosecond speed PLZT optical switch [40].

5.6.3 Lithium Tantalate ($LiTaO_3$)

Lithium tantalate ($LiTaO_3$) is a ferroelectric crystal isostructural with $LiNbO_3$. It is widely used in electro-optic [101, 103, 325] and integrated optics [34, 104, 326–328] devices. Many of the applications depend on the shaping of antiparallel (180°) ferroelectric domains [102, 329, 330]. This has driven the recent interest in the study of domain processes in this material. The low T_c (=610 °C) of lithium tantalate precludes most metal indiffusion techniques for waveguide fabrication in quasi-phase-matched devices. In the past, several research groups have demonstrated quasi-phase-matched SHG of blue and ultraviolet light in $LiTaO_3$ waveguides periodically poled by patterned proton exchange [27–29, 105]. SHG of blue light has also been reported in waveguides poled by application of periodic fields with interdigital electrodes [30].

Different types of electro-optic devices have been developed using $LiTaO_3$ crystals and one of them is described here.

5.6.3.1 Second Harmonic Generator, EO Lens, and EO Scanner

An integrated optical device was designed by Chiu et al. [34] in z-cut $LiTaO_3$ that contains the following three functional parts: a quasi-phase-matched SHG grating, an electro-optic (EO) lens, and an electro-optic scanner. Figure 5.27 shows the schematic of the device. The first section is a quasi-phase-matched SHG grating. The grating consists of periodically inverted ferroelectric domains with a period of $\Lambda = 3.5$ μm. Channel waveguides, 4.5 μm wide and perpendicular to the grating, are used to confine the light and increase the power density and conversion efficiency. The phase matching wavelength λ of the input infrared (IR) is related to the grating period by $\lambda = 2\Lambda(n_e^{2\omega} - n_e^{\omega})$ where n_e is the mode index in the waveguide at the indicated frequencies. After the channel waveguide in the SHG section, the light enters a planar waveguide region where the EO lenses and EO prism scanners are fabricated. The lenses collimate the rapidly diverging beam from the channel waveguides [331, 332] and can also be used for focus control in an optical disk system. The collimated beam then enters an EO prism scanner and the output angle can be controlled by an applied voltage [102, 333]. Both the lens

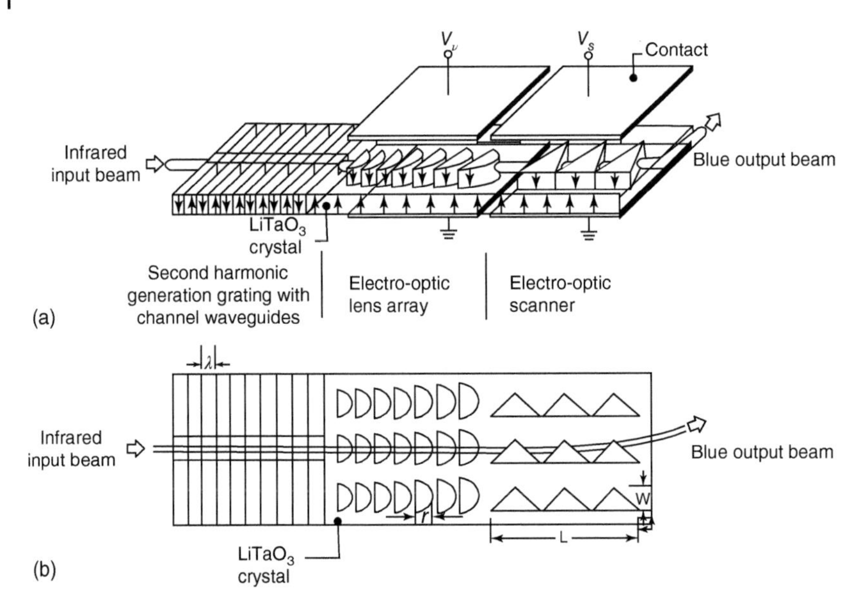

Figure 5.27 Device schematic (a) top view and (b) perspective view. Source: Chiu et al. 1999 [34]. Reproduced with permission of Journal of Lightwave Technology.

and the scanner consist of photolithographically defined and domain-inverted regions in the substrate. There are in total 43 semicircular inverted domains for the lens stack. They are spaced 5 μm apart and the radii (r) range from 75 to 195 μm. The scanner is divided into seven smaller triangular domains with 1 mm bases and 0.48 mm heights. The total dimensions of the scanner are, therefore, 7 mm (L) × 0.48 mm (W). When a uniform voltage is applied across the thickness of the device, the lens and prism regions have different indices of refraction from the surrounding areas. Therefore, the focusing power or scanning angle can be controlled by varying the applied voltage.

The final device configuration and test setup are shown in Figure 5.28. Waveguide characteristics were tested in both channel and planar waveguides by m-line spectroscopy. To test the SHG performance, a Ti-sapphire laser was used as the input IR source. A color filter was used in front of the detector to block the IR light when measuring blue light. The measured IR power was 60 mW. The SHG tuning curve is shown in Figure 5.29. The matching peaks of the two phases were due to multiple modes in the channel waveguides.

The performance of the lens and scanner sections was characterized by imaging the output facet of the device with an objective lens onto a charge-coupled device (CCD) camera. Since the domain inversion was carried out through the thickness of the substrate, the lens and scanner can still function as a wafer device, instead of a waveguide device.

Figure 5.30a shows the operation of the lenses for various applied voltages. As can be seen in these images, the focusing power of the lens increases with the applied voltage due to the increased index difference. Figure 5.30b shows the

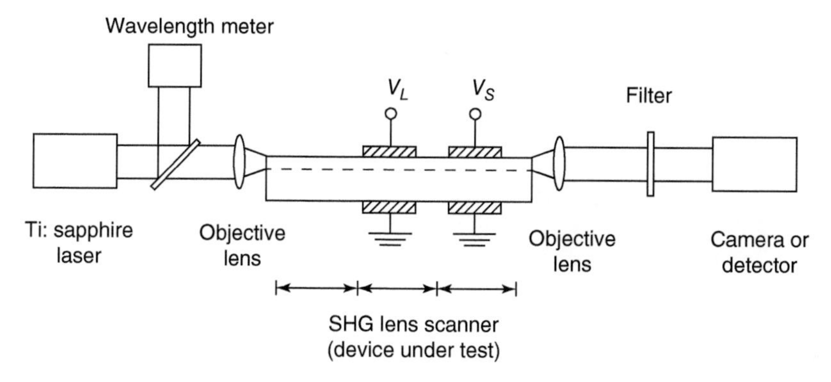

Figure 5.28 Device configuration and test setup. Source: Chiu et al. 1999 [34]. Reproduced with permission of Journal of Lightwave Technology.

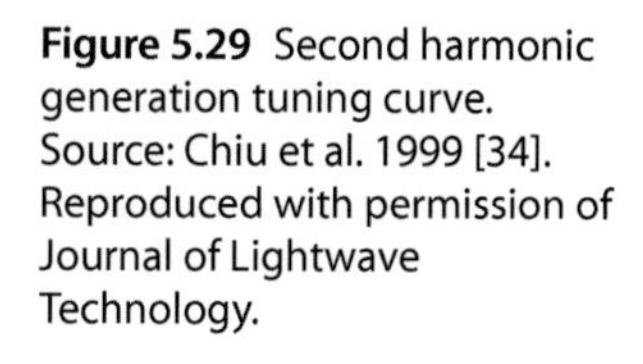
Figure 5.29 Second harmonic generation tuning curve. Source: Chiu et al. 1999 [34]. Reproduced with permission of Journal of Lightwave Technology.

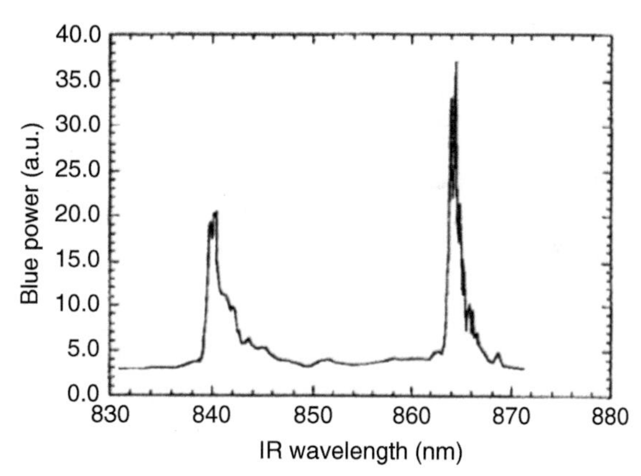

intensity profiles integrated over the thickness of the sample to reduce the scattering seen in Figure 5.30a. The measured $1/e^2$ beam diameter in another device was compared to an (fast fourier transform) FFT-based beam-propagation method simulation [334] and the result is shown in Figure 5.31. The agreement between the measured and calculated values is good. Figure 5.30b also shows beam profiles with voltages applied to both the lenses and the scanner.

The voltage on the lenses was −1200 V and that on the scanner was 1000 V. The measured scanning sensitivity was 17 mrad/kV, compared to the calculated value inside the wafer $\theta = (\Delta n/n)L/W = 15$ mrad/kV for scanner geometry ($L = 7$ mm, $W = 0.48$ mm, $d = 150$ μm).

Compared to the device of [103] where the SHG process takes place in the bulk crystal, the SHG grating in the channel waveguide has higher conversion efficiency. The EO lens stack is used to collimate the light from the channel waveguide. All three components are fabricated in waveguides so that the light remains confined in the entire device. This is the first demonstration of such an integrated device.

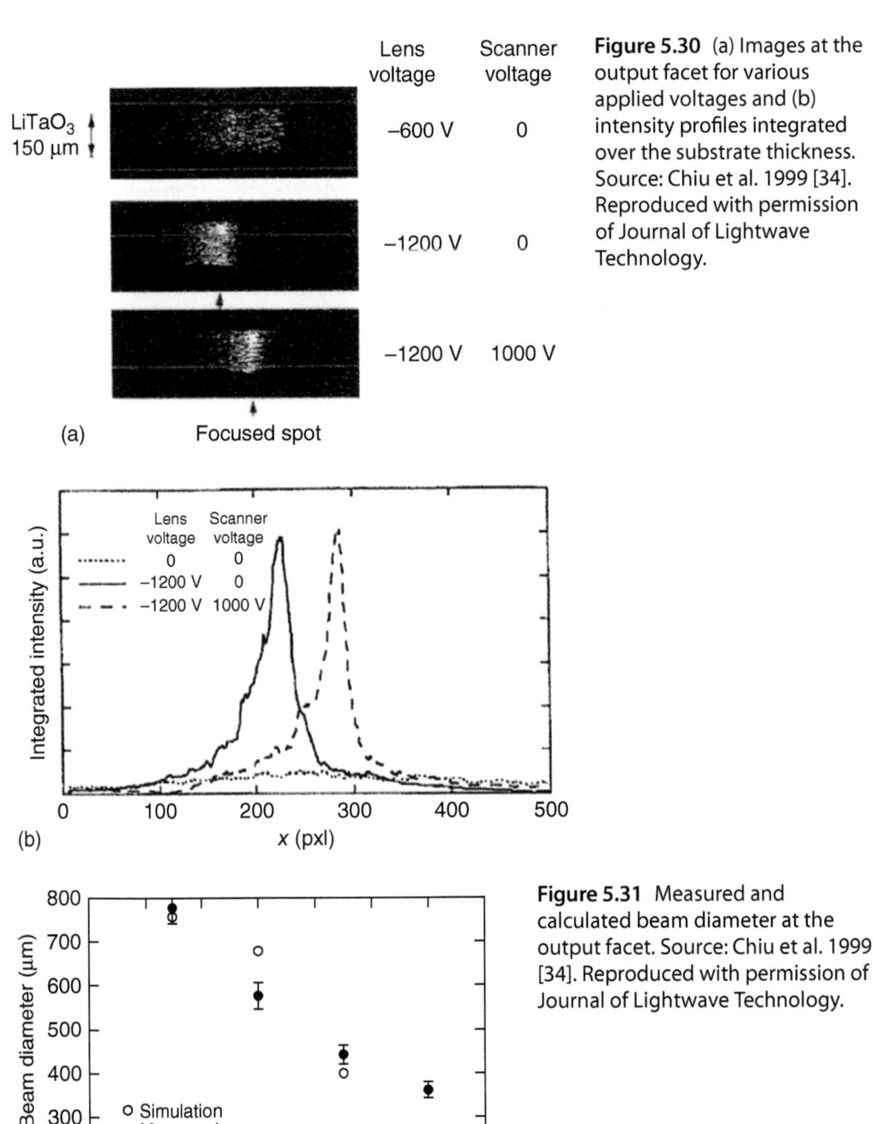

Figure 5.30 (a) Images at the output facet for various applied voltages and (b) intensity profiles integrated over the substrate thickness. Source: Chiu et al. 1999 [34]. Reproduced with permission of Journal of Lightwave Technology.

Figure 5.31 Measured and calculated beam diameter at the output facet. Source: Chiu et al. 1999 [34]. Reproduced with permission of Journal of Lightwave Technology.

5.6.4 Lithium Niobate ($LiNbO_3$)

$LiNbO_3$ has become a very attractive material for integrated optical applications because of its excellent electro-optical and nonlinear optical properties. The fabrication of waveguides in $LiNbO_3$ crystals by means of proton exchange (PE) dates back to 1982 [26] and since then extensive work has been carried out on its many possible implementations [117–123]. The PE process has proved to be a simple and effective method of fabricating low-loss waveguides in $LiNbO_3$ crystals that has the advantage, with respect to the Ti-indiffusion method [21, 110–115], of a much higher photorefractive damage threshold [116] and much lower process temperatures. Schmidt and Kaminow [124] designed metal-diffused optical waveguides in $LiNbO_3$. Thomson et al. [125] designed a z-cut $LiNbO_3$ optical waveguide using femtosecond pulses in the low repetition rate regime. P. Rabiei and P. Günter [126] designed submicrometer $LiNbO_3$ slab waveguides by crystal ion slicing and wafer bonding. R. Geiss et al. [127] designed $LiNbO_3$ photonic crystal waveguide. H. Lu et al. [128] designed $LiNbO_3$ photonic crystal wire optical waveguide on a smart-cut thin film.

$LiNbO_3$ waveguides are already widely used in many functional electro-optic devices [24, 31, 32, 129–146, 148, 335]. In recent years, they have also been emerging as the most promising candidates for the development of all-optical devices based on quadratic nonlinearities, due to substantial technological advances in ferroelectric domain microstructuring (the "poling" technology) that provided an effective way to implement the quasi-phase-matching (QPM) principle [336]. Progress in waveguide and poling technologies has increased the efficiency of quadratic ($\chi^{(2)}$) interactions in $LiNbO_3$ by more than a factor of 10^4 since the initial demonstrations of blue and green light generation [149, 150]. Nowadays, the best nonlinear optical devices in $LiNbO_3$ [151–154, 337] have reached efficiencies in excess of 3000%/W or 100%/W/cm^2. The variety of applications of $\chi^{(2)}$ devices has also increased: periodically poled $LiNbO_3$ (PPLN) waveguides can not only provide coherent sources ranging from the ultraviolet [155] through to the MIR [338–340] but also signal processing devices for optical communications [341]. Recently, a UV laser direct writing method for channel waveguide fabrication has been proposed [342] along with detailed characterization [343, 344]. More specifically, a 55% increase in the r_{33} coefficient was observed compared to the bulk material properties in UV-written $LiNbO_3$ waveguides that have been subjected to polling inhibition [345, 346].

5.6.4.1 Application of $LiNbO_3$

The electro-optic effect describes how the index of refraction is changed with the application of an electric field. In a bulk optics application, a Pockels cell rotates the polarization of the light when a voltage is applied. This device is used in Q-switching Nd:YAG lasers. A schematic of such a laser cavity is shown in Figure 5.32. With no applied voltage, the $LiNbO_3$ (LN) cell does not affect the polarization state of the light because the light propagates along the optic axis.

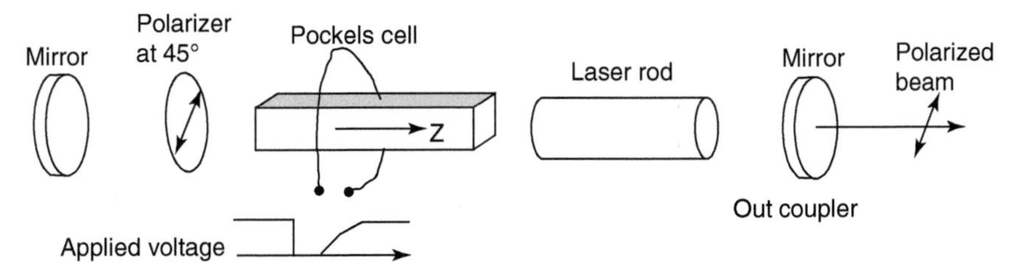

Figure 5.32 Schematic of laser cavity using $LiNbO_3$ as Q-switch. Source: Reproduced with permission of Crystal Technology [147].

The laser will operate continuously with output polarization aligned with the polarizer as shown in the Figure 5.32. When the voltage is applied, slight changes in refractive index are induced for light polarized along and perpendicular to the electric field. For the correct voltage, the relative phase retardation of the two polarization components for a double-pass through the cell is 180°. In that case, light exiting the polarizer and propagating to the right will return at a polarization state that is 90° rotated and thus completely rejected by the polarizer, making laser oscillation impossible. In a typical application, this voltage is applied while the flash lamps pump the laser medium. At the time of maximum gain in the laser rod, the electrodes are shorted, dropping the voltage to zero and enabling the laser radiation to build up so that all the stored energy is extracted from the laser rod into the short pulse.

With the increased demand for bandwidth on fiber optic communication lines, $LiNbO_3$ has gained importance in various integrated optic components [347, 348]. Among those, the high-speed electro-optic modulator, fabricated on an X- or a Z-wafer, is the most prevalent. The advent of erbium fiber amplifiers has increased the distance between electrical–optical converters and has also enabled wavelength division multiplexing (WDM) where dozens of digital signal streams are launched into the same fiber by using slightly different wavelengths for each of them. The $LiNbO_3$ modulators are well matched to the requirements for WDM systems because they enable fast (40 GHz or more) modulation speed at reasonable voltage levels with very low pulse chirp [349].

The basic layout of an intensity modulator is shown in Figure 5.33. Light is coupled from an incoming fiber (from a continuously emitted laser at a particular wavelength) into the waveguide. At the first Y-junction, the energy is split evenly into the waveguide paths. At the second Y-junction, the energy is recombined into the output waveguide and to the fiber. With no voltage applied, the two waves experience the same optical path length in the two arms of the interferometer and the recombination is perfect so that all the light is coupled into the receiving fiber. If a voltage is applied to the electrodes, the electrical field inside the crystal at the waveguide location causes an index change and the path length for that arm is changed, causing a relative phase difference of the two optical waves at the second Y-junction. If the voltage applied is V_{π}, the induced phase change is π and the two fields from the waveguides are out of phase so that the energy gets radiated into the substrate and no light is coupled into the waveguide or the fiber.

High-speed operation of the modulators requires velocity matching of the electric radio frequency (RF) wave with the propagating optical wave. The dielectric constant in $LiNbO_3$ is large (~30), so the electrodes need to be electroplated to

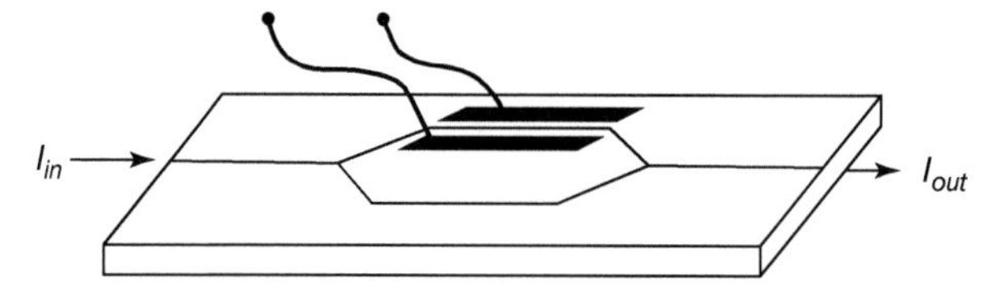

Figure 5.33 Schematic of integrated optical modulator based on a Mach–Zehnder interferometer. Source: Reproduced with permission of Crystal Technology [147].

have a significant amount of energy traveling in air, speeding up the traveling microwave. Actual designs can be quite complex, requiring various processing steps such as buffer layer deposition or reactive $LiNbO_3$ etching [350].

Over the years, a number of highly integrated electro-optic devices have been designed using $LiNbO_3$ thin films, and some of them are described here.

$LiNbO_3$ Modulators for Space Environment Photonics systems, subsystems, and components are found in an increasing number of applications of many high technology industry segments. This remark applies particularly to space-embarked systems, which rely on the versatility and reliability of photonic systems to realize many of the critical functions needed to ensure their safe and durable operation.

Many embarked space photonic systems use light modulators as a key component to achieve intensity or phase modulation of various light sources at different operating wavelengths. In particular, the electro-optic $LiNbO_3$ modulators offer a unique combination of performance that makes them prime candidates not only to satisfy the optical system specifications but also to meet the tough requirements of space operation.

The $LiNbO_3$ based modulator is one of the many optical modulators that have been developed in recent years. Other materials (e.g. InP, GaAs) based modulators have been also used to make external light modulators. Initially, the development of these modulators was driven by the fiber-optic telecommunication market, which needed ever-increasing modulation speeds. Today, EO modulators are used in a large number of both telecom and non-telecom applications. For space applications, the accumulated number of hours of operation and the proven reliability of $LiNbO_3$ modulators make them a very attractive choice compared with products issued from competing technologies.

In addition, the $LiNbO_3$ modulators, beside their long-standing proven record of use in many applications, and their many comprehensive successful qualifications (e.g. Telcordia) offer both a large optical bandwidth ranging from 780 to 2500 nm and a very broad electro-optic modulation bandwidth (>40 GHz). $LiNbO_3$ modulators are used in very diverse space applications that include navigation, measure–countermeasure, telecommunications, sensing, etc.

Fiber-optic gyroscopes (FOGs) are high-performance sensors used in demanding navigation systems. It is now proved that they can overtake the performance of classical laser gyroscopes, with the benefit of a smaller footprint, lower weight, and lower cost. Systems with accuracy better than 0.001 °C/h are commercially available, and FOG-based navigation systems have been used in satellites since 2010. These FOG modules integrate custom-designed $LiNbO_3$ phase modulators.

Figure 5.34 shows a three axis ASTRIX fiber-optic inertial measurement unit from AIRBUS DEFENCE & SPACE (formerly ASTRIUM). This unit is designed for long lifetime mission. Each axis uses a $LiNbO_3$ modulator. All opto-electronics and opto-components are fully compliant with HiRel Telecom satellite standard (SCC-B or equivalent).

Free space optical (FSO) communication (Figure 5.35) has been implemented between satellites since the 1990s using directly modulated high-power laser diodes at 820–850 nm. The emergence of fiber lasers in the near infrared and the

Figure 5.34 Three axis ASTRIX fiber-optic inertial measurement unit. Source: Reproduced with permission of Photoline. https://photonics.ixblue.com/files/files/pdf/ArticleSpaceModulators.pdf.

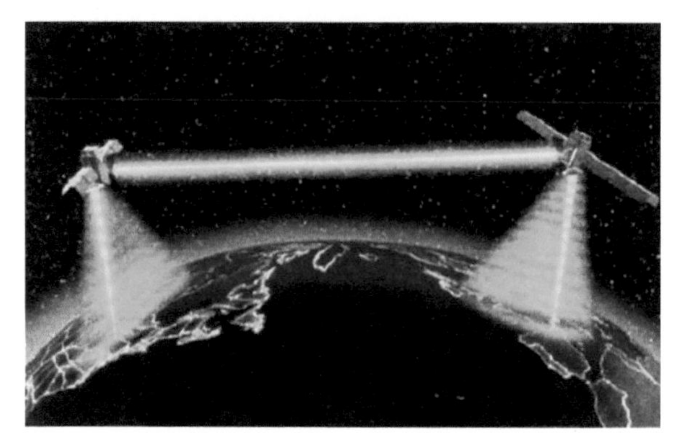

Figure 5.35 Artist's view of FSO (free space optics) communication demonstrations. Source: Reproduced with permission of Photoline. https://photonics.ixblue.com/files/files/pdf/ArticleSpaceModulators.pdf.

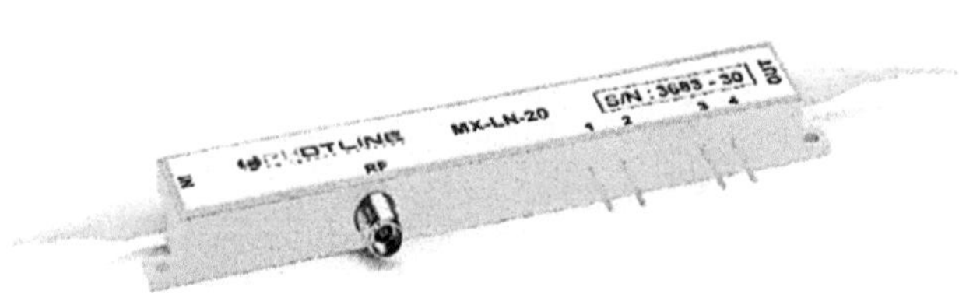

Figure 5.36 Space compatible 20 GHz intensity modulator. Source: Reproduced with permission of Photoline. https://photonics.ixblue.com/files/files/pdf/ArticleSpaceModulators.pdf.

availability of $LiNbO_3$ modulators in this band have made possible space optical links using more efficient modulation formats and offering improved data rates and BER (bit error rate/ratio).

Over the years, intensive design works and test programs have been conducted at IXBlue–Photline to improve the environmental performance of the modulators (Figure 5.36) and to make them compatible with space applications.

Recently, the most comprehensive range of commercial $LiNbO_3$ electro-optic intensity and phase modulators, from low frequencies to 40 Gbps/40 GHz and for a broad range of wavelength windows including 800, 1060, 1300, 1550 nm, and 2 μm, have been designed by iXBlue Photonics (https://photonics.ixblue.com/products-and-applications/lithium-niobate-electro-optic-modulator). When they are matched with the family of RF drivers, these modulators serve

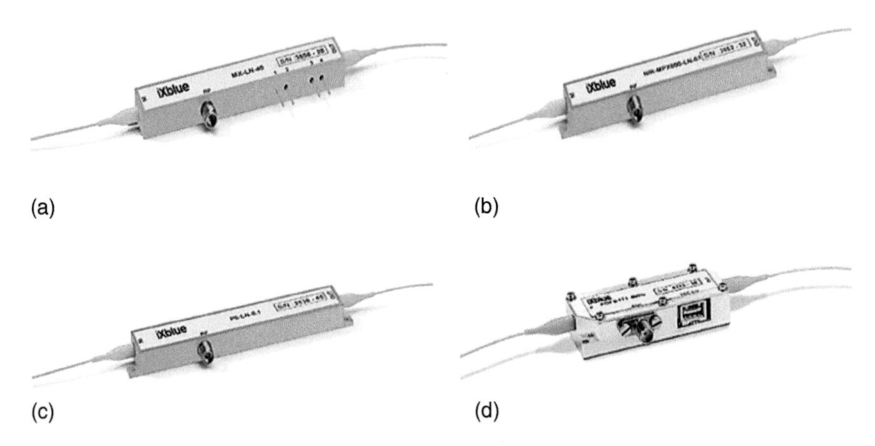

Figure 5.37 (a) Intensity modulator, (b) phase modulator, (c) polarization modulator, and (d) specialty $LiNbO_3$ modulator. Source: Reproduced with permission of Photoline. https://photonics.ixblue.com/products-and-applications/lithium-niobate-electro-optic-modulator.

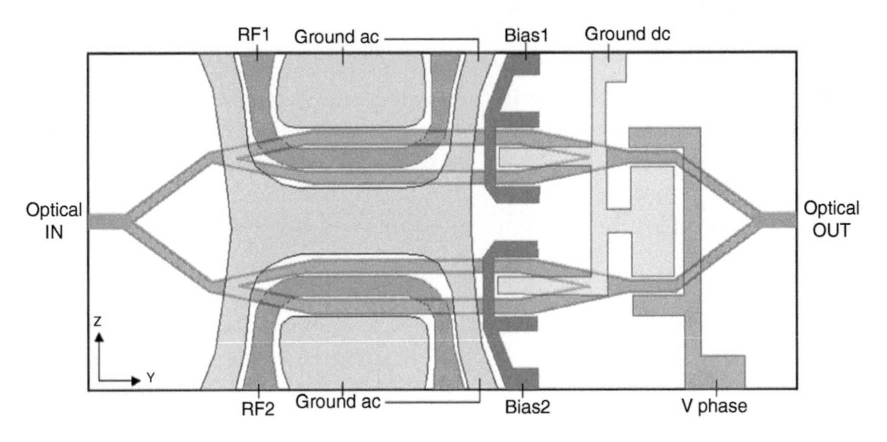

Figure 5.38 Schematic of the integrated $LiNbO_3$ quadrature modulator. Source: Kaplan et al. [156]. Reproduced with permission of CeLight.

all applications, from laboratory experiments to demanding industrial systems. Figure 5.37 shows the design of intensity modulator, phase modulator, polarization modulator, and specialty $LiNbO_3$ modulator.

***$LiNbO_3$* Quadrature Modulator** The x-cut highly integrated $LiNbO_3$ optical modulator and the hybrid receiver were developed for coherent QPSK (quadrature phase shift keying) fiber-optical communications, as well as other applications [156]. The main advantage of the proposed optimized design is the high integration level of electro-optical components within a single $LiNbO_3$ crystal wafer [351, 352]. A quadrature modulator with the structure is shown in Figure 5.38.

The single x-cut $LiNbO_3$ chip consists of

- two passive 3 dB Y-junctions for separating and combining the optical signals;
- two MZ (Mach–Zehnder) electro-optical "push–pull" modulators;

- two "push–pull" phase shifters for the MZ modulator's bias adjustments;
- one phase shifter for phase difference introduction between the two modulated signals before combining.

The device was designed for a 4″ $LiNbO_3$ wafer, using titanium indiffused technology. The waveguides were designed for single-mode TE operation on x-cut, Y-propagation crystal. Diffusion time and temperature were 10 hours and 1000 °C, respectively. The substrate was coated with a buffer layer of silicon dioxide. A thick gold-plated coplanar-waveguide (CPW) traveling-wave type electrode structure was formed on the buffer layer. SiO_2 and gold thickness were optimized for velocity matching of the optical and electrical signals. Each MZ is driven by RF signals applied to the on-chip CPWs. The CPWs were designed for 50 Ω impedance. The lengths of the RF electrodes were about 40 mm. The integrated phase modulator package was designed for at least tens of Giga symbol/s quadrature or binary phase shift keying (BPSK) modulation. The actual device was optimized for 12 GHz. Two MZ type modulators are combined using two 3-dB Y-junctions at the input and output. The MZ modulator is an optical switch that splits the incoming light wave into two optical beams and then combines them after an appropriate distance. Separate bias pads are utilized to optimize the dc bias point of each MZ modulator. An additional V-phase pad has been added to obtain quadrature phase difference between the two MZ modulators. The modulator is pigtailed with polarization maintaining fibers (PMFs). The half wavelength voltages were designed to be 5 V at dc and 7 V at 10 GHz. The optical insertion loss, including waveguide propagation loss, branch loss, and connection loss between the waveguide and fibers was optimized to be less than 8 dB.

Balanced 90° $LiNbO_3$ Coherent Receiver (Hybrid) The architecture of the integrated balanced phase diversity receiver [353, 354] comprising the optical hybrid is shown in Figure 5.39a [156]. The x-cut $LiNbO_3$, titanium diffused integrated 3″ wafer technology was used. The single hybrid chip includes four tunable 3 dB couplers for mixing the optical signals and two electro-optical phase shifters.

The design is optimized for use with balanced photodetectors for coherent detection of QPSK and BPSK signals.

The design includes the following features:

- Optical "I" (in-phase) and "Q" (quadrature) outputs
- External dc biasing capability of optical hybrid for optical amplitude and phase compensation
- Separate optical input for local oscillator (LO) laser to heterodyne (or homodyne) with signal on chip
- Package includes the option for thermoelectric cooler (TEC) and thermistor for temperature control and monitoring.

The input signal is fed into input S and split by tunable 3 dB coupler VH6. The phase between the signals at the upper ports of VH4 and VH3 can be tuned by the phase shifter VH5. The local oscillator signal is fed into input LO and split by VH1. If the phase between the LO signals at the lower ports of VH4 and VH3

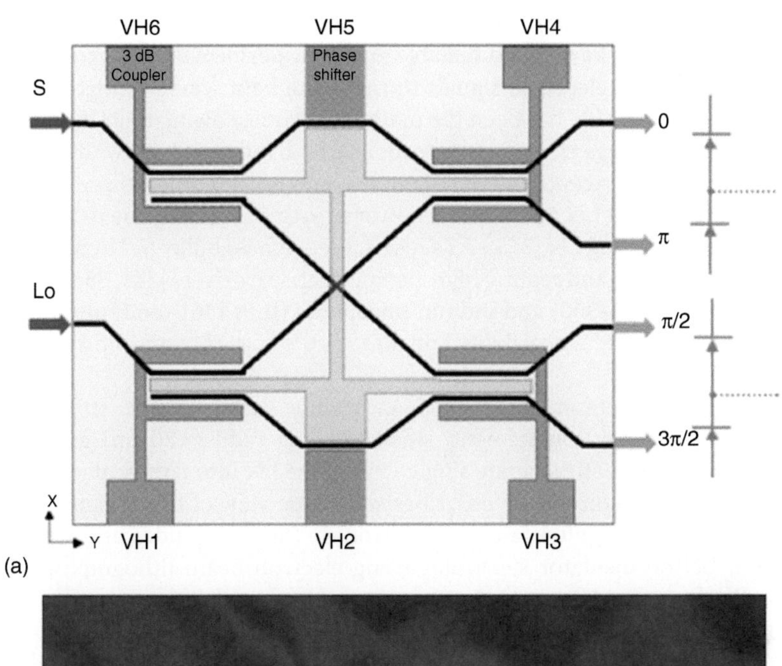

(a)

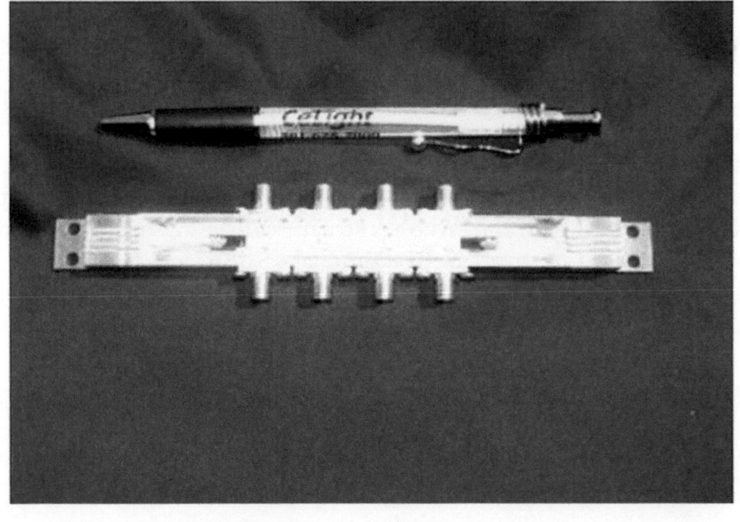

(b)

Figure 5.39 (a) Schematic of the integrated $LiNbO_3$ coherent receiver and (b) $LiNbO_3$ coherent receiver package. Source: Kaplan et al. [156]. Reproduced with permission of CeLight.

is adjusted to 90°, the signals at the device output ports will show the desired phase relation of the 90° hybrid: {180, 90, 270} with respect to the first output. Figure 5.39b shows the actual packaged $LiNbO_3$ coherent receiver [156].

These highly integrated devices will enable low manufacturing cost and high signal-to-noise ratio (SNR) in coherent optical communication systems. The integration of all components in a single $LiNbO_3$ chip dramatically reduces the cost, improves performance, and provides better stability and control to increase communication distance.

Nanophotonic $LiNbO_3$ Modulator Data centers, metropolitan, and long-haul data communication networks demand scalable and high-performance electro-optic modulators to convert electrical signals to modulate light waves at high speed [349, 355]. For decades, LN has been the material of choice owing to its excellent properties – namely, large electro-optic response, high intrinsic bandwidth, wide transparency window, exceptional signal quality, and good temperature stability [349, 355, 356]. Existing LN modulators, however, are not scalable due to the difficulty in nanostructuring LN [355]. As a result, they remain bulky (~10 cm long), discrete, and expensive, and require high-power electrical drivers [349, 355]. Integrated silicon (Si) [357–360] and indium phosphide (InP) [361–363] photonics are promising solutions for scalability but come at the cost of compromised performance [357–363].

Wang et al. [61] demonstrated single-crystalline LN photonic structures with submicron optical confinement, small bending radii (<20 μm) and low propagation loss by directly shaping single-crystalline LN into nanoscale waveguides (Figure 5.40a). Figure 5.40b describes schematic view of the device layout with thin film LN waveguides and RF electrodes. The waveguides are defined on thin-film LN-on-insulator substrates using electron beam lithography and subsequently dry etched in Ar^+ plasma using a deposited Si hard mask. The index contrast between the LN core and the silicon dioxide (SiO_2) cladding is $\Delta n = 0.67$, over an order of magnitude higher than ion-diffused LN waveguides. Figure 5.41a,b show a range of fabricated nanophotonic LN devices including nano-waveguides, ring resonators, racetrack resonators, and MZIs. The typical propagation loss of these structures is ~3 dB/cm, which is limited by etching roughness [180] and can be further improved by at least 1 order of magnitude [181, 364]. The resulting MZI and racetrack structures have low on-chip insertion loss of ~2 and ~1 dB respectively (with additional ~5 dB/facet coupling loss).

The highly confined optical mode allows to maximize electro-optic modulation efficiency by placing gold micro-RF electrodes close to the LN waveguide

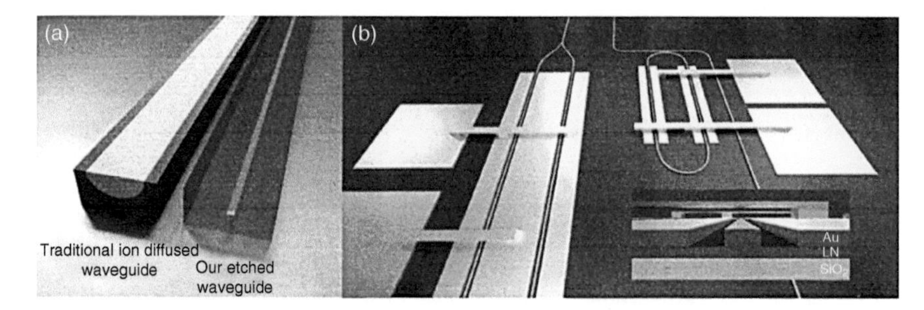

Figure 5.40 Device design. (a) Schematics of a traditional ion-diffused LN waveguide (left) and the etched LN waveguide embedded in SiO_2 (right), roughly to scale. The blue region indicates the approximate waveguiding core in each circumstance. The larger index contrast in etched waveguides allows for much stronger light confinement. (b) Schematic view of the device layout with thin-film LN waveguides and RF electrodes. Metal bias and bridges are fabricated to achieve modulation on both sides of the devices. Inset shows a schematic of the device cross section with an overview of the metal bridge. Source: Wang et al. 2017 [61]. Reproduced with permission of OSA Publishing.

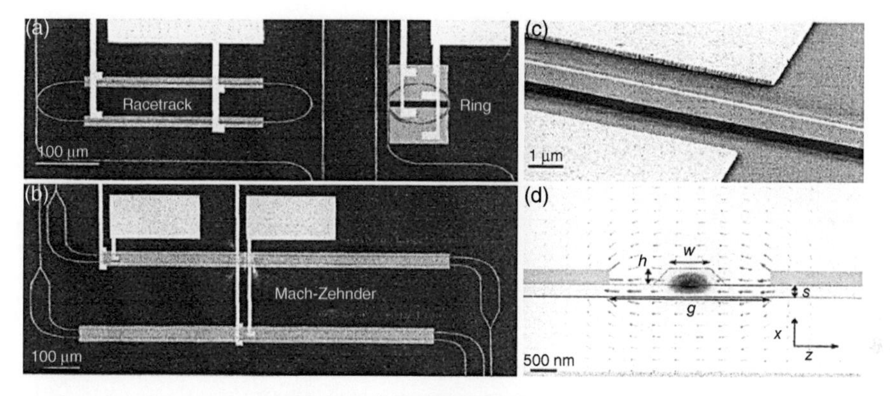

Figure 5.41 Fabricated optical devices and electrical contacts (a, b). (a) False-color scanning electron microscope (SEM) images of the fabricated racetrack and ring resonator based modulators and (b) Mach–Zehnder interferometer based modulators. (c) Cross-section view of the simulated optical TE mode profile (E_z-component) and RF electrical field (shown by arrows). The x-cut LN used here is most sensitive to the horizontal component of the electric field (E_z). h, LN waveguide height; w, waveguide width; s, LN slab thickness; g, metal electrode gap. (d) Close-up SEM image of the metal electrodes and the optical waveguide. Source: Wang et al. 2017 [61]. Reproduced with permission of OSA Publishing.

(Figure 5.41c,d). The devices make use of an x-cut LN configuration, where transverse-electric (TE) optical modes and in-plane electric fields (E_z) interact through the highest electro-optic tensor component (r_{33}) of LN. Wang et al. [61] designed the waveguide geometry and the micro-RF electrode positions to achieve optimal overlap between the optical and electric fields, while minimizing the bending loss and the metal-induced absorption loss. Figure 5.41c shows the numerically simulated overlap between the corresponding optical and electric fields. The optical waveguides have a top width $w = 900$ nm, rib height $h = 400$ nm, and a slab thickness $s = 300$ nm (Figure 5.41c). To maximize the in-plane electric field (E_z), they sandwiched the optical waveguide between the signal and ground electrodes with a gap of $g = 3.5$ μm. A SiO_2 cladding layer is used to further enhance this overlap by increasing the dielectric constant of the surrounding media to match the high dielectric constant of LN (~28) [356].

Wang et al. [61] showed efficient and linear electro-optic tuning in a racetrack modulator and a micro-MZI modulator. Figure 5.42a shows a typical transmission spectrum of a racetrack resonator with a loaded quality (Q) factor ~50 000. When a voltage is applied, the change of refractive index modifies the effective optical path length of the resonator, resulting in a resonance frequency shift. The electrical fields on the two racetrack arms are aligned to the same direction so that the modulation on the two arms adds up (Figure 5.40). The measured electro-optic efficiency is 7.0 pm/V with good linearity and no observable changes in resonance extinction ratio and linewidth (Figure 5.42a,b). The MZI modulator is a balanced interferometer with two 50 : 50 Y-splitters and two optical paths. The applied voltage induces a phase delay on one arm and a phase advance on the other, which in turn change the output intensity at the Y-combiner by interference. The minimum voltage that is needed to completely switch the output between "on" and "off" is defined as the half-wave voltage (V_π).

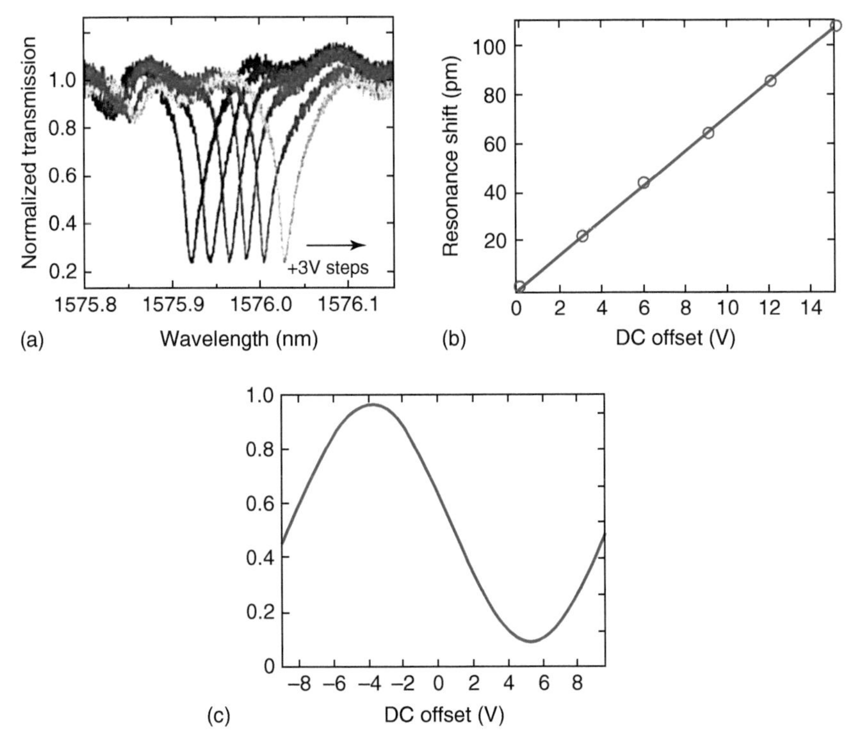

Figure 5.42 DC electrical and optical characterization. (a) Measured transmission spectra of a high Q (~50 000) racetrack resonator exhibiting large frequency shift with applied dc voltages. (b) Linear resonant wavelength shift as a function of dc voltage with error bars. The measured tuning efficiency is 7.0 pm/V. (c) Optical transmission of a 2 mm long MZI modulator versus dc voltage applied, indicating a half-wave voltage (V_π) of 9 V and a voltage–length product of 1.8 V.cm. Source: Wang et al. 2017 [61]. Reproduced with permission of OSA Publishing.

Wang et al. [61] measured a V_π of 9 V from a 2 mm long MZI modulator, with 10 dB extinction ratio (Figure 5.42c). This translates to a voltage–length product of 1.8 V.cm, an order of magnitude better than bulk LN devices [181, 364] and significantly higher than the previously reported LN thin-film devices because of the highly confined electro-optic overlap [186, 365–369].

The miniaturized LN devices exhibit high electro-optic bandwidths (S_{21} parameter), which is characterized using a network analyzer and a high-speed photodiode (Figure 5.43a). For a racetrack resonator modulator featuring a Q factor of 8000, the measured electro-optic 3 dB bandwidth is 30 GHz (Figure 5.43b). This value is limited by the cavity-photon lifetime of the resonator (~6 ps). Wang et al. [61] confirmed the lifetime limited bandwidth by testing additional resonators with Q values of 5700 and 18 000. The resulting 3 dB bandwidths are 40 and 11 GHz, respectively. The Q factors are engineered from the intrinsic value by controlling the distance between the RF electrodes and the optical waveguide. The intrinsic RC bandwidth limit of the racetrack modulator is estimated to be over 100 GHz. For the 2 mm long MZI device with direct capacitive modulation, the measured electro-optic 3 dB bandwidth is ~15 GHz (Figure 5.43c).

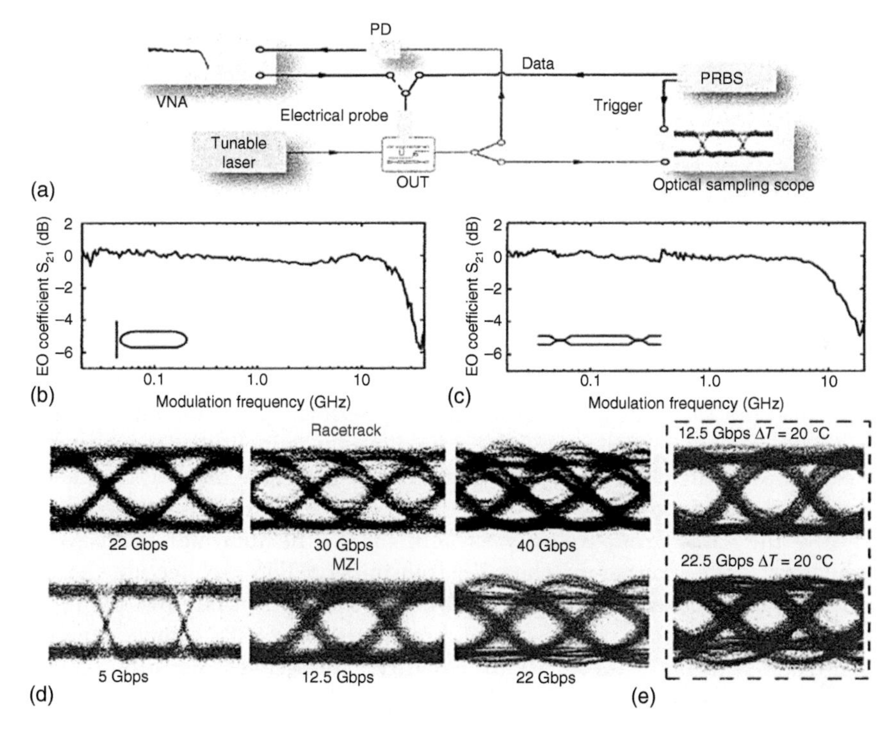

Figure 5.43 Bandwidth and high-speed data operation. (a) Simplified experimental setup for testing eye diagrams. Dashed lines indicate the signal path for electro-optic bandwidth measurement. (b, c) Electro-optic bandwidths (S_{21} parameter) of a racetrack resonator with $Q \sim 8000$ (a) and a 2 mm long MZI (b). The corresponding 3 dB bandwidths are 30 and 15 GHz respectively. (d) Eye diagrams of the racetrack (blue) and MZI (purple) modulator with data rates up to 40 and 22 Gbps. All eye diagrams are measured with 2^7-1 PRBS in a non-return-to-zero scheme with a 5.66 V_{pp} electrical drive. (e) Eye diagrams of the MZI modulator at 12.5 and 22 Gbps with the device heated up by 20 °C. The extinction ratios in (d) and (e) are 3 and 8 dB for racetrack resonator and MZI respectively. VNA, vector network analyzer; PD, photodiode; PRBS, pseudo-random binary sequence; DUT, device under test. Source: Wang et al. 2017 [61]. Reproduced with permission of OSA Publishing.

This is limited by the RC constant due to a larger capacitance (~0.2 pF) induced by the longer RF electrode used. Note that since the on-chip electrical resistance is rather small (<10 Ω), the measured bandwidth is currently limited by the 50 Ω impedance of the network analyzer drive.

The platform supports data transmission rates as high as 40 Gbps. Figure 5.43d displays non-return-to-zero (NRZ) open eye diagrams for both racetrack and MZI modulators at various data rates, obtained with 2^7-1 (pseudo-) random binary sequence at 5.66 V_{pp}. Because of the high signal quality, the devices can generally operate at data rates 1.5 times their 3 dB bandwidth, which translates to 40 and 22 Gbps for the racetrack and MZI devices respectively. The measured extinction ratios of the modulators are 3 and 8 dB with power consumptions ($CV^2/4$) of 240 fJ/bit and 1.6 pJ/bit respectively.

Wang et al. [61] confirmed that the MZI modulator maintains the stable thermal properties of their bulk counterparts, due to the low thermo-optic coefficient of LN ($3.9 \times 10^{-5}\,K^{-1}$) [356]. They varied the temperature of the chip within a $\Delta T = 20\,°C$ range, limited by the setup, and recorded the eye diagrams. They found that the MZI modulator is able to maintain an open eye diagram at the maximum data rate of 22 Gbps without any feedback to compensate for temperature drifts (open loop configuration) (Figure 5.43e).

The micrometer scale LN modulators feature high band width, excellent linearity, low voltage, and good temperature stability. The monolithic LN photonic platform that was demonstrated here could lead to a paradigm shift in electro-optic modulator scalability, performance, and design. For nearly a decade, existing LN modulator performance has been capped due to the non-ideal phase matching between the RF and optical fields [355]. The high dielectric constant of LN ($\varepsilon_{RF} \sim 28$) dictates that RF fields in LN propagate much slower than optical fields ($\varepsilon_{opt} \sim 4$) resulting in performance trade-off between bandwidth and driving voltage [355]. In the thin-film monolithic LN approach, instead, phase matching can be much better achieved since the electrical field primarily resides in the low dielectric SiO_2 ($\varepsilon_{opt} \sim 4$) and readily propagates at nearly the same group velocity as light [355]. The thin-film micro-MZI modulators, with a phase-matched RF transmission line architecture, could simultaneously achieve ultrahigh bandwidth (>60 GHz) and low modulation voltage (~1 V), and therefore are directly drivable with complementary metal-oxide-semiconductor (CMOS) circuitry. The active micro-resonators and low-loss waveguides could enable a chip-scale photonic circuit densely integrated with novel switches, filters, and nonlinear wavelength sources and operates in a wide wavelength range (from visible to mid-IR). Furthermore, the ultracompact footprint (as small as $30 \times 30\,\mu m^2$) of micro-ring modulators is attractive for data center applications where real estate is at a premium. This high-performance monolithic LN nanophotonic platform could become a practical cost-effective solution to meet the growing demands of next-generation data center, and metropolitan and long-haul optical telecommunications.

Electro-optic Field Sensors In high voltage or impulse discharge research, many phenomena are related to very high amplitudes of electric field, which is usually in the MV/m or higher range. For example, the average breakdown electric field of air gaps is in the order of 10^6 V/m. Owing to the considerable size of the sensor head and metal part in the traditional sensor, it is difficult to measure precisely such kinds of intensive transient electric field, especially in nonuniform fields and narrow space measurement situations. Moreover, most traditional electric field sensors may distort the original field and cannot measure frequency and phase information, which limits their application in transient electric field measurements [370].

Therefore, sensors using optical technique to overcome the above problems have been developed by many researchers [371–373]. Basically, the optical electric field sensors can be classified into two main types: active and passive. The active sensor utilizes a laser diode or a light emitting diode to convert the electric signal to optical signal and transmit the measured signal. This kind of active sensor has the distinct advantage of increased sensitivity because of the signal amplification in the sensor. Meanwhile, the passive sensor converts the electric field signal by an optical modulator using electro-optical effect crystals such as $LiNbO_3$ and $LiTaO_3$ [374]. This kind of passive sensor does not need a power source for the sensor head and performs much better for its reduced influence on the original field, well insulated, and wide enough operating bandwidth from "dc" (direct current) to GHz measurement.

So, during the past 30 years, different types of passive sensors based on electro-optic effect have been developed [375]. One kind is using the electro-optic effect crystal in high voltage measurements [376]; some research has been performed to measure very high electrical fields, such as the electric field near high voltage transmission lines. However, in order to couple the light into the sensor head, this kind of measurement system must be constructed with many independent optical components, which makes this kind of system so sophisticated that it limits its application in practice.

Recently, more and more researchers are interested in integrated electro-optic sensors for their small size, simple structure, and system-on-chip characteristics. But, research on integrated electro-optic sensors mainly focuses on how to increase the sensitivity of the sensors for near-field measurement [158]. Therefore, this kind of sensors are widely used in the detection of weak electromagnetic signals [377]. In the area of high-voltage measurements, some researchers used this sensor to acquire weak discharge or small voltage signals. The history of a variety of electro-optic methods for high-voltage measurements and electric field measurements in Laplacian and Poissonian fields is presented in detail [378]. The electro-optic modulator was applied to acquire the signal of partial discharges in high-voltage power transmission cables [379, 380]. The MZI was used in conjunction with a capacitive divider to measure high voltages such as those encountered in SF_6 gas bus ducts [381] and to measure the propagation characteristics of electromagnetic pulses (EMPs) induced by electrostatic discharge (ESD) [382]. A fiber-optic sensor was developed using the MZI for partial discharges in oil-filled power transformers using a 3×3 coupler in the passive homodyne scheme to improve the stability of the temperature [383].

The configuration of the electric field measurement system is illustrated in Figure 5.44. The laser source is a high-resolution stable power source, which was connected with a 50 m long PMF. The output of the laser source is a linear polarized optical beam with a wavelength of 1550 nm, which is widely used in telecommunication systems. This laser source was connected with the sensor head and the light in the sensor was modulated while passing through the outside electric field. A 50 m long mono-mode fiber connects the sensor to the photodetector (PD), which was used to convert the modulated light into an electric signal. The length of the two fibers ensures a safe distance between high-voltage devices and the observers.

Using MZI [131], one of the typical structures of the sensor is illustrated in Figure 5.45. The input light is divided into two equal parts by the left Y-coupling and then transmitted through two horizontal waveguides, waveguide 1 and waveguide 2. The light through waveguide 1 is modulated by the electric field, which is imposed by the electrode and antenna. And these two parts of light interfere with each other by passing through the right Y-coupling. So, when an electric field is applied to the dipole antenna, a voltage is induced across the

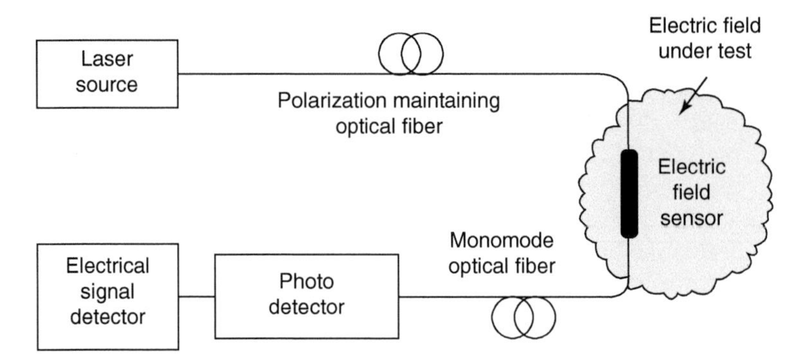

Figure 5.44 The configuration of the measurement system. Source: Alferness 1982 [131]. Reproduced with permission of IEEE.

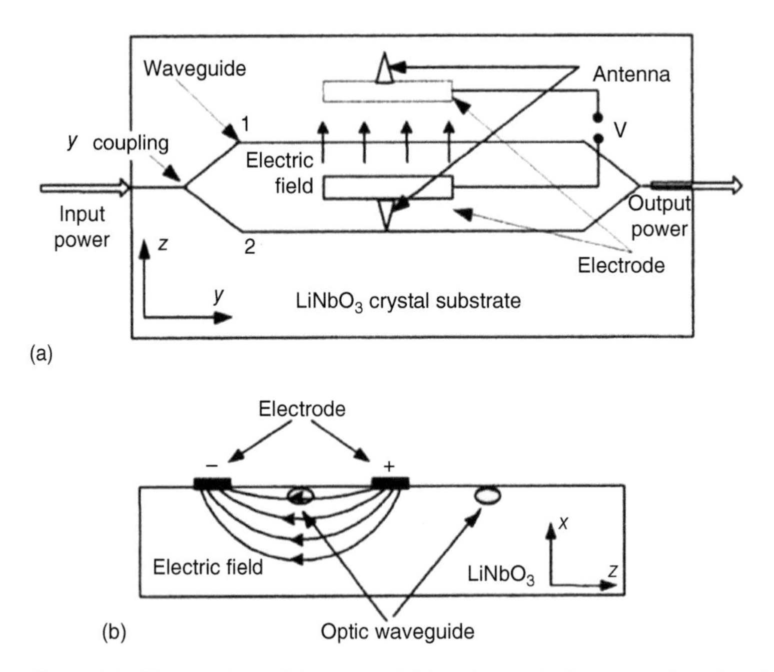

Figure 5.45 The structure of the sensor: (a) top view and (b) cross-section view. Source: Alferness 1982 [131]. Reproduced with permission of IEEE.

electrodes of the modulator. Then the output power of the sensor outputs an optical signal, which is in proportion to the applied electric field strength. The amplitude of the optical signal is then measured by the optical receiver. A laser diode is used as the optical source and a photodiode such as a PIN photodiode or avalanche photodiode is used as the optical receiver. The modulated light wave is converted to an electrical signal, which is detected by a photodiode, suitable for analysis with a spectrum analyzer, oscilloscope, or other signal processors.

The fundamental principle of electro-optic modulation can be found in other papers [131, 384]. The refractive index of electro-optic crystals, such as $LiNbO_3$, will change when an electric field is applied on it. The phase shift of the light wave during propagation is given by

$$\varphi = \frac{\pi n^3 r L \Gamma}{\lambda d} V \tag{5.38}$$

where n is the intrinsic refractive index of the crystal, r is the electro-optic coefficient corresponding to the polarization of the electric field E, L is the length of the electrode, λ is the wavelength, V is the voltage across the gap d, and Γ (<1) is the overlap factor of the electric field with the light wave in the crystal.

For simplicity, the half-wave electric field E_π is defined as

$$E_\pi = \frac{\lambda}{\Gamma n^3 r L} \tag{5.39}$$

According to the fundamental principle of Mach–Zehnder modulators, the modulator output of optical power can be expressed as

$$P_0 = \frac{P_i}{2}[1 + \cos(\Delta\varphi_0 + \varphi)] \tag{5.40}$$

where P_0 and P_i are the output and input laser power, respectively and $\Delta\varphi_0$ is the intrinsic phase difference between two waveguides. When the intrinsic phase difference between the two optical paths is $\pi/2$, the output of optical power will be

$$P_0 = \frac{P_i}{2}[1 - \sin\varphi] \tag{5.41}$$

If $\varphi \ll 1$, Eq. (5.41) reads

$$P_0 = \frac{P_i}{2}(1 - \varphi) \tag{5.42}$$

Obviously, the voltage between the electrodes, which is induced by the antenna, is linear with the environmental electric field. Thus, the output laser power of the sensor is linear with the measured electric field when the sensor is working near the static working point. In other words, the value of the electric field can be determined by measuring the output laser power.

Based on the theory mentioned above, the integrated electro-optic electric field sensors were designed by R. Zeng et al. [385] (Figure 5.46). The fabricated sensors have E_π ranges from ~10 to ~100 kV/m and a flat frequency response curve from the power frequency to greater than 100 MHz. For the design shown in Figure 5.46a, two electrodes with a distance of 40 μm are connected to the vertical dipole antennas with a length of 2 mm and the E_π is approximately 600 kV/m. For the sensor illustrated in Figure 5.46b, dipole antennas are combined with the electrodes. The gap between the electrodes is enlarged to 100 μm and the length of the antenna is shortened to 10 mm, leading to an E_π approximately 70 kV/m.

To expand the dynamic range, a sensor with a mono-shielding electrode was developed and is illustrated schematically in Figure 5.46c. The E-field distribution around the waveguide is altered by the mono-shielding electrode, resulting in unbalanced modulation on the two arms of the MZI. The mono-shielding gold electrode was fabricated by photolithography, with a thickness of 500 nm and a width of 100 μm. The fabricated sensors have an E_π range from 2000 to

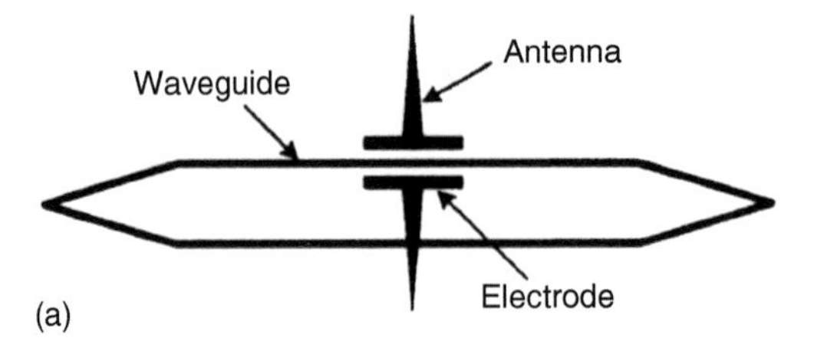

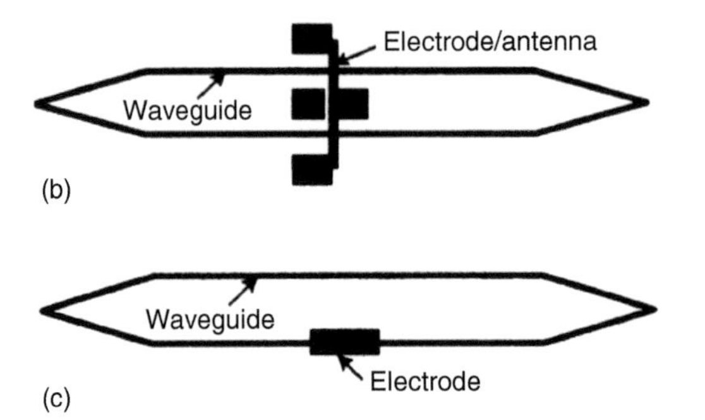

Figure 5.46 The structure of the sensors: (a) two electrodes connected with vertical dipole antennas design; (b) horizontal dipole antennas combined with electrodes design; and (c) mono-shield electrode design. Source: Alferness 1982 [131]. Reproduced with permission of IEEE.

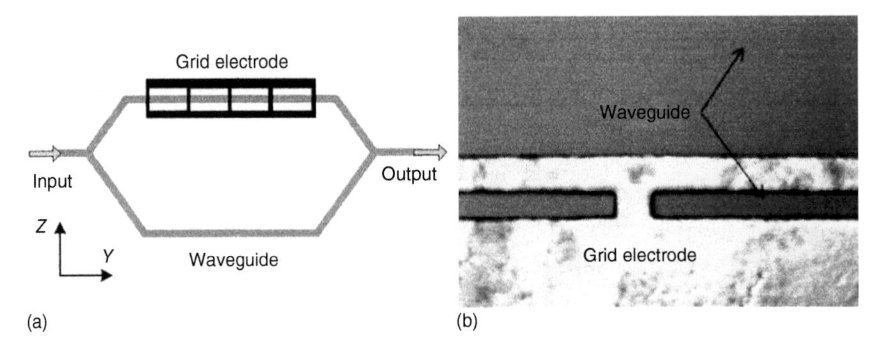

Figure 5.47 Mono-shielding electrode optimized as a grid type. (a) Schematic of the sensor; (b) micrograph of the fabricated electrode [388].

8000 kV/m; the E_π is altered by changing the length of the electrode (5–20 mm) and meets the requirement for air discharge field measurement.

A buffer layer (Si or SiO_2) should be added between the waveguide and electrode to eliminate the influence of the metal on the waveguide. However, the buffer layer deteriorates the frequency response [386] and the stability [387]. The mono-shielding electrode was optimized as a grid type, shown in Figure 5.47 [388]. This improvement ameliorates the stability while maintaining the electrical characteristics of the sensor. The reduction of metal material may also generate less interference to the original field. The MZI type is the most widely studied integrated electro-optic electric field sensor; it has a simple geometrical structure with an E_π ranging from ~1 V/m to ~1 MV/m and is tuned using different types of antennas and electrodes.

A coupler interferometer (CI) based integrated optical electric field sensor (IOES) can improve the controllability of φ, but it has a complicated transfer

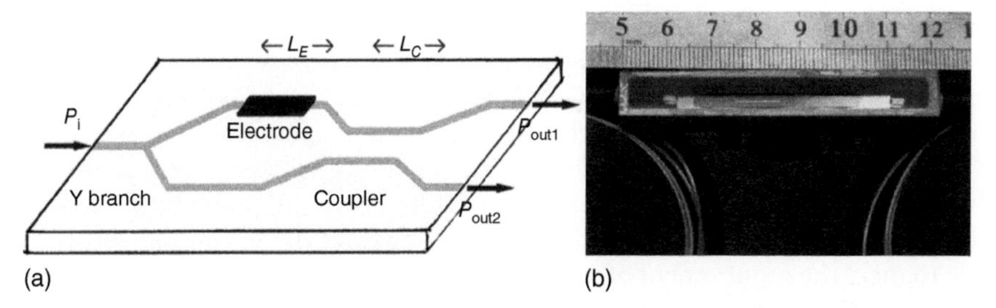

Figure 5.48 (a) Schematic of CI-based IOES. (b) Sensor after encapsulation. Source: Zeng et al. 2011 [390]. Reproduced with permission of SPIE.

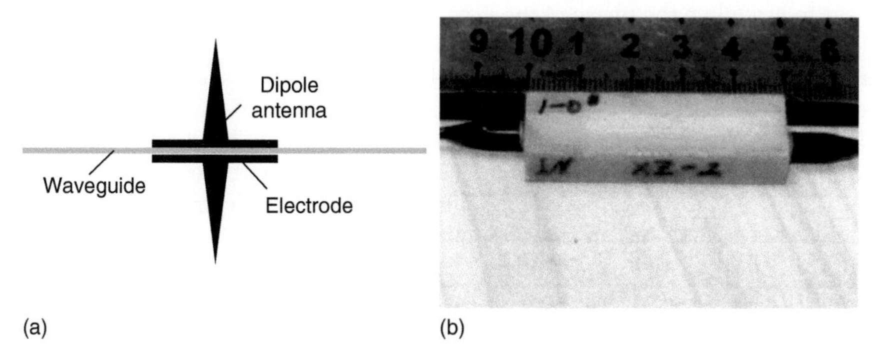

Figure 5.49 The CPI-based IOES with dipole antenna and electrode. (a) Schematic of the sensor; (b) fabricated sensor with dimensions of $5 \times 1.2 \times 0.5\,\text{cm}^3$ [391].

function [389]. A new CI-based IOES has been proposed that has a simple sinusoidal transfer function and a more favorable optical bias; this IOES has the advantages of both the MZI and conventional CI types [390]. The structure of the novel sensor is shown in Figure 5.48a. The width of the waveguide was designated to be 6 μm to support a fundamental mode only. The distance of the coupler was selected as 7 μm and the interaction length L was chosen as 2.8 mm. The fabricated sensor is shown in Figure 5.48b and has dimensions of $7 \times 1 \times 1\,\text{cm}^3$. However, the problem of temperature stability of the operating point remains unsolved.

The potential advantage of the common path interferometer (CPI) has drawn much attention in the area of voltage measurement and intense electric field detection. To improve the sensitivity, R. Zeng et al. [391] have designed antennas and electrodes around the waveguide, as shown in Figure 5.49a. The Ti-indiffusion waveguide has a length of 2 mm and a width ranging from 10 to 12 μm; the antenna has a length of 2 mm and the electrode has variable dimensions for different E_π. The IOES after encapsulation is shown in Figure 5.49b. This type of sensor could be shorter than the MZI or CI types. Because the waveguide with a favorable optical bias in the CPI could have a length on the order of millimeters, the Y branch of the MZI or CI limits the reduction of the waveguide length.

The fabricated sensors have an E_π of approximately 2500 kV/m and the optical biases deviate from the ideal values within 8°. The operating point varies slowly with temperature and the drift rate is approximately 0.015 °C/mm (the waveguide length is 20 mm). This value is similar to the 0.014 °C/mm value reported by University of British Columbia (UBC) [388], and both demonstrate that the CPI-based IOES has much better temperature stability. The CPI-based IOES has overcome the key problems of the MZI or CI types: controllability of the optical bias and temperature stability of the operating point. By designing dipole antennas and electrodes, the dynamic range of the optimized sensor is able to meet the requirement of intense electric field measurement.

In 2012, H. S. Jung [172] has fabricated an electric field sensor (Figure 5.50a) that integrated Ti:$LiNbO_3$ MZI and conventional lumped push–pull electrodes (Figure 5.50b) with a plate-type probe antenna, which utilizes the electro-optic effect to modulate the phase of the light propagating in each arm of the device. The sensor has a small device size of $46 \times 7 \times 1\ mm^3$ and operates at a wavelength of 1.3 μm. The output characteristic of the interferometer shows the modulation depth of 100% and 75%, and V_π voltage of 6.6 and 6.6 V at the 200 Hz and 1 kHz, respectively. The minimum detectable electric field is ~1.84, ~3.28, and ~11.6 V/m at the frequencies 500 kHz, 1 MHz, and 5 MHz, respectively. The sensor exhibits linear-like response tendency for the applied electric field intensity from 0.293 to 23.2 V/m, corresponding to a dynamic range of about ~22, ~17, and ~6 dB at frequencies of 500 kHz, 1 MHz, and 5 MHz, respectively.

In 2012, a $LiNbO_3$-based integrated optical sensor with Mach–Zehnder optical waveguide interferometer and tapered antenna array was designed and fabricated by Sun et al. [392] for electric field detection. The schematic configuration of $LiNbO_3$-based integrated optical sensor system is shown in Figure 5.51. It consists of three main modules, namely, laser diode and polarization controller module, photonic probe module, and photodetector and spectrum analyzer module. An optical beam from a 1.31 μm laser diode is transmitted to a polarization controller to generate a TE polarized mode, and then it is coupled into the photonic probe by using a PMF. The output beam is detected by a photodetector and analyzed by a spectrum analyzer or an oscillator, after passing a single-mode fiber. As the most important module of the proposed system, the photonic probe is made of $LiNbO_3$ crystals.

As shown in Figure 5.51, the M–Z optical waveguide interferometer is designed and fabricated by using annealed proton-exchange technology on x-cut y-propagation $LiNbO_3$ substrate. The proton source is benzoic acid and with a small amount of lithium benzoate. Because the annealed proton-exchange process results in the extraordinary index are noticeably increased and the ordinary index is unnoticeably decreased, the waveguides fabricated on the x-cut y-propagation $LiNbO_3$ slide only support TE polarization. After coating a SiO_2 buffer layer by radio frequency magnetic sputtering, the antenna array contains six pairs of tapered antenna on one of the two arms of M–Z interferometer by sputtering metals of Cr and Au and then electroplating Au. The M–Z modulator's parameters are 5 dB insertion loss, 5 V half-wave voltage, and 10 dB extinction ratio.

On the other hand, Figure 5.52 schematically shows the pattern of the tapered antenna array. This configuration improves the frequency response of the photonic probe from two aspects. Firstly, the resistance of the tapered pattern

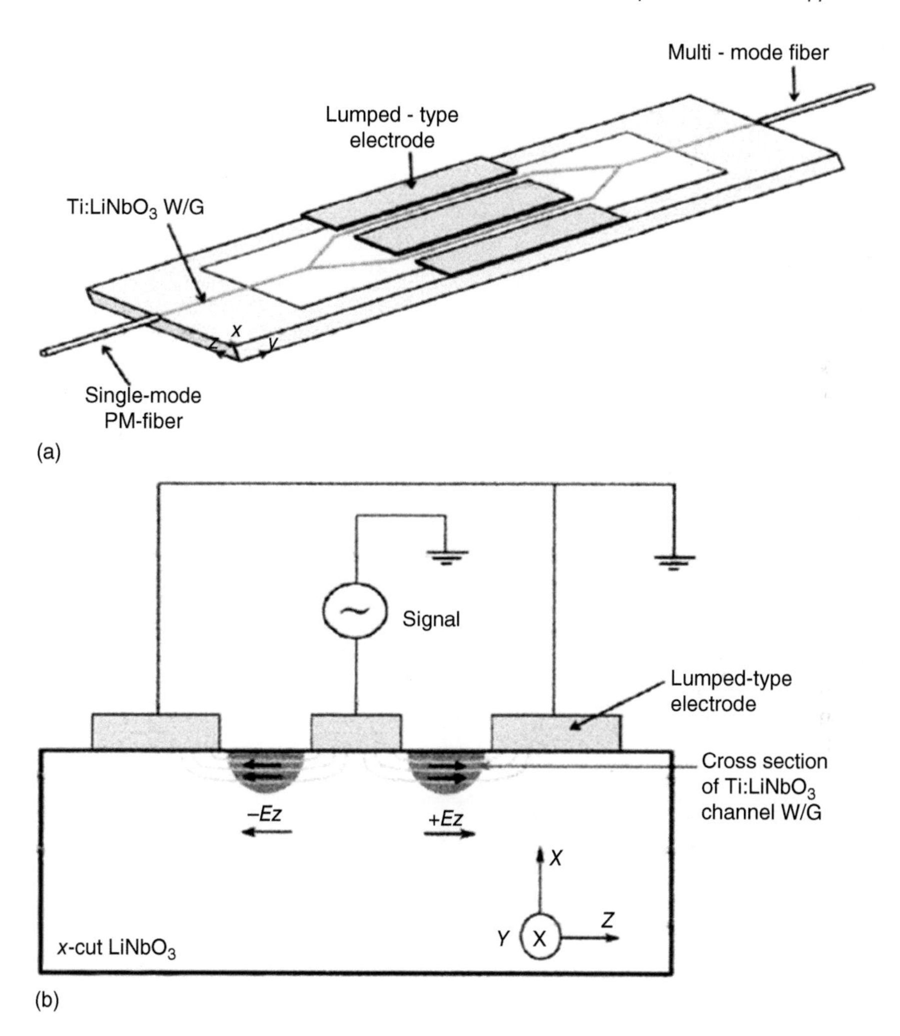

Figure 5.50 Schematic diagram of a symmetric Mach–Zehnder interferometer with metal electrode: (a) Perspective view of the electrode placement along the channel waveguides and crystal orientation. (b) Cross section through the interferometer arms showing the electric field generated by a voltage applied to the electrode. Source: Jung 2012 [172]. Reproduced with permission of OSA Publishing.

element increases from the bottom to the tip of the pattern, the reflection current at the tip can be reduced or eliminated, and the formation of standing waves can be avoided. Secondly, the effective capacitance of the modulating electrodes can be reduced by utilizing segmented electrodes structure, which has been proved to be an effective pattern for the trade-off between the bandwidth and the sensitivity of the electro-optic photonic probe based on the M–Z interferometer structure [393]. Considering that this photonic probe is designed for the detection of EMP signal, which usually does not need high sensitivity, only six pairs of tapered antenna are used in this design. Moreover, in view of the frequency range in which the sensitivity is constantly increasing with decreasing

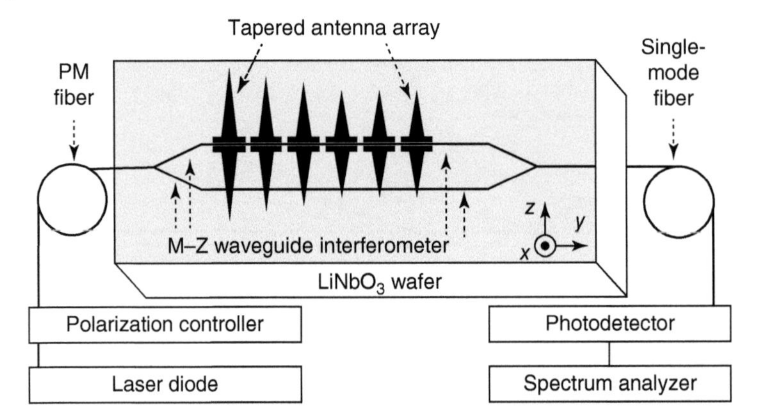

Figure 5.51 Schematic configuration of the $LiNbO_3$-based integrated optical sensor system. Source: Sun et al. 2012 [392]. Reproduced with permission of IEEE.

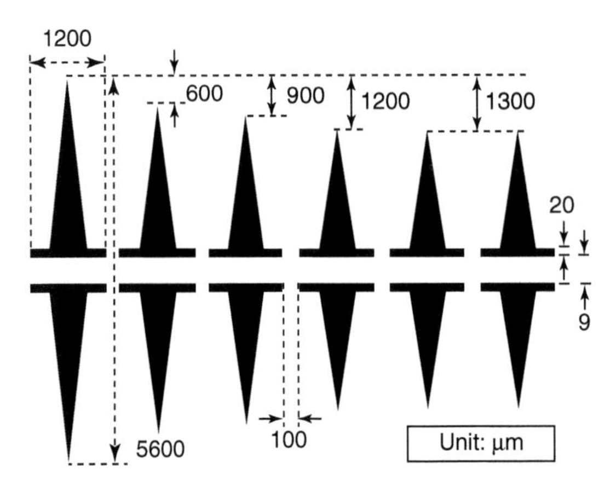

Figure 5.52 Geometry of tapered antenna array. Source: Sun et al. 2012 [392]. Reproduced with permission of IEEE.

element length of the tapered antenna array, the element of tapered antenna array has different length so as to further improve the frequency of the photonic probe. The structural parameters of the tapered antenna array are also shown in Figure 5.52.

When a detected pulsed electric field is received by the tapered antenna array, an induced voltage, which depends on the intensity of the received pulsed electric field, is applied to the modulating electrodes of the M–Z interferometer of the probe that are constructed by the tapered antennas. A laser beam, which travels through the M–Z interferometer, will be modulated by the induced voltage signal. The information on the detected pulsed electric field can be extracted by a photodetector to obtain this modulated optical signal and an oscillator to analyze the output signal from the photodetector. Experimental results reveal that this sensor has a frequency response deviation less than ±10 dB within the range of 10 kHz–18 GHz, and its minimum detectable electric field intensity is 0.4 V/m. Moreover, a pulsed electric field with nanoseconds width and electric field intensity of 103 V/m has been detected by this sensor.

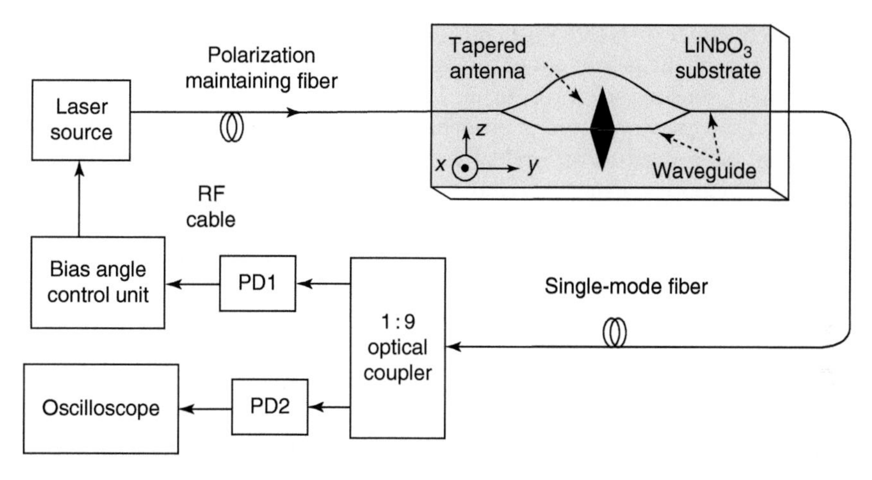

Figure 5.53 Configuration of the integrated optical waveguide *E*-field measurement system. Source: Zhang et al. 2014 [176]. Reproduced with permission of Springer Nature.

In 2014, a $LiNbO_3$ based integrated optical *E*-field sensor (Figure 5.53) with an optical waveguide MZI and a tapered antenna (Figure 5.54) has been designed and fabricated by Zhang et al. [176] for the measurement of the pulsed electric field. The minimum detectable *E*-field of the sensor was 10 kV/m. The sensor showed a good linear characteristic while the input *E*-fields varied from 10 to 370 kV/m. Furthermore, the maximum detectable *E*-field of the sensor, which could be calculated from the sensor input/output characteristic, was

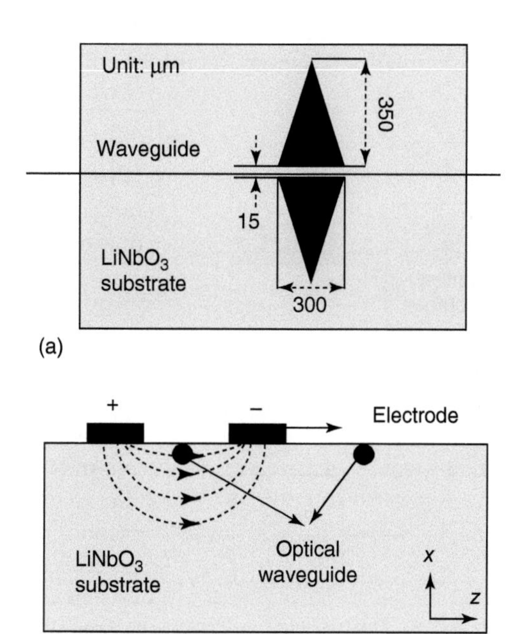

Figure 5.54 Schematic of the sensor: (a) geometry of the tapered antenna and (b) cross-section view. Source: Zhang et al. 2014 [176]. Reproduced with permission of Springer Nature.

approximately equal to 1000 kV/m. All these results suggest that such sensors can be used for the measurement of the lighting impulse electric field.

Electric field sensing schemes, using $LiNbO_3$ electro-optical retarders, are being studied since several years. A birefringent $LiNbO_3$ electro-optic waveguide (BEOW) has been used as an optical retarder and simultaneously as electric field sensor. The optical delay acts as the carrier of the sensed electric field. The modulated optical delay is transmitted to a measurement receiver, which is based on a second optical retarder acting as demodulator. The sensed electric field can be detected only when the sensor and demodulator are optically matched. An important issue when using BEOWs is their inherent sensitivity to temperature, which manifests as dc-drift at the output of the sensing-detection process and limits practical applications. A novel electric field sensing scheme, using $LiNbO_3$ unbalanced MZIs instead of BEOWs is designed by Gutiérrez-Martínez et al. [179]. This new scheme contributes to three main aspects related to electro-optic sensing:

(a) sensing-detection of electric fields using matched optical delays;
(b) proposes a simple technique for linearizing the sensing-detection process;
(c) demonstrates a strong reduction of the dc-drift phenomenon.

The contributing aspects are novel and become very attractive for implementing practical electric field sensing schemes using $LiNbO_3$ integrated optics devices. In this basic scheme shown in Figure 5.55, an optical path difference d_{m0} is generated and simultaneously modulated by the sensed electric field. The sensed signal is transmitted through an optical channel. At the receiver, the sensed electric field can be recuperated as an optical intensity variation only when a second optical delay is introduced and is optically matched to the sensor's one. An attractive feature of using optical delays as signal carriers is that several integrated optics retarders can be cascaded for introducing successive

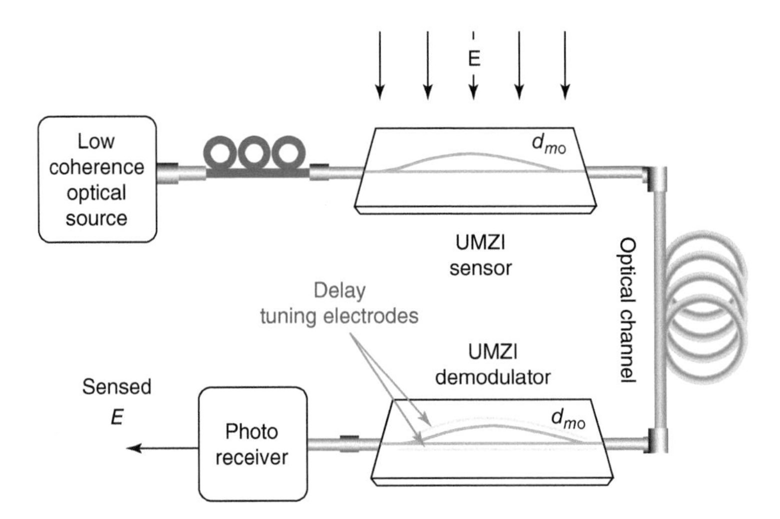

Figure 5.55 Electric field sensing scheme using UMZIs as optical retarders. Source: Gutiérrez-Martínez et al. 2017 [179]. Reproduced with permission of Springer Nature.

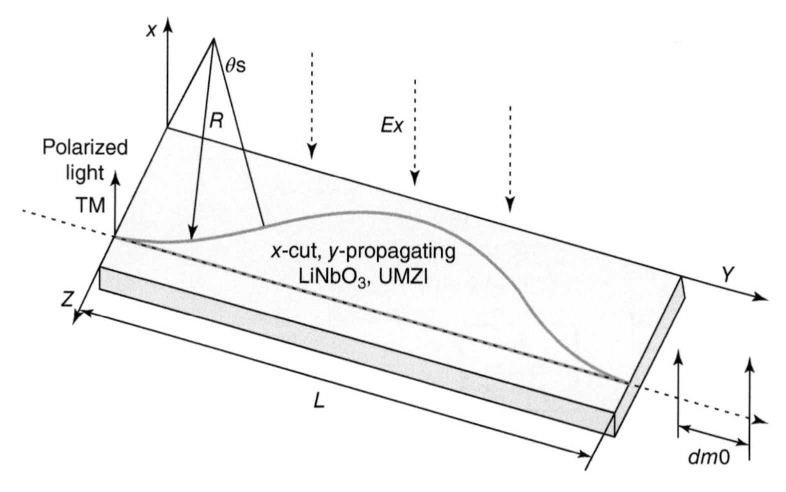

Figure 5.56 Unbalanced Mach–Zehnder electric field sensor. Source: Gutiérrez-Martínez et al. 2017 [179]. Reproduced with permission of Springer Nature.

longer delays. Successive optical delays can potentially be used for configuring multipoint sensing schemes, in either series or parallel architectures [174, 394].

An unbalanced integrated optics $LiNbO_3$ Mach–Zehnder sensor is shown in Figure 5.56. The unbalanced Mach-Zehnder interferometer (UMZI) is fabricated on an x-cut, y-propagating crystal by standard technology for titanium (Ti) indiffusion on a $LiNbO_3$ crystal. The diffused Ti, at high temperature, creates the optical waveguides of the electro-optic sensor. The physical dimensions of the UMZI chips are $54 \times 2 \times 0.5\,\text{mm}^3$ and, as depicted in Figure 5.56, the chip contains only one UMZI with one input and one output waveguide. The chips are provided of PMF at the input and output waveguides. According to Figure 5.56, the UMZI structure is composed of straight and curved waveguides. The straight arm has a physical length $L = 40\,\text{mm}$. The curved arm is implemented by four complementary arc sections [395] with curvature radius $R = 97.8\,\text{mm}$; $\theta_s = 5.9°$ is the angle of the arc section.

From the conceptual sensing-detection scheme depicted in Figure 5.55, the modulated path difference coming out from the UMZI retarder sensor travels through an optical fiber channel. At the receiver, a second UMZI, used as demodulator, is optically matched to ensure the detection of the sensed electric field. The second retarder detects the transmitted delay and generates a fringe pattern only when its own delay is d_{m0}. If the electric field variations are small when compared to the half-wave electric field E_π, the detected electric field is linearly recuperated as [179]

$$I_r = \frac{I_i}{4}\left(1 + \frac{\pi}{2}\frac{E_x}{E_\pi}\right) \tag{5.43}$$

As observed in Figure 5.55, in the proposed sensing scheme, a novel aspect is that the demodulating UMZI is provided of "tuning" electrodes, which ensure a linear sensing-detection by setting the receiving optical delay to $d_{m0} + \lambda_0/4$.

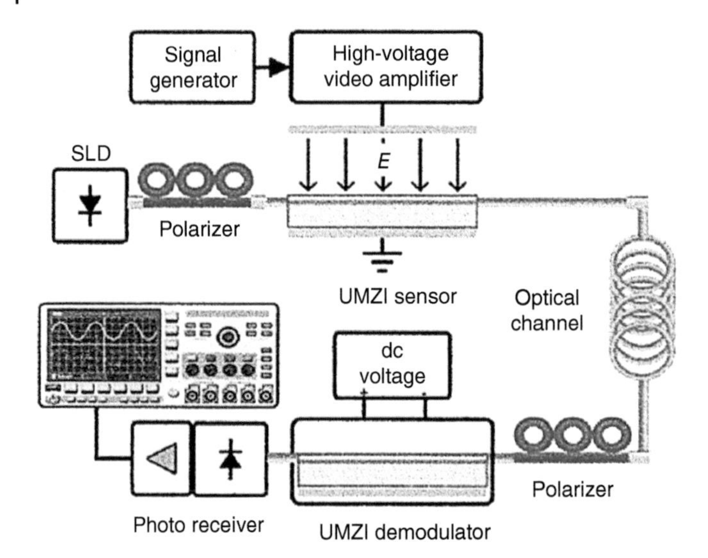

Figure 5.57 Experimental electric field sensing scheme using matched UMZIs. Source: Gutiérrez-Martínez et al. 2017 [179]. Reproduced with permission of Springer Nature.

To test the UMZI-based electric field sensing scheme, an experimental setup was implemented as illustrated in Figure 5.57. The experimental scheme has been tested just in laboratory. The optical source is a low-coherence superluminescent diode (SLD) (Qphotonics QSDM-1300), emitting a Gaussian spectrum at $\lambda_0 = 1310\,\text{nm}$ and power of 1 mW, showing a coherence length of 70 μm. Integrated optics x-cut, y-propagating $LiNbO_3$ UMZIs are used for the sensing-detection process. The sensed electric field has been generated by a signal generator and a high-voltage video amplifier (0–5 MHz-Leysop 250 series). The high voltage is limited by the output voltage range of the video amplifier (10–100 V_{pp}; gain 80). The high voltage is applied to a set of parallel plates covering the surfaces of the crystals ($45 \times 5\,\text{mm}^2$). When the parallel plates are in contact with the crystal surfaces, the maximum voltage produces an internal electric field of 200 kV_{pp}/m. The sensing-detection process was conducted under such a condition. When the plates are not in contact with the crystals, the internally sensed electric field (E_{LiNbO_3}) drops drastically by the boundary conditions $E_{LiNbO_3} = (\varepsilon_{air}/\varepsilon_{LiNbO_3})E_{air}$ [396]. For experimental testing, the SLD optical beam is firstly polarized along the x-axis for a TM propagating mode. The UMZI sensor introduces a static optical delay, which is simultaneously modulated by the x-oriented electric field. The scheme senses only the E_x electric field. Other electric field components E_z and E_y could be measured by designing 3D sensing configurations, such as the one proposed in Ref. [397]. In this setup, at the receiver the matched demodulator UMZI allows the detection of the sensed electric field and its response is linearized by applying a dc voltage to its associated electrodes. The sensed electric field is converted to an output voltage by a non-optimized photoreceiver, which is based on a PIN photodiode (Thorlabs D400) with a load resistance of 10 kΩ and 1-MHz bandwidth low-gain home-made electronic amplifier. The average received optical power at the input of the photodetector is 30 μW (−15 dBm). Three complementary tests were conducted, in order to determine the performance of this sensing scheme, in the frame of the main objectives of this proposal.

The experimental scheme has been characterized by three main aspects: the first is the test of matched optical delays for electric field sensing; the second aspect is the linearization of the sensing-detection process; the third issue is the dc-drift performance. Regarding the sensing-detection process, as the UMZI sensor is not biased at the quadrature point of its optical transfer function, the response is electrically linearized at the demodulator. Related to the dc-drift, the proposed scheme minimizes it, which becomes a very attractive feature. Finally, the proposed sensing scheme can be used in practical measurements in hard electromagnetic environments as found in industry and electric power facilities.

Since the 1980s, the rapid growth of integrated optics technology has led to an increase in worldwide research on integrated optical sensors. Time domain electric and magnetic field measurements are key research topics for scientific work in areas such as high-voltage engineering, high-power EMPs, and high-energy physics. Until recently, in these research areas, only IOESs were able to satisfy most of the application requirements, which include small size (millimeter), broadband response (dc ~ GHz), high signal amplitude (kV/m ~ MV/m), and appropriate insulation (MV). In the search for clear advantages, researchers have focused on the IOES advancements that have been achieved over the past 37 years. These researchers have developed a wide variety of IOES designs, and some have been implemented and applied in practice.

***$LiNbO_3$* Microresonator** High-Q crystalline microresonators have attracted tremendous attention for their broad range of applications in quantum electrodynamics, sensing, and optical signal processing [364]. Recently, several groups have demonstrated realization of freestanding on-chip microresonators in LN crystal of Q factors on the order of 10^6 [59, 60, 180, 181]. Among them, the maskless processing technique combining femtosecond laser direct writing with FIB milling allows rapid prototyping of LN microresonators of various sizes and geometries [59, 60, 182, 183]. Besides, the femtosecond laser direct writing technique alone can also provide functionality of straightforward integration of various functional microcomponents in a single substrate [184], which are highly in demand by either scientific research or industrial applications. For most applications utilizing a microresonator, wavelength tuning is of vital importance [185]. The large electro-optic coefficient of LN has enabled fast and efficient tuning of hybrid silicon and LN ring microresonators with typically loaded Q factors on the order of 10^3–10^4 [186, 187]. Electro-optic (EO) tuning of freestanding LN microresonators has recently been demonstrated by Wang et al. with external indium-tin-oxide (ITO) electrodes on top of the microresonator [181]. Nevertheless, fully integrated wavelength tunable freestanding LN microresonators have not been demonstrated due to the difficulty in fabricating the tiny microelectrodes in the close vicinity of the microresonators.

In 2017, Wang et al. [188] demonstrated electro-optic tuning of an on-chip $LiNbO_3$ microresonator with integrated in-plane microelectrodes. The procedure of fabricating the integrated sample is schematically illustrated in

Figure 5.58a–d. First, to realize the electro-optic tuning of the microresonator, two trenches with a depth of 6 μm were first inscribed on the LN substrate surface using a 20× objective lens of NA 0.4 at an average power of 0.42 mW and a scan speed of 160 μm/s. Then, a pair of parallel embedded microelectrodes separated by 75 μm was produced by filling copper into the trenches using femtosecond assisted selective electroless copper plating. The details of the selective metallization can be found in Ref. [398]. The sample was then completely cleaned

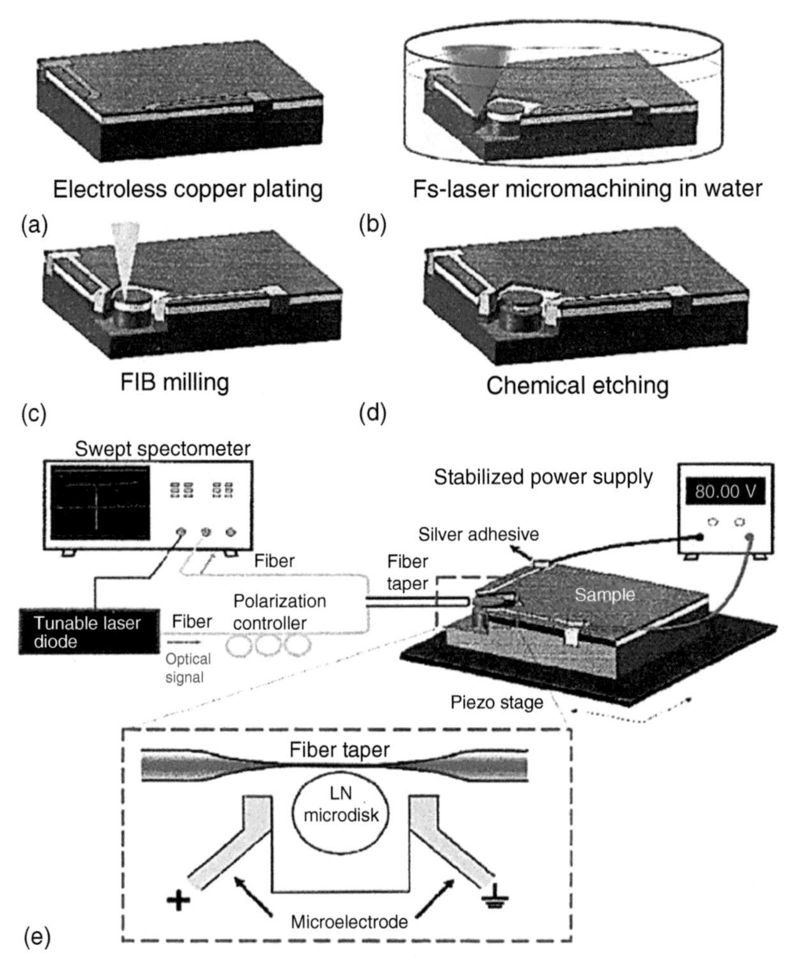

Figure 5.58 The processing flow of fabricating an on-chip electro-optic tunable LN microresonator with integrated in-plane microelectrodes is illustrated in (a)–(d): (a) formation of two microelectrodes using femtosecond laser assisted selective electroless copper plating; (b) water-assisted femtosecond laser ablation in LN substrate to form a cylindrical post; (c) focused ion beam (FIB) milling to smooth the periphery of the cylindrical post; (d) chemical wet etching of the fabricated sample to form the freestanding LN microdisk resonator. (e) Experimental setup for measuring the electro-optic tuning of the resonate wavelength in the $LiNbO_3$ microresonator. Source: Wang et al. 2017 [188]. Reproduced with permission of OSA Publishing.

with distilled water for removal of the electroless plating solution. The structure of the microelectrodes is schematically illustrated in Figure 5.58a.

Next, a cylindrical post was fabricated between the two planar electrodes by ablating the LN substrate immersed in water using femtosecond laser pulses focused with a 100X objective lens (NA 0.8), as shown in Figure 5.58b. The height of the cylindrical post is 13 μm. The periphery of the cylindrical post was then smoothed by FIB milling, as shown in Figure 5.58c. Finally, chemical wet etching, which selectively removes the silica underneath the LN thin film to form a freestanding LN microdisk resonator, was conducted by immersing the fabricated structure in a solution of 2% hydrofluoric acid (HF) diluted with water, as shown in Figure 5.58d. The diameter of the LN microdisk was 26.34 μm. More details of the process flow of fabricating the LN microresonator can be found in Refs. [59, 60].

The on-chip EO tuning of the LN microresonator was demonstrated using a setup schematically shown in Figure 5.58e. A narrow-band continuous-wave tunable diode laser (New Focus, Model 688-LN) was used to excite the whispering gallery mode (WGM) of the LN microresonator via a fiber taper fabricated by pulling a section of SMF-28 single-mode fiber to a diameter of ~1 μm. By using an online fiber polarization controller, WGM modes in the microdisk with certain polarization were excited. A transient optical power detector (Lafayette, Model 4650) was used to measure the transmission spectra at the output end of the fiber taper. The measurement system could record the light signal over a 100 nm wavelength span with 0.5 pm wavelength resolution and 0.015 dB power accuracy in less than one second. As illustrated in the bottom inset of Figure 5.58e, the fiber taper was placed close to the microdisk boundary by adjusting the XYZ-piezo stage, coupling the laser light into and out of the LN microresonator. During the EO tuning, the fiber taper was placed in contact with the top surface of the LN microdisk to enhance the stability of the measurement. Selective excitation of a certain spatial mode can be achieved by adjusting the position of the fiber taper with respect to the microresonator. An open-loop piezo controller (Thorlabs, Model MDT693B) was used as the voltage supply for microelectrodes, which provided a variable voltage ranging from 0 to 150 V with a resolution of 0.1 V.

Figure 5.59a shows the top view of the fabricated device consisting of a high-Q LN microresonator and two microelectrodes. Figure 5.59b presents the scanning electron microscope (SEM) image of the freestanding microdisk. The sidewall of the microdisk is very smooth; however, a taper angle of approximately 10° can be observed, which is caused by FIB milling.

Characterization of the optical mode spectrum of the microresonator was accomplished with the fiber taper coupling method [399]. First, a coarse scan was performed in the range from 1560 to 1617 nm with a wavelength resolution of 5 pm to obtain the transmission spectrum, as shown in Figure 5.59c. FSR was measured to be approximately 15.8 nm. Next, the Q factor of a mode around 1582.6 nm was measured by employing a fine scan at a wavelength resolution of 2 pm. The Q factor of the mode was estimated to be 1.83×10^5 as indicated by the Lorentz fitting (red colour solid line) in Figure 5.59d, which is consistent with the results [59, 60, 182].

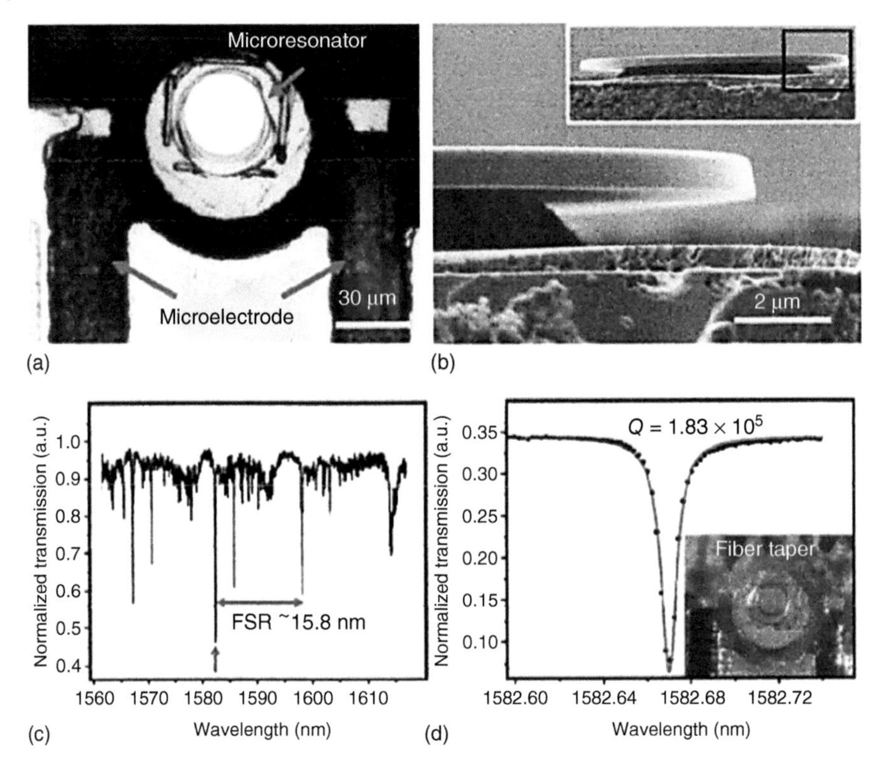

Figure 5.59 (a) The top view of the EO tunable LN microresonator integrated with in-plane microelectrodes. (b) The SEM image of the freestanding LN microdisk with smooth boundary. Inset of (b) is the overview of the LN microdisk. (c) Normalized transmission spectrum of the LN microresonator measured under critical coupling condition without the applied voltage. (d) The Lorentz fitting (red curve) of the dip under over-coupling condition showing a Q-factor of 1.83×10^5. The inset of (d) is the top-view optical image of the LN microresonator coupled with a fiber taper. Source: Wang et al. 2017 [188]. Reproduced with permission of OSA Publishing.

The performance of the fully integrated on-chip EO tunable LN microresonator was examined by applying different voltages to the planar microelectrodes. The voltage was tuned from 0 to 150 V. Figure 5.60a shows eight transmission spectra of a particular mode. The resonant wavelengths were measured as 1582.621 57, 1582.636 78, 1582.657 15, 1582.669 64, 1582.717 16, 1582.771 98, 1582.8258, and 1582.89 872 nm, with corresponding applied voltages of 0, 33.2, 56.8, 65.7, 100.6, 112.0, 135.0, and 150.5 V, respectively. The experimental results indicate a maximum 277.15 pm wavelength shift when the voltage was increased from 0 to 150.5 V.

Figure 5.60b plots the wavelength shift as a function of the applied voltage. It reveals that resonant wavelength of the WGM possesses a nonlinear dependence on the voltage between the microelectrodes (i.e. the electric field strength in the microdisk). The microresonator shows a relatively low linear tuning coefficient of 0.826 pm/V at low voltages (i.e. <80 V), which can be attributed to Pockels effect with linear EO relation. However, when the applied voltage is above 80 V, the resonant wavelength shifts drastically with increasing voltage, where

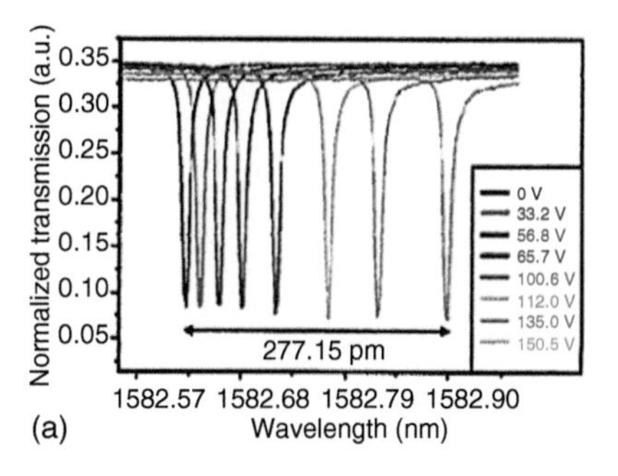

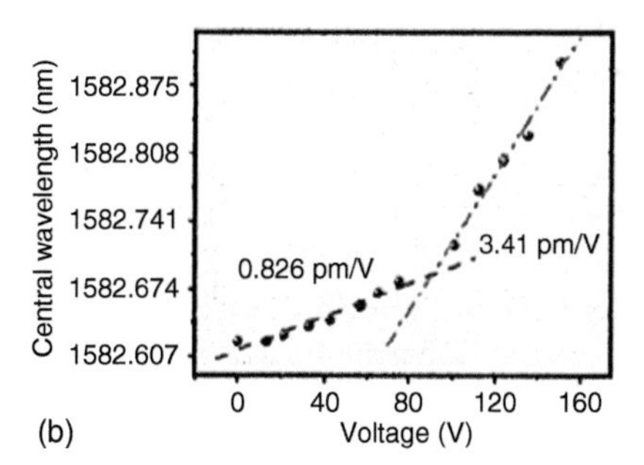

Figure 5.60 (a) The resonant wavelength in the fabricated microresonator as a function of the applied voltage. (b) The resonant wavelength shifts as a function of the voltage with a nonlinear relationship. At low voltages (i.e. 0–80 V), the tuning efficiency is 0.826 pm/V according to the linear fitting plotted by the blue dashed line. At high voltages (i.e. above 80 V), the tuning coefficient is 3.41 pm/V according to the linear fitting plotted by the green dash dot line. Source: Wang et al. 2017 [188]. Reproduced with permission of OSA Publishing.

the second-order Kerr effect dominates due to the small interelectrode separation [400]. The tuning coefficient is estimated to be 3.41 pm/V by the linear fitting indicated by the green dashed dot line in Figure 5.60b, which is significantly higher than the tuning coefficient of 0.826 pm/V achieved at low voltages. The result shows that the integrated in-plane microelectrodes can provide efficient EO tuning due to the strong electric field generated between two microelectrodes.

The present technique relies on consecutive patterning of the microelectrodes and microresonators with high-precision femtosecond laser direct writing, which leads to high spatial precision sufficient for accurately assembling the isolated components into the fully integrated microdevice. It is also noteworthy that the electrodes are bulk metallic structures buried within the LN substrate, which can produce strong bonding between the microelectrodes and the substrates. Furthermore, the electric field is more uniformly distributed in the plane perpendicular to the LN microdisk in comparison with microelectrodes formed by patterning a metal thin film coated on the surface of LN substrate. The demonstration of electro-optic tuning of a single microresonator can be readily applied to fabricate a coupled multiple microresonators whose resonant wavelengths can be tuned independently. In addition, the technique developed in this work allows straightforward integration of freestanding thin film waveguides with the microresonator. This technique shows promising potential for applications ranging from classical optical filters and modulators to cavity quantum electrodynamics.

5.6.5 Barium Strontium Titanate (BST)

$Ba_{1-x}Sr_xTiO_3$ (BST) is one of the most interesting thin-film ferroelectric materials due to its high dielectric constant, composition-dependent Curie temperature, and high optical nonlinearity. The composition-dependent T_c enables a maximum infrared response to be obtained at room temperature. The BST thin films in the paraelectric phase have characteristics such as good chemical and thermal stability and good insulating properties; they are often considered the most suitable capacitor dielectrics for successful fabrication of high-density gigabit scale dynamic random-access memories (DRAMs).

In 2000, Li et al. [189] investigated the electro-optic effects in $Ba_{1-x}Sr_xTiO_3$ (BST) thin films deposited on LAO (001) substrates. Both the Pockels and Kerr effects were observed and the electro-optic coefficients were quite high. From then, BST thin films were considered as promising candidates not only in microwave but also in electro-optic applications. In 2003, Kim et al. [190, 191] reported an extremely large QEO coefficient ($R_c = 1.0 \times 10^{-15}$ m^2/V^2) resulting in BST thin films grown on MgO (001) substrates, which implies that BST thin films are applicable for electro-optic modulators and optical waveguides devices. Wang et al. [192, 193] observed large linear ($r_c = 125.0 \times 10^{-12}$ m/V) and QEO coefficients ($R_c = 10.0 \times 10^{-18}$ m^2/V^2) in $B_{0.7}S_{0.3}TiO_3$ thin films, which is very promising for rib waveguides and Mach–Zehnder electro-optic modulators. In 2009, Takeda et al. [194] studied the birefringence and electro-optic effect in epitaxial BST thin films and the Kerr coefficient of BST film was determined to be 3.44×10^{-17} m^2/V^2. Bain et al. [401, 402] fabricated a number of planar optical waveguides with BST ferroelectric thin films and studied the optical (refractive index) property of the BST ferroelectric thin films in the 1450–1580 nm wavelength range at room temperature. The average value of the refractive index was found to be 1.985, which is considered to be important for optoelectronic device applications.

5.6.6 Lead Magnesium Niobate-Lead Titanate (PMN–PT)

PMN–PT $(1-x)Pb(Mg_{1/3}Nb_{2/3})O_3$–$xPbTiO_3$, one of the most widely investigated relaxor ferroelectric with a perovskites structure, was first synthesized in the late 1950s [195]. A study of the optical, electrical, and electro-optical properties of PMN–PT ceramics as a function of the stoichiometries is of interest both for possible insight into the physical nature of relaxor ferroelectrics and for making practical extension of its several present applications to include usage in electro-optical devices.

Present applications take advantages of the singularly excellent dielectric, low thermal expansion, and high electrostrictive properties of PMN–PT [403]. From the viewpoint of crystal chemistry, the substitution of Ti^{4+} ions for the complex $(Mg_{1/3}Nb_{2/3})^{4+}$ ions on the B-site of the perovskite structure in the $(1-x)Pb(Mg_{1/3}Nb_{2/3})O_3$–$xPbTiO_3$ (PMN–PT) system leads to the outstanding properties of the PMN–PT ceramics that exhibit excellent electrical and electro-optical performance, which make them promising for applications in multilayer capacitors, piezoelectric transducers and actuators, and optical devices [404–408]. Its electro-optic effect is about 2–5 times higher than that of PLZT and nearly 100 times higher than that of $LiNbO_3$ at room temperature. To reach the same index change (or phase shift), the voltage applied on PMN–PT is much lower than on PLZT and $LiNbO_3$. The remarkably good transparency over a wide wavelength range of 500–77 000 nm makes PMN–PT ceramic best suited for almost all the visible to MIR optical applications, such as optical limiter [196], polarization retarder [197], optical switch [198, 199], optical waveguides [200, 201], electro-optic devices [39, 202], and integrated photonic devices [203].

The excellent optical and electro-optic properties of PMN–PT ceramics are ideal for dynamic photonic device applications. Several types of dynamic photonic devices have been developed using PMN–PT ceramic materials and are described here.

5.6.6.1 Electro-optic Tunable Filter

The characteristic of electro-optic tunable filer is based on the electro-optic effect. This device operates by selectively coupling principal polarizations in a birefringent crystal at a phase-matched wavelength by means of a spatially periodic refractive index perturbation. It is tuned by varying the spatial period dc electric field via an array of separately addressable finger electrodes. The main advantages of this filter are its very low power consumption and its versatility in pass band programming by virtue of separately addressable voltages under microprocessor control.

With the increasing demand for higher bandwidths and larger data throughout in optical networks, tunable optical filters play a key role in next generation's high-speed communication systems. In WDM systems tunable optical filters are needed to manipulate or select a desired wavelength from the band of available channels. With the advance to dense wavelength division multiplexing (DWDM) a suitable tunable optical filter needs both a large tuning range and a narrow bandwidth.

There are two kinds of electro-optic tunable filters that have been widely used: tunable Lyot filter (TLF) and tunable Fabry–Pérot filter (TFPF). TLF has a wider and more stable tuning range but higher optical loss since it requires polarizers. TFPF usually has narrower bandwidth, larger aperture, better transmission, but narrower tuning range. They can be used separately or jointly. Both TLF and TFPF are fabricated using opto-ceramic PMN–PT as the dynamic tuning component [409].

Figure 5.61a shows a four-stage TLF where each stage consists of a polarizer and a tunable wave plate of PMN–PT. The tunable wave plates are in harmonic lengths. Figure 5.61b shows a picture of a fiber-version four-stage TLF (uncovered) with driving circuit. Characteristics of the TLF are shown in Figure 5.62a, where three stages were powered. The transmitted band position of TLF is determined by the total birefringence and the tuning is realized by voltage-induced birefringence. That is, when the birefringence varies, the peak wavelength moves.

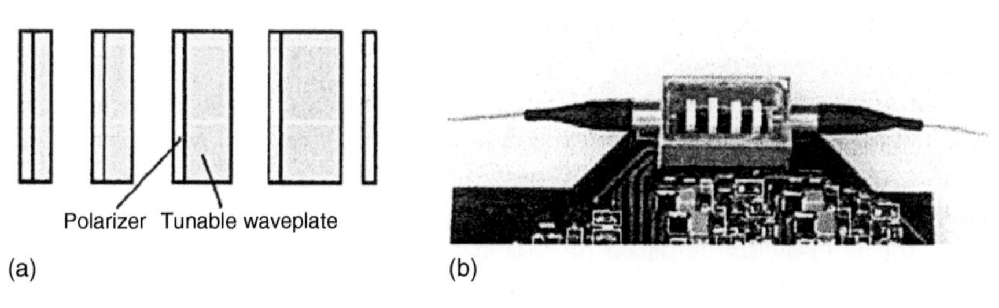

Figure 5.61 (a) A schematic design of TFL and (b) a fiber-version four-stage TLF with driving circuit. Source: Jiang et al. 2005 [409]. Reproduced with permission of SPIE.

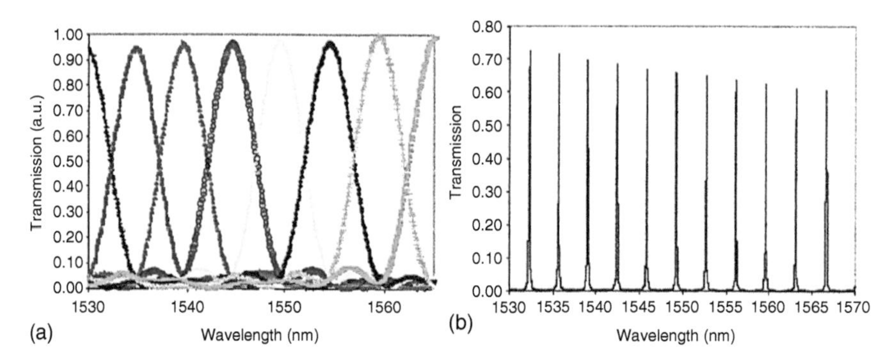

Figure 5.62 (a) TLF responses when three stages were powered and (b) transmission spectra of a TFPF. Source: Jiang et al. 2005 [409]. Reproduced with permission of SPIE.

The TFPF is constructed by sandwiching an opto-ceramic PMN–PT with both sides coated with transparent electrodes, between two partial reflection mirrors. ITO was used as the transparent electrode while SiO_2/TiO_2 multilayer was used as the dielectric mirror. In order to operate the TFPF in the entire specified wavelength range, the tunability of a TFPF must be such that the tuning range of each transmission peak is greater than or equal to the FSR of the etalon so that the adjacent peak could continue to cover the next wavelength window by a range of another FSR.

Figure 5.62b shows the transmission spectrum of a TFPF (150 μm thick) characterized by using an Ando 6317 optical spectra analyzer. The finesse was determined to be 33, the insertion loss (IL) 1.6 dB, and the FSR 3.5 nm, respectively. About 2.1 nm wavelength shift was obtained by applying a 1.0 V/μm electric field, which is very close to the estimated value (2.3 nm) according to the measured Δn. A tuning range of 6 nm was demonstrated by applying a higher electric field. The tuning speed measured from another setup on similar opto-ceramic structure was about 500 ns.

By combining two TFPFs with slightly different FSRs, the tuning can be very adaptive and efficient (lower voltage required). For instance, when a transmission peak of TFPF 1 coincides with a transmission peak of TFPF 2, light of this wavelength is transmitted. Figure 5.63 shows the experimental results of tuning range of a dual-cavity TFPF when only one TFPF is tuned. With moderate electric field, the FSR shown in Figure 5.63 is about 70 nm. When both TFPFs are tuned, the wavelength selection range is wider.

5.6.6.2 Electro-optic Q-Switches

Electro-optic Q-switching is one of the most common schemes to generate pulsed laser. The name Q-switch originates from the working principle of Q-switch laser where the laser output is turned off by periodically spoiling the resonator quality factor *Q* with the help of a modulated absorber inside the resonator. A Q-switch is based on loss switching. Because the pump continuously delivers constant power at all times, energy is stored in the atoms in the form of an accumulated population difference during off-times. When the losses are reduced during on-times, the large accumulated population difference is

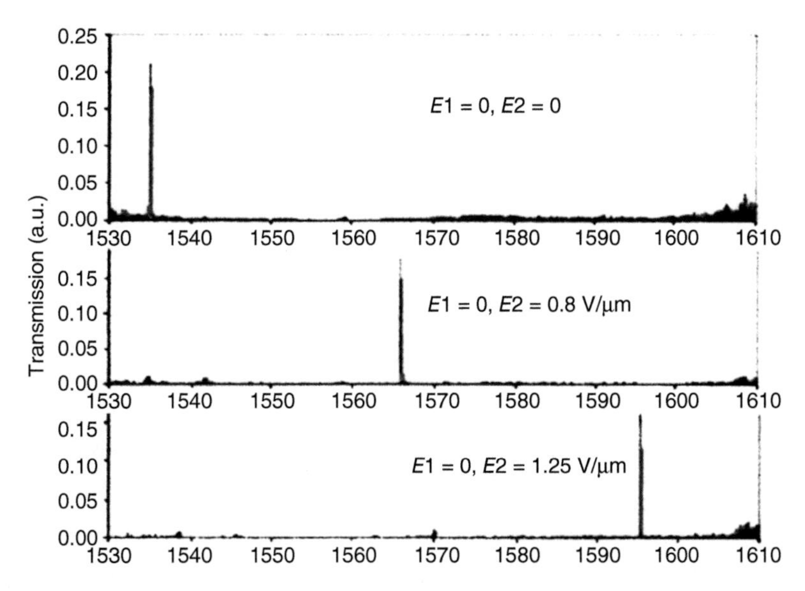

Figure 5.63 Measured outputs of a composite TFPF with different applied voltages. Source: Jiang et al. 2005 [409]. Reproduced with permission of SPIE.

released, generating intense short pulses of light. Among the existing *Q*-switches [410, 411], the electro-optic version is attractive in terms of size and agility [412, 413]. Typical electro-optic *Q*-switches are made from KDP and $LiNbO_3$. Owing to the small modulation efficiency of these materials, a high driving voltage (>1000 V) is usually required.

In contrast, opto-ceramic PMN–PT has very high electro-optic coefficients; therefore, much lower voltage is needed for *Q*-switching. Low voltage is very important to achieve *Q*-switched lasers with high repetition rate. An opto-ceramic *Q*-switch was developed with driving voltage as low as 48 V, which allowed us to reach a high repetition rate over 200 kHz [409]. In order to evaluate the *Q*-switch, a *Q*-switched diode pumped solid-state (DPSS) laser was built, as shown in Figure 5.64. The components used for constructing the DPSS include a 807 nm laser diode with a polarizer, a gradient-index (GRIN) lens to focus the

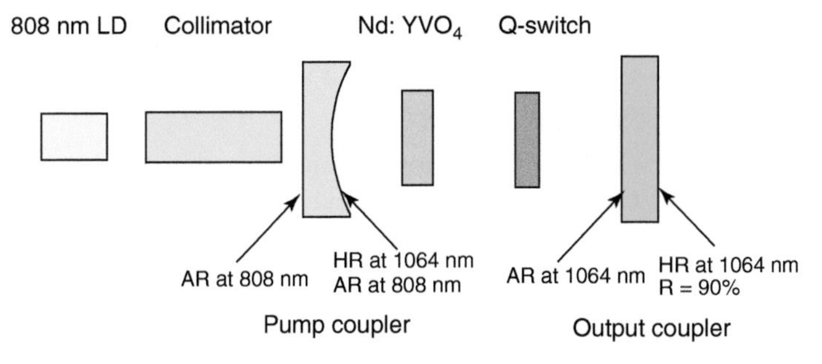

Figure 5.64 Configuration of a Q-switched DPSS laser. Source: Jiang et al. 2005 [409]. Reproduced with permission of SPIE.

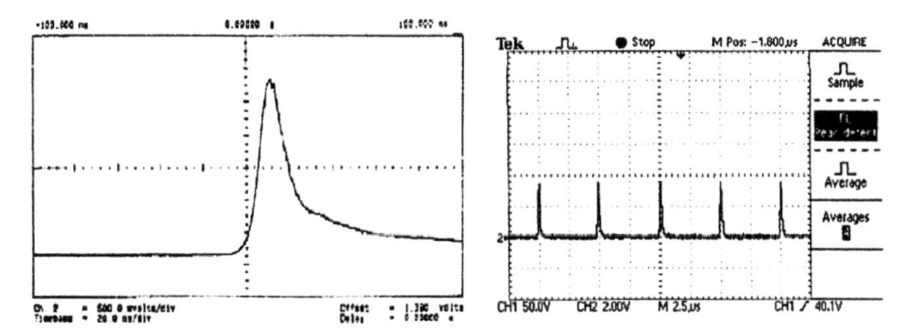

Figure 5.65 Pulse width and repetition rate of a Q-switch made from an opto-ceramic PMN–PT. Source: Jiang et al. 2005 [409]. Reproduced with permission of SPIE.

pump laser into the laser crystal, a plano-concave lens as the pump coupler, a Nd-doped YVO_4 laser crystal, a PMN–PT *Q*-switch with its optical axis 45° aligned to the YVO_4 *c*-axis, and a 90% plane mirror as the output coupler.

Various laser pulses with different durations were demonstrated from 10 to 100 ns, with driving voltage ranging from 48 to 106 V. It is also found that the low repetition rate tended to give shorter pulses. Figure 5.65 shows the pulse width and repetition rate of a *Q*-switch with an aperture of $1.8 \times 0.5\ mm^2$ made from an opto-ceramic PMN–PT. This small aperture *Q*-switch was specially designed for fiber lasers. Other large aperture *Q*-switches have also been developed for free space lasers.

5.6.6.3 Variable Optical Attenuator

Electronically controllable VOA play a crucial role in controlling optical signal levels throughout the network. The main network applications for VOA are as follows:

(1) *Pre-emphasis*: In DWDM transmission, the optical power between all channels needs to be equalized before they are combined into a single optical fiber.
(2) *Channel balancing:* At add/drop network nodes, channel equalization is imperative because optical signals arrive independently from different points in a network.
(3) *Optical automatic gain control:* Simultaneous attenuation of multiple wavelengths between stages in erbium-doped fiber amplifiers (EDFAs) enables tuning of amplifier gain for specific span lengths. This capability optimizes power levels throughout the transmission line and reduces spectral gain tilt induced in EDFAs by changing optical power.

The VOAs can be combined with other functionalities for new uses. Beyond network uses, demand for VOA from instrumentation manufacturers has dramatically increased. For example, VOAs contribute to tunable laser instrumentation by maintaining constant output power during a sweep across a range of wavelengths.

The opto-ceramic (OC) materials, especially transparent electro-optic PMN–PT ceramics, enable the design of devices using a non-guided, free space configuration. In the design, light travels perpendicular to the surface

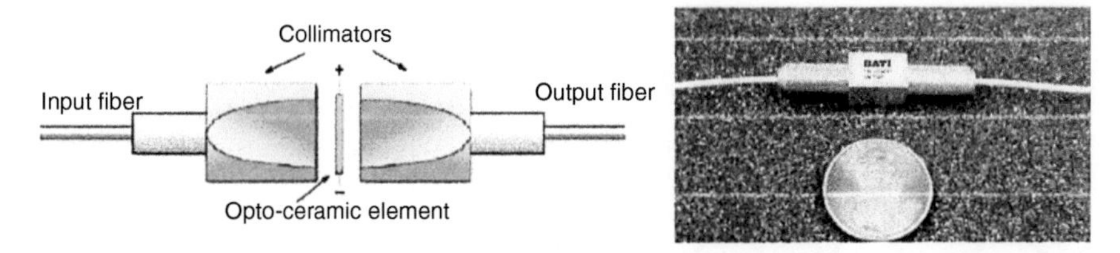

Figure 5.66 Schematic of the VOA construction and an Eclipse™ VOA device. Source: Jiang et al. 2005 [409]. Reproduced with permission of SPIE.

Table 5.2 Summary of the VOAs performance.

Attributes	Typical	Achieved
Insertion loss (IL)	≤0.6 dB	≤0.3 dB
Insertion loss (IL)	≥25 dB	≥35 dB
Insertion loss (IL)		≤0.25 dB
PDL at 1550 nm and 15 dB	≤0.25 dB	0.1 dB
Response time	<30 μs	<0.1 μs
Input optical power	500 mW	2.7 W
Return loss	≥55 dB	
Operating temperature range	0–70 °C	
Storage temperature range	−40 to 85 °C	
Dimensions		$10 \times 6.5 \times 6$ mm^3

Source: Jiang et al. 2005 [409]. Reproduced with permission of SPIE.

of the OC material, enabling low insertion loss and polarization-insensitive operation. Light is introduced via an input collimator; it passes through an OC element and exits by way of an output collimator. Attenuation control is realized by adjusting the electric field within the OC element – modification of the material's refractive index changes the level of light output. A schematic of the construction as well as an Eclipse™ VOA device is shown in Figure 5.66. This compact VOA has a footprint of only 65 mm^2; therefore, they are readily arrayed up to cover a group of channels in a straightforward manner.

The typical dynamic range of the VOAs is 25 dB and the better ones can reach 35 dB. Typical wavelength-dependent loss (WDL) and polarization-dependent loss (PDL) are 0.25 dB (0.1 dB achievable) at 15 dB attenuation with insertion loss less than 0.6 dB (0.3 dB achievable). The fundamental response time of these VOAs is well under 1 μs; such high speed permits the implementation of a real-time closed loop control system, which has the benefits of precise attenuation settings based on real-time light intensities and very small temperature dependence. Table 5.2 summarizes the performance parameters of VOAs made from electro-optic PMN–PT ceramics.

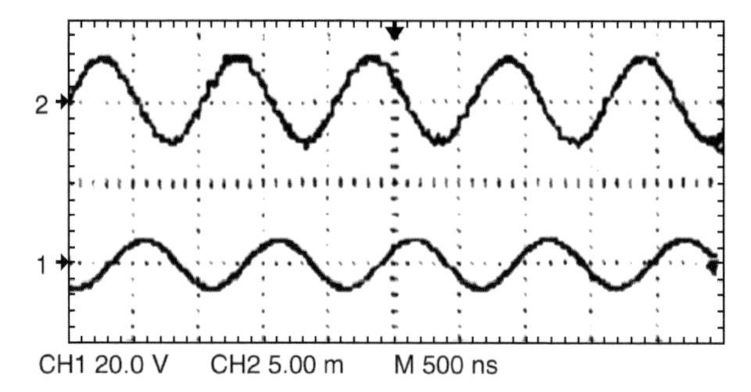

Figure 5.67 Optical modulation of a VOA at 1 MHz. Source: Jiang et al. 2005 [409]. Reproduced with permission of SPIE.

The intrinsic fast speed of these VOAs makes it possible for modulation at sub-megahertz range, which is suitable for channel supervising. Figure 5.67 shows the optical modulation of a VOA at 1 MHz. With the advent of high-speed VOA technology, network designers are now able to increase the flow of control information between network nodes without requiring the expense and high insertion loss involved in optical signal de-multiplexing. By using a VOA that has an intrinsically fast material response speed, both signal level and modulation functionalities can be implemented without component count.

5.6.6.4 Polarization Controller (PC)

Light is a polarized electromagnetic field. At any given instant it has distinct orientation in space and its propagation through medium depends on the polarization direction. Polarization-related impairments, such as PDL and polarization mode dispersion (PMD), have become the major obstacles to the increased transmission rates in DWDM systems. The existence of PDL is the difference between the maximum and minimum insertion losses for all possible input states of polarization (SoP), while PMD is a time delay or differential group delay (DGD) when the two polarization components travel with different speeds. At the receiving end, the delay between the arrivals of the two modes of the optical signal is interpreted as dispersion, which makes it difficult or impossible to interpret the optical signal. PDL may weaken signal strength while PMD may cause signal bit errors, especially when the bandwidth is high. Therefore, polarization management is a major task in controlling optical communication systems. A dynamic polarization controller is identified as the most important element for overcoming polarization-related impairments.

Two dynamic retardation plates (also called retarders) placed 45° with respect to each other would be capable of changing the input SoP to any polarization; one adjusts the polarization component along the longitude direction while another adjusts the same along the latitude direction on the Poincare sphere [414]. If an SoP is parallel or perpendicular to the fast (or slow) axis of the plate, such plate would have no effect on the SoP. However, when two plates are placed 45° with respect to each other, there would be no singularity point on the entire Poincare sphere since there is no circumstance that the incoming SoP would simultaneously align to the fast or slow axis of the plates. Therefore, two dynamic retarders would be sufficient to change any incoming SoP to any output SoP.

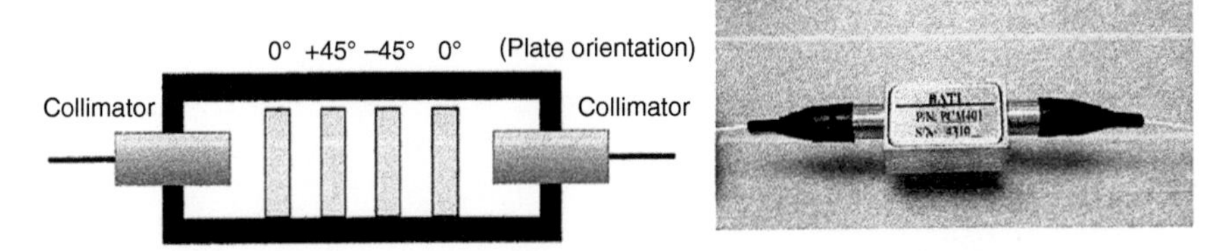

Figure 5.68 Schematics of a polarization controller design and an Acrobat™ polarization controller. Source: Jiang et al. 2005 [409]. Reproduced with permission of SPIE.

Opto-ceramic PMN–PT is ideal to play the role of dynamic retarders since its refractive index is adjustable under an electric field. The large electro-optic effect of opto-ceramic material requires a small electric field to achieve π phase shift, which minimizes the electric field induced scattering from the material; in turn, it produces negligible activation loss (AL). The major application of polarization controllers is for network impairment correction, where continuous operation is required. In circumstances where one retardation-plate reaches π phase shift, there are two options for subsequent operations: either further applying voltage beyond V_π to maintain system compensation or reducing the voltage to zero and starting over again. Applying higher than V_π voltage would be at the price of higher activation loss while rewinding to zero voltage would leave the system an uncontrolled period, which is not acceptable. We add another dynamic retarder to help the rewinding process; that is, when the voltage on the rewinding plate is reduced, the extra plate would be biased up to realize a seamless and endless control.

Figure 5.68 shows the design of a polarization controller and an actual Acrobat™ PC device. Four opto-ceramic plates are used to rotate the income SoP and they are oriented at 45°, −45°, and 0°, respectively, with respect to the first plate. The functioning of the first three plates has been described in the previous paragraph; however, the additional one was added for simplifying the control algorithm of system design where more than V_π voltage is required. Table 5.3 lists the key parameters of the device. Fast speed, low PDL, low insertion loss, and low activation loss featured by the Acrobat PC are all critical parameters in evaluating a dynamic polarization controller.

Figure 5.69 shows the redrawing of screen prints of a polarimetry measurement of an Acrobat polarization controller; rotations along the longitude, latitude, and angular directions are controlled by three dynamic retardation plates independently. By adjusting the retardation on two plates (or three plates if the incoming polarization is coincidently parallel to one of the other plates), the SoP can be controlled anywhere on the Poincare sphere. These polarization controllers can be used for PMD compensation, polarization scrambling, polarization multiplexing, polarization generation, and other polarization management functions.

5.6.6.5 Variable Gain Tilt Filters (VGTF) and Dynamic Gain Flattening Filters (DGFF)

In WDM optical links, it is important to keep all channels at the same power levels in order to avoid SNR degradation. A major cause of dynamic power variations

Table 5.3 Key parameters of polarization controllers.

Attributes	Performance
Insertion loss	≤0.6 dB
Polarization-dependent loss	≤0.1 dB
Polarization mode dispersion	≤0.01 ps
Input power	≤2.7 W
Activation loss at V_π	≤0.02 dB
Speed	≤30 μs
Return loss	≥55 dB
Power consumption	600 mW
Dimensions	$22.3 \times 11 \times 7.8\ \text{mm}^3$

Source: Jiang et al. 2005 [409]. Reproduced with permission of SPIE.

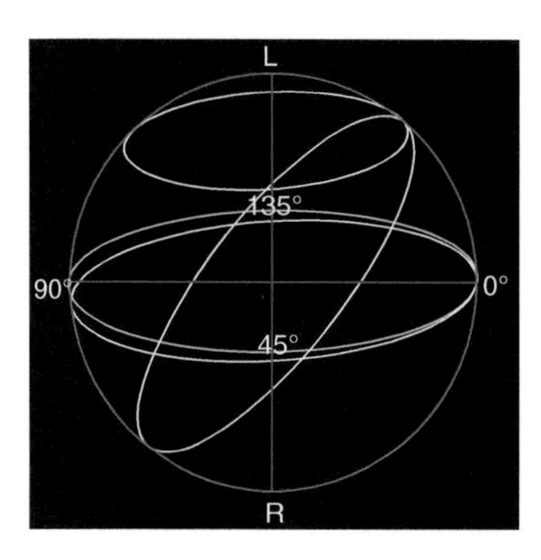

Figure 5.69 A screen print of the measurement of a PC. Source: Jiang et al. 2005 [409]. Reproduced with permission of SPIE.

is the power and wavelength-dependent gain characteristics of optical amplifiers [415]. Equinox™ dynamic gain flattering filter (DGFF) is based on opto-ceramic harmonic elements. The principle behind this is that any smooth spectrum curve can be represented by a sum of Fourier curves A_i or raised on the logarithm scale

$$H(\omega) = \sum \log A_i(\omega) \tag{5.44}$$

where ω is the harmonic frequency or a set of frequencies selected to best fit the spectrum curve. The expression of A_i is in the form of a sinusoidal wave

$$A_i(\varphi, \theta, \omega) = 1 - \frac{1}{2}\sin^2(\varphi)[1 + \cos(\theta, \omega)] \tag{5.45}$$

where variable φ controls the amplitude of the sinusoidal wave and θ adjusts the phase of the filter, respectively. The FSR of the sinusoidal wave was determined by the harmonic frequency ω.

Figure 5.70a is a schematic of variable gain tilt filters (VGTFs), also called a sinusoidal filter, which consists of a pair of collimators, an amplitude tuning element, and a phase tuning element. Both the amplitude and phase tuning elements are

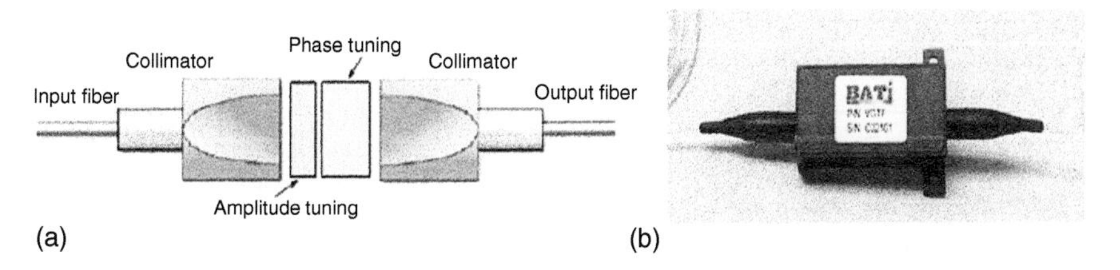

Figure 5.70 Schematic of a sinusoidal filter and (b) Equinox™ VGTF. Source: Jiang et al. 2005 [409]. Reproduced with permission of SPIE.

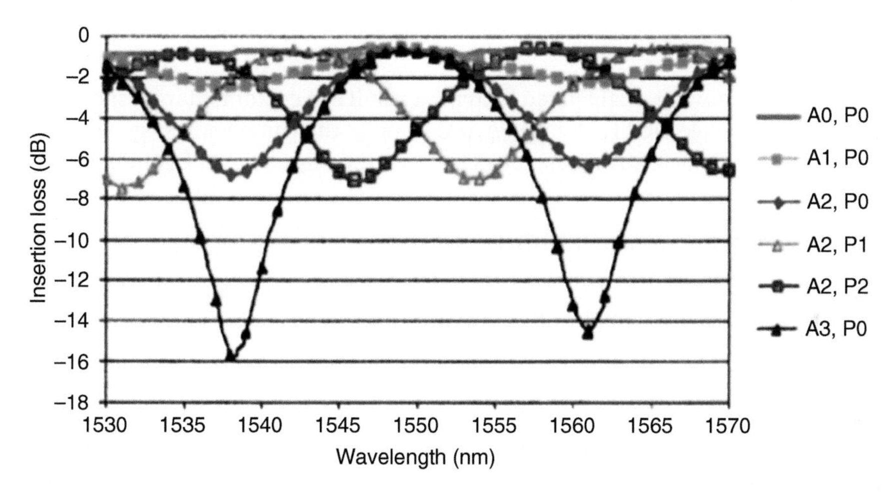

Figure 5.71 Measured results of a 24 nm sine filter. Source: Jiang et al. 2005 [409]. Reproduced with permission of SPIE.

Figure 5.72 Equinox™ DGFF. Source: Jiang et al. 2005 [409]. Reproduced with permission of SPIE.

composed from opto-ceramic PMN–PT materials and the corresponding passive optical components.

Figure 5.71 is a typical tuning curve of a sinusoidal wave filter with an FSR = 24 nm. The device started with a flat spectrum and at minimum insertion loss state. Both amplitude (A) and phase (P) can be tuned individually, where A and P indicate the amplitude and phase tuning positions, respectively. For simple spectra manipulation, such as changing the slope of the spectra, a sinusoidal filter with proper frequency-response-shaped (FRS) would be sufficient.

In order to flatten an arbitrary gain profile, multiple sinusoidal filters with different FSRs will be needed to form a DGFF. Figure 5.72 is an Equinox DGFF consisting of five sinusoidal filters and a VOA. Such devices can also be composed from individual sinusoidal filters and VOA.

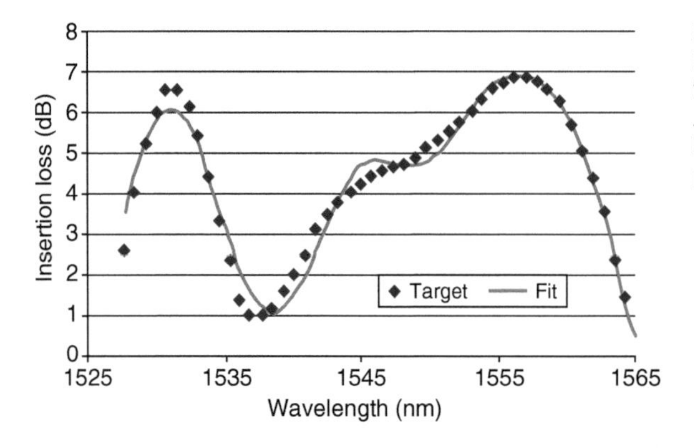

Figure 5.73 Fitting of a DGFF consists of five sine filters and a VOA. Source: Jiang et al. 2005 [409]. Reproduced with permission of SPIE.

Figure 5.73 shows the calculated flattening result of a customer target spectrum using the five-stage DGFF. The ripple in C-band of the compensated spectrum is less than 0.5 dB. Such fitting result has been verified by using an Equinox DGFF.

5.6.7 Potassium Niobate ($KNbO_3$)

Potassium niobate ($KNbO_3$) is a perovskite–structure material that shows ferroelectricity below 435 °C. It belongs to the most promising ferroelectric oxides with excellent optical properties [231]. Recently, potassium niobate has attracted much attention from the scientific community, due to its many applications in several technological fields. A large electro-optic coefficient [204–212] and a nonlinear optical coefficient [213–229] and excellent photorefractive sensitivity [230–235] make the crystal a particularly attractive candidate for applications such as optical waveguide [221, 223, 227, 236–256], electro-optic modulators [236, 262], SHG (also called frequency doubling) [22, 23, 213, 225, 248, 250, 251, 263–270], optical parametric oscillation [270, 277–279], and holographic storage medium [280–282]. The material also has a highly promising use in surface acoustic wave devices [206, 207, 416, 417], since its electromechanical coupling coefficient is higher than that of $LiNbO_3$, which is the most manufactured material for this application. These properties rank $KNbO_3$ at the top of Tuttle's FoM comparison [283] and Holman's device-level analysis [284] of materials in this class including $BaTiO_3$, PLZT, and $LiNbO_3$.

A permanent multimode waveguide in $KNbO_3$ single domain substrate crystal were produced for the first time by implantation of 2 MeV He^+ ions in 1988 by Bremer et al. [239]. Ridged channel waveguides in $KNbO_3$ were produced using He^+ ion implantation [227]. A minimum loss of 2 dB/cm was measured at a wavelength of 0.633 μm. A continuous-wave second-harmonic output power of 14 mW at 438 nm was obtained with an in-coupled fundamental power of 340 mW in a 0.73 cm long waveguide, yielding a normalized internal conversion efficiency of 25% W^{-1}/cm^2 (13% W^{-1}). A planar optical waveguide in $KNbO_3$ was fabricated with a minimum loss of $(1.0 \pm 0.2)\,cm^{-1}$ after implantation of 2 MeV ions at a dose of $5 \times 10^{14}\,cm^{-2}$ [241]. Permanent optical channel waveguides in single crystals of $KNbO_3$ were formed with low-dose MeV He ion irradiation, and propagation losses as low as 1.0 dB/cm were measured without the need for any annealing [246]. A minimum propagation loss of less than $0.2\,cm^{-1}$ (1 dB/cm) has been found in He^+ ion-implanted $KNbO_3$ waveguide at a wavelength of 0.633 μm [254]. Planar optical waveguides in single crystals of $KNbO_3$ were produced by 1 and 2 MeV He ion implantation with doses between $5 \times 10^{13}\,cm^{-2}$ and $5 \times 10^{14}\,cm^{-2}$ [242], and the possibility of producing nonleaky waveguides by ultralow dose implantation promises the creation of complicated channel waveguide structures within minutes.

In the recent past, thin films of potassium niobate found applications in integrated optoelectronic devices such as SHG [257, 258, 271–276], optical switching device [257], optical waveguide [257–261], and acoustic wave device [418–421]. Various deposition techniques such as epitaxial growth by MOCVD [208, 220, 222, 258, 275, 422–427], pulsed laser deposition (PLD) [260, 271, 428–431], rf-sputtering or ion beam [205, 219, 259, 261, 274, 432, 433], sol–gel [276, 434–436], polymeric precursor [229], and physical vapor deposition (PVD) [257] have been used for this compound.

Recently, the nanostructures of $KNbO_3$ have attracted great interest from the scientific community, mainly because of their promising applications as nanoelectromechanical systems (NEMSs). The nanostructure materials are ideal for studies of size-dependent properties; however, characterization of such properties of $KNbO_3$ is scare and mostly limited to studies on thin films and in superlattice structures where strain is an important issue. The nanoneedles/rods and wires of $KNbO_3$ have been synthesized by the templated crystallization of precursor gel [437, 438] and hydrothermal [41, 54, 439–445] methods. The $KNbO_3$ nanoneedles exhibited clearly SHG response [54] and showed potential application in nano-optical devices as an efficient nanoscale second harmonic light source. Nakayama et al. [41] developed an electrode-free, continuously tunable coherent visible light source compatible with physiological environments, from individual potassium niobate ($KNbO_3$) nanowires. These wires exhibit efficient SHG, and act as frequency converters, allowing the local synthesis of a wide range of colors via sum and difference frequency generation that is sufficient for in situ scanning and fluorescence microscopy.

5.6.8 Potassium Tantalate Niobate (KTN)

$KTa_{1-x}Nb_xO_3$ (KTN) is a ferroelectric material with a perovskite-like (ABO_3) structure. It is a compatible solid solution of $KTaO_3$ and $KNbO_3$, the phase structures and the properties can be modulated by controlling the composition, and the Curie temperature of paraelectric–ferroelectric phase transition varies with Ta/Nb ratio [285, 446]. KTN has received a great deal of attention due to its remarkable ferroelectric, dielectric, photorefractive, electro-optic, and nonlinear optical properties [285, 286, 289, 447–468]. In last decades, KTN single crystals have been widely investigated owing to their excellent potential for optical device performance as their large QEO effect. The KTN single crystal is a transparent isotropic material with very large electro-optic coefficients of about 600 pm/V, which is 20 times larger than that of conventional $LiNbO_3$ [469]. Currently, the KTN crystals are widely used in opto-electronic devices such as electro-optic modulator [36, 37, 285, 470–472], electro-optic deflector (EOD) [290], electro-optic sensor [473], optical waveguide [474–482], and electro-optic beam scanner [43, 292, 293, 483]. A total internal reflection diffraction device has been constructed using a QEO material KTN [484]. This device exhibits the best zero-order extinction voltage of 40 V, and a capacitance of 630 pf at room temperature (20 °C above Curie temperature). Very recently, a fast varifocal lens has been developed based on the KTN single crystal that responds as fast as 1 μs, which is 1000 times faster than the conventional commercial devices [44, 288, 483].

The KTN thin films appear more suitable for optical devices as well as some other applications, since not only the problems of compositional gradient and inhomogeneity but also the process of cutting and polishing can be avoided [485]. Many methods have been used for preparing KTN films, such as sol–gel deposition [486–488], chemical solution deposition (CSD) [489, 490], MOCVD [422, 491], magnetron sputtering [492], and PLD [485, 493–498]. Pure or dominating perovskite phase and oriented KTN films have been successfully grown on Si [490, 499], (100) MgO [489, 495, 496], (100) and (110) $SrTiO_3$ [485, 486, 488], platinum coated silicon [492], and sapphire substrates [496]. The study about optical properties of KTN films is also one of the investigative hotspots and the optical and nonlinear properties have been reported previously [459, 493–495, 498, 500, 501]. Since the past two decades, the KTN thin films are widely used for nonlinear optical applications, which include electro-optic modulator [493], integrated optics [494], and surface acoustic wave infrared/thermal detector [501]. Recently, the KTN nanoparticles and nanorods have been synthesized through solvothermal [502] and PLD [503] methods, which have potential interest for nonlinear optical devices.

Several types of electro-optic devices have been developed using KTN crystals and some are described here.

5.6.8.1 Electro-optic Phase Modulator

High-speed optical modulators with nanosecond operation are important components in systems designed to send/receive large amounts of information such as screen and/or moving image data. They are useful as transponders in metropolitan networks, which can send/receive modulated signals. Many kinds of modulator have already been reported that employ materials with electro-optic (EO) effects [504–506]. EO modulators that use $LiNbO_3$ (LN) are the most widely investigated. The electrode length for modulation is 5 cm although the driving voltage is less than 1 V [504]. There is a trade-off between driving voltage and modulation length, and materials with high EO coefficients are needed to improve these two aspects of performance simultaneously. Thus, certain problems remain as regards the modulator due to the intrinsic properties of their materials. $KTa_{1-x}Nb_xO_3$ (KTN) crystals exhibit large quadratic EO effects [507].

In 2004, Toyoda et al. [36] successfully fabricated a KTN buried waveguide on a 2″ KTN crystal. The core and cladding layers were formed using liquid phase epitaxy. The cross section and near-field pattern of the buried waveguide are shown in Figure 5.74. They confirmed that the KTN waveguides were capable

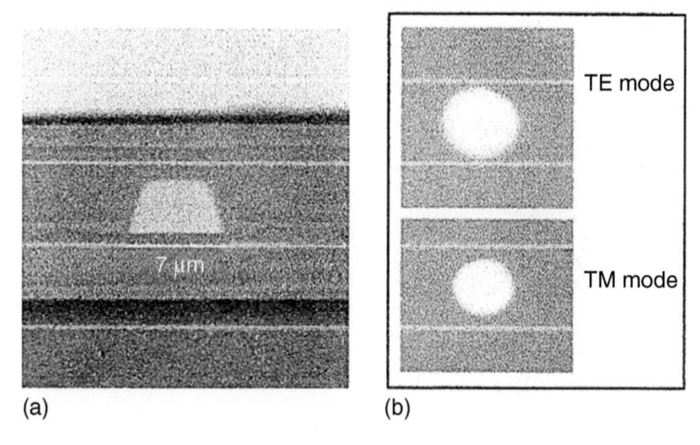

Figure 5.74 Cross section of fabricated KTN buried waveguide, and near-field pattern of KTN buried waveguide. (a) Cross section of fabricated KTN buried waveguide and (b) near-field pattern of KTN buried waveguide. Source: Toyoda et al. 2004 [36]. Reproduced with permission of IEEE.

of single-mode beam propagation. The mode field diameter was about 7.6 μm for both the TE and TM modes. This core size provides with a lower coupling loss between optical fibers and the KTN phase modulator than between optical fibers and a semiconductor modulator. Toyoda et al. [36] measured the propagation loss using optical low coherence reflectometry techniques and it was less than 0.5 dB/cm. They also measured the PDL with a PDL meter (JDS Fitel PS3) and found it to be less than 0.1 dB/cm. This low PDL value results from the slight core ridge roughness generated during fabrication.

Figure 5.75 shows the experimental setup for realizing EO phase modulators by retardation techniques. The polarized light, which was inclined at 45° to the optical axis of the KTN waveguide, was launched into the phase modulator using a polarizer and polarization-maintaining and absorption-reducing (PANDA) fiber. The second polarizer was positioned at the output. These two polarizers were arranged orthogonally in relation to each other to extinguish the input light. The modulated output light intensity was detected by supplying a voltage to the phase modulator.

The output light was modulated in accordance with the following equation:

$$I_{out} = I_{in}(1 - \cos\phi) \tag{5.46}$$

Here, I_{in}, I_{out}, and ϕ, respectively, are the input light intensity, output light intensity, and retardation (phase difference between the TE and TM modes) caused by EO effects. The retardation ϕ in KTN crystals is expressed by the following equation:

$$\phi = -\frac{\pi}{\lambda_0} n_0^3 (s_{11} - s_{12}) \left(\frac{V}{d}\right)^2 l \tag{5.47}$$

Here, λ_0, n_0, V, d, l, and s denote the input light wavelength, refractive index, applied voltage, electrode gap, electrode length, and EO constant, respectively,

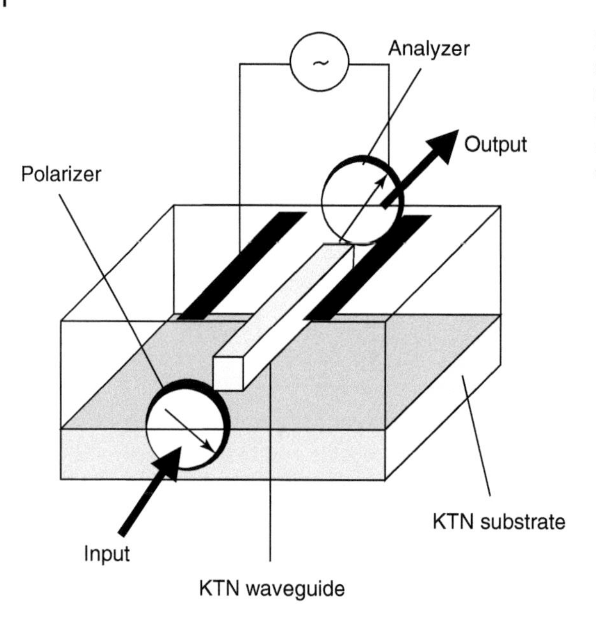

Figure 5.75 Experimental setup for realizing EO phase modulators by retardation techniques. Source: Toyoda et al. 2004 [36]. Reproduced with permission of IEEE.

and parameter s is expressed by the following equation:

$$s = g\varepsilon_0^2\varepsilon_r^2 \tag{5.48}$$

Here, g, ε_0, and ε_r denote a constant, the dielectric constant in a vacuum, and the dielectric constant in KTN crystals. The l and d values are 6 mm and 16 μm, respectively.

Figure 5.76 shows the optical output intensity against applied voltage. It is clearly seen that the output light modulation depends on the applied voltage. V_π is about 4.0 V. The local minimum was found around 6.5 V where the phase shift reaches 2π and the local maximum was around 9.6 V where the phase shift reaches 3π. This means that the V_π value can be reduced to 2.5 V by applying a biased potential of 4.0 V in advance. The $V_\pi \times L$ value is an important performance index for optical phase modulators. The value of $V_\pi \times L$ is 1.5 cm V,

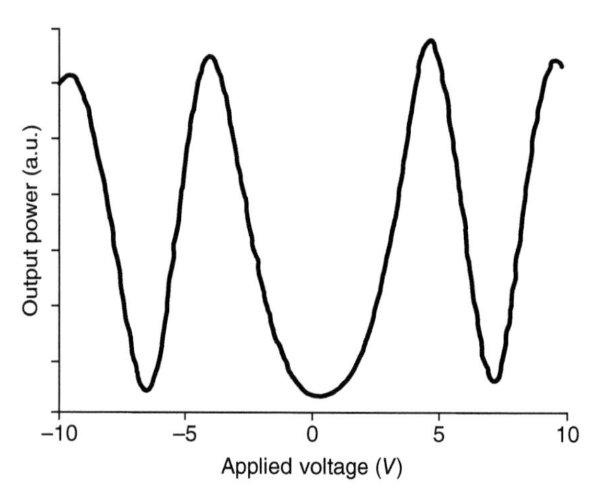

Figure 5.76 Optical output intensity against applied voltage. Source: Toyoda et al. 2004 [36]. Reproduced with permission of IEEE.

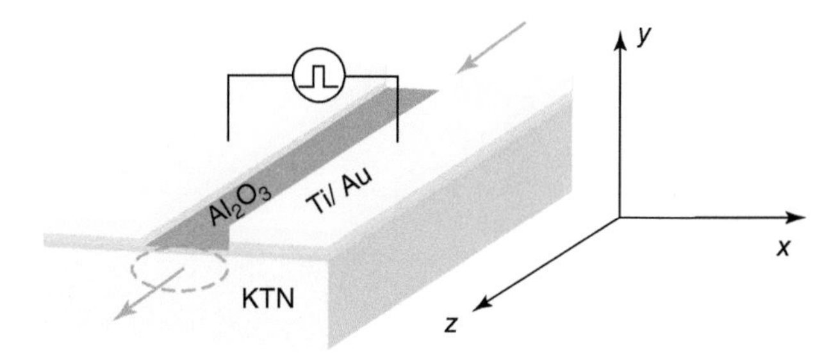

Figure 5.77 An illustration of configuration of dynamic optical waveguide based on KTN electro optical crystal. Source: Chang et al. 2012 [482]. Reproduced with permission of OSA Publishing.

which is less than one-third that of LN. Moreover, the electric field drop of the center core to the surface electric field was estimated by using a static electric field calculation and found to be about 0.34. The Kerr constant, $(s_{11} - s_{12})$, is estimated to be $2.6 \times 10^{-15} \left(\frac{m}{v}\right)^2$ from (5.47), which is a value comparable to that of the bulk crystal [37].

These results suggest that an optimized electrode and waveguide configuration will lead to a further decrease in the $V_\pi \times L$ value (about 0.5 cm V). It confirms the fast response of the EO phase modulator. The rise time was 187 ps at 3 GHz, which is comparable to that of the input electric signals. The modulation depth did not change between 1 and 3 GHz. This shows that KTN-based EO devices are useful for optical communication applications including transponders, packet switching, and optical interconnection.

In 2012, Chang et al. [482] designed a field induced dynamic waveguide based on KTN electro-optic crystals. Figure 5.77 illustrates the schematic configuration of the fabricated device. Two metal contacts with a gap of 50 μm were located on top of the crystal, which generated an electric field perpendicular to the z-axis. It was also assumed that the incident wave propagates along the z-axis. To quantitatively analyze the field induced index change of the proposed dynamic optical waveguide, the electric field distribution was calculated via finite element method (FEM).

Figure 5.78 shows the calculated field distribution on the x–y cross-section plane when the voltage difference between two contacts is 100 V. The arrows denote the vector of the electric field, $\mathbf{E} = E_x(x, y)\hat{x} + E_y(x, y)\hat{y}$, at points, (x, y), where the lengths proportionally present the magnitude of the electric field. One can observe that the field, parallel to x-axis, is dominated near the gap center, which is also the area of interest of the device. Furthermore, the gradient of field in x-direction, $E_x(x, y)$, is also displayed within the rainbow scale in units of volts per meter. The rainbow plot reveals that a strong electric field along x-axis direction exists near the central region of the electrode gap, which can induce significant refractive index change and enable the guiding effect.

The refractive index distributions and the corresponding light field distributions for the H- and V-polarized light beams were quantitatively computed under different levels of applied external electric fields. The simulation results

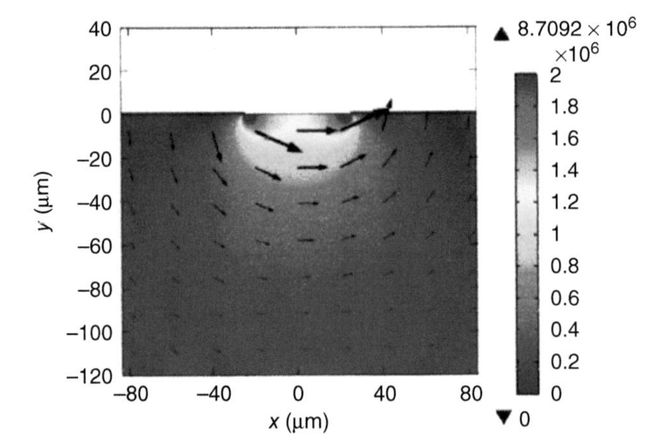

Figure 5.78 The simulated electric field distribution is plotted in *x*–*y* cross-sectional plane when the bias voltage is 100 V. The black arrows and the rainbow color, respectively, indicate the proportional field vector, $\vec{E}(x, y)$, and the field magnitude in *x*-component, $E_x(x, y)$ (V/m). Source: Chang et al. 2012 [482]. Reproduced with permission of OSA Publishing.

showed that the V-polarized light could be guided within the regions between two electrodes. The simulation results also indicated that not only the total intensity but also the distribution of the light fields could be controlled by adjusting the magnitude of the applied external electric field, which enabled the dynamic optical waveguide. The dynamic nature of the optical waveguide allowed to quickly control the guiding effect in optical waveguides, which could be a great benefit for many tunable optical devices such as VOAs and dynamic gain equalizers.

Furthermore, a multimode dynamic waveguide was fabricated and its performance including output light intensity distribution with and without applied external electric field was quantitatively evaluated. It could be observed that the experimental results agreed relatively well with the theoretical models, which confirmed the feasibility of the proposed dynamic optical waveguide. A fast response time (tens of nanoseconds) was also experimentally observed, which was much faster than the recently reported KTN crystal based optical beam scanner because the effect of slow electron injection was avoided via the Al_2O_3 barrier layer [37]. The tuning speed could also be further increased by reducing the RC time constant of the system.

In 2013, Chang et al. [508] designed a lower driving energy/power EO modulator based on a nanodisordered KTN crystal. They fabricated a transverse KTN based EO modulator, as shown in Figure 5.79a,b, respectively. The KTN crystal used in the transverse EO modulator had a composition of $KTa_{0.65}Nb_{0.35}O_3$ and a dimension of $10 \times 10 \times 0.5\,mm^3$. Gold electrodes with a width of 1.75 mm were coated on both top and bottom surfaces, as illustrated in Figure 5.79a. After the electrode coating, the optical aperture dimension of the modulator became $L = 10\,mm$, $w = 6.5\,mm$, and $t = 0.5\,mm$. Since both L and w were much larger than t, the electric field was quite uniform within the crystal.

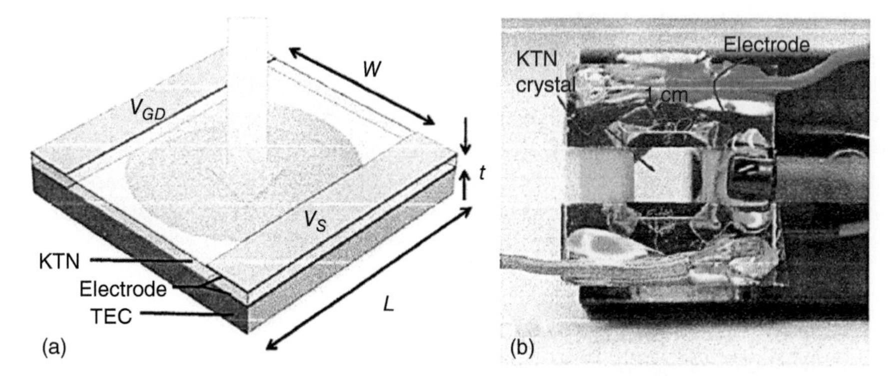

Figure 5.79 (a) A schematic sketch of transverse EO modulator; (b) picture of fabricated transverse EO modulator. Source: Chang et al. 2013 [508]. Reproduced with permission of OSA Publishing.

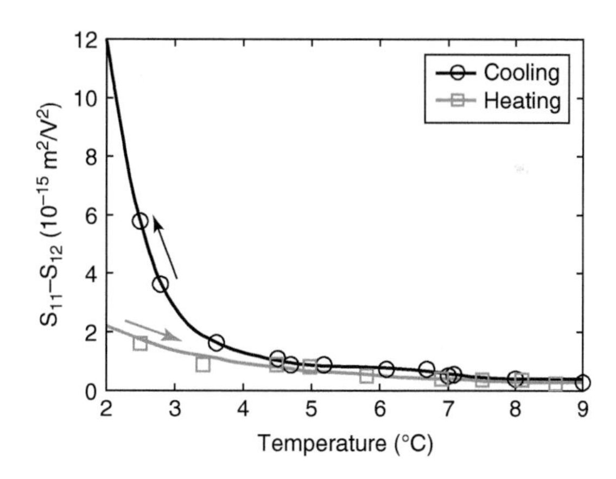

Figure 5.80 The experimentally measured transmission quadratic EO coefficient as a function of temperature. Circle line: decreasing temperature history; square line: increasing temperature history. Source: Chang et al. 2013 [508]. Reproduced with permission of OSA Publishing.

Figure 5.80 shows the experimentally measured QEO coefficients $(s_{11} - s_{12})$ as a function of temperature for different temperature cycling histories. The circle and square lines denote the temperature decreasing and temperature increasing processes, respectively. One can clearly see that the quadratic EO coefficients $(s_{11} - s_{12})$ were larger when the sample had a decreased temperature cycling history due to the formation of a nanodisordered KTN crystal. Also, the difference in EO coefficient between the decreasing and increasing temperature cycling histories becomes larger and larger when the temperature approaches the Curie point. For example, $(s_{11} - s_{12}) = 5.78 \times 10^{-15} \mathrm{m}^2/\mathrm{V}^{***\mathrm{kerr}2}$ with a decreasing temperature history at 2.5 °C. On the other hand, $(s_{11} - s_{12}) = 1.64 \times 10^{-15} \mathrm{m}^2/\mathrm{V}^2$ with an increasing temperature history at the same 2.5 °C measurement temperature. Thus, there is a 3.5 fold increase in quadratic EO coefficient, which results in reduction by a factor of 1.87 in the required driving voltage. This will be greatly beneficial for the construction of a broadband large field-of-view EO modulator due to the reduced driving energy and driving power.

Furthermore, Chang et al. [508] also experimentally demonstrated that this kind of EO modulator was also broadband (GHz range) and had a large (±30°)

field of view. Since the KTN crystal is highly transparent within 400–4000 nm spectral range [509] and the optical attenuation except the Fresnel surface reflection loss is less than a few percent, the modulation bandwidth could be further increased by taking advantage of the broad transmittance of the KTN crystal (400–4000 nm) such as via WDM technology [509]. The presented EO modulator could play an important role in a variety of applications such as modulating retro reflectors for broadband FSO communications, laser *Q*-switches, laser pulse shaping, and high-speed optical shutters.

5.6.8.2 KTN Fast Varifocal Lenses

NTT Photonics laboratories has recently developed much faster varifocal lenses by using single crystals of KTN ($KTa_{1-x}Nb_xO_3$). KTN is a perovskite-type oxide material that has a very strong electro-optic (EO) effect [44]. The EO effect is a phenomenon whereby the refractive index of a substance is modulated by an externally applied electric field. The common conventional material is LN. For LN, the effect is linear, which means that the index modulation is proportional to the electric field. KTN has a very strong quadratic effect where the index modulation is proportional to the square of the electric field. With an appropriate bias electric field, the effect is a few orders of magnitude stronger than that for LN.

The large and fast EO effect of KTN can be applied to varifocal lenses. With a simple device structure, Imai et al. [44] were able to obtain fairly large changes in focal length, from infinity to 1.6 m, in addition to quick responses, as fast as 1 μs, which is 3 orders of magnitude faster than the conventional commercial varifocal lenses. This high speed will lead to KTN varifocal lenses being applied to new technical fields such as laser manufacturing.

The structure of a KTN varifocal lens is schematically shown in Figure 5.81a. It consists simply of a KTN single-crystal block together with four metal electrode stripes. A light beam is incident on the top face of the block between a pair of parallel electrodes. The beam leaves the crystal from the bottom face where another electrode pair is formed symmetrically.

When a voltage is applied between the top and bottom electrode pairs, an electric field is generated in the crystal. This field is mainly formed in the two regions between the top and bottom electrodes. However, it also spreads into the gap region between the left and right electrodes through which the light travels. The field is stronger closer to the electrodes, so according to the EO effect, refractive index changes are larger near the electrodes than near the central optical axis. This index modulation, which was calculated from an electric field distribution obtained by the FEM, is shown in Figure 5.81b. The horizontal axis is the x-coordinate shown in Figure 5.81a with the origin at the optical axis. The vertical axis shows the optical path length obtained by integrating the refractive index along the z-axis. The KTN varifocal lens exhibits different behaviors for different light polarizations because of the crystallographic symmetry. For a light beam with its vibrating electric field parallel to the x-axis, the optical path length curve has a peak at the center, as shown by the red line. This means that the KTN device works as a convex converging lens. For a light beam with its vibrating electric field parallel to the y-axis, the curve has a trough at the center, as shown by the blue line, which means that the device works as a concave diverging lens. Imai

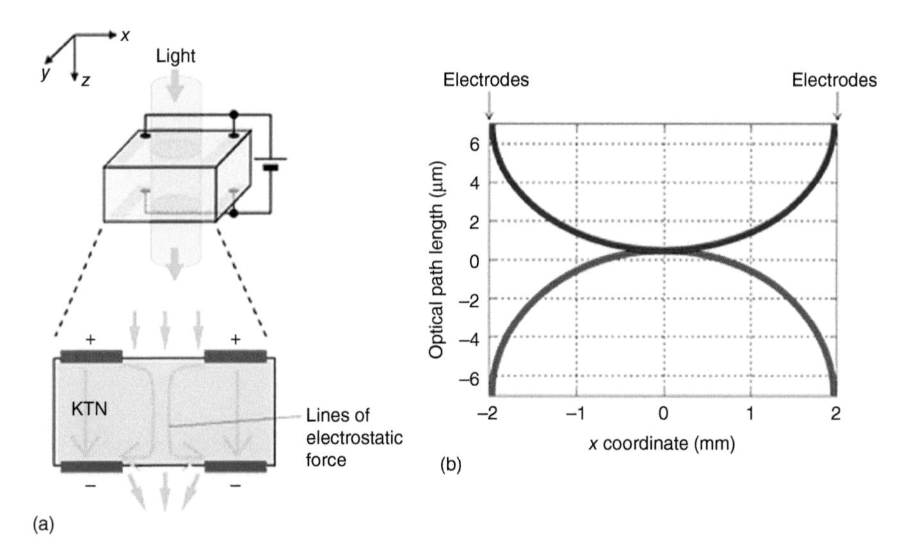

Figure 5.81 Structure and functions of KTN varifocal lens. (a) Structure of KTN varifocal lens and (b) calculated optical path length distributions. Source: Imai et al. 2009 [44]. Reproduced with permission of NTT Technical Review.

et al. [44] treated only light with its electric field parallel to the x-axis and thus converging lenses. For this polarization, the refractive index and the optical path length become lower near the electrodes than near the center region. Because light tends to be attracted to regions with high refractive indices, light rays are bent toward the center, and the KTN works as a converging lens.

Actually, this device works not as an ordinary spherical lens but as a cylindrical lens. A cylindrical lens focuses light onto a straight line because it converges light in only one direction, while an ordinary lens focuses light onto a spot because it converges light both vertically and horizontally. The KTN varifocal lens corresponds to a cylindrical lens because it converges light only in the x-direction in Figure 5.81a. However, a lens can be constructed that focuses in two dimensions, corresponding to an ordinary spherical lens, by assembling two KTN lenses as shown in Figure 5.82. The half-wave plate between the lenses rotates the polarization of the light so that the second KTN lens converges the light as well as the first lens rather than diverging it.

5.6.8.3 Electro-optic Deflectors

An EOD relies on the change of the refractive index n of a material as a result of an electric field E applied to an optically transparent material [510–512]. The latter is achieved by applying an electric voltage over the medium. The change in the refractive index is caused by electromagnetic forces that perturb the position, orientation, or shape of atom or molecule structure in the material. EODs refract a laser beam by introducing a phase delay across the cross section of the laser beam [289]. Two types of electro-optic effects are distinguished. If the refractive index varies linearly with the electric field, such as in $LiNbO_3$, $LiTaO_3$ and $KTiOPO_4$ crystals, this electro-optic effect is referred to as the Pockels effect [512–514].

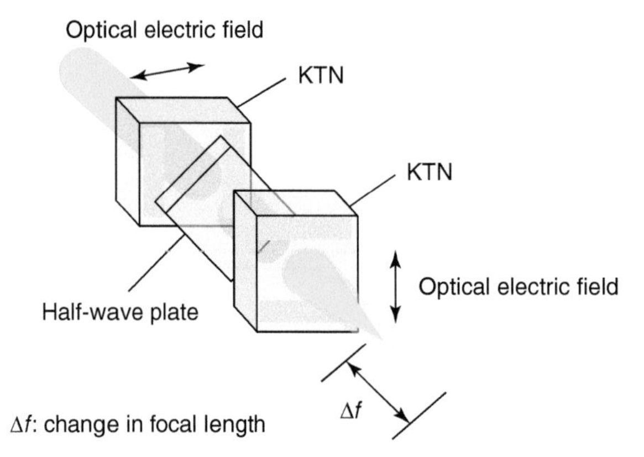

Figure 5.82 Two-dimensional convergence by two KTN lenses. Source: Imai et al. 2009 [44]. Reproduced with permission of NTT Technical Review.

If it varies quadratically with the field strength, as with E^2, it is referred to as the Kerr effect or QEO effect. All materials exhibit the Kerr effect, but it is usually much weaker than the Pockels effect. An exception is potassium tantalate niobate ($KTa_{1-x}Nb_xO_3$, or short KTN), which shows a large electro-optic effect at a particular temperature, which depends on the ratio of tantalate and niobate [285, 457]. Therefore, this crystal is to be actively controlled to a temperature just above the Curie temperature (typically chosen via the Ta_{1-x}:Nb_x ratio to be approximately 35 °C), at which the Kerr constant is at its maximum [457, 515]. Then, at a field strength of 500 V/mm the variation of the refractive index in a KTN crystal equals 1.52×10^{-2}, whereas at this field strength in a $LiNbO_3$ crystal (Pockels effect) the variation is only 9.0×10^{-5} [483]. Therefore, EODs based on KTN crystals show a larger maximum deflection angle as well as deflection velocity than EODs based on the Pockels effect.

Figure 5.83a shows an EOD based on refraction at the interface(s) of an optical prism, equipped with two electrodes and a driver/amplifier, inducing an electric field perpendicular to the electrodes. The angle of deflection θ_p is proportional to the width l of the base and inversely proportional to the height w of the prism [512].

$$\theta_p = \frac{\Delta n}{n}\frac{l}{w} \tag{5.49}$$

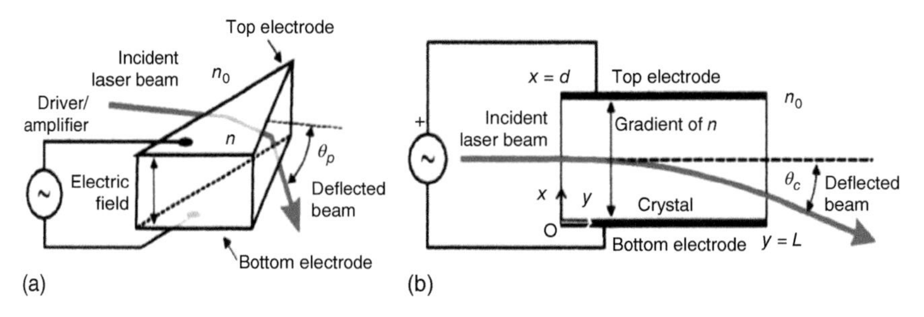

Figure 5.83 Typical configuration of an electro-optic deflector. (a) EOD based on refraction at the interface(s) of an optical prism. Adapted from [285] (b) EOD based on refraction by an index gradient that exists perpendicular to the direction of the propagation of the laser beam. Source: Adapted from Yagi 2009 [483].

where n denotes the refractive index of the crystal and Δn the difference between the refractive index of the crystal and the surrounding material. By placing several prisms in sequence, the maximum deflection angle can be increased. Then the number of prisms and the dimensions of the individual prisms are optimized for large angle of deflection, high damage threshold, and minimized reflection losses [510, 511]. Note that in this configuration, refraction (and therefore deflection) takes place at the interfaces of electro-optic and embedding material only. Therefore, maximum deflection angles are limited to a few degrees. Figure 5.83b shows an EOD based on refraction by an index gradient that exists perpendicular to the direction of the propagation of the laser beam. Besides the electro-optic crystal, it consists of two electrodes and a driver/amplifier. In this configuration, the laser beam is cumulatively refracted as it propagates through the crystal. Deflectors based on KTN crystals are of this type.

The fact that KTN shows larger deflection than other materials, such as $BaTiO_3$, $SrTiO_3$, and $Sr_xBa_{1-x}Nb_2O_6$, at the same field strength, is attributed to a specific distribution of the electric field perpendicular to the beam propagation, not observed in other Kerr cells [516]. This specific distribution is attributed to "trapped charges" inside the crystal, as a result of electrons injected into the crystal from the electrodes [517]. This effect was referred to as the *space-charge effect* and is superimposed on the Kerr effect [43, 289, 515]. Assuming the laser beam enters the crystal along the optical axis of the crystal (see Figure 5.83b) the angle of deflection due to a voltage V over the KTN crystal equals [483, 515]

$$\theta_e = -0.153 n^3 \varepsilon_0^2 \varepsilon_r^2 \frac{V^2}{d^3} L \tag{5.50}$$

where ε_o denotes the permittivity of vacuum, $\varepsilon_r = 3 \times 10^4$ the relative permittivity, and $n = 2.2$, the refractive index of KTN at $V = 0$, L denotes the propagation length of the laser beam in the crystal and finally d denotes the thickness of the crystal. The negative sign in Eq. (5.50) indicates that the laser beam is deflected toward the cathode. Hence, controlling the voltage allows to control the beam deflection angle. Substituting typical dimensions ($d = 1$ mm and $L = 6$ mm) and voltage $V = \pm 400$ V shows that the typical maximum deflection angle reaches $\theta_e = \pm 128$ mrad ($\pm 7.3°$), which is indeed large compared to an EOD based on the Pockels effect.

To enhance the deflection angle and beam size of the KTN deflector, NTT [55] has developed beam – shaping optics consisting of two cylindrical concave lenses as shown in Figure 5.84b. Controlling the beam by placing a lens in front of the deflector is essential for the efficient utilization of the deflector aperture. The lens on the output side expands the deflection angle while maintaining the beam at the same size as the incident beam. This new improved arrangement enabled to fabricate a swept light source for high resolution and deep imaging range OCT (optical coherence tomography) for a wide range of applications in the medical sector.

This product operates at a wavelength sweep speed of 200 kHz, double the speed of conventional products. In addition to the high-speed sweep at 200 kHz, the product has the following performance features: wavelength sweep span larger than 100 nm, average light output of 15 mW, and coherence length longer than 7 mm. This allows instantaneous imaging, thereby reducing examination

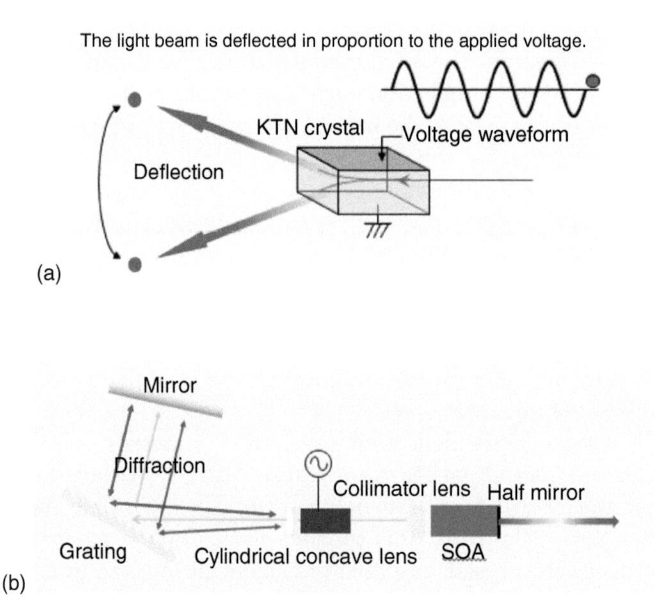

Figure 5.84 (a) Space-charge-controlled high-speed KTN deflector and (b) structure of wavelength swept light sources [55].

time. This high-speed operation also makes it possible to take high-resolution images with 4000 voxels at a rate of 50 fps, allowing an OCT system using this light source to produce a high-resolution 3D image of a living body.

In order to develop a high-resolution, high-speed multidimensional KTN deflector, Zhu et al. [518] studied the deflection behavior of KTN deflectors in the case of coexisting pre-injected space charge and composition gradient. They found that such coexistence can enable new functionalities of KTN crystal based EODs.

In particular, when the direction of the composition gradient is perpendicular to the direction of the external electric field (Figure 5.85), two-dimensional beam scanning can be achieved by harnessing only one single piece of KTN crystal, which can result in a compact, high-speed two-dimensional deflector. These new functionalities can expedite the usage of KTN deflection in many applications such as high-speed 3D printing, high-speed, high-resolution imaging, and free space broadband optical communication.

5.6.8.4 KTN Optical Beam Scanner

Using KTN crystal, Naganuma et al. [43] made a deflector for light rather than electrons, as shown in Figure 5.86. Since then, they have managed to find an explanation for this odd phenomenon: electrons are injected into the crystal through the cathode, forming a space charge distribution that, in turn, modifies the electric field between the electrodes so that the field becomes uneven. Such a field, combined with the electro-optic (EO) effect, which the KTN crystal exhibits, causes the light path to be bent [519].

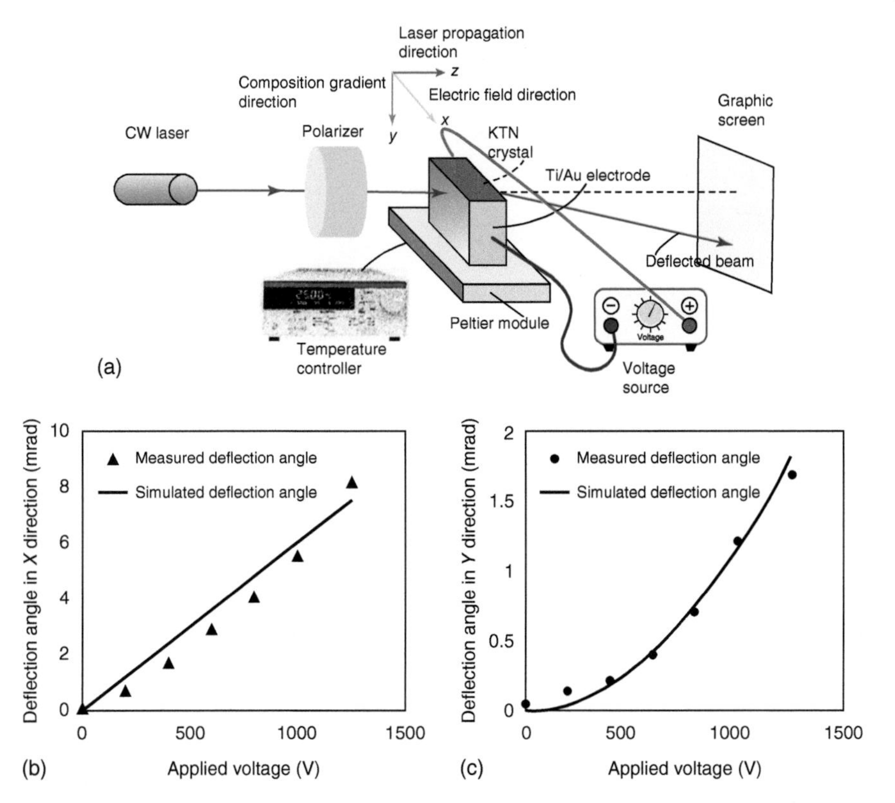

Figure 5.85 (a) The experimental system used to measure the deflection angle as a function of applied voltage. The electric field is perpendicular to the direction of the composition gradient. (b) The measured deflection angle and the computed deflection angle in the *x*-direction. (c) The measured deflection angle and the computed deflection angle in the *y*-direction. Source: Zhu et al. 2017 [518]. Reproduced with permission of AIP.

KTN meets both these requirements simultaneously, and it is almost unique among crystals in manifesting the optical deflector function based on such space-charge-controlled EO effect. Thus, the KTN optical scanner is a kind of EO device. Accordingly, it is fast and might be able to follow a sub-gigahertz (GHz) control voltage, as has been demonstrated for KTN modulators [37]. On the other hand, the refractive index gradation is induced all over the crystal and the bending effect accumulates along the entire light path in the crystal, so a fairly large deflection can be obtained. In this regard, the KTN scanner is quite different from the prism-based devices [512], where the bending effect is merely localized at the air–crystal interfaces.

In the KTN scanner, the optical beam should be focused onto the crystal with a small spot diameter of less than the crystal thickness d, so that the beam can pass through the narrow gap between the electrodes. Diffraction causes such a beam to inevitably have a fairly large divergence angle, as shown in Figure 5.87. Provided that the change in deflection angle caused by the applied voltage is smaller than the inherent divergence angle $\Delta\theta$, the deflection angle change is hardly discernible because it is overwhelmed by divergence. From this viewpoint, the total

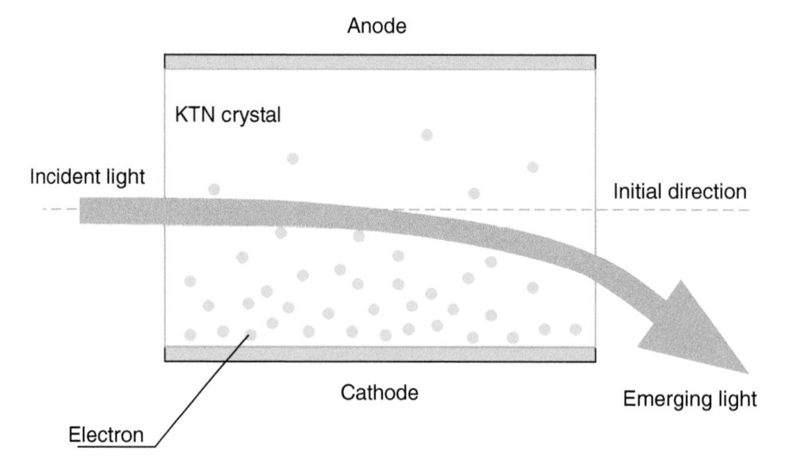

Figure 5.86 KTN optical scanner and its mode of operation. Source: Naganuma et al. 2009 [43]. Reproduced with permission of NTT Technical Review.

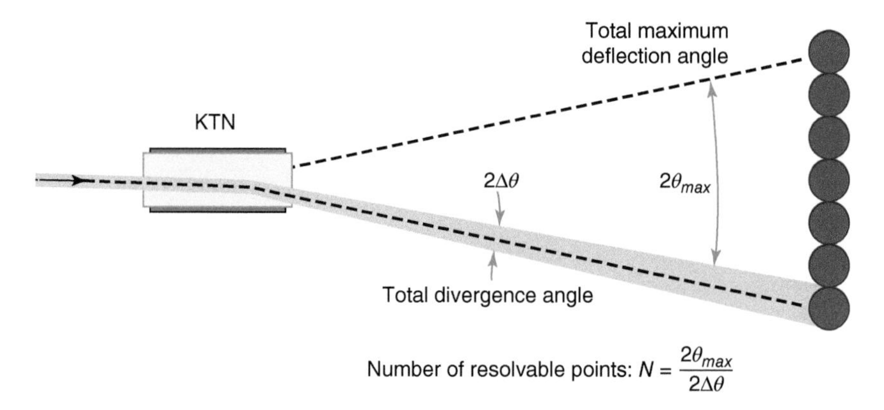

Figure 5.87 Number of resolvable points for the optical scanner. Source: Naganuma et al. 2009 [43]. Reproduced with permission of NTT Technical Review.

divergence angle $2\Delta\theta$ is like a granularity of the deflection angle for an optical scanner. Furthermore, the total range of deflection angle $2\theta_{max}$ normalized by the divergence angle, i.e. $N = 2\theta_{max}/2\Delta\theta$, represents how many points can be distinguished, namely, the number of resolvable points.

An analysis of the space-charge-controlled EO effect gives the following formula for the maximum deflection angle θ_{max}

$$\theta_{max} = \frac{1}{2}n^3 g_{11} E_{max}^2 \varepsilon^2 \frac{L}{d} \tag{5.51}$$

Here, d denotes the distance between the electrodes, i.e. the crystal thickness; n, ε, and g_{11} respectively express the refractive index, permittivity, and second-order electro-optic coefficient of the KTN crystal; and E_{max} represents the maximum electric field that can be imposed on the crystal. Since the electric field reaches its maximum value of 1.5 times its average of V/d right in front of the anode when the space charge is formed, it can be translated into a value $V_{max} = (2/3)E_{max}d$ for the maximum voltage that can be applied to the crystal.

On the other hand, for a light beam of wavelength λ, the divergence angle accompanying a radius of beam-waist w is expressed as $\Delta\theta = \lambda/(\pi w)$. This, along with Eq. (5.51), gives the following expression for the number of resolvable points N.

$$N = \frac{\pi}{2\lambda} n^3 g_{11} E_{\max}^2 \frac{w}{d} \varepsilon^2 L \tag{5.52}$$

As for the material constants of KTN, the refractive index n is 2.29 [285] at a wavelength $\lambda = 633$ nm, while the second-order electro-optic coefficient has been reported to be $g_{11} = 0.136\,\text{m}^4/\text{C}^2$ [286]. The electric field upper-bound E_{max} is known empirically to be around 600 V/mm. By plugging these values into Eq. (5.52) and using permittivity of 15 000, $N = 25.7\ (w/d)L$, where size factors w, d, and L are all assumed to be expressed in millimeters.

The beam diameter $2w$ cannot exceed the crystal thickness d by nature. Moreover, in this case, for an optical scanner, it is essential to leave room for the deflected beam. A simple geometric consideration tells that to maximize N, the optimal beam diameter should be half the crystal thickness, namely, $w/d = 1/4$. Taking this into account the devised rule of thumb: $N = 6.4\ L$. For instance, the number of resolvable points that can be expected with a crystal with $L = 5$ mm is at the most 32.

From the above discussion, it is shown that to increase the number of resolvable points either the crystal length L, or more precisely, the length of the path along which the electric field interacts with the light beam needs to be increased. Instead of actually lengthening the crystal of the KTN-crystal scanner, Naganuma et al. [43] tried to improve its resolution by using a multi-pass scheme (Figure 5.88) in which the light beam passes through the crystal multiple times. The KTN crystal they used had a thickness d of 1 mm and length L of 6 mm (of which the length of the interaction region was 5 mm). After depositing antireflection coatings on its light incident surfaces, Naganuma et al. [43]

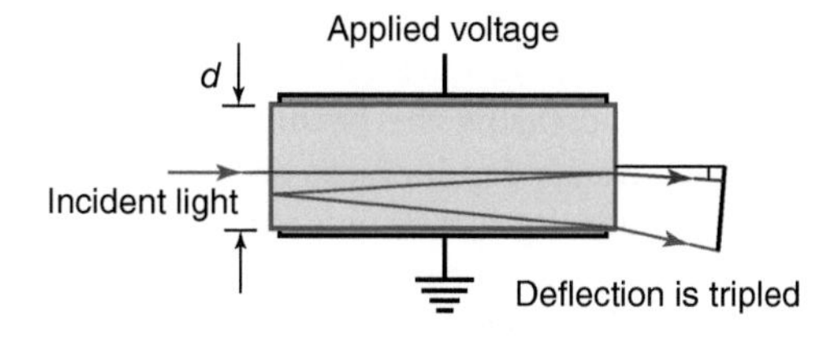

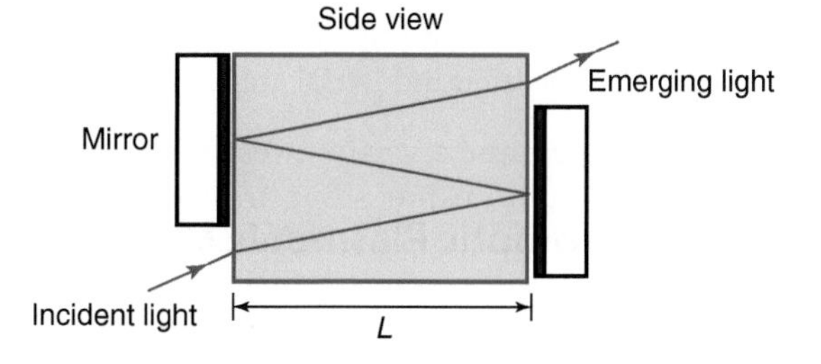

Figure 5.88 Multi-pass scheme for KTN optical scanner. Source: Naganuma et al. 2009 [43]. Reproduced with permission of NTT Technical Review.

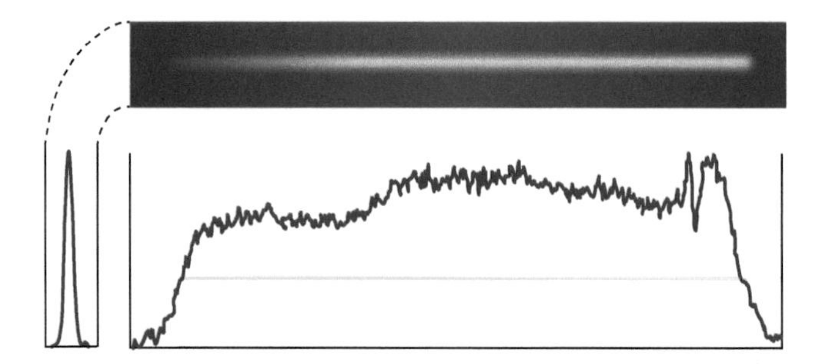

Figure 5.89 On-screen scanning trace giving an estimation of N. Source: Naganuma et al. 2009 [43]. Reproduced with permission of NTT Technical Review.

reflected light back into the crystal by placing mirrors close to its surfaces. In this way, they made a three-pass configuration. As the side view shows, the incident beam enters the crystal with a slight upward inclination, and after three trips through the crystal, it reaches the portion on the output surface that has no mirror and leaves as the emerging beam. A scan trace drawn by a deflected laser beam of wavelength $\lambda = 633$ nm on a screen placed on the output side is shown in Figure 5.89. The beam was focused onto the KTN crystal with a spot diameter of 300 μm, and the applied voltage was varied in the range of ±350 V. This arrangement tripled the deflection angle compared with the single-pass configuration, and the number of resolvable points was also tripled to become 61, as estimated from the length-to-width ratio of the trace.

A grown KTN crystal usually has some composition gradation along the growth direction. Composition variation manifests as a difference in phase-transition temperature T_c, which in turn causes a change in permittivity ε, leading to a larger change in Kerr constant that is proportional to ε^2. In short, the performance of the KTN scanner is critically dependent on the composition variation. Thus, the multi-pass scheme, which achieves deflection equivalent to a longer crystal, is particularly beneficial in the case of KTN.

To further improve the resolution, either raise the geometrical upper bound by increasing the crystal thickness d or boost the number of resolvable points per length L by increasing the permittivity ε. The latter approach is especially promising because a steep increase according to ε^2 is expected. For instance, if the relative permittivity ε_r is doubled to 30 000, one can expect to get 360 resolvable points, which rivals the resolution of current POSs (point-of sales) scanners. Thus, a large improvement in resolution can be expected with more homogenous crystals, which will be attainable through further refinements in the KTN growth technique. The resulting improvements in resolution will make KTN scanners applicable to a broader industrial arena.

5.7 Electro-optic Plasmonic Materials and Applications

Plasmonics enables the merging between two major technologies: nanometer-scale electronics and ultrafast photonics [520]. Metal–dielectric interfaces can support the waves known as SPPs that are tightly coupled to the interface, and allow manipulation of light at the nanoscale, overcoming the diffraction limit. Plasmonic technologies can lead to a new generation of fast, on-chip, nanoscale devices with unique capabilities [521, 522]. To provide the basic nanophotonic circuitry functionalities, elementary plasmonic devices such as waveguides, modulators, sources, amplifiers, and photodetectors are required. Various designs of plasmonic waveguides have been proposed to achieve the highest mode localization and the lowest propagation losses [522].

In addition to waveguides, modulators are the most fundamental components for digital signal encoding and are paramount to the development of nanophotonic circuits. In this regard, opto-electronic modulators can be designed to achieve ultrafast operational speeds in tens of gigahertz. Many plasmonic waveguide and modulator structures have been proposed and experimentally verified, but most of these structures use metals such as gold or silver, which are not CMOS compatible [45, 46, 50, 52, 523–537].

The promising development of chip-scale plasmonic devices with traditional noble metals is hindered by challenges such as high losses, continuous thin-film growth, and non-tunable optical properties. Moreover, noble metals as plasmonic building blocks are not compatible with the established semiconductor manufacturing processes. This limits the ultimate applicability of such structures for future consumer devices. Recently, there have been efforts toward addressing the challenge of developing CMOS compatible material platforms for integrated plasmonic devices [538]. Similar to the advances in silicon technologies that led to the information revolution worldwide, the development of new CMOS-compatible plasmonic materials with adjustable/tunable optical properties could revolutionize the field of hybrid photonic/electronic devices. This technology would help address the needs for faster, smaller, and more efficient photonic systems, renewable energy, nanoscale fabrication, and biotechnologies. These new materials can bring exciting new functionalities that cannot be achieved with traditional metals.

Recently, efforts have been directed toward developing new material platforms for integrated plasmonic devices [538–545]. Transparent conducting oxides (TCOs), for instance, ITO and aluminum-doped zinc oxide (AZO), are oxide semiconductors conventionally used in optoelectronic devices such as flat panel displays and in photovoltaics. The optical response of TCOs is governed by free electrons, whose density is controlled through the addition of n-type dopants. The free carrier concentration in TCOs can be high enough so that TCOs exhibit metal-like behavior in the near-infrared (NIR) and MIR ranges and can be exploited for subwavelength light manipulation (Figure 5.90). Moreover, in contrast to the optical properties of noble metals, which cannot be tuned or changed, the permittivity of TCOs can be adjusted via doping and/or fabrication process [539, 544, 547] (Figure 5.91), providing certain advantages for designing various plasmonic and nanophotonic devices.

The plasmonic properties of TCOs have been intensively studied over the past decade. The shortlist of successful experimental demonstrations includes (i) possibility of high doping [547, 548] and achieving negative permittivity for ITO [539, 549], AZO [539, 549–552], gallium-doped zinc oxide (GZO) [539, 550, 553, 554], and indium oxide doped with both tin and zinc (zinc–indium–tin oxide

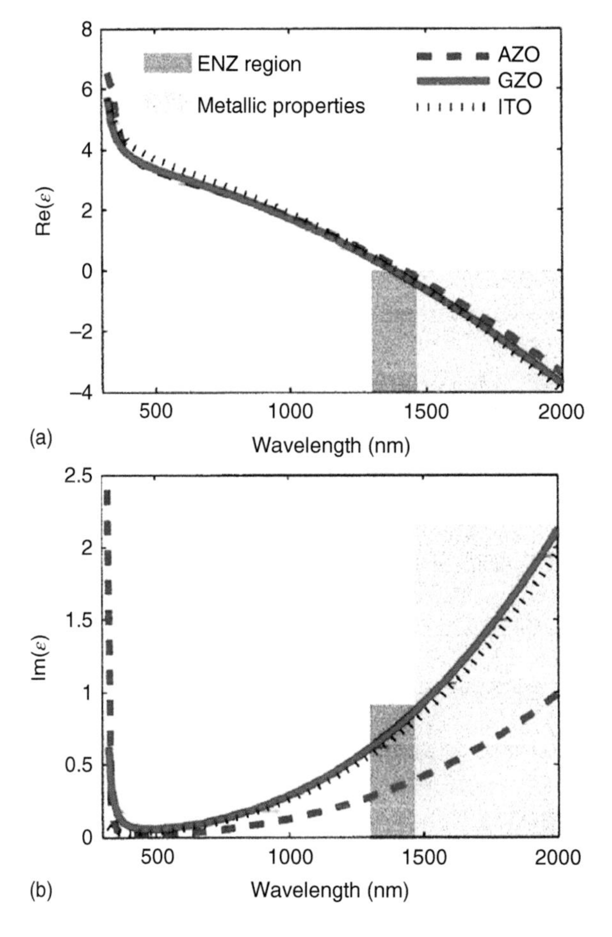

Figure 5.90 Optical properties of highly doped transparent conducting oxides: (a) real and (b) imaginary parts of permittivity. Shaded parts correspond to wavelengths where TCOs are metallic and allow plasmonic resonances. Crossover wavelength, that is ENZ region, is approximately at 1400 nm. Real values of permittivity of AZO, GZO, and ITO are similar. However, AZO possesses the lowest losses [546]. AZO, aluminum-doped zinc oxide; GZO, gallium-doped zinc oxide; ITO, indium tin oxide.

[ZITO], indium-tin-zinc oxide [ITZO]) [549]; (ii) SPP guiding in ITO [549, 555–557], AZO, and GZO [557]; (iii) tunability of localized surface plasmon resonance (LSPR) in colloidal nanocrystals of ITO [558, 559], AZO [560], and indium-doped cadmium oxide (ICO) [561]; (iv) LSPR in ITO nanorods and an increase in the electron concentration in the conduction band [562, 563]; (v) LSPR [557, 564] or SPP guiding [565] with pattern films; and (vi) negative refraction [566] as well as near-perfect absorption in multilayer structures with ZnO components [567]. Theoretical studies of GZO [568] and first-principle theoretical calculations of optical properties of AZO [569] were reported as well.

Importantly, TCOs possess a unique property of having a small negative real part of the permittivity at telecommunication wavelengths [539, 570]. Structures that exhibit such near-zero permittivity (epsilon-near-zero [ENZ]; [571, 572]) in the optical range can have a plasmonic resonance accompanied by slow light propagation. Thus, conditions for enhanced light–matter interactions are created [573].

An initial study of planar structures with silver layers showed that a plasmonic mode in such conditions experiences an unexpected giant modal gain. The

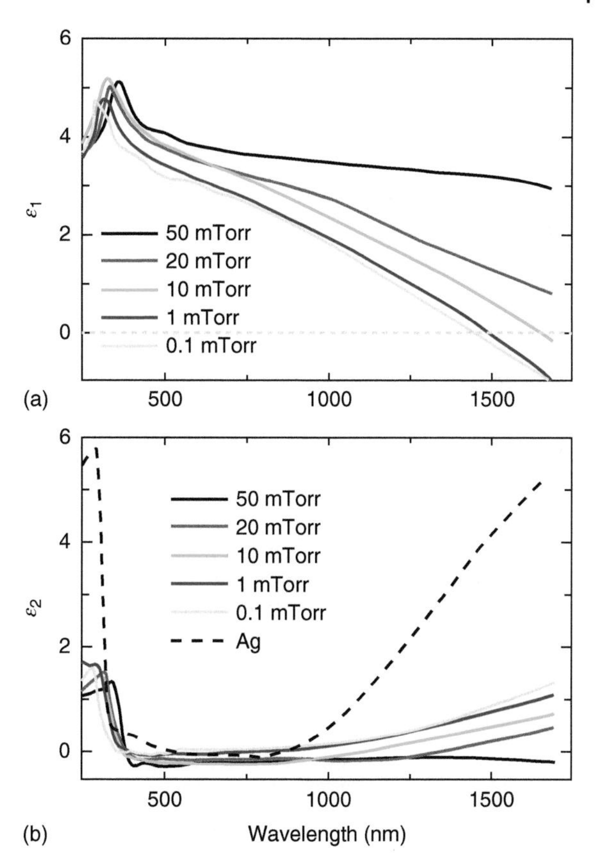

Figure 5.91 Strong dependence of AZO permittivity (real (a) and imaginary (b) parts) on oxygen pressures upon deposition (0.1–50 mTorr). The value can be changed up to several units. The film thickness is approximately 120 nm. Losses in AZO is several times lower than losses of silver (Ag) at telecommunication wavelengths. Source: Kim et al. 2013 [547]. Reproduced with permission of AIP.

physical origin of such an increase is the slowing down of light propagation in the structure [574, 575]. However, this effect can only be realized at $\lambda = 540$ nm, close to the crossover point of the silver permittivity. It cannot be in any way tuned to the NIR range, where the real part of the silver permittivity is too high in magnitude. In contrast to conventional noble metals, TCO materials provide conditions for plasmonic resonances and slow light propagation [573, 576]. Implementing a TCO layer in such multilayer structures instead of silver opens up a possibility to achieve slow light propagation at technologically important telecommunication wavelength $\lambda = 1.55\,\mu$m (Figure 5.92). Consequently, ENZ materials can drastically facilitate plasmonic nanolaser [579, 580] operation. According to numerical calculations, the extreme nonlinear dispersion and, correspondingly, extraordinarily high group indices can be achieved with only a moderate increase in the optical losses [581]. It allows designing a cavity-free plasmonic nanolaser with an almost zero threshold and operational sizes far beyond the diffraction limit [582].

In addition to linear properties, nonlinearities at the ENZ point are of interest and are propelled by the pronounced local-field enhancement facilitated by metal nanostructures [583–586]. Recently, the enhancement of nonlinear properties in a hybrid TCO-gold structure due to the plasmon-induced hot carrier injection and tight-field confinement in the vicinity of a plasmonic nanostructure has

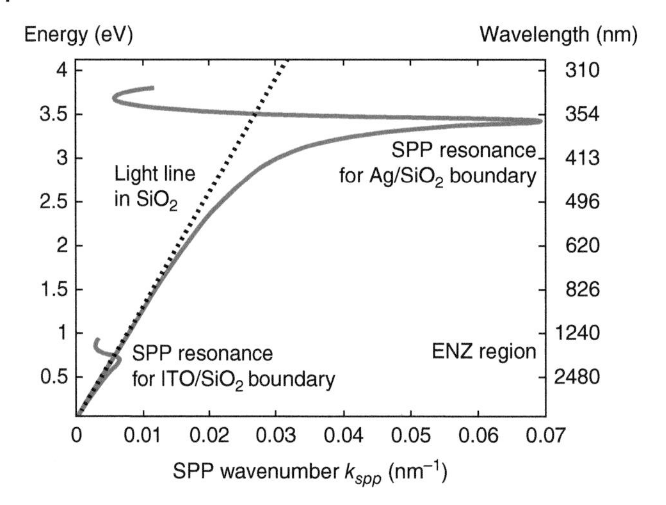

Figure 5.92 Dispersion of SPPs on silver/silica and ITO/silica interfaces. For the case of silver, the SPP resonance is in the ultraviolet range and cannot be tuned (detailed analysis of resonant properties can be found in Refs. [577, 578]). For the ITO film, the resonance is on the NIR wavelength and can be used for telecom applications [546]. ITO, indium tin oxide; NIR, near-infrared: SPP, surface plasmon polaritons.

been demonstrated [587–589]. The rapid energy transfer in the TCO produces a localized rise in the carrier temperature and increased carrier mobility, thus, providing a stronger nonlinear response. This ability to dramatically change the optical properties of TCOs from dielectric-like to metallic-like under an applied bias triggers new approaches for the realization of photonic modulators [48, 590].

To modulate and switch plasmonic signals in the waveguiding configuration, changes can be induced in either the real or imaginary parts of the permittivity by applying a bias. Traditional photonic approaches rely on small refractive index changes in the material forming the waveguides core. Consequently, it requires long propagation distances to accumulate the sufficient phase or absorption changes, which result in increase in the size of devices (up to hundreds of micrometers). In contrast, TCOs are promising candidates for adding electro-optical capabilities to ultracompact plasmonic devices. A large increase in the carrier concentration within a10 nm thick ITO film of a metal-oxide-semiconductor (MOS) stack is demonstrated in Ref. [46]. Furthermore, additional functionalities can be gained by tuning a small absolute value of the TCO permittivity in and out of the plasmonic resonance [49]. The theoretically predicted extinction ratio (ER) is 50 dB/μm [49]. Such a high value allows for the 3 dB modulation depth in an extremely compact, only 60 nm long device. The effect is based on the carrier concentration changes, and thus, the associated timescale is on the order of picoseconds. Therefore, frequency modulation speed of around 100 GHz or even higher is expected [46].

In general, for plasmonic modulators, two classes of active materials were suggested based on two different physical mechanisms for the refractive index control: carrier concentration change (used in e.g. silicon [45, 51, 591–597] and TCOs [46, 49, 52, 53, 534, 598–606]) and structural phase transitions (e.g. gallium [607–610] and vanadium dioxide [535, 611–615]). In addition, plasmonic modulators based on thermo-optic [531, 532, 616, 617] or nonlinear polymer [47, 58], graphene [618–620], indium phosphide [536, 621], barium titanate [622], bismuth ferrite [623], and germanium [624] were also reported.

The refractive index changes that accompany nanoscale material phase transitions are much higher than those caused by the carrier concentration change. In particular, an ER up to 2.4 dB/μm was demonstrated for a hybrid plasmonic modulator based on the metal–insulator phase transition in vanadium dioxide [612]. However, neither structural transitions nor thermo-optic responses provide the necessary bit rate, so only megahertz modulation operations are expected due to microsecond timescales of the aforementioned alternative processes.

An experimental study performed for a plasmonic latching switch, whose design is similar to the nanophotonic modulators based on conventional silicon waveguides, revealed the possibility of the creation and elimination of a conductive path in a gold/silicon dioxide/ITO layered structure with a switching speed on the order of a few milliseconds [625, 626]. Similar to processes in an a-silicon layer, changes in the optical transmission are attributed to the formation and annihilation of nanoscale metallic filaments [627].

The carrier concentration change in the accumulation layer makes faster structural phase transitions, and thus optoelectronic modulators can theoretically achieve ultrafast operational speeds in the hundreds of gigahertz range, being limited only by the RC time constant. It was shown that some plasmonic-based designs even outperform conventional silicon-based modulators [46, 592].

Effective modulation performance was obtained with horizontally arranged metal–insulator–silicon–insulator metal slot waveguides [51, 596, 597]. The governing principle is based on inducing a high free electron density in the infinitesimal vicinity of the SiO_2/Si interface (Figure 5.93). Such electro-absorption CMOS-compatible modulation was characterized, and a 3-dB broadband modulation with just 3 μm length of the device at approximately 6.5 V bias was reported [596]. Propagation losses varied between 1 and 2.9 dB/μm, which correspond to an FoM $f = 1.9$. The advantage of this configuration is that it can be readily integrated in standard Si circuits. However, its high mode localization requires a very high aspect ratio of the waveguide core, which is challenging to fabricate.

Ultracompact efficient plasmonic modulators based on strong light localization in vertically arranged metal–silicon–insulator–metal waveguides were reported in Ref. [45]. The PlasMOStor consists of the semiconductor core with Si and SiO_2 layers, sandwiched between two silver plates [45]. It supports two modes: photonic and plasmonic, which interfere while propagating. The modulation principle is based on cutting off the photonic mode and thereby changing the integral transmittance. For the photonic mode, propagation losses can be changed from 2.4 to 28 dB/μm, providing the possibility to operate with high $f = 11$.

In contrast to silicon, changing the carrier concentration in TCOs can result in a much higher change in the effective refractive index. In particular, a striking unity order change $\Delta n = 1.5$ was experimentally demonstrated [590], which is 3 orders of magnitude higher than the maximal change in the effective refractive index in a silicon waveguide: $\Delta n = 10^{-3}$ [628]. Thus, TCOs can be efficiently

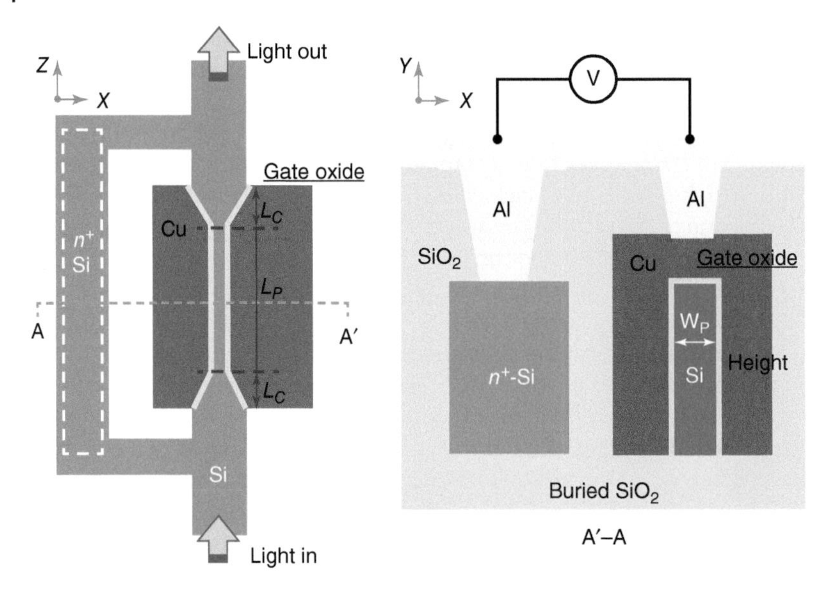

Figure 5.93 Schematic view of the horizontally arranged silicon-based plasmonic electro-absorption modulator. Source: Zhu et al. 2011 [596]. Reproduced with permission of AIP.

exploited as active materials for light modulation providing extraordinary tuning of their complex refractive indices by changing the carrier concentration with the application of an electric field.

5.7.1 Transparent Conducting Oxides

The first experimental demonstration of the carrier concentration change was made for a relatively thick TCO layer (thickness 300 nm) in a metal–insulator–metal (MIM) structure. The carrier concentration growth from 1×10^{21} to 2.83×10^{22} cm^{-3} was achieved in a 5 nm thick accumulation layer [590].

The general concept of the carrier concentration change is shown in Figure 5.94. Applied bias causes either charge accumulation or its depletion in the TCO layer, depending on the direction of the electric field. In turn, variations in the carrier concentration $n_{c.c.}$ affect the TCO plasma frequency ω_{pl}, and consequently, its permittivity ε_{acum}, which follows the Drude model $\varepsilon_{acum}(\omega) = \varepsilon_\infty - \omega_{pl}^2/\omega(\omega + i\gamma)$, where $\omega_{pl}^2 = n_{c.c.}e^2/\varepsilon_0 m_{eff}$, ε_∞ is the background permittivity, γ is the collision frequency, e is the charge of an electron, m_{eff} is the effective mass of an electron, and ε_0 is the vacuum permittivity. The process can be described by the Thomas–Fermi screening theory [52, 534]. One can either find the spatial distribution of free carriers directly from the Thomas–Fermi screening theory or use instead its exponential approximation with the Thomas–Fermi screening length (Figure 5.95a). Evidently, the TCO permittivity can reach relatively low values (Figure 5.95b).

Figure 5.98 shows that utilizing an HfO_2 layer, which possesses a high-static dielectric constant $k = 25$, instead of SiO_2 with ordinary $k = 4.5$, is very beneficial.

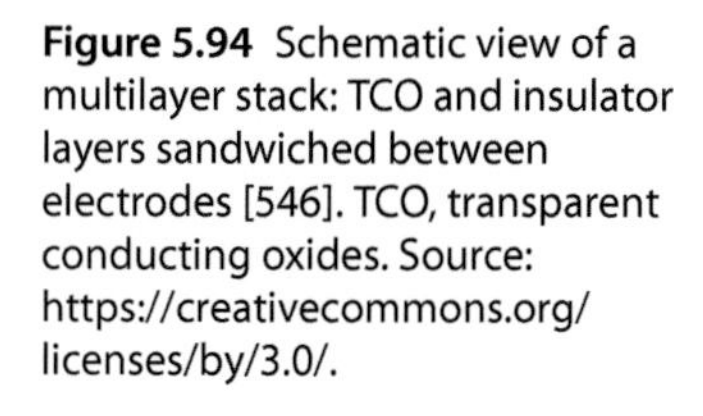

Figure 5.94 Schematic view of a multilayer stack: TCO and insulator layers sandwiched between electrodes [546]. TCO, transparent conducting oxides. Source: https://creativecommons.org/licenses/by/3.0/.

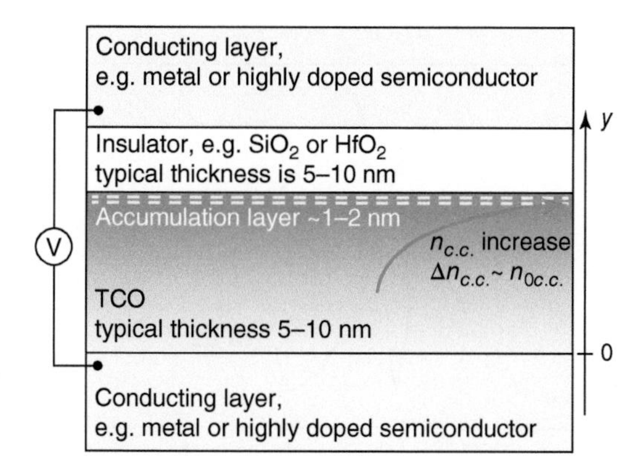

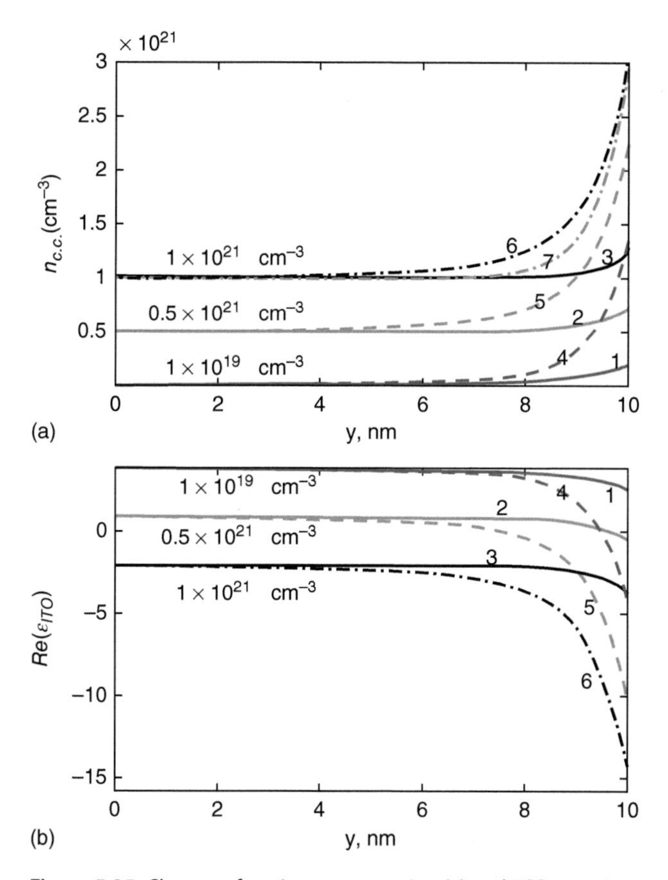

Figure 5.95 Change of carrier concentration (a) and TCO permittivity (b) according to the Thomas–Fermi screening theory. 2.5 V per 5 nm thick SiO_2 or HfO_2 layer was taken for the calculations. 10^{19}, 5×10^{20}, and 10^{21} cm^{-3} correspond to different initial concentrations of carriers in the TCO layer. Solid curves 1, 2, and 3 correspond to SiO_2 and dashed curves 4, 5, and 6 correspond to HfO_2. In (a), curve 7 shows approximation with the exponential decay defined by the Thomas–Fermi screening length [546]. TCO, transparent conducting oxides. Source: https://creativecommons.org/licenses/by/3.0/.

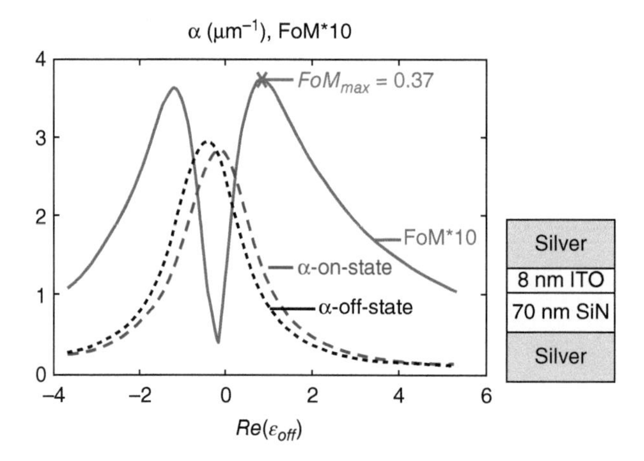

Figure 5.96 Influence of ITO permittivity on the absorption coefficient and FoM of the metal–insulator–metal–waveguide-based modulator. To obtain the complex propagation constant, the dispersion equation was solved for a four-layer structure, that is, 8 nm thick ITO and 70 nm thick silicon nitride insulation embedded in silver claddings, where the real part of the ITO permittivity was varied. ITO, indium tin oxide; FoM, figure of merit. Source: Babicheva and Lavrinenko 2012 [599]. Reproduced with permission of Elsevier.

For the same applied voltage, the carrier concentration induced in the accumulation layer can be several times higher than that in the case with SiO_2, providing larger change in the permittivity value. As the effect occurs only in a 5 nm accumulation layer, the rest of the material reacts passively for the applied voltage. It does not bring additional functionality, but rather increases optical losses significantly. Thus, the focus of the study shifted later toward ultrathin (8–10 nm) active films [46, 534, 600]. Since the TCO's properties depend strongly on fabrication conditions such as the doping level, annealing environment, pressure, and temperature [539, 544], it allows one to adjust TCO's permittivity, and hence, to tailor the material for a specific device design [599], as shown in Figure 5.96.

5.7.2 Ultracompact Plasmonic Modulators

MIM configurations, where the dielectric core is sandwiched between two metal plates, were shown as one of the most compact and efficient waveguide layouts [629–632]. In such waveguides, light is localized in the gap with typical sizes of approximately 100 nm and less, which facilitate manipulation of light on the subwavelength scale. The SPP propagation length in the MIM waveguide is not extended and can be up to 10 μm only. However, the tight MIM mode confinement provides a base for designing extremely fast and ultracompact plasmonic devices, including modulators, photodetectors, lasers, and amplifiers, where the long propagation length is not required.

In the MIM configurations, an additional layer such as an ultrathin TCO film allows for control of the dispersion in the waveguide, and thus, a higher efficiency plasmonic modulator in comparison to other switchable materials. In addition to the high speed and compactness of the MIM configurations, the metal plates can serve as intrinsic electrodes for electrical pumping of the active material, offering a simplified design.

One of the first proposed concepts of the TCO-based modulator includes layers of oxide and semiconductor, which are embedded in metal. For example, ITO was implemented in an MIM waveguide structure to demonstrate a subwavelength plasmonic modulator (Figure 5.97a,b), in which a 5% change in the average carrier density (from 9.25×10^{20} cm^{-3} to 9.7×10^{20} cm^{-3}) was assumed [534]. The authors employed a one-dimensional, four-layer structure model to find the SPP dispersion and the Thomas– Fermi screening theory to determine the carrier density distribution. Numerical simulations with the FEM were used to model the performance of a modulator with realistic dimensions. The structure supports an SPP having resonance conditions close to 1.55 μm due to the ITO layer, which has

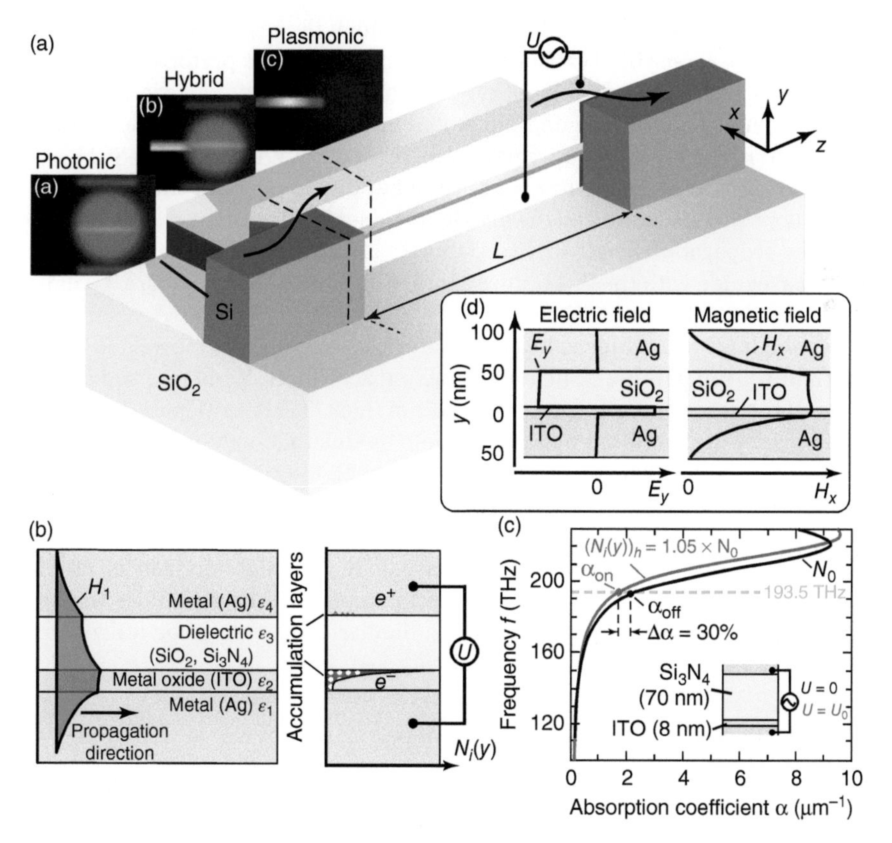

Figure 5.97 (a) SPPAM: the modulator section consists of a stack of silver, ITO, and silica layers, and can be integrated by means of a directional coupler with a silicon waveguide. The insets show how a photonic mode in the silicon waveguide excites the MIM waveguide mode in SPPAM via a hybrid mode in the directional coupler. (b) The absorption coefficient of the SPP is modulated by applying a voltage between two silver electrodes. (c) The MIM waveguide possesses a plasmonic resonance and is strongly affected by changes in the permittivity of the active layer. ITO, indium tin oxide; MIM, metal–insulator–metal; SPPAM, surface plasmon polariton absorption modulator. Source: Melikyan et al. 2011 [534]. Reproduced with permission of OSA Publishing.

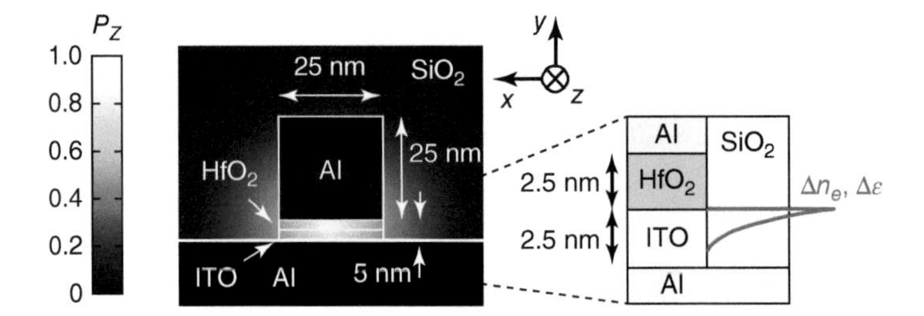

Figure 5.98 Design of nanowire-based modulator with the size 25 × 25 nm^2: SPP mode in a nanowire-MIM waveguide and a zoom of the active region with the change of electron density Δn_e. MIM, metal–insulator–metal; SPP, surface plasmon polaritons. Source: Krasavin and Zayats 2012 [52]. Reproduced with permission of APS.

small permittivity values in the NIR range, although the resonance is broadened because of high losses in ITO. Consequently, this decreases performance of the device but increases the operational bandwidth at the same time. A similar structure based on a silicon-waveguide-integrated multilayer stack was fabricated and an ER up to $r = 0.002$ dB/μm was achieved. Theoretically, the Ag-ITO-Si_3N_4-Ag structure can have an ER up to $r = 2$ dB/μm, but due to the high-mode confinement in the MIM structure and the high losses associated with the metal and ITO layers, the propagation length in this system is rather limited: losses are 24 dB/μm for the waveguide with the Si_3N_4-core and 9 dB/μm with the SiO_2-core. Overall, such a modulator is extremely compact.

A very high ER was achieved in a modulator based on an MIM waveguide with an ultrathin 5 nm gap [52]. Such an ultracompact design leads to a significant change of the carrier concentration in a 2.5 nm thick ITO layer (Figure 5.98). The total length of the proposed nano-plasmonic modulator, including the coupling tapers, can be below 500 nm, providing a very small footprint acceptable for the on-chip integration.

The suggested plasmonic modulator can be a part of a MIM waveguide or coupled to another type of a plasmonic waveguide with high mode localization. In addition, there are various possibilities to couple MIM waveguides to conventional photonic waveguides [633–635]. Similar concepts can be utilized for plasmonic waveguide modulator structures.

5.7.3 Silicon Waveguide-Based Modulators

The MIM design provides a very high ER, and consequently allows an ultracompact layout. However, device fabrication, its integration, and characterization are challenging. While Melikyan et al. [534] proposed the MIM design, proof-of-concept demonstration was performed with a conventional silicon waveguide-based structure, where a MOS stack was deposited on top of a silicon waveguide (Figure 5.99).

A similar approach was taken in Refs. [46, 598, 600, 601] where silicon photonic modulators used the TCO film as a dynamic layer. An ER $r = 1$ dB/μm was demonstrated for a plasmonic modulator utilizing a metal-oxide ITO stack on top of a silicon photonic waveguide ([46], Figure 5.100). Under applied

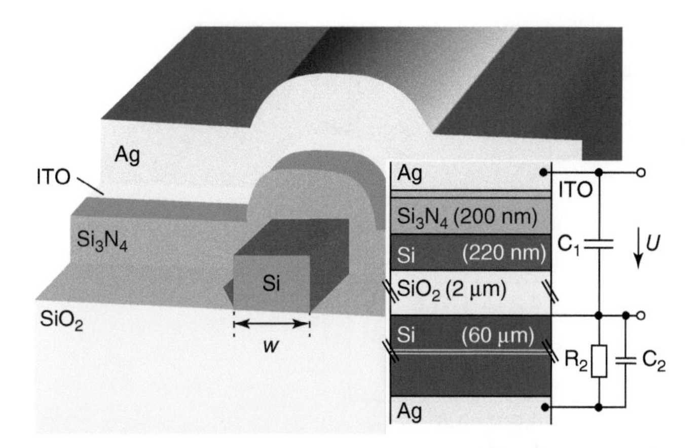

Figure 5.99 Hybrid design of plasmonic/photonic absorption modulator. Source: Melikyan et al. 2011 [534]. Reproduced with permission of OSA Publishing.

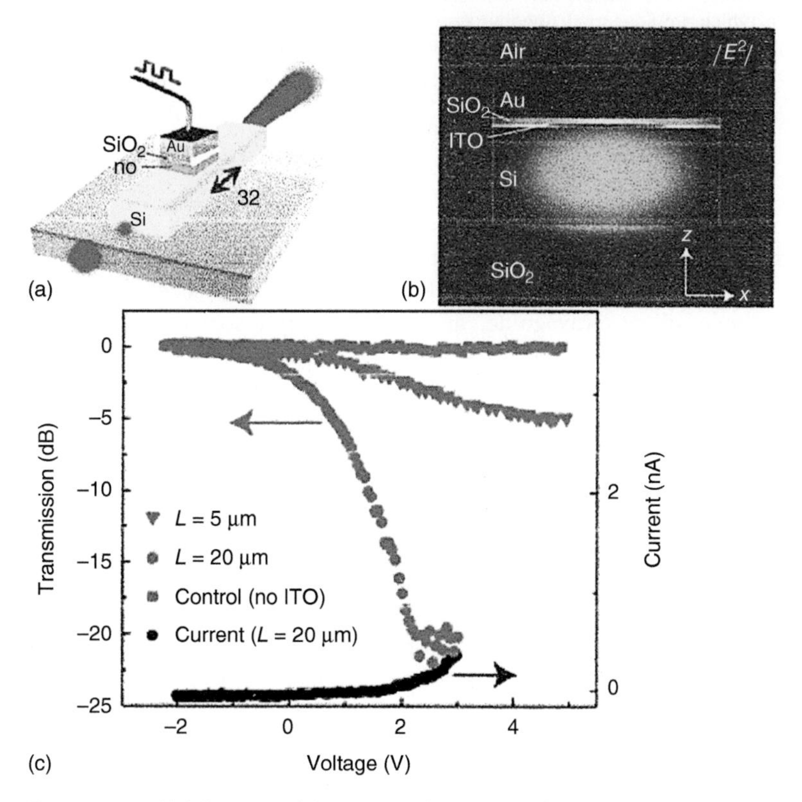

Figure 5.100 (a) Schematic of the waveguide-integrated silicon-based nanophotonic modulator. (b) Electric field density across the active MOS region of the modulator. (c) Experimentally demonstrated ER up to 1 dB/μm. MOS, metal–oxide–semiconductor. Source: Sorger et al. 2012 [46]. Reproduced with permission of Nanophotonics.

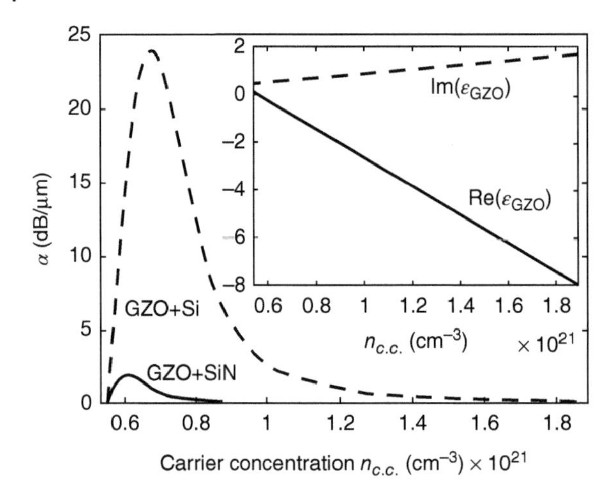

Figure 5.101 Increase of the absorption coefficient in a multilayer waveguide for the carrier concentration, which provides small negative permittivity (see inset), required for acquiring the plasmonic resonance at the working frequencies. Multilayer structures with the high-index claddings, for example, silicon, give much higher values of changes in the absorption coefficient, i.e. ER, in comparison to the low-index claddings. Source: Babicheva et al. 2013 [49]. Reproduced with permission of OSA Publishing.

bias, the carrier concentration is increased from 1×10^{19} to $6.8\times10^{20}\,\text{cm}^{-3}$, reducing the propagation length from 34 to 1.3 μm. Such waveguide-integrated plasmonic modulators are relatively easy to fabricate, their size is comparable to a single silicon waveguide, and propagation losses originate only from one metal interface. The configuration possesses low insertion loss (1 dB) and a broad operational band due to the non-resonant MOS-mode propagation.

One of the most important steps in developing TCO modulators is utilizing its small value of permittivity (Figure 5.101). It was pointed out in the first paper by Melikyan et al. [534] (Figure 5.97c) and the optimal operation was studied further [599] (Figure 5.96). However, because of the high initial carrier concentration in TCOs, the structure did not give high performance. Later, it was theoretically predicted that a very high ER (up to $r = 20$ dB/μm) can be achieved utilizing ENZ properties of AZO [598]. For a pure photonic system (e.g. [598, 600], Figure 5.102), the ENZ condition $\varepsilon = 0$ causes a resonance. Owing to the small absolute value of the permittivity, a large portion of the electric field is localized within the TCO layer providing efficient modulation. If plasmonic elements are involved, the resonance condition shifts toward negative values $\varepsilon = -4, \ldots, 0$. The effect was discussed in detail in Ref. [49]. Later, in [601], it was mentioned that the ENZ point and absorption maximum do not coincide, which happens because metal layers (TiN and Cu) are put on a silicon waveguide (Figure 5.103). In general, recently proposed designs of silicon waveguide-integrated TCO modulators are similar but include a different number of additional layers, for example Refs. [601] and [605].

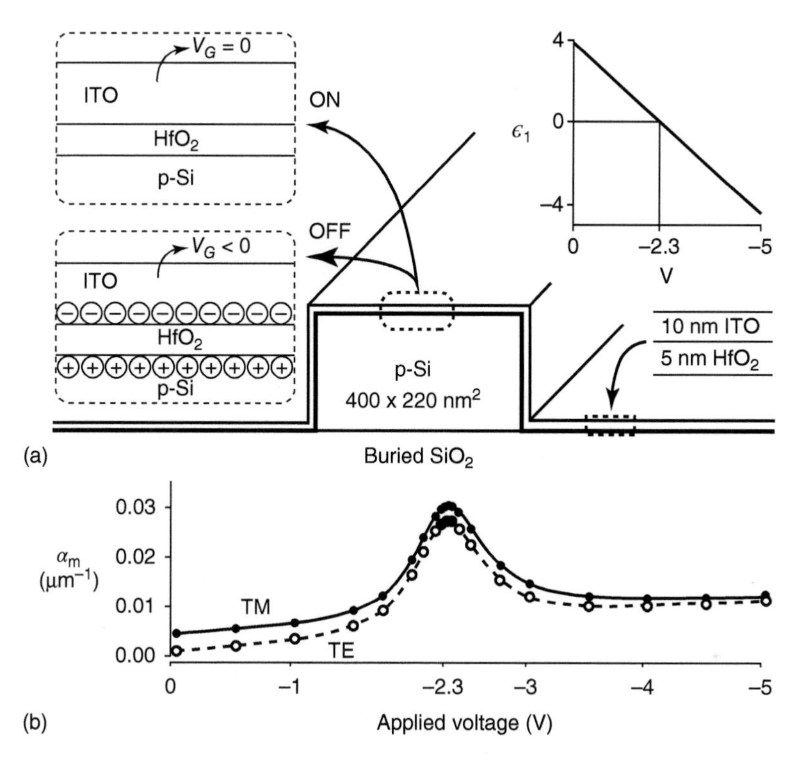

Figure 5.102 (a) One of the simplest yet efficient structures includes only two additional layers (HfO_2 and ITO). Inset: real part of the TCO permittivity in the accumulation layer (approximately 1 nm) under different voltage. (b) Absorption coefficient reaches maximum at the ENZ point. ENZ, epsilon-near-zero; ITO, indium tin oxide; TCO, transparent conducting oxides. Source: Vasudev et al. 2013 [600]. Reproduced with permission of OSA Publishing.

5.7.4 CMOS Compatibility

Numerous plasmonic waveguides and modulators have been proposed and experimentally verified, but most of these structures incorporate noble metals such as gold or silver as plasmonic building blocks. However, noble metals are not compatible with the well-established CMOS manufacturing processes and pose challenges in fabrication and device integration. This fact limits the ultimate applicability of plasmonic structures based on noble metals in future optoelectronic devices. Thus, the next step toward practical implementation of plasmonic technologies requires the development of fully CMOS-compatible passive and active integrated-optics components. Recently, CMOS-compatible materials with adjustable/tailorable optical properties have been proposed as alternatives to noble metals in plasmonic devices [541, 544, 545, 637]. New configurations open up a way for dynamic switching and modulation capabilities offering low cost and highly stable nonlinear properties. Similar to the advances in silicon technologies that led to information revolution worldwide, the development of new CMOS compatible plasmonic materials could revolutionize the field of hybrid photonic/electronic devices.

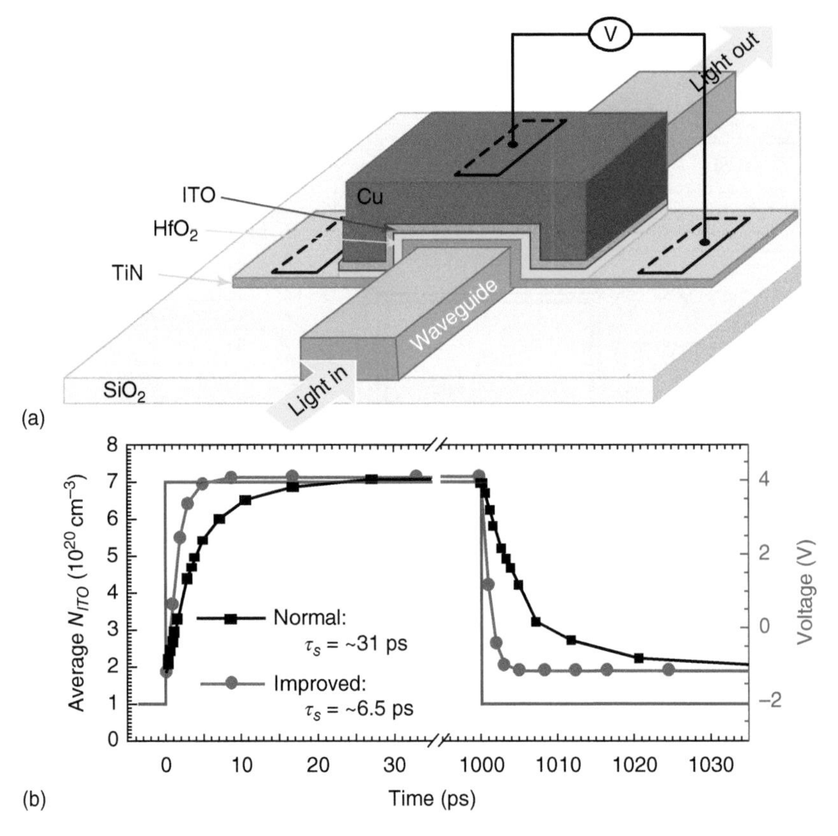

Figure 5.103 (a) Modulator design [636] includes two additional metal layers: TiN and Cu. (b) Numerical modeling of the change in the carrier concentration. Cutoff frequency is approximately 11–54 GHz. Source: Zhu et al. 2014 [601]. Reproduced with permission of OSA Publishing.

Plasmonic ceramics, such as titanium nitride and zirconium nitride, are among the best candidates that can replace conventional plasmonic metals [638, 639]. Titanium nitride is CMOS compatible and provides higher mode confinement in comparison to gold [638]. Furthermore, titanium nitride is thermally and chemically stable, extremely hard (one of the hardest ceramics), and biocompatible [636, 640–642]. It can be grown epitaxially on many substrates including c-sapphire and [578] silicon, forming ultra-smooth and ultrathin layers [643]. It was theoretically shown that a TCO modulator integrated with a long-range TiN-strip waveguide can provide a high ER (up to $r = 46$ dB/μm) as well as low losses in the transmittive state ([49], Figure 5.104).

5.7.5 Perspectives of EO-Plasmonic Materials

The MIM-modulators offer high compactness (3 dB modulation can be achieved on a sub-100 nm scale), but the overall performance (e.g. FoM) is lower than that of silicon waveguide-based devices. However, because conventional dielectric

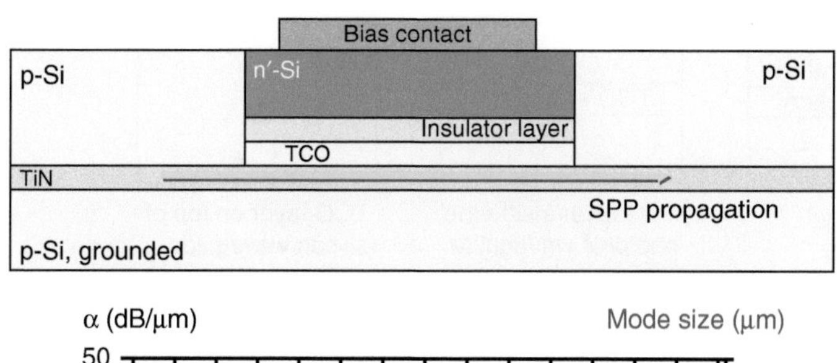

(a)

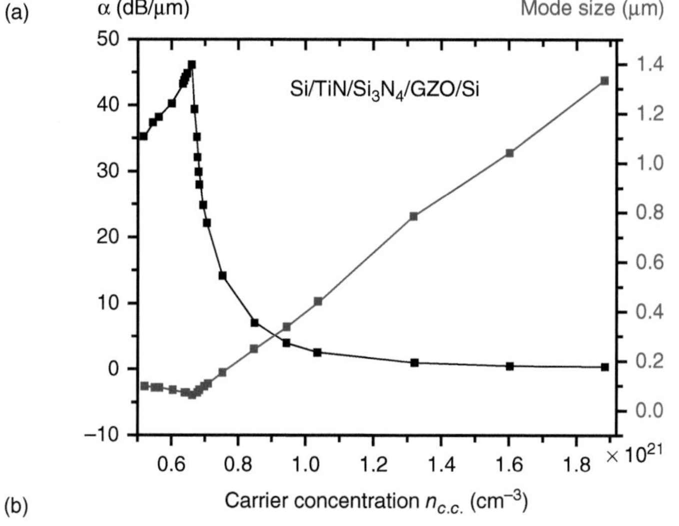

(b)

Figure 5.104 (a) TCO plasmonic modulator along with integration scheme: the input/output interface with low-loss plasmonic strip interconnects. (b) High $r = 46$ dB/μm and small mode size in the range 0.05–1.3 μm can be achieved. TCO, transparent conducting oxides. Source: Babicheva et al. 2013 [49]. Reproduced with permission of OSA Publishing.

devices exploit photonic modes, the miniaturization level of such devices is limited. New configurations of hybrid plasmonic waveguides, such as slot based or long-range dielectric-loaded plasmonic waveguides (Figure 5.105), provide the best trade-off between compactness and propagation losses. So, hybrid photonic waveguides can serve as the most promising platform for plasmonic electro-optical modulators. Recently, a plasmonic slot waveguide, which is filled with low-doped TCO, was suggested (Figure 5.106a) [603]. Such a waveguide has reasonably low losses and a small mode size forming a promising direction for further optimization. An ER $r = 2.7$ μm/dB and insertion losses 0.45 dB/μm are measured for a 300×200 nm^2 slot (Figure 5.106b).

Furthermore, a more sophisticated design, which is based on directional coupling to conventional silicon waveguides, was suggested [602]. Another approach is to introduce a resonant structure along the waveguide [53, 644]. While higher ER can be achieved by utilizing directional coupling or resonant features, the operation bandwidth of the modulator is essentially limited. Similar to designs with ring resonators, these devices suffer from high sensitivity to fabrication imperfections and temperature-dependent functionality.

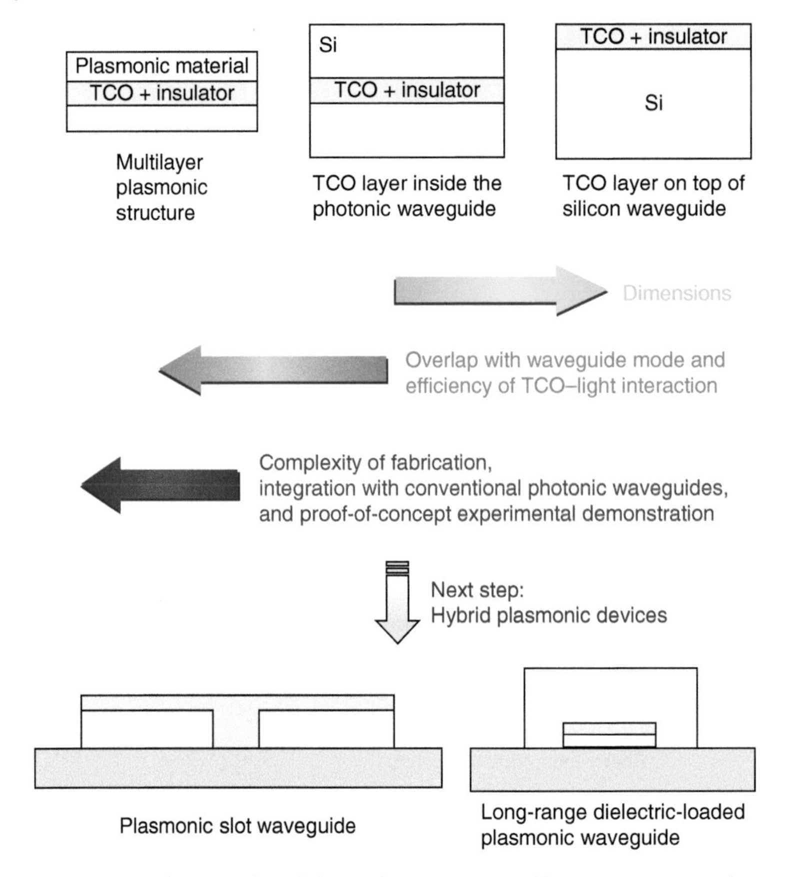

Figure 5.105 Suggested modulators designs are varied by compactness and ease of integration with low-loss waveguide (arrows show direction of more profound influence). MIM, metal–insulator–metal; TCO, transparent conducting oxides. Source: Babicheva et al. [546]. Reproduced with permission of Nanophotonics.

Plasmonics enables the merger between two major technologies: nanometer-scale electronics and ultrafast photonics. As alternatives to conventional noble metals, new plasmonic materials (such as highly doped oxide semiconductors and plasmonic ceramics) offer many appealing advantages, including low-intrinsic loss, semiconductor-based design, compatibility with standard nanofabrication processes, and tunability. The demonstration of the plasmonic (metal-like) response of TCOs in the optical range was an important milestone, which opened up perspectives in employing this material for plasmonic applications. Implementing TCO layers in a waveguide pushes plasmonic resonances to lower frequencies, which are of interest for telecom applications.

It is expected that developing and optimizing highly tunable materials such as TCOs will have a significant impact on nanophotonics technology providing CMOS-compatible passive and active devices. TCO-based plasmonic

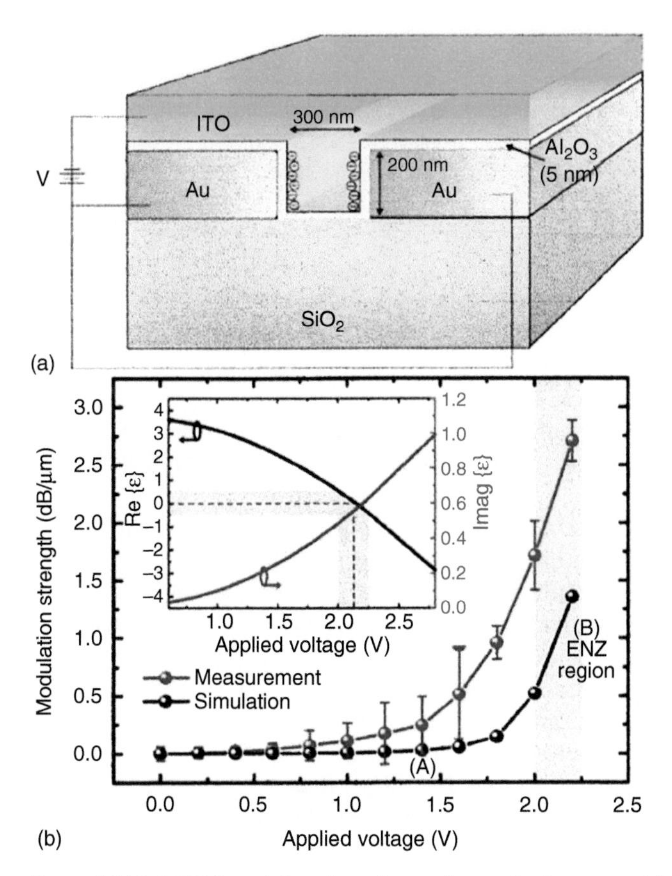

Figure 5.106 (a) TCO modulator based on a plasmonic slot waveguide. (b) High modulation strength is achieved because of the good overlap of the propagating plasmonic mode and the ITO layer. ITO, indium tin oxide; TCO, transparent conducting oxides. Source: Lee et al. 2014 [603]. Reproduced with permission of ACS.

devices could also play an important role in footprint reduction and energy consumption, as well as increasing bandwidth, which are one of the main criteria in research in photonics technologies. Up to now, the lowest estimated power consumption of an un-optimized TCO based plasmonic modulator is about 4 fJ/bit [603], which can be further decreased by engineering device design, for example, improving mode confinement, choosing more optimal geometry, and appropriate constituent materials. Thus, more experimental studies investigating the switching properties, that is, modulating strength and characteristic timescales, are required as well as the continued search for TCO materials with lower losses. The envisioned picture of new materials employment is complex hybrid circuitry bridging existing nanophotonic and nanoelectronics platforms and eventually, setting new standards in the levels of integration.

References

1 Kerr, J. (1875). A new relation between electricity and light: dielectrified media birefringent. *Philos. Mag.* 50 (332): 337–348.
2 Röntgen, W.C. (1883). Über die durch elektrische Kräfte erzeugte Anderung der Doppelbrechung des Quarzes. *Ann. Phys. Chem.* 18: 213–228. (Germany).
3 Röntgen, W.C. (1883). Bemerkung zu der Abhandlung des Hrn. A. Kundt: Uber des optisches Verhalten des Quarzes im elektrischen Felds. *Ann. Phys. Chem.* 19: 319–323. (Germany).
4 Kundt, A. (1883). Ober des optisches Verhalten des Quarzes im elektrischen Felds. *Ann. Phys. Chem.* 18: 228–233. (Germany).
5 Pockels, F. (1894). Ober den Einfluss des elektrostatischen Feldes auf das optische Verhalten piezoelektrischer Klistalle. *Abh. Gott.* 39: 1–204. (Germany).
6 Pockels, F. (1906). *Lehrbuch der Kristalloptik*. Leipzig: B. G. Teubner.
7 Mueller, H. (1935). Properties of Rochelle salt. *Phys. Rev.* 47: 175–191.
8 Mueller, H. (1940). Properties of Rochelle salt, IV. *Phys. Rev.* 58: 805–811.
9 Valasek, J. (1922). Properties of Rochelle salt related to the piezoelectric effect. *Phys. Rev.* 20: 639–664.
10 Anistratov, A.T. and Aleksandrov, K.S. (1967). Electro-optical properties of Rochelle salt. *Sov. Phys. Crystallogr.* 12: 392–396.
11 Jacquerod, A. (1942). The Kerr effect in potassium dihydrogen phosphate and potassium dihydrogen arsenate. *Helv. Phys. Acta* 15: 324–325.
12 Zwicker, B. and Scherrer, P. (1943). Electro-optical behaviour of KH_2PO_4 and KD_2PO_4 crystals. *Helv. Phys. Acta* 16: 214–216.
13 Zwicker, B. and Scherrer, P. (1944). Electro-optical properties of the signette-electric crystals KH_2PO_4 and KD_2PO_4. *Helv. Phys. Acta* 17: 346–373.
14 Billings, B.H. (1947). A tunable narrow-band optical filter. *J. Opt. Soc. Am.* 37: 738–746.
15 Miller, R.C. (1964). Optical harmonic generation in single crystal $BaTiO_3$. *Phys. Rev.* 134 (5A): 1313–1319.
16 Kaminow, I.P. (1965). Barium titanate light phase modulator. *Appl. Phys. Lett.* 7: 123.
17 Land, C.E. (1967). *Electron Devices Meeting*, vol. 13, 94–96.
18 Ballman, A.A., et al. (1970). $LiTaO_3$ electro-optic modulator. US Patent 3,506,929A, filed 14 April 1970.
19 Haertling, G.H. (1970). Hot pressed ferroelectric PLZT ceramics for electro-optical application. *Am. Ceram. Soc. Bull.* 49 (6): 564–567.
20 Miyazawa, S. and Uchida, N. (1975). Light intensity modulation based on guided-to-radiation mode coupling in heterostructure waveguides. *Opt. Quantum Electron.* 7: 451.
21 Fukuma, M., Noda, J., and Iwasaki, H. (1978). Optical properties in titanium-diffused $LiNbO_3$ strip waveguides. *J. Appl. Phys.* 49 (7): 3693.
22 Günter, P., Asbeck, P.M., Choy, M.M., and Kurtz, S.K. (1980). Second-harmonic generation with $Ga_{1-x}Al_x$ as lasers and $KNbO_3$ crystals. *Ferroelectrics* 28: 379, Appl. Phys. Lett., 35, 461 (1979).

23 Kato, K. (1979). High-efficiency second-harmonic generation at 4250–4680 Å in $KNbO_3$. *IEEE J. Quantum Electron.* 15 (6): 410–411.

24 Leonberger, F.J. (1980). High-speed operation of $LiNbO_3$ electro-optic interferometric waveguide modulators. *Opt. Lett.* 5 (7): 312–314.

25 Bulmer, C.H., Burns, W.K., and Moeller, R.P. (1980). Linear interferometric waveguide modulator for electromagnetic-field detection. *Opt. Lett.* 5 (5): 176–178.

26 Jackel, J.J., Rice, C.E., and Veselka, J.J. (1982). Proton exchange for high index waveguides in $LiNbO_3$. *Appl. Phys. Lett.* 41: 607–608.

27 Ahlfedt, H., Webjorn, J., and Arvidsson, G. (1991). Periodic domain inversion and generation of blue light in lithium tantalite. *Photonics Technol. Lett.* 3: 638–639.

28 Yamamoto, K., Mizuuchi, K., and Taniuchi, T. (1991). Milliwatt-order blue-light generation in a periodically domain-inverted $LiTaO_3$ waveguide. *Opt. Lett.* 16 (15): 1156–1158.

29 Mizuuchi, K., Yamamoto, K., and Taniuchi, T. (1991). Second harmonic generation of blue light in $LiTaO_3$ waveguides. *Appl. Phys. Lett.* 58: 2732–2734.

30 Matsumoto, S., Lim, E.J., Hertz, H.M., and Fejer, M.M. (1991). Second harmonic generation of blue light in a periodically poled $LiTaO_3$ waveguide. In: *Integrated Photonics Research*, Technical Digest Series 97. Washington, DC: Optical Society of America.

31 Dolfi, D.W. and Ranganath, T.R. (1992). 50 GHz velocity-matched, broad wavelength $LiNbO_3$ modulator with multimode active section. In: *Proceedings of Integrated Photonics Research Conference*, post-deadline paper PD-2.

32 Murphy, T.O., Murphy, E.J., and Irvin, R.W. (1994). An 8 × 8 Ti: $LiNbO_3$ polarization independent photonic switch. In: *Proceedings of Photonics in Switching/ECOC*, 174–176. Florence, Italy.

33 Gill, D.M., Conrad, C.W., Ford, G. et al. (1997). Thin-film channel waveguide electro-optic modulator in epitaxial $BaTiO_3$. *Appl. Phys. Lett.* 71 (13): 1783–1785.

34 Chiu, Y., Gopalan, V., Kawas, M.J. et al. (1999). Integrated optical device with second harmonic generator, electro-optic lens, electro-optic scanner in $LiTaO_3$. *J. Lightwave Technol.* 17 (3): 462.

35 Chen, D., Murphy, T.E., Chakrabarti, S., and Phillips, J.D. (2004). Optical wave guiding in $BaTiO_3/MgO/Al_xO_y/GaAs$ heterostructures. *Appl. Phys. Lett.* 85 (22): 5206–5208.

36 Toyoda, S., Fujiura, K., Sasaura, M. et al. (2004). KTN-crystal-waveguide-based electro-optic phase modulator with high performance index. *Electron. Lett.* 40 (13): 830.

37 Toyoda, S., Fujiura, K., Sasaura, M. et al. (2004). Low-driving-voltage electro-optic modulator with novel $KTa_{1-x}Nb_xO_3$ crystal waveguides. *Jpn. J. Appl. Phys.* 43 (8B): 5862–5866.

38 Tang, P., Towner, D., Hamano, T. et al. (2004). Electro-optic modulation up to 40 GHz in a barium titanate thin film waveguide modulator. *Opt. Express* 12 (24): 5962–5967.

39 Jiang, H., Zou, Y.K., Chen, Q. et al. (2005). Transparent electro-optic ceramics and devices. *Proc. SPIE Optoelectron. Devices Integr.* 5644: 380. http://dx.doi.org/10.1117/12.582105.

40 Nashimoto, K. (2006). High speed PLZT optical switches, EpiPhotonics production manual, Optical Network Test Beds Workshop 3 (ONT3), Tokyo, Japan. http://epiphotonics.com/products2.htm.
41 Nakayama, Y., Pauzauskie, P.J., Radenovic, A. et al. (2007). Tunable nanowire nonlinear optical probe. *Nature* 447: 1098–1101.
42 Dicken, M.J., Sweatlock, L.A., Pacifici, D. et al. (2008). Electro-optic modulation in thin film barium titanate plasmonic interferometers. *Nano Lett.* 8 (11): 4048–4052.
43 Naganuma, K., Miyazu, J., and Yagi, S. (2009). High-resolution KTN optical beam scanner. *NTT Tech. Rev.* 7 (12): 1–6.
44 Imai, T., Yagi, S., Toyoda, S., and Sasaura, M. (2009). Fast varifocal lens based on $KTa_{1-x}Nb_xO_3$ (KTN) single crystals. *NTT Tech. Rev.* 7 (12): 1–5.
45 Dionne, J.A., Diest, K., Sweatlock, L.A., and Atwater, H.A. (2009). PlasMOStor: a metal-oxide-Si field effect plasmonic modulator. *Nano Lett.* 9 (2): 897–902.
46 Sorger, V.J., Lanzillotti-Kimura, N.D., Ma, R.-M., and Zhang, X. (2012). Ultra-compact silicon nanophotonic modulator with broadband response. *Nanophotonics* 1: 17–22.
47 Melikyan, A. et al. (2014). High-speed plasmonic phase modulators. *Nat. Photonics* 8: 229–233.
48 Shi, K., Haque, R.R., Zhao, B. et al. (2014). Broadband electro-optical modulator based on transparent conducting oxide. *Opt. Lett.* 39: 4978–4981.
49 Babicheva, V.E., Kinsey, N., Naik, G.V. et al. (2013). Towards CMOS-compatible nanophotonics: ultra-compact modulators using alternative plasmonic materials. *Opt. Express* 21: 27326–27337.
50 Cai, W., White, J.S., and Brongersma, M.L. (2009). Compact, high-speed and power-efficient electro-optic plasmonic modulators. *Nano Lett.* 9: 4403.
51 Thomas, R., Ikonic, Z., and Kelsall, R.W. (2012). Electro-optic metal–insulator–semiconductor–insulator–metal Mach–Zehnder plasmonic modulator. *Photonics Nanostruct. Fundam. Appl.* 10: 183–189.
52 Krasavin, A.V. and Zayats, A.V. (2012). Photonic signal processing on electronic scales: electro-optical field-effect nanoplasmonic modulator. *Phys. Rev. Lett.* 109: 053901-1–053901-5.
53 Ye, C., Pickus, S., Liu, K. et al. (2014). High performance graphene and ITO-based electro-optic modulators and switches. In: *Proceedings of OECC/ACOFT 2014*, Melbourne, Australia (6–10 July 2014), 404–406.
54 Wang, Y., Chen, Z., Ye, Z., and Huang, J.Y. (2012). Synthesis and second harmonic generation response of $KNbO_3$ nanoneedles. *J. Cryst. Growth* 341 (1): 42–45.
55 New Medical Light Source using NTT's Communication Laser Technology, NTT Advanced Technology Corporation & Hamamatsu Photonics K.K. (Press Release, January 31, 2013).
56 Zografopoulos, D.C. and Beccherelli, R. (2013). Design of a vertically coupled liquid-crystal longrange plasmonic optical switch. *Appl. Phys. Lett.* 102: 101103.
57 Wang, Y. et al. (2014). Plasmonic switch based on composite interference in metallic strip waveguides. *Laser Photonics Rev.* 8 (4): L47–L51.

58 Krasavin, A.V. and Zayats, A.V. (2010). Electro-optic switching element for dielectric-loaded surface plasmon polariton waveguides. *Appl. Phys. Lett.* 97: 041107.
59 Lin, J.T., Xu, Y.X., Fang, Z.W. et al. (2015). Second harmonic generation in a high-Q lithium niobate microresonator fabricated by femtosecond laser micromachining. *Sci. China Phys. Mech. Astron.* 58: 114209.
60 Lin, J., Xu, Y., Fang, Z. et al. (2015). Fabrication of high-Q lithium niobate microresonators using femtosecond laser micromachining. *Sci. Rep.* 5: 8072.
61 Wang, C., Zhang, M., Stern, B. et al. (2017). Nanophotonic lithium niobate electro-optic modulators. *Opt. Express* 26 (2): 1547. https://arxiv.org/pdf/1701.06470.pdf.
62 Moretti, P., Thevenard, P., Godefroy, G. et al. (1990). Waveguides in barium titanate by helium implantation. *Phys. Status Solidi A* 117: K85.
63 Youden, K.E., James, S.W., Eason, R.W. et al. (1992). Photorefractive planar waveguides in $BaTiO_3$ fabricated by ion-beam implantation. *Opt. Lett.* 17: 1509.
64 Eknoyan, O., Taylor, H.F., Tang, Z. et al. (1992). Strain induced optical waveguides in lithium niobate, lithium tantalate, and barium titanate. *Appl. Phys. Lett.* 60 (4): 407–409.
65 Gill, D.M., Block, B.A., Conrad, C.W. et al. (1996). Thin film channel waveguides fabricated in metalorganic chemical vapor deposition grown $BaTiO_3$ on MgO. *Appl. Phys. Lett.* 69: 2968.
66 Beckers, L., Schubert, J., Zander, W. et al. (1998). Structural and optical characterization of epitaxial waveguiding $BaTiO_3$ thin films on MgO. *J. Appl. Phys.* 83: 3305.
67 Petraru, A., Schubert, J., Schmid, M., and Buchal, C. (2002). Ferroelectric $BaTiO_3$ thin film optical waveguide modulators. *Appl. Phys. Lett.* 81 (8): 1375.
68 Zhifu, L., Lin, P.-T., Wessels, B.W. et al. (2007). Nonlinear photonic crystal waveguide structures based on barium titanate thin films and their optical properties. *Appl. Phys. Lett.* 90 (20): 201104–2011043.
69 Tang, P., Towner, D., Meier, A., and Wessels, B. (2007). $BaTiO_3$ thin film waveguides and related modulator devices. US Patent 7,224,878, filed 29 May 2007.
70 Kim, S.S., Towner, D., Teren, A., et al. (2001). Integrated Photonics Research Topical Meeting, IWB1.
71 Petraru, A., Schubert, J., Schmid, M., and Buchal, Ch. Epitaxial and polycrystalline ferroelectric $BaTiO_3$ thin films used for optical switching. https://www.ecio-conference.org/wp-content/uploads/2016/05/2003/2003_ThP20.pdf.
72 Seneviratne, D.A. (2007). Materials and devices for optical switching and modulation of photonic integrated circuits. http://dspace.mit.edu/handle/1721.1/39539.
73 Teren, A.R., Kim, S.-S., Ho, S.-T., and Wessels, B.W. (2001). Erbium-doped barium titanate thin film waveguides for integrated optical amplifiers. *Mater. Res. Soc. Symp. Proc.* 688: http://dx.doi.org/10.1557/PROC-688-C9.7.1.

74 Petraru, A., Schubert, J., Schmid, M. et al. (2003). Integrated optical Mach Zehnder modulator based on polycrystalline $BaTiO_3$. *Opt. Lett.* 28 (24): 2527.

75 Wessels, B.W. (2007). Ferroelectric epitaxial thin films for integrated optics. *Annu. Rev. Mater. Res.* 37: 659–679.

76 Klein, M.B. and Valley, G.C. (1986). Characteristics of $BaTiO_3$ for electro-optic devices. In: *Proceedings of SPIE 0567, Advances in Materials for Active Optics,* 116.

77 Zgonik, M., Bernasconi, P., Duelli, M. et al. (1994). Dielectric, elastic, piezoelectric, electro-optic, and elasto-optic tensors of $BaTiO_3$ crystals. *Phys. Rev. B* 50: 5941.

78 Wessels, B.W. and Block, B.A. (2000). Rare earth doped barium titanate thin film optical working medium for optical devices. US Patent 6,122,429, filed 19 September 2000.

79 Tang, P., Towner, D.J., Meier, A.L., and Wessels, B.W. (2004). Low-voltage, polarization-insensitive, electro-optic modulator based on a polydomain barium titanate thin film. *Appl. Phys. Lett.* 85 (20): 4615–4617.

80 Tang, P., Towner, D.J., Meier, A.L., and Wessels, B.W. (2005). Velocity matching of BaTiO3 thin film electro-optic waveguide modulators, *Conference Paper: Integrated Photonics Research and Applications,* San Diego, CA.

81 Zhifu, L., Bo, L., Guoyang, X. et al. (2005). Barium titanate thin film electro-optic modulator low half-wave voltage at 1310 nm. In: *Conference on Lasers and Electro-Optics (CLEO),* vol. 3, 1870–1872.

82 Sun, D.-G., Liu, Z., Huang, Y. et al. (2005). Performance simulation for ferroelectric thin-film based waveguide electro-optic modulators. *Opt. Commun.* 255: 319–330.

83 Tang, J., Yang, S., and Bhatranand, A. (2007). Strain induced waveguide electro-optic modulators in barium titanate crystal. In: *Quantum Electronics and Laser Science Conference, QELS '07,* 1–2.

84 Zhifu, L., Lin, P.-T., and Wessels, B.W. (2008). Cascaded Bragg reflectors for a barium titanate thin film electro-optic modulator. *J. Opt. A: Pure Appl. Opt.* 10 (1): 015302.

85 Tang, J., Shujun, Y., and Bhatranand, A. (2009). Electro-optic barium titanate waveguide modulators with transparent conducting oxide electrodes. In: *Lasers and Electro Optics and The Pacific Rim Conference on Lasers and Electro-Optics (CLEO/PACIFIC RIM '09),* 1–2.

86 Lee, V.C., Sis, S.A., Zhu, X., and Mortazawi, A. (2010). Intrinsically switchable interdigitated barium titanate thin film contour mode resonators. In: *Microwave Symposium Digest (MTT),* 1. IEEE MTT-S International.

87 Bowman, D. and Bhandarkar, S. (2012). Sol–gel processing of $BaTiO_3$ for electro-optic waveguide devices. In: *Advances in Photonic Materials and Devices,* vol. 163 (ed. S. Bhandarkar). Hoboken, NJ: Wiley.

88 Sun, D.-G., Fu, X., Hall, T.J., and Jiang, H. (2013). Analysis for the nonlinear electro-optic modulation effect of $BaTiO_3$ crystal thin-film waveguide modulators. *J. Appl. Phys.* 113: 184502.

89 Lu, H.A., Wills, L.A., Wessels, B.W. et al. (1993). Second-harmonic generation of poled $BaTiO_3$ thin films. *Appl. Phys. Lett.* 62 (12): 1314–1316.

90 Lu, H.A. et al. (1993). Electron beam induced poling of $BaTiO_3$ thin films. *Appl. Phys. Lett.* 63 (70): 874–876.

91 Wang, W., Yang, G., Chen, Z. et al. (2003). Nonlinear optical properties of Fe/$BaTiO_3$ composite thin films prepared by two-target pulsed-laser deposition. *J. Opt. Soc. Am. B* 20: 1342–1345.

92 Haertling, G.H. (1971). Improved hot-pressed electro-optic ceramic in the (Pb,La)(Zr,Ti)O_3 system. *J. Am. Ceram. Soc.* 54 (6): 303–309.

93 Haertling, G.H. (1986). Piezoelectric and electro-optic ceramics. In: *Ceramic Materials for Electronics* (ed. R.C. Buchanan), 135–225. New York: Marcel Dekker.

94 Haertling, G.H. (1988). Electro-optic ceramics and devices. In: *Electronic Ceramics* (ed. L.M. Levinson), 371–492. New York: Marcel Dekker.

95 Cutchen, J.T. et al. (1973). Electro-optic devices utilizing quadratic PLZT ceramic elements. Paper presented at the 1973 Western Electronic show and convention (Wescon), session 30, on ferroelectric ceramic electro-optic devices.

96 Baudrant, A., Vial, H., and Daval, J. (1978). Liquid phase epitaxial growth of $LiNbO_3$ thin films. *J. Cryst. Growth* 43: 197.

97 Land, C.E. (1989). Longitudinal electro-optic effects and photosensitivities of lead zirconate titanate thin films. *J. Am. Ceram. Soc.* 72: 2059.

98 Bazylenko, M. and Mann, I.K. (2002). Electro-optic waveguide structure. US 2002/0186948 A1, 12 December 2002.

99 Levine, J.D. and Essaian, S. (2002). Electro-optical waveguide switching method and apparatus. US 2002/0154852 A1, 24 October 2002.

100 Swartz, S.L. (1990). Topics in electronic ceramics. *IEEE Trans. Dielectr. Electr. Insul.* 25: 935.

101 Li, J., Cheng, H.C., Kawas, M.J. et al. (1996). Electro-optic wafer beam deflector in $LiTaO_3$. *IEEE Photonics Technol. Lett.* 8 (11): 1486–1488.

102 Chen, Q., Chiu, Y., Lambeth, D.N. et al. (1994). Guided-wave electro-optic beam deflector using domain reversal in $LiTaO_3$. *J. Lightwave Technol.* 12: 1401–1404.

103 Gopalan, V., Kawas, M.J., and Gupta, M.C. (1996). Integrated quasi-phase-matched second-harmonic generator and electro-optic scanner on $LiTaO_3$ single crystals. *IEEE Photonics Technol. Lett.* 8 (12): 1704–1706.

104 Baldi, P., De Micheli, M.P., Hadi, K.E. et al. (1998). Proton exchanged waveguides in $LiNbO_3$ and $LiTaO_3$ for integrated lasers and nonlinear frequency converters. *Opt. Eng.* 37 (4): 1193–1202.

105 Kiminori, M., Kazuhisa, Y., and Makoto, K. (1997). Generation of ultraviolet light by frequency doubling of a red laser diode in a first-order periodically poled bulk $LiTaO_3$. *Appl. Phys. Lett.* 70 (10): 1201–1203.

106 Champert, P.A., Popov, S.V., Taylor, J.R., and Meyn, J.P. (2000). Efficient second-harmonic generation at 384 nm in periodically poled lithium tantalate by use of a visible Yb–Er-seeded fiber source. *Opt. Lett.* 25 (17): 1252.

107 Gao, Z.D., Zhu, S.N., and Kung, A.H. (2006). Monolithic red-green-blue laser light source based on cascaded wavelength conversion in periodically poled stoichiometric lithium tantalite. *Appl. Phys. Lett.* 89: 181101.

108 Lobino, M., Marshall, G.D., Xiong, C. et al. (2011). Correlated photon-pair generation in a periodically poled MgO doped stoichiometric lithium tantalate reverse proton exchanged waveguide. *Appl. Phys. Lett.* 99: 081110.
109 Okazaki, M., Yoshimoto, S., Chichibu, T., and Suhara, T. (2015). Electro-optic spatial light modulator using periodically-poled MgO:s-$LiTaO_3$ waveguide. *IEEE Photonics Technol. Lett.* 27 (15): 1646–1648.
110 Stulz, L.W. (1979). Titanium in-diffused $LiNbO_3$ optical waveguide fabrication. *Appl. Opt.* 18 (12): 2041–2044.
111 Miyazawa, S. (1979). Ferroelectric domain inversion in Ti-diffused $LiNbO_3$ optical waveguide. *J. Appl. Phys.* 50: 4599.
112 Fukuma, M. and Noda, J. (1980). Optical properties in titanium-diffused $LiNbO_3$ strip waveguides and their coupling to a fibre characteristics. *Appl. Opt.* 19 (4): 591–597.
113 Korotky, S.K., Minford, W.J., Buhl, L.L. et al. (1982). Modesize and method for estimating the propagation constant of single-mode Ti:$LiNbO_3$ waveguides. *IEEE Trans. Microw. Theory Tech.* 30 (10): 1784.
114 Griffiths, G. and Esdaile, R. (1984). Analysis of titanium diffused planar optical waveguides in lithium niobate. *IEEE J. Quantum Electron.* 20 (2): 149–159.
115 Hofmann, D., Schreiber, G., Hasse, C. et al. (1999). Continuous wave mid infrared optical parametric oscillators with periodically poled Ti:$LiNbO_3$ channel waveguides. *Opt. Lett.* 24: 696.
116 Fujiwara, T., Srivastava, R., Cao, X., and Ramaswamy, R.V. (1993). Comparison of photorefractive index change in proton-exchanged and Ti-indiffused $LiNbO_3$ waveguides. *Opt. Lett.* 18: 346.
117 Yi-Yan, A. (1983). Index instabilities in proton-exchanged $LiNbO_3$ waveguides. *Appl. Phys. Lett.* 42 (8): 633–635.
118 Yi-Yan, A., Primot, J., Burgeat, J., and Guglielmi, R. (1983). Proton-exchanged $LiNbO_3$ waveguides: an X-ray analysis. In: *Proceedings of 2nd ECIO*, 17–18. Florence, Italy.
119 Suchoski, P.G., Findakly, T.K., and Leonberger, J. (1988). Stable low-loss proton-exchanged $LiNbO_3$ waveguide devices with no electro-optic degradation. *Opt. Lett.* 13: 1050–1052.
120 Bortz, M.L., Field, S.J., Feyer, M.M. et al. (1994). Noncritical quasi-phasematched 2nd harmonic generation in an annealed proton exchanged $LiNbO_3$ waveguide. *IEEE J. Quantum Electron.* 30: 2953–2960.
121 Chen, W.L., Chen, R.S., Lee, J.H., and Wang, W.S. (1995). Lithium niobate ridge waveguides by nickel diffusion and proton-exchanged wet etching. *IEEE Photonics Technol. Lett.* 7: 1318–1320.
122 Baldi, P., De Micheli, M.P., El Hadi, K. et al. (1998). Proton exchange waveguides in $LiNbO_3$ and $LiTaO_3$ for integrated lasers and nonlinear frequency converters. *Opt. Eng.* 37 (4): 1193.
123 Gallo, K., Gawith, C.B.E., Prawiharjo, J., et al. (2004). UV-written channel waveguides in proton-exchanged lithium niobate. *Paper CTuF3 in the Technical Digest of the Conference on Lasers and Electro-Optics*, San Francisco.
124 Schmidt, R.V. and Kaminow, I.P. (1974). Metal-diffused optical waveguides in $LiNbO_3$. *Appl. Phys. Lett.* 25: 458–460.

125 Thomson, R.R., Campbell, S., Blewett, I.J. et al. (2006). Optical waveguide fabrication in *z*-cut lithium niobate ($LiNbO_3$) using femtosecond pulses in the low repetition rate regime. *Appl. Phys. Lett.* 88: 111109.

126 Rabiei, P. and Günter, P. (2004). Optical and electro-optical properties of submicrometer lithium niobate slab waveguides prepared by crystal ion slicing and wafer bonding. *Appl. Phys. Lett.* 85 (20): 4603–4605.

127 Geiss, R., Diziain, S., Iliew, R. et al. (2010). Light propagation in a free-standing lithium niobate photonic crystal waveguide. *Appl. Phys. Lett.* 97 (13): 131109.

128 Lu, H., Sadani, B., Courjal, N. et al. (2012). Enhanced electro-optical lithium niobate photonic crystal wire waveguide on a smart-cut thin film. *Opt. Express* 20 (3): 2974.

129 Schmidt, R.V. and Alferness, R.C. (1979). Directional coupler switches, modulators, and filters using alternating techniques. *IEEE Trans. Circuits Syst.* CAS-26: 1099–1108.

130 Alferness, R.C. and Buhl, L.L. (1980). Electro-optic waveguide TE $\leftrightarrow$ TM mode converter with low drive voltage. *Opt. Lett.* 5 (11): 473–475.

131 Alferness, R.C. (1982). Waveguide electro-optic modulators. *IEEE Trans. Microwave Theory Tech.* MTT-30: 1121–1137.

132 Becker, R.A. (1983). Comparison of guided-wave interferometric modulators fabricated on $LiNbO_3$ via Ti indiffusion and proton exchange. *Appl. Phys. Lett.* 43: 131–133.

133 Noguchi, K., Mitomi, O., Kawano, K., and Yanagibashi, M. (1993). Highly efficient 40-GHz bandwidth Ti: $LiNbO_3$ optical modulator employing ridge structure. *IEEE Photonics Technol. Lett.* 5: 52–54.

134 Gopalakrishnan, G., Burns, W.K., McElhanon, R.W. et al. (1994). Performance and modeling of broadband $LiNbO_3$ traveling wave optical intensity modulators. *J. Lightwave Technol.* 12: 1807–1819.

135 Noguchi, K., Mitomi, O., Miyazawa, H., and Seki, S. (1995). A broadband Ti: $LiNbO_3$ optical modulator with a ridge structure. *J. Lightwave Technol.* 13: 1164–1168.

136 Noguchi, K., Mitomi, O., and Miyazawa, H. (1996). Low voltage and broadband Ti: $LiNbO_3$ optical modulator operating in the millimeter wavelength region. In: *Technology DIG Conference Optical Fiber Communications*, 205–206. San Francisco, CA, paper ThB2.

137 Asano, M. and Ejiri, T. (1997). Optical waveguide device having substrate made of ferroelectric crystals. US Patent 5,621,839, filed 15 April 1997.

138 Moyer, R.S., Grencavich, R., Judd, F.F. et al. (1998). Design and qualification of hermetically packaged lithium niobate optical modulator. *IEEE Trans. Compon. Packag. Manuf. Technol. Part B* 21: 130–135.

139 Maack, D. (1999). Reliability of lithium niobate Mach–Zehnder modulators for digital optical fiber telecommunication systems. In: *Proceedings of SPIE Critical Reviews: Reliability of Optical Fibers and Optical Fiber Systems*, 197–230. Boston, MA.

140 Chang, S.-J., Tsai, C.-L., Lin, Y.-B. et al. (1999). Improved electro-optic modulator with ridge structure in *x*-cut $LiNbO_3$. *J. Lightwave Technol.* 17: 843–847.

141 Hallemeier, P., Kissa, K., McBrien, G., and Horton, T. (1999). Next generation 10 Gb/s lithium niobate modulator components for RZ based transmission techniques. In: *Proceedings of National Fiberoptic Engineers Conference*, Chicago, IL, September 26–30, 1999.
142 Burns, W., Howerton, M., Moeller, R., et al. (1999). Low drive voltage, 40 GHz $LiNbO_3$ modulators. *OFC '99*, San Diego, CA (February 1999), paper ThT1.
143 Shur, V.Y., Rumyantsev, E.L. et al. (2000). Regular ferroelectric domain array in lithium niobate crystals for nonlinear optic applications. *Ferroelectrics* 236: 129–144.
144 Ramadan, T.A., Levy, M., and Osgood, R.M. Jr., (2000). Electro-optic modulation in crystal-ion-sliced *z*-cut $LiNbO_3$ thin films. *Appl. Phys. Lett.* 76 (11): 1407–1409.
145 Rabiei, P. and Steier, W.H. (2005). Lithium niobate ridge waveguides and modulators fabricated using smart guide. *Appl. Phys. Lett.* 86 (16): 161115.
146 Janner, D., Tulli, D., Belmonte, M., and Pruneri, V. (2008). Waveguide electro-optic modulation in micro-engineered $LiNbO_3$. *J. Opt. A: Pure Appl. Opt.* 10: 104003. (6 pp.).
147 Jundt, D.H. *Niobium Compounds in Acoustics and Electro-optics*. Palo Alto, CA: Crystal Technology, Inc.
148 Minakata, M., Kawaguchi, T., and Imaeda, M. (1999). Optical waveguide substrate, optical waveguide device, second harmonic generation device and process of producing optical waveguide substrate. US Patent 5,991,067, 23 November 1999.
149 Lim, E.J., Fejer, M.M., Byer, R.L., and Kozlovsky, W.J. (1989). Blue light generation by frequency doubling in periodically poled lithium niobate channel waveguides. *Electron. Lett.* 25: 731.
150 Amin, J., Pruneri, V., Webjorn, J. et al. (1996). Blue light generation in a periodically poled Ti:$LiNbO_3$ channel waveguide. *Opt. Commun.* 135 (1–3): 41–44.
151 Gawith, C.B.E., Shepherd, D.P., Abernethy, J.A. et al. (1999). Second-harmonic generation in a direct-bonded periodically ploed $LiNbO_3$ buried waveguide. *Opt. Lett.* 24 (7): 481–483.
152 Hofmann, D., Schreiber, G., Haase, C. et al. (1999). Quasi-phasematched difference frequency generation in periodically poled Ti:$LiNbO_3$ channel waveguides. *Opt. Lett.* 24 (13): 896–898.
153 Parameswaran, K.R., Route, R.K., Kurz, J.R. et al. (2002). Highly efficient second-harmonic generation in buried waveguides formed by annealed and reverse proton exchange in periodically poled lithium niobate. *Opt. Lett.* 27: 179.
154 Lee, Y.L., Suche, H., Min, Y.H. et al. (2003). Wavelength- and time-selective all-optical channel dropping in periodically poled Ti: $LiNbO_3$ channel waveguides. *IEEE Photonics Technol. Lett.* 15: 978.
155 Mizuuchi, K., Morikawa, A., Sugita, T., and Yamamoto, K. (2003). Efficient second-harmonic generation of 340 nm light in a 1.4 μm periodically poled bulk MgO:$LiNbO_3$. *Jpn. J. Appl. Phys.*, part 2 42: L90.

156 Kaplan, A., Achiam, K., Greenblatt, A. et al. *$LiNbO_3$ Integrated Optical QPSK Modulator and Coherent Receiver*. Silver Spring, MD: CeLight Inc.
157 Catherine, H. (1988). Bulmer, integrated optical sensors in lithium niobite. *Opt. News* 14 (2): 20–23.
158 Kuwabara, N., Tajima, K., Kobayashi, R., and Amemiya, F. (1992). Development and analysis of electric field sensor using $LiNbO_3$ optical modulator. *IEEE Trans. Electromagn. Compat.* 34 (4): 391–396.
159 Lee, S.S. et al. (1993). Integrated optical high voltage sensor using a z-cut $LiNbO_3$ modulator. *Photon. Technol. Lett.* 5: 9.
160 Yim, Y.-S., Shin, S.-Y., Shay, W.-T., and Lee, C.-T. (1998). Lithium niobate integrated-optic voltage sensor with variable sensing ranges. *Opt. Commun.* 152 (4–6): 225–228.
161 Rao, Y.J., Gnewuch, H., Pannell, C.N., and Jackson, D.A. (1999). Electro-optic electric field sensor based on periodically poled $LiNbO_3$. *Electron. Lett.* 35 (7): 596–597.
162 Cecelja, F., Bordovsky, M., and Balachandran, W. (2001). Lithium niobate sensor for measurement of DC electric fields. *IEEE Trans. Instrum. Meas.* 50 (2): 465–469.
163 Zaldívar-Huerta, I. and Rodríguez-Asomoza, J. (2004). Electro-optic E-field sensor using an optical modulator. In: *Proceedings of the 14th International Conference on Electronics, Communications and Computers*, 220.
164 Lee, T.H., Wu, P.I., and Lee, C.T. (2007). Intergrated $LiNbO_3$ electro-optical electromagnetic field sensor. *Microwave Opt. Technol. Lett.* 49 (9): 2312–2314.
165 Lee, T.H., Wu, P.I., and Lee, C.T. (2007). Intergraded $LiNbO_3$ electro-optical electromagnetic field sensor. *Microwave Opt. Technol. Lett.* 49: 2312–2314.
166 Gutiérrez-Martínez, C. and Santos-Aguilar, J. (2008). Electric field sensing scheme based on matched $LiNbO_3$ electro-optic retarders. *IEEE Trans. Instrum. Meas.* 57 (7): 1362–1368.
167 Gutiérrez-Martínez, C. (2009). Electric field sensing schemes using low coherence light and $LiNbO_3$ optical retarders. In: *Optical Fibre, New Developments*, Chapter 6 (ed. C. Lethien). Croatia: In-Tech.
168 Gutiérrez-Martínez, C. and Santos-Aguilar, J. (2009). Modeling the electro-optic response of dielectric $LiNbO_3$ electric field sensors. In: *12th International Symposium on Microwave and Optical Technology*. New Delhi, India.
169 Tulli, D., Janner, D., Garcia-Granda, M. et al. (2011). Electrode-free optical sensor for high voltage using a domain-inverted lithium niobate waveguide near cut-off. *Appl. Phys. B* 103: 399–403.
170 Lee, C.T. and Huang, W.H. (2011). Intergrated azimuthal lithium niobate electromagnetic field sensor with Mach–Zehnder waveguide modulator and micromultiannular antenna. *Microwave Opt. Technol. Lett.* 53 (3): 565–567.
171 Gutiérrez-Martínez, C., Santos-Aguilar, J., Ochoa-Valiente, R. et al. (2011). Modelling and experimental electro-optic response of dielectric lithium niobate waveguides used as electric field sensors. *Meas. Sci. Technol.* 22 (3).
172 Jung, H.-S. (2012). Electro-optic electric field sensor utilizing $Ti:LiNbO_3$ symmetric Mach–Zehnder interferometers. *J. Opt. Soc. Korea* 16 (1): 47–52.

173 Chen, L. and Reano, R.M. (2012). Compact electric field sensors based on indirect bonding of lithium niobate to silicon microrings. *Opt. Express* 20 (4): 4032.

174 Gutiérrez-Martínez, C., Santos-Aguilar, J., and Morales-Díaz, A. (2013). On the design of video-bandwidth electric field sensing systems using dielectric $LiNbO_3$ electro-optic sensors and optical delays as signal carriers. *IEEE Sens. J.* 13 (11): 4196–4203.

175 Zhang, J., Chen, F., Sun, B., and Li, C. (2013). A novel bias angle control system for $LiNbO_3$ photonic sensor using wavelength tuning. *IEEE Photonics Technol. Lett.* 25 (20): 1993–1995.

176 Zhang, J., Chen, F., Sun, B., and Chen, K. (2014). Integrated optical waveguide sensor for lighting impulse electric field measurement. *Photonic Sens.* 4 (3): 215–219.

177 Yang, Q., Sun, S., Han, R. et al. (2015). Intense transient electric field sensor based on the electro-optic effect of $LiNbO_3$. *AIP Adv.* 5: 107130.

178 Jung, H. (2016). Ti:$LiNbO_3$ integrated optic electric-field sensors based on electro-optic effect. *Fiber Integr. Opt.* 35 (4): 161–180.

179 Gutiérrez-Martínez, C., Santos-Aguilar, J., Meza-Pérez, J., and Morales-Díaz, A. (2017). Novel electric field sensing scheme using integrated optics $LiNbO_3$ unbalanced Mach–Zehnder interferometers and optical delay-modulation. *J. Lightwave Technol.* 35 (1): 27–33.

180 Wang, C., Burek, M.J., Lin, Z. et al. (2014). Integrated high quality factor lithium niobate microdisk resonators. *Opt. Express* 22 (25): 30924–30933.

181 Wang, J., Bo, F., Wan, S. et al. (2015). High-Q lithium niobate microdisk resonators on a chip for efficient electro-optic modulation. *Opt. Express* 23 (18): 23072–23078.

182 Lin, J.T., Xu, Y.X., Ni, J.L. et al. (2016). Phase-matched second-harmonic generation in an on-chip $LiNbO_3$ microresonator. *Phys. Rev. Appl.* 6 (1): 014002.

183 Wang, M., Lin, J.T., Xu, Y.X. et al. (2016). Fabrication of high-Q microresonators in dielectric materials using a femtosecond laser: principle and applications. *Opt. Commun.* In press.

184 Liao, Y., Xu, J., Cheng, Y. et al. (2008). Electro-optic integration of embedded electrodes and waveguides in $LiNbO_3$ using a femtosecond laser. *Opt. Lett.* 33 (19): 2281–2283.

185 Armani, D., Min, B., Martin, A., and Vahala, K. (2004). Electrical thermo-optic tuning of ultrahigh-Q microtoroid resonators. *Appl. Phys. Lett.* 85 (22): 5439–5441.

186 Guarino, A., Poberaj, G., Rezzonico, D. et al. (2007). Electro-optically tunable microring resonators in lithium niobate. *Nat. Photonics* 1 (7): 407–410.

187 Chen, L., Wood, M.G., and Reano, R.M. (2013). 12.5 pm/V hybrid silicon and lithium niobate optical microring resonator with integrated electrodes. *Opt. Express* 21 (22): 27003–27010.

188 Wang, M., Xu, Y., Fang, Z. et al. (2017). On-chip electro-optic tuning of a lithium niobate microresonator with integrated in-plane microelectrodes. *Opt. Express* 25 (1): 124–129.

189 Li, J., Duewer, F., Gao, C. et al. (2000). Electro-optic measurements of the ferroelectric–paraelectric boundary in $Ba_{1-x}Sr_xTiO_3$ materials chips. *Appl. Phys. Lett.* 76: 769.

190 Kim, D.-Y., Moon, S.E., Kim, E.-K. et al. (2003). Electro-optic characteristics of (001)-oriented $Ba_{0.6}Sr_{0.4}TiO_3$ thin films. *Appl. Phys. Lett.* 82 (9): 1455–1457.

191 Kim, D.-Y., Moon, S.E., Kim, E.-K. et al. (2003). High electro-optic coefficient of $Ba_{0:6}Sr_{0:4}TiO_3$/MgO(001) dielectric thin film. *J. Korean Phys. Soc.* 42: S1347–S1349.

192 Wang, D. (2006). Barium Strontium Titanate Thin Films for Electro-optic Applications, ProQuest Dissertations and Theses; Ph.D. Thesis – Hong Kong Polytechnic University, Hong Kong 9780542691010.

193 Wang, D.Y., Wang, J., Chan, H.L.W., and Choy, C.L. (2007). Linear electro-optic effect in $Ba_{0.7}Sr_{0.3}TiO_3$ thin film grown on LSAT (001) substrate. *Integr. Ferroelectr.* 88 (1): 12–20.

194 Takeda, K., Muraishi, T., Hoshina, T. et al. (2009). Birefringence and electro-optic effect in epitaxial BST thin films. *Mater. Sci. Eng. B, Solid-State Mater. Adv. Technol.* 161 (1–3): 61–65.

195 Smolenskii, G.A. and Agranovskaya, A.I. (1958). Dielectric polarization and losses of some complex compounds. *Sov. Phys. Tech. Phys.* 3 (7): 1380–1382.

196 Lu, Y.L. and Gao, C. (2008). Optical limiting in lead magnesium niobate–lead titanate multilayers. *Appl. Phys. Lett.* 92 (12): 121109.

197 Lim, B.C., Phua, P.B., Lai, W.J., and Hong, M.H. (2008). Fast switchable electro-optic radial polarization retarder. *Opt. Lett.* 33 (9): 950–952.

198 Zou, Y.K., Chen, Q.S., Zhang, R. et al. (2005). Low voltage, high repetition rate electro-optic Q-switch. In: *Conference on Lasers and Electro-Optics/Quantum Electronics and Laser Science and PhotonicApplications Systems Technologies*. Technical Digest (CD) (Optical Society of America, paper CTuZ5.

199 Qiao, L., Ye, Q., Gan, J.L. et al. (2011). Optical characteristics of transparent PMNT ceramic and its application at high speed electro-optic switch. *Opt. Commun.* 284 (16–17): 3886–3890.

200 Cheng, K.-C. (2004). Dielectric and electro-optic properties of PMN–PT single crystals and thin films. http://hdl.handle.net/10397/3361.

201 Lu, Y., Jin, G.-H., Cronin-Golomb, M. et al. (1998). Fabrication and optical characterization of $Pb(Mg_{1/3}Nb_{2/3})O_3$–$PbTiO_3$ planar thin film optical waveguides. *Appl. Phys. Lett.* 72 (23): 2927.

202 Li, K.K. et al. (2004). Electro-optic ceramic material and device, PMN–PT. US Patent 6,746,618.

203 Makarov, V. and Stadler, B.J.H. Integrated photonic device fabrication using PMN–PT. http://www.ece.umn.edu/~stadler/Research/pdf/Vlad%202007.pdf.

204 Günter, P. (1974). Electro-optical properties of $KNbO_3$. *Opt. Commun.* 11: 285–290.

205 Graettinger, T.M., Rou, S.H., Ameen, M.S. et al. (1991). Electro-optic characterization of ion beam sputter-deposited $KNbO_3$ thin films. *Appl. Phys. Lett.* 58 (18): 1964–1966.

206 Zgonik, M., Schlesser, R., Biaggio, I. et al. (1993). Materials constants of $KNbO_3$ relevant for electro- and acousto-optics. *J. Appl. Phys.* 74 (2): 1287–1297.

207 Zgonik, M., Schlesser, R., Biaggio, I., and Günter, P. (1994). Electro-and acousto-optic properties of $KNbO_3$ crystals. *Ferroelectrics* 158: 217–222.

208 Nystrom, M.J. and Wessels, B.W. (1996). The effects of domain structure on the electro-optic response of potassium niobate thin films. *Mater. Res. Soc. Symp. Proc.* 453: 259.

209 Hoerman, B.H., Nichols, B.M., Nystrom, M.J., and Wessels, B.W. (1999). Dynamic response of the electro-optic effect in epitaxial $KNbO_3$. *Appl. Phys. Lett.* 75: 2707.

210 Hoerman, B.H. et al. (1999). Dynamic response of the electro-optic effect in epitaxial ferroelectric thin films. *Mater. Res. Soc. Symp. Proc.* 597: 157.

211 Hoerman, B.H., Nichols, B.M., and Wessels, B.W. (2002). Dynamic response of the dielectric and electro-optic properties of epitaxial ferroelectric thin films. *Phys. Rev. B* 65: 224110.

212 Sastry, P.U. (2002). Linear electro-optical properties of orthorhombic $KNbO_3$. *Solid State Commun.* 122: 41.

213 Fukuda, T. and Uematsu, Y. (1972). Preparation of $KNbO_3$ single crystal for optical applications. *Jpn. J. Appl. Phys.* 11: 163–169.

214 Wiesendanger, E. (1973). Dielectric, mechanical and optical properties of orthorhombic $KNbO_3$. *Ferroelectrics* 6 (1): 263–281.

215 Uematsu, Y. (1974). Nonlinear optical properties of $KNbO_3$ single crystal in the orthorhombic phase. *Jpn. J. Appl. Phys.* 13: 1362–1368.

216 Günter, P., Flückiger, U., Huignard, J.P., and Micheron, F. (1976). Optically induced refractive index changes in $KNbO_3$:Fe. *Ferroelectrics* 13 (1): 297–299.

217 Zysset, B., Biaggio, I., and Günter, P. (1992). Refractive indices of orthorhombic $KNbO_{3.}$ I. dispersion and temperature dependence. *J. Opt. Soc. Am. B* 9: 380.

218 Biaggio, I., Kerkoc, P., Wu, L.-S. et al. (1992). Refractive indices of orthorhombic $KNbO_{3.}$ II. Phase-matching configurations for nonlinear-optical interactions. *J. Opt. Soc. Am. B* 9: 507.

219 Graettinger, T.M. et al. (1993). Growth, microstructures and optical properties of $KNbO_3$ thin films. *Mater. Res. Soc. Symp. Proc.* 310: 301.

220 Wessels, B.W. et al. (1995). Epitaxial niobate thin films and their nonlinear optical properties. *Mater. Res. Soc. Symp. Proc.* 401: 211.

221 Pliska, T., Mayer, F., Fluck, D. et al. (1995). Nonlinear optical investigation of the optical homogeneity of $KNbO_3$ bulk crystals and ion-implanted waveguides. *J. Opt. Soc. Am. B* 12: 1878.

222 Nystrom, M.J., Wessels, B.W., Chen, J. et al. (1995). Deposition of potassium niobate thin films by metalorganic chemical vapor deposition and their nonlinear optical properties. *Mater. Res. Soc. Symp. Proc.* 392: 183.

223 Fluck, D., Pliska, T., Kupfer, M., and Gunter, P. (1995). Depth profile of the nonlinear optical susceptibility of ion-implanted $KNbO_3$ waveguides. *Appl. Phys. Lett.* 67 (6): 748–750.

224 Nichols, B.M. et al. (1998). Epitaxial $KNbO_3$ and its nonlinear optical properties. *Mater. Res. Soc. Symp. Proc.* 541: 741.
225 Lu, C.-H., Lo, S.-Y., and Lin, H.-C. (1998). Hydrothermal synthesis of nonlinear optical potassium niobate ceramic powder. *Mater. Lett.* 34 (3–6): 172–176.
226 Xue, D. and Zhang, S. (1998). Linear and nonlinear optical properties of $KNbO_3$. *Chem. Phys. Lett.* 291: 401.
227 Pliska, T., Fluck, D., Günter, P. et al. (1998). Linear and nonlinear optical properties of $KNbO_3$ ridge waveguides. *J. Appl. Phys.* 84: 1186.
228 Alford, W.J. and Smith, A.V. (2001). Wavelength variation of the second-order nonlinear coefficients of $KNbO_3$, $KTiOPO_4$, $KTiOAsO_4$, $LiNbO_3$, $LiIO_3$, β-BaB_2O_4, KH_2PO_4, and LiB_3O_5 crystals: a test of Miller wavelength scaling. *J. Opt. Soc. Am. B* 18: 524.
229 Simões, A.Z. et al. (2005). Optical properties of potassium niobate thin films prepared by the polymeric precursor method. *Mater. Lett.* 59: 598.
230 Günter, P. and Micheron, F. (1978). Photorefractive effects and photocurrents in $KNbO_3$:Fe. *Ferroelectrics* 18: 27.
231 Günter, P. (1982). Holography, coherent light amplification and optical phase conjugation with photorefractive materials. *Phys. Rep.* 93: 199–299.
232 Zha, M.Z. and Günter, P. (1985). Nonreciprocal optical transmission through photorefractive $KNbO_3$:Mn. *Opt. Lett.* 10 (4): 184–186.
233 Reeves, R.J., Jani, M.G., Jassemnejad, B. et al. (1991). Photorefractive properties of $KNbO_3$. *Phys. Rev. B* 43: 71–82.
234 Buse, K., Hesse, H., Stevendaal, U. et al. (1994). Photorefractive properties of ruthenium-doped potassium niobate. *Appl. Phys. A Solids Surf.* 59: 563.
235 Buchal, C. (2000). Ion implantation for photorefractive devices and optical emitters. *Nucl. Instrum. Methods Phys. Res., Sect. B* 166–167 (2): 743–749.
236 Baumert, J.-C., Walther, C., Buchmann, P. et al. (1985). $KNbO_3$ electro-optic induced optical waveguide/cut-off modulator. *Appl. Phys. Lett.* 46: 1018.
237 Baumert, J.-C., Walther, C., Buchmann, P. et al. (1985). Electro-optically induced optical waveguide in $KNbO_3$. *Integr. Opt.* 48: 44–48.
238 Holman, R.L., Johnson, L.M.A., and Skinner, D.P. (1987). Desirability of electro-optic materials for guided-wave optics. *Opt. Eng.* 26 (2): 262134.
239 Bremer, T., Heiland, W., Hellermann, B. et al. (1988). Waveguides in $KNbO_3$ by He^+ implantation. *Ferroelectr. Lett. Sect.* 9 (1): 11–14.
240 Zhang, L., Chandler, P.J., and Townsend, P.D. (1990). Detailed index profiles of ion implanted waveguides in $KNbO_3$. *Ferroelectr. Lett. Sect.* 11: 89.
241 Strohkendl, F.P., Günter, P., Buchal, C., and Irmscher, R. (1991). Formation of optical waveguides in $KNbO_3$ by low dose MeV He^+ implantation. *J. Appl. Phys.* 69: 84.
242 Strohkendl, F.P., Fluck, D., Günter, P. et al. (1991). Nonleaky optical waveguides in $KNbO_3$ by ultralow dose MeV He ion implantation. *Appl. Phys. Lett.* 59: 3354–3356.
243 Fluck, D., Günter, P., Irmscher, R., and Buchal, C. (1991). Optical strip waveguides in $KNbO_3$ formed by He ion implantation. *Appl. Phys. Lett.* 59: 3213.

244 Moretti, P., Thevenard, P., Wirl, K. et al. (1992). Proton implanted waveguides in $LiNbO_3$, $KNbO_3$ and $BaTiO_3$. *Ferroelectrics* 128 (1): 13–18.
245 Fluck, D., Irmscher, R., Buchal, C., and Günter, P. (1992). Tailoring of optical planar waveguides in $KNbO_3$ by MeV He ion implantation. *Ferroelectrics* 128: 79.
246 Fluck, D., Günter, P., Fleuster, M., and Buchal, C. (1992). Low-loss optical channel waveguides in $KNbO_3$ by multiple energy ion implantation. *J. Appl. Phys.* 72: 1671.
247 Fluck, D., Jundt, D.H., Günter, P. et al. (1993). Modeling of refractive index profiles of He^+ ion-implanted $KNbO_3$ waveguides based on the irradiation parameters. *J. Appl. Phys.* 74 (10): 6023–6031.
248 Fluck, D., Zha, M., Günter, P. et al. (1994). Second harmonic generation and two-wave mixing in ion-implanted $KNbO_3$ waveguides. *Ferroelectrics* 151: 205.
249 Brüulisauer, S., Fluck, D., Solcia, C. et al. (1995). Nondestructive waveguide loss-measurement method using self-pumped phase conjugation for optimum end-fire coupling. *Opt. Lett.* 20 (17): 1773–1775.
250 Pliska, T., Jundt, D.H., Fluck, D. et al. (1995). Low-temperature annealing of ion-implanted $KNbO_3$ waveguides for second-harmonic generation. *J. Appl. Phys.* 77: 6114.
251 Fluck, D., Pliska, T., Gunter, P. et al. (1996). Blue-light second-harmonic generation in ion-implanted $KNbO_3$ channel waveguides of new design. *Appl. Phys. Lett.* 69 (27): 4133–4135.
252 Pliska, T., Solcia, C., Fluck, D. et al. (1997). Radiation damage profiles of the refractive indices of He^+ ion-implanted $KNbO_3$ waveguides. *J. Appl. Phys.* 81: 1099.
253 Pliska, T., Fluck, D., Günter, P. et al. (1998). Birefringence phase-matched blue light second-harmonic generation in a $KNbO_3$ ridge waveguide. *Appl. Phys. Lett.* 72: 2364.
254 Pliska, T., Fluck, D., Günter, P. et al. (1998). Mode propagation losses in He^+ ion-implanted $KNbO_3$ waveguides. *J. Opt. Soc. Am. B* 15 (2): 628–639.
255 Dittrich, P., Montemezzani, G., Bernasconi, P., and Günter, P. (1999). Fast, reconfigurable light-induced waveguides. *Opt. Lett.* 24 (21): 1508–1510.
256 Fluck, D. and Gunter, P. (2000). Second-harmonic generation in potassium niobate waveguides. *IEEE J. Sel. Topics Quantum Electron.* 6 (1): 122–131.
257 Kawai, H., Hashima, H., and Nakajima, S. (2001). Thin film of potassium niobate process for producing the thin film and optical device using the thin film. US Patent 6,203,860 B1, files 20 March 2001.
258 Romanov, M.V., Korsakov, I.E., Kaul, A.R. et al. (2004). MOCVD of $KNbO_3$ ferroelectric films and their characterization. *Chem. Vap. Deposition* 10 (6): 318–324.
259 Graettinger, T.M., Lichtenwalner, D.J., Chow, A.F. et al. (1995). Processing thin films of $KNbO_3$ for optical waveguides. *Integr. Ferroelectr.* 6 (1–4): 363–373.
260 Beckers, L., Zander, W., Schubert, J. et al. (1996). Epitaxial $BaTiO_3$ and $KNbO_3$ thin films on various substrates for optical waveguide applications. *Mater. Res. Soc. Symp. Proc.* 441.

261 Schwyn Thöny, S. et al. (1992). Sputter deposition of epitaxial waveguiding $KNbO_3$ thin films. *Appl. Phys. Lett.* 61: 373.
262 Pliska, T., Gamper, E., and Fluck, D. Electro-optical waveguide modulator on $KNbO_3$. http://www.nlo.ethz.ch/dl/jb1994/JB94-3.10.pdf.
263 Yu, N., Kurimura, S., Kitamura, K. et al. (2004). Efficient broadband second harmonic generation in periodically-poled potassium niobate. In: *Conference on Lasers and Electro-Optics*, San Francisco, CA.
264 Denev, S.A., Lummen, T.T.A., Barnes, E. et al. (2011). Probing ferroelectrics using optical second harmonic generation. *J. Am. Ceram. Soc.* 94: 2699.
265 Schmidt, N., Voigt, P., Hellermann, B. et al. (1991). Phase matching for second harmonic generation in $KNbO_3$:Ta crystals with Nd:YAG- and GaAs-laser wavelengths. *J. Appl. Phys.* 69: 3426.
266 Günter, P. (1979). Near-infrared noncritically phase-matched second-harmonic generation in $KNbO_3$. *Appl. Phys. Lett.* 34: 650.
267 Kolinsky, P.V. (1992). New materials and their characterization for photonic device applications. *Opt. Eng.* 31 (8): 1676–1684.
268 Wu, L.S., Looser, H., and Günter, P. (1990). High-efficiency intracavity frequency doubling of Ti:Al_2O_3 lasers with $KNbO_3$ crystals. *Appl. Phys. Lett.* 56: 2163.
269 Kapphan, S., Schmidt, N., Voigt, P. et al. (1991). Frequency doubling of Nd:YAG- and GaAs-lasers by $KNbO_3$: Ta crystals. *Radiat. Eff. Defects Solids* 119–121: 281.
270 Polzik, E.S. and Kimble, H.J. (1991). Frequency doubling and optical parametric oscillation with potassium niobate. In: *Proceedings of SPIE 1561, Inorganic Crystals for Optics, Electro-Optics, and Frequency Conversion*, vol. 143.
271 Gopalan, V. and Raj, R. (1996). Domain structure and phase transitions in epitaxial $KNbO_3$ thin films studied by in situ second harmonic generation measurements. *Appl. Phys. Lett.* 68 (10): 1323.
272 Gopalan, V. and Raj, R. (1997). Electric field induced domain rearrangement in potassium niobate thin films studied by in situ second harmonic generation measurements. *J. Appl. Phys.* 81: 865.
273 Gopalan, V. and Raj, R. (1996). Domain structure-second harmonic generation correlation in potassium niobate thin films deposited on a strontium titanate substrate. *J. Am. Ceram. Soc.* 79: 3289.
274 Chow, A.F., Lichtenwalner, D.J., Auciello, O., and Kingon, A.I. (1995). Second harmonic generation in potassium niobate thin films. *J. Appl. Phys.* 78: 435.
275 Murzina, T.V., Savinov, S.A., Ezhov, A.A. et al. (2006). Ferroelectric properties in $KNbO_3$ thin films probed by optical second harmonic generation. *Appl. Phys. Lett.* 89 (6): 062907.
276 Xu, Y., Cheng, C.-H., and Mackenzie, J.D. (1996). Epitaxial KNbO3 and Fe-doped KNbO3 thin films prepared by the sol–gel technique. *Mater. Res. Soc. Symp. Proc.* 433.
277 Bosenberg, W.R. and Jarman, R.H. (1993). Type-II phase-matched $KNbO_3$ optical parametric oscillator. *Opt. Lett.* 18 (16): 1323–1325.

278 Urschel, R., Fix, A., Wallenstein, R. et al. (1995). Generation of tunable narrow-band midinfrared radiation in a type I potassium niobate optical parametric oscillator. *J. Opt. Soc. Am. B* 12: 726.

279 Petrov, V. and Noack, F. (1996). Mid-infrared femtosecond optical parametric amplification in potassium niobate. *Opt. Lett.* 21 (19): 1576–1578.

280 Günter, P. and Krumins, A. (1980). High-sensitivity read-write volume holographic storage in reduced $KNbO_3$ crystals. *Appl. Phys.* 23 (2): 199–207.

281 Montemezzani, G. and Gunter, P. (1990). Thermal hologram fixing in pure and doped $KNbO_3$ crystals. *J. Opt. Soc. Am. B: Opt. Phys.* 7 (12): 2323–2328.

282 Tong, X., Zhang, M., Yariv, A., and Agranat, A. (1996). Thermal fixing of volume holograms in potassium niobate. *Appl. Phys. Lett.* 69: 3966.

283 Tuttle, B. (1987). Electronic ceramic thin films: trends in research and development. *MRS Bull.* 40–45.

284 Holman, R., Johnson, L.A., and Skinner, D. (1986). *Proceedings of the Sixth IEEE International Symposium on Applied of Ferroelectrics*, 32–41. New York: IEEE.

285 Chen, F.S., Geusic, J.E., Kurtz, S.K. et al. (1966). Light modulation and beam deflection with potassium tantalate-niobate crystals. *J. Appl. Phys.* 37 (1): 388–398.

286 Geusic, J.E., Kurtz, S.K., Van Uitert, L.G., and Wemple, S.H. (1964). Electro-optic properties of ABO_3 perovskites in the paraelectric phase. *Appl. Phys. Lett.* 4 (8): 141–143.

287 Imai, T., Sasaura, M., Nakamura, K., and Fujiura, K. (2007). Crystal growth and electro-optic properties of $KTa_{1-x}Nb_xO_3$. *NTT Tech. Rev.* 5: 1–8.

288 Imai, T. et al. (2012). Fast response varifocal lenses using $KTa_{1-x}Nb_xO_3$ crystals and a simulation method with electrostrictive calculations. *Appl. Opt.* 51: 1532.

289 Nakamura, K., Miyazu, J., Sasaura, M., and Fujiura, K. (2006). Wide-angle, low voltage electro-optic beam deflection based on space-charge-controlled mode of electrical conduction in $KTa_{1-x}Nb_xO_3$. *Appl. Phys. Lett.* 89 (13): 131115.

290 Yagi, S., Naganuma, K., Imai, T. et al. (2011). A mechanical-free 150-kHz repetition swept light source incorporated a KTN electro-optic deflector. In: *Proceedings of SPIE 7889, Optical Coherence Tomography and Coherence Domain Optical Methods in Biomedicine XV, 78891J*.

291 Miyazu, J., Sasaki, Y., Naganuma, K. et al. (2010). 400 kHz beam scanning using $KTa_{1-x}Nb_xO_3$. In: *Conference on Lasers and Electro-Optics/Quantum Electronics and Laser Science and Photonic Applications Systems Technologies, Technical Digest (CD)*. Optical Society of America, paper CTuG5.

292 Foshee, J., Tang, S., Tang, Y. et al. (2007). A novel high-speed electro-optic beam scanner based on KTN crystals. In: *Proceedings of SPIE 6709, Free-Space Laser Communications VII, 670908*.

293 Tang, Y., Wang, J., Wang, X. et al. (2008). KTN-based electro-optic beam scanner. In: *Proceedings of SPIE 7135, Optoelectronic Materials and Devices III, 713538* (18 November 2008).

294 Gramotnev, D.K. and Bozhevolnyi, S.I. (2010). Plasmonics beyond the diffraction limit. *Nat. Photonics* 4 (2): 83–91.

295 Ye, C., Li, Z., Liu, K. et al. (2014). A compact plasmonic silicon-based electro-optic 2×2 switch. *Nanophotonics*, 4(3): 261–268.
296 Saleh, B.E.A. and Teich, M.C. (1991). *Fundamentals of Photonics*. New York: Wiley.
297 Miller, R.C., Kleinman, D.A., and Savage, A. (1963). Quantitative studies of optical harmonic CdS, $BaTiO_3$ and KH_2PO_4 type crystals. *Phys. Rev. Lett.* 11 (4): 146–152.
298 Wang, J., Zhang, T., Xiang, J., and Zhang, B. (2008). High-tunability and low-leakage current of the polycrystalline compositionally graded (Ba,Sr)TiO_3 thin films derived by a sol–gel process. *Mater. Chem. Phys.* 108: 445.
299 Ihlefeld, J. (2008). Enhanced dielectric tunability in barium titanate thin films with boron additions. *Scr. Mater.* 58: 549.
300 Kobayashi, Y. (2008). Fabrication and dielectric properties of the $BaTiO_3$-polymer nano-composite thin films. *J. Eur. Ceram. Soc.* 28: 117.
301 Song, S. (2007). Effects of buffer layer on the dielectric properties of $BaTiO_3$ thin films prepared by sol–gel processing. *Mater. Sci. Eng., B* 145: 28.
302 Wu, C. and Lu, F. (2001). Electrochemical deposition of barium titanate films using a wide electrolytic voltage range. *Thin Solid Films* 398: 621.
303 Ianculescu, A. (2007). Structure–properties correlations for barium titanate thin films obtained by rf-sputtering. *J. Eur. Ceram. Soc.* 27: 1129.
304 Zhu, J.S., Lu, X.M., Jiang, W. et al. (1997). Optical study on the size effects in $BaTiO_3$ thin films. *J. Appl. Phys.* 81: 1392.
305 Huang, H. and Yao, X. (2004). Preparation of $BaTiO_3$ thin films by mist plasma evaporation on MgO buffer layer. *Ceram. Int.* 30: 1535.
306 Tan, C. (2008). Growth and dielectric properties of $BaTiO_3$ thin films prepared by the microwave-hydrothermal method. *Thin Solid Films* 516: 5545.
307 Tan, C. (2007). Growth and dielectric properties of solvothermal $BaTiO_3$ polycrystalline thin films. *Thin Solid Films* 515: 6572.
308 Huang, Y. (2006). Preparation and photoluminescence properties of ZnO/amorphous-$BaTiO_3$ thin-films by sol–gel process. *J. Mater. Lett.* 60: 3818.
309 Silvan, M. (2002). $BaTiO_3$ thin films obtained by sol–gel spin coating. *Surf. Coat. Technol.* 151: 118.
310 Yuk, J. (2003). Sol–gel $BaTiO_3$ thin film for humidity sensors. *Sens. Actuators, B* 94: 290.
311 Kumazava, H. (1999). Fabrication of barium titanate thin films with a high dielectric constant by a sol–gel technique. *Thin Solid Films* 353: 144.
312 Sharma, H. (1998). Electrical properties of sol–gel processed barium titanate films. *Thin Solid Films* 330: 178.
313 Zhang, H.X. (2000). Optical and electrical properties of sol–gel derived $BaTiO_3$ films on ITO coated glass. *Mater. Chem. Phys.* 63: 174.
314 Ashiri, R. and Helisaie, A. (2011). Preparation and characterization of optical properties of barium titanate nanothin films. *Metal*, Brno, Czech Republic, EU 5: 18–20.
315 Guo, H., Liu, L., Chen, Z. et al. (2006). Structural and optical properties of $BaTiO_3$ ultrathin films. *Europhys. Lett.* 73 (1): 110–115.

316 Tang, P., Towner, D.J., Meier, A.L., and Wessels, B.W. (2003). Polarisation-insensitive Si_3N_4 strip-loaded $BaTiO_3$ thin-film waveguide with low propagation losses. *Electron. Lett.* 39: 1651–1652.
317 Towner, D.J., Ni, J., Marks, T.J., and Wessels, B.W. (2003). Effects of two-stage deposition on the structure and properties of heteroepitaxial $BaTiO_3$ thin films. *J. Cryst. Growth* 255: 107–113.
318 Chowdhury, A. and McCaughan, L. (2001). Figure of merit for near-velocity-matched traveling-wave modulators. *Opt. Lett.* 26: 1317–1319.
319 Pernice, W.H.P., Xiong, C., Walker, F.J., and Tang, H.X. (2014). Design of a silicon integrated electro-optic modulator using ferroelectric $BaTiO_3$ film. *IEEE Photonics Technol. Lett.* 26 (13): 1344.
320 Abel, S. et al. (2013). A strong electro-optically active lead-free ferroelectric integrated on silicon. *Nat. Commun.* 4: 1671.
321 Gu, X. et al. (2009). Commercial molecular beam epitaxy production of high quality $SrTiO_3$ on large diameter Si substrates. *J. Vac. Sci. Technol., B, Microelectron. Nanometer Struct.* 27 (3): 1195–1199.
322 Schneider, S., Mono, T., Albrethsen-Keck, B. et al. (1998). Dry-etching of barium-strontium-titanate thin films. In: *Proceedings of 11th IEEE International Symposium of Applications of Feroelectrics (ISAF 98)*, 51–54.
323 Trapani, G.B. (1980). Rigid electro-optic device using a transparent ferroelectric ceramic element. US Patent 4,201,450, filed 6 May 1980.
324 AC Photonics, Inc. Santa Clara, CA: http://www.acphotonics.com/.
325 Tangonan, G.L., Persechini, D.L., Lotspeich, J.F., and Barnoski, M.K. (1978). Electro-optic diffraction modulation in Ti-diffused $LiTaO_3$. *Appl. Opt.* 17 (20): 3259–3263.
326 Byer, R.L. (1997). Quasi-phasematched nonlinear interactions and devices. *J. Nonlinear Opt. Phys. Mater.* 6: 549–592.
327 Findakly, T., Suchoski, P., and Leonberger, F. (1988). High-quality $LiTaO_3$ integrated-optical waveguides and devices fabricated by the annealed-proton-exchange technique. *Opt. Lett.* 13 (9): 797–798.
328 Kushibiki, J., Miyashita, M., and Chubachi, N. (1996). Quantitative characterization of proton-exchanged layers in $LiTaO_3$ optoelectronic devices by line-focus-beam acoustic microscopy. *IEEE Photonics Technol. Lett.* 8 (11): 1516–1518.
329 Gahagan, K.T., Gopalan, V., Robinson, J.M. et al. (1999). Integrated electro-optic lens/scanner in a $LiTaO_3$ single crystal. *Appl. Opt.* 38: 1186–1190.
330 Batchko, R.G., Shur, V.Y., Fejer, M.M., and Byer, R.L. (1999). Backs witch poling in lithium niobate for high-fidelity domain patterning and efficient blue light generation. *Appl. Phys. Lett.* 75: 1673.
331 Yamada, M., Saitoh, M., and Ooki, H. (1996). Electric-field induced cylindrical lens, switching and deflection devices composed of the inverted domains in $LiNbO_3$ crystals. *Appl. Phys. Lett.* 69 (24): 3659–3661.
332 Kawas, M.J., Stancil, D.D., Schlesinger, T.E., and Gopalan, V. (1997). Electro-optic lens stacks on $LiTaO_3$ by domain inversion. *J. Lightwave Technol.* 15: 1716–1719.

333 Lotspeich, J.F. (1968). Electro-optic light beam deflection. *IEEE Spectr.* 5: 45–53.

334 Feit, M.D. and Fleck, J.A. (1978). Light propagation in graded-index optical fibers. *Appl. Opt.* 17 (24): 3990–3998.

335 Poberaj, G., Koechlin, M., Sulser, F. et al. (2009). Ion-sliced lithium niobate thin films for active photonic devices. *Opt. Mater.* 31 (7): 1054–1058.

336 Armstrong, J.A., Bloembergen, N., Ducuing, J., and Pershan, P.S. (1962). Interactions between light waves in a nonlinear dielectric. *Phys. Rev.* 127: 1918.

337 Chanvillard, L., Aschieri, P., Baldi, P. et al. (2000). Soft proton exchange on periodically poled $LiNbO_3$: a simple waveguide fabrication process for highly efficient nonlinear interactions. *Appl. Phys. Lett.* 76: 1089.

338 Petrov, K.P., Roth, A.P., Patterson, T.L., and Bamford, D.J. Efficient channel waveguide device for difference-frequency mixing of diode lasers. In: *Technical Digest of the Conference on Lasers and Electro-Optics*, Baltimore, MD (23–28 May 1999).

339 Hoffman, D. et al. Mid-infrared difference frequency generation in periodically poled Ti:$LiNbO_3$ channel waveguides. In: *ECIO'99 Turin.*

340 Schreiber, G., Ricken, R., and Scholer, W. Near infrared second harmonic and difference frequency generation in periodically poled Ti:$LiNbO_3$ waveguides. In: *ECIO'99 Turin.*

341 Chou, M.-H., Parameswaran, K.P., Fejer, M.M., and Brener, I. (2000). Optical signal processing and switching with second-order nonlinearities in waveguides. *IEICE Trans. Electron.* E83-C: 869.

342 Maillis, S., Riziotis, C., Wellington, I.T. et al. (2003). Direct UV writing of channel waveguides in congruent lithium niobate single crystals. *Opt. Lett.* 28: 1433–1435.

343 Sones, C.L., Ganguly, P., Ying, Y.J. et al. (2009). Spectral and electro-optic response of UV-written waveguides in $LiNbO_3$ single crystals. *Opt. Express* 17: 23755–23764.

344 Ganguly, P., Sones, C.L., Ying, Y.J. et al. (2009). Determination of refractive indices from the mode profiles of UV-written channel waveguides in $LiNbO_3$-crystals for optimization of writing conditions. *IEEE J. Lightwave Technol.* 27: 3490–3497.

345 Sones, C.L., Muir, A.C., Ying, Y.J. et al. (2008). Precision nanoscale domain engineering of lithium niobate via UV laser induced inhibition of poling. *Appl. Phys. Lett.* 92: 072905.

346 Sones, C.L., Ganguly, P., Ying, Y.J. et al. (2010). Poling-inhibition ridge waveguides in lithium niobate crystals. *Appl. Phys. Lett.* 97: 151112.

347 Bosso, S.S. (1999). Applications of lithium niobate integrated optic in telecommunication systems. *Proc. SPIE, Integr. Opt. Devices III* 3620: 34–37.

348 Fritz, D., McBrien, G., and Suchoski, P.G. (2000). Practical applications for integrated optics in communications systems. *Proc. SPIE, Integr. Opt. Devices IV* 3936: 218–221.

349 Wooten, E.L., Kissa, K.M., Yu-Yan, A. et al. (2000). A review of lithium niobate modulators for fiber-optic communications systems. *IEEE J. Sel. Top. Quantum Electron.* 6: 69–82.

350 Noguchi, K., Mitomi, O., Kawano, K., and Yanagibashi, M. (1993). Highly 40-GHz bandwidth Ti:$LiNbO_3$ optical modulator employing ridge structure. *IEEE Photonics Technol. Lett.* 5: 52–54.

351 Shimotsu, S. et al. (2001). Single side-band modulation performance of a $LiNbO_3$ integrated modulator consisting of four-phase modulator waveguides. *IEEE Photonics Technol. Lett.* 13 (4): 364–366.

352 Griffin, R.A. et al. 10 Gb/s optical differential quadrature phase shift key (DQPSK) transmission using GaAs/AlGaAs integration. In: *Proceedings OFC-2002,* Anaheim, CA, post-deadline paper.

353 Ghirardi, F. et al. (1994). InP-based 10 GHz bandwidth polarization diversity heterodyne photoreceiver with electro-optical adjustability. *IEEE Photonics Tech. Lett.* 6 (7): 814–816.

354 Hoffmann, D. et al. (1989). Integrated optics eight-port 90° hybrid on $LiNbO_3$. *IEEE J. Lightwave Technol.* 7 (5): 794–798.

355 Janner, D., Tulli, D., Garca-Granda, M. et al. (2009). Micro-structured integrated electro-optic $LiNbO_3$ modulators. *Laser Photonics Rev.* 3: 301.

356 Nikogosyan, D.N. (2005). *Nonlinear Optical Crystals: A Complete Survey.* New York: Springer-Science.

357 Xu, Q., Schmidt, B., Pradhan, S., and Lipson, M. (2005). Micrometre-scale silicon electro-optic modulator. *Nature* 435: 325.

358 Timurdogan, E., Sorace-Agaskar, C.M., Sun, J. et al. (2014). An ultralow power athermal silicon modulator. *Nat. Commun.* 5: 4008.

359 Zhang, C., Morton, P.A., Khurgin, J.B. et al. (2016). Ultralinear heterogeneously integrated ring-assisted Mach–Zehnder interferometer modulator on silicon. *Optica* 3: 1483.

360 Sun, C., Wade, M.T., Lee, Y. et al. (2015). Single-chip microprocessor that communicates directly using light. *Nature* 528: 534.

361 Rolland, C., Moore, R.S., Shepherd, F., and Hillier, G. (1993). 10 Gbit/s, 1.56μm Multiquantum well InP/InGaAsP Mach–Zehnder Optical Modulator. *Electron. Lett.* 29: 471.

362 Aoki, M., Suzuki, M., Sano, H. et al. (1993). InGaAs/InGaAsP MQW electroabsorption modulator integrated with a DFB laser fabricated by band-gap energy control selective area MOCVD. *IEEE J. Quantum Electron.* 29: 2088.

363 Kikuchi, N., Yamada, E., Shibata, Y., and Ishii, H. (2012). *IEEE Compound Semiconductor Integrated Circuit Symposium (CSICS),* 1–4.

364 Ilchenko, V.S., Savchenkov, A.A., Matsko, A.B., and Maleki, L. (2004). Nonlinear Optics and Crystalline Whispering Gallery Mode Cavities *Phys. Rev. Lett.* 92: 043903.

365 Chen, L., Xu, Q., Wood, M.G., and Reano, R.M. (2014). Hybrid silicon and lithium niobate electro-optical ring modulator. *Optica* 1: 112.

366 Rao, A., Patil, A., Chiles, J. et al. (2015). Heterogeneous microring and Mach–Zehnder modulators based on lithium niobate and chalcogenide glasses on silicon. *Opt. Express* 23: 22746.

367 Jin, S., Xu, L., Zhang, H., and Li, Y. (2016). $LiNbO_3$ thin-film modulators using silicon nitride surface ridge waveguides. *IEEE Photonics Technol. Lett.* 28: 736.

368 Rao, A., Patil, A., Rabiei, P. et al. (2016). High-performance and linear thin-film lithium niobate Mach–Zehnder modulators on silicon up to 50 GHz. *Opt. Lett.* 41: 5700.
369 Cai, L., Kang, Y., and Hu, H. (2016). Electric-optical property of the proton exchanged phase modulator in single-crystal lithium niobate thin film. *Opt. Express* 24: 4640.
370 Pfaff, W.R., Feser, K., and Lutz, M. (1989). Potential-free special sensor for field strength measurement in NEMP research and testing. In: *Proceedings of eighth Internship Zurich Symposium on EMC*, 35–40.
371 Ogawa, O., Sowa, T., and Ichzono, S. (1999). A guided wave optical electric field senor with improved temperature stability. *J. Lightwave Technol.* 17: 823–830.
372 Santos, J.C., Taplamacioglu, M.C., and Hidaka, K. (2000). Pockels high voltage measurement system. *IEEE Trans. Power Delivery* 15: 8–13.
373 Liang, W.L., Shay, W., Huang, M., and Tseng, W. (2003). Progress of photonic sensor development measure the RF electric and magnetic fields. In: *Asia-Pacific Conference on Environmental Electromagnetics (CEEM)*, Hangzhou, China, 609–612.
374 Masterson, K.D., Driver, L.D., and Kanda, M. (1989). Photonic probes for the measurement of electromagnetic fields over broad bandwidths. In: *IEEE International Symposium on Electromagnetic Compatibility (EMC)*, 1–6.
375 Hidaka, K. (1996). Progress in Japan of space charge field measurement in gaseous dielectrics using a Pockels sensor. *IEEE Electr. Insul. Mag.* 12 (1): 17–28.
376 Santos, J.C., Taplamacioglu, M.C., and Hidaka, K. (1999). Optical high voltage measurement using Pockels microsingle crystal. *Rev. Sci. Instrum.* 70: 3271–3276.
377 Chang, J., Vittitoe, C.N., Neyer, B.T., and Ballard, W.P. (1985). An electro-optical technique for intense microwave measurements. *J. Appl. Phys.* 57: 4843–4848.
378 Hidaka, K. (1999). Electric field and voltage measurement by using electro-optic sensor. In: *11th International Symposium on High Voltage Engineering*, 1–14.
379 Pommerenke, D. and Masterson, K. (2000). A novel concept for monitoring partial discharge on EHV-cable system accessories using no active components at the accessories. In: *Eighth International Conference on Dielectric Materials, Measurements and Applications*, 145–149.
380 Tian, Y., Lewin, P.L., Pommerenke, D. et al. (2004). Partial discharge on-line monitoring for HV cable systems using electro-optic modulators. *IEEE Trans. Dielectr. Electr. Insul.* 11: 861–869.
381 Jaeger, N.A.F. and Young, L. (1989). High-voltage sensor employing an integrated optics Mach–Zehnder interferometer in conjunction with a capacitive divider. *J. Lightwave Technol.* 1: 229–235.
382 Tajima, K., Masugi, M., and Kuwabara, N. (1999). Propagation characteristics of ESD-induced electromagnetic pulses measured using optical *E*-field sensor. In: *International Symposium on Electromagnetic Compatibility*, 142–144.

383 Mac Alpine, M., Zhao, Z., and Demokan, M.S. (2002). Development of a fiber-optic sensor for partial discharges in oil-filled power transformers. *Electr. Power Syst. Res.* 63: 27–36.

384 Zeng, R., Zhang, Y., and Chen, W. (2008). Mesurement of electrical field distribution along composite insulators by integrated optical electric field sensor. *IEEE Trans. Dielectr. Electr. Insul.* 15: 302–310.

385 Zeng, R., Wang, B., Yu, Z.Q., and Chen, W.Y. (2011). Design and application of an integrated electro-optic sensor for intensive electric field measurement. *IEEE Trans. Dielectr. Electr. Insul.* 18: 312–319.

386 Bull, J.D., Jaeger, N.A.F., and Rahmatian, F. (2005). A new hybrid current sensor for high-voltage applications. *IEEE Trans. Power Delivery* 20: 32–38.

387 Korotky, S.K. and Veselka, J.J. (1996). An RC network analysis of long term Ti: $LiNbO_3$ bias stability. *J. Lightwave Technol.* 14: 2687–2697.

388 Rong, Z., Zhanqing, Y., Bo, W. et al. (2009). Integrated optical sensor with grid electrode for intense electric field measurement. CN 200920143317.7.

389 Thaniyavarn, S. (1986). Modified 1×2 directional coupler waveguide modulator. *Electron. Lett.* 22: 941–942.

390 Zeng, R., Wang, B., Yu, Z.-Q. et al. (2011). Integrated optical *E*-field sensor based on balanced Mach–Zehnder interferometer. *Opt. Eng.* 50: 11404.

391 Zeng, R., Wang, B., Yu, Z.-Q. et al. (2012). Development and application of integrated optical sensors for intense *E*-field measurement. *Sensors* 12: 11406–11434.

392 Bao, S., Chen, F., Chen, K. et al. (2012). Integrated optical electric field sensor from 10 kHz to 18 GHz. *IEEE Photonics Technol. Lett.* 24 (13): 1106–1108.

393 Meier, T., Kostrzewa, C., Petermann, K., and Schueppert, B. (1994). Integrated optical *E*-field probes with segmented modulator electrodes. *J. Lightwave Technol.* 12 (8): 1497–1503.

394 Brooks, J.L., Wenworth, R.H., Youngquist, R.C. et al. (1985). Coherence multiplexing of fiber-optic interferometric sensors. *J. Lightwave Technol.* LT-3 (5): 1062–1072.

395 Hauden, J., Porte, H., and Goedgebuer, J.P. (1994). Quasi-polarization independent coherence modulator/demodulator integrated in *Z*-propagating lithium niobate. *IEEE J. Quantum Electron.* 30 (10): 2325–2331.

396 Gutiérrez-Martínez, C., Sánchez-Rinza, B., Rodríguez-Asomoza, J., and Pedraza-Contreras, J. (2000). Automated measurement of optical coherence lengths and optical delays in coherence-modulated optical transmissions. *IEEE Trans. Instrum. Meas.* 49 (1): 32–36.

397 Zhang, J., Chen, F., Sun, B. et al. (2014). 3D integrated optical *E*-field sensor for lightning electromagnetic impulse measurement. *IEEE Photonics Technol. Lett.* 26 (23): 2353–2356.

398 Xu, J., Liao, Y., Zeng, H. et al. (2007). Selective metallization on insulator surfaces with femtosecond laser pulses. *Opt. Express* 15 (20): 12743–12748.

399 Serpengüzel, A., Arnold, S., and Griffel, G. (1995). Excitation of resonances of microspheres on an optical fiber. *Opt. Lett.* 20 (7): 654–656.

400 Melnichuk, M. and Wood, L.T. (2010). Direct Kerr electro-optic effect in non-centrosymmetric materials. *Phys. Rev. A* 82 (1): 013821.

401 Bain, A.K., Jackson, T.J., Koutsonas, Y. et al. (2007). Optical properties of barium strontium titanate (BST) ferroelectric thin films. *Ferroelectr. Lett.* 34 (5–6): 149–154.

402 Bain, A.K. and Chand, P. (2013). Optical properties of ferroelectrics and measurement procedures. In: *Advances in Ferroelectrics*. Croatia: InTech publication. ISBN: 980-953-307-657-2.

403 Jang, S.J., Uchino, K., Nomura, S., and Cross, L.E. (1980). Electrostrictive behavior of lead magnesium niobate based ceramic dielectrics. *Ferroelectrics* 27 (1): 31–34. https://doi.org/10.1080/00150198008226059.

404 Chen, Y.H., Uchino, K., Shen, M., and Viehland, D. (2001). Substituent effects on the mechanical quality factor of $Pb(Mg_{1/3}Nb_{2/3})O_3$–$PbTiO_3$ and $Pb(Sc_{1/2}Nb_{1/2})O_3$–$PbTiO_3$ ceramics. *J. Appl. Phys.* 90 (3): 1455–1458. https://doi.org/10.1063/1.1379248.

405 Kong, L.B., Ma, J., Zhu, W., and Tan, O.K. (2002). Rapid formation of lead magnesium niobate-based ferroelectric ceramics via a high-energy ball milling process. *Mater. Res. Bull.* 37 (3): 459–465. https://doi.org/10.1016/S0025-5408(01)00823-6.

406 Gomi, M., Miyazawa, Y., Uchino, K. et al. (1982). Optical stabilizer using a bistable optical device with a PMN electrostrictor. *Appl. Opt.* 21 (14): 2616–2619.

407 Nomura, L.S., Gomi, M., and Uchino, K. (1985). Bistable optical device with a PMN-based ceramic electrostrictor. *Ferroelectrics* 63 (1): 209–216.

408 Kawamura, S., Kaneko, J.H., Fujimoto, H. et al. (2006). Possibility of using a PMN–PT single crystal as a neutron optical device. *Physica B* 385–386, Part 2, 15: 1277–1279.

409 Jiang, H., Zou, Y.K., Chen, Q. et al. (2005). Transparent electro-optic ceramics and devices. http://www.bostonati.com/whitepapers/SPIE04paper.pdf.

410 Yin, X., Zhang, S., and Wang, J. (2004). Multual action of optical activity and electro-optic effect and its influence on the electro-optic Q-switch. *Opt. Rev.* 11: 328–331.

411 Chen, Y.H. and Huang, Y.C. (2003). Actively Q-switched Nd-YVO_4 laser using an electro-optic periodically poled lithium niobate crystal as a laser Q-switch. *Opt. Lett.* 28: 1460.

412 Ozolinsh, M. et al. (1997). Q-switching of Er:YAG (2.9 μm) solid state laser by PLZT electro-optic modulator. *IEEE J. Quantum Electron.* 33: 10.

413 Wang, G., Jiang, H., and Zhao, J. (1998). Low voltage electro-optic Q-switching of 1.06 μm microlasers by PLZT. In: *CLEO '98*, vol. 485.

414 Poole, C.D. and Wagner, R.E. (1986). Phenomenological approach to polarisation dispersion in long single-mode fibres. *Electron. Lett.* 22: 1029.

415 Keiser, G. (1999). A review of WDM technology and applications. *Opt. Fiber Technol.* 5: 3.

416 Nakamura, K., Oshiki, M., and Kitazume, H. (1998). SH-mode SAW and its acousto-optic interaction in $KNbO_3$. In: *Ultrasonics Symposium Proceedings*, vol. 2, 1305–1308. IEEE,.

417 Yamanouchi, K., Wagatsuma, Y., Odagawa, H., and Cho, Y. (2001). Single crystal growth of $KNbO_3$ and application to surface acoustic wave devices. *J. Eur. Ceram. Soc.* 21 (15): 2791–2795.

418 Wu, M.-S. and Shih, W.-C. (1997). Propagation characteristics of surface acoustic waves in $KNbO_3/SrTiO_3/Si$ layered structures. *Jpn. J. Appl. Phys.* 36: 2192.
419 Odagawa, H., Kotani, K., Cho, Y., and Yamanouchi, K. (1999). Observation of ferroelectric polarization in $KNbO_3$ thin films and surface acoustic wave properties. *Jpn. J. Appl. Phys.* 38: 3275–3278.
420 Nayak, R., Gupta, V., and Sreenivas, K. (2001). SAW and AO propagation characteristics of $KNbO_3$/spinel thin film layered structure. In: *Ultrasonics Symposium*, vol. 1, 269–272. IEEE.
421 Nayak, R., Gupta, V., and Sreenivas, K. (2003). Influence of spinel substrate and over-layer for enhanced SAW and AO properties with $KNbO_3$ thin film. *IEEE Trans. Ultrason. Ferroelectr. Freq. Control* 50 (6): 577–584.
422 Onoe, A. et al. (2001). Epitaxial growth of orientation-controlled $KNbO_3$ crystal films on MgO using $KTa_xNb_{1-x}O_3$ intermediate layer by metalorganic chemical vapor deposition. *Appl. Phys. Lett.* 78: 49.
423 Nystrom, M.J. et al. (1997). The effects of substrate thermal mismatch on the domain structure of MOCVD-derived potassium niobate thin films. *MRS Proc.* 474: 31.
424 Onoe, A. et al. (1996). Heteroepitaxial growth of $KNbO_3$ single-crystal films on $SrTiO_3$ by metalorganic chemical vapor deposition. *Appl. Phys. Lett.* 69: 167.
425 Nystrom, M.J. et al. (1996). Microstructure of epitaxial potassium niobate thin films prepared by metalorganic chemical vapor deposition. *Appl. Phys. Lett.* 68: 761.
426 Mantese, J.V. et al. (1992). Characterization of potassium tantalum niobate films formed by metalorganic deposition. *J. Appl. Phys.* 72: 615.
427 Nichols, B.M., Wessels, B.W., Belot, J.A., and Marks, T.J. (1998). Epitaxial $KNbO_3$ and its nonlinear optical properties. *MRS Proc.* 541.
428 Yang, R. et al. (2008). Pulsed laser deposition of stoichiometric $KNbO_3$ films on Si (100). *Thin Solid Films* 516: 8559.
429 Yang, R. et al. (2007). Composition controlling of $KNbO_3$ thin films prepared by pulsed laser deposition. *Mater. Lett.* 61: 2658.
430 Rousseau, A., Laur, V., Guilloux-Viry, M. et al. (2006). Pulsed laser deposited $KNbO_3$ thin films for applications in high frequency range. *Thin Solid Films* 515: 2353.
431 Zaldo, C. et al. (1994). Growth of $KNbO_3$ thin films on MgO by pulsed laser deposition. *Appl. Phys. Lett.* 65: 502.
432 Kakio, S. et al. (2008). Fabrication and evaluation of highly (110)-oriented potassium niobate thin films prepared by RF-magnetron sputtering. *Jpn. J. Appl. Phys.* 47: 3802.
433 Graettinger, T.M., Morris, P.A., Roshko, A. et al. (1994). Growth of epitaxial $KNbO_3$ thin films. *MRS Proc.* 341.
434 Derderian, G.J. et al. (1994). Epitaxial growth of (110) $KNbO_3$ films on (100) MgO substrates via sol–gel processing. *MRS Proc.* 341: 277.
435 Derderian, G.J. et al. (1994). Microstructure/process relations in sol–gel-prepared $KNbO_3$ thin films on (100) MgO. *J. Am. Ceram. Soc.* 77: 820.

436 Derderian, G.J. et al. (1993). Microstructural changes due to process conditions in sol–gel derived $KNbO_3$ thin films. *MRS Proc.* 310: 339.

437 Pribošič, I., Makovec, D., and Drofenik, M. (2005). Formation of nanoneedles and nanoplatelets of $KNbO_3$ perovskite during templated crystallization of the precursor gel. *Chem. Mater.* 17(11): 2953–2958.

438 Lin, Y., Yang, H., Zhu, J. et al. (2008). Low-temperature synthesis of nanocrystalline $KNbO_3$. *Mater. Manuf. Processes* 23: 796.

439 Vasco, E., Magrez, A., Forró, L., and Setter, N. (2005). Growth kinetics of one-dimensional $KNbO_3$ nanostructures by hydrothermal processing routes. *J. Phys. Chem. B* 109: 14331.

440 Liu, J.F., Li, X.L., and Li, Y.D. (2002). Novel synthesis of polymorphous nanocrystalline $KNbO_3$ by a low temperature solution method. *J. Nanosci. Nanotechnol.* 2: 617. J. Cryst. Growth, 247, 419 (2003).

441 Magrez, A., Vasco, E., Seo, J.W. et al. (2006). Growth of Single-Crystalline $KNbO_3$ Nanostructures *J. Phys. Chem. B* 110: 58.

442 Suyal, G., Colla, E., Gysel, R. et al. (2004). Piezoelectric response and polarization switching in small anisotropic perovskite particles. *Nano Lett.* 4: 1339.

443 Wang, G., Yu, Y., Grande, T., and Einarsrud, M.A. (2009). Synthesis of $KNbO_3$ nanorods by hydrothermal method. *J. Nanosci. Nanotechnol.* 9 (2): 1465–1469.

444 Wang, G., Selbach, S.M., Yu, Y. et al. (2009). Hydrothermal synthesis and characterization of $KNbO_3$ nanorods. *CrystEngComm* 11: 1958–1963.

445 Paula, A.J., Parra, R., Zaghete, M.A., and Varela, J.A. (2008). Synthesis of $KNbO_3$ nanostructures by a microwave assisted hydrothermal method. *Mater. Lett.* 62: 2581–2584.

446 Triebwasser, S. (1959). Study of ferroelectric transitions of solid-solution single crystals of $KNbO_3$-$KTaO_3$. *Phys. Rev.* 114: 63.

447 Chen, F.S. (1967). A Laser · Induced Inhomogeneity of Refractive Indices in KTN. *J. Appl. Phys.* 38: 3418.

448 Haas, W. and Johannes, R. (1967). Linear electro-optic effect in potassium tantalate niobate crystals. *Appl. Opt.* 6: 2007.

449 Fay, H. (1967). Characterization of potassium tantalate-niobate crystals by electro-optic measurements. *Mater. Res. Bull.* 2: 499.

450 van Raalte, J.A. (1967). Linear electro-optic effect in ferroelectric KTN. *J. Opt. Soc. Am.* 57: 671.

451 Fox, A.J. and Whipps, P.W. (1971). Longitudinal quadratic electro-optical effect in KTN. *Electron. Lett.* 7 (5–6): 139–140.

452 Boatner, L.A., Krätzig, E., and Orlowski, R. (1980). KTN as a holographic storage material. *Ferroelectrics* 27: 247.

453 Loheide, S., Riehemann, S., Mersch, F. et al. (1993). Refractive indices, permittivities, and linear electrooptic coefficients of tetragonal potassium tantalate-niobate crystals. *Phys. Status Solidi A* 137: 257.

454 Vor der Linde, D., Glass, A.M., and Rodgers, K.F. (1975). High-sensitivity optical recording in KTN by two-photon absorption. *Appl. Phys. Lett.* 26: 22.

455 Pattnaik, R. and Toulouse, J. (1997). New dielectric resonances in mesoscopic ferroelectrics. *Phys. Rev. Lett.* 79: 4677.

456 Ishai, P.B., de Oliveira, C.E.M., Ryabov, Y. et al. (2004). Glass-forming liquid kinetics manifested in a KTN:Cu crystal. *Phys. Rev. B* 70: 132104.
457 Wang, J.Y., Guan, Q.C., Wei, J.Q., and Liu, Y.G. (1992). Growth and characterization of cubic $KTa_{1-x}Nb_xO_3$ crystals. *J. Cryst. Growth* 116: 27.
458 Buse, K., Havermeyer, F., Glabasnia, L. et al. (1996). Quadratic polarization-optic coefficients of cubic $KTa_{1-x}Nb_xO_3$ crystals. *Opt. Commun.* 131: 339–342.
459 Zhang, H.Y., He, X.H., Shih, Y.H. et al. (1997). Optical and nonlinear optical study of $KTa_{0.52}Nb_{0.48}O_3$ epitaxial film. *Opt. Lett.* 22: 1745.
460 Agranat, A., Leyva, V., and Yariv, A. (1989). Voltage-controlled photorefractive effect in paraelectric $KTa_{1-x}Nb_x0_3$:CuV. *Opt. Lett.* 14 (18): 1017–1019.
461 Sun, H.-G., Zhou, Z.-X., Yuan, C.-X. et al. (2012). Structural, electronic and optical properties of $KTa_{0.5}Nb_{0.5}O_3$ surface: a first-principles study. *Chin. Phys. Lett.* 29 (1): 017303.
462 Zhao, X.Y., Wang, Y.H., Zhang, M. et al. (2011). *Chin. Phys. Lett.* 28: 0607101.
463 Gong, S., Wang, Y.H., Zhao, X.Y. et al. (2011). Structural, Electronic and Optical Properties of $BiAl_xGa_{1-x}O_3$ (x = 0, 0.25, 0.5 and 0.75) *Chin. Phys. Lett.* 28: 087402.
464 Shen, Y.Q. and Zhou, Z.X. (2008). Structural, electronic, and optical properties of ferroelectric $KTa_{1/2}Nb_{1/2}O_3$ solid solutions. *J. Appl. Phys.* 103: 074113.
465 Wang, Y., Shen, Y.Q., and Zhou, Z.X. (2011). Structures and elastic properties of paraelectric and ferroelectric $KTa_{0.5}Nb_{0.5}O_3$ from first-principles calculation *Physica B* 406: 850.
466 Shen, Y.Q. and Zhou, Z.X. (2008). The effect of local ordering on the structure phase transition in disordered $KTa_{1/2}Nb_{1/2}O_3$ from first principles studies *Comput. Mater. Sci.* 42: 434.
467 Ishai, P.B., Agranat, A.J., and Feldman, Y. (2006). Confinement kinetics in a KTN:Cu crystal: experiment and theory *Phys. Rev. B* 73: 104.
468 Yang, L., Jun, L., Zhongxiang, Z. et al. (2013). Low-frequency–dependent electro-optic properties of potassium lithium tantalate niobate single crystals. *Europhys. Lett.* 102 (3): 37004.
469 Wang, X.P., Wang, J.Y., Zhang, H.J. et al. (2008).Thermal properties of cubic $KTa_{1-x}Nb_xO_3$ crystals *J. Appl. Phys.* 103: 033513.
470 Milek, J.T. and Neuberger, M. (1972). Potassium tantalate niobate (KTN). In: *Linear Electro-optic Modular Materials*, 223–245.
471 Fox, A.J. (1975). Nonlinear longitudinal KTN modulator. *Appl. Opt.* 14 (2): 343–352.
472 Kazuo, F., Seiji, T., Masahiro, S. et al. (2004). High-efficiency electro-optic phase modulator with KTN crystal waveguides. IEIC Technical Rep., Vol. 103, No. 616 (OPE2003 231–242), 25–30.
473 Shinagawa, M. et al. (2013). Sensitive electro-optic sensor using $KTa_{1-x}Nb_xO_3$ crystal. *Sens. Actuators, A* 192: 42.
474 Bohac, P. and Kaufmann, H. (1986). KTN optical waveguides grown by liquid-phase epitaxy. *Electron. Lett.* 22 (16): 861–862.

475 Fujiura, K. and Nakamura, K. (2004). KTN crystal with a large electro-optic effect and the potential for improving optical device performance. *Microwave Photonics, MWP'04. IEEE International Topical Meeting* (4–6 October 2004).

476 Fujiura, K. and Nakamura, K. (2005). KTN optical waveguide technologies with a large electro-optic effect, lasers and electro-optics. In: *CLEO/Pacific Rim 2005. Pacific Rim Conference* (30–02 August 2005), 69–70.

477 Fujiura, K. and Nakamura, K. (2005). KTN optical waveguide devices with an extremely large electro-optic effect. In: *Proceedings of SPIE 5623, Passive Components and Fiber-based Devices*, 518 (19 January 2005.

478 Chang, Y.-C. and Yin, S. (2012). Dynamic and tunable optical waveguide based on KTN electro-optic crystals. In: *Proceedings of SPIE 8497, Photonic Fiber and Crystal Devices: Advances in Materials and Innovations in Device Applications VI, 84970J*. (15 October 2012).

479 Wang, L.-L., Wang, X.-L., Chen, F. et al. (2007). Optical properties of $KTa_xNb_{1-x}O_3$ waveguides formed by carbon and proton implantation. *Jpn. J. Appl. Phys.* 46 (9A): 5885.

480 Imai, T., Fujiura, K., Shimokozono, M. et al. (2008). Optical waveguide material and optical waveguide. US Patent 7,340,147 B2, filed 4 March 2008 and issued 27 January 2005.

481 Koji, E. (2004). Optical waveguide devices using electro-optic KTN crystals. *J. Ceram. Soc. Jpn.* 39 (7): 552–554.

482 Chang, Y.C., Lin, C.M., Yao, J. et al. (2012). Field induced dynamic waveguides based on potassium tantalate niobate crystals. *Opt. Express* 20 (19): 21126–21136.

483 Yagi, S. (2009). KTN crystals open up new possibilities and applications. *NTT Tech. Rev.* 7 (12): 1–5.

484 Scibor-Rylski, M.T.V. (1976). A total internal reflection device using KTN. *Opt. Quantum Electron.* 8 (3): 197–201.

485 Yilmaz, S., Venkatesan, T., and Gerhar-Multhaupt, R. (1991). Pulsed laser deposition of stoichiometric potassium-tantalate-niobate films from segmented evaporation targets. *Appl. Phys. Lett.* 58: 2479.

486 Lu, C.J. and Kuang, A.X. (1997). Preparation of potassium tantalate niobate through sol-gel processing. *J. Mater. Sci.* 32: 4421.

487 Nazeri, A. (1994). Crystallization of sol-gel deposited potassium-tantalate-niobate thin films on platinum *Appl. Phys. Lett.* 65: 295.

488 Kuang, A.X., Lu, C.J., Huang, G.Y., and Wang, S.M. (1995). Preparation of $KTa_{0.65}Nb_{0.35}O_3$ thin films by a sol-gel process. *J. Cryst. Growth* 149: 80.

489 Suzuki, K., Sakamoto, W., Yogo, T., and Hirano, S. (1999). Processing of oriented $K(Ta,Nb)O_3$ films using chemical solution deposition. *J. Am. Ceram. Soc.* 82: 1463.

490 Bursik, J., Zelezny, V., and Vanek, P. (2005). Preparation of potassium tantalate niobate thin films by chemical solution deposition and their characterization *J. Eur. Ceram. Soc.* 25: 2151.

491 Nichols, B.M., Hoerman, B.H., Hwang, J.-H. et al. (2003). Phase stability of epitaxial $KTa_xNb_{1-x}O_3$ thin films deposited by metalorganic chemical vapor deposition *J. Mater. Res.* 18: 106.

492 Sashital, S.R., Krishnakumar, S., and Esener, S. (1993). Synthesis and characterization of rf-planar magnetron sputtered $KTa_xNb_{1-x}O_3$ thin films *Appl. Phys. Lett.* 62: 2917.

493 Yilmaz, S., Gerhard-Multhaupt, R., Bonner, W.A. et al. (1994). Electro-optic potassium-tantalate-niobate films prepared by pulsed laser deposition from segmented pellets. *J. Mater. Res.* 9 (05): 1272–1279.

494 Knauss, L.A., Harshavardhan, K.S., Christen, H.-M. et al. (1998). Growth of nonlinear optical thin films of $KTa_{1-x}Nb_xO_3$ on GaAs by pulsed laser deposition for integrated optics. *Appl. Phys. Lett.* 73: 3806.

495 Rousseau, A., Guilloux-Viry, M., Dogheche, E. et al. (2007). Growth and optical properties of $KTa_{1-x}Nb_xO_3$ thin films grown by pulsed laser deposition on MgO substrates. *J. Appl. Phys.* 102: 093106.

496 Rousseau, A., Laur, V., Deputier, S. et al. (2008). Influence of substrate on the pulsed laser deposition growth and microwave behaviour of $KTa_{0.6}Nb_{0.4}O_3$ potassium tantalate niobate ferroelectric thin films. *Thin Solid Films* 516: 4882.

497 Peng, W., Guilloux-Viry, M., Deputier, S. et al. (2007). Structural improvement of PLD grown $KTa_{0.65}Nb_{0.35}O_3$ films by the use of $KNbO_3$ seed layers *Appl. Surf. Sci.* 254: 1298.

498 Yang, W. et al. (2012). Effect of oxygen atmosphere on the structure and refractive index dispersive behavior of $KTa_{0.5}Nb_{0.5}O_3$ thin films prepared by PLD on Si(001) substrates. *Appl. Surf. Sci.* 258: 3986.

499 Ma, W.D., Zhao, Z.S., Wang, S.M. et al. (1999). Preparation of Perovskite Structure $K(Ta_{0.65}Nb_{0.35})O_3$ Films by Pulsed Laser Deposition on Si Substrates. *Phys. Status Solidi A* 176: 985.

500 Yang, W., Zhou, Z., Yang, B. et al. (2011). Structure and refractive index dispersive behavior of potassium niobate tantalate films prepared by pulsed laser deposition. *Appl. Surf. Sci.* 257: 7221–7225.

501 Korsah, K., Kisner, R., Boatner, L. et al. (2005). Preliminary investigation of KTN as a surface acoustic wave infrared/thermal detector. *Sens. Actuators, A* 119 (2): 358–364.

502 Zheng, K.-Y., Wei, N., Yang, F.-X. et al. (2007). $KTa_{0.4}Nb_{0.6}O_3$ nanoparticles synthesized through solvothermal method. *Front. Phys. China* 2 (4): 436–439.

503 Simon, Q., Dorcet, V., Boullay, P. et al. (2013). Nanorods of potassium tantalum niobate tetragonal tungsten bronze phase grown by pulsed laser deposition. *Chem. Mater.* https://doi.org/10.1021/cm401018k.

504 Fetterman, H., Udupa, A., Chan, D. et al. Polymer modulators with bandwidth exceeding 100 GHz. In: *ECOC'98*, vol. 3, 501–502. Madrid, Spain.

505 Sugiyama, M., Doi, M., Taniguchi, S. et al. (2002). Driver-less 40 Gb/s $LiNbO_3$ modulator with sub-1 V drive voltage. *Proceedings of Optical fiber communications Conference (OFC'02)*, Anaheim, CA, Paper PD paper FB6-1.

506 Watanabe, K., Kikuchi, N., Arimoto, H. et al. Ultra-compact 40-Gbit/s optical transmitter with a novel hybrid integration technique of electro-absorption modulation on driver-IC. In: *ECOC'03*, 448–451. Rimini, Italy, Paper We.2.5.1.

507 Rytz, D. and Scheel, H.J. (1982). Crystal growth of $KTa_{1-x}Nb_xO_3$ $(0 < x < 0.04)$ solid solutions by a slow-cooling method. *J. Cryst. Growth* 59: 468–484.
508 Chang, Y.-C., Wang, C., Yin, S. et al. (2013). Kovacs effect enhanced broadband large field of view electro-optic modulators in nanodisordered KTN crystals. *Opt. Express* 21 (15): 17760.
509 Wang, X., Wang, J., and Liu, B. (2011). Growth and properties of cubic potassium tantalate niobate crystals. *Adv. Mat. Res* 306–307: 352–357.
510 Gottlieb, M., Ireland, C., and Ley, J. (1983). *Electro-optic and Acousto-optic Scanning and Deflection*. New York: Marcel Dekker Publishing.
511 Maldonado, T.A. (1995). *Handbook of Optics. Volume II: Devices, Measurements and Properties*. Chapter 13, 13.1–13.35.
512 Scrymgeour, D.A., Barad, Y., Gopalan, V. et al. (2001). Large-angle electro-optic laser scanner on $LiTaO_3$ fabricated by in situ monitoring of ferroelectric-domain micropatterning. *Appl. Opt.* 40: 6236–6241.
513 Djukic, D., Roth, R., Yardley, J.T., and Osgood, R.M. Jr., (2004). Low-voltage planar-waveguide electro-optic prism scanner in crystal-ion-sliced thin-film $LiNbO_3$. *Opt. Express* 12: 6159–6164.
514 Chiu, Y., Stancil, D.D., Schlesinger, T.E., and Risk, W.P. (1996). Electro-optic beam scanner in $KTiOPO_4$. *Appl. Phys. Lett.* 69: 3134–3136.
515 Nakamura, K., Miyazu, J., Sasaki, Y. et al. (2008). Space-charge-controlled electro-optic effect: optical beam deflection by electro-optic effect and space-charge-controlled electrical conduction. *J. Appl. Phys.* 104: 013105.
516 Huang, C., Sasaki, Y., Miyazu, J. et al. (2014). Trapped charge density analysis of KTN crystal by beam path measurement. *Opt. Express* 22: 7783–7789.
517 Miyazu, J., Imai, T., Toyoda, S. et al. (2011). New beam scanning model for high-speed operation using $KTa_{1-x}Nb_xO_3$ crystals. *Appl. Phys Express* 4: 111501.
518 Zhu, W., Chao, J.-H., Chen, C.-J. et al. (2017). New functionalities of potassium tantalate niobate deflectors enabled by the coexistence of pre-injected space charge and composition gradient. *J. Appl. Phys.* 122: 133111.
519 Nakamura, K. (2007). Optical beam scanner using Kerr effect and space charge-controlled electrical conduction in $KTa_{1-x}Nb_xO_3$ crystal. *NTT Tech. Rev.* 5 (9): https://www.ntt-review.jp/archive/ntttechnical.php?contents=ntr200709sp2.html.
520 Dionne, J.A. and Atwater, H.A. (2012). Plasmonics: metal-worthy methods and materials in nanophotonics. *MRS Bull.* 37: 717–724.
521 Brongersma, M.L. and Shalaev, V.M. (2010). Applied physics: the case for plasmonics. *Science* 328: 440.
522 Sorger, V.J., Oulton, R.F., Ma, R.-M., and Zhang, X. (2012). Toward integrated plasmonic circuits. *MRS Bull.* 37: 728–738.
523 Brongersma, M.L. and Kik, P.G. (eds.) (2007). *Surface Plasmon Nanophotonics*. Netherlands: Springer.
524 Bozhevolnyi, S. (ed.) (2008). *Plasmonic Nanoguides and Circuits*. Pan Stanford Publishing.

525 Berini, P. (2009). Long-range surface plasmon polaritons. *Adv. Opt. Photonics* 1: 484–588.

526 Boltasseva, A., Nikolajsen, T., Leosson, K. et al. (2005). Integrated optical components utilizing long-range surface plasmon polaritons. *J. Lightwave Technol.* 23: 413–422.

527 Charbonneau, R., Lahoud, N., Mattiussi, G., and Berini, P. (2005). Demonstration of integrated optics elements based on long ranging surface plasmon polaritons. *Opt. Express* 13: 977–984.

528 Charbonneau, R., Scales, C., Breukelaar, I. et al. (2006). Passive integrated optics elements based on long-range surface plasmon polaritons. *J. Lightwave Technol.* 24: 477–494.

529 Sorger, V.J., Ye, Z., Oulton, R.F. et al. (2011). Experimental demonstration of low-loss optical waveguiding at deep sub-wavelength scales. *Nat. Commun.* 2: 331-1–331-5.

530 Volkov, V.S., Han, Z., Nielsen, M.G. et al. (2011). Long-range dielectric-loaded surface plasmon polariton waveguides operating at telecommunication wavelengths. *Opt. Lett.* 36: 4278–4280.

531 Nikolajsen, T., Leosson, K., and Bozhevolnyi, S.I. (2004). Surface plasmon polariton based modulators and switches operating at telecom wavelengths. *Appl. Phys. Lett.* 85: 5833–5835.

532 Nikolajsen, T., Leosson, K., and Bozhevolnyi, S.I. (2005). In-line extinction modulator based on long-range surface plasmon polaritons. *Opt. Commun.* 244: 455–459.

533 MacDonald, K.F. and Zheludev, N.I. (2010). Active plasmonics: current status. *Laser Photon. Rev.* 4: 562.

534 Melikyan, A., Lindenmann, N., Walheim, S. et al. (2011). Surface plasmon polariton absorption modulator. *Opt. Express* 19: 8855–8869.

535 Sweatlock, L.A. and Diest, K. (2012). Vanadium dioxide based plasmonic modulators. *Opt. Express* 20: 8700–8709.

536 Babicheva, V.E., Kulkova, I.V., Malureanu, R. et al. (2012). Plasmonic modulator based on gain-assisted metal–semiconductor–metal waveguide. *Photonics Nanostruct. Fundam. Appl.* 10: 389.

537 Melikyan, A., Alloatti, L., Muslija, A. et al. (2013). Surface plasmon polariton high-speed modulator. In: *CLEO: OSA Technical Digest*, paper CTh5D.2.

538 Boltasseva, A. and Atwater, H.A. (2011). Low-loss plasmonic metamaterials. *Science* 331: 290–291.

539 West, P.R., Ishii, S., Naik, G. et al. (2010). Searching for better plasmonic materials. *Laser Photonics Rev.* 4: 795–808.

540 Naik, G.V. and Boltasseva, A. (2010). Semiconductors for plasmonics and metamaterials. *Phys. Status Solidi RRL* 4: 295–297.

541 Naik, G.V., Kim, J., and Boltasseva, A. (2011). Oxides and nitrides as alternative plasmonic materials in the optical range. *Opt. Mater. Express* 1: 1090–1099.

542 Naik, G.V. and Boltasseva, A. (2011). A comparative study of semiconductor-based plasmonic metamaterials. *Metamaterials* 5: 1–7.

543 Khurgin, J.B. and Boltasseva, A. (2012). Reflecting upon the losses in plasmonics and metamaterials. *MRS Bull.* 37 (8): 768–779.

544 Naik, G., Shalaev, V.M., and Boltasseva, A. (2013). Alternative plasmonic materials: beyond gold and silver. *Adv. Mater.* 25: 3264–3294.

545 Boltasseva, A. (2014). Empowering plasmonics and metamaterials technology with new material platforms. *MRS Bull.* 39: 461.

546 Babicheva, V.E., Boltasseva, A., and Lavrinenko, A.V. (2015). Transparent conducting oxides for electro-optical plasmonic modulators. *Nanophotonics* 4: 165–185.

547 Kim, H., Osofsky, M., Prokes, S.M. et al. (2013). Optimization of Al-doped ZnO films for low loss plasmonic materials at telecommunication wavelengths. *Appl. Phys. Lett.* 102: 171103.

548 Look, D.C., Droubay, T.C., and Chambers, S.A. (2012). Stable highly conductive ZnO via reduction of Zn vacancies. *Appl. Phys. Lett.* 101 (10): 102101.

549 Noginov, M.A. et al. (2011). Transparent conductive oxides: plasmonic materials for telecom wavelengths. *Appl. Phys. Lett.* 99: 021101.

550 Bodea, M.A., Sbarcea, G., Naik, G.V. et al. (2012). Negative permittivity of ZnO thin films prepared from aluminium and gallium doped ceramics via pulsed-laser deposition. *Appl. Phys. A* 110 (4): 929–934.

551 Frölich, A. and Wegener, M. (2011). Spectroscopic characterization of highly doped ZnO films grown by atomic-layer deposition for three-dimensional infrared metamaterials [Invited]. *Opt. Mater. Express* 1: 883–889.

552 Pradhan, A.K. et al. (2014). Extreme tunability in aluminium doped zinc oxide plasmonic materials for near-infrared applications. *Sci. Rep.* 4: 6415.

553 Sadofev, S., Kalusniak, S., Schäfer, P., and Henneberger, F. (2013). Molecular beam epitaxy of *n*-Zn(Mg)O as a low-damping plasmonic material at telecommunication wavelengths. *Appl. Phys. Lett.* 102: 181905.

554 Sadofev, S., Kalusniak, S., Schäfer, P. et al. (2015). Free-electron concentration and polarity inversion domains in plasmonic (Zn,Ga)O. *Phys. Status Solidi B* 252 (3): 607–611.

555 Rhodes, C., Franzen, S., Maria, J.-P. et al. (2006). Surface plasmon resonance in conducting metal oxides. *J. Appl. Phys.* 100: 054905.

556 Michelotti, F., Dominici, L., Descrovi, E. et al. (2009). Thickness dependence of surface plasmon polariton dispersion in transparent conducting oxide films at 1.55 μm. *Opt. Lett.* 34: 839–841.

557 Kim, J., Naik, G.V., Emani, N.K. et al. (2013). Plasmonic resonances in nanostructured transparent conducting oxide films. *IEEE J. Sel. Top. Quantum Electron.* 19: 4601907.

558 Garcia, G. et al. (2011). Dynamically modulating the surface plasmon resonance of doped semiconductor nanocrystals. *Nano Lett.* 11: 4415–4420.

559 Kanehara, M., Koike, H., Yoshinaga, T., and Teranishi, T. (2009). Indium tin oxide nanoparticles with compositionally tunable surface plasmon resonance frequencies in the near-IR region. *J. Am. Chem. Soc.* 131 (49): 17736–17737.

560 Buonsanti, R., Llordes, A., Aloni, S. et al. (2011). Tunable infrared absorption and visible transparency of colloidal aluminum-doped zinc oxide nanocrystals. *Nano Lett.* 11 (11): 4706–4710.

561 Gordon, T.R., Paik, T., Klein, D.R. et al. (2013). Shape-dependent plasmonic response and directed self-assembly in a new semiconductor building block, indium-doped cadmium oxide (ICO). *Nano Lett.* 13 (6): 2857–2863.

562 Li, S.-Q. et al. (2014). Plasmonic-photonic mode coupling in indium-tin-oxide nanorod arrays. *ACS Photonics* 1: 163–172.

563 Tice, D.B., Li, S.-Q., Tagliazucchi, M. et al. (2014). Ultrafast modulation of the plasma frequency of vertically aligned indium tin oxide rods. *Nano Lett.* 14: 1120–1126.

564 Abb, M., Wang, Y., Papasimakis, N. et al. (2014). Surface-enhanced infrared spectroscopy using metal oxide plasmonic antenna arrays. *Nano Lett.* 14 (1): 346–352.

565 Santiago, K., Mundle, R., Samantaray, C.B. et al. (2012). Nanopatterning of atomic layer deposited Al:ZnO films using electron beam lithography for waveguide applications in the NIR region. *Opt. Mater. Express* 2: 1743–1750.

566 Naik, G.V., Liu, J., Kildishev, A.V. et al. (2012). Demonstration of Al:ZnO as a plasmonic component of near infrared metamaterials. *Proc. Natl. Acad. Sci.* 109 (23): 8834–8838.

567 Zhang, Y., Wei, T., Dong, W. et al. (2013). Near perfect infrared absorption from dielectric multilayer of plasmonic aluminum-doped zinc oxide. *Appl. Phys. Lett.* 102: 213117.

568 Kimetal, J. (2013). Optical properties of gallium-doped zinc oxide – a low-loss plasmonic material: first-principles theory and experiment. *Phys. Rev.* X3: 041037.

569 Calzolari, A., Ruini, A., and Catellani, A. (2014). Transparent conductive oxides as near-IR plasmonic materials: the case of Al-doped ZnO derivatives. *ACS Photonics* 1: 703–709.

570 Traviss, D., Bruck, R., Mills, B. et al. (2013). Ultrafast plasmonics using transparent conductive oxide hybrids in the epsilon near-zero regime. *Appl. Phys. Lett.* 102: 121112.

571 Engheta, N. (2013). Pursuing near-zero response. *Science* 340 (6130): 286–287.

572 Silveirinha, M. and Engheta, N. (2006). Tunneling of electromagnetic energy through subwavelength channels and bends using epsilon-near-zero materials. *Phys. Rev. Lett.* 97: 157403.

573 Tsakmakidis, K.L. and Hess, O. (2012). Extreme control of light in metamaterials: complete and loss-free stopping of light. *Phys. B Conden. Matter* 407: 4066–4069.

574 Li, D.B. and Ning, C.Z. (2009). Giant modal gain, amplified surface plasmon-polariton propagation, and slowing down of energy velocity in a metal-semiconductor-metal structure. *Phys. Rev. B* 80: 153304.

575 Li, D.B. and Ning, C.Z. (2010). Peculiar features of confinement factors in a metal-semiconductor waveguide. *Appl. Phys. Lett.* 96: 181109.

576 Tsakmakidis, K.L., Boardman, A.D., and Hess, O. (2007). Trapped rainbow storage of light in metamaterials. *Nature* 450: 397–401.

577 Dionne, J.A., Sweatlock, L.A., Atwater, H.A., and Polman, A. (2005). Planar metal plasmon waveguides: frequency-dependent dispersion, propagation, localization, and loss beyond the free electron model. *Phys. Rev. B* 72: 075405.

578 Dionne, J.A., Sweatlock, L.A., Atwater, H.A., and Polman, A. (2006). Plasmon slot waveguides: towards chip-scale propagation with subwavelength-scale localization. *Phys. Rev. B* 73: 035407.

579 Lu, Y.-J. et al. (2012). Plasmonic nanolaser using epitaxially grown silver film. *Science* 337: 450.

580 Oulton, R.F. et al. (2009). Plasmon lasers at deep subwavelength scale. *Nature* 461: 629–632.

581 Tsakmakidis, K., Hamm, J., Pickering, T.W., and Hess, O. (2012). Plasmonic nanolasers without cavity, threshold and diffraction limit using stopped light. *Frontiers in Optics*, Laser Science XXVIII, OSA Technical Digest (online) (Optical Society of America), paper FTh2A.2.

582 Tsakmakidis, K.L., Pickering, T.W., Hamm, J.M. et al. (2014). Completely stopped and dispersion less light in plasmonic waveguides. *Phys. Rev. Lett.* 112: 167401.

583 Kauranen, M. and Zayats, A. (2012). Nonlinear plasmonics. *Nat. Photonics* 6: 737–748.

584 Aouani, H., Rahmani, M., Navarro-Cía, M., and Maier, S.A. (2014). Third harmonic-up conversion enhancement from a single semiconductor nanoparticle coupled to a plasmonic antenna. *Nat. Nanotechnol.* 9: 290–294.

585 Metzger, B. et al. (2014). Doubling the efficiency of third harmonic generation by positioning ITO nanocrystals into the hot-spot of plasmonic gap-antennas. *Nano Lett.* 14 (5): 2867–2872.

586 Abb, M., Wang, Y., de Groot, C.H., and Muskens, O.L. (2014). Hot spot mediated ultrafast nonlinear control of multifrequency plasmonic nano-antennas. *Nat. Commun.* 5: 4869.

587 Abb, M., Albella, P., Aizpurua, J., and Muskens, O.L. (2011). All-optical control of a single plasmonic nanoantenna – ITO hybrid. *Nano Lett.* 11: 2457–2463.

588 Abb, M. and Muskens, O.L. (2012). Ultrafast plasmonic nanoantenna-ITO hybrid switches. *Inter. J. Opt.* 132542.

589 Abb, M., Sepúlveda, B., Chong, H.M.H., and Muskens, O.L. (2012). Transparent conducting oxides for active hybrid metamaterial devices. *J. Opt.* 14: 114007.

590 Feigenbaum, E., Diest, K., and Atwater, H.A. (2010). Unity-order index change in transparent conducting oxides at visible frequencies. *Nano Lett.* 10 (6): 2111–2116.

591 Dionne, J.A., Sweatlock, L.A., Sheldon, M.T. et al. (2010). Silicon-based plasmonics for on-chip photonics. *IEEE J. Sel. Top. Quantum Electron.* 16: 295–306.

592 Wassel, H.M.G., Dai, D., Tiwari, M. et al. (2012). Opportunities and challenges of using plasmonic components in nanophotonic architectures. *IEEE J. Emerging Sel. Top. Circuits Syst.* 2: 154–168.

593 Hirata, T., Kajikawa, K., Tabei, T., and Sunami, H. (2008). Proposal of a metal–oxide–semiconductor silicon optical modulator based on inversion-carrier absorption. *Jpn. J. Appl. Phys.* 47: 2906.

594 Tabei, T., Hirata, T., Kajikawa, K., and Sunami, H. (2009). Potentiality of metal–oxide–semiconductor silicon optical modulator based on free carrier absorption. *Jpn. J. App. Phys.* 48: 114501.

595 Tabei, T. and Yokoyama, S. (2012). Proposal of a silicon optical modulator based on surface plasmon resonance. *Proceedings of SPIE8431*.

596 Zhu, S., Lo, G.Q., and Kwong, D.L. (2011). Electro-absorption modulation in horizontal metal-insulator-silicon-insulator metal nano-plasmonic slot waveguides. *Appl. Phys. Lett.* 99: 151114.
597 Zhu, S., Lo, G.Q., and Kwong, D.L. (2010). Theoretical investigation of silicon MOS-type plasmonic slot waveguide based MZI modulators. *Opt. Express* 18: 27802.
598 Zhaolin, L., Zhao, W., and Shi, K. (2012). Ultra-compact electro absorption modulators based on tunable epsilon-near zero-slot waveguides. *IEEE Photonics J.* 4: 735–740.
599 Babicheva, V.E. and Lavrinenko, A.V. (2012). Plasmonic modulator optimized by patterning of active layer and tuning permittivity. *Opt. Commun.* 285: 5500–5507.
600 Vasudev, A.P., Kang, J.-H., Park, J. et al. (2013). Electro-optical modulation of a silicon waveguide with an "epsilon-near-zero" material. *Opt. Express* 21: 26387–26397.
601 Zhu, S., Lo, G.Q., and Kwong, D.L. (2014). Design of an ultra-compact electro-absorption modulator comprised of a deposited TiN/HfO_2/ITO/Cu stack for CMOS back end integration. *Opt. Express* 22: 17930–17947.
602 Kim, J.T. (2014). Silicon optical modulators based on tunable hybrid plasmonic directional couplers. *IEEE J. Sel. Top. Quantum Electron.* 21 (4): 3300108.
603 Lee, H.W.H., Papadakis, G., Burgos, S.P. et al. (2014). Nanoscale conducting oxide PlasMOStor. *Nano Lett.* 14 (11): 6463–6468.
604 Amemiya, T., Murai, E., Gu, Z. et al. (2014). GaInAsP/InP-based optical modulator consisting of gap surface-plasmon-polariton waveguide: theoretical analysis. *J. Opt. Soc. Am. B* 31: 2908–2913.
605 Zhao, H., Wang, Y., Capretti, A. et al. (2015). Broad band electro-absorption modulators design based on epsilon-near-zero indium tin oxide. *IEEE J. Sel. Top. Quantum Electron.* 21 (4): https://doi.org/10.1109/JSTQE.2014.2375153.
606 Ye, C., Khan, S., Li, Z.R. et al. (2014). λ-Size ITO and graphene-based electro-optic modulators on SOI. *IEEE J. Sel. Top. Quantum Electron.* 20 (4): 3400310.
607 Krasavin, A.V. and Zheludev, N.I. (2004). Active plasmonics: controlling signals in Au/Ga waveguide using nanoscale structural transformations. *Appl. Phys. Lett.* 84: 1416.
608 Krasavin, A.V., MacDonald, K.F., Zheludev, N.I., and Zayats, A.V. (2004). High-contrast modulation of light with light by control of surface plasmon polariton wave coupling. *Appl. Phys. Lett.* 85: 3369–3371.
609 Krasavin, A.V., Zayats, A.V., and Zheludev, N.I. (2005). Active control of surface plasmon–polariton waves. *J. Opt. A: Pure Appl. Opt.* 7: S85–S89.
610 Zhao, W. and Lu, Z. (2011). Nanoplasmonic optical switch based on Ga–Si_3N_4–Ga waveguide. *Opt. Eng.* 50 (7): 074002.
611 Kruger, B.A., Joushaghani, A., and Poon, J.K.S. (2012). Design of electrically driven hybrid vanadium dioxide (VO_2) plasmonic switches. *Opt. Express* 20: 23598–23609.

612 Joushaghani, A., Kruger, B.A., Paradis, S. et al. (2013). Sub-volt broadband hybrid plasmonic vanadium dioxide switches. *Appl. Phys. Lett.* 102: 061101.

613 Ooi, K.J.A., Bai, P., Chu, H.S., and Ang, L.K. (2013). Ultracompact vanadium dioxide dual-mode plasmonic waveguide electro-absorption modulator. *Nanophotonics* 2: 13–19.

614 Choe, J.-H. and Kim, J.T. Design of vanadium dioxide based plasmonic modulator for both TE and TM polarization mode. *IEEE Photonics Technol. Lett.* https://doi.org/10.1109/LPT.2014.2384020.

615 Kim, J.T. (2014). CMOS-compatible hybrid plasmonic modulator based on vanadium dioxide insulator-metal phase transition. *Opt. Lett.* 39: 3997–4000.

616 Gosciniak, J. and Bozhevolnyi, S.I. (2013). Performance of thermo-optic components based on dielectric-loaded surface plasmon polariton waveguides. *Sci. Rep.* 3: 1803.

617 Gosciniak, J., Bozhevolnyi, S.I., Andersen, T.B. et al. (2010). Thermo-optic control of dielectric-loaded plasmonic waveguide components. *Opt. Express* 18 (2): 1207–1216.

618 Gosciniak, J. and Tan, D.T.H. (2013). Graphene-based waveguide integrated dielectric-loaded plasmonic electro-absorption modulators. *Nanotechnology* 24: 185202.

619 Lao, J., Tao, J., Wang, Q.J., and Huang, X.G. (2014). Tunable graphene-based plasmonic waveguides: nano-modulators and nano-attenuators. *Laser Photonics Rev.* 8 (4): 569–574.

620 Yang, L. et al. (2014). Ultracompact plasmonic switch based on graphene-silica metamaterial. *Appl. Phys. Lett.* 104: 211104.

621 Babicheva, V.E., Malureanu, R., and Lavrinenko, A.V. (2013). Plasmonic finite-thickness metal-semiconductor-metal waveguide as ultra-compact modulator. *Photonics Nanostruct. Fundam. Appl.* 11: 323–334.

622 Babicheva, V.E. and Lavrinenko, A.V. (2013). Plasmonic modulator based on metal-insulator-metal waveguide with barium titanate core. *Photonics Lett. Poland* 5: 57–59.

623 Babicheva, V.E., Zhukovsky, S.V., and Lavrinenko, A.V. (2014). Bismuth ferrite as low-loss switchable material for plasmonic waveguide modulator. *Opt. Express* 22: 28890–28897.

624 Abadía, N., Bernadin, T., Chaisakul, P. et al. (2014). Low-power consumption Franz–Keldysh effect plasmonic modulator. *Opt. Express* 22: 11236–11243.

625 Hoessbacher, C., Fedoryshyn, Y., Emboras, A. et al. (2014). The plasmonic memristor: a latching optical switch. *Optica* 1: 198–202.

626 Emboras, A., Hoessbacher, C., Fedoryshyn, Y. et al. (2014). Plasmonic switches. In: *Proceedings of OECC/ACOFT2014*, 730–732. Melbourne.

627 Emboras, A., Goykhman, I., Desiatov, B. et al. (2013). Nanoscale plasmonic memristor with optical read out functionality. *Nano Lett.* 13 (12): 6151–6155.

628 Reed, G.T., Mashanovich, G., Gardes, F.Y., and Thomson, D.J. (2010). Silicon optical modulators. *Nat. Photonics* 4: 518–526.

629 Zia, R., Selker, M.D., Catrysse, P.B., and Brongersma, M.L. (2004). Geometries and materials for subwavelength surface plasmon modes. *J. Opt. Soc. Am.* A21: 2442–2446.

630 Verhagen, E., Dionne, J.A., Kuipers, L.(.K.). et al. (2008). Near-field visualization of strongly confined surface plasmon polaritons in metal–insulator–metal waveguides. *Nano Lett.* 8 (9): 2925–2929.

631 Miyazaki, H.T. and Kurokawa, Y. (2006). Squeezing visible light waves into a 3-nm-thick and 55-nm-long plasmon cavity. *Phys. Rev. Lett.* 96: 097401.

632 Kurokawa, Y. and Miyazaki, H.T. (2007). Metal-insulator-metal plasmon nanocavities: analysis of optical properties. *Phys. Rev. B* 75: 035411.

633 Emboras, A., Briggs, R.M., Najar, A. et al. (2012). Efficient coupler between silicon photonic and metal-insulator-silicon-metal plasmonic waveguides. *Appl. Phys. Lett.* 101 (25): 251117.

634 Yang, R., Abushagur, M.A.G., and Lu, Z. (2008). Efficiently squeezing near infrared light into a 21 nm-by-24 nm nano spot. *Opt. Express* 16: 20142–20148.

635 Yang, R. and Lu, Z. (2011). Silicon-on-insulator platform for integration of 3D nanoplasmonic devices. *IEEE Photonics Technol. Lett.* 23: 1652–1654.

636 Guler, U., Shalaev, V.M., and Boltasseva, A. (2015). Nanoparticle plasmonics: going practical with transition metal nitrides. *Mater. Today* 18 (4): 227–237.

637 Kinsey, N., Ferrera, M., Shalaev, V.M., and Boltasseva, A. (2015). Examining nanophotonics for integrated hybrid systems: are view of plasmonic interconnects and modulators using traditional and alternative materials [Invited]. *J. Opt. Soc. Am. B.* 32: 121–142.

638 Naik, G.V., Schroeder, J.L., Ni, X. et al. (2012). Titanium nitride as a plasmonic material for visible and near-infrared wavelengths. *Opt. Mater. Express* 2 (4): 478–489.

639 Kinsey, N., Ferrera, M., Naik, G.V. et al. (2014). Experimental demonstration of titanium nitride plasmonic interconnects. *Opt. Express* 22: 12238–12247.

640 Guler, U., Naik, G.V., Boltasseva, A. et al. (2012). Performance analysis of nitride alternative plasmonic materials for localized surface plasmon applications. *Appl. Phys.* B107 (2): 285–291.

641 Guler, U., Ndukaife, J.C., Naik, G.V. et al. (2013). Local heating with lithographically fabricated plasmonic titanium nitride nanoparticles. *Nano Lett.* 13 (12): 6078–6083.

642 Guler, U., Boltasseva, A., and Shalaev, V.M. (2014). Refractory plasmonics. *Science* 344: 263–264.

643 Naik, G.V., Saha, B., Liu, J. et al. (2014). Epitaxial superlattices with titanium nitride as a plasmonic component for optical hyperbolic metamaterials. *Proc. Natl. Acad. Sci. U.S.A.* 111 (21): 7546–7551.

644 Lee, H.W.H., Burgos, S.P., Papadakis, G., and Atwater, H.A. (2013). Nanoscle conducting oxide plasmonic slot waveguide modulator. In: *Frontiers in Optics* (ed. I. Kang, D. Reitze, N. Alic and D. Hagan). Optical Society of America (OSA) Technical Digest (online).

6

Photorefractive Effect

6.1 Introduction

In 1966, A. Ashkin et al. at Bell Laboratories were experimenting with the inorganic crystals $LiNbO_3$ and $LiTaO_3$ for use in frequency doubling [1]. During their testing, they found what they termed "optical damage." The effect was interesting but highly detrimental to the optics of nonlinear devices that are based on these crystals. Shortly afterwards, experiments with a new material potassium tantalate niobate (KTN) showed that it exhibited the same type of optical damage reported earlier with $LiNbO_3$ and $LiTaO_3$. The major difference was that the optical damage occurred only in the presence of an electric field. Since most Physicists and Engineers adore problems or things that they do not understand, experiments were performed to explain the behaviors observed in nature. What was discovered has been given the name *photorefractive effect.*

Photorefractive effect can show very high light sensitivity, permitting the observation of such effects at very low light powers (<μW). On the other hand, detailed understanding of the photorefractive effects helped develop materials with negligible light sensitivity, which can be then used without "damage" for electro-optics and nonlinear optics. Such materials are also of benefit for the nonconventional (band-to-band or inter-band) regime.

6.2 Photorefractive Effect

Photorefractive effect is a spatial modulation of the refractive index caused by variations in optical radiation (light) intensity. These variations in light intensity occur when two beams of light create an interference pattern.

The effect can be used to store temporary, erasable holograms and is useful for holographic data storage. It can also be used to create a phase conjugate mirror or an optical spatial soliton.

The specific processes required for the photorefractive effect are the following:

(a) Photogeneration of charge carriers
(b) Transport of mobile carriers
(c) Trapping of mobile carriers based upon destructive interference of light
(d) Change of refractive index due to space charge set up by mobile charge carriers.

Crystal Optics: Properties and Applications, First Edition. Ashim Kumar Bain.

(a) *Generation of charge carriers:* The first step in the photorefractive process is the generation of charge carriers with light. One of the great discoveries of this century was that light exists as quanta of energy ($h\nu$) – photons. When light is exposed on a material, energy from the light can be imparted to the material if the energy of light ($h\nu$) is larger than the bandgap energy level of material. For our purposes, it will be easier to think of the material as having no extra electrons in the conduction band initially. In reality, there may already be electrons in the conduction band during equilibrium but this small difference will not change important behavior. While light is exposed on a photoconductive material the energy of the light is enough to raise electrons to the conduction band. If we were to now apply an electric field across the material, electron current would flow. With our assumption that there was no electron in the conduction band previously, without the light no current would have flowed with the electric field. Thus, we have photogeneration of charge carriers as shown in Figure 6.1.

(b) *Charge transport:* In the picture (Figure 6.2), we are exposing two beams of light into a material at angle $\pm\theta$. Because light has wave properties this creates

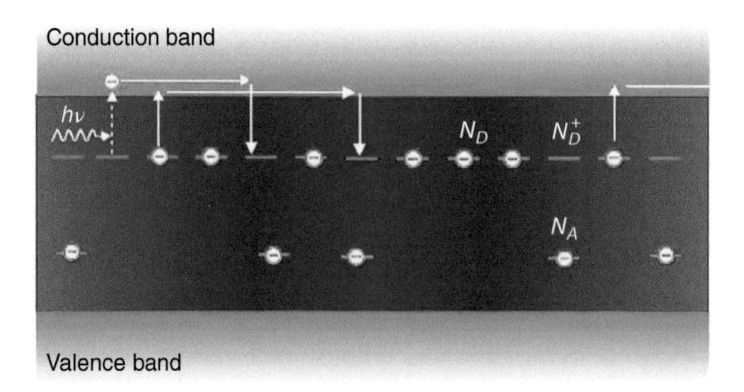

Figure 6.1 Band diagram behavior.

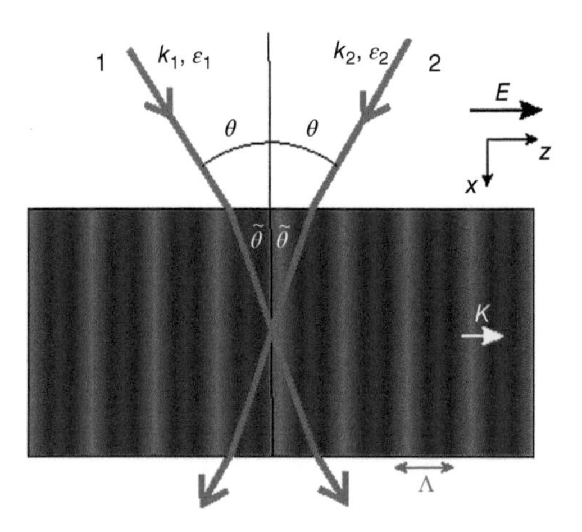

Figure 6.2 Light application to media.

an interference pattern where the light has a pattern of greater intensity in one area and less intensity in another. The exact intensity is a function of the interference itself and can therefore be controlled. When a variation in light intensity occurs within a photorefractive material, charge carriers are generated as a function of light intensity. Where the interference pattern causes the optical radiation (light) to be more intense charge carriers are generated. Where the light interference pattern causes the light to be less intense, little or no charge carriers are generated. The abundance of charge carriers (electrons) in the conduction band generated by the intense light are free to move about the material due to diffusion, charge density gradients, or drift in an externally applied field. It is important to note that the ions that are created (ionized donors) are not free to move about the material and must stay in the position that existed before the material received light. It is also important to note that the location of immobile ions where the radiation intensity is high is different than where the light intensity is less. The result is the creation of an inhomogeneous space-charge distribution that can remain in place for a period of time after the light is removed.

(c) *Trapping of mobile carriers:* The charge carriers (electrons) in the conduction band generated by the intense light are free to move about the material due to diffusion, charge density gradients, or drift in an externally applied field. This is only true for a short time, specifically the recombination time of electron with acceptors. We know that the photorefractive material contains donor atoms or impurities that can give up an electron to the conduction band. What is not stated until now is that the material also contains acceptor atoms looking for free electrons. Therefore, as free electrons move from the area of high light intensity to an area of low light intensity they can become attached to an acceptor atom looking for an electron. Once the electron is attached to the acceptor atom the free charge carrier is trapped. The trapped charge will remain trapped even after the light has been removed. This gives us another ion to match the one formed when the charge carrier was freed by the light energy.

(d) *Change of refractive index:* Because of the light intensity differences in the material, we now have a separation of fixed ions. These fixed ions are not free to move about in the material. Between the separated ions there is an electric field or space charge.

Photorefractive materials also exhibit an electro-optic effect. Materials exhibiting an electro-optic effect alter their refraction of index in response to an electric field. The field distorts the structure of the material and the distortion alters the refraction of index. The space charge gradients create a gradient in the refractive index of the material. The most exciting part is that the refractive index gradients are a phase grating or hologram that can diffract a beam of light.

The material can be brought back to its original state (erased) by illumination with uniform light or by heating. Thus, the material can be used to record and store image, much like a photographic emulsion stores an image. The complete process is illustrated in Figure 6.3 for doped lithium niobate ($LiNbO_3$). Important photorefractive materials include barium titanate ($BaTiO_3$), bismuth silicon

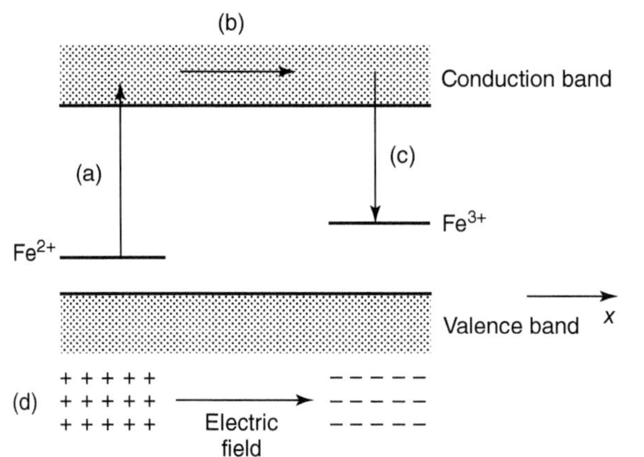

Figure 6.3 Energy-level diagram of $LiNbO_3$ illustrating the processes of (a) photoionization, (b) diffusion, (c) recombination, and (d) space–charge formation and electric-field generation. Fe^{2+} impurity centers act as donors, becoming Fe^{3+} when ionized, while Fe^{3+} centers act as traps, becoming Fe^{2+} after recombination.

oxide ($Bi_{12}SiO_{20}$), lithium niobate ($LiNbO_3$), potassium niobate ($KNbO_3$), gallium arsenide (GaAs), and strontium barium niobate (SBN).

6.2.1 Conventional Model of Photorefractive Effect

The conventional model of photorefractive effect was elaborated by Kukhtarev et al. [2–4] considering only photoexcitation and recombination of one species of charge carriers between a single donor level and the corresponding conduction band. This single-level band scheme, along with the involved physical mechanisms, is depicted in Figure 6.4. Such a simplification is justifiable for light illumination with photon energies much smaller than the bandgap energy. For laser light with higher photon energies, additional impurity levels have to be taken into account. The involved processes can be described by the following set of equations:

$$\frac{\partial N_D^+}{\partial t} = (S_e I + \beta_e)(N_D - N_D^+) - \gamma_e n N_D^+ \tag{6.1}$$

$$\frac{\partial n}{\partial t} = \frac{\partial N_D^+}{\partial t} + \frac{1}{e}\nabla J_e \tag{6.2}$$

$$J_e = en\mu_e E + k_B T \mu_e \nabla n + eS_e I(N_D - N_D^+)L_{ph} \tag{6.3}$$

$$\nabla . E_{sc} = \frac{e}{\varepsilon_0 \varepsilon_{eff}}(N_D^+ - n - N_A) \tag{6.4}$$

where the photoexcited charges are assumed to be electrons. In the above equations n is the free electron concentration in the conduction band, N_D is the donor concentration, N_D^+ is the concentration of ionized donors, N_A is the concentration of ionized donors (acceptors) in the dark, J_e is the electron current density, E is the total electric field, E_{sc} is the space charge field, I is the light intensity, L_{ph} is the photogalvanic transport length vector, S_e is the photoionization constant for electrons, β_e is the thermal (dark) generation rate for electrons, γ_e is the recombination constant for electrons, μ_e is the electron mobility tensor, ε_o is the vacuum dielectric permittivity, ε_{eff} is the effective dielectric constant for the chosen configuration [5], e is the absolute value of

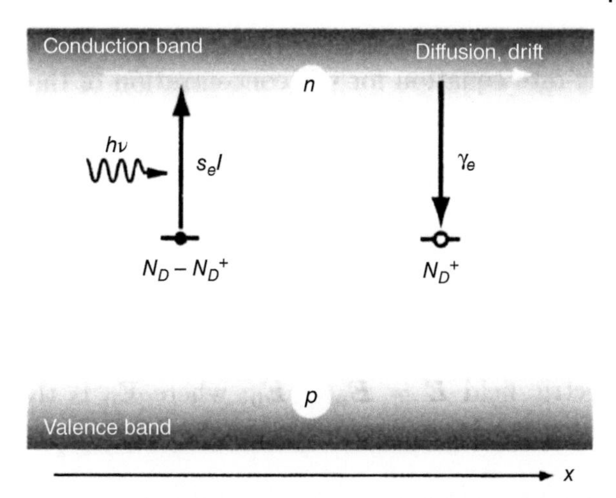

Figure 6.4 Single-level band scheme in the conventional model at low intensities. An electron is photoexcited to the conduction band from a mid-bandgap level. Diffusion or drift results in a displacement of the electron. Near a trap center, the electron can be retrapped.

the elementary charge, k_B is the Boltzmann constant, and T is the absolute temperature, respectively.

Equation (6.1) is the rate equation for the concentration of the ionized donors. The first term describes the photoionization process (s_eI) and the thermal excitation of electrons from the donor level. The second term takes into account the recombination of the electrons (γ_e) into traps, in the case of ionized donors. The second equation (Eq. (6.2)) is the continuity equation for the electron density. The additional term with respect to Eq. (6.1) describes the divergence of the electron current density. Equation (6.3) describes the different contributions to the electron current density. The first term gives the drift current in the total electric field $E = E_{sc} + E_0$, where E_0 is the applied field. The second term describes the diffusion process of the electrons generated by the electron concentration gradient. Thus, this term only gives a contribution for inhomogeneous illumination. The last term gives the photogalvanic current, if present.

The last equation (Eq. (6.4)) is the Poisson equation for the electric field. It describes the spatially modulated part of the electric field generated by the nonuniform distribution of the charge carriers in the crystal. These four equations are valid in this form only for isotropic photoexcitation, i.e. if s_e is independent of the light polarization and intensity or if the intensity and polarization of the exciting light wave is invariant, i.e. constant with respect to its propagation direction, as in the case of a plane wave.

6.2.1.1 Photorefractive Index Gratings

Homogeneously irradiating a photorefractive material will not generate a space charge field or refractive index grating but only change some of the bulk material properties such as conductivity or absorption. In contrast, for an inhomogeneous illumination the charges are locally redistributed and will result in a space charge field. An effective way to produce an intensity pattern is given by the interference of two coherent plane waves (Figure 6.6).

Considering two plane waves with wave vectors $\mathbf{k}_1$ and $\mathbf{k}_2$ and electric field amplitudes A_1 and A_2, respectively, we obtain a sinusoidal intensity pattern of

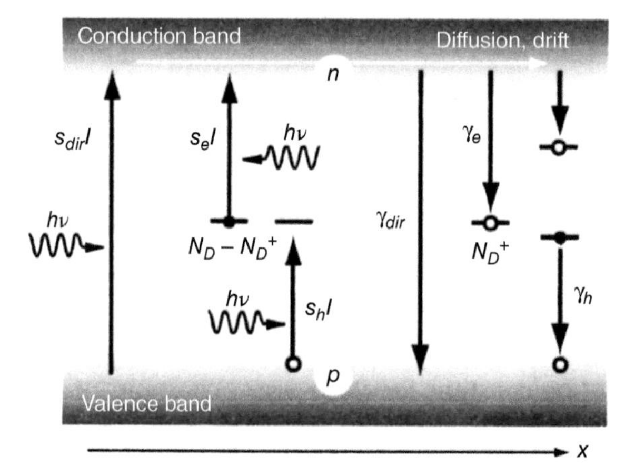

Figure 6.5 Simplified band scheme of possible charge transitions in a photorefractive material with the concentration of non-ionized donors $N_D - N_D^+$ and traps N_D^+. Excitation of charge carriers may occur via band-to-band (with a generation rate $s_{dir}I$), trap-to-band (s_eI), and band-to-trap transitions (s_hI). Excited charges are displaced by diffusion and drift. They can recombine through band-to-band electron–hole recombination (with a recombination rate γ_{dir}), or be trapped in mid-bandgap levels (γ_e and γ_h). The arrows show the movements of the electrons, and thermal excitation is not depicted.

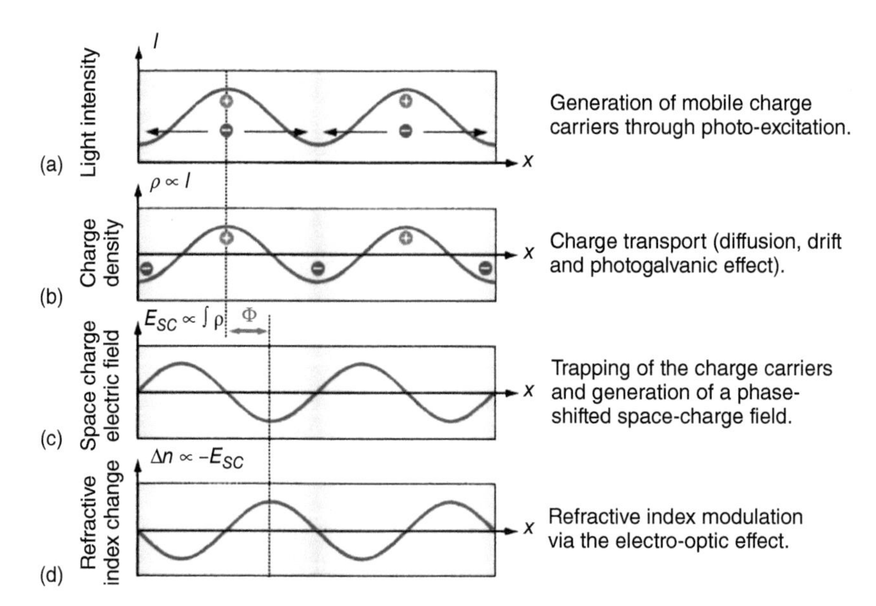

Figure 6.6 Mechanisms involved in the photorefractive effect, illustrated here for the case of electron charge transport. Λ is the grating constant and ϕ is the spatial shift between light intensity and refractive index modulation. For charge transport dominated by diffusion, one gets $\phi = \Lambda/4$.

the form

$$I(x) = I_0 + I_1 \exp(i\boldsymbol{K} \cdot \boldsymbol{x}) \tag{6.5}$$

where $\mathbf{K} = \mathbf{k}_1 - \mathbf{k}_2$ is the grating vector, $\boldsymbol{x}$ is the position vector, $I_0 = \frac{1}{2}\sqrt{\varepsilon_0/\mu_0}(|A_1|^2 + |A_2|^2)$ is the mean intensity, and $I_1 = \sqrt{\varepsilon_0/\mu_0}(|A_1||A_2|)$is the intensity modulation. Assuming a small modulation depth of $m = I_1/I_0$, we can follow the usual linearization procedure and obtain for the quantities n, N_D^+, E, and J_e a sinusoidal expression of the type

$$F(x) = F_0 + F_1 \exp(iK \cdot x) \tag{6.6}$$

where only the real part corresponds to the value of the physical property. Since F_1 is a complex value, the modulation of the different quantities can exhibit a phase shift ϕ with respect to the light fringes. This implies a spatial shift of the maxima by ϕ/K. The most important quantity for describing the photorefractive effect is the electric field and we will denote in the following the amplitude of the first spatial Fourier component (E_1) of the space charge field with E_{sc}.

In an electro-optic material, the presence of a periodic space charge field E_{sc} generates a refractive index grating that can be investigated by diffraction experiments of a third plane wave, eventually at a different wavelength. For non-centrosymmetric materials that exhibit the linear electro-optic effect (*Pockels effect*), the change of the dielectric permittivity tensor is defined as

$$\Delta\varepsilon^{-1} = r_{eff}(\boldsymbol{K})E_{sc} \tag{6.7}$$

where ε is the dielectric permittivity tensor of the material at light wavelength and $r_{eff}(\boldsymbol{K})$ is a 3×3 matrix describing the electro-optic effect, which considers the unclamped electro-optic tensor and the additional contributions from the piezoelectric effect [6]. Since in Eq. (6.7) and alike in the following formula for Δn, the oscillating spatial term is not included, the components on the left-hand side of this equation are the complex first spatial Fourier components.

For a certain geometry and light polarization, the amplitude of the spatially periodic refractive index change Δn can be expressed as

$$\Delta n \approx -\frac{1}{2} n^3 r_{eff} E_{sc} \tag{6.8}$$

where n is the refractive index seen by the beam and r_{eff} is the effective electro-optic coefficient relevant for this geometry.

6.2.1.2 Space Charge Field for Sinusoidal Illumination

A solution of Eqs. (6.1)–(6.4) can be found in the case where the material is illuminated with a periodic intensity distribution as described by Eq. (6.5). The space charge field is parallel to the grating vector and its amplitude is $|E_{sc}|$. The complex amplitude of the space charge field E_{sc} in the case of small modulation depth m is given by [3]

$$E_{sc} = -im\frac{E_q(E_d - iE_0)}{E_q + E_D - iE_0} \tag{6.9}$$

where i is the imaginary unit. For a better physical interpretation of E_{sc} we have introduced the diffusion field E_D and the trap limited field E_q. The first is defined as

$$E_D = \frac{k_B T}{e} K \tag{6.10}$$

and corresponds to the field amplitude of a sinusoidal electric field that exactly counteracts the effect of the charge diffusion process. The trap limited field is

$$E_q = \frac{e}{\varepsilon_{eff} \varepsilon_0 K} N_{eff} \tag{6.11}$$

and gives a limit for the maximum E_{sc} that can be generated with the available traps. In Eq. (6.11) N_{eff} is the effective trap density defined as

$$N_{eff} = \frac{N_{D0}^+ (N_D - N_{D0}^+)}{N_D} \tag{6.12}$$

where $N_{D0}^+ = n_0 + N_A$ is the spatial average of ionized donors with illumination switched "on." Although n_0, the spatial average of electrons in the conduction band when the light is on, is dependent on the intensity I_0 of the light, n_0 is in most cases much smaller than N_A, so that $N_{D0}^+ \simeq N_A$, and N_{eff} and E_q are independent of the light intensity in the conventional model. In case no electric field is applied, Eq. (6.9) can be simplified to

$$E_{sc} = -im \frac{E_q E_D}{E_q + E_D} \tag{6.13}$$

An important conclusion can be deduced from Eq. (6.13), which shows that if one of the two fields E_q or E_D is much smaller than the other, the total space–charge field is limited by this field. These limiting cases can be observed experimentally by varying the grating constant K. Furthermore, we see from Eq. (6.13) that without externally applied field the space-charge field is purely imaginary and the grating is shifted by $\phi_0 = \pi/2$ compared to the illumination grating. For large grating spacing $\Lambda \gg 1\,\mu m$ the buildup time of such photorefractive gratings is defined by the dielectric time constant

$$\tau_D = \frac{\varepsilon_0 \varepsilon_{eff}}{\sigma} = \frac{\varepsilon_0 \varepsilon_{eff}}{e \mu_e n_0} \tag{6.14}$$

Here, $\sigma = e\mu_e n_0$ is the total conductivity of the material and depends on the light intensity through n_0. Solving Eq. (6.1) considering only the spatial average values gives $n_0 \propto I_o$ and therefore

$$\tau_D \propto \frac{1}{I_0} \tag{6.15}$$

for the conventional photorefractive effect. The correction factor in the buildup time for smaller grating spacing is in the order of 1 and is independent on the light intensity.

6.2.2 Inter-band Photorefractive Effect

The conventional model only considers photoexcitation of charges from one mid-gap impurity level to one of the conduction bands. Accordingly, the absorption constant is small and the excitation process rather slow (milliseconds to seconds). As shown in Figure 6.5, other transitions are possible as well. Under light illumination with photon energies larger than the bandgap, band-to-band transitions occur and dominate over impurity-to-band transitions, particularly when the impurity concentration is low. Because of a much higher absorption constant in this regime, many more free charge carriers are produced, thus leading to a much faster excitation process. This effect is known as the "*inter-band photorefractive effect.*"

The most common set of equations describing the charge dynamics is given in [7] and includes one trap level in addition to the direct band-to-band excitation. No thermal excitations and no photogalvanic current are considered. The following equations describe the processes illustrated in Figure 6.5:

$$\frac{\partial N_D^+}{\partial t} = s_e I(N_D - N_D^+) + \gamma_h p(N_D - N_D^+) - \gamma_e n N_D^+ \tag{6.16}$$

$$\frac{\partial n}{\partial t} = s_{dir} I(N_V - p) + s_e I(N_D - N_D^+) - \gamma_e n N_D^+ - \gamma_{dir} np + \frac{1}{e}\nabla J_e \tag{6.17}$$

$$\frac{\partial p}{\partial t} = s_{dir} I(N_V - p) + s_h I(N_V - p) N_D^+ - \gamma_h p(N_D - N_D^+) - \gamma_{dir} np - \frac{1}{e}\nabla J_e \tag{6.18}$$

$$J_e = en\mu_e E_{sc} + k_B T \mu_e \nabla n \tag{6.19}$$

$$J_h = ep\mu_h E_{sc} - k_B T \mu_h \nabla p \tag{6.20}$$

$$\nabla E_{sc} = \frac{e}{\varepsilon_0 \varepsilon_{eff}}(N_D^+ + p - n - N_A) \tag{6.21}$$

The symbols have the same meaning as in Eqs. (6.1)–(6.4) whereas the newly added ones are defined as follows: p is the free hole concentration in the valence band, N_V is the density of electrons close enough to the top of the valence band to be photoexcited, J_h the hole current density, s_{dir} the photoexcitation constant for direct band-to-band phototransitions, γ_h the recombination constant for the hole–donor interaction, and μ_h the hole mobility tensor.

The mathematical complexity of this set of equations is already for one impurity level so high that no closed solution has yet been found. Analytic solutions for E_{sc} were found by applying some simplifications [7, 8]. Here we present only the solution for pure inter-band regime, i.e. without considering any trap level ($N_D = N_D^+ = N_A = 0$). This limit well describes the effect for high light intensities, where the contribution from trap levels becomes negligible. In this limitation, the hole and electron densities are equal and are given by

$$p_0 = n_0 = \sqrt{\frac{gI_o}{\gamma_{dir}}} \tag{6.22}$$

where the constant g is related to the absorption constant α_{dir} for band-to-band excitation divided by the photon energy $h\nu$:

$$g \equiv s_{dir} N_V = \frac{\alpha_{dir}}{h\nu} \tag{6.23}$$

The resulting amplitude of the space charge field E_{sc} for a continuous sinusoidal illumination with a small light intensity modulation m is given by [7]

$$E_{sc} = -im\frac{E_{qf}[E_D(E_{Rh} - E_{Re}) - iE_0(E_{Rh} + E_{Re})]}{(E_D + E_{Re} + E_{Rh})(E_D + 2E_{qf}) + E_0^2 + iE_0(E_{Re} - E_{Rh})} \tag{6.24}$$

where $E_D = Kk_BT/e$ represents the diffusion field, E_{qf} the free career-limited field

$$E_{qf} = \frac{e}{\varepsilon_0\varepsilon_{eff}K}\sqrt{\frac{gI_0}{\gamma_{dir}}} \tag{6.25}$$

and $E_{Re,Rh}$, the electron (hole) recombination fields

$$E_{Re,Rh} = \frac{1}{K\mu_{e,h}}\sqrt{gI_0\gamma_{dir}} \tag{6.26}$$

The recombination field can be interpreted as the average electric field needed to drift one electron or hole by a distance $K^{-1} = \Lambda/2\pi$ before a direct band-to-band recombination takes place.

As in the conventional photorefractive effect, in the case of pure diffusion charge transport mechanism ($E_0 = 0$) the space–charge field is purely imaginary, i.e. $\pi/2$ phase shifted with respect to the light fringes. Equation (6.24) is a little bit more complex than Eq. (6.9), valid for the conventional single level model. Different regimes dominated by one of the fields E_{qf}, E_D, E_{Rh}, and E_{Re} are possible.

In the inter-band photorefractive effect, the steady state of the space–charge field depends not only on the grating spacing Λ but also on the light intensity I_0. For high intensities ($E_{qf} \gg E_D$) and small grating spacing Λ, the E_{sc} will grow proportional to $\sqrt{I_0}$.

The dynamics of the buildup is given by a double exponential function [8] and the buildup time constants are proportional to

$$\tau_{ib} \propto \frac{1}{\sqrt{I_0}} \tag{6.27}$$

in the pure inter-band regime. The different intensity dependence of the buildup time for conventional and inter-band effects, compared with Eq. (6.15), can be used to easily identify different regimes.

6.3 Applications of Photorefractive Effect

Photorefractive effect refers to the spatial modulation of the refractive index (Δn) of materials that possess both photoconductivity and optical nonlinearity simultaneously [9, 10]. Under the nonhomogeneous illumination formed by the interference of two coherent laser beams, a spatially modulating space-charge field is formed because of the redistribution of photo charges, which subsequently modulate the refractive index of material via an electro-optic effect. The attractive features of the photorefractive effect are that (i) a large Δn (up to the order of 10^{-2}) can be induced by a low-power laser on the order of milliwatts and (ii) the photorefractive grating formed can be completely erased by the illumination of single laser beam. Consequently, photorefractive material has been considered a promising recording medium, especially for reversible optical storage and information processing in real-time applications [9, 11, 12].

Currently, photorefractive effect is widely used in holography, multiplexing, phase encoding, beam coupling, two waves mixing, four waves mixing, phase conjugation, beam fanning, novelty filters, etc. Here, some of the applications are described in more detail.

6.3.1 Holographic Storage

Holographic data storage is one promising application for photorefractive materials [11–15]. It has been extensively applied to optical data storage because a large amount of data can be stored via various multiplexing methods and optical information can be retrieved with a high data-transfer rate due to the parallelism of optics [13]. Figure 6.7a shows the interference pattern of the two light waves. Figure 6.7b shows the application of the interference into a material that can store the interference pattern. Figure 6.7c shows the interference pattern permanently in the holographic material.

In holography, usually one of the light beams acts as a reference to the second light beam, which contains the data. The intensity of the sum of two waves forms a sinusoidal interference pattern, which is recorded in the photorefractive crystal in the form of a refractive index variation. The crystal then serves as a volume phase hologram. To construct the stored object wave, the crystal is illuminated with the reference wave. Acting as a volume diffraction grating, the crystal reflects the reference wave and reproduces the object wave.

Figure 6.8 helps provide an understanding of how to make use of the stored interference pattern. If we apply either the reference or conjugate beam to the interference pattern, we recover the original data light beam that was used to cause the interference. The photorefractive effect observed by A. Ashkin et al. [1] at Bell Laboratories has progressed to a means to store vast amounts of information in the form of refractive index gratings.

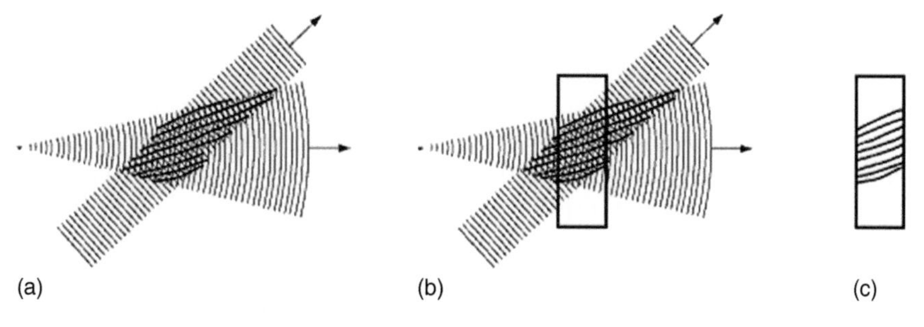

Figure 6.7 Hologram basics.

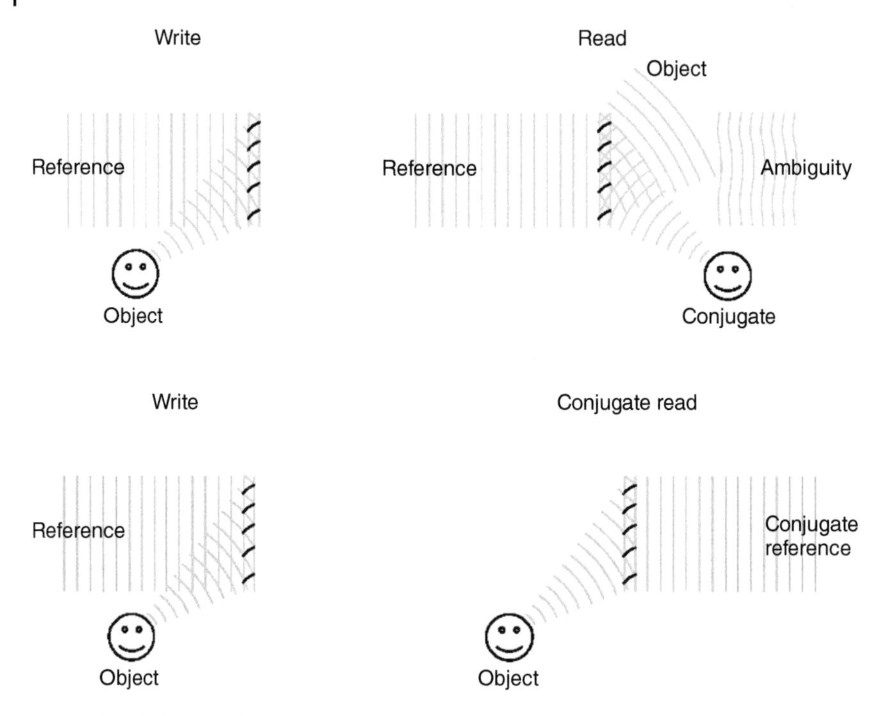

Figure 6.8 Hologram data recovery.

6.3.2 Two Waves Mixing

Since the recording process is relatively fast, the processes of recording and reconstruction can be carried out simultaneously. The object and reference waves travel together in the medium and exchange energy via reflection from the created grating. This process is called *two waves mixing or two beam coupling*. As shown in Figure 6.9, waves 1 and 2 interfere and form a volume grating. Wave 1 reflects from the grating and adds to wave 2; wave 2 reflects from the grating and adds to wave 1. Thus, the two waves are coupled together by the grating they create in the medium. Consequently, the transmission of wave 1 through the medium is controlled by the presence of wave 2 and vice versa. For example, wave 1 may be amplified at the expense of wave 2. The mixing of two (or more) waves also occurs in other nonlinear optical materials with light-dependent optical properties.

If we go back to a setup similar to Figure 6.9 and shine two beams of light through the medium with diffraction grating we find that we can control the energy transfers between the two beams of light. Owing to the nonlocality of the grating, one diffracted beam interferes destructively with its companion beam. The direction of the energy transfer depends upon the sign of the electro-optic coefficient and the sign of the charge career. This means that energy transfer does not necessarily have to go from stronger beam to weaker beam or vice-versa. The ability to transfer energy provided by beam coupling allows us to amplify signals or images. Wave mixing has numerous applications in optical data

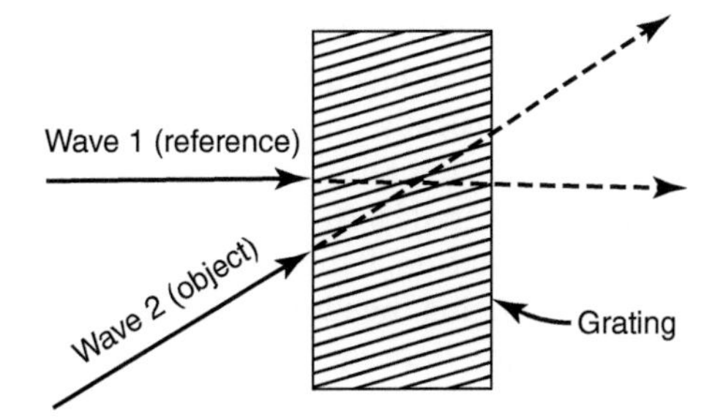

Figure 6.9 Two-wave mixing is a form of dynamic holography.

processing, including image amplification, the removal of image aberrations, cross-correlation of images, and optical interconnections.

6.3.3 Light-Induced Waveguides

In photoconducting electro-optic crystals reconfigurable waveguides are produced by inter-band photorefraction [16] as explained in the following and as illustrated in Figure 6.10. A homogeneous external electric field E that is applied to an electro-optic crystal induces a uniform change in the refractive index:

$$\Delta n = -\frac{n^3}{2} r_{eff} E \tag{6.28}$$

where r_{eff} is the effective electro-optic coefficient for the chosen configuration and n is the refractive index for the corresponding wavelength and polarization. If a small portion of such a biased photorefractive crystal is nonuniformly illuminated, free charge carriers are produced in the bright regions, which screen the

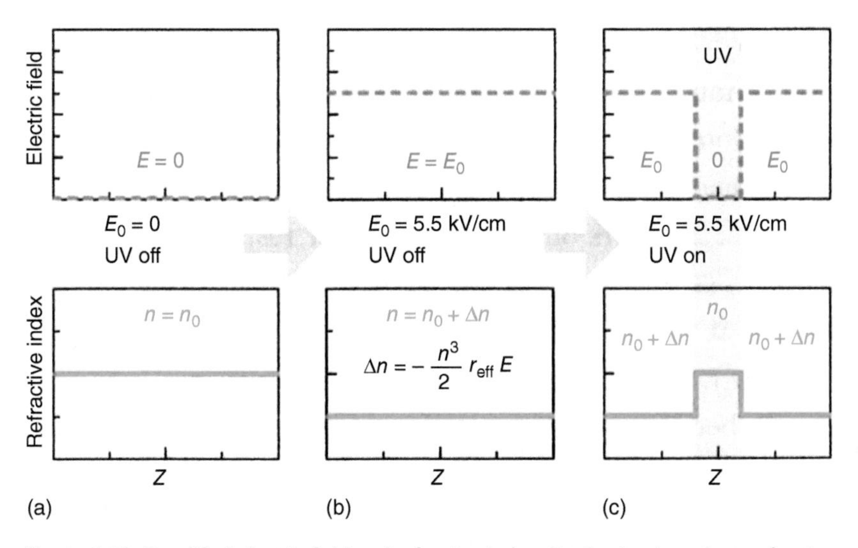

Figure 6.10 Simplified electric field and refractive index distribution in a photorefractive crystal for the generation of light-induced waveguides by band-to-band excitation: (a) Unperturbed state with refractive index n_0. (b) External field E_0 decreases the refractive index homogeneously by Δn. (c) UV-excited charges screen the external field and produce a waveguide.

applied field due to charge transport and trapping in dark regions. As a result, one gets a strongly reduced electric field in the bright regions, while in dark regions the field is basically unchanged. Via the electro-optic effect, this field distribution generates the refractive index profile, where the refractive index in the dark regions is lowered by Δn. Using a proper field direction and a suitable distribution of the illumination, light-induced waveguide structures can be generated in this way. By changing the illumination, the waveguide structure can be reconfigured.

By choosing light with photon energy larger than the bandgap energy of the material, charges can be excited directly from one band to the other, i.e. by inter-band excitation [17]. This process is much more effective in terms of use of the incident photons with respect to a conventional photorefractive effect, where the photoexcitation occurs from dopant or impurity energy levels within the material bandgap. In conventional photorefraction, charges are excited from mid-bandgap levels. When the photon energy of the interfering beams is larger than the bandgap energy of the material, charges can be excited directly from one band to the other. This inter-band photorefractive effect produces a holographic grating which, in a simplified way, consists of two grating types [18]. Near the surface, where the writing intensity is higher, we obtain gratings for which band-to-band recombination is dominant. At greater depth, where the light intensity is smaller, recombination into traps dominates.

In $LiTaO_3$ inter-band excitation leads to 3 orders of magnitude faster response compared to conventional trap-level excitation [19]. Owing to the strong light absorption at inter-band wavelengths ($\alpha_z = 690 \pm 40\,\mathrm{cm}^{-1}$ at $\lambda = 257\,\mathrm{nm}$ in SLT:Mg, which was measured using a thin plate of thickness 100 μm [20]), the screened regions are just underneath the surface. Thus, the waveguides were probed directly beneath the surface, which was possible due to sharp polished crystal edges. Inter-band photorefractive effects in SLT:Mg have been previously studied in longitudinal geometry, where the writing beam and the readout beam enter the crystal from the same surface [19] and the diffracted beam is a mixture of diffraction at different grating compositions along the beam path. The depth structure of an inter-band photorefractive grating can be determined in the transverse readout geometry for which the readout beam is propagating parallel to the input surface of the recording beams. In this geometry, the readout beam interacts over the whole crystal length with the same grating composition. Therefore, the diffraction efficiency, which is defined as the ratio between diffracted and incident light intensities inside the crystal, reflects the grating composition at the depth of propagation and was determined by measuring the diffracted power after the crystal, taking into account Fresnel losses.

6.4 Photorefractive Materials and Devices

The photorefractive effect is a phenomenon in which the local index of refraction is changed by the spatial variation of the light intensity. Such an effect was first discovered in 1966 [1]. The spatial index variation leads to the distortion of the wave front and such an effect was referred to as "optical damage." The photorefractive effect has since been observed in many electro-optic crystals, including

$LiNbO_3$, $BaTiO_3$, SBN, bismuth silicon oxide (BSO), BGO, GaAs, Indium phosphide (InP), etc. Photorefractive materials are, by far, the most efficient media for the recording of dynamic/static holograms [2, 4, 21–32]. In these media, information can be stored, retrieved, and erased by the illumination of light. In addition to the holographic properties, energy coupling occurs between the recording beams and also between the reading beam and the diffracted beam. Photorefractive media provide a promising candidate for information systems because of their unique properties, such as low-intensity operation, massive storage capacity, directional energy transfer, real-time response, and large dynamic range. These features make them attractive materials for volume holographic data storage (VHDS), image processing, optical interconnections, computing, and neural networks [21–23, 33–38]. Recently, the rapid development of optical fiber communication systems has stimulated the advancement of photorefractive devices. Photorefractive materials have been proposed for and used in fiber-optic devices such as modulators, switches, dispersion compensators, filters, and wavelength division multiplexers/demultiplexers.

Although the phrase *"photorefractive effect"* has been traditionally used for such effects in electro-optic materials, new materials, including photopolymers and photosensitive glasses, have been developed in recent years and are playing increasingly important roles in optical fiber communication systems. Photopolymers in combination with liquid crystal (LC) are ideal materials for wavelength selective tunable devices. The improved optical quality and large dynamic range of photopolymers make them promising materials for holographic recording. Holographic gratings recorded in photopolymers can be employed as distributed Bragg reflectors (DBR). The large birefringence of liquid crystals can be used to tune the index of refraction to cover a large wavelength range. In addition, the combination of photopolymer and liquid crystal also leads to a new material known as holographic polymer dispersed liquid crystal (H-PDLC), which provides a medium for switchable holograms. Photonic devices made of these materials can be easily incorporated into an optical fiber system because of the low index of refraction of polymers and liquid crystals and their relatively easy processing techniques. Besides photopolymers, photosensitive glasses are also promising for applications in fiber-optic systems. Fiber Bragg gratings (FBG) in photosensitive glass fibers have been used as band pass filters and dispersion compensators.

Here, we describe several applications of various photorefractive materials in information storage, processing, and communication systems. Specifically, we first briefly review the applications of the traditional electro-optic photorefractive crystals in optical information processing and VHDS, then discuss the recent works on the applications of photopolymers, H-PDLC, and photosensitive glasses in photonic devices for optical fiber communications. We describe a flat-topped tunable wavelength division multiplexing (WDM) filter incorporating DBR mirrors recorded holographically in photopolymers, demonstrate a wavelength selective switch based on switchable gratings in H-PDLC, and construct a multichannel dispersion slope compensator using a novel sampled FBG with post recording exposures.

6.4.1 Electro-optic Photorefractive Crystals

The photorefractive effect is generally believed to arise from optically generated charge carriers that migrate when the crystal is exposed to a spatially varying pattern of illumination with photons having sufficient energy. Migration of the charge carriers due to drift, diffusion, and the photovoltaic effect produces a space–charge separation, which then gives rise to a strong space–charge field. Such a field induces a refractive index change via the electro-optic (Pockels) effect. This simple picture of the photorefractive effect can be employed to explain several interesting optical phenomena in these media. Nonlinear optical processes [33] including two-wave mixing, four waves mixing, and phase conjugation are typical characteristics of the photorefractive effect. Since its discovery, the photorefractive effect [14, 36–45] has been applied in optical data storage and information processing.

In recent years, the research on the applications of photorefractive materials has been focused on VHDS. VHDS is becoming competitive [46–51] due to its large storage capacity (TBits/cm^3) and fast access rate (GBits/s). Multiple holograms can be recorded in a photorefractive crystal or a photopolymer sequentially by using an exposure schedule that equalizes the diffraction efficiency or the bit-error rate (BER). Different reference beam angles or wavelengths or phase distributions can be chosen for different exposures, known as wavelength multiplexing, angle multiplexing, and phase multiplexing respectively. Once information is stored in a volume holographic medium, it can be retrieved and can serve as a library for pattern recognition or other processing. A unique benefit of such a system is the parallel nature of the readout where an input reference beam can read out an entire page of information or alternatively, an input object can be compared with all the stored images simultaneously to achieve high-speed pattern recognition.

In the 1990s, a series of efforts have been made to push the holographic data storage technology to its maturity [52]. A brief historical review of various holographic data storage systems (HDSS) developed was summarized by Sincerbox in the book "Holographic Data Storage" [52]. Here, we merely mention a few recent representative works and provide links to websites of current research. At Stanford, Hesselink's group demonstrated the first multi-page digital holographic storage system [47, 53]. At Caltech and Holoplex, a group led by Psaltis et al. [45, 54] has pioneered a broad range of research on holographic memories. In 1994 DARPA sponsored two major programs, photorefractive information storage materials (PRISM) and HDSS, involving many companies and universities. As part of this joint endeavor, IBM [55, 56] developed two systems: PRISM [57, 58] and Demodulation of envelope modulation on noise (DEMON) (for demonstration) [59, 60]. PRISM is a material tester to measure and compare the BER introduced by various materials and configurations for holographic storage and DEMON is a complete holographic recording and retrieval platform built to evaluate various coding and signal processing techniques. At Rockwell, efforts were focused on building compact systems of read-only memory and holographic correlator for avionics applications [61–63]. At the same time, a team at Lucent

Technology was developing new materials (photopolymers) for VHDS [64–66]. Later on, this team formed a start-up company, InPhase Technologies [67] in Colorado.

A unique feature of the VHDS technology is its natural ability to be a content addressable memory or to perform associative retrieval. As we know, two beams, one object and the other reference, are involved in the recording of a hologram. During the reconstruction process, if the reference beam is used to read out the object beam, it is the regular data retrieval process. Instead, if the object beam (or part of the object beam) is used to read out the reference beam, the reconstructed wave will be the particular reference beam associated with the object during recording. Using this ability, we can search the address for a given object or content. If partial content is used during the search, the retrieved address can in turn be used to obtain the entire information stored on that page. This associative retrieval is very useful in database (or internet) search as well as pattern recognition. Using conventional computers with localized bit storage, it is difficult and slow to implement associative retrieval. With holographic memory, since each input object pattern is compared with all patterns stored in the same volume simultaneously, it is ideal for high-speed address search. When the holographic memory system is used for content addressable memory or associative retrieval, it is usually called a holographic correlator. Volume holographic correlators have been built for pattern recognition and database search applications, for example, the finger print recognition system built at Caltech/Holoplex [68], the query-by-image content (QBIC) database developed at IBM Almaden Research Center [69], and the compact correlator constructed at Rockwell Science Center [70]. This feature of content addressable memory remains the unique advantage of VHDS.

Materials have always been the main obstacle for commercial applications of photorefractive holographic storage. Although iron-doped $LiNbO_3$ is the mainstay of holographic data storage efforts, several shortcomings, especially the low response speed, impede it from becoming a commercial recording medium. The recent advances in the photorefraction of doped lithium niobate crystals are reviewed [71], especially the photorefractive performance of LN:Hf, LN:Zr, LN:Sn, LN:Hf,Fe, LN:Zr,Fe, LN:Zr,Fe,Mn, LN:Zr,Cu,Ce, LN:Ru, LN:V, and LN:Mo. Among them, Zr, Fe, and Mn triply doped $LiNbO_3$ shows excellent nonvolatile holographic storage properties, and V and Mo monodoped $LiNbO_3$ has fast response and multi-wavelength storage characteristics. Some of the recently designed photorefractive devices based on electro-optic photorefractive crystals are described in more detail here.

6.4.1.1 Photorefractive Waveguides

Optical waveguides are one of the fundamental building blocks for integrated optical applications and devices such as modulators, filters, switches, and couplers [72]. In nonlinear optical applications such as harmonic generation or parametric mixing, guiding the light allows maintaining a high intensity over a much longer interaction length, leading to larger conversion efficiencies as in bulk materials. There are many techniques for fabricating permanent waveguides including ion in-diffusion, ion exchange, ion implantation, fs laser ablation, photolithography, and epitaxial thin film deposition. By using these techniques, a permanent change of the refractive index in the waveguide region is obtained. This precludes an easy reconfiguration of the waveguide structure.

Recently, several techniques based on light-induced refractive index changes in photorefractive materials have been developed [16, 73–78]. Compared with waveguides fabrication by traditional methods, optically induced waveguides possess many advantages. For example, they can be fabricated solely by laser beam illumination with milliwatt power level, both surface and buried waveguides can be easily formed, and the waveguide structures can be easily erased or fixed, or even modified. So far, various materials have been investigated to fabricate waveguides by light irradiation, including $LiNbO_3$ [75, 79, 80], SBN [81], $KNbO_3$ [16], $LiTaO_3$ [77] crystals, etc.

Recently, F. Juvalta et al. have fabricated an array of planar waveguides in $Sn_2P_2S_6$ crystals for dynamic waveguide applications [82]. $Sn_2P_2S_6$ is a semiconducting ferroelectric material with interesting optical and nonlinear optical properties: high photorefractive efficiency in the infrared up to the telecommunication wavelength 1.55 μm [83–85] and a large electro-optic coefficient (r_{111} = 174 pm/V at 633 nm [86]). Furthermore, the photorefractive response of $Sn_2P_2S_6$ in the near-infrared is very fast, more than 2 orders of magnitude faster than in any other photorefractive ferroelectric crystal such as Rh-doped $BaTiO_3$ [87].

Figure 6.11 shows the schematic diagram of the fabricated technique for an array of planar waveguides in $Sn_2P_2S_6$ crystals. The crystal sample was nominally pure, to minimize the possibility of deep level trapping [7]. The use of inter-band light provides a faster effect, but on the other hand also a higher absorption for the controlling light. The absorption in $Sn_2P_2S_6$ is $\alpha = 490\,\text{cm}^{-1}$ at the controlling wavelength of $\lambda_{CL} = 514\,\text{nm}$ [88]. Therefore, the waveguide reaches a depth of only a few tens of micrometers below the surface and the crystal needs sharp edges for in- and out-coupling of the guided light. Since the structures are written between the bands, readout at sub-bandgap wavelengths such as red or telecommunication wavelengths does not disturb the waveguide structures.

The illumination of the crystal is shown in Figure 6.11a. The controlling light (argon ion laser at 514 nm) homogeneously illuminated a mask that was imaged

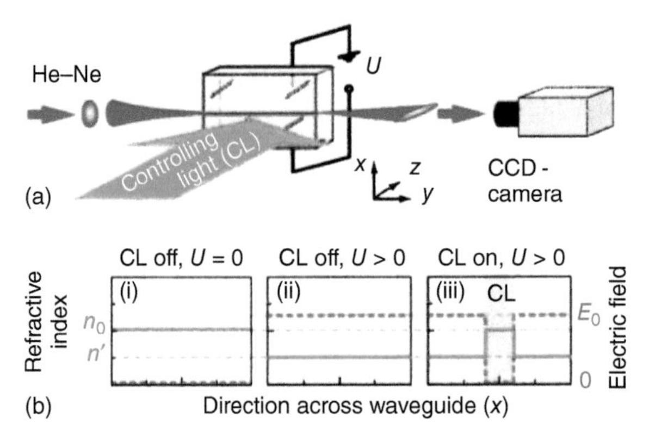

Figure 6.11 (a) Arrangement for recording light-induced waveguide structures. (b) Simplified electric field (dashed light grey) and refractive index (solid dark grey) distribution in a photorefractive crystal during the formation of the waveguides. Source: Juvalta et al. 2009 [88]. Reproduced with permission of OSA Publishing.

onto the crystal z-surface by appropriate optics. The probe beam (He–Ne at 633 nm) traveled along the crystalline y-direction and was focused onto the input face of the crystal by a spherical lens ($f = 40$ mm) to a diameter of $2\omega_0 = 22$ μm. An out-coupling lens imaged the output face onto a charge-coupled device (CCD) camera. In order to excite an eigenmode, the readout beam was polarized along the dielectric 3-axis (d_1, d_2, d_3) at an angle of $\psi = 43°$ with respect to the x-axis according to the orientation of the indicatrix in $Sn_2P_2S_6$ [85, 86]. The controlling light was polarized in the x-direction and an electric field was applied along the x-direction as well. For this configuration, an effective electro-optic coefficient of $r_{eff} = 183$ pm/V at 633 nm was obtained using the coordinate system as defined in [85]. Uniform background illumination at 514 nm produced a homogeneous conductivity that is needed for better confinement of the waveguides [77].

The basic process for inducing a step index profile in photorefractive crystals is schematically shown in Figure 6.11b, where (i) is the unperturbed state with a uniform refractive index n_0. In a first step (ii), an electric field E_0 is applied, which homogeneously decreases the refractive index via the electro-optic effect to a value of $n' = n_0 - \frac{1}{2}n_0^3 r_{eff} E_0$. Finally, the controlling light is switched on (iii) and electrons are excited to the conduction band. Free charges, electrons in the conduction band, and holes in the valence band drift and screen the applied electric field in the illuminated region. This results in an electric field pattern that is correlated to the pattern of the controlling light. Thus, a refractive index structure is produced, which has its maximum in the illuminated regions.

Figure 6.12a shows the output of a 6.8 mm long pure $Sn_2P_2S_6$ crystal. A straight slit was imaged onto the crystal z-surface to a width of 17 μm in the x-direction. In the experiments, the applied electric field was $E = 900$ V/cm, which resulted in a refractive index change of $\Delta n = 2.3 \times 10^{-4}$ for the He–Ne probe beam. The profile of the guided light perfectly matches a $\cos^2$-function, for the first waveguide mode, with an FWHM of 12 μm, which is in good agreement with the expected FWHM of 11 μm for the given index profile. The buildup times τ_b of light-induced

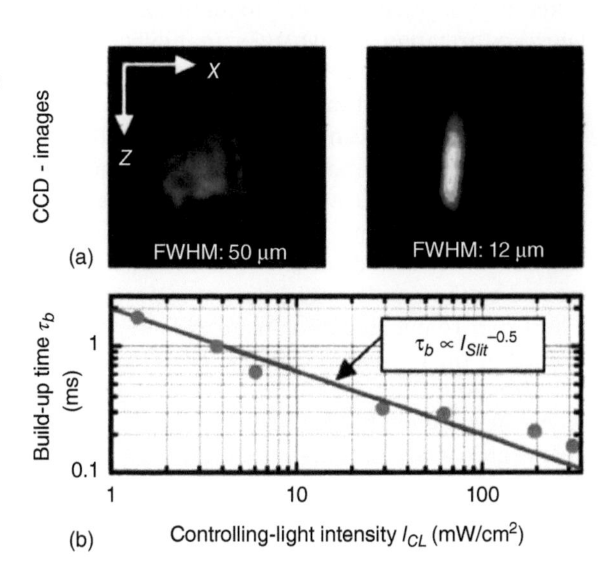

Figure 6.12 (a) CCD images of the output face of a 6.8 mm long Sodium polystyrene sulfonate (SPS) crystal without (left) and with (right) a photoinduced waveguide. (b) Buildup times τ_b of the light-induced waveguide as a function of the controlling light intensity. Source: Juvalta et al. 2009 [88]. Reproduced with permission of OSA Publishing.

waveguides in $Sn_2P_2S_6$ as a function of the controlling light intensity are shown in Figure 6.12b. The response is very fast, with $\tau_b < 200\,\mu s$ for intensities above $0.1\,W/cm^2$. This is two times faster than the buildup times observed in $KNbO_3$ [16] and more than 1 order of magnitude faster than the buildup times measured in $LiTaO_3$ [77] for recording with the same intensity but in the UV. The buildup times were determined by recording the temporal evolution of the peak intensity of the output light. This was measured using a photodiode and a $100\,\mu m$ pinhole in the image plane of the out-coupling lens. The square root intensity dependence of the buildup times confirms the inter-band nature of the structure formation [7].

Fabricating arrays of three-dimensional waveguides (3D-WGs) in bulk media has always been a challenge. Recently, light-induced 3D-WGs in photosensitive materials, e.g. various glasses [89, 90] and photorefractive crystals [79, 91], have attracted much research interest. At the same time, several optical approaches for creating arrays of 3D-WGs have been proposed in bulk photorefractive crystals [80, 92]. Matoba et al. have proposed generating array of 3D-WGs in $LiNbO_3$ by illumination of a four-plane wave interference pattern [80] and the technique has been improved to obtain large amplitudes of index variations [92]. Zhao et al. have fabricated an array of planar waveguides in an iron doped $LiNbO_3$ by exposure to a two-beam interference pattern [93]. Chen and Martin have demonstrated an array of 3D-WGs employing pixel-like solitons in an SBN crystal [94]. These light-induced photorefractive waveguide arrays can be created at milliwatt or even lower power levels and can guide probe beams with higher power level at a less photosensitive wavelength. The output power of a guided signal beam can be dynamically controlled by another guided beam [95] or by an introduced control beam [94]. This may be used as adaptive interconnections. Additionally, the waveguide structures can be erased by uniform light illuminations or by heating the crystal and can be fixed thermally [96] or electrically [97]. Thus, both dynamic and permanent waveguide arrays can be fabricated.

Recently, P. Zhang et al. proposed and demonstrated a novel approach to fabricate an array of 3D-WGs in $LiNbO_3$ employing illuminations of two mutually incoherent or coherent interferograms formed by two plane waves [98]. Figure 6.13 shows a schematic diagram of the fabrication technique for an array of 3D-WGs in a $LiNbO_3$ crystal. Coherent collimated writing beams 1 and 2, which can be mutually incoherent or coherent, propagate through Fresnel biprisms BP1 and BP2 to illuminate the crystal. If the writing beams are mutually coherent, they are switched on in turn. Otherwise, they may or may not be switched on at the same time. The angles of the grating vectors formed by the interference patterns with respect to the c-axis of the crystal are adjustable. To obtain approximately identical amplitudes of the index variations formed by the two patterns, the angles are adjusted to be equal, both defined as θ (indicated in Figure 6.13) [99]. The periods of the interference fringes Λ_1 and Λ_2 are determined by the deflection angles and the wavelengths of the writing beams propagating through the Fresnel biprisms, which are denoted by α_1 and α_2, λ_1 and λ_2. Thus, the intensity patterns propagating along the x-axis and the

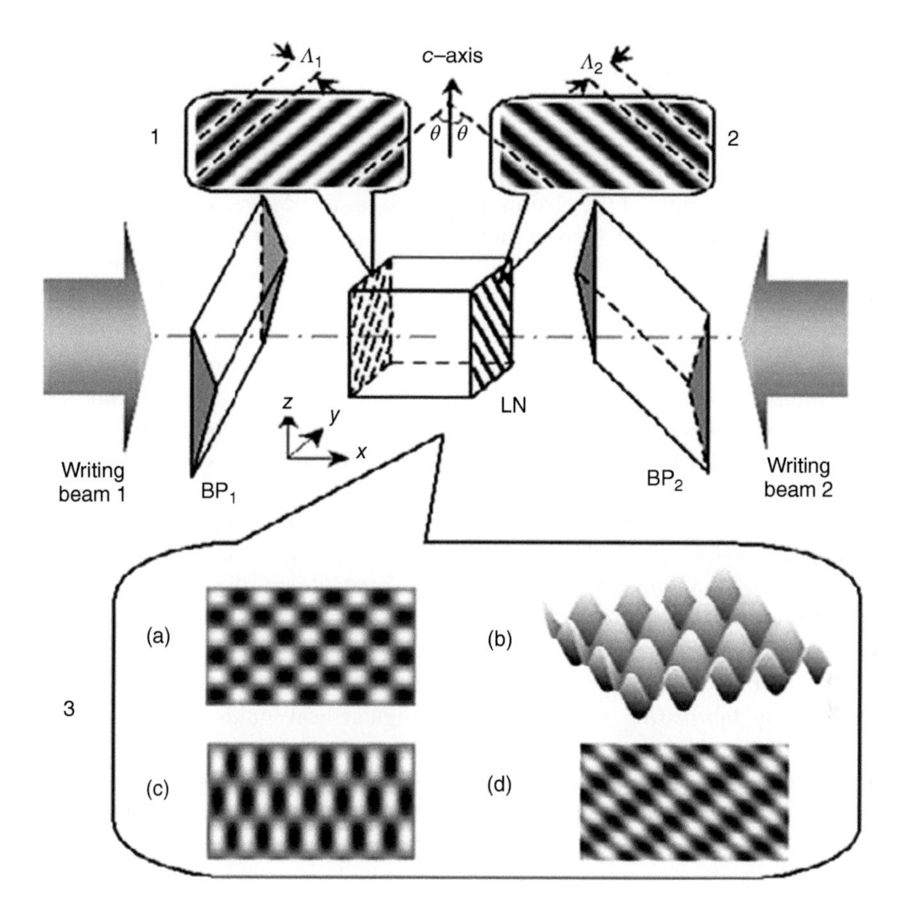

Figure 6.13 Schematic diagram of the fabricating technique for array of 3D-WGs [98]. BP, Fresnel's biprism; LN, $LiNbO_3$ crystal. Insets 1 and 2 show the intensity patterns propagating along the *x*- and −*x*-axis, respectively. Inset 3 depicts the index variations in the waveguide array; (a) and (b) for $\Lambda_1 = \Lambda_2, \theta = 45°$; (c) for $\Lambda_1 = \Lambda_2, \theta = 20°$; (d) for $\Lambda_1 = 2\Lambda_2, \theta = 45°$.

negative x-axis can be respectively described as

$$I_1(y,z) = I_{01}\cos^2\left[\frac{2\pi}{n_1\lambda_1}(y\cos\theta - z\sin\theta)\sin\alpha_1\right] \tag{6.29}$$

$$I_2(y,z) = I_{02}\cos^2\left[\frac{2\pi}{n_2\lambda_2}(y\cos\theta + z\sin\theta)\sin\alpha_2\right] \tag{6.30}$$

where I_{01} and I_{02} are the amplitudes of intensity distributions of the two interferograms, and n_1 and n_2 are the crystal indices at wavelengths of λ_1 and λ_2. The insets 1 and 2 in Figure 6.13 depict the examples of the intensity patterns I_1 and I_2, respectively. In the $LiNbO_3$ crystal, the intensity patterns will induce the corresponding space–charge fields and modulate the index distribution of the crystal through the linear electro-optic effect. According to the simplified model described in Ref. [100], the distribution of index change for extraordinary light is given by

$$\Delta n(y,z) = -\gamma I(y,z) \tag{6.31}$$

where γ is a constant that depends on the crystal properties. Owing to the linear superposition principle of the electric fields, the total index changes induced by the two interferograms can be described by

$$\Delta n(y,z) = -\gamma I_{total} = -\gamma[I_1(y,z) + I_2(y,z)] \tag{6.32}$$

where I_{total} is the total light intensity distribution to illuminate the crystal sample. It is important to note that $I(y,z)$ is uniform along the direction of the x-axis, so the index distribution along the x-axis is uniform, without considering the absorption of the crystal. The insets 3(a) and 3(b) in Figure 6.13 depict the 2D and 3D illustrations of the index distribution of the light-induced waveguide array, under the condition of $\theta = 45°$ and $\Lambda_1 = \Lambda_2$. By adjusting Λ_1 and Λ_2 by changing α_1 and α_2 or λ_1 and λ_2, different waveguide arrays can be obtained. By further altering θ, the arrays of square, rectangular, or parallelogram waveguides can be created. For $\Lambda_1 = \Lambda_2$, $\theta = 20°$ and $\Lambda_1 = 2\Lambda_2$, $\theta = 45°$, illustrations of 2D maps of index distributions of the waveguide arrays are shown in the insets 3(c) and 3(d) in Figure 6.13, respectively.

Z. Peng et al. experimentally demonstrate the feasibility of the present technique by fabricating an array of rectangular waveguides (Figure 6.14). An expanded and collimated ordinary beam emitted from a diode-pumped solid-state laser (DPL) with wavelength of 532 nm propagates through a Fresnel biprism to illuminate a $LiNbO_3$:Fe crystal with 0.07 wt% dopant and dimensions of $5 \times 5 \times 5\,mm^3$. The period of the interference pattern in the crystal

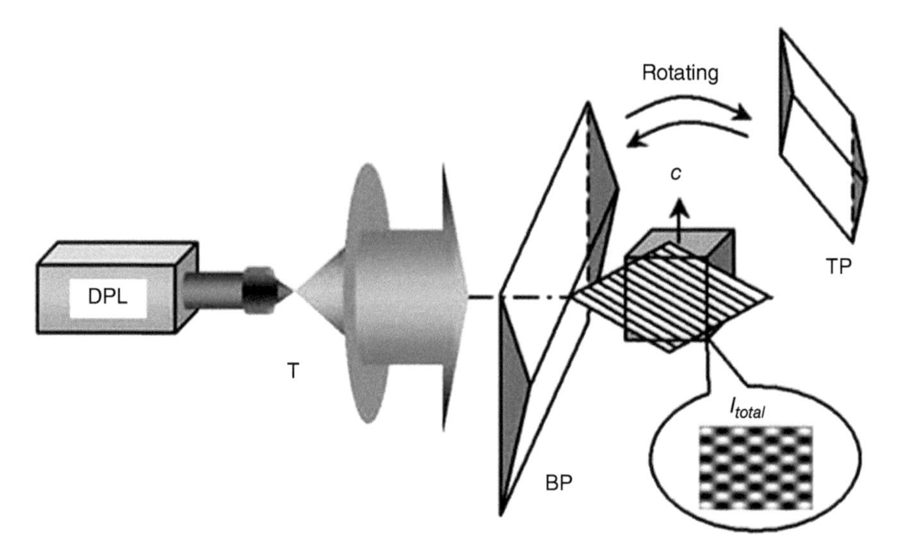

Figure 6.14 Experimental setup for fabricating an array of rectangular waveguides in $LiNbO_3$:Fe [98]. DPL, diode-pumped solid-state laser; T, telescope; BP, Fresnel's biprism; I_{total}, total intensity for illuminating the crystal.

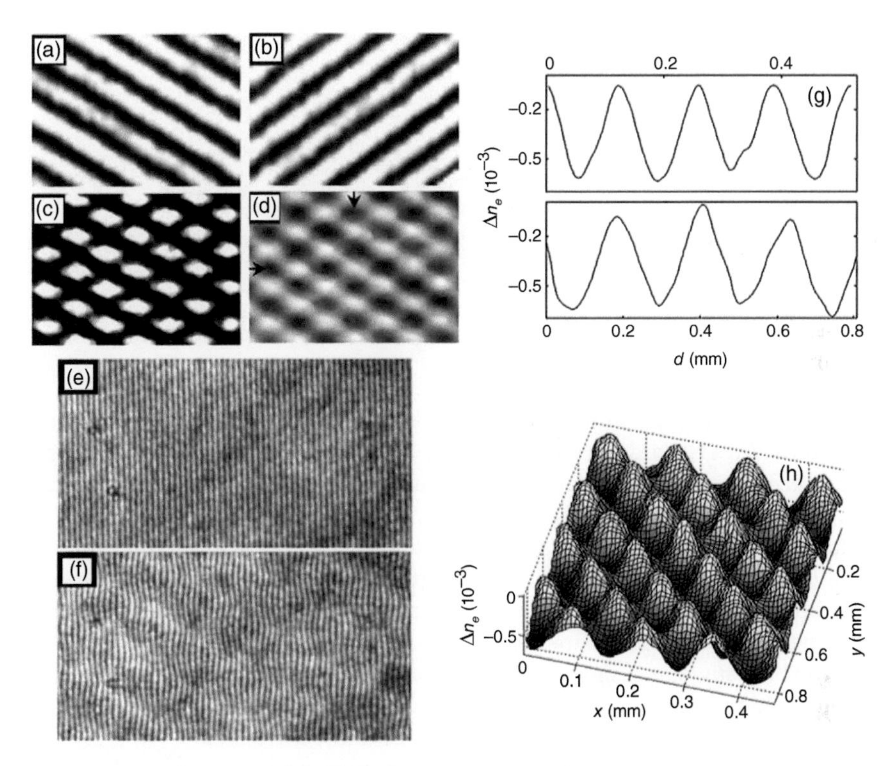

Figure 6.15 Experimental results for fabricating an array of rectangular waveguide [98]. (a, b) the intensity patterns for the first and second exposures, respectively; (c) the near-field pattern of the waveguide array; (d) 2D index map of the waveguide array; (e, f) the holograms that are used for index profile measurement, before and after waveguide fabrication, respectively; (g) the index profiles along the horizontal and the vertical arrows in (d), top for vertical, bottom for horizontal; (h) 3D display of the 2D index distribution of the waveguide array.

sample is approximately 100 μm. Rotating the Fresnel biprism, two exposures of the crystal with $\theta = 60°$ orientation (insets 1 and 2 of Figure 6.13) can be realized.

Figure 6.15a,b shows the interferograms for the double exposures. For each one, illumination is switched on for 30 minutes and the amplitude of the power intensity of the writing beam is approximately 20 mW/cm^2. After double exposure, the near-field pattern of the waveguide array under illumination by an expanded beam is shown in Figure 6.15c. The index distribution of the waveguide array is measured by using digital holography [101]. Figure 6.15d shows a 2D map of index distribution of the illuminated region. The holograms, which are used for index profile measurement, before and after waveguide fabrication are shown in Figure 6.15e,f, respectively. The index profiles along the horizontal and the vertical arrows in Figure 6.15d are depicted in Figure 6.15g, where both the index profiles have large amplitudes of index variations and the periods of the array are about 200×115 μm^2 (horizontal × vertical), and in addition the size of each channel is approximately 100×58 μm^2. A 3D display of the 2D index distribution of the waveguide array is shown in Figure 6.15h.

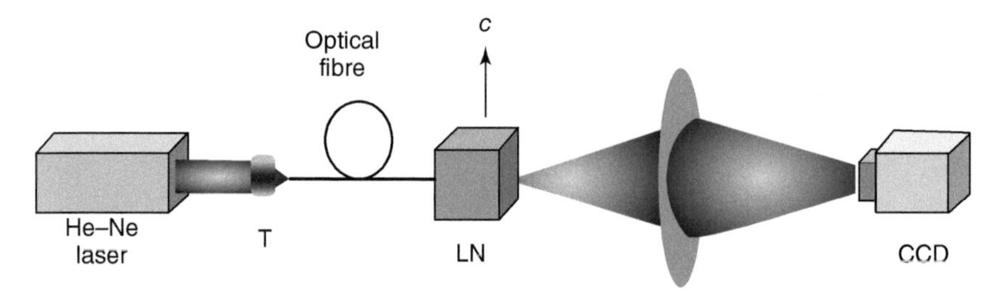

Figure 6.16 Experimental setup for guiding tests of the waveguide array. T, telescope; LN, $LiNbO_3$:Fe crystals; CCD, charge-coupled device camera. Source: Zhang et al. 2004 [98]. Reproduced with permission of IOP.

The experimental setup for guiding tests of the waveguide array is depicted in Figure 6.16. An extraordinary He–Ne laser beam with wavelength of 632.8 nm and power intensity in submicrowatts is coupled into a single mode fiber by an objective lens. Then the fiber is directly coupled with the waveguide array. The rear face of the crystal sample is imaged onto the target plane of a CCD camera. Figure 6.17a,b shows the light intensity patterns monitored by the CCD camera without and with guidance. The normalized light intensity profiles along the white lines in Figure 6.17a,b are depicted in Figure 6.17c, from which the probe beam can be trapped in a single waveguide of the array. Shifting the optical fiber, the guided phenomena are periodically repeated. From the results, it can be seen that an array of rectangular waveguides is successfully fabricated.

For using the technique proposed here, one must notice that with decreasing angle θ, the amplitude of the index variations in the waveguide array will be decreased, due to the decrease in the space–charge field along the c-axis [99]. The double exposure fabrication is suitable to create waveguide arrays in photorefractive materials with long decay time, such as $LiNbO_3$. However, for materials with short decay time, e.g. SBN, the writing beams 1 and 2 must be switched on simultaneously. The present technique, compared to that employing interference of four plane waves and soliton arrays, has several advantages. Apart from the fact that the optical system can be implemented more easily, the most important advantage is the better adjustability. By adjusting several experimental parameters, an array of various 3D-WGs with different periods, e.g. square, rectangular, and parallelogrammical waveguides, can be fabricated. Waveguide arrays generated by this optical method may find versatile applications in the field of massively parallel optical interconnections, neural network systems, and nonlinear discrete systems.

6.4.1.2 Photorefractive Tunable Filters

With the increasing demand for higher bandwidths and larger data throughput in optical networks, tunable optical filters play a key role in next generation's high-speed communication systems. In WDM systems tunable optical filters are needed to manipulate or select a desired wavelength from the band of available channels. With the advance to dense wavelength division multiplexing (DWDM), a suitable tunable optical filter needs both a large tuning range and a narrow bandwidth.

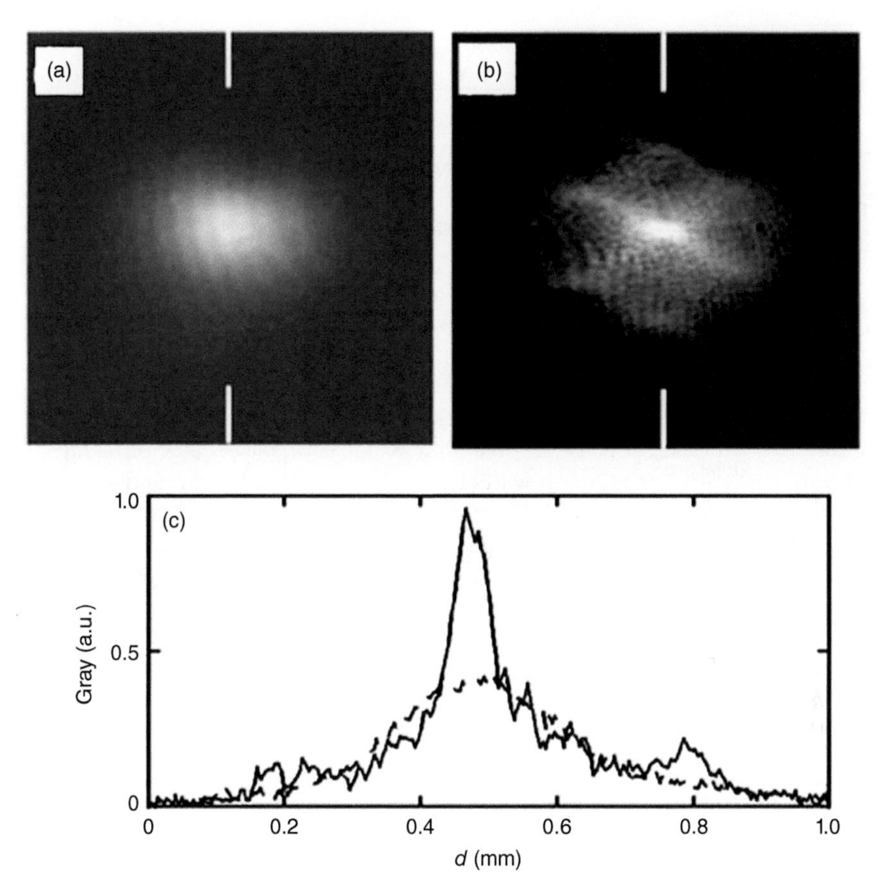

Figure 6.17 Experimental results for guiding tests of the waveguide array. (a, b) Light intensity patterns without and with guidance, respectively; (c) light intensity profiles along the white lines in (a) and (b), dashed line for (a), solid line for (b). Source: Zhang et al. 2004 [98]. Reproduced with permission of IOP.

Various approaches are employed for the development of dynamic optical filters, such as filters based on Fabry–Perot interferometers, micro machined devices, FBGs, ring resonator tunable filters, acousto-optical or electro-optical tunable filters, and volume holograms. Volume holograms offer the important advantage of a very high wavelength selectivity of the filter. In the past, fixed narrow-band interference filters have been developed in lithium niobate ($LiNbO_3$) using volume holography [102, 103]. Fine-tuning was achieved by varying the filter temperature or by applying an electric field [104]. Wavelength demultiplexing with superimposed fixed volume-phase holograms has been shown in $LiNbO_3$ [105, 106]. Wavelength-selective photonic switching with fixed holographic gratings by the application of a uniform electric field has also been demonstrated in $LiNbO_3$ [107] and in potassium lithium tantalate niobate [108].

An optically tunable optical transmission filter at the wavelength 1560 nm with a bandwidth of about 5 nm and a switching speed of about one second has been demonstrated in barium titanate [109]. The tuning of this filter was achieved by

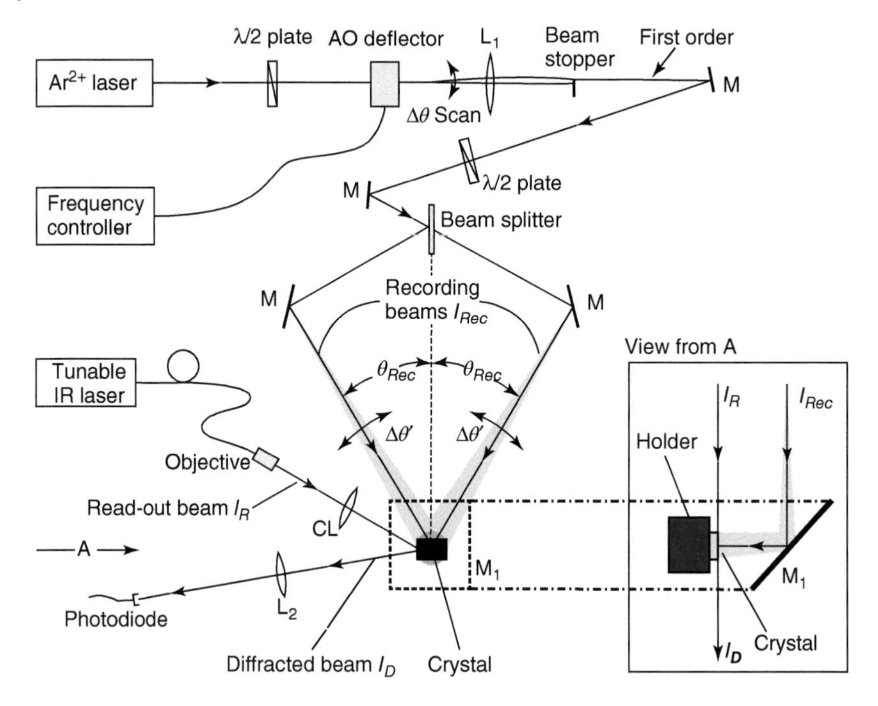

Figure 6.18 Schematic top view of the experimental setup. The recording beams are directed to the crystal from the top by means of a 45° inclined mirror, as can be better seen by the view from direction *A* shown in the inset. Source: Dittrich et al. 2002 [110]. Reproduced with permission of Elsevier.

changing the angle of one writing beam by rotating a mirror. This approach has the disadvantage that the direction of the grating vector is also changed and that the output beam position shifts significantly. To record fast dynamic holographic filters, however, different recording mechanisms and materials are required. One such possible mechanism is the fast inter-band photorefractive effect [7], where a refractive index grating is created via the linear electro-optic effect by a spatially modulated light distribution that induces band-to-band photoexcitation of charge carriers.

Figure 6.18 shows the schematic diagram of the fabricated technique used to record and read out the light-induced tunable filters [110]. Light of an Ar^{2+}-laser ($\lambda_{Rec} = 363.8$ nm) is used as the grating recording beam I_{Rec}. It is deflected by an acousto-optic (AO) deflector made of SiO_2 (from Neos Technologies). The zeroth order is blocked in the focal plane of L_1, while the first deflection order passes. The deflection angle $\theta_c = 0.28°$ of the first diffraction order can be changed ($\pm\Delta\theta$) by shifting the frequency f that is applied to the AO deflector from its center frequency at $f_c = 80$ MHz, covering a bandwidth of $\Delta f = \pm 15$ MHz. The angular deviation $\Delta\theta$ from the central deflection angle θ_c is directly proportional to the deviation of the frequency Δf from the center frequency f_c applied to the acousto-optical cell and is given by $\Delta\theta = (\lambda_{Rec}\Delta f)/v_s$ where $v_s = 5960$ m/s is the speed of the acoustic wave in the deflector crystal.

The output plane of the deflector is imaged by the lens L_1 ($f = 200$ mm) and the spot size enlarged by a factor of $M \approx 6$ onto the crystal surface. After passing a beam splitter the two deflected beams interfere at the top crystal surface and induce the reflection grating by the inter-band photorefractive effect. The inset in the right bottom of Figure 6.18 shows in a side view from direction A how the beams are redirected by a 45° inclined mirror and hit the crystal from the top through the surface, oriented normal to the a-axis of $KNbO_3$ (transverse diffraction geometry). The crystal is a pure $KNbO_3$ crystal of dimensions $2.4 \times 8.4 \times 10.3\ mm^3$ ($a \times b \times c$-axis of the orthorhombic system). It is fixed onto a holder that permits fine adjustment of all six translational and rotational degrees of freedom. The imaging by lens L_1 ensures that by varying the deflection angle by $\Delta\theta$ only the mutual angle of the two recording beams is changed by $-2\,\Delta\theta'$, where $\Delta\theta' = \Delta\theta/M$, but not their positions on the crystal surface. This allows a precise tuning of the grating spacing $\Lambda = \lambda_{Rec}/(2\sin\theta_{Rec})$, where θ_{Rec} is the incident angle of the recording beams. By increasing the deflection angle after the AO deflector, the angle θ_{Rec} is decreased according to $\theta_{Rec} = \theta_c' - \Delta\theta'$, where θ_c' is the incident angle for $f = f_c$. The central mutual angle $2\theta_c'$ between the two recording beams is about 60.6° and corresponds to a central grating spacing of $\Lambda_c = \lambda_{Rec}/(2\sin\theta_c') = 0.361\ \mu m$. Two $\lambda/2$ plates are used to control the polarization of the recording beam, which is horizontal before entering the AO deflector and parallel to the c-axis of the crystal after the last inclined mirror.

The readout beam I_R is provided by a tunable IR laser (Santec, TSL-220) at the telecommunication wavelength λ_R near 1550 nm and with a power of $P_R = 2.0$ mW. An objective is used to collimate the fiber output of this laser. The beam is then focused vertically by the cylindrical lens (CL) ($f = 120$ mm) onto the surface normal to the c-axis of the crystal with a vertical beam width of about 40 μm. Its vertical center is positioned about 40 μm below the edge of the surface where the grating is recorded (surface perpendicular to the a-axis). The polarization of the readout beam is vertical, i.e. parallel to the a-axis of the crystal. At this wavelength and for this polarization, the refractive index of $KNbO_3$ is $n_a = 2.20$ [111]. Since the dispersion is very small at this wavelength, n_a is considered to be constant in the experiment. The diffracted beam I_D is collected by lens L_2 ($f = 200$ mm) and its intensity measured by a photodiode.

To characterize the filter, the grating configuration and angle convention used in this work are given in more detail in Figure 6.19, which shows a top view of the bc-plane of the $KNbO_3$ crystal. The grating vector $\boldsymbol{K}$ is oriented perpendicular to the fringe planes and is of magnitude $K = 2\pi/\Lambda$. $\boldsymbol{k}_R$ and $\boldsymbol{k}_D$ are the wave vectors inside the crystal of the readout beam and of the diffracted beam, respectively. The corresponding angles $\alpha_R \cong 12.6^\circ$ and $\alpha_D \cong 171^\circ$ between the wave vectors and the c-axis of the crystals are internal to the crystal. All $\boldsymbol{k}$-vectors are lying in the bc-plane of the $KNbO_3$ crystal. The c-axis of the crystal and the grating vector are not parallel but slightly rotated by $\phi \cong 3.3^\circ$ in order to spatially separate the diffracted signal and the Fresnel reflection of the readout beam.

Considering isotropic diffraction with s-polarization, the lengths of the vectors $\boldsymbol{k}_R$ and $\boldsymbol{k}_D$ are equal, if the Bragg condition

$$\cos(\alpha_R - \phi) = \frac{K}{2k_R} \tag{6.33}$$

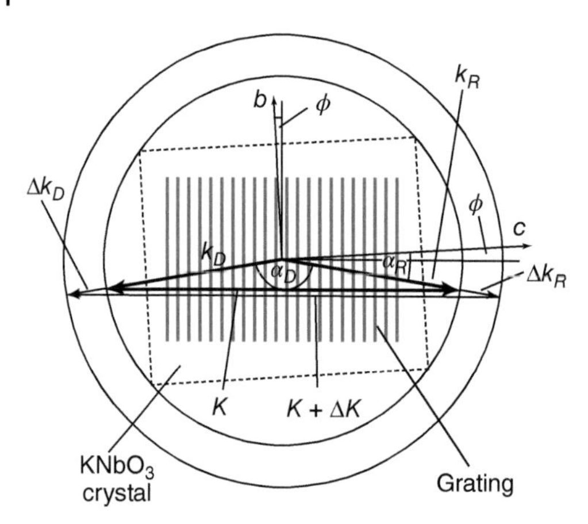

Figure 6.19 Grating configuration and angle convention used in this work. Top view at the *bc*-plane of the $KNbO_3$ crystal; all ***k***-vectors are lying in this plane. The angles are internal to the crystal. If ***K*** is changed by $\Delta \boldsymbol{K}$ by tuning the grating; $\boldsymbol{k}_R$ has to change by $\Delta \boldsymbol{k}_R$ in order that the Bragg condition is still fulfilled. Source: Dittrich et al. 2002 [110]. Reproduced with permission of Elsevier.

is fulfilled. The three vectors k_R, k_D and K are then in the symmetric configuration indicated in Figure 6.19 (inner circle). If the length K of the grating vector $\boldsymbol{K}$ is now enlarged by ΔK (by tuning the filter), the length of the wave vectors $\boldsymbol{k_R}$ and $\boldsymbol{k_D}$ also has to increase by $\Delta k_R = \Delta k_D$ in order that the Bragg condition is still fulfilled, as indicated by the outer circle. Considering that $(\alpha_R - \phi)$ is small, in a first-order approximation, the change of magnitude of the grating and the wave vectors are

$$\Delta k_R \cong \Delta k_D \cong \frac{\Delta K}{2} \tag{6.34}$$

On the other hand, if the magnitude of only one vector, either K or k_R, is changed and the Bragg condition is no longer fulfilled, one introduces a phase mismatch that can be expressed accordingly by either ΔK or by Δk_R. Equation (6.34) is still valid in this case.

As already mentioned, the length of the grating vector $\boldsymbol{K}$ is controlled by the frequency f applied to the AO deflector. Considering that the incident angle of the recording beam is decreased for Δf being positive, the grating mismatch ΔK is related to the frequency detuning Δf by

$$\Delta K = \frac{4\pi}{Mv_s}\cos\theta_{Rec}\Delta f \cong -\frac{4\pi}{Mv_s}\cos\theta'_c\Delta f \equiv C_1\Delta f \tag{6.35}$$

where $\Delta\theta'$ is considered to be very small. The relationship between the Bragg wavelength λ_{Bragg} of an induced grating and the frequency deviation Δf from the center frequency f_c that is applied to the AO deflector can be calculated as

$$\lambda_{Bragg} = \lambda_{Rec}\frac{n_a\cos(\alpha_R - \phi)}{\sin\theta_{Rec}} = \lambda_{Rec}\frac{n_a\cos(\alpha_R - \phi)}{\sin(\theta'_c - C_2\Delta f)} \tag{6.36}$$

where $C_2 \equiv \frac{\lambda_{Rec}}{Mv_s} = \frac{\Delta\theta'}{\Delta f} =$ constant.

Neglecting absorption gratings and light absorption, the diffraction efficiency η (defined as the ratio between diffracted and incident light intensities) for the reflection gratings is derived from coupled wave theory [112, 113] and is given by

$$\eta = -\frac{\sin^2(\sqrt{\nu^2+\xi^2})}{1-\sin^2(\sqrt{\nu^2+\xi^2})+\left(\frac{\xi^2}{\nu^2}\right)} \tag{6.37}$$

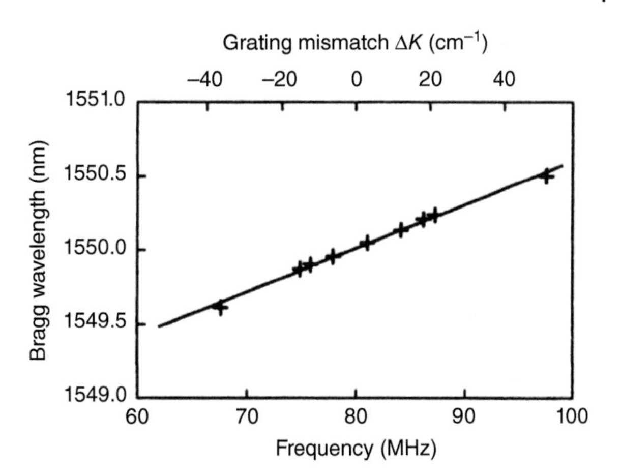

Figure 6.20 Bragg wavelength as a function of the frequency applied to the acousto-optic deflector and of the corresponding grating mismatch ΔK for the central wavelength. Source: Dittrich et al. 2002 [110]. Reproduced with permission of Elsevier.

where, $\nu \equiv i\pi\Delta n d/(\lambda_R\sqrt{\cos\alpha_R\cos(\pi-\alpha_D)})$ is the coupling coefficient, $\xi \equiv \Delta k_R d/2$ is the off-Bragg parameter describing the phase mismatch Δk_R according to Eq. (6.34), Δn is the amplitude of the refractive index modulation, d is the effective thickness of the grating, and λ_R is the vacuum wavelength of the diffracted light. Note that ν is purely imaginary and therefore $\nu^2 < 0$. Equation (6.37) is identical to the diffraction efficiency formula derived by Kogelnik [112].

In Figure 6.20 the tuning of the Bragg wavelength of the filter is depicted as a function of the frequency that is applied to the acousto-optic deflector. A tuning range of about $\Delta\lambda = 1$ nm was obtained with good linearity, indicated by a regression coefficient of $R = 0.997$. On the top axis of Figure 5.70, the corresponding induced grating mismatch ΔK with respect to a central Bragg wavelength of 1550 nm at 80 MHz is indicated.

The wavelength selectivity of the dynamic grating filter is shown in Figure 6.21. The induced grating was subsequently set to three different grating spacing, separated by about $\Delta K = 40\,\text{cm}^{-1}$ and corresponding to three different Bragg

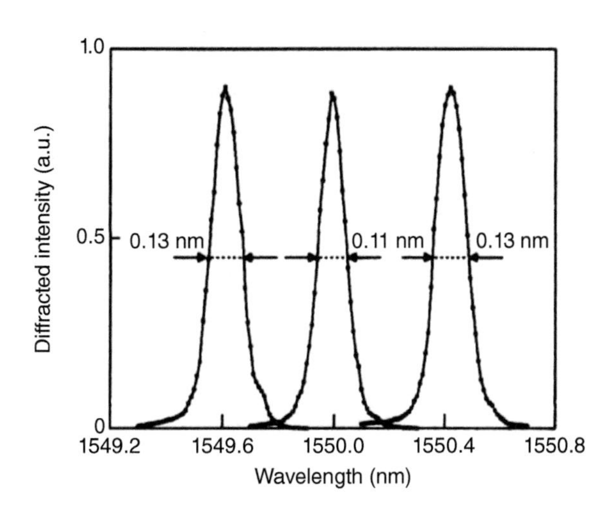

Figure 6.21 Wavelength selectivity scan for three different grating vectors separated by 0.4 nm (50 GHz). The three peaks are obtained using three different frequencies applied to the acousto-optic deflector mutually separated by 13.4 MHz, corresponding to a grating mismatch of $\Delta K = 40\,\text{cm}^{-1}$. The solid lines are guides to the eye. Source: Dittrich et al. 2002 [110]. Reproduced with permission of Elsevier.

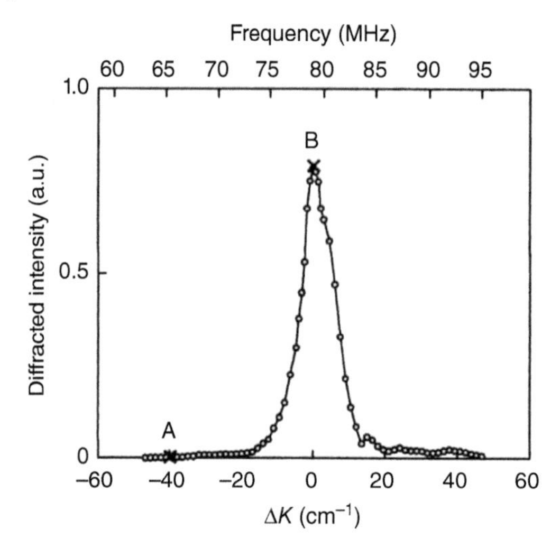

Figure 6.22 Measured diffraction signal as a function of the grating spacing at fixed readout wavelength $\lambda_R = 1550$ nm. To determine the tuning times of the optical filter the diffraction signal was recorded while tuning the grating from point A to point B and vice versa. The solid line is a guide to the eye. Source: Dittrich et al. 2002 [110]. Reproduced with permission of Elsevier.

wavelengths, separated by approximately 0.4 nm or 50 GHz. For each grating, a wavelength scan was performed by changing the wavelength of the readout beam in steps of 10 pm and the signal of the diffracted beam was measured. The average total intensity of the writing beams at the crystal surface was 15 mW/cm^2 in each case. The measured bandwidth of the filter (FWHM) is 0.13 nm or about 16 GHz. The crosstalk at the center wavelength between two neighboring reflection peaks is −18 dB. The diffraction efficiency of the filter is in the order of 0.1%.

The selectivity of the reflection grating as a function of the grating spacing at a fixed readout wavelength $\lambda_R = 1550$ nm is shown in Figure 6.22. The corresponding frequency that is applied to the AO deflector is given at the top axis. The tuning times of the holographic reflection grating were determined by measuring the diffracted signal while changing the grating's Bragg wavelength. The response time of the AO deflector is about 1 μs. "Tuning-in" means that the Bragg wavelength of the filter was set equal to the used readout wavelength, i.e. the diffraction grating was tuned from point A in Figure 6.22 by $\Delta K = 40$ cm^{-1} to the point of maximal diffraction, as indicated by point B. "Tuning-out" denotes the reversed process of de-tuning the grating from point B to point A.

Figure 6.23 depicts the time development of the diffraction signal for tuning-in and tuning-out and the tuning time of the filter as a function of the average recording beam intensity at the crystal surface: (a) shows a typical measurement (here for $I_R = 3.9$ mW/cm^2) of the diffraction efficiency η for tuning-in and tuning-out, which is then quite well described by a simple exponential increase

$$\eta \cong n_0(1 - e^{-t/\tau_1}) \tag{6.38}$$

or decrease

$$\eta = \eta_0 e^{-t/\tau_2} \tag{6.39}$$

respectively, where n_0 is the diffraction efficiency at point B in Figure 6.22. It is found that the tuning times τ_1 and τ_2 are proportional to $I^{-0.8}$ for both processes, which is shown in Figure 6.23b.

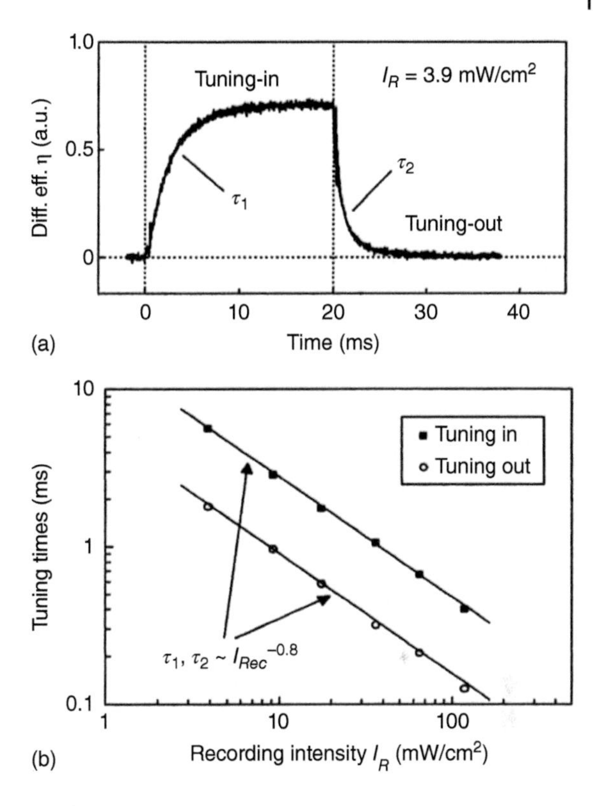

Figure 6.23 (a) Time development of the diffraction efficiency for tuning-in and tuning-out for $I_R = 3.9\,\text{mW/cm}^2$. (b) Tuning time of the holographic reflection grating as a function of the average total recording intensity at the crystal surface. Source: Dittrich et al. 2002 [110]. Reproduced with permission of Elsevier.

For practical implementations the diffraction efficiency of the proposed tunable filter should be increased. Significant higher diffraction efficiencies (potentially approaching 100%), while not changing any of the other characteristics of the filter, should be feasible by employing anisotropic diffraction geometries exploiting larger effective electro-optic coefficients [114, 115]. Antireflection coatings can be used to further minimize losses. Since the tuning of the filter is governed by the time constants of the inter-band photorefractive effect, one can expect to reach even faster tuning times (in the order of a few microseconds) by increasing the intensity of the recording beams further [7]. It is also important to note that sub-nanosecond optical response was already demonstrated for gratings recorded in $KNbO_3$ with pulsed light (fluence ~ $100\,\mu\text{J/cm}^2$) using inter-band transitions [116].

6.4.1.3 Photorefractive Switches

Intensive research has been going on in the field of fast and efficient optical switching. The most popular development is the optical switch based on Pockels effect using a $LiNbO_3$ crystal [117]. This type of optical switch can achieve very high speed due to the fast Pockels effect. Another type of switch is the one using absorption by carriers in semiconductor optical modulator, which seems promising in waveguide optics [118]. A switch using the Kerr nonlinear photonic crystals has also been proposed to achieve efficient optical switch [119].

Photorefractive crystal is an important nonlinear optical medium. Since the photorefractive effect is capable of efficient beam coupling even at a very low level of power and has various promising applications in optical phase conjugate wave generation, optical processing, optical interconnection, and switching, etc., it has attracted much interest among researchers in the past decades [33, 120]. However, a crucial disadvantage of the photorefractive effect is its rather slow temporal response, which is usually larger than the order of seconds. The slow response time greatly limits its applications, particularly in fast optical switching.

Recently, P. Xie and coworkers have proposed and experimentally demonstrated that this problem of slow speed of switching can be overcome by applying an electric field to control the efficiency of photorefractive four-wave mixing [121, 122]. They have experimentally studied the fast switching of single beam propagation through a cerium-doped $BaTiO_3$ photorefractive crystal as well as the consequent phase conjugate generation [123]. The switching is achieved by applying an external voltage. The system is rather simple and is very easy to align in practice. The high efficiency of switching with very weak input power and with the applied voltage smaller than 100 V makes it rather promising in practical applications.

The experimental setup is schematically shown in Figure 6.24. An extraordinary beam of wavelength $\lambda = 514.5$ nm is incident upon the $BaTiO_3$:Ce crystal with an angle of θ_0 outside the crystal, where the angle is defined positive for a clockwise rotation from the surface normal. The doped cerium can enhance the photorefractive effect of $BaTiO_3$, maybe due to the increased density of acceptor impurity. The dimension of the crystal is $6.1 \times 4.0 \times 3.8\ \text{mm}^3$ ($L = 3.8$ mm), with c-axis along the 6.1 mm edge. To apply an electric field to the crystal a pair of indium tin oxide (ITO) transparent electrodes is coated onto the input surface of the crystal and its opposite surface.

At first, P. Xie et al. considered the case when there is no external electric field. Owing to the strong photorefractive effect the fanning beam is amplified at the

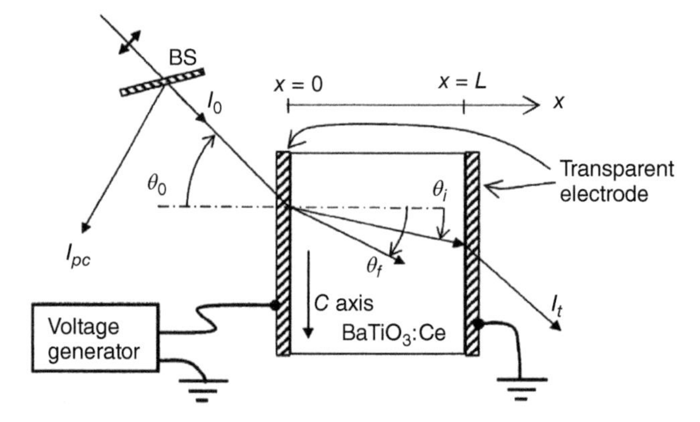

Figure 6.24 Schematic diagram of the experimental setup. BS represents a beam splitter, I_0 is the input power, I_t is the power of the transmitted beam, I_{pc} is the output power of the phase conjugate beam, θ_0 is the incident angle outside the crystal, and θ_i and θ_f are propagation angle of the incident beam and fanning beam relative to the x-axis, respectively, inside the crystal. Source: Xie and Mishima 2005 [123]. Reproduced with permission of Elsevier.

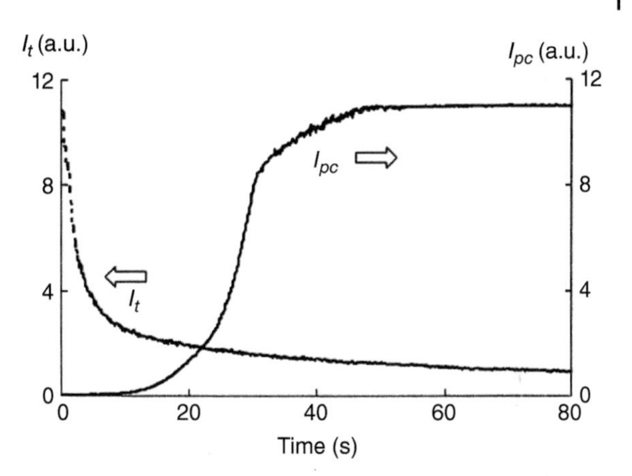

Figure 6.25 Temporal evolutions of the transmitted power I_t and of the output phase conjugate power I_{pc} under no applied electric field. The input power I_0 is turned on at $t = 0$. Source: Xie and Mishima 2005 [123]. Reproduced with permission of Elsevier.

expense of the incident beam power and consequently the counter-propagating phase conjugate beam is generated [124, 125]. Figure 6.25 shows a typical temporal evolution of the power of the transmitted beam at $x = L$ and the output phase conjugate power at $x = 0$ for incident angle $\theta_0 = 30°$, where the input beam is turned on at $t = 0$ and the input power at $x = 0$ is $I_0 = 9.1$ mW. The steady-state output phase conjugate power is $I_{pc} = 1.24$ mW and the steady-state power of the transmitted beam is $I_t = 3.8$ μW. (Owing to the poor transparency of the electrodes, the phase conjugate power inside the crystal is in fact larger than that measured here.) It is seen that due to the slow speed of the photorefractive grating formation the transmitted power (or the phase conjugate power) takes more than 40 seconds to reach its steady state.

After the power reaches the steady state, an electric field E_x is applied to the crystal along the x-axis, as shown in Figure 6.24. Under the field, the refractive indices for the incident and the fanning beams change due to Pockels effect, as follows [121, 122]:

$$\frac{1}{n'^2_\alpha} = \frac{\sin^2\theta_a}{n_0^2} + \frac{\cos^2\theta_a}{n_e^2} - r_{42}\sin 2\theta_a E_x \quad (\alpha = i \text{ or } f) \tag{6.40}$$

where θ_i and θ_f are propagation angles of the incident beam and the fanning beam relative to the x-axis, respectively; inside the crystal, n_i' and n_f' are the refractive indices for the incident beam and the fanning beam in the presence of electric field, respectively, while n_e and n_o are the two principal refractive indices of the $BaTiO_3$:Ce crystal and r_{42} is the component of its electro-optic tensor. Here, for simplicity of analysis, the fanning beam is considered with only one propagation direction, θ_f, which corresponds to its Fourier plane-wave component of the largest amplified intensity. It should be emphasized that owing to the fast speed of Pockels effect (which is usually faster than the order of nanoseconds [126]) the response time of the refractive index change is much faster than that of the photorefractive grating formation/erasure, which is slower than the order of seconds, as can be seen in Figure 6.24. During a short period of time, which is justified in the fast optical switching, the built-in photorefractive grating can be considered to be unchanged. On the other hand, the change in refractive index induced by the applied electric field via Pockels effect, as shown in Eq. (6.40), can result in changes in wave vectors of the incident and fanning beams inside the crystal. The changes in the wave vectors induce the phase mismatch

$$\Delta \boldsymbol{k} = \boldsymbol{k}_g - (\boldsymbol{k}_f{}' - \boldsymbol{k}_i{}') \tag{6.41}$$

for the diffraction, where $\boldsymbol{k}_g = \boldsymbol{k}_f - \boldsymbol{k}_i$ is the built-in grating and $\boldsymbol{k}_i{}'(\boldsymbol{k}_i)$ and $\boldsymbol{k}_f{}'(\boldsymbol{k}_f)$ are the wave vectors of the incident and the fanning beams in the presence (absence) of the external field E_x, respectively, the magnitudes of which are approximately derived as

$$k'_\alpha = k_\alpha \left\{ 1 + \frac{r_{42} \sin 2\theta_\alpha E_x}{2\left[\frac{\sin^2\theta_\alpha}{n_0^2} + \frac{\cos^2\theta_\alpha}{n_e^2}\right]} \right\} \quad (\alpha = i \text{ or } f) \tag{6.42}$$

From Eqs. (6.41) and (6.42) it is easy to see that for the large value of r_{42} in $BaTiO_3$:Ce crystal, when the difference between θ_f and θ_i is large a mediate external field can induce a phase mismatch to reduce diffraction to a great extent. Thus, efficient switching of the transmitted beam and consequent switching of the phase conjugate beam can be achieved. The switching time is the same as the response time of the Pockels effect of the crystal. The switching time cannot be limited by the time of the grating formation/erasure.

Figures 6.26 and 6.27 show some examples of the measured temporal evolution of the power of the transmitted beam and that of the phase conjugate beam with an external *ac* voltage. The experimental conditions are the same as those in Figure 6.25, i.e. with the same incident angle $\theta_0 = 30°$ and input power $I_0 = 9.1$ mW. Figure 6.26 plots the power of the transmitted beam and Figure 6.27 plots the output power of the phase conjugate beam. In Figure 6.26a, at $t < 0.525$ seconds the voltage is off. At $t \approx 0.526$ seconds a 50 Hz square-form voltage is turned on, the high and the low values of which are 0 and −114 V, respectively. The voltage of −114 V corresponds to the electric field of −30 V/mm. In Figure 6.26b, at $t \approx 0.842$ seconds a similar voltage as in Figure 6.26a is turned on but with the low and the high values being 0 and +114 V, respectively. From Figure 6.26a,b it is seen that since the grating has been previously formed under no external field, the transmitted power I_t increases to a larger value when the voltage changes from zero to the other values.

Figure 6.26c, which is the elongation of Figure 6.26b along the time axis, shows that the switching time of the transmitted power is about 30 μs, which is exactly the same as the switching time of the applied voltage that is determined by the voltage generator, not by the time of the grating formation/erasure. This is because the switching time of the voltage generator is slower than the response time of the Pockels effect of the crystal, which is estimated to be of the order of 0.1 ns [126] and therefore, the switching time of the transmitted power is only determined by the voltage generator in the experiment.

Similarly, Figure 6.27a shows the temporal evolution of the phase conjugate output power with a 50 Hz square-form voltage. The high and the low values of the voltage are 0 and −114 V, respectively. Figure 6.27b is with a similar voltage, but with the low and the high values being 0 and +114 V, respectively. Contrary to the transmitted power as shown in Figure 6.26, the phase conjugate power

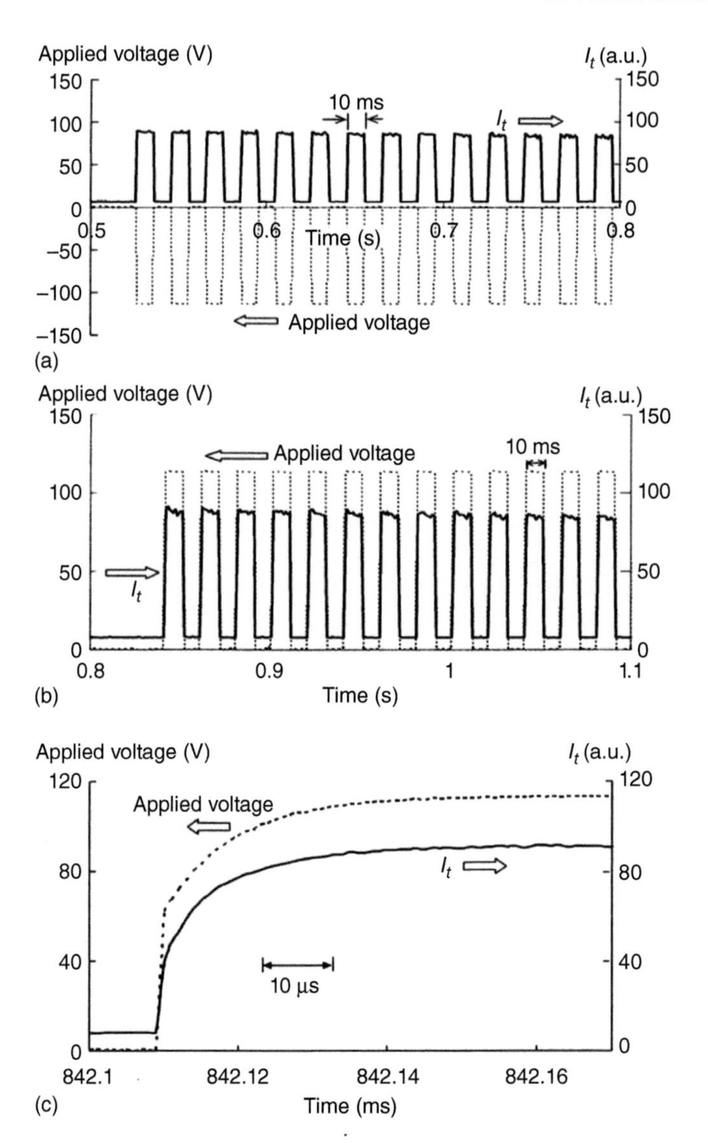

Figure 6.26 (a, b) Temporal evolution of the transmitted power I_t with an *ac* square-form external voltage applied to the crystal after the system reaches the steady state. (c) is the elongation of (b) along the time axis. Source: Xie and Mishima 2005 [123]. Reproduced with permission of Elsevier.

decreases from a high value $I_{pc}(0)$ to a very low value $I_{pc}(\mathrm{V})$ simultaneously as the voltage changes from 0 to ± 114 V as shown in Figure 6.27a,b. The power increases from a very low value $I_{pc}(\mathrm{V})$ to a high value $I_{pc}(0)$ simultaneously as the voltage changes from ± 114 to 0 V, also as shown in Figure 6.27a,b.

To see the efficiency of the switching, $I_t(\mathrm{V})/I_t(0)$ and $I_{pc}(\mathrm{V})/I_{pc}(0)$ are shown as a function of the applied voltage in Figure 6.28a,b, respectively, for various incident angles θ_0. Using the same input power $I_0 = 9.1$ mW for all θ_0 in Figure 6.28

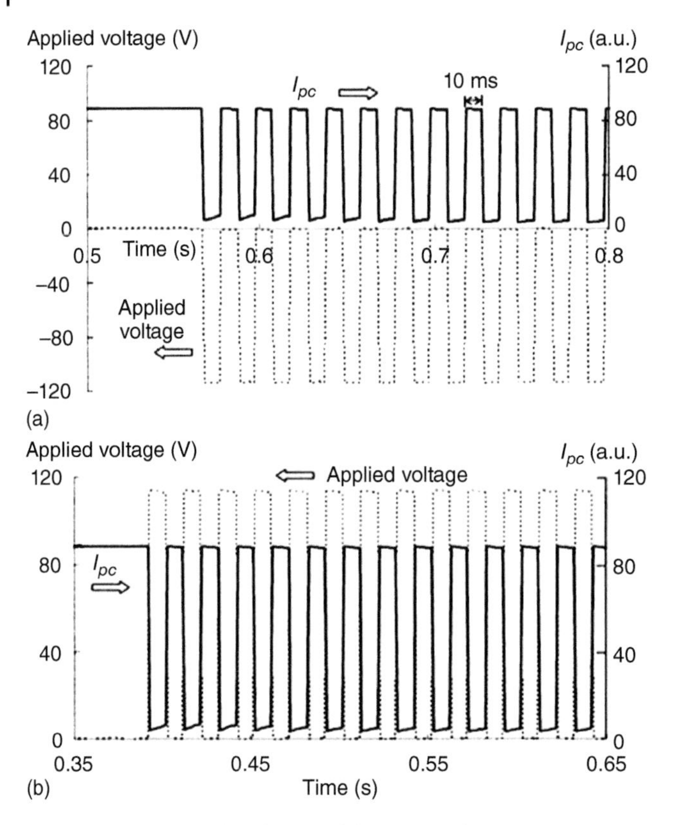

Figure 6.27 Temporal evolution of the output phase conjugate power I_{pc} with an *ac* square-form external voltage applied to the crystal after the system reaches its steady state. Source: Xie and Mishima 2005 [123]. Reproduced with permission of Elsevier.

the measured steady-state output powers of the phase conjugate beam without applied field are $I_{pc}(0) = 1.24\,\text{mW}$ for $\theta_0 = 30°$, $195\,\mu\text{W}$ for $\theta_0 = 2.5°$, and $61\,\mu\text{W}$ for $\theta_0 = -30°$. From Figure 6.28a,b it is seen that the efficiency of switching is nearly symmetric for positive and negative voltages and it increases with increase in the magnitude of the voltage, as can be noted from Eqs. (6.41) and (6.42). From Figure 6.28b it is seen that the switching is more effective for a smaller θ_0 than for a larger θ_0. This can be understood as follows: as θ_0 decreases, the difference between angles θ_f and θ_i increases since in $BaTiO_3$:Ce crystal the energy is mostly transferred to the plane-wave component of the fanning beam with a smaller angle relative to the *c*-axis [124]. From Eqs. (6.41) and (6.42) it is noted that increase in this difference leads to increase in the phase mismatch, which results in more effective switching. From Figure 6.28b it is seen that to reduce the phase conjugate power by 90% the applied voltage required is only about 70 V for $\theta_0 = -30°$, which can be easily implemented in practice as a fast switching device.

6.4.1.4 Holographic Interferometers

Holographic interferometry [127] is a powerful optical method for surface analysis in the field of nondestructive testing, being extremely useful in applications

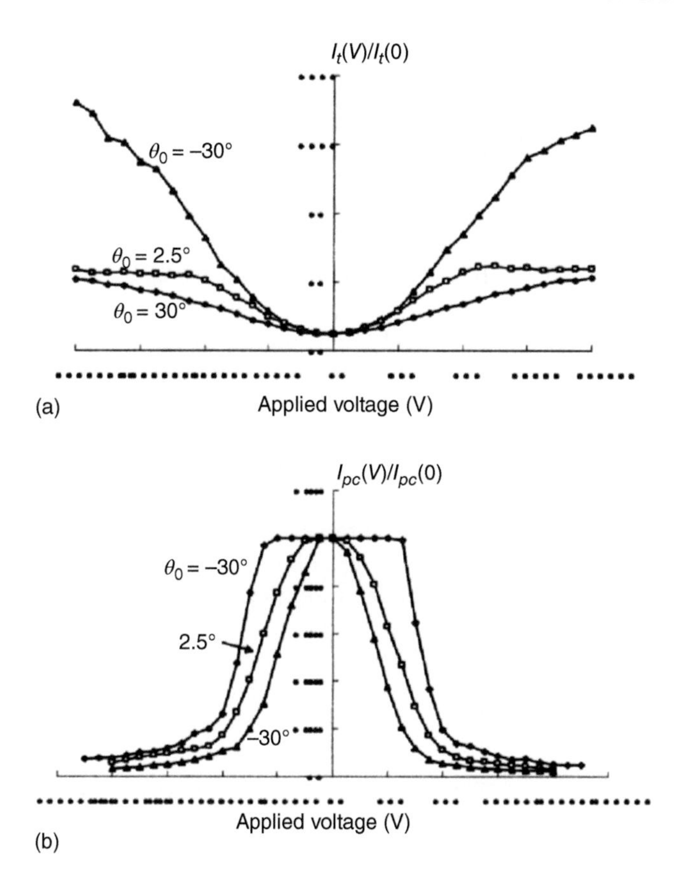

Figure 6.28 (a) $I_t(V)/I_t(0)$ and (b) $I_{pc}(V)/I_{pc}(0)$ as a function of the applied voltage V for incident angle $\theta_0 = 30°$ (lines with filled rhombuses), $\theta_0 = 2.5°$ (lines with unfilled squares), and $\theta_0 = -30°$ (lines with filled triangles). Source: Xie and Mishima 2005 [123]. Reproduced with permission of Elsevier.

in basic research, biomedical, and technological areas. These techniques present advantages in relation the conventional techniques: therefore, do not present any contact with the surfaces, it guarantees absolute reliability to them; besides presenting high accuracy; and also, it allows to make qualitative analyses through visual inspection. However, holographic interferometry techniques in real time with conventional materials (silver halide emulsions, photo-thermoplastics, and others) present serious difficulties – for example, the chemical processing necessary for the holographic film. Photorefractive crystals present as an attractive recording medium. The phenomenon that characterizes these crystals is the photorefractive effect [128, 129]. Such effect consists of refractive index modulation through photoinduction and linear electro-optic effect (Pockels effect), which allows the register volume holograms. These crystals also present advantages as in situ self-proceedings of recording medium and its indefinite reusability, or it does not present fatigue. Therefore, the processes are dynamic and reversible. These properties indicate these crystals for real-time holographic interferometry (RTHI). The photorefractive crystal that was used in this work was $Bi_{12}SiO_{20}$ (BSO) of the sillenite family (as $Bi_{12}TiO_{20}$ [BTO] and $Bi_{12}GeO_{20}$ [BGO]) [129], and presents such properties as good sensitivity and diffraction efficiency, short response time, birefringence, optical activity, and other parameters, besides being indicated for works in diffusive regimen with configuration exhibiting diffraction anisotropy. Many works involving holographic interferometry using photorefractive crystals had been studied through the last decades [130–132].

In RTHI [130, 132], where a hologram is registered, in the reading process the object is illuminated and an interferogram resulting from the overlapping of the diffracted wave front for the hologram and the incoming object is directly observed. Each variation of the object, when this is stimulated, is observed directly in real time, and used in measurements in static processes (static micro-displacements, deformations, tensions and shipments, detention of defects in surfaces, etc.) and dynamic processes (dynamic micro-displacements, dynamic variation of temperature, deformations, dynamic tensions and shipments, etc.), with sequential reading of holographic interferograms. In RTHI with photorefractive BSO crystals, the writing–reading hologram process is made in real time in diffusive regimen, with the configuration exhibiting diffraction anisotropy.

In this work, Gesualdi et al. used a spatial phase measurements technique with phase shift, described by Creath [133] as a phase-shifting technique. The phase of the interferogram was calculated using holographic interferogram intensities that were obtained with the holographic interferometer. The many interferograms captured have had the intensities changed due to phase shift using piezoelectric linear micro-displacements (PZL). If the phase shift is time continuous it will be called phase-shifting; Otherwise, if the phase shift is discrete it will be called phase-stepping. Gesualdi et al. used the second technique, so that a sequence of interferograms is captured, and between each pair of interferograms they introduced a known change in the reference beam, to help in phase determination of each point of the interferogram. In accordance with the amount of interferograms captured they have a phase shift technique. Here, they used the four frames phase shift technique to determine the phase of one point (i,j). In this technique they used the following values of phase shift (θ): 0, $\pi/2$, π, $3\pi/2$. The term that brings the information is the term of the interference, and thus, the values of intensity $I_n(i,j)$ in each point (i,j) of the interferogram for each change of phase can be written as

$$I_n(i,j) = I_0(i,j)\cos\left[\phi(i,j) + \frac{(n-1)\pi}{2}\right] \quad \text{where } n = 1, 2, 3, \text{and } 4 \tag{6.43}$$

Using the trigonometric relations and combining the intensities to obtain the phase $\Phi(i,j)$

$$\phi(i,j) = \arctan\left[\frac{I_4(i,j) - I_2(i,j)}{I_1(i,j) - I_3(i,j)}\right] \tag{6.44}$$

When the phase $\Phi(i,j)$ of each point (i,j) can build a 2D graphic where the (i,j)-coordinate identifies a point on the xy-plane and each value of phase as a gray level intensity in point on the xy-plane, changing between white and black.

The black corresponds to $-\pi$ and white to $+\pi$, and the gray levels in between correspond to values between $-\pi$ and $+\pi$, having 256 gray levels for a system of images of 8 bits. To remove the phase wrapping $\Phi(i,j)$ of the phase map, for which the values of the tangent function are between $-\pi$ and $+\pi$, it is not possible to distinguish the values of phase that exceed this range. This wrapping appears in the phase map at areas where the changes of phase are presented. It appears like a border from white to black or black to white. This change corresponds to a variation from $+\pi$ to $-\pi$, or $-\pi$ to $+\pi$. This change is known as phase jump. When this wrapping occurs, it is called the wrapped phase; else, when the wrapping is removed it is called the unwrapped phase. This process of removing the wrapping is called unwrapping. There are many works on unwrapping phase techniques [134]. Ghiglia et al. used the cellular automate technique [135], which consist in removal of wrapping for phase integration of the near points. This technique is path independent and not much sensitive to noise.

In the experimental setup to determine some characteristic figures of merit of photorefractive BSO crystals used in this work, Gesualdi et al. evaluated the conditions under which BSO crystals presented better performance for real-time holographic interferometer in diffusive regimen, as shown in Figure 6.29 [136]. Thus, they analyzed the results, and the best ones resulted with measurements in holographic gratings for angles in the range 40° and 50°; the ratio of intensities between reference and object beams was $I_R/I_S = 6.0$, which was well characterized by Gesualdi [137]. So, with the PZT calibrated and interferometer optimized, the experimental procedure used was to capture the holographic interferograms with the software Global Lab and Captura; calculate the phase map and remove the wrapping of phase map with software of calculate built with software Borland

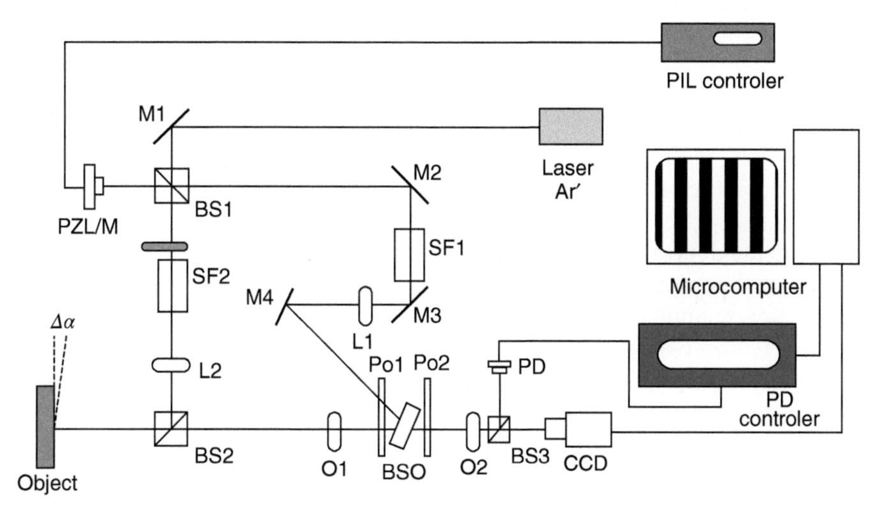

Figure 6.29 Real-time holographic interferometry with photorefractive BSO crystals setup where the light source is an argon laser ($\lambda = 5145$ nm); M1, M2, M3, M4 are mirrors; BS1, BS2, BS3 are beam splitters; SF1, SF2 are spatial filters; L1, L2 are lens; O1, O2 are objectives; Po1, Po2 are polarizers; PZL/M is the piezoelectric translator + mirror; PD is the photodetector; BSO is the photorefractive crystal; and CCD is the camera for capture of images. Source: Gesualdi et al. 2003 [136]. Reproduced with permission of Annals of Optics.

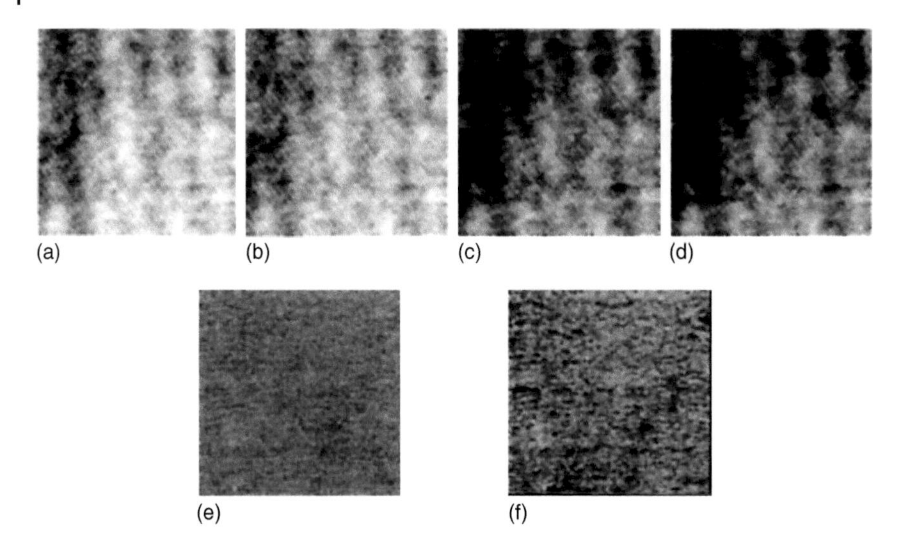

Figure 6.30 Holographic interferograms of a mirror static: (a) $\theta = 0$, (b) $\theta = \pi/2$, (c) $\theta = \pi$, (d) $\theta = 3\pi/2$. Phase map: (e) wrapped and (f) unwrapped. Source: Gesualdi et al. 2003 [136]. Reproduced with permission of Annals of Optics.

C^{++} and the functions of the Data Translation Global Lab; and plot and analyze the surfaces with software Mathcad. To calculate the phase map and to execute the unwrapping process, Gesualdi et al. used the programs of microcomputer that were developed by Soga [138].

Gesualdi et al. presented the results obtained from measurements on the surfaces of opaque objects; in each figure, there are the four holographic interferograms digitalized, the phase map, and the unwrapped phase map. In Figure 6.30, they studied a mirror static ($\Delta\alpha \cong 0$) that presents a phase map without any jump phase that was expected. In Figure 6.31, they studied a mirror slopping ($\Delta\alpha \cong 0.4$ mrad), where the phase map presents some phase jump that shows the variations of phase; the phase jump appears in the vertical direction as the axis of the slope is in the vertical direction. The phase map obtained is consistent with the displacements done on mirrors and coins.

Zamiri et al. reported a low-cost and high-accuracy interferometric technique for detecting nanometer vibrations by using a photorefractive crystal interferometer based on two-wave mixing within a $Bi_{12}SiO_{20}$ (BSO) crystal [139]. The results of small displacement detection on the sample (1–2 nm) and comparison between the sensitivity of Michelson and photorefractive adaptive interferometers were presented.

In the experiments, ultrasound waves with different frequencies and amplitudes are generated with a piezoceramic transducer (PZO, $2 \times 5 \times 5\,mm^3$). The transducer has a resonance frequency of 500 kHz and a maximum stroke of 3 μm/150 V. As sample, a small circular mirror of diameter 4 mm connected to PZO is used. In Figure 6.32, an image of the PZO structure can be seen.

By applying sinusoidal voltages (0.05–5 V) on the PZO, displacements in the order of nanometers with frequencies between 1 Hz and 70 kHz are generated.

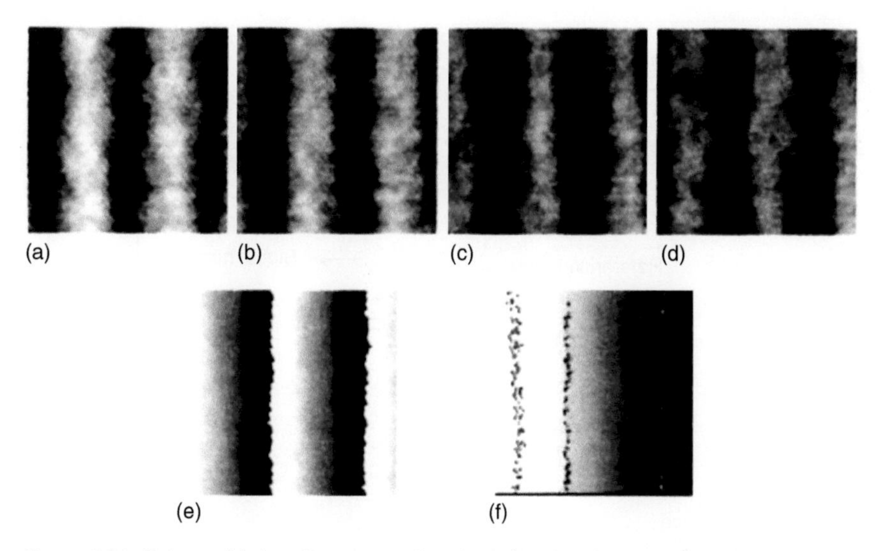

Figure 6.31 Holographic interferograms of a mirror slopping: (a) $\theta = 0$, (b) $\theta = \pi/2$, (c) $\theta = \pi$, (d) $\theta = 3\pi/2$. Phase map: (e) wrapped and (f) unwrapped. Source: Gesualdi et al. 2003 [136]. Reproduced with permission of Annals of Optics.

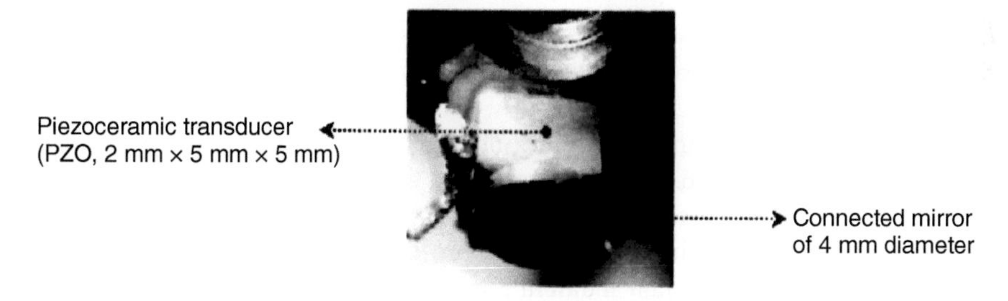

Figure 6.32 Image of PZO and the connected mirror. Source: Zamiria et al. 2010 [139]. Reproduced with permission of Elsevier.

To measure the vibrations, Zamiri et al. used a Michelson and a photorefractive crystal interferometer. For the latter, a 532 nm detection laser beam is divided into reference and signal beams [140]. The signal beam is focused on the sample surface. By surface vibrations, the reflected beam gets modulated and interferes with the reference beam in a BSO (bismuth silicon oxide: $Bi_{12}SiO_{20}$) photorefractive crystal with a [97] crystallographic axis. The crystal has a size of $5 \times 5 \times 5\,mm^3$. An external electric field is applied in the [001] direction by evaporated gold contacts. It is notable that reversing the applied electric field direction will change the sign of the photorefractive gain. Silver paste was used to contact wires from a high voltage power supply to the gold electrodes on both surfaces. Zamiri et al. used this BSO crystal because of its high photorefractive gain, its fast response time, and reasonable cost.

Owing to the photorefractive grating generated in the crystal, the reference beam gets diffracted in the direction of the signal beam. By interfering both beams on a photodiode, one can measure the phase shifts via amplitude

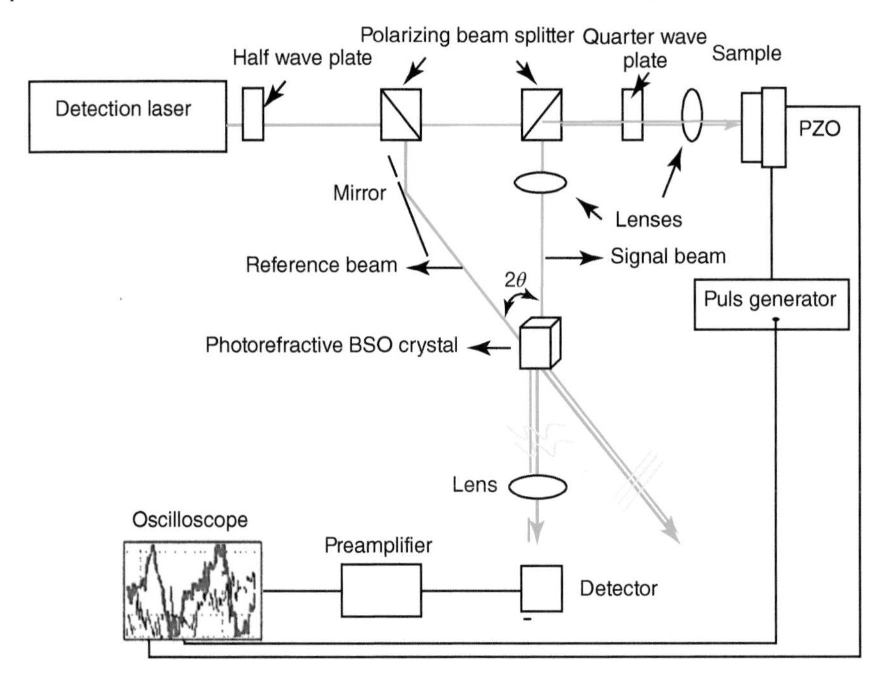

Figure 6.33 Schematic of the photorefractive interferometer based on the two-wave mixing in a BSO crystal. Source: Zamiria et al. 2010 [139]. Reproduced with permission of Elsevier.

modulations of the laser intensity and thus the ultrasonic displacements on the sample surface. In Figure 6.33, a simple schematic of the interferometer is shown. They optimized this interferometer by changing the signal and reference beam intensities ratio ($R = 10$), their beam diameter on the crystal ($A_{signal\text{-}in} = 0.03\,\text{cm}^2$, $A_{reference\text{-}in} = 0.2\,\text{cm}^2$), and their incident angle ($2\theta = 10°$). It is worth noting that all the physical parameters (beam spot size on the sample, laser beam power on the detector, detector type, and object beam power) for the photorefractive interferometer and the Michelson interferometer are chosen similarly.

Zamiri et al. found a flat frequency response for Michelson interferometer. In the case of the adaptive interferometer, the response frequency is approximately constant for high frequencies (1–40 kHz) while for frequencies lower than 1 kHz (f_{cut}) the sensitivity is decreased. By using the adaptive interferometer and applying a high voltage of 1.5–2 kV across the BSO crystal, they found a minimum detectable displacement (sensitivity) of 1–2 nm. In Figures 6.34 and 6.35, one can see the PRC frequency response and dependence of the SNR on the vibration amplitudes at 20 kHz for both types of interferometers, respectively.

The photorefractive crystal interferometer shows in comparison to the classical homodyne (Michelson) interferometer with the same physical parameters chosen, four to five times less sensitivity for mirror-like surfaces. However, for highly rough industrial specimens the sensitivity of the photorefractive crystal interferometer is much higher than that of the Michelson interferometer. By using a simple photorefractive crystal interferometer small displacements and vibrations of the order of a few nanometers on the surface of specimens were detected.

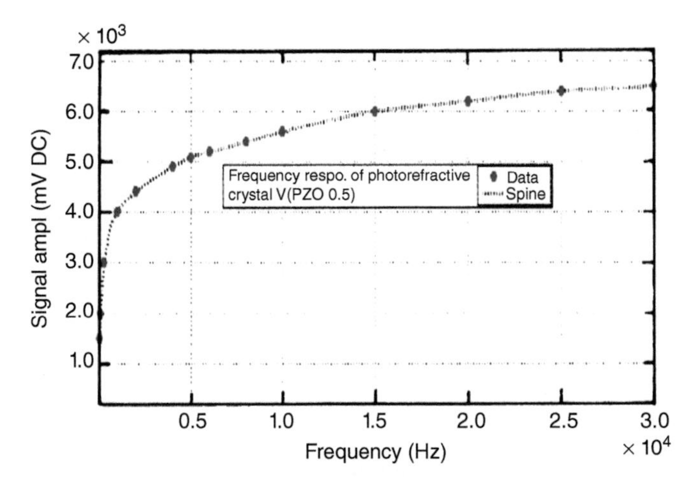

Figure 6.34 Frequency response of the BSO photorefractive crystal at a laser intensity of 0.1 W/cm^2. Source: Zamiria et al. 2010 [139]. Reproduced with permission of Elsevier.

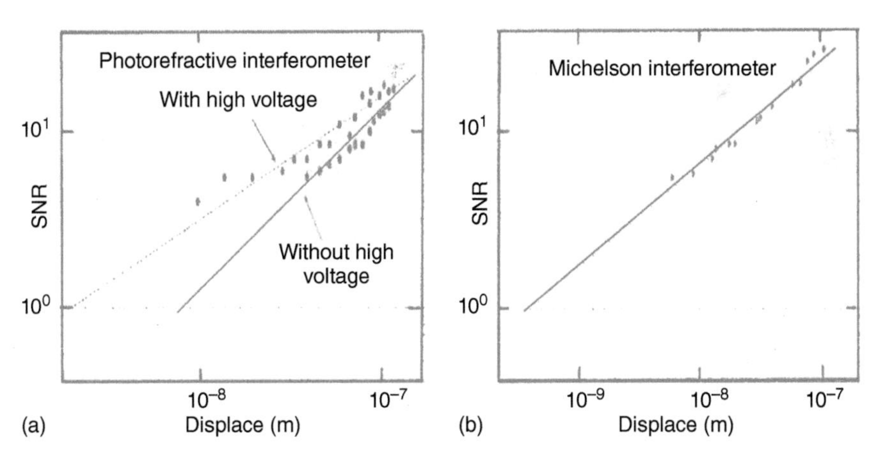

Figure 6.35 Comparison of the sensitivity of a photorefractive (a) and Michelson interferometer (b) on a mirror surface at 20 kHz. Source: Zamiria et al. 2010 [139]. Reproduced with permission of Elsevier.

The sensitivity of the interferometer can be improved by applying moderate external electric fields of 4 kV/cm along the crystal. These results show the potential of such interferometers for nondestructive testing in industrial applications.

6.4.2 Photorefractive Polymers

The photorefractive effect in polymers was first discovered in 1991 [141] and since then there has been a considerable amount of research dedicated to these materials. Photorefractive (PR) polymers or photopolymers [141–144] are a promising media for providing updatable or rewritable holograms due to high photosensitivity, low cost, and high processability. Thus, using PR polymers, updatable real-time three-dimensional holograms can be recorded and simultaneously reconstructed or replayed. Some important applications of PR polymers are described below.

6.4.2.1 Holographic Display

Holographic displays have intrigued scientists and the public alike with vivid three-dimensional images appearing to hover in mid-air. Applications range from artistic installments and eye-catching advertisement platforms to holographic television and displays. The flexibility of the photorefractive polymer system lends itself to large display sizes and response characteristics tunable for each proposed application. Holographic recording and display requires a photorefractive film that has a high sensitivity and fast response time in order to record the wave front as quickly and efficiently as possible. While the device must be optimized to exhibit high diffraction efficiency for bright image reconstruction, photorefractive gain is not an important parameter in these applications. The performance of the PR material can be adjusted to match the given application. For video rate displays, short image persistence and pulse laser response are key metrics. For static or slowly changing applications, image brightness and long persistence are the critical metrics.

Long persistence holograms are useful in cases where the same image or scene will be viewed for minutes to hours. Increasing the glass transition temperature of the photorefractive compound is one way to achieve high-efficiency systems that are stable on this time frame, requiring that the temperature of the media be raised above this temperature for recording and then be subsequently cooled. The application of an external field after cooling is also not required to maintain the overall alignment of the chromophores. Li and Wang [145] recorded diffraction grating persistence times of more than two hours in their unplasticized system based on a TPD:carbazole:DCST functionalized matrix doped with 7-DCST. With a T_g of 43 °C, local preheating of the film for 1.5 minutes using a CO_2 laser allowed them to update the film, achieving a peak diffraction efficiency of 92% and higher than 72% over the first five minutes. Tay et al. [146] recorded a holographic stereogram into a 4″ × 4″ plasticized PATPD-CAAN:FDCST:ECZ film at room temperature T_g. They were able to achieve both fast recording and hours of persistence by writing the hogels at 9 kV and then dropping the voltage to 4 kV in a process called voltage "kick-off." The higher writing electric field enhances the recording dynamics of the composite, while the lower reading voltage maintains the grating without contributing as significantly to its decay. This process allowed them to increase the diffraction efficiency from the 1.5% they obtained by recording for 0.5 seconds at 4 kV to 55% for the same recording time at 9 kV.

For applications in which high-speed updating is necessary, the ability to quickly record and rewrite the grating is necessary. Tsutsumi et al. [147] fabricated a 1.8 cm^2 area device based on PVK:7-DCST:CzEPA:TNF. This film exhibited 68% diffraction efficiency at 45 V/μm and a weighted average rise time of 59 ms (1.5 W/cm^2 writing power). They recorded real object holograms as well as a two-dimensional image using an SLM displaying an image, requiring one second per exposure. Tsujimura et al. [148] also recorded a two-dimensional image displayed by an SLM into a PDAS:FDCST:ECZ:PCBM film. While the steady-state diffraction efficiency of their system only reached 35% at 45 V/μm, they were able to refresh their image in only 50 ms.

Figure 6.36 Multicolor holographic stereograms [149, 150].

Blanche et al. [149] continued the work by Tay et al. adding a sensitizer to the PATPD-CAAN composition (PATPD-CAAN:FDC ST:ECZ:PCBM) and implementing a 6 ns pulsed laser that delivered 200 mJ per pulse at a repetition rate of 50 Hz. Each hogel was written with a single pulse, shortening the writing time of the entire 4 × 4″ display to two seconds. This high-speed writing enabled the integration of a "telepresence" system, taking live feed from 16 cameras of a speaker and transmitting it to the holographic printer. Additionally, multicolor reconstruction was performed by angularly multiplexing separate holograms written for each wavelength, the results of which are shown in Figure 6.36. Recent developments in their work have been the exhibition of a 12 × 12″ active area device, three-color holograms with a color gamut exceeding that of the standard HDTV, and display brightness of 2500 cd/m^2 [151].

Computer generated holograms (CGHs) have recently been written in a photorefractive polymer device by Jolly [152]. Instead of holographically recording the necessary views as in holographic stereography, CGHs are calculated by computationally interfering an object and reference beam and recording the resulting fringe pattern directly. This is performed through a process known as direct fringe writing in which the fringes are displayed on an SLM and imaged onto the film where the intensity distribution is recorded. The benefit of this method is that instead of approximating the object wave front by a series of perspectives, the CGH is able to directly reproduce the object wave front within the limitations of the recording resolution. To display the grating with sufficient resolution to capture an analog approximation to the phase grating requires many pixels on the SLM and demagnification of its image onto the PR film to achieve the necessary diffraction angles. The limited number of available pixels requires a balance between the fidelity of the grating and the extent of the image.

6.4.2.2 Holographic Autocorrelator

Laser pulses on the order of femtoseconds, called ultrashort pulses, are difficult to characterize as their time evolution occurs significantly faster than standard semiconductor detecting devices can respond. One method by which the temporal extent of these pulses can be measured is by using a technique called autocorrelation. In a photorefractive autocorrelator, the signal beam is split into two paths and one arm is sent through a variable delay device so that when the two beams are recombined in the photorefractive media, their peak intensities are temporally offset. The effect of this delay is that the modulation of the grating, and therefore the efficiency with which the reading beam is diffracted, becomes a function of delay time and the autocorrelation function of the pulse can be recorded [153, 154].

Nau et al. [155] implemented such a device using an NIR sensitized PVK:DMNPAA/MNPAA:ECZ:TNFM photorefractive film at voltages up to 50 V/μm. They successfully measured the temporal characteristics of 130 and 13 fs pulses centered at 800 nm with diffraction efficiencies of around 2%. Since the temporal confinement of a bandwidth-limited pulse is inversely related to the range of optical frequencies or the bandwidth it carries, short pulses are composed of many wavelengths, each of which records its own grating within the media. In testing the spectral characteristics of the recorded hologram, this group verified the ability of the photorefractive polymer to accurately record pulses in this time range.

6.4.2.3 Laser Ultrasonic Receiver

Laser ultrasound is an optical metrology technique that is used for nondestructive testing and is implemented in many industrial settings. It is especially useful in the interrogation of rough surfaces, a case in which many other optical based interrogation methods have difficulties due to scatter. The ultrasound in these systems is often created by high peak power Nd:YAG or Excimer pulse lasers causing localized thermal expansion or ablation. The ultrasound then travels along the surface and through the bulk of the material under test until it reaches a defect or edge, at which point it scatters or reflects. The ultrasound is measured optically using a laser-based interferometric system to map displacements of the material surface.

Zamiri et al. [156] implemented a PVK:AODCST:BBP:PCBM photorefractive film device as an adaptive beam combiner in a homodyne interferometer configuration. The gain of the device was on the order of 70 cm^{-1} at 50 V/μm with a response time of 60 ms. Using this system, they were able to measure surface displacements as small as 0.2 nm caused by back surface and defect reflections of the ultrasound traveling at thousands of meters per second. Based on the time of travel from each of these locations and the characteristic surface and bulk acoustic wave velocities of the material, their locations relative to the ultrasound generation and the detection beam location can be accurately determined.

6.4.2.4 Ultrasound-Modulated Optical Tomography

Time reversed ultrasonically encoded (TRUE) optical focusing is a technique developed by Xu et al. [157] to dynamically direct light inside scattering media. An ultrasonic beam is focused inside the sample to converge the incident irradiation to the point of interest within the media. Interaction with the modulated ultrasonic waves of frequency f_a alters the initial optical frequency of the scattered beam (f_o) by frequency shifting it to ($f_o + f_a$) and ($f_o - f_a$). By pre-shifting the sample beam by f_a, upon transmission part of the scattered beam will return to the original laser frequency. At this point, the scattered light is focused onto a photorefractive medium along with a reference beam from the same laser and a hologram is recorded. A counter-propagating probe beam is used to read the

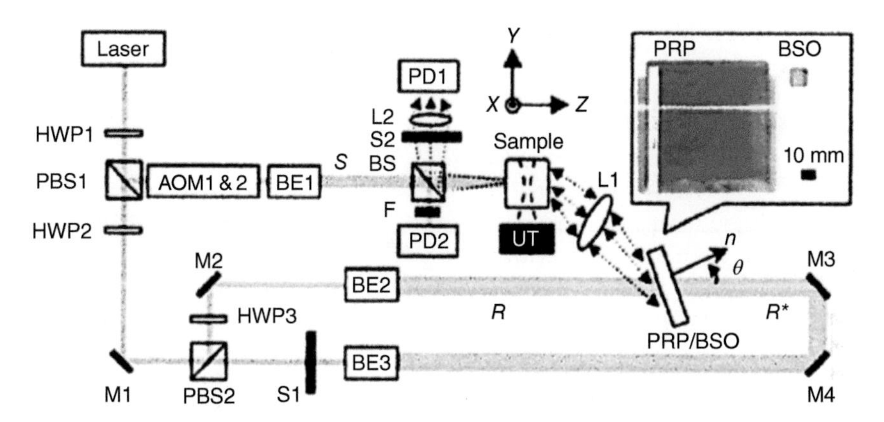

Figure 6.37 Schematic of the optical tomography testbed. The inset image compares the size of a photorefractive polymer (PRP) device to a BSO crystal. Source: Suzuki et al. 2012 [158]. Reproduced with permission of Journal of Biomedical Optics.

hologram, and the signal is sent back through the sample in the same path that it initially traveled and is detected on the other side. The sample is translated such that the focus moves through the sample and a depth resolved image is formed of the media. The initial implementation by Xu et al. implemented a photorefractive crystal, $Bi_{12}SiO_2$ (BSO), as the phase conjugation device but the small sizes of these crystals limited the etendue of the capture system.

Suzuki et al. [158] integrated a PATPD-CAAN:FDCST:ECZ photorefractive polymer into this system, with device performance characteristics of 90% diffraction efficiency at 4 kV and 200 cm^{-1} of gain at 5 kV. They were able to increase the focused optical energy within the media 40-fold over the BSO crystal system. A representative measurement system is illustrated in Figure 6.37 as well as a comparison between the active areas of the photorefractive polymer device (PRP) and the BSO crystal. In Figure 6.38, the sample under test is shown along with the TRUE imaging results. The DC data is the signal without the ultrasonic focusing or time reversal and the TRDC signal is that without the ultrasonic focusing but incorporating time reversal (phase conjugation). This system was able to image through media up to 120 mean free photon paths, or 120 times the distance that an average photon travels through the media unperturbed [158, 159].

A subsequent implementation by Suzuki et al. [160] removed the time reversal component of the system, instead directly amplifying the signal beam using two-beam coupling gain. This system, while limited to imaging distance of ~94 mean free paths, was significantly less complex. Lai et al. [161] incorporated a fiber-optic bundle to receive the scattered signal, allowing the signal gain and detection to be easily removed from the sample location.

6.4.2.5 Holographic Optical Coherence Imaging

Coherence gated holographic imaging is a medical imaging tool that is used to create depth maps of biological tissues and features. This type of system uses the short coherence length of ultrashort pulse lasers in order to selectively record holographic images throughout the depth of the sample. When the probe beam

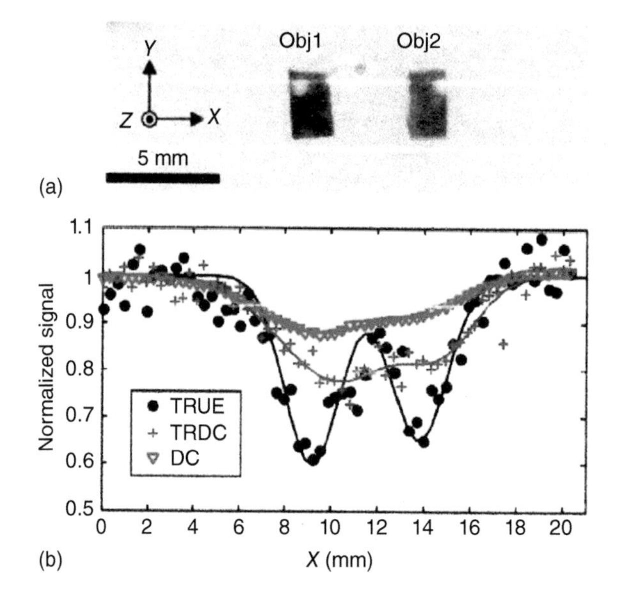

Figure 6.38 (a) Image of the sample under test. (b) The normalized signals detected as a function of spatial position within the sample. TRDC refers to the case in which the reference and object beams are of the same optical frequency, and DC is the direct signal without frequency shifting or amplification. Source: Suzuki et al. 2012 [158]. Reproduced with permission of Journal of Biomedical Optics.

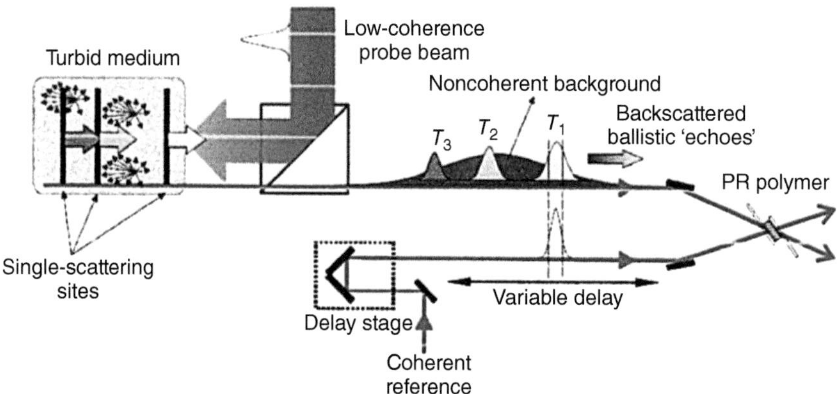

Figure 6.39 Illustration of the concept of coherence gated holographic imaging using a PR device as a coherence filter. Source: Salvador et al. 2009 [162]. Reproduced with permission of OSA Publishing.

backscatters from an entity within the medium under test, it can undergo additional scattering events, at which point it loses its coherence, or it can exit the material. The time that it takes for the probe beam to exit the sample is related to the distance traveled and is used to correlate the measured signal to the depth from which it was taken. In order to selectively measure the signal from a specific location, the scattered beam is interfered with a coherent reference beam in a holographic recording medium as shown in Figure 6.39. By delaying the pulse of the reference beam, the signals from various depths will be holographically recorded.

Salvador et al. [162] implemented a TPD:DMNPAA/MNPAA:PCBM photorefractive film device as the recording media. This film was characterized by a near

100% steady-state diffraction efficiency with *cw* illumination and an initial rise time 250 times faster than their PVK:TNFM comparison sample. By imaging the diffracted reading beam onto a CCD camera to record the spatial map and varying the delay of the reference beam, they were able to image a 600 μm thick tumor with 12 μm axial resolution and 8 μm lateral resolution. One of the main problems they encountered was the small coherent beam overlap region within the sample, as the size of this region is related to the coherence length of the signal pulse and the inter-beam angle on the holographic recording media. This limits the field of view that can be recorded in the plane of incidence to a fraction of that which is recorded perpendicular to the plane of incidence. A method by which they were able to mitigate this effect, increasing the horizontal field of view by a factor of eight, was by optically demagnifying the image beam prior to interference in the photorefractive polymer device [163].

6.4.2.6 Surface Waveguide

Guiding high intensities of light along the surface of a media is often desired in applications such as surface spectroscopy and microscopy or other surface waveguide applications. Photorefractive materials have the ability to self-waveguide along a surface, experiencing total internal reflection (TIR) at the electrode surface and self-bending within the bulk of the media as the light changes the refractive index. The efficiency with which this process can occur depends not only on the gain of the material, but also on its ability to amplify scattering, or beam fanning. Ren et al. [164] performed a theoretical analysis of this type of surface wave and found that the depth that the optical field penetrates into the bulk decreases with increasing applied electric field as well as for increasing refractive index modulation magnitude. While the self-guiding effect has been found in photorefractive crystals, this externally controllable wave front profile is unique to the photorefractive polymer implementation.

Fujihara et al. [165] demonstrated surface waveguiding in photorefractive polymers in PVK:PDCST:BBP:C60, as it had been previously shown to exhibit efficient beam fanning, or amplified scattering. The polymer film was placed between two electrodes as usual but in this implementation, the refractive index of the substrate on one side matched that of the polymer to facilitate light injection while the other electrode substrate had a lower index to provide the necessary surface for TIR [165]. Measurements of the buildup of the waveguide effect with respect to pump power show faster and stronger waveguiding with increasing pump power. The results of these measurements are shown in Figure 6.40 as is an image of the waveguide output in the figure inset.

6.4.3 Holographic Polymer Dispersed Liquid Crystal (H-PDLC)

An H-PDLC is a photopolymer mixed with LC. During the grating formation inside the photopolymer–LC mixture, photopolymerization occurs in the bright regions faster than in the dark regions. While the monomer diffuses to the bright regions, the LC molecules diffuse to dark regions [167]. After the final uniform curing, the H-PDLC composite system consists of alternating layers of polymer planes and LC-rich droplet planes. If the refractive index of the polymer is

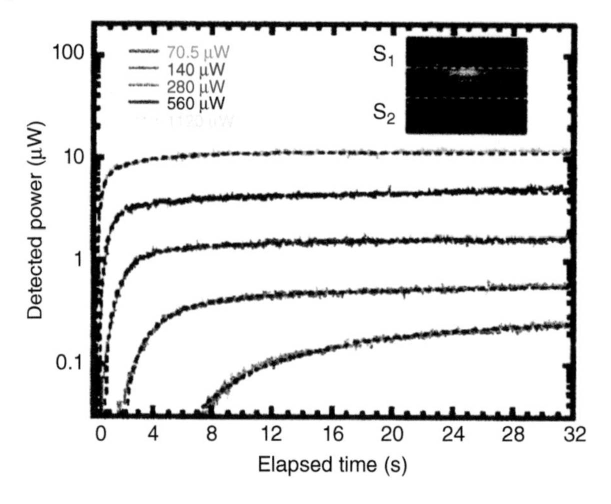

Figure 6.40 Detected power at the end of the waveguide as a function of time for various pump powers. Inset shows an image of the waveguide end with the dotted lines illustrating the interfaces between layers. Source: Fujihara et al. 2012 [166]. Reproduced with permission of OSA Publishing.

matched with one of the principal refractive indices (n_o or n_e) of the LC but not the other, the grating recorded inside the H-PDLC can be switched on or off by an electrical field that changes the orientation of the LC molecules. Using this technique, different types of devices have been designed, and some of them are described here.

6.4.3.1 Wavelength Switch

Optical switches with wavelength selectivity are important and useful in WDM systems, especially in network reconfiguration. One example of such devices is a switchable add/drop filter which is capable of switching between all-through state and wavelength adding (or dropping) state. The building block of these switches is a 2×2 switch. Most of the existing 2×2 switches, such as those involving two prisms operating in TIR mode or total transmission mode, require light being coupled in and out of the fiber resulting in large device size and high insertion loss.

In 2003, C. Gu et al. have designed and demonstrated a novel wavelength-selective 2×2 switch by recording electrically switchable holographic gratings in a layer of H-PDLC sandwiched between two-side polished fibers [168]. This device provides in-line operation capability and is particularly suitable for WDM network reconfiguration. A schematic drawing of the basic structure of the 2×2 wavelength switch is shown in Figure 6.41. Two fibers are partially cut through their claddings by side polishing. The two polished sides are coated with ITO

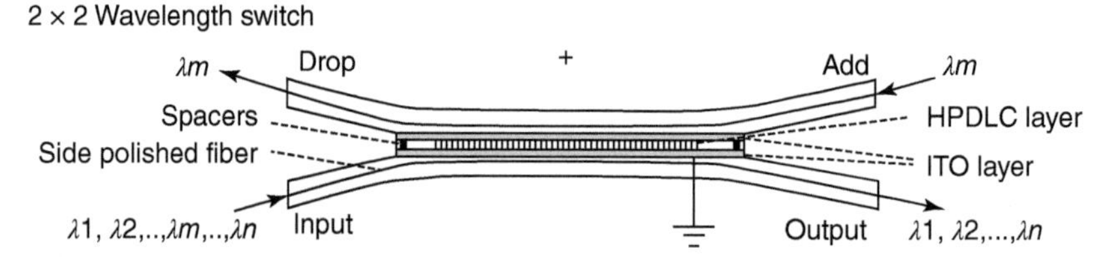

Figure 6.41 Structure of the 2×2 wavelength switch. Source: Gu et al. 2003 [168]. Reproduced with permission of Elsevier.

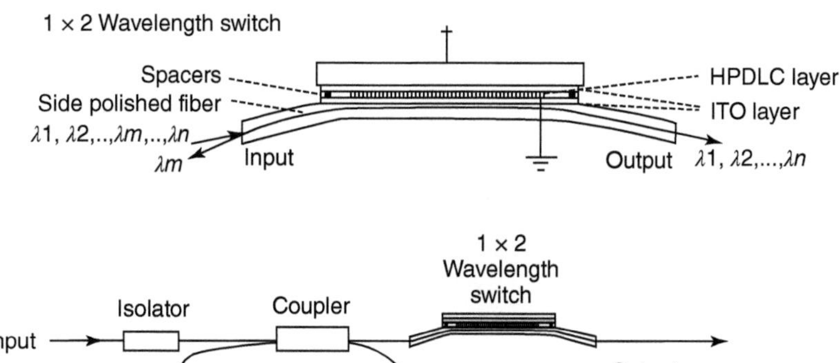

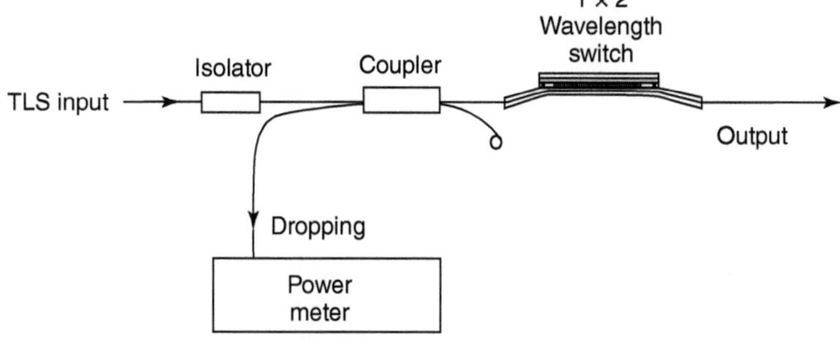

Figure 6.42 The 1 × 2 wavelength switch and the experimental setup for demonstrating the device operation. Source: Gu et al. 2003 [168]. Reproduced with permission of Elsevier.

electrodes. Spacers are placed between the two side-polished fibers to form a cell that is then filled with H-PDLC. A holographic grating can be recorded in the H-PDLC layer by interfering two plane waves from a laser [169]. At the field-on state, the grating is switched off and light will propagate through each fiber without any coupling, which provides the all-through state. When there is no electric field, the grating is switched on and the incident light at the INPUT port will be reflected back and come out from the DROP port, which provides the wavelength dropping state. Meanwhile, the incident light from the ADD port will be reflected into the OUTPUT port, which provides the wavelength adding state.

To provide a proof of concept demonstration, C. Gu et al. first used one side polished fiber with a H-PDLC cell built on top of it. The structure is shown in Figure 6.42. The grating in the H-PDLC layer is recorded by interfering two Ar beams. The angle between the two interfering beams (2θ) is calculated according to $\sin\theta = \overline{n}\lambda_{Ar}/\lambda$, where λ_{Ar} is the wavelength of the Ar laser (488 nm), λ is the wavelength to be reflected, and $\overline{n}$ is the mode index of the corresponding wavelength. To measure the reflected wave, they placed a coupler at the input side as shown in Figure 6.42.

The results of the experiment are shown in Figure 6.43. By adjusting the angle between the two writing beams during the holographic recording they obtained a reflection peak at 1548 nm at the field off state. The 3 dB bandwidth of the reflection peak is about 3 nm. Adjusting the angle between the two writing beams can control the peak wavelength of the reflection. When they apply the electric field, the holographic grating is switched off and an extinction ratio of more than 25 dB has been achieved.

Further optimization of the experimental condition is expected to improve the results significantly in terms of diffraction efficiency and bandwidth. By reducing the refractive index of the H-PDLC layer, the insertion loss can be minimized.

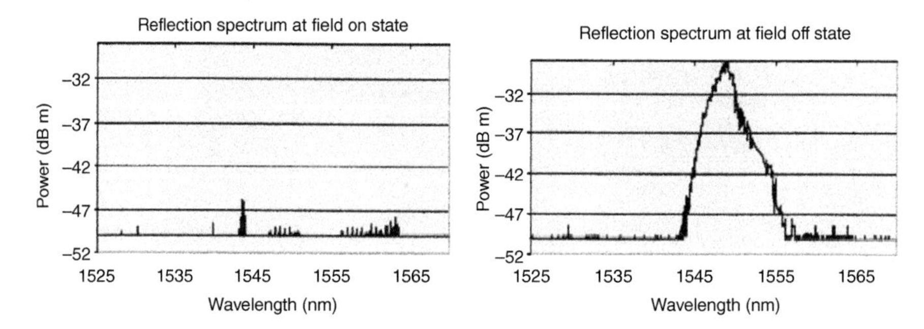

Figure 6.43 Reflection spectrum measured at field-on and field-off states of the switch. Source: Gu et al. 2003 [168]. Reproduced with permission of Elsevier.

All of the side-polished fibers have a polishing length of 10 mm. After it is polished to about 5 μm near the core, by using the special polishing technique, the insertion loss is often less than 0.2 dB. By applying a low refractive index material on it, it should be possible to compensate the loss at the polished spot. In addition, by adjusting the grating parameters, such as grating length and apodization, etc., the reflection spectrum can be modified to achieve the required bandwidth and line shape.

6.4.3.2 Electrically Switchable Cylindrical Fresnel Lens

Many approaches have been developed to fabricate Fresnel lenses by different materials; among them the electrically controllable liquid crystal Fresnel lens [170–175] is the most promising because of the good electro-optical properties of liquid crystals due to its molecular orientation in an external electric field. By combining liquid crystals with polymers and employing holographic techniques, a new composite material called H-PDLCs [176] has emerged, and electrically switchable holographic diffractive gratings [177–180] for various applications, using photopolymerizable liquid crystals polymer composite, have been demonstrated. Recently, Jashnsaz et al. have reported holographic techniques to fabricate an electrically switchable LC/polymer Fresnel lens using Fresnel pattern resulting from the superposition of a plane wave and a spherical wave front [181, 182]. Also, they proposed and fabricated all optical switchable holographic Fresnel lens using these techniques [183]. In this work, Jashnsaz et al. [183] fabricated and characterized an electrically switchable cylindrical Fresnel lens with focusing property in only one direction using an astigmatism lens in one path of Michelson interferometer. This is the first study to propose and fabricate an electrically switchable one-dimensional Fresnel lens.

The consequence of superposition of plane wave and spherical wave front is a set of concentric bright and dark rings of Fresnel pattern, and irradiating an LC/polymer composite film with this pattern yields a concentric Fresnel lens that focuses light in two dimensions and results in a point image (Figure 6.44a) as discussed completely in the works [181, 182]. If one interferes a plane wave with a cylindrical wave front instead of spherical wave front, the result of interference is a cylindrical (one-dimensional) Fresnel pattern instead of concentric (two-dimensional) bright and dark rings, as shown in Figure 6.44b.

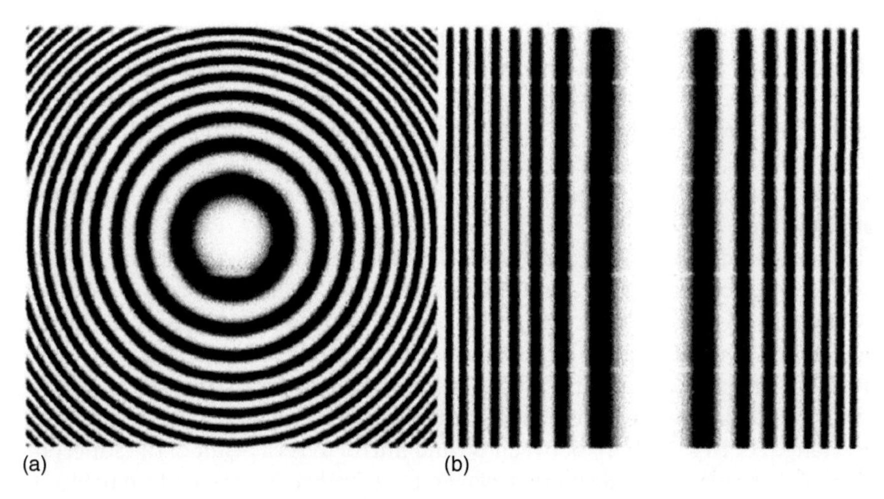

Figure 6.44 Fresnel zone plates; (a) concentric zone plate acts as a two-dimensional Fresnel lens and focuses light to a point image and (b) cylindrical zone plate acts as a one-dimensional Fresnel lens and focuses light to a line image [181, 182].

When a photocurable homogeneous LC/prepolymer composite film is exposed to this pattern of light, photopolymerization occurs in the bright regions faster than in the dark regions, and the monomer diffuses to the brighter regions while the LC molecules diffuse to the darker regions and form LC droplets. As a result of the photopolymerization-induced phase separation (PIPS), the concentration and size of liquid crystal droplets within the dark regions exceed those existing in the bright regions and hardened polymer layers form in the bright regions, so that alternating layers of polymer and LC-droplet rich regions form in the bright and dark regions, respectively. By choosing the polymer in such a manner as to match its refractive index with one of the principal refractive indices (n_o or n_e) of the LC but not the other one, this composite film acts as a cylindrical Fresnel lens because of the refractive index mismatch between the adjacent zones in one dimension, and it focuses light in one dimension only, resulting in a line image. When an *ac* electric field is applied across the film, the LC molecules are reoriented and the polymer network remains almost unchanged. Therefore, the relative phase difference between the zones $\Delta\phi = 2\pi d(n_{LC} - n_{\text{polymer}})/\lambda$ alters. The focusing efficiency of the lens reaches its maximum value when the phase difference is equal to an odd multiple of π while the film becomes isotropic for an even multiple of π.

Jashnsaz et al. utilized materials formulation similar to that reported in the work [182]. The prepolymer–liquid crystal composite was made of commercially available materials. The materials used to fabricate the prepolymer mixture were trimethylolpropane triacrylate (TMPTA) as the main monomer, *N*-vinylpyrrollidone (NVP) as the crosslinking monomer, rose bengal (RB) as the photoinitiator, *N*-phenylglycine (NPG) as the co-initiator, and S-271 POE sorbitan monooleate as the surfactant, all from Aldrich. The ratio of TMPTA/NVP/S-271/NPG/RB was 80/10/7/2/1 by weight. The refractive index of the cured polymer was 1.522 at 632.8 nm. After mixing prepolymer composites, the liquid crystal E7 with positive optical anisotropy from Merck and with the ordinary refractive index of $n_o = 1.521$ and optical birefringence of $\Delta n = 0.225$ was added to the mixture. The prepolymer and LC were mechanically blended at 35/65 weight ratio, respectively, at 30 °C. Readymade homogeneous (HG) cells formed by two ITO coated glass plates separated by $d \approx 9.9\,\mu m$ thick spacer were filled with the photosensitive prepolymer/LC homogenous mixture by means of the capillary flow.

Figure 6.45 depicts the experimental setup of the Michelson interferometer, used to produce Fresnel pattern to cure the sample. Light from a *cw* argon ion laser with wavelength of $\lambda = 514\,nm$ passes from a beam expander, and is then divided into two paths of the interferometer by a beam splitter (BS). Light propagates through path 2 as a planner wave, while it changes to an elliptical (approximately cylindrical) wave front in path 1 by placing an astigmatism lens (with first focal length of F1 ≈ 100 cm) between BS and mirror 1. As a result of the superposition of these planner and elliptical waves, the elliptical Fresnel pattern forms in the plane perpendicular to the propagation axis in the output of the interferometer, and the central ellipse's width (2a) and length (2b) alter by a displacement toward or away from the BS according to the $a = (2\lambda d_1)^{1/2}$ and $b = (2\lambda d_2)^{1/2}$ [183], where $\lambda = 514\,nm$, and the focal lengths of the astigmatism lens L3 is responsible for d_1 and d_2. The variable a is one-half the length of the central ellipse's major axis; b is one-half the length of the minor axis. This gives a binary phase Fresnel lens with two different focal lengths $f_1 \equiv f(x)$ and $f_2 \equiv f(y)$ in the direction of x and y, respectively, and they are related to the innermost zone diameters a and b as $f_1 \equiv f(x) = a^2/\lambda$ and $f_2 \equiv f(y) = b^2/\lambda$, where λ is the wavelength of the incident probe beam. Thus, various Fresnel lenses with different focal lengths can be

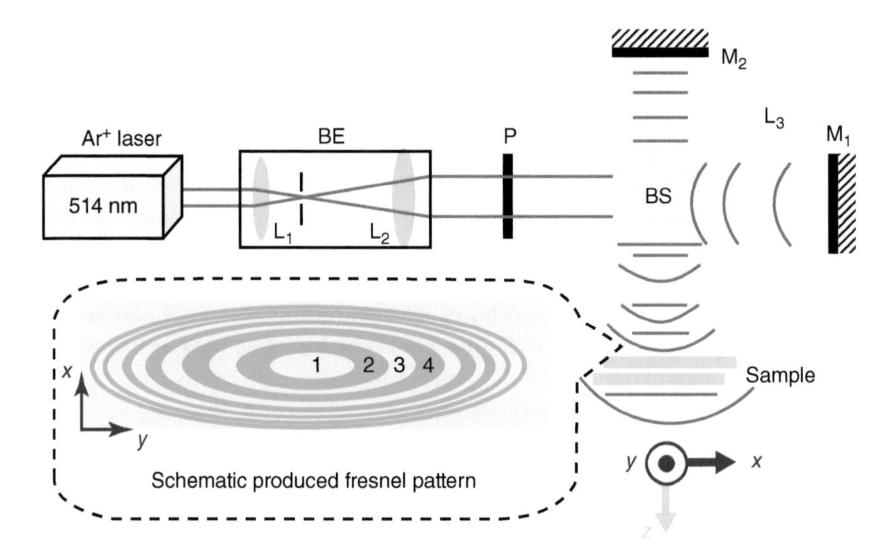

Figure 6.45 Michelson interferometer; experimental arrangement for fabricating the holographic Fresnel lens; BE, beam expander; BS, beam splitter; L1, L2, lenses; L3, astigmatism lens (approximately cylindrical lens); M1 and M2, mirrors; P, polarizer. Source: Jashnsaz et al. 2010 [184]. Reproduced with permission of Cornell University Library.

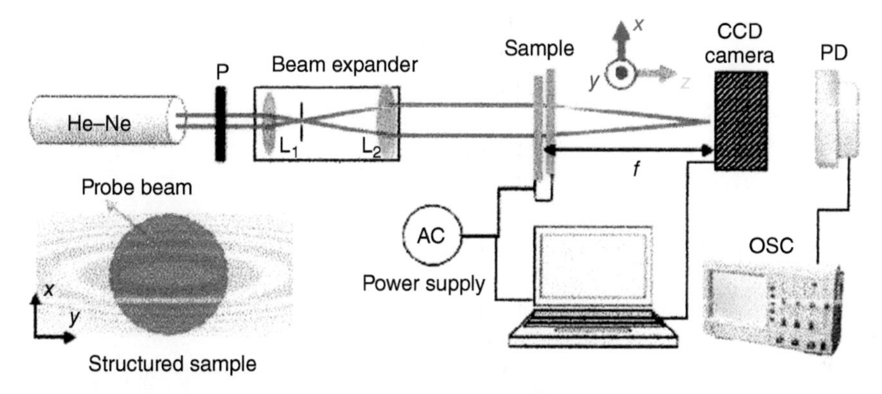

Figure 6.46 Experimental arrangement for studying the light focusing behavior and electro-optical properties of the fabricated lens; L1 and L2, lenses; P, polarizer; PD, photodiode detector; schematic structure of patterned sample and its cross area with probe beam as shown in *xy*-plane. Source: Jashnsaz et al. 2010 [184]. Reproduced with permission of Cornell University Library.

fabricated by this arrangement only by moving the sample toward or away from the BS. Also, different astigmatism Fresnel lenses with different ratio of f_1/f_2 can be fabricated by using different astigmatism lenses (different F1/F2) in the place of L3. Polarizer P determines the polarization state of the curing beam. The power of laser was 120 mW over the sample, the exposure time was four minutes, and the time of the final uniform UV curing was about three minutes. The experiment is carried out on a vibration-isolated optical bench at room temperature. The central zone was dark, so the odd and even rings shown in the Figure 6.45 became liquid crystal and polymer domain layers, respectively.

Figure 6.46 depicts the experimental setup used to study the performance and electro-optical properties of the fabricated lens. Expanded He–Ne laser beam with wavelength of 632.8 nm covers the area of the structured sample in a manner to experience Fresnel zone plate in x-direction only, but not in y-direction and in a rather good approximation, it can be regarded as Fig. 6.44(b) for incident probe beam, and it acts as a one-dimensional Fresnel lens. To demonstrate its light focusing behavior, a CCD camera and a photodiode detector (PD) were set at the focal length of the lens. The data were analyzed by a computer and the output of the PD monitored on an oscilloscope (OSC). In order to study the electro-optical behavior of the sample, a 5 kHz *ac* electric field was applied across the sample. Polarizer P sets the polarization state of the probe beam in x–direction, which is the direction of the surface rubbing of the HG cell.

Optical polarizing microscope images of the structured sample are shown in Figure 6.47. The polymer zones form an isotropic network that cannot rotate the polarization of the incident light; however, in the LC-droplet rich layers, nematic molecules form an anisotropic region and show birefringence behavior. Therefore, the polarization of the light will be rotated after passing through LC layers. Thus, under the crossed-polarizer optical microscope, the odd and even zones look bright and black, which corresponds to the anisotropic LC and isotropic polymer domain layers, respectively as shown in Figure 6.47a.

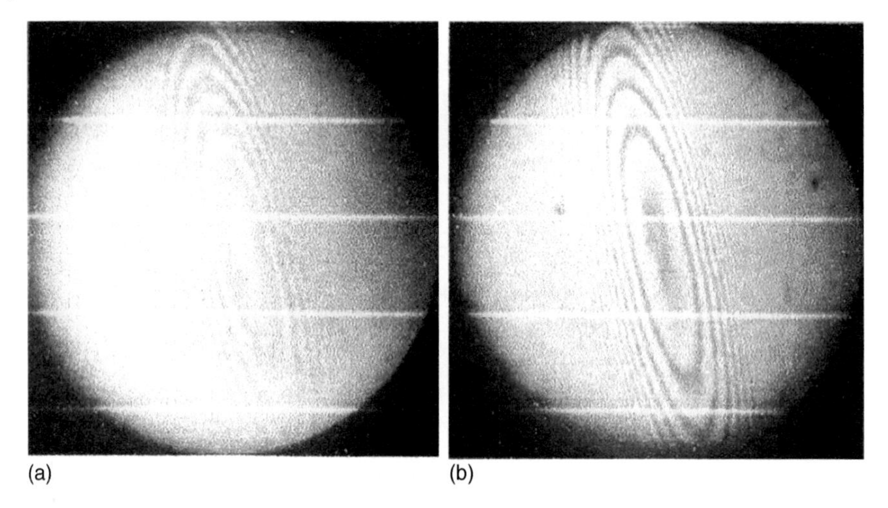

Figure 6.47 Polarizing microscope images of the patterned sample with (a) crossed polarizers, (b) parallel polarizers. Source: Jashnsaz et al. 2010 [184]. Reproduced with permission of Cornell University Library.

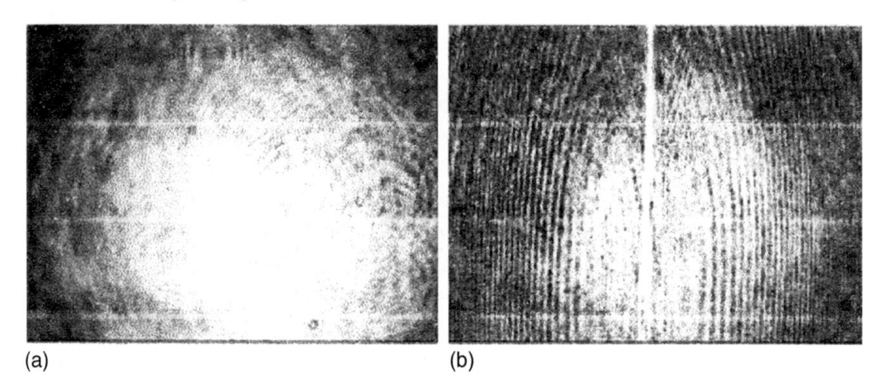

Figure 6.48 The observed laser beam images (a) without LC sample, (b) with LC sample, using a CCD camera at focal point at $\lambda = 623.8$ nm. Source: Jashnsaz et al. 2010 [184]. Reproduced with permission of Cornell University Library.

The reverse is true if the microscope is under the parallel-polarizer condition as shown in Figure 6.47b. The central ellipse shaped zone is estimated to be about 0.5 mm width, $2a = 0.5$ mm.

Figure 6.48 shows the focusing property of the fabricated lens using a CCD camera set at focal point. The incident probe laser beam profile is shown in Figure 6.48a in the absence of sample. When the sample is present, focusing behavior occurs in one dimension, in x-direction, as shown in Figure 6.48b, which is captured by a CCD camera set at about 8 cm from the sample at its focal point. This is due to different refractive index of the LC and polymer domain layers for the incident probe light, which acts as a one-dimensional Fresnel zone plate because of refractive index mismatch between adjacent layers. So, it focuses light in the x-direction, resulting in a line image. These results indicate that the sample indeed behaves like a cylindrical lens.

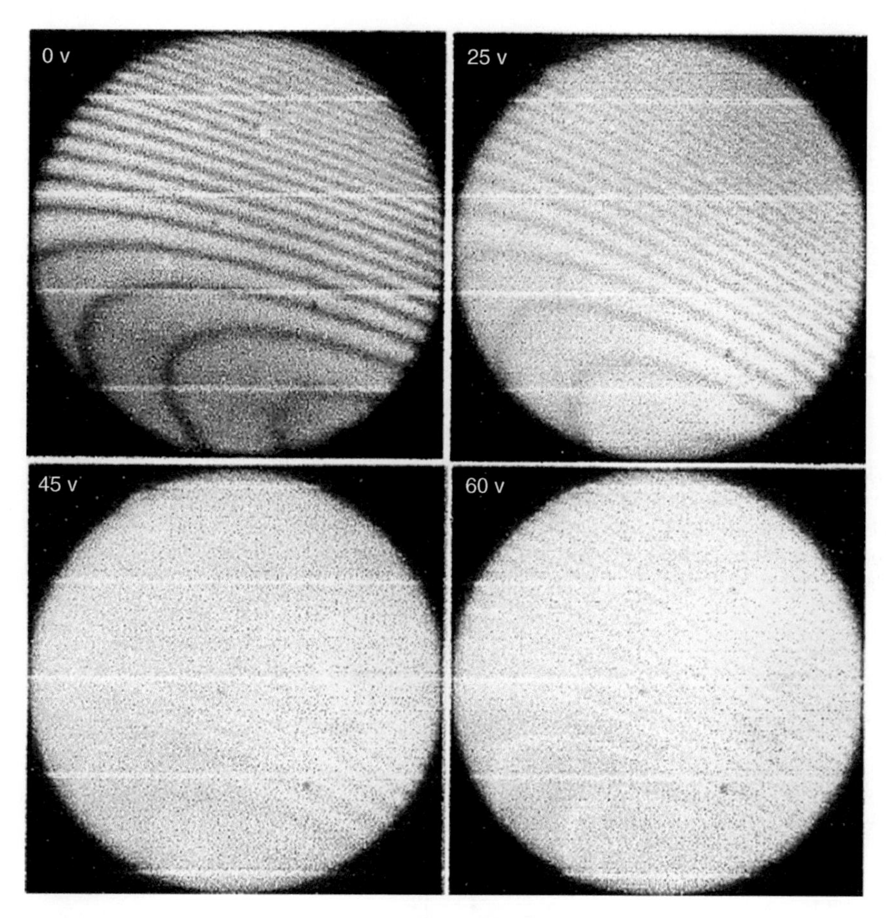

Figure 6.49 Microscopic images of the patterned sample at various applied voltages (Polarizer||Analyzer). Source: Jashnsaz et al. 2010 [184]. Reproduced with permission of Cornell University Library.

Figure 6.49 shows the optical polarizing microscope images of the sample with parallel polarizers at various applied voltages. It appears that by increasing the applied electric field across the film, the dark layers in which the concentration of LCs is high get brighter where there is no serious change in the brightness of the bright isotropic polymeric layers, and the intensity contrast between the dark and bright regions decreases and eventually at about 60 V_{rms} applied voltage across the sample the initially dark layers completely gets bright and the sample becomes isotropic. The contrast between the dark and bright regions corresponds to the refractive index mismatch between adjacent LC rich and polymer layers.

Figure 6.50 shows the intensity profiles of the outgoing focused beam using a CCD camera set at focal point of the sample at different applied voltages. The CCD camera captured picture without applied voltage is given in the inset the figure and the profiles at various voltages all have yield along the indicated horizontal dotted lines (analyzed in Origin Pro.).

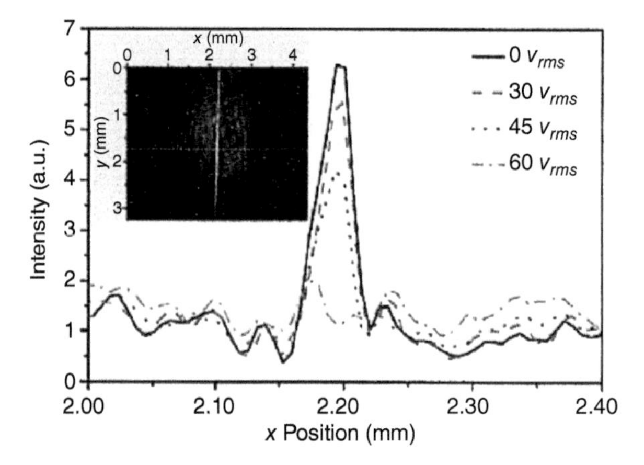

Figure 6.50 Intensity profiles of the outgoing focused (in one direction, *x*-direction) beam by a CCD camera set at focal point of the sample at various applied voltages. Picture captured by the CCD camera without applied voltage to sample is given in the inset of the figure and profiles at various voltages all have yield along indicated horizontal dotted lines. Source: Jashnsaz et al. 2010 [184]. Reproduced with permission of Cornell University Library.

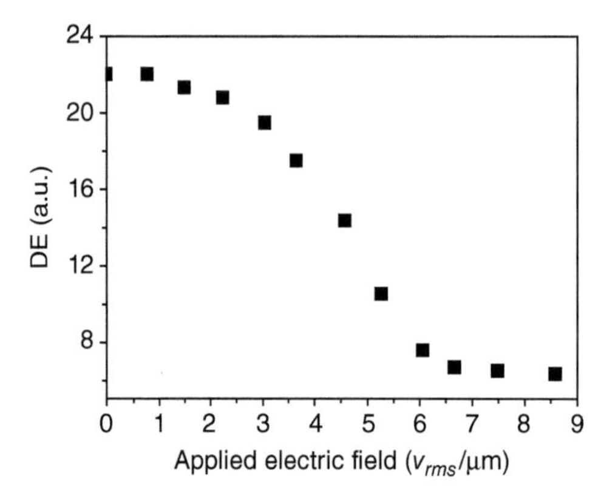

Figure 6.51 Diffraction efficiency at various applied electric fields measured by a photodiode detector set at primary focal point of the sample. Source: Jashnsaz et al. 2010 [184]. Reproduced with permission of Cornell University Library.

Figure 6.51 shows the diffraction efficiency at various applied electric fields to the sample by a photodiode detector set at the focal point of the fabricated lens. By increasing the applied electric field across the sample, the focusing property of the sample decreases because of the refractive index matching between the LC and polymer layers, and eventually the diffraction efficiency is eliminated at 6 V_{rms}/μm, which corresponds to the switching voltage. This is the state in which the uniaxial rod-like LC molecules have reoriented in the direction of the applied electric field (perpendicular to the cell surface), so transversely polarized light passing through the LC layers experiences the ordinary (n_o) refractive index of LC, which is equal to the refractive index of the polymer layers.

Jashnsaz et al. [184] also have measured the response times of the fabricated Fresnel lens by switching the applied electric field (60 V_{rms} across the cell) "on"

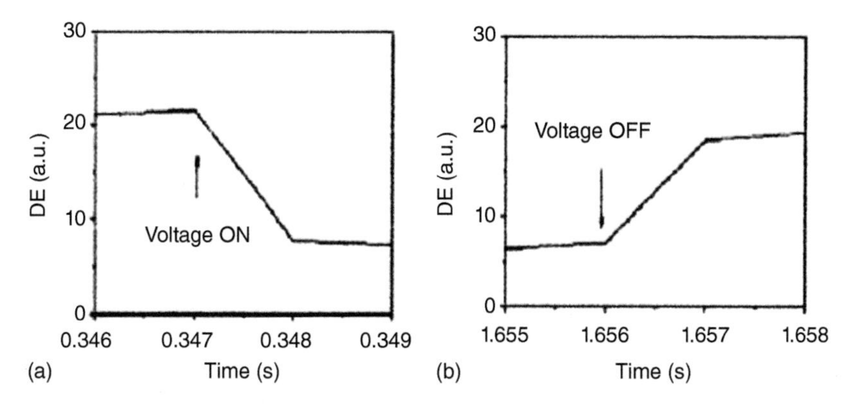

Figure 6.52 Electro-optical response time of the fabricated Fresnel lens monitored on an OSC using a photodiode detector at the focal point, at (a) the rise time and (b) the decay time, by switching "on" and "off" an electric field of $E = 6\,V_{rms}/\mu m$ across the sample. Source: Jashnsaz et al. 2010 [184]. Reproduced with permission of Cornell University Library.

and "off" for rise and decay times, respectively. To perform this, a photodiode detector was set at the focal point and its output monitored on an OSC; the results have been depicted in the Figure 6.52. It is found that at the applied electric field of $E = 6\,V_{rms}/\mu m$ across the sample, the focusing property of the sample disappears with a response of about 1 ms in a reversible manner.

6.4.4 Photosensitive Glass

Photosensitivity in glass was first discovered in optical fibers by Ken Hill et al. in 1978 [185], followed by a breakthrough by Meltz et al. in 1989 [186] reporting on the holographic writing of gratings using single-photon absorption at 244 nm. Since the last few decades, there have been many excellent individuals and institutions involved in the research, development, and application of these very interesting materials. In 2003, Z. Chen et al. has demonstrated the self-trapping of light in an organic PR glass, along with an effective approach for switching between self-focusing and self-defocusing [187]. Applications proposed with solitons in inorganic PR crystals [188] and in liquid crystals [189] might also be realized in organic PR glasses, while taking advantage of such materials for their low cost and structure flexibility compared with inorganic crystals and for their amenable absorption and scattering compared with liquid crystals. For instance, it might be possible to develop directional couplers and efficient nonlinear frequency converters in soliton-induced waveguides based on organic materials [188, 190]. In 2003, Corning Inc. has developed a new PR glass that can be used as a direct-write holographic medium for a variety of high-performance filtering applications [191]. In 2005, M. Asaro et. al has demonstrated optical wave guiding of a probe beam at 980 nm by a soliton beam at 780 nm in an organic photorefractive monolithic glass [192]. Both planar and circular waveguides induced by one- and two-dimensional spatial solitons formed as a result of orientationally enhanced photorefractive nonlinearity are produced in the organic glass. In 2012, N. F. Borrelli et Al. [193] developed a new photorefractive glass composition consisting of weight percent ("wt%") of 70–73 SiO_2, 13–17% B_2O_3, 8–10% Na_2O, 2–4% Al_2O_3, 0.005–0.1% CuO, <0.4% Cl, 0.1–0.5% Ag, and 0.1–0.3% Br. In another embodiment the composition consists essentially of 70–77% SiO_2, 13–18% B_2O_3, 8–10% Na_2O, 2–4% Al_2O_3, 0.005–0.1% CuO, <0.4% Cl, 0.1–0.5% Ag, and 0.1–0.3% Br. The glass can be used to make articles or elements that can exhibits both the photorefractive effect and the polarizing effect within a single element or article and can be used to make a variety of optical elements including Bragg gratings, filtering elements, beam shaping elements and light collection elements for use in display, security, defense, metrology, imaging, and communications applications.

Photosensitivity is investigated in silica-based materials because it allows the realization of several optical devices, mainly based upon the principle of FBG [194–196]. Exposure to laser radiation, typically in the ultraviolet (UV) region, causes a permanent change in refractive index in silica-based photorefractive materials [194]. This change can be selectively induced in a pattern to create a periodic or chirped grating or an optical channel waveguide [197, 198], finally obtaining filtering or guiding functions in a fiber or in a planar film [194, 197]. Some of the requirements for photosensitive materials are (i) photoinduced refractive index change of at least 10^{-4}; (ii) thermal stability at few 10^2 °C of the refractive index change; (iii) compatibility with silica-based optical telecommunication technology. At the present moment, Ge-doped silica Ge > 15 mol%, possibly impregnated with hydrogen, is the most investigated photosensitive silica-based material [195, 196]. Two of the mechanisms are believed to be involved in the formation of index gratings in germanosilicate fibers: the formation of color centers that changes the index of refraction via Kramers–Kronig relationship and the densification that occurs inside glass fibers upon illumination by UV light.

Recently sampled FBG has attracted attention because of its unique properties, such as multiple reflection/transmission peaks with very precise spacing and relatively easy fabrication procedures compared with long chirped grating writing. These properties make FBG ideal for WDM applications. The development of WDM technology has had a profound effect on increasing the bandwidth of fiber-optic networks over the last decade. Practically all WDM devices/subsystems incorporate wavelength selective filters employing various physical phenomena to achieve the required wavelength selectivity. The VBG (volume Bragg gratings) technology combines and perhaps exceeds the best characteristics of the TFF (thin film filters) and FBG filters.

The volume Bragg grating technology developed at PD-LD Inc. [199] is based on a proprietary photorefractive glass that changes index of refraction in areas exposed to UV light. This characteristic can be utilized for direct writing of periodic structures such as gratings to create filters with desirable properties. Filter shape, center wavelength, pass bandwidth, side band suppression, and other characteristics are determined by the combination of recording and processing procedures and the composition of glass. Physically, VBG elements are small glass cubes or parallelepipeds, 2–5 mm on a side, robust and easy to handle, well suited for high temperature processing (soldering, brazing, glass fritting, etc.) and automated manufacturing. Since the glass is silica based, the produced filters are physically and chemically very stable, different from filters recorded in photorefractive plastics or crystals, which usually encounter problems with dimensional stability and permanence of recording, when subjected to typical telecom operating and storage temperatures.

Figures 6.53 and 6.54 depicts VBG filters operating either in transmission or in reflection mode. In reflection mode, the selected wavelength is reflected by FBG filters, while express channels go through with only a minimal loss. A receiving collimator-fiber assembly, just like in TFF based devices, easily picks up the reflected beam. Hence FBG-like performance combined with TFF-like

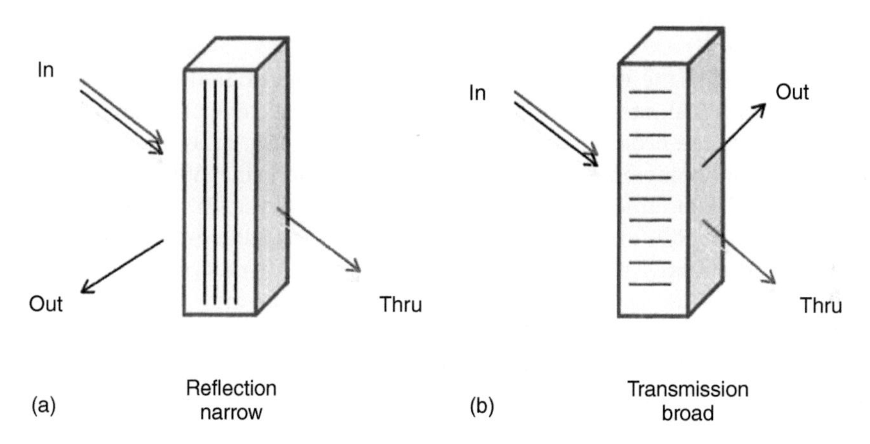

Figure 6.53 Principles of operation of VBG filters – thin lines denote planes with changed index of refraction. In a reflection type, all wavelengths except the selected one pass through the filter with a minimal loss. The reflected wavelength is picked up by a fiber-collimator assembly positioned at the proper angle. In a transmission-type VBG, all wavelengths pass through the filter with a minimal loss and only the selected wavelength is deflected to be picked up by a fiber-collimator assembly positioned at the proper angle. Note that a VBG element can contain more than one grating in the same volume to select more than one wavelength from the multi-wavelength stream. Note also that a different wavelength(s) can be selected by angular tuning of the VBG element. Source: Volodin et al. [199]. Reproduced with permission of Photonics Online.

Figure 6.54 Several VBG filters can be overlapped or multiplexed in the same volume. This allows fabrication of multi-port, multifunctional elements in a very compact volume. Source: Volodin et al. [199]. Reproduced with permission of Photonics Online.

In

Thru

packaging avoids the need for circulators to extract the selected wavelength. Reflection-type VBGs are suitable for DWDM applications with channel spacing of 25–200 GHz.

In a transmission-type VBG all channels go through the filter but the selected wavelength is deflected from the beam, to be picked up by the properly placed collimator-fiber assembly. Transmission-type VBGs are suitable for CWDM (coarse wavelength division multiplexing) applications, where pass bands are several nanometers wide. Another interesting and unique characteristic of VBG elements is that a single element can contain multiple filters in the same volume. Thus, one element can manage several, e.g. four selected wavelengths. This is a distinct advantage over conventional TFF or FBG-based approaches that would require four distinct and separate filters to accommodate four distinct wave lengths. This reduction in number of filters employed translates into huge cost and space savings.

Manufacturing of the VBG filters and devices is a complex, multi-step process that requires vertical integration of the company. Several steps of the manufacturing sequence are proprietary and unique to PD-LD (e.g. glass composition, glass processing, holographic recording, filter shape modeling, VBG characterization, etc.). Photorefractive glasses developed and used at PD-LD are oxide glasses, based on SiO_2 and containing numerous other additives. It is believed that the photorefractive action is based on a redox reaction of silver initiated by the UV exposure. The elemental silver particles serve as nucleation centers for the growth of a second phase during the thermal development process, at $\sim$500 °C for several hours. This second phase is thought to be Na and K halogenide rich, thus resulting in a material with somewhat lower index of refraction than in the unexposed areas.

The produced glass must be of optical quality with excellent compositional uniformity and absence of bubbles, striations, and other imperfections. It also must be free of chemical impurities that might strongly influence the extent and nature of photoinduced reactions. Subsequent to melting and annealing, the glass boule is cut into wafers, which are then polished. These wafers are then subjected to holographic recording and development.

In the recording of the VBG filter(s), a wafer of the photorefractive glass is illuminated by two mutually coherent laser beams. These beams create a standing wave pattern with sinusoidal variation of the light intensity. This pattern is imprinted in the bulk of the material, penetrating its entire thickness. At this stage, the photochemical changes induced by light are not complete (i.e. they are latent) and the index of refraction is unchanged. Only bringing the material to the elevated temperature sufficient for the rapid growth of the second phase as mentioned above completes the process. During this step, the latent grating is transformed into the grating of the refractive index or the true VBG. These gratings are completely stable at temperatures as high as 150–200 °C. The recording of the VBG wafer can be done by using truly "holographic" techniques or alternatively by using a phase mask, much like the ones used for the recording of the FBG. In either method, the spectral shape of the filter can be constructed (such as, for example, the flat-top, steep roll-off filters) for the fiber-optic communication networks. After development, wafers are diced into individual VBG

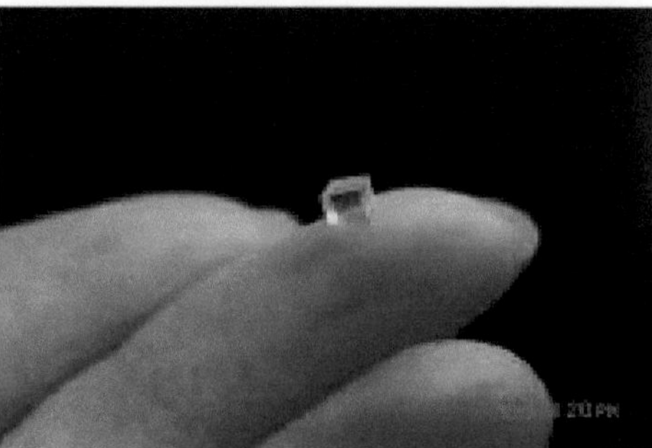

Figure 6.55 Photograph of a diced wafer and a single VBG element. Manufacturing VBG elements is an efficient process, since the whole wafer is recorded in one step, to produce many VBG elements at a time. Source: Volodin et al. [199]. Reproduced with permission of Photonics Online.

filter elements, which might be AR coated. The finished filter elements undergo QC and characterization and are ready for application. Figure 6.55 shows a diced wafer and a single VBG element.

The VBG filter can reflect (or deflect) the selected wavelength of light at a convenient angle to be picked up by a receiving collimator or sent to some other suitable target (e.g. photodiode). The very important implication of this is that, unlike with an FBG, VBG requires no circulators to operate. All other wavelengths of light that are outside of the passband are transmitted through the VBG unaffected. The amplitude and phase envelopes of the VBG determine the spectral shape of the filter. With proper design, steep filters are obtained having, for example, 0.5 dB width for 0.2 nm and 25 dB width for <0.7 nm. For comparison of filter shapes Figure 6.56 shows spectral curves for VBG filters: (a) 50 GHz DWDM filters, operating in reflection. (b) Unapodized 12 nm bandpass CWDM filter, operating in transmission.

In the case of the narrow filter, the light at the peak wavelength is reflected by the grating practically completely, with 55+ dB attenuation, making it also an excellent notch filter. As mentioned previously, multiple VBG filters can be recorded in the same volume, thus reducing the number of filter elements in a given device. This means that very compact, integrated multifunctional subsystems can be constructed using this technology. Moreover, since the peak wavelength of the VBG depends somewhat on the angle of incidence, VBG filters can be tuned by rotation if tunability is required for a particular application.

As the above description has shown, VBG filters have a significant potential for application in numerous WDM devices and subsystems due to their excellent performance characteristics, combined with small size, robustness, ease of application, and efficient manufacturing techniques. At this point, PD-LD Inc. is concentrating on products based on the transmission-type VBGs suitable for applications with wide wavelength separation. It offers discrete VBG elements themselves as well as the universal broadband WDM combiners described below. It has also demonstrated the capability to build hybrid modules in which active devices (lasers and detectors) are combined with VBG filters to produce unique components and subsystems.

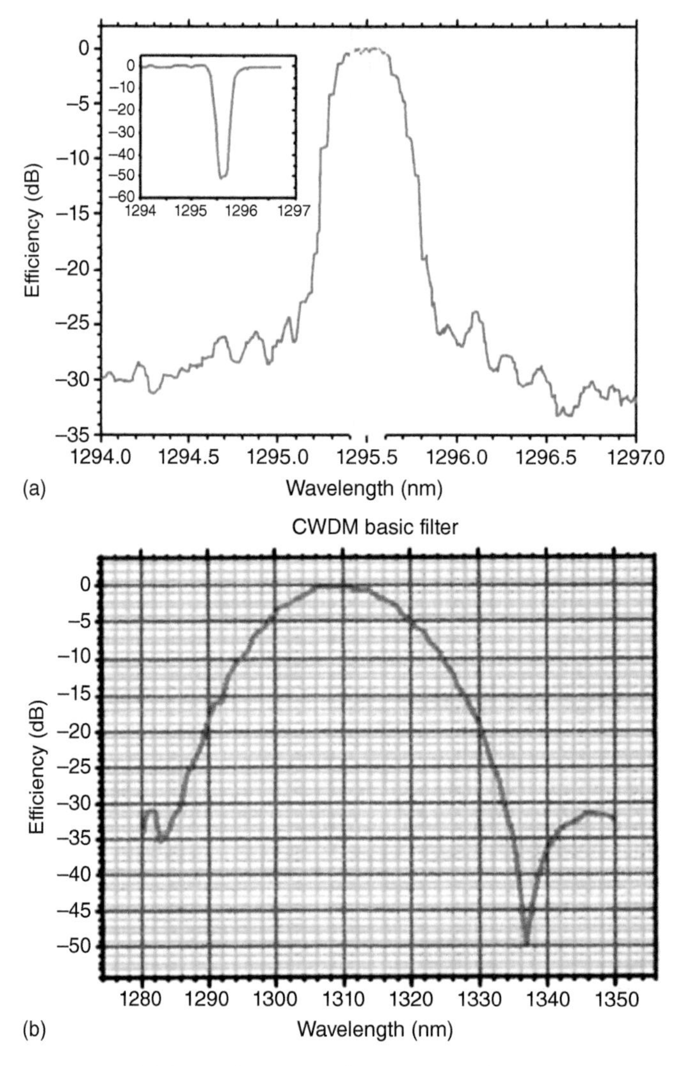

Figure 6.56 (a) Spectral response of a 50 GHz VBG filter for DWDM applications (reflection type) [199]. (b) Spectral response of a 12 nm wide CWDM VBG filter (transmission type). Source: Volodin et al. [199]. Reproduced with permission of Photonics Online.

6.4.4.1 VBG Discrete Filter Elements

The essential characteristics of VBG filters and methods for manufacturing them have been discussed previously. The main advantages of VBG filters are the following:

- FBG performance combined with TFF simplicity
- More than one filter in the same volume
- Small size, robustness, durability
- Filter shape control, excellent performance characteristics
- Amenable to automated packaging
- Low cost.

Figure 6.57 Low-cost package for a four-channel WDM combiner. In this package one uses very wide VBG filters for combining four lasers from the near IR bands. Source: Volodin et al. [199]. Reproduced with permission of Photonics Online.

6.4.4.2 Universal WDM Combiners/Splitters

The WDM combiners/splitters for combining two wavelengths such as 1310 and 1550 nm are well-known devices. They are based on fused fiber (low isolation units) or TFF technology (high isolation units). WDM combiners/splitters based on VBG technology can combine up to four wavelengths with excellent isolation and low losses.

The initial product of PD-LD Inc. combines three wavelength bands, 850–1100, 1300, and 1550 nm bands, respectively. Other combinations are possible, e.g.:

1310, 1490, 1550 nm – for Cable television (CATV) applications,
1275, 1350, 1490, 1550 nm – for low-cost CWDM applications,
440, 534, 630 nm – applications in displays, laser shows, pointers, etc.

Figure 6.57 depicts a universal four-channel WDM combiner. In comparison with other four-channel multiplexers, this design offers a large advantage in price per channel and is only a fraction in size in comparison with competing devices.

6.4.4.3 Integrated Combiner Modules

PD-LD Inc. has demonstrated a four-channel, 1300 nm band CWDM transmitter module integrating DFB lasers with a compact VBG-based multiplexer in a unit designed to perform to the LX 4 standard, specified for use in 10 Gbit Ethernet transmission (Figure 6.58). Figure 6.59 shows the output of this unit. Note high coupled power of 1 mW (0 dBm) for all channels in contrast with – 7.5 dBm or less, when a standard 4×1 fused coupler is used. Thus, VBG technology enables compact component integration into various multi-source 4D transceiver footprints. The advantages of the integrated approach are as follows:

- Compact size
- High coupled power (>0 dBm/channel, making reaches of over 50 km feasible)
- Utilizes standard hermetically packaged coaxial un-cooled lasers
- Flexible, other wavelengths available on a custom basis.

6.4.4.4 Other VBG-Based Devices and Subsystems

In addition to devices described in the previous section, VBG elements can be used in most devices where either TFF or FBG elements are used. The following list describes some but not all of the devices that might benefit from the VBG technology.

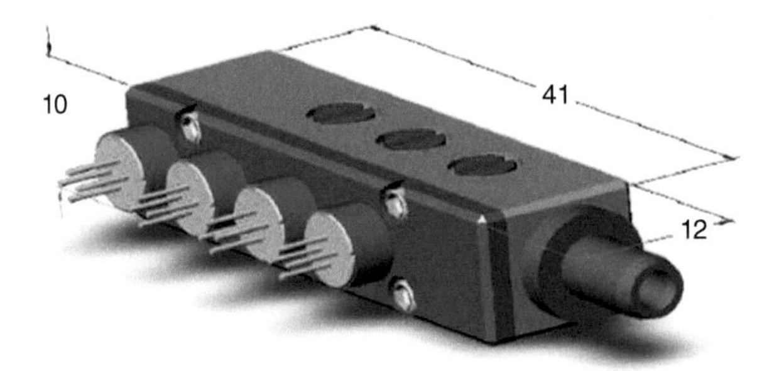

Figure 6.58 An integrated module based on VBG technology. This module contains four CWDM DFB lasers and VBG elements for multiplexing these lasers into a single mode fiber. Note that this unit couples over 1 mW per channel, thus avoiding the 7.5 dB loss associated with commonly used 4 × 1 fused fiber couplers. The unit is sufficiently small to fit into a Xenpak package specified for 10 Gbit Ethernet networks. Source: Volodin et al. [199]. Reproduced with permission of Photonics Online.

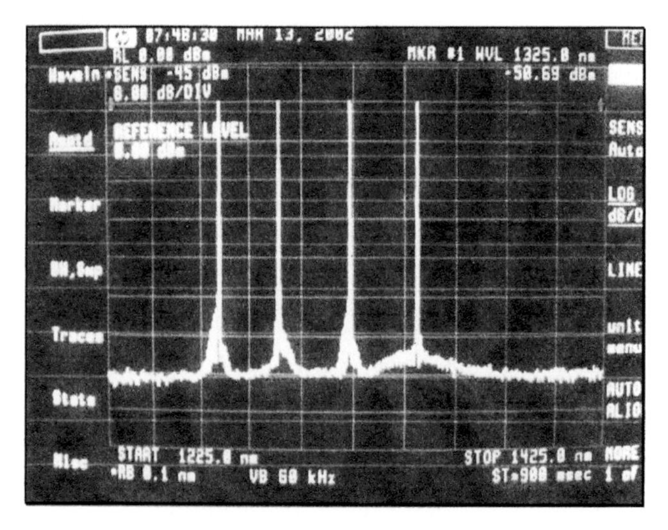

Figure 6.59 Spectral output of a four-channel CWDM transmitter based on VBG elements. Note the high coupling efficiency (0 dBm) for the long-range reach. Source: Volodin et al. [199]. Reproduced with permission of Photonics Online.

- Low-cost pump combiner
- Optical Add-Drop Multiplexer (OADMs) for DWDM networks
- Wavelength sensitive tap for monitoring
- Wavelength locker
- Tunable devices – slight angular motion of a VBG element provides wavelength tuning.

The PD-LD Inc. has developed a new technology of VBG. Characteristics such as compact size, physical toughness and durability, and high volume manufacturability, combined with superior optical characteristics such as low insertion loss, high wavelength selectivity, and free space operation promise to make VBG elements essential parts of many future components used in communication networks. We thus believe that VBG elements represent a technological and commercial breakthrough development for numerous applications in photonic industry.

References

1 Ashkin, A., Boyd, G.D., Dziedzic, J.M. et al. (1966). Optically-induced refractive index inhomogeneities in $LiNbO_3$ and $LiTaO_3$. *Appl. Phys. Lett.* 9: 72.
2 Kukhtarev, N.V., Markov, V.B., Odulov, S.G. et al. (1979). Holographic storage in electro-optic crystals. II. Beam coupling-light amplification. *Ferroelectrics* 22 (3–4): 961–964.
3 Kukhtarev, N.V., Markov, V.B., Odulov, S.G. et al. (1979). Holographic storage in electro-optic crystals. I. Steady state. *Ferroelectrics* 22 (3–4): 949–960.
4 Vinetskii, V.L., Kukhtarev, N.V., Odulov, S.G., and Soskin, M.S. (1979). Dynamic self-diffraction of coherent light beams. *Sov. Phys. Usp.* 22 (9): 742–756.
5 Zgonik, M., Bernasconi, P., Duelli, M. et al. (1994). Dielectric, elastic, piezoelectric, electro-optic and elasto-optic tensors of $BaTiO_3$ crystals. *Phys. Rev. B* 50 (9): 5941–5949.
6 Günter, P. and Zgonik, M. (1991). Clamped-unclamped electro-optic coefficient dilemma in photorefractive phenomena. *Opt. Lett.* 16 (23): 1826–1828.
7 Montemezzani, G., Rogin, P., Zgonik, M., and Günter, P. (1994). Interband photorefractive effects: theory and experiments in $KNbO_3$. *Phys. Rev. B* 49 (4): 2484–2502.
8 Bernasconi, P. (1998). Physics and applications of ultraviolet light induced photorefractive gratings. PhD thesis, ETH Zürich.
9 Günter, P. and Huignard, J.-P. (2010). *Photorefractive Materials and Their Applications*, vol. I. Berlin: Springer-Verlag, Vol. II, 1989.
10 Moerner, W.E. and Silence, S.M. (1994). Polymeric photorefractive materials. *Chem. Rev.* 94: 127.
11 Agullo-Lopez, F., Cabrera, J.M., and Agullo-Rueda, F. (1994). *Electro-optics.* London: Academic.
12 Zilker, S.J. (2000). Materials design and physics of organic photorefractive systems. *Chem. Phy. Chem.* 1: 72.
13 Hariharan, P. (1996). *Optical Holography*, 2ee. Cambridge, England: Cambridge University Press.
14 Van Heerden, P.J. (1963). Theory of optical information storage in solids. *Appl. Opt.* 2: 393.
15 Duelli, M., Montemezzani, G., Zgonik, M., and Günter, P. (2000). *Nonlinear Optical Effects and Materials; Chapters 4 and 5.* Berlin: Springer-Verlag.
16 Dittrich, P., Montemezzani, G., Bernasconi, P., and Günter, P. (1999). Fast, reconfigurable light-induced waveguides. *Opt. Lett.* 24 (21): 1508–1510.
17 Günter, P. and Huignard, J.-P. (2006). *Photorefractive Materials and their Applications 1: Basic Effects.* Berlin: Springer Verlag.
18 Bernasconi, P., Montemezzani, G., and Günter, P. (1999). Off-Bragg-angle light diffraction and structure of dynamic inter band photorefractive gratings. *Appl. Phys. B* 68 (5): 833–842.

19 Dittrich, P., Koziarska-Glinka, L., Montemezzani, G. et al. (2004). Deep-ultraviolet interband photorefraction in lithium tantalate. *J. Opt. Soc. Am. B: Opt. Phys.* 21 (3): 632–639.
20 Mosimann, R., Haertle, D., Jazbinšek, M. et al. (2006). Determination of the absorption constant in the inter band region by photocurrent measurements. *Appl. Phys. B* 83 (1): 115–119.
21 Yeh, P. and Gu, C. (eds.) (1995). *Landmark Papers on Photorefractive Nonlinear Optics.* New Jersey: World Scientific Publishing.
22 Yeh, P. and Gu, C. (eds.) (1994). *Photorefractive Materials, Effects and Applications.* SPIE Critical Reviews.
23 Yeh, P., Gu, C., Khoo, I.C. et al. (eds.) (1994). *Nonlinear Optics and Optical Physics,* 341, (Chapter 12). New Jersey: World Scientific Publishing.
24 Feinberg, J., Heiman, D., Tanguay, A.R. Jr., and Hellwarth, R. (1980). Photorefractive effects and light-induced charge migration in barium titanate. *J. Appl. Phys.* 51: 1297.
25 Ducharme, S. and Feinberg, J. (1984). Speed of the photorefractive effect in a $BaTiO_3$ single crystal. *J. Appl. Phys.* 56: 839.
26 Mullen, R.A. and Hellwarth, R.W. (1985). Optical measurement of the photorefractive parameters of $Bi_{12}SiO_{20}$. *J. Appl. Phys.* 58: 40.
27 Yeh, P. (1989). Two-wave mixing in nonlinear media. *IEEE J. Quantum Electron.* 25: 484.
28 Markov, V., Odulov, S., and Soskin, M. (1979). Dynamic holography and optical image processing. *Opt. Laser Technol.* 11: 95.
29 Marrakchi, A., Huignard, J.P., and Günter, P. (1981). Diffraction efficiency and energy transfer in two-wave mixing experiments with $Bi_{12}SiO_{20}$ crystals. *Appl. Phys. A* 24: 131.
30 Feinberg, J. and Hellwarth, R.W. (1980). Phase-conjugating mirror with continuous-wave gain. *Opt. Lett.* 5: 519.
31 Feinberg, J. (1982). Asymmetric self-defocusing of an optical beam from the photorefractive effect. *J. Opt. Soc. Am. A:* 72: 46.
32 (a) Yeh, P. (1992). Photorefractive phase conjugators. *IEEE Proc.* 80: 436. (b) Zel'dovich, B.Y., Pilipetsky, N.F., and Shkunov, V.V. (1985). *Principles of Phase Conjugation.* New York: Springer Verlag.
33 Yeh, P. (1993). *Introduction to Photorefractive Nonlinear Optics.* New York: Wiley.
34 Yeh, P. (1987). Fundamental limit of the speed of photorefractive effect and its impact on device applications and material research. *Appl. Opt.* 26: 602, 26, 3190 (1987).
35 Gu, C. and Yeh, P. (1991). Reciprocity in photorefractive wave mixing. *Opt. Lett.* 16: 455.
36 Yeh, P., Gu, C., and Neurgaonkar, R.R. (eds.) (1994). *Materials for Optoelectronic Applications.* Switzerland: Trans Tech Publishing.
37 Gu, C. and Yeh, P. (1994). Application of photorefractive volume holography in optical computing. *Int. J. Nonlinear Opt. Phys.* 3: 317.
38 Yeh, P., Chiou, A.E., Hong, J. et al. (1989). Photorefractive nonlinear optics and optical computing. *Opt. Eng.* 28: 328.
39 Chiou, A.E.T. and Yeh, P. (1989). Parallel image subtraction using a phase-conjugate Michelson interferometer. *Opt. Lett.* 11: 306.

40 Chang, T.Y., Beckwith, P.H., and Yeh, P. (1988). *Opt. Lett.* 13: 586.
41 Yeh, P., Chiou, A.E.T., and Hong, J. (1988). Optical interconnection using photorefractive dynamic holograms. *Appl. Opt.* 27: 2093.
42 Chiou, A.E.T. and Yeh, P. (1990). Energy efficiency of optical interconnections using photorefractive holograms. *Appl. Opt.* 29: 1111.
43 Lee, H., Gu, X.G., and Psaltis, D. (1989). Volume holographic interconnections with maximal capacity and minimal cross talk. *J. Appl. Phys.* 65: 2191.
44 Psaltis, D., Gu, X.G., Brady, D. et al. (eds.) (1990). *An Introduction to Neural and Electronic Networks*. New York: Academic Press.
45 Psaltis, D., Brady, D., Gu, X.G., and Lin, S. (1990). Holography in artificial neural networks. *Nature* 343: 325.
46 Psaltis, D. and Mok, F. (1995). *Sci. Am.* 273: 70.
47 Heanue, J.F., Bashaw, M.C., and Hesselink, L. (1994). Holographic memories. *Science* 265: 749.
48 Ashley, J., Bernal, M.-P., Blaum, M., et al. (1996). Holographic storage promises high data density. *Laser Focus World* 32: 81.
49 Curtis, K. (1997). OSA Annual Meeting, ThR1.
50 Hong, J.H. and Psaltis, D. (1996). Dense holographic storage promises fast access. *Laser Focus World* 32(4): 119–122.
51 Gu, C., Yeh, P., Yi, X., and Hong, J. (2000). *Holographic Data Storage* (ed. H. Coufal, D. Psaltis and G. Sincerbox), 63. New York: Springer Verlag.
52 Sincerbox, T., Coufal, H., Psaltis, D., and Sincerbox, G. (eds.) (2000). *Holographic Data Storage*, 3. New York: Springer Verlag.
53 Hesselink, L., Coufal, H., Psaltis, D., and Sincerbox, G. (eds.) (2000). *Holographic Data Storage*, 383. New York: Springer Verlag.
54 Psaltis, D. Optical Information Processing Group Publications [Online]. Available: http://optics.caltech.edu/Publications/publications.htm.
55 Ashley, J., Bernal, M.-P., Burr, G.W. et al. (2000). Holographic data storage. *IBM J. Res. Dev.* 44 (3): 341.
56 Holographic data storage, IBM site. http://www.research.ibm.com/journal/rd/443/ashley.html.
57 Bernal, M.-P., Coufal, H., Grygier, R.K. et al. (1996). A precision tester for studies of holographic optical storage materials and recording physics. *Appl. Opt.* 35: 2360.
58 Shelby, R.M., Hoffnagle, J.A., Burr, G.W. et al. (1997). Pixel-matched holographic data storage with megabit pages. *Opt. Lett.* 22: 1509.
59 Burr, G.W., Ashley, J., Coufal, H. et al. (1997). Modulation coding for pixel-matched holographic data storage. *Opt. Lett.* 22: 639.
60 Sanford, J.L., Greier, P.F., Yang, K.H. et al. (1998). A one-megapixel reflective spatial light modulator system for holographic storage. *IBM J. Res. Dev.* 42 (3/4): 411.
61 Hong, J.H., McMichael, I., Chang, T.Y. et al. (1995). Volume holographic memory systems: techniques and architectures. *Opt. Eng.* 34: 2193.
62 McMichael, I., Christian, W., Pletcher, D. et al. (1996). Compact holographic storage demonstrator with rapid access. *Appl. Opt.* 35: 2375.
63 Ma, J., Chang, T., Choi, S. et al. (eds.) (2000). *Holographic Data Storage*, 409. New York: Springer Verlag.

64 Dhar, L., Hale, A., Katz, H.E. et al. (1999). *Opt. Lett.* 24: 487.

65 Wilson, W.L., Curtis, K., Tackitt, M. et al. (2000). *Opt. Quantum Electron.* 32: 393.

66 Physical Sciences - Holographic Data Storage - What Is It? Physical Sciences Research. Bell Laboratories, 2000. Web. 09 Dec. 2011. http://www.bell-labs.com/org/physicalsciences/projects/hdhds/2.html.

67 InPhase Technologies [Online]. Available: http://www.inphase-technolo-gies.com/.

68 Zhou, G., Qiao, Y., Mok, F., and Psaltis, D. (1996). *Opt. Photonics News* 43.

69 Flickner, M., Sawhney, H., Niblack, W. et al. (1995). *Computer* 28: 23.

70 Rockwell Scientific (2000–2003). Photonic processing [Online]. Available: http://www.rockwellscientific.com/html/imaging-optics.html#PhotonicProcessing.

71 Kong, Y., Liu, S., and Xu, J. (2012). Recent advances in the photorefraction of doped lithium niobate crystals. *Materials* 5: 1954–1971. www.mdpi.com/journal/materials.

72 Günter, P. (2000). *Nonlinear optical effects and materials*, vol. 72 of Springer Series in Optical Science. Berlin, Heidelberg, New York: Springer.

73 Christodoulides, D.N., Lederer, F., and Silberberg, Y. (2003). Discretizing light behaviour in linear and nonlinear waveguide lattices. *Nature* 424 (6950): 817–823.

74 Fleischer, J.W., Segev, M., Efremidis, N.K., and Christodoulides, D.N. (2003). Observation of two-dimensional discrete solitons in optically induced nonlinear photonic lattices. *Nature* 422 (6928): 147–150.

75 Matoba, O., Inujima, T., Shimura, T., and Kuroda, K. (1998). Segmented photorefractive waveguides in $LiNbO_3$:Fe. *J. Opt. Soc. Am. B: Opt. Phys.* 15 (7): 2006–2012.

76 Zhang, P., Zhao, J.L., Yang, D.X. et al. (2003). Optically induced photorefractive waveguides in KNSBN:Ce crystal. *Opt. Mater.* 23 (1-2): 299–303.

77 Juvalta, F., Koziarska-Glinka, B., Jazbinšek, M. et al. (2006). Deep UV light induced, fast reconfigurable and fixed waveguides in Mg doped $LiTaO_3$. *Opt. Express* 14 (18): 8278–8289.

78 Petter, J. and Denz, C. (2001). Guiding and dividing waves with photorefractive solitons. *Opt. Comm.* 188 (1-4): 55–61.

79 Itoh, K., Matoba, O., and Ichioka, Y. (1994). Fabrication experiment of photorefractive three-dimensional waveguides in lithium niobite. *Opt. Lett.* 19: 652.

80 Matoba, O., Itoh, K., and Ichioka, Y. (1996). Array of photorefractive waveguides for massively parallel optical interconnections in lithium niobite. *Opt. Lett.* 21: 122.

81 Bekker, A., Pedael, A., Berger, N.K. et al. (1998). Optically induced domain waveguides in $Sr_xBa_{1-x}Nb_2O_6$ crystals. *Appl. Phys. Lett.* 72: 3121.

82 Juvalta, F., Mosimann, R., Jazbinšek, M., and Günter, P. (2009). Fast dynamic waveguides and waveguide arrays in photorefractive $Sn_2P_2S_6$. *Opt. Express* 17: 379–384.

83 Shumelyuk, A., Odoulov, S., Oleynik, O. et al. (2007). Spectral sensitivity of nominally undoped photorefractive $Sn_2P_2S_6$. *Appl. Phys. B* 88 (1): 79–82.

84 Mosimann, R., Marty, P., Bach, T. et al. (2007). High-speed photorefraction at telecommunication wavelength 1.55 μm in $Sn_2P_2S_6$:Te. *Opt. Lett.* 32: 3230–3232.
85 Grabar, A.A., Jazbinšek, M., Shumelyuk, A. et al. (2007). Photorefractive effects in $Sn_2P_2S_6$. In: *Photorefractive Materials and Their Applications 2* (ed. Peter Günter and Jean-Pierre Huignard), 327–362. Springer.
86 Haertle, D., Caimi, G., Haldi, A. et al. (2003). Electro-optical properties of $Sn_2P_2S_6$. *Opt. Commun.* 215 (4-6): 333–343.
87 Bach, T., Jazbinšek, M., Günter, P. et al. (2005). Self pumped optical phase conjugation at 1.06 μm in Te-doped $Sn_2P_2S_6$. *Opt. Express* 13 (24): 9890–9896.
88 Juvalta, F., Mosimann, R., Jazbinšek, M., and Günter, P. (2009). Fast dynamic waveguides and waveguide arrays in photorefractive $Sn_2P_2S_6$ induced by visible light. *Opt. Express* 17 (2): 379–384.
89 Miura, K., Qiu, J.R., Inouye, H., and Mitsuyu, T. (1997). Photowritten optical waveguides in various glasses with ultrashort pulse laser. *Appl. Phys. Lett.* 71: 3329.
90 Gao, R.X., Zhang, J.H., Zhang, L.G. et al. (2002). Femtosecond Laser Induced Optical Waveguides and Micro-mirrors Inside Glasses. *Chin. Phys. Lett.* 19: 1424.
91 Zhang, P., Zhao, J.L., Yang, D.X. et al. (2003). *Appl. Opt.* 42: 4208.
92 Matoba, O., Kuroda, K., and Itoh, K. (1998). Fabrication of a two-dimensional array of photorefractive waveguides in $LiNbO_3$:Fe using non-diffracting checkered pattern. *Opt. Commun.* 145: 150.
93 Zhao, J.L., Li, B.L., Yang, D.X. et al. (2003). Experiment of light writing planar waveguides array in $LiNbO_3$:Fe Crystal. *Acta Photon. Sin.* 32: 421.
94 Chen, Z.G. and Martin, H. (2003). Waveguides and waveguide arrays formed by incoherent light in photorefractive materials. *Opt. Mater.* 23: 235.
95 Matoba, O., Ikezawa, K., Itoh, K., and Ichioka, Y. (1995). Modification of photorefractive waveguides in lithium niobate by guided beam for optical dynamic interconnection. *Opt. Rev.* 2: 438.
96 Park, N., Yang, B., Lee, H.S., and Lee, B. (2000). Photorefractive waveguide formation with refractive index reversal by use of thermal fixing. *Electron. Lett* 36: 429.
97 Wesner, M., Herden, C., and Kip, D. (2001). Electrical fixing of waveguide channels in strontium-barium niobate crystals. *Appl. Phys. B* 72: 733.
98 Zhang, P., Yang, D.X., Zhao, J.L. et al. (2004). Light-induced array of three-dimensional waveguides in lithium niobate by employing two-beam interference field. *Chin. Phys. Lett.* 21 (8): 1558.
99 Chen, F.S. (1969). Optically induced change of refractive indices in $LiNbO_3$ and $LiTaO_3$. *J. Appl. Phys.* 40: 3389.
100 Song, Q.W., Zhang, C., and Talbot, P.J. (1993). Self-defocusing, self-focusing, and speckle in $LiNbO_3$ and $LiNbO_3$:Fe crystals. *Appl. Opt.* 32: 7266.
101 Zhao, J.L., Zhang, P., Zhou, J.B. et al. (2003). Visualizations of light-induced refractive index changes in photorefractive crystals employing digital holography. *Chin. Phys. Lett.* 20: 1748.

102 Müller, R., Santos, M.T., Arizmendi, L., and Cabrera, J.M. (1993). A narrow-band interference filter with photorefractive $LiNbO_3$. *J. Phys. D: Appl. Phys.* 27: 241.

103 Rakuljic, G.A. and Leyva, V. (1993). Volume holographic narrow-band optical filter. *Opt. Lett.* 18: 459.

104 Müller, R., Alvarez-Bravo, J.V., Arizmendi, L., and Cabrera, J.M. (1994). Tuning of photorefractive interference filters in $LiNbO_3$. *J. Phys. D: Appl. Phys.* 27: 1628.

105 Breer, S. and Buse, K. (1998). Wavelength demultiplexing with volume phase holograms in photorefractive lithium niobite. *Appl. Phys. B* 66: 339.

106 Boffi, P., Ubaldi, M.C., Piccini, D. et al. (2000). 1550-nm volume holography for optical communication devices. *IEEE Photonics Technol. Lett.* 12 (10): 1355.

107 Petrov, V.M., Denz, C., Shamray, A.V. et al. (2001). Electrically controlled volume $LiNbO_3$ holograms for wavelength demultiplexing systems. *Opt. Mater.* 18: 191.

108 Agranat, A.J., Secundo, L., Golshani, N., and Razvag, M. (2001). Wavelength-selective photonic switching in paraelectric potassium lithium tantalate niobate. *Opt. Mater.* 18: 195.

109 James, R.T.B., Wah, C., Iizuka, K., and Shimotahira, H. (1995). Optically tunable optical filter. *Appl. Opt.* 34: 8230.

110 Dittrich, P., Montemezzani, G., and Günter, P. (2002). Tunable optical filter for wavelength division multiplexing using dynamic interband photorefractive gratings. *Opt. Commun.* 214: 363–370.

111 Zysset, B., Biaggio, I., and Günter, P. (1992). Refractive indices of orthorhombic $KNbO_3$. I. Dispersion and temperature dependence. *J. Opt. Soc. Am. B: Opt. Phys.* 9: 380.

112 Kogelnik, H. (1969). Coupled wave theory for thick hologram gratings. *Bell Syst. Tech. J.* 48: 2909.

113 Montemezzani, G. and Zgonik, M. (1997). Light diffraction at mixed phase and absorption gratings in anisotropic media for arbitrary geometries. *Phys. Ref. E* 55: 1035.

114 Montemezzani, G. (2000). Optimization of photorefractive two-wave mixing by accounting for material anisotropies: $KNbO_3$ and $BaTiO_3$. *Phys. Rev. A* 62: 053803.

115 Zgonik, M., Schlesser, R., Biaggio, I., and Günter, P. (1994). Electro-and acousto-optic properties of $KNbO_3$ crystals. *Ferroelectrics* 158: 217.

116 Ewart, M., Zgonik, M., and Günter, P. (1997). Nanosecond optical response to pulsed UV excitation in $KNbO_3$. *Opt. Commun.* 141: 99.

117 Yariv, A. (1991). *Optical Electronics*, 4ee. Philadelphia, (Chapters 9 and 18): Saunders College Publishing.

118 Ogawa, I., Ebisawa, F., Hanawa, F. et al. (1998). Hybrid integrated four-channel SS-SOA array module using planar light wave circuit platform. *Electron. Lett* 34: 361.

119 Xie, P. and Zhang, Z.-Q. (2004). Dynamical control of light propagation in nonlinear photonic crystals by applied electric fields. *J. Appl. Phys.* 95: 1630.

120 Günter, P. and Huignard, J.-P. (eds.) (1988). *Photorefractive Materials and Their Applications I, II*, Vols. 61 and 62 of Topics in Applied Physics. Berlin: Springer-Verlag, 1989.

121 Taj, I.A., Xie, P., and Mishima, T. (2001). Fast switching of photorefractive output by applied electric field. *Opt. Commun.* 187: 7.

122 Mishima, T., Xie, P., Koyanagi, K., and Taj, I.A. (2003). Demonstration of fast switching of photorefractive four-wave mixing via Pockels effect. *Opt. Commun.* 225: 211.

123 Xie, P. and Mishima, T. (2005). Fast switching of light propagation in a photorefractive crystal via Pockels effect. *Opt. Commun.* 246 (1–3): 29–34.

124 Xie, P., Dai, J.H., Wang, P.Y., and Zhang, H.J. (1996). Mechanism of self-pumped phase conjugation in photorefractive crystals. *Appl. Phys. Lett.* 69: 4005.

125 Xie, P., Dai, J.H., Wang, P.Y., and Zhang, H.J. (1997). Self-pumped phase conjugation in photorefractive crystals: reflectivity and spatial fidelity. *Phys. Rev. A* 55: 3092.

126 Yariv, A. (1975). *Quantum Electronics*, 2ee. John Wiley & Sons, Inc., (Ch. 14).

127 Vest, C.M. (1979). *Holographic Interferometry*. Willey and Sons. Inc.

128 Yeh, P. (1993). *Introduction to Photorefractive Nonlinear Optics*. Willey and Sons. Inc.

129 Günter, P. and Huinard, J.P. (1988). *Photorefractive Effects and Materials I e Photorefractive Materials and Their Applications II, Topics in Applied Physics*, vol. 61–62. Springer-Verlag.

130 Frejlich, J. and Garcia, P.M. (2000). Advances in real-time holographic interferometry for the measurements of vibrations and deformations. *Opt. Lasers Eng.* 32: 515–521.

131 Dirksen, D. and von Bally, G. (1994). Phase shifting holographic double exposure interferometry with fast photorefractive crystals. *J. Opt. Soc. Am. A:* B11: 1858–1864.

132 Georges, M.P. and Lemaire, P.C. (1999). Real-time holographic interferometry using sillenite photorefractive crystals. Study and optimization of a transportable set-up for quantified phase measurements on large objects. *Appl. Phys.* B68: 1073–1083.

133 Creath, K. (1988). *Phase-Measurement Interferometry Techniques, Progress in Optics*, vol. XXVI. Elsevier Science Publishers B.V.

134 Judge, T.R. and Bryanston-Cross, P.J. (1994). A review of phase unwrapping techniques in fringe analysis. *Opt. Lasers Eng.* 24: 199–239.

135 Ghiglia, D.C., Mastin, G.A., and Romero, L.A. (1987). Cellular-automata method for phase unwrapping. *J. Opt. Soc. Am. A:* 4: 267–280.

136 Gesualdi, M.R.R., Soga, D., and Muramatsu, M. (2003). Real-time holographic interferometry using photorefractive $Bi_{12}SiO_{20}$ crystals and their applications in surfaces analysis with phase-stepping technique. *Ann. Opt.* 5.

137 Gesualdi, M.R.R. (2000). Técnicas de Interferometria Holográfica Usando Cristais Fotorrefrativos da Família das Silenitas do Tipo Bi12SiO20 (BSO). Universidade de São Paulo, Instituto de Física.

138 Soga, D. (2000). Medida de Topografia de Superfície Usando a Técnica de Deslocamento de Fase. Universidade de São Paulo, Instituto de Física.
139 Zamiria, S., Reitinger, B., Berer, T. et al. (2010). Determination of nanometer vibration amplitudes by using a homodyne photorefractive crystal interferometer. *Procedia Eng.* 5: 299–302.
140 Zamiri, S., Reitinger, B., Berer, T. et al. (2010). Laser ultrasonic measurements of metal thickness using a photorefractive crystal. Microelectronics Conference, ME10, Vienna, Austria.
141 Ducharme, S., Scott, J.C., Twieg, R.J., and Moerner, W.E. (1991). Observation of the photorefractive effect in a polymer. *Phys. Rev. Lett.* 66: 1846.
142 Ostroverkhova, O. and Moerner, W.E. (2004). Organic photorefractives: mechanism, materials and applications. *Chem. Rev.* 104: 3267–3314.
143 Meerholz, K., Volodin, B.L., Kippelen, B., and Peyghambarian, N.A. (1994). Photorefractive polymer with high optical gain and diffraction efficiency near 100%. *Nature* 371: 497–500.
144 Kippelen, B., Meerholz, K., and Peyghambarian, N. (1996). *Nonlinear Optics of Organic Molecules and Polymers* (ed. H.S. Nalwa and S. Miyata) Chapter 8,, 507–623. Boca Raton, FL, USA: CRC.
145 Li, G. and Wang, P. (2010). Efficient local fixing of photorefractive polymer hologram using a laser beam. *Appl. Phys. Lett.* 96: 111109.
146 Tay, S., Blanche, P.-A., Voorakaranam, R. et al. (2008). An updatable holographic three-dimensional display. *Nature* 451: 694–698.
147 Tsutsumi, N., Kinashi, K., Nonomura, A., and Sakai, W. (2012). Quickly updatable hologram images using Poly(*N*-vinyl Carbazole) (PVCz) photorefractive polymer composite, materials. *Materials* 5: 1477–1486.
148 Tsujimura, S., Kinashi, K., Sakai, W., and Tsutsumi, N. (2012). High-speed photorefractive response capability in triphenylamine polymer-based composites. *Appl. Phys Express* 5: 064101.
149 Blanche, P.-A., Bablumian, A., Voorakaranam, R. et al. (2010). Holographic three-dimensional telepresence using large-area photorefractive polymer. *Nature* 468: 80–83.
150 Onsager, L. (1934). Deviations from Ohm's law in weak electrolytes. *J. Chem. Phys.* 2: 599–615.
151 Lynn, B., Blanche, P.-A., Bablumian, A. et al. (2013). Recent advancements in photorefractive holographic imaging. *J. Phys. Conf. Ser.* 415: 012050.
152 Jolly, S., Smalley, D.E., Barabas, J., Bove, V.M. (2013). Errata: Direct fringe writing architecture for photorefractive polymer-based holographic displays: analysis and implementation. *Opt. Eng.* 52: 055801.
153 Kolner, B.H. (1994). Space-time duality and the theory of temporal imaging. *IEEE J. Quantum Electron.* 30: 1951–1963.
154 Fernandez-Ruiz, M.R., Li, M., and Azaña, J. (2013). Time-domain holograms for generation and processing of temporal complex information by intensity-only modulation processes. *Opt. Express* 21: 10314–10323.
155 Nau, D., Christ, A., Giessen, H. et al. (2009). Femtosecond properties of photorefractive polymers. *Appl. Phys. B: Lasers Opt.* 95: 31–35.
156 Zamiri, S., Reitinger, B., Portenkirchner, E. et al. (2014). Laser ultrasonic receivers based on organic photorefractive polymer composites. *Appl. Phys. B* 114(4): 509.

157 Xu, X., Liu, H., and Wang, L.V. (2011). Time-reversed ultrasonically encoded optical focusing into scattering media. *Nat. Photonics* 5: 154–157.
158 Suzuki, Y., Xu, X., Lai, P., and Wang, L.V. (2012). Energy enhancement in time-reversed ultrasonically encoded optical focusing using a photorefractive polymer. *J. Biomed. Opt.* 17: 080507.
159 Suzuki, Y., Lai, P., Xu, X., and Wang, L.V. (2013). *Photons Plus Ultrasound: Imaging and Sensing 2013* (ed. A.A. Oraevsky and L.V. Wang), 85811G. Bellingham, Washington, United States: SPIE.
160 Suzuki, Y., Lai, P., Xu, X., and Wang, L.V. (2013). High-sensitivity ultrasound-modulated optical tomography with a photorefractive polymer. *Opt. Lett.* 38: 899–901.
161 Lai, P., Suzuki, Y., Xu, X., and Wang, L.V. (2013). *Photons Plus Ultrasound: Imaging and Sensing 2013* (ed. A.A. Oraevsky and L.V. Wang), 85812X. Bellingham, WA: SPIE.
162 Salvador, M., Prauzner, J., K, ober, S. et al. (2009). Three-dimensional holographic imaging of living tissue using a highly sensitive photorefractive polymer device. *Opt. Express* 17: 11834–11849.
163 Salvador, M., Prauzner, J., Kober, S., and Meerholz, K. (2011). Beam walk-off suppression in photorefractive polymer-based coherence domain holography. *Appl. Phys. B: Lasers Opt.* 102: 803–807.
164 Ren, X.K., Yang, D.Y., Zhang, T.H. et al. (2010). Polymeric photorefractive surface waves. *Opt. Commun.* 283: 3792–3797.
165 Fujihara, T., Sassa, T., Muto, T. et al. (2009). Surface waves in photorefractive polymer films. *Opt. Express* 17: 14150–14155.
166 Fujihara, T., Umegaki, S., Hara, M., and Sassa, T. (2012). Formation speed and formation mechanism of self-written surface wave-based waveguides in photorefractive polymers. *Opt. Mater. Express* 2: 849–855.
167 Bowley, C. and Crawford, G.P. (2000). Diffusion kinetics of formation of holographic polymer-dispersed liquid crystal display materials. *Appl. Phys. Lett.* 76: 2235.
168 Gu, C., Xu, Y., Liu, Y. et al. (2003). Applications of photorefractive materials in information storage, processing and communication. *Opt. Mater.* 23: 219–227.
169 Sutherland, R.L., Natarajan, L.V., Tondiglia, V.P. et al. (1996). Switchable volume hologram materials and devices, US Patent 5,942,157.
170 Cudney, R., Ríos, L., and Escamilla, H. (2004). Electrically controlled Fresnel zone plates made from ring-shaped 180° domains. *Opt. Express* 12: 5783–5788.
171 Patel, J.S. and Rastani, K. (1991). Electrically controlled polarization-independent liquid crystal Fresnel lens arrays. *Opt. Lett.* 16: 532–534.
172 Fan, Y.H., Ren, H., and Wu, S.T. (2003). Switchable Fresnel lens using polymer-stabilized liquid crystals. *Opt. Express* 11: 3080–3086.
173 Fan, Y.H., Ren, H., and Wu, S.T. (2005). Electrically switchable Fresnel lens using a polymer separated composite film. *Opt. Express* 13: 4141–4147.
174 Ren, H., Fan, Y.H., and Wu, S.T. (2003). Tunable Fresnel lens using nanoscale polymer dispersed liquid crystals. *Appl. Phys. Lett.* 83: 1515–1517.

175 Domash, L.H., Chen, T., Gomatam, B.N. et al. (1996). Switchable-focus lenses in holographic polymer-dispersed liquid crystal. *Proc. SPIE* 2689: 188.
176 Liu, Y.J. and Sun, X.W. (2008). Holographic polymer-dispersed liquid crystals: materials, formation, and applications. *Adv. Opto Electron.* 2008: 1–52.
177 Woo, J.Y., Kim, E.H., Shim, S.S., and Kim, B.K. (2008). High dielectric anisotropy compound doped transmission gratings of HPDLC. *Opt. Commun.* 281: 2167–2172.
178 Liu, Y.J., Zhang, B., Jia, Y., and Xu, K. (2003). Improvement of the diffraction properties in holographic polymer dispersed liquid crystal bragg gratings. *Opt. Commun.* 218: 27–32.
179 Xin, Z., Cai, J., Shen, G. et al. (2006). A dynamic gain equalizer based on holographic polymer dispersed liquid crystal gratings. *Opt. Commun.* 268: 79–83.
180 Beev, K., Criante, L., Lucchetta, D.E. et al. (2006). Total internal reflection holographic gratings recorded in polymer-dispersed liquid crystals. *Opt. Commun.* 260: 192–195.
181 Nemati, H., Mohajerani, E., Moheghi, A. et al. (2009). A simple holographic technic for fabricating a LC/polymer switchable Fresnel lens. *EPL* 87: 64001.
182 Jashnsaz, H., Mohajerani, E., Nemati, H. et al. (2011). Electrically switchable holographic liquid crystal/polymer Fresnel lens using a Michelson interferometer. *Appl. Opt.* 50: 2701–2707.
183 Jashnsaz, H., Nataj, N.H., Mohajerani, E., and Khabbazi, A. (2011). All-optical switchable holographic Fresnel lens based on azo-dye doped polymer-dispersed liquid crystals. *Appl. Opt.* 50: 4295–4301.
184 Jashnsaz, H., Mohajerani, E., and Hosain Nataj, N. (2010). Electrically switchable cylindrical Fresnel lens based on holographic polymer-dispersed liquid crystals using a Michelson interferometer. *IJOP* 4 (2): 121–126.
185 Hill, K.O., Fujii, Y., Johnson, D.C., and Kawasaki, B.S. (1978). Photosensitivity in optical fiber waveguides: Application to reflection filter fabrication. *Appl. Phys. Lett.* 32, 647.
186 Meltz, G., Morey, W.W., and Glenn, W.H. (1989). Formation of Bragg gratings in optical fibers by a transverse holographic method. *Opt. Lett.* 14: 823.
187 Chen, Z., Asaro, M., Ostroverkhova, O. et al. (2003). Self-trapping of light in an organic photorefractive glass. *Opt. Lett.* 28 (24): 2509.
188 Lan, S., DelRe, E., Chen, Z. et al. (1999). Second-harmonic generation in waveguides induced by photorefractive spatial solitons. *Opt. Lett.* 24: 1145.
189 Assanto, G., Peccianti, M., and Conti, C. (2003). Nematicons: Optical spatial solitons in nematic liquid crystals. *Opt. Photonics News* 44.
190 Singer, K.D., Sohn, J.E., and Lalama, S.J. (1986). Second harmonic generation in poled polymer films. *Appl. Phys. Lett.* 49: 248.
191 Corning display Glass (2003). http://www.corning.com/emea/en/products/display-glass.html.
192 Asaro, M., Sheldon, M., Chen, Z. et al. (2005). Soliton-induced waveguides in an organic photorefractive glass. *Opt. Lett.* 30 (5): 519–521.
193 Borrelli, N.F., Schroeder, J.F., and Seward, T.P. (2012). Polarizing photorefractive glass. US Patent 8,179,595 B2, filed 15 May 2012.

194 Poumellec, B. and Kherbouche, F. (1996). The photorefractive Bragg gratings in the fibers for telecommunications. *J. Phys. III France* 6: 1595.

195 Niay, P., Douay, M., Bernage, P. et al. (1999). Does photosensitivity pave the way towards the fabrication of miniature coherent light sources in inorganic glass waveguides? *Opt. Mater.* 11: 115.

196 Canning, J., Sommer, K., Englund, M., and Huntington, S. (2001). Direct evidence of two types of UV-induced glass changes in silicate-based optical fibers. *Adv. Mater.* 13: 970.

197 Othonos, A. and Kalli, K. (1999). *Fiber Bragg Gratings*. Boston: Artech House.

198 Potter, B.G. Jr., and Simmons-Potter, K. (2000). Photosensitive point defects in optical glasses: Science and applications. *Nucl. Instrum. Methods Phys. Res., Sect. B* 771: 166–167.

199 Volodin, B.L., Dolgy, S.V., Melnik, E.D., and Ban, V.S. *Volume Bragg Gratings^{TM}, A New Platform Technology for WDM Applications*. Pennington, NJ: PD-LD Inc.

Index

Crystal Optics: Properties and Applications, First Edition. Ashim Kumar Bain.

d

e

q